Lecture Notes in Computer Science 6300

Volume Editors

Andreas Blass
University of Michigan, Mathematics Department
Ann Arbor, MI 48109-1043, USA
E-mail: ablass@umich.edu

Nachum Dershowitz
Tel Aviv University, School of Computer Science
Ramat Aviv, Tel Aviv 69978, Israel
E-mail: nachumd@post.tau.ac.il

Wolfgang Reisig
Humboldt-Universität zu Berlin, Institut für Informatik
Unter den Linden 6, 10099 Berlin, Germany
E-mail: reisig@informatik.hu-berlin.de

About the Cover
The cover illustration is a "Boaz" Plate created by Maurice Ascalon's Pal-Bell Company circa 1948. Image and artwork copyright Ascalon Studios, Inc. Used by permission. The Hebrew legend is from the Book of Psalms 126:5, "They that sow in tears shall reap in joy."

Credits
The frontispiece photograph was taken by Bertrand Meyer at the Eidgenössische Technische Hochschule (ETH) in Zürich, Switzerland on May 16, 2004. Used with permission.

Library of Congress Control Number: 2010931832

CR Subject Classification (1998): F.3, D.2, D.3, C.2, F.2, F.1

LNCS Sublibrary: SL 2 – Programming and Software Engineering

ISSN 0302-9743
ISBN-10 3-642-15024-1 Springer Berlin Heidelberg New York
ISBN-13 978-3-642-15024-1 Springer Berlin Heidelberg New York

springer.com

Printed in Germany

Typesetting: Camera-ready by author, data conversion by Scientific Publishing Services, Chennai, India
Printed on acid-free paper 06/3180

Andreas Blass   Nachum Dershowitz
Wolfgang Reisig (Eds.)

Fields of Logic and Computation

Essays Dedicated to Yuri Gurevich on the Occasion of His 70th Birthday

Springer

Dedicated to

Yuri Gurevich

in honor of his 70th birthday,
with deep admiration and affection.

They that sow in tears shall reap in joy.

הזרעים בדמעה ברנה יקצרו.

(Psalms 126:5)

Wishing him many,
many happy returns.

Yuri Gurevich
(b. 1940)

Preface

Yuri Gurevich has played a major role in the discovery and development of applications of mathematical logic to theoretical and practical computer science. His interests have spanned a broad spectrum of subjects, including decision procedures, the monadic theory of order, abstract state machines, formal methods, foundations of computer science, security, and much more.

In May 2010, Yuri celebrated his 70th birthday. To mark that occasion, on August 22, 2010, a symposium was held in Brno, the Czech Republic, as a satellite event of the 35th International Symposium on Mathematical Foundations of Computer Science (MFCS 2010) and of the 19th EACSL Annual Conference on Computer Science Logic (CSL 2010). The meeting received generous support from Microsoft Research.

In preparation for this 70th birthday event, we asked Yuri's colleagues (whether or not they were able to attend the symposium) to contribute to a volume in his honor. This book is the result of that effort. The collection of articles herein begins with an academic biography, an annotated list of Yuri's publications and reports, and a personal tribute by Jan Van den Bussche. These are followed by 28 technical contributions. These articles – though they cover a broad range of topics – represent only a fraction of Yuri's multiple areas of interest.

Each contribution was reviewed by one or two readers. In this regard, the editors wish to thank several anonymous individuals for their assistance.

We offer this volume to Yuri in honor of his birthday and in recognition of his grand contributions to the fields of logic and computation.

June 20, 2010

Andreas Blass
Nachum Dershowitz
Wolfgang Reisig

Table of Contents

On Yuri Gurevich

Technical Papers

Yuri, Logic, and Computer Science

Andreas Blass[1], Nachum Dershowitz[2], and Wolfgang Reisig[3]

[1] Mathematics Department, University of Michigan, Ann Arbor, MI 48109–1043, U.S.A.
[2] School of Computer Science, Tel Aviv University, Ramat Aviv 69978, Israel
[3] Humboldt-Universität zu Berlin, Institut für Informatik, Unter den Linden 6, 10099 Berlin, Germany

Yuri Gurevich was born on May 7, 1940, in Nikolayev, Ukraine, which was a part of Soviet Union at the time. A year later, World War II reached the Soviet Union, and Yuri's father was assigned to work in a tank body factory near Stalingrad. So that's where Yuri spent the second year of his life, until the battle of Stalingrad forced the family, except for his father, to flee. Their home was destroyed by bombing only hours after they left. But fleeing involved crossing the burning Volga and then traveling in a vastly overcrowded train, in which many of the refugees died; in fact, Yuri was told later that he was the only survivor among children of his age. His mother decided that they had to leave the train, and the family lived for two years in Uzbekistan. In May 1944, the family reunited in Chelyabinsk, in the Ural Mountains, where the tank body factory had moved in the meantime, and that is where Yuri attended elementary and high school.

An anecdote from his school days (recorded in [123][1]) can serve as a premonition of the attention to resources that later flowered in Yuri's work on complexity theory. To prove some theorem about triangles, the teacher began with "Take another triangle such that ... ." Yuri asked, "Where does another triangle come from? What if there are no more triangles?" (commenting later that shortages were common in those days). For the sake of completeness, we also record the teacher's answer, "Shut up."

After graduating from high school, Yuri spent three semesters at the Chelyabinsk Polytechnik. Because of a dissatisfaction with the high ratio of memorization to knowledge in the engineering program, Yuri left after a year and a half and enrolled in Ural State University to study mathematics.

Yuri obtained four academic degrees associated with Ural State University: his master's degree in 1962, his candidate's degree (equivalent to the Western Ph.D.) in 1964, his doctorate (similar to habilitation, but essentially guaranteeing an appointment as full professor) in 1968, and an honorary doctorate in 2005. At Ural State University, Yuri ran a flourishing logic seminar, and he founded a Mathematical Winter School that is still functioning today. It should also be noted that the four-year interval between the candidate's and doctor's degrees was unusually short.

[1] Numerical references are to the annotated bibliography in this volume; they match the numbering on Yuri's web site.

A. Blass, N. Dershowitz, and W. Reisig (Eds.): Gurevich Festschrift, LNCS 6300, pp. 1–6, 2010.

That four-year interval contained some very important non-academic events. Yuri and Zoe were married in March 1965, and their twin daughters, Hava and Naomi, were born in October 1966.

As mentioned above, once he had his doctorate (in the Russian sense), Yuri would ordinarily get a professorship, but a glance at his curriculum vitae shows that he overshot a bit, becoming not only a professor but chair of the Mathematics Department at the National Economy Institute in Sverdlovsk in 1969. This does not indicate a great enthusiasm for administrative work; in fact, though he's very good at administration, Yuri is not very fond of it. Nor does it indicate great interest in economics. What made this appointment extremely attractive is that it provided an apartment – a very important benefit in the Soviet Union, especially for a man with a young family.

Because of the political situation in the Soviet Union, Yuri and Zoe decided to move to Israel. That rather dangerous journey took them to Krasnodar and then to Tbilisi, the capital of beautiful and hospitable Georgia. Eventually they emigrated to Israel in October 1973.

During the period from his master's degree until his departure from the Soviet Union, Yuri established himself as a first-rate algebraist and logician. Already in his master's thesis [1], he solved an open problem in group theory, but his most important work from this period, at the interface of logic and algebra, concerned ordered abelian groups. His 1964 thesis for the candidate's degree [3] proved the decidability of the first-order theory of these groups; later he obtained decidability of the richer theory that includes quantification not only over elements but also over convex subgroups. This richer theory includes essentially all the questions that had attracted the attention of researchers in ordered abelian groups. It is fair to say that this work [19,25] of Yuri's subsumed that entire field.

In addition to this work on ordered abelian groups, Yuri made fundamental contributions to the decision problem for first-order logic. In particular, he completed [6,7,13] the analysis of decidability of classes of first-order formulas of the form "specified quantifier prefix and specified vocabulary." Yuri's work from that period contains other contributions, for example the undecidability of the first-order theory of lattice-ordered abelian groups [9], but there is also an important but non-technical contribution that must be mentioned here.

We quote part of a toast offered by Yuri's first Ph.D. student, Alexander Livchak, at a 2010 anniversary celebration of the Faculty of Mathematics of the Ural State University:

> Gurevich taught us to think freely. It was helpful that his specialty was logic – the science of proofs. He tried unobtrusively to impress upon us that the final judgment is ours and not that of the Central Committee of the Communist Party or that of Marx–Engels.
>
> It all started with a seminar on axiomatic set theory. The idea of a winter school was born there. The schedule of the Winter Math School included not only studies but also mandatory daily skiing and various entertainment activities. For example, Gurevich liked debates à la medieval scholastic disputes. He would volunteer to argue any ridiculous

> and obviously false thesis of our choice in order to demonstrate the art of arguing. Therein lay his secret "counter-revolutionary Zionist" (in the terminology of the time) plot: to teach us to argue, doubt, prove, refute. In general to teach us to think independently.

Yuri lived in Israel from 1973 to 1981, teaching at Ben-Gurion University of the Negev in Beer-Sheva, except for leaves of absence spent at Simon Fraser University in Vancouver, the Hebrew University of Jerusalem, and Bowling Green State University in Ohio. Very soon after his arrival in Israel, he impressed people by solving several problems posed by Saharon Shelah. This work and other results from his Israeli period concerned the monadic theory of linear orders, either in general or in the context of specific linear orders like the real line. One of these results is that, if the continuum hypothesis holds, the countability of subsets of the real line can be defined in monadic second-order logic. That work led to the first of numerous deep joint papers with Shelah. It also led to connections with the theory of ordered (non-abelian) groups, especially groups of automorphisms of linear orders.

Another major contribution from Yuri's Israeli period (although the paper [40] was prepared and published after Yuri was in the U.S.) is the Gurevich-Harrington theorem. This theorem concerns the existence of winning strategies in certain infinite games. In various contexts (mostly in topology), people had considered strategies that look only at the opponent's immediately previous move (rather than the whole history of the play) or a fixed number of previous moves. Yuri and Leo Harrington showed that, for many games, the winning player has a strategy that remembers, at any stage, only finitely much information from the past, though there is no bound on how long ago that information might have appeared. They used this result to greatly simplify the hardest part of Michael Rabin's proof of the decidability of the monadic theory of two successor functions.

Yuri spent the academic year 1981–82 as a visiting professor at Bowling Green State University in Ohio, which was at that time a major center of research on ordered groups; Andrew Glass and Charles Holland were on the faculty there. In addition to research on ordered groups, Yuri resumed thinking about computer science, which he had already been interested in even in the Soviet Union. He sought a computer science position in Israel or the U.S.

In 1982, Yuri accepted an appointment as professor of computer science[2] at the University of Michigan.[3] Yuri took his conversion to computer science seriously. He did not just write mathematics papers with a computer science facade. Although he finished various mathematical projects, he immediately began

---

[2] Technically, he was professor of Computer and Communication Sciences, since that was the name of the department. A subsequent reorganization put him and his fellow computer scientists into the engineering college, in the Computer Science and Engineering Division of the Department of Electrical Engineering and Computer Science. He now holds the title of Professor Emeritus of Electrical Engineering and Computer Science.

[3] An observation by Andreas Blass: The angriest I've ever seen Andrew Glass is when he talked about Bowling Green's failure to make a serious effort to keep Yuri.

thinking deeply about computational issues and making significant contributions to his new field. Furthermore, with the enthusiasm of a new convert, he began the difficult project of trying to convert Andreas Blass to computer science. The project didn't entirely succeed; Blass still claims to be a set-theorist, but he certainly learned a great deal of computer science from Yuri.

Yuri's contributions to computer science span a vast part of that field, and we can mention only a few of them here. First, there are many results in complexity theory, but to appreciate this part of Yuri's work it is necessary to take into account the great variety of topics that fall under this heading. There is traditional complexity theory, largely directed toward the P vs. NP question but also covering many other aspects of polynomial-time computation. But there is also a strong connection to probabilistic issues, both in connection with the use of randomness in computation and in connection with average-case (as opposed to worst-case) complexity of algorithms. Yuri made important contributions in all these areas, including good explanations of Leonid Levin's theory of average-case complexity and natural complete problems for this theory. He also investigated far stricter resource bounds, including linear time, and in a joint paper [82] with Shelah showed that the notion of "linear times polylog time" is remarkably robust across different models of computation as long as one excludes ordinary Turing machines.

A second broad area of Yuri's research in computer science is connections between computation and logic, and this, too, spans several sub-areas. A particularly important contribution is Yuri's emphasis on the computational relevance of finite structures. Classical logic is greatly changed by restricting attention to finite structures, mainly because the compactness theorem, one of the chief traditional tools, becomes false in this context. Although there had certainly been earlier work on finite structures, Yuri's papers [60] and [74] led to a major increase of interest and activity in this field. Yuri also formulated in [74] the main open problem about the connection between logic and complexity, namely his conjecture that there is no logic that exactly captures polynomial time computability on unordered structures. (Part of the contribution here is making the conjecture precise by saying what should be meant by a "logic" in this context.)

Yuri's contributions to the interface between logic and computer science also include studies of Hoare logic and (motivated by better compatibility with Hoare logic) existential fixed-point logic. Another of Yuri's contributions is the introduction, in joint work with Erich Grädel [109], of metafinite model theory, in which models are primarily finite but are allowed to have a secondary, infinite part so that such operations as counting can be accommodated in a natural way.

We should also mention here Yuri's "Logic in Computer Science" column in the *Bulletin of the European Association for Theoretical Computer Science.* Here we find not only a great number of interesting columns written by Yuri himself and exploring the most diverse areas that could fit under the "logic and computer science" heading, often in the form of a Socratic dialogue with his friend and disciple, "Quisani," but also columns that he solicited from other experts. The columns, later collected along with other *BEATCS* material in three books titled *Current Trends in Theoretical Computer Science*,[4] make fascinating reading. They are not all just surveys either; Yuri sometimes used the column to present his new results. An outstanding example is [131], in which he proves that, if we could compute, in polynomial time, a complete isomorphism invariant for graphs, then we could also compute in polynomial time, from any graph as input, a standard representative (a canonical form) of its isomorphism class.

In terms of subsequent impact, Yuri's biggest achievement during his Michigan period was the invention of abstract state machines[5] (ASMs) and the ASM thesis. ASMs are an extraordinarily clean and general model of computation. Here "clean" means that ASMs have a simple, unambiguous semantics; "general" means that they can easily simulate a huge variety of computations, ranging from high-level algorithms down to hardware. Yuri proposed the "ASM thesis" that *every* algorithm can be faithfully represented, at its natural level of abstraction, by an ASM. Initial support for the thesis came from numerous case studies, in which Yuri, his students, and others gave ASM descriptions of a wide variety of software and hardware systems as well as abstract algorithms. Later, support came from rigorous proofs, but this is getting ahead of the story.

In 1998, Yuri joined Microsoft Research as a senior researcher, with the job of bringing ASMs into the real world of large-scale software development. The move from Michigan to Microsoft happened remarkably fast; the decision was made in August and involved getting a leave of absence from Michigan for the fall semester, which begins in early September. Fortunately, the relevant administrators at Michigan granted the leave, fully expecting that Yuri would soon return, especially because his job at Microsoft involved building a new research group from scratch. But Yuri handled his new administrative duties well, hiring several first-rate researchers, and he really enjoyed (and continues to enjoy) the knowledge that his work is having an impact on real computing. So, after two years on leave from Michigan, he officially retired. He has now been a professor emeritus for ten years.

Among Yuri's many contributions while at Microsoft, we describe just a few, of which several involve ASMs. Perhaps the most surprising is his discovery that, in certain contexts, the ASM thesis can actually be proved. In [141], Yuri presented some simple, natural postulates about sequential, non-interactive algorithms; argued that they are satisfied by anything that one would intuitively consider to be such an algorithm; and then proved that anything satisfying the postulates is equivalent, in a very strong sense, to an ASM (of a particular sort). That result has since been extended to parallel algorithms in [157-1,157-2], to interactive sequential algorithms in [166,170,171,176,182], and to a notion of equivalence that keeps track of exactly which part of the state is relevant at each step in [201].

A second, quite different, use of ASMs occurs in describing "choiceless polynomial time" computation. Here, the input to a computation is an unordered structure, and the computation is allowed to use parallelism and essentially

[4] Edited by G. Rozenberg, A. Salomaa, and (for the last two) G. Paun; published by World Scientific in 1993, 2001, and 2004.

[5] Originally called "dynamic structures" and subsequently "evolving algebras."

arbitrary data structures, but it is not allowed to make arbitrary choices of elements (or, equivalently, to linearly order the input structure). This concept, originally introduced with quite a complicated definition by Shelah, turned out to be equivalent to computation by a rather standard sort of ASM over a structure consisting of the original input plus all hereditarily finite sets over it. The hereditarily finite sets capture the arbitrary data structures, and the ASMs take care of the rest of the computational issues. By itself, choiceless polynomial time is rather weak; it can't even count [120]. But when extended by counting, it is a surprisingly strong logic [150] and in fact is one of very few logics that might possibly capture polynomial-time computation on unordered structures (though that seems very unlikely).

Yet another use of ASMs is the proof of Church's thesis [188] on the basis of natural assumptions about computability.

Another aspect of Yuri's contribution to computer science is that he accurately assesses the quality of people's work and acts on the basis of that assessment. We omit the negative examples, to avoid unnecessary controversy, but describe one positive example, as seen by Blass. Ben Rossman, while taking a year off after finishing his undergraduate degree, solved an old problem of Yuri's and sent him the solution. Many people, getting a rather difficult-to-read manuscript, out of the blue, from someone with no credentials, would be inclined to ignore it, but Yuri read it carefully, decided it was correct, and told Blass about it enthusiastically. Not long afterward, Yuri, Rossman, and Blass were all at a LICS conference, and Rossman, who had met Blass a few months earlier at another conference but had not yet met Yuri, mentioned to Blass that he was looking for something interesting to do in the following summer. Blass immediately thought of Microsoft Research, but not of Yuri's group, which was at that time heavily engaged in very applied work, far from Rossman's theoretical interests. But there might be a possibility of an internship with Microsoft's Theory Group, so Blass suggested that Rossman check with Yuri about that possibility. Yuri promptly offered Rossman a visiting position in his group, and this unusual investment in theory paid off in several significant contributions by Rossman to the group's work [169,176,182].

Among Yuri's other recent technical contributions are work on efficient file transfer [183,190], on software testing [154,160,163,173], on security assessment [202], and on decentralized authorization [191,198,200].

There is a great deal more to be said about Yuri's work, in both mathematics and computer science, his generosity toward colleagues and students, and his amazing energy level. But this is being written while the rest of this volume is ready to go to the publisher, so we'll stop here. Some more information can be found in Jan Van den Bussche's contribution to this volume, and an indication of the community's admiration for Yuri can be inferred from the size of this volume.

Annotated List of Publications of Yuri Gurevich

The following list of publications and annotations is derived from Yuri Gurevich's website,[1]

`http://research.microsoft.com/en-us/um/people/gurevich/annotated.htm`.

Abbreviations:

BEATCS	=	Bulletin of the European Association for Theoretical Computer Science
JSL	=	Journal of Symbolic Logic
LNCS	=	Lecture Notes in Computer Science
MSR-TR-Y-N	=	Microsoft Research Technical Report number N of year Y
ACM TOCL	=	ACM Transactions of Computation Logic
Doklady	=	Doklady Akademii Nauk SSSR (Proceedings of the USSR Academy of Sciences)

0. Egon Börger, Erich Grädel, Yuri Gurevich: The Classical Decision Problem. Springer Verlag, Perspectives in Mathematical Logic, 1997. Second printing, Springer Verlag, 2001. Review in Journal of Logic, Language and Information 8:4 (1999), 478–481. Review in ACM SIGACT News 35:1 (March 2004), 4–7

   The classical decision problem is (in its modern meaning) the problem of classifying fragments of first-order logic with respect to the decidability and complexity of the satisfiability problem as well as the satisfiability problem over finite domains. The results and methods employed are used in logic, computer science and artificial intelligence.

   The book gives the most complete and comprehensive treatment of the classical decision problem to date, and includes an annotated bibliography of 549 items. Much of the material is published for the first time in book form; this includes the classifiability theory, the classification of the so-called standard fragments, and the analysis of the reduction method. Many proofs have been simplified and there are many new results and proofs.

1. Yuri Gurevich: Groups covered by proper characteristic subgroups. Trans. of Ural University 4:1 (1963), 32–39 (Russian, Master's thesis)

2. Yuri Gurevich, Ali I. Kokorin: Universal equivalence of ordered abelian groups. Algebra and Logic 2:1 (1963), 37–39 (Russian)

   We prove that no universal first-order property distinguishes between any two ordered abelian groups.

[1] The editors thank Zoe Gurevich for her help.

A. Blass, N. Dershowitz, and W. Reisig (Eds.): Gurevich Festschrift, LNCS 6300, pp. 7–48, 2010.

3. Yuri Gurevich: Elementary properties of ordered abelian groups. Algebra and Logic 3:1 (1964), 5–39 (Russian, Ph.D. thesis)

We classify ordered abelian groups by first-order properties. Using that classification, we prove that the first-order theory of ordered abelian groups is decidable; this answers a question of Alfred Tarski.

3a. Yuri Gurevich: Elementary properties of ordered abelian groups. AMS Translations 46 (1965), 165–192

This is an English translation of [3].

4. Yuri Gurevich: Existential interpretation. Algebra and Logic 4:4 (1965), 71–85 (Russian)

We introduce a method of existential interpretation, and we use the method to prove the undecidability of fragments of the form $\exists^r\forall^*$ of various popular first-order theories.

5. Yuri Gurevich: On the decision problem for pure predicate logic. Doklady 166 (1966), 1032–1034 (Russian)

The $\forall\exists\forall\exists^*$ fragment of pure predicate logic with one binary and some number k of unary predicates is proven to be a conservative reduction class. Superseded by [6].

5a. Yuri Gurevich: On the decision problem for pure predicate logic. Soviet Mathematics 7 (1966), 217–219

This is an English translation of [5].

6. Yuri Gurevich: The decision problem for predicate logic. Doklady 168 (1966), 510–511 (Russian)

The $\forall\exists\forall\exists^*$ fragment of pure predicate logic with one binary and no unary predicates is a conservative reduction class and therefore undecidable for satisfiability and for finite satisfiability. This completes the solution of the classical decision problem for pure predicate logic: the prefix-vocabulary classes of pure predicate logic are fully classified into decidable and undecidable. See a more complete exposition in [7].

6a. Yuri Gurevich: The decision problem for predicate logic. Soviet Mathematics 7 (1966), 669–670

This is an English translation of [6].

7. Yuri Gurevich: Recognizing satisfiability of predicate formulas. Algebra and Logic 5:2 (1966), 25–35 (Russian)

This is a detailed exposition of the results announced in [6].

8. Yuri Gurevich: The word problem for some classes of semigroups. Algebra and Logic 5:2 (1966), 25–35 (Russian)

The word problem for finite semigroups is the following decision problem: given some number n of word pairs $(u_1, v_1), ..., (u_n, v_n)$ and an additional word pair (u_0, v_0), decide whether the n equations $u_1 = v_1, ..., u_n = v_n$ imply the additional equation $u_0 = v_0$ in all finite semigroups. We prove that the word problem for finite semigroups is undecidable. In fact, the undecidability result holds for a particular premise $E = (u_1 = v_1$ and ... and $u_n = v_n)$. Furthermore, this particular E can be chosen so that the following are recursively inseparable:

– $\{(u_0, v_0) :$ E implies $u_0 = v_0$ in every periodic semigroup$\}$,
– $\{(u_0, v_0) :$ E fails to imply $u_0 = v_0$ in some finite semigroup$\}$.

The paper contains some additional undecidability results.

9. Yuri Gurevich: Hereditary undecidability of the theory of lattice-ordered abelian groups. Algebra and Logic 6:1 (1967), 45–62 (Russian)

Delimiting the decidability result of [3] for linearly ordered abelian groups and answering Malcev's question, we prove the theorem in the title.

10. Yuri Gurevich: Lattice-ordered abelian groups and K-lineals. Doklady 175 (1967), 1213–1215 (Russian)

10a. Yuri Gurevich: Lattice-ordered abelian groups and K-lineals. Soviet Mathematics 8 (1967), 987–989

This is an English translation of [10].

11. Yuri Gurevich: A new decision procedure for the theory of ordered abelian groups. Algebra and Logic 6:5 (1967), 5–6 (Russian)

12. Yuri Gurevich: The decision problem for some algebraic theories. Doctor of Physico-Mathematical Sciences Thesis, Sverdlovsk, USSR, 1968 (Russian)

13. Yuri Gurevich: The decision problem for logic of predicates and operations. Algebra and Logic 8 (1969), 284–308 (Russian)

The article consists of two chapters. In the first part of the first chapter, the author rediscovers well-partial-orderings and well-quasi-orderings, which he calls *tight* partial orders and tight quasi-orders, and develops a theory of such orderings. (In this connection, it may be appropriate to point out Joseph B. Kruskal's article "The theory of well-quasi-ordering: A frequently discovered concept" in J. Comb. Theory A, vol. 13 (1972), 297–305.) To understand the idea behind the term "tight", think of a boot: you cannot move your foot far down or sidewise – only up. This is similar to tight partial orders where infinite sequences have no infinite descending subsequences, no infinite antichains, but always have infinite ascending subsequences.

In the second part of the first chapter, the author applies the theory of tight orders to prove a classifiability theorem for prefix-vocabulary classes of first-order logic. The main part of the classifiability theorem is that the partial order of prefix-vocabulary classes (ordered by inclusion) is tight. But there is an additional useful part of the classifiability theorem, about the form of the minimal classes outside a downward closed collection, e.g. the minimal classes that are undecidable in one way or another.

In the second chapter, the author completes the decision problem for (the prefix-vocabulary fragments of) pure logic of predicates and functions, though the treatment of the most difficult decidable class is deferred to [18]. In particular, the classes $[\forall^2,(0,1),(1)]$ and $[\forall^2,(1),(0,1)]$ are proved to be conservative reduction classes. (This abstract is written in January 2006.)

13a. Yuri Gurevich: The decision problem for logic of predicates and operations. Algebra and Logic 8 (1969), 160–174 (English)

This is an English translation of [13].

14. Yuri Gurevich: The decision problem for decision problems. Algebra and Logic 8 (1969), 640–642 (Russian)

Consider the collection D of first-order formulas α such that the first-order theory with axiom α is decidable. It is proven that D is neither r.e. nor co-r.e. (The second part had been known earlier.)

14a. Yuri Gurevich: The decision problem for decision problems. Algebra and Logic 8 (1969), 362–363 (English)

This is an English translation of [14].

15. Yuri Gurevich: Minsky machines and the $\forall\exists\forall\&\exists^*$ case of the decision problem. Trans. of Ural University 7:3 (1970), 77–83 (Russian)

An observation that Minsky machines may be more convenient than Turing machines for reduction purposes is illustrated by simplifying the proof from [7] that some $[\forall\exists\forall\&\exists^*,(k,1)]$ is a reduction class.

16. Yuri Gurevich, Igor O. Koriakov: A remark on Berger's paper on the domino problem. Siberian Mathematical Journal, 13 (1972), 459–463 (Russian)

Berger proved that the decision problem for the unrestricted tiling problem (a.k.a. the unrestricted domino problem) is undecidable. We strengthen Berger's result. The following two collection of domino sets are recursively inseparable:
(1) those that can tile the plane periodically (equivalently, can tile a torus) and
(2) those that cannot tile the plane at all.
It follows that the collection of domino sets that can tile a torus is undecidable.

16a. Yuri Gurevich, Igor O. Koriakov: A remark on Berger's paper on the domino problem. Siberian Mathematical Journal 13 (1972), 319–321 (English)

This is an English translation of [16].

17. Yuri Gurevich, Tristan Turashvili: Strengthening a result of Suranyi. Bulletin of the Georgian Academy of Sciences 70 (1973), 289–292 (Russian)

18. Yuri Gurevich: Formulas with one universal quantifier. In: *Selected Questions of Algebra and Logic*, Volume dedicated to the memory of A.I. Malcev, Publishing house Nauka – Siberian Branch, Novosibirsk, (1973), 97–110 (Russian)

The main result, announced in [9], is that the $\exists^*\forall\exists^*$ class of first-order logic with functions but without equality has the finite model property (and therefore is decidable for satisfiability and finite satisfiability). This result completes the solution in [9] for the classical decision problem for first-order logic with functions but without equality.

19. Yuri Gurevich: The decision problem for the expanded theory of ordered abelian groups. Soviet Institute of Scientific and Technical Information (VINITI), 6708:73 (1974), 1–31 (Russian)

20. Yuri Gurevich: The decision problem for first-order logic. Manuscript (1971), 124 pages (Russian)

This was supposed to be a book (and eventually it became the core of the book [0]), but the publication of the original Russian book was aborted when the author left USSR. A German translation of the manuscript can be found in Universitätsbibliothek Dortmund (Ostsprachen-Übersetzungsdienst) and Technische Informationsbibliothek und Universitätsbibliothek Hannover.

21. Yuri Gurevich: The decision problem for standard classes. JSL 41 (1976), 460–464

The classification of prefix-signature fragments of (first-order) predicate logic with equality, completed in [7], is extended to first-order logic with equality and functions. One case was solved (confirming a conjecture of this author) by Saharon Shelah.

22. Yuri Gurevich: Semi-conservative reduction. Archiv für Math. Logik und Grundlagenforschung 18 (1976), 23–25

23. Ilya Gertsbakh, Yuri Gurevich: Constructing an optimal fleet for a transportation schedule. Transportation Science 11 (1977), 20–36

A general method for constructing all optimal fleets is described.

24. Yuri Gurevich: Intuitionistic logic with strong negation. Studia Logica 36 (1977), 49–59

    Classical logic is symmetric with respect to True and False but intuitionistic logic is not. We introduce and study a conservative extension of first-order intuitionistic logic that is symmetric with respect to True and False.

25. Yuri Gurevich: Expanded theory of ordered abelian groups. Annals of Mathematical Logic 12 (1977), 193–228

    The first-order theory of ordered abelian groups was analyzed in [3]. However, algebraic results on ordered abelian groups in the literature usually cannot be stated in first-order logic. Typically they involve so-called convex subgroups. Here we introduce an expanded theory of ordered abelian groups that allows quantification over convex subgroups and expresses almost all relevant algebra. We classify ordered abelian groups by the properties expressible in the expanded theory, and we prove that the expanded theory of ordered abelian groups is decidable. Curiously, the decidability proof is simpler than that in [3]. Furthermore, the decision algorithm is primitive recursive.

26. Yuri Gurevich: Monadic theory of order and topology, I. Israel Journal of Mathematics 27 (1977), 299–319

    We disprove two of Shelah's conjectures and prove some more results on the monadic theory of linearly orderings and topological spaces. In particular, if the Continuum Hypothesis holds then there exist monadic formulæ expressing the predicates "X is countable" and "X is meager" over the real line and over Cantor's Discontinuum.

27. Yuri Gurevich: Monadic theory of order and topology, II. Israel Journal of Mathematics 34 (1979), 45–71

    Assuming the Continuum Hypothesis, we interpret the theory of (the cardinal of) the continuum with quantification over constructible (monadic, dyadic, etc.) predicates in the monadic (second-order) theory of real line, in the monadic theory of any other short non-modest chain, in the monadic topology of Cantor's Discontinuum and some other monadic theories. We exhibit monadic sentences defining the real line up to isomorphism under some set-theoretic assumptions. There are some other results.

28. Yuri Gurevich: Modest theory of short chains, I. JSL 44 (1979), 481–490

    The composition (or decomposition) method of Feferman-Vaught is generalized and made much more applicable.

29. Yuri Gurevich, Saharon Shelah: Modest theory of short chains, II. JSL 44 (1979), 491–502

    We analyze the monadic theory of the rational line and the theory of the real line with quantification over "small" subsets. The results are in some sense the best possible.

30. Yuri Gurevich: Two notes on formalized topology. Fundamenta Mathematicae 57 (1980), 145–148

31. Yuri Gurevich, W. Charles Holland: Recognizing the real line. Transactions of American Math. Society 265 (1981), 527–534

    We exhibit a first-order statement about the automorphism group of the real line that characterizes the real line among all homogeneous chains.

32. Andrew M. W. Glass, Yuri Gurevich, W. Charles Holland, Saharon Shelah: Rigid homogeneous chains. Math. Proceedings of Cambridge Phil. Society 89 (1981), 7–17

33. Andrew M. W. Glass, Yuri Gurevich, W. Charles Holland, Michèle Jambu-Giraudet: Elementary theory of automorphism groups of doubly homogeneous chains. Springer Lecture Notes in Mathematics 859 (1981), 67–82

34. Yuri Gurevich: Crumbly spaces. Sixth International Congress for Logic, Methodology and Philosophy of Science (1979) North-Holland (1982), 179–191

    Answering a question of Henson, Jockush, Rubel and Takeuti, we prove that the rationals, the irrationals and the Cantor set are all elementarily equivalent as topological spaces.

35. Stal O. Aanderaa, Egon Börger, Yuri Gurevich: Prefix classes of Krom formulas with identity. Archiv für Math. Logik und Grundlagenforschung 22 (1982), 43–49

36. Yuri Gurevich: Existential interpretation, II. Archiv für Math. Logik und Grundlagenforschung 22 (1982), 103–120

37. Yuri Gurevich, Saharon Shelah: Monadic theory of order and topology in ZFC. Annals of Mathematical Logic 23 (1982), 179–198

    In the 1975 Annals of Mathematics, Shelah interpreted true first-order arithmetic in the monadic theory of order under the assumption of the continuum hypothesis. The assumption is removed here.

38. Ilya Gertsbakh, Yuri Gurevich: Homogeneous optimal fleet. Transportation Research 16B (1982), 459–470

39. Yuri Gurevich: A review of two books on the decision problem. Bulletin of the American Mathematical Society 7 (1982), 273–277

40. Yuri Gurevich, Leo Harrington: Automata, trees, and games. 14th Annual Symposium on Theory of Computing, ACM (1982), 60–65

    We prove a forgetful determinacy theorem saying that, for a wide class of infinitary games, one of the players has a winning strategy that is virtually memoryless: the player has to remember only boundedly many bits of information. We use forgetful determinacy to give a transparent proof of Rabin's celebrated result that the monadic second-order theory of the infinite tree is decidable.

41. Yuri Gurevich, Harry R. Lewis: The inference problem for template dependencies. Information and Control 55 (1982), 69–79

    Answering a question of Jeffrey Ullman, we prove that the problem in the title is undecidable.

42. Andreas Blass, Yuri Gurevich: On the unique satisfiability problem. Information and Control 55 (1982), 80–88

    Papadimitriou and Yannakakis were interested whether Unique Sat is hard for $\{L - L' : L, L' \in NP\}$ when NP differs from co-NP (otherwise the answer is obvious). We show that this is true under one oracle and false under another.

43. Edmund M. Clarke, Nissim Francez, Yuri Gurevich, A. Prasad Sistla: Can message buffers be characterized in linear temporal logic? Symposium on Principles of Distributed Computing, ACM (1982), 148–156

    In the case of unbounded buffers, the negative answer follows from a result in [28].

44. Yuri Gurevich: Decision problem for separated distributive lattices. JSL 48 (1983), 193–196

It is well known that for all recursively enumerable sets X_1, X_2 there are disjoint recursively enumerable sets Y_1, Y_2 such that $Y_i \subseteq X_i$ and $(Y_1 \cup Y_2) = (X_1 \cup X_2)$. Alistair Lachlan called distributive lattices satisfying this property *separated.* He proved that the first-order theory of finite separated distributive lattices is decidable. We prove here that the first-order theory of all separated distributive lattices is undecidable.

45. Yuri Gurevich, Menachem Magidor, Saharon Shelah: The monadic theory of ω_2. JSL 48 (1983), 387–398

In a series of papers, Büchi proved the decidability of the monadic (second-order) theory of ω_0, of all countable ordinals, of ω_1, and finally of all ordinals $< \omega_2$. Here, assuming the consistency of a weakly compact cardinal, we prove that, in different set-theoretic worlds, the monadic theory of ω_2 may be arbitrarily difficult (or easy).

46. Yuri Gurevich, Saharon Shelah: Interpreting second-order logic in the monadic theory of order. JSL 48 (1983), 816–828

Under a weak set-theoretic assumption, we interpret full second-order logic in the monadic theory of order.

47. Yuri Gurevich, Saharon Shelah: Rabin's Uniformization Problem. JSL 48 (1983), 1105–1119

The negative solution is given.

48. Yuri Gurevich, Saharon Shelah: Random models and the Gödel case of the decision problem. JSL 48 (1983), 1120–1124

We replace Gödel's sophisticated combinatorial argument with a simple probabilistic one.

49. Andrew M. W. Glass, Yuri Gurevich: The word problem for lattice-ordered groups. Transactions of American Math. Society 280 (1983), 127–138

The problem is proven to be undecidable.

50. Yuri Gurevich: Critiquing a critique of Hoare's programming logics. Communications of ACM (May 1983), 385 (Tech. communication)

51. Yuri Gurevich: Algebras of feasible functions. 24th Annual Symposium on Foundations of Computer Science, IEEE Computer Society Press, 1983, 210–214

We prove that, under a natural interpretation over finite domains,
(i) a function is primitive recursive if and only if it is logspace computable, and
(ii) a function is general recursive if and only if it is polynomial time computable.

52. Yuri Gurevich, Peter H. Schmitt: The theory of ordered abelian groups does not have the independence property. Trans. of American Math. Society 284 (1984), 171–182

53. Yuri Gurevich, Harry R. Lewis: The word problem for cancellation semigroups with zero. JSL 49 (1984), 184–191

In 1947, Post showed the word problem for semigroups to be undecidable. In 1950, Turing strengthened this result to cancellation semigroups, i.e. semigroups satisfying the cancellation property

(1) if $xy = xz$ or $yx = zx$ then $y = z$.
No semigroup with zero satisfies (1). The cancellation property for semigroups with zero and identity is
(2) if $xy = xz \neq 0$ or $yx = zx \neq 0$ then $y = z$.
The cancellation property for semigroups with zero but without identity is the conjunction of (2) and
(3) if $xy = x$ or $yx = x$ then $x = 0$.
Whether or not a semigroup with zero has an identity, we refer to it as a cancellation semigroup with zero if it satisfies the appropriate cancellation property. It is shown in [8] that the word problem for finite semigroups is undecidable. Here we show that the word problem is undecidable for finite cancellation semigroups with zero; this holds for semigroups with identity and also for semigroups without identity. (In fact, we prove a stronger effective inseparability result.) This provides the necessary mathematical foundation for [41]).

54. Yuri Gurevich, Larry J. Stockmeyer, Uzi Vishkin: Solving NP-hard problems on graphs that are almost trees, and an application to facility location problems. Journal of the ACM 31 (1984), 459–473

Imagine that you need to put service stations (or MacDonald's restaurants) on roads in such a way that every resident is within, say, 10 miles of the nearest station. What is the minimal number of stations and how does one find an optimal placement? In general, the problem is NP hard; however in important special cases there are feasible solutions.

55. Andreas Blass, Yuri Gurevich: Equivalence relations, invariants, and normal forms. SIAM Journal on Computing 13 (1984), 682–689

For an equivalence relation E on the words in some finite alphabet, we consider the following four problems.

Recognition. Decide whether two words are equivalent.

Invariant. Calculate a function constant on precisely the equivalence classes.

Normal form. Calculate a particular member of an equivalence class, given an arbitrary member.

First member. Calculate the first member of an equivalence class, given an arbitrary member.

A solution for any of these problems yields solutions for all earlier ones in the list. We show that, for polynomial time recognizable E, the first member problem is always in the class Δ_2^P (solvable in polynomial time with an oracle for an NP set) and can be complete for this class even when the normal form problem is solvable in polynomial time. To distinguish between the other problems in the list, we construct an E whose invariant problem is not solvable in polynomial time with an oracle for E (although the first member problem is in $\mathrm{NP}^E \cap \text{co-NP}^E$), and we construct an E whose normal form problem is not solvable in polynomial time with an oracle for a certain solution of its invariant problem.

56. Andreas Blass, Yuri Gurevich: Equivalence relations, invariants, and normal forms, II. Springer LNCS 171 (1984), 24–42

We consider the questions whether polynomial time solutions for the easier problems of the list for [55] yield NP solutions for the harder ones, or vice versa. We show that affirmative answers to several of these questions are equivalent to natural principles like NP = co-NP, (NP $\cap$ co-NP) = P, and the shrinking principle for NP sets. We supplement known oracles with enough new ones to show that all questions considered have negative answers relative to some oracles. In other words, these questions cannot be answered affirmatively by means of relativizable polynomial-time Turing reductions. Finally, we show that the analogous questions in the framework where Borel sets play the role of polynomial time decidable sets have negative answers.

57. Yuri Gurevich, Saharon Shelah: The monadic theory and the 'next world'. Israel Journal of Mathematics 49 (1984), 55–68

Let r be a Cohen real over a model V of ZFC. Then the second-order $V[r]$-theory of the integers (even the reals if V satisfies CH) is interpretable in the monadic V-theory of the real line. Contrast this with the result of [79].

58. Warren D. Goldfarb, Yuri Gurevich, Saharon Shelah: A decidable subclass of the minimal Gödel case with identity. JSL Logic 49 (1984), 1253–1261

59. Yuri Gurevich, Harry R. Lewis: A logic for constant depth circuits. Information and Control 61 (1984), 65–74

We present an extension of first-order logic that captures precisely the computational complexity of (the uniform sequences of) constant-depth polynomial-time circuits.

60. Yuri Gurevich: Toward logic tailored for computational complexity. In: M. Richter et al. (eds.) Computation and Proof Theory, Springer Lecture Notes in Math. 1104 (1984), 175–216

The pathos of this paper is that classical logic, developed to confront the infinite, is ill prepared to deal with finite structures, whereas finite structures, e.g. databases, are of so great importance in computer science. We show that famous theorems about first-order logic fail in the finite case, and discuss various alternatives to classical logic. The message has been heard.

60.5. Yuri Gurevich: Reconsidering Turing's thesis (toward more realistic semantics of programs). Technical report CRL-TR-36-84 University of Michigan, September 1984

The earliest publication on the abstract state machine project.

61. John P. Burgess, Yuri Gurevich: The decision problem for linear temporal logic. Notre Dame JSL 26 (1985), 115–128

The main result is the decidability of the temporal theory of the real order.

62. Yuri Gurevich, Saharon Shelah: To the decision problem for branching time logic. In: P. Weingartner and G. Dold (eds.) Foundations of Logic and Linguistics: Problems and their Solutions, Plenum (1985), 181–198

63. Yuri Gurevich, Saharon Shelah: The decision problem for branching time logic. JSL 50 (1985), 668–681

Define a tree to be any partial order satisfying the following requirement: the predecessors of any element x are linearly ordered, i.e. if ($y < x$ and $z < x$) then ($y < z$ or $y = z$ or $y > z$). The main result of the two papers [62,63] is the decidability of the theory of trees with additional unary predicates and quantification over nodes and branches. This gives the richest decidable temporal logic.

64. Yuri Gurevich: Monadic second-order theories. In: J. Barwise and S. Feferman (eds.) Model-Theoretical Logics, Springer-Verlag, Perspectives in Mathematical Logic (1985), 479–506

In this chapter we make a case for the monadic second-order logic (that is to say, for the extension of first-order logic allowing quantification over monadic predicates) as a good source of theories that are both expressive and manageable. We illustrate two powerful decidability techniques here. One makes use of automata and games. The other is an offshoot of a composition theory where one composes models as well as their theories. Monadic second-order logic appears to be the most natural match for the composition theory.

Undecidability proofs must be thought out anew in this area; for, whereas true first-order arithmetic is reducible to the monadic theory of the real line R, it is nevertheless not interpretable in the monadic theory of R. A quite unusual undecidability method is another subject of this chapter.

In the last section we briefly review the history of the methods thus far developed and mention numerous results obtained using the methods.

64.5. Yuri Gurevich: A new thesis. Abstracts, American Mathematical Society 6:4 (August 1985), p. 317, abstract 85T-68-203

The first announcement of the "new thesis", later known as the *Abstract State Machine thesis.*

65. Andreas Blass, Yuri Gurevich, Dexter Kozen: A zero-one law for logic with a fixed-point operator. Information and Control 67 (1985), 70–90

The zero-one law, known to hold for first-order logic but not for monadic or even existential monadic second-order logic, is generalized to the extension of first-order logic by the least (or iterative) fixed-point operator. We also show that the problem of deciding, for any π, whether it is almost-sure is complete for exponential time, if we consider only π's with a fixed finite vocabulary (or vocabularies of bounded arity) and complete for double-exponential time if π is unrestricted.

66. Andreas Blass, Yuri Gurevich: Henkin quantifiers and complete problems. Annals of Pure and Applied Logic 32 (1986), 1–16

We show that almost any non-linear quantifier, applied to quantifier-free first-order formulas, suffices to express an NP-complete predicate; the remaining non-linear quantifiers express exactly co-NL predicates (NL is Nondeterministic Logspace).

67. Larry Denenberg, Yuri Gurevich, Saharon Shelah: Definability by constant-depth polynomial-size circuits. Information and Control 70 (1986), 216–240

We investigate the expressive power of constant-depth polynomial-size circuit models. In particular, we construct a circuit model whose expressive power is precisely that of first-order logic.

68. Amnon Barak, Zvi Drezner, Yuri Gurevich: On the number of active nodes in a multicomputer system. Networks 16 (1986), 275– 282

Simple probabilistic algorithms enable each active node to find estimates of the fraction of active nodes in the system of n nodes (with a direct communication link between any two nodes) in time $o(n)$.

69. Yuri Gurevich: What does $O(n)$ mean? SIGACT NEWS 17:4 (1986), 61–63

70. Yuri Gurevich, Saharon Shelah: Fixed-point extensions of first-order logic. Annals of Pure and Applied Logic 32 (1986), 265–280

We prove that the three extensions of first-order logic by means of positive, monotone and inflationary inductions have the same expressive power in the case of finite structures. An extended abstract of the above, in Proc. 26th Annual Symposium on Foundation of Computer Science, IEEE Computer Society Press (1985), 346–353, contains some additions.

71. Yuri Gurevich, Saharon Shelah: Expected computation time for Hamiltonian Path Problem. SIAM J. on Computing 16:3 (1987), 486–502

Let $G(n,p)$ be a random graph with n vertices and the edge probability p. We give an algorithm for Hamiltonian Path Problem whose expected run-time on $G(n,p)$ is $cn/p+o(n)$ for any fixed p. This is the best possible result for the case of fixed-edge probability. The expected run-time of a slightly modified version of the algorithm remains polynomial if $p = p(n) > n - c$ where c is positive and small.

The paper is based on a 1984 technical report.

72. Miklós Ajtai, Yuri Gurevich: Monotone versus positive. Journal of the ACM 34 (1987), 1004–1015

A number of famous theorems about first-order logic were disproved in [60] in the case of finite structures, but Lyndon's theorem on monotone vs. positive resisted the attack. It is defeated here. The counterexample gives a uniform sequence of constant-depth polynomial-size (functionally) monotone boolean circuits not equivalent to any (however nonuniform) sequence of constant-depth polynomial-size positive boolean circuits.

73. Andreas Blass, Yuri Gurevich: Existential fixed-point logic. In: E. Börger (ed.) Logic and complexity, Springer LNCS 270 (1987), 20–36

The purpose of this paper is to draw attention to existential fixed-point logic (EFPL). Among other things, we show the following.

- If a structure A satisfies an EFPL formula φ then A has a finite subset F such that every structure that coincides with A on F satisfies φ.
- Using EFPL instead of first-order logic removes the expressivity hypothesis in Cook's completeness theorem for Hoare logic.
- In the presence of a successor relation, EFPL captures polynomial time.

74. Yuri Gurevich: Logic and the challenge of computer science. In: E. Börger (ed.) Current Trends in Theoretical Computer Science, Computer Science Press (1988), 1–57

The chapter consists of two quite different parts. The first part is a survey (including some new results) on finite model theory. One particular point deserves a special attention. In computer science, the standard computation model is the Turing machine whose inputs are strings; other algorithm inputs are supposed to be encoded with strings. However, in combinatorics, database theory, etc., one usually does not distinguish between isomorphic structures (graphs, databases, etc.). For example, a database query should provide information about the database rather than its implementation. In such cases, there is a problem with string presentation of input objects: there is no known, easily computable string encoding of isomorphism classes of structures. Is there a computation model whose machines do not distinguish between isomorphic structures and compute exactly PTIME properties? The question is intimately related to a question by Chandra and Harel in "Structure and complexity of relational queries", J. Comput. and System Sciences 25 (1982), 99–128. We formalize the question as the question whether there exists a logic that captures polynomial time (without presuming the presence of a linear order) and conjecture the negative answer. The first part is based on lectures given at the 1984 Udine Summer School on Computation Theory and summarized in the technical report "Logic and the Challenge of Computer Science", CRL-TR-10-85, Sep. 1985, Computing Research Lab, University of Michigan, Ann Arbor, Michigan.

In the second part, we introduce a new computation model: evolving algebras (later renamed *abstract state machines*). This new approach to semantics of computations and, in particular, to semantics of programming languages emphasizes dynamic and resource-bounded aspects of computation. It is illustrated on the example of Pascal. The technical report mentioned above contained an earlier version of part 2. The final version was written in 1986.

75. Yuri Gurevich: Algorithms in the world of bounded resources. In: R. Herken (ed.) The universal Turing machine – a half-century story, Oxford University Press (1988), 407–416

In the classical theory of algorithms, one addresses a computing agent with unbounded resources. We argue in favor of a more realistic theory of multiple addressees with limited resources.

76. Yuri Gurevich: Average case completeness. J. Computer and System Sciences 42:3 (June 1991), 346–398 (a special issue with selected papers of FOCS'87)

We explain and advance Levin's theory of average case complexity. In particular, we exhibit the second natural average-case-complete problem and prove that deterministic reductions are inadequate.

77. Yuri Gurevich, James M. Morris: Algebraic operational semantics and Modula-2. CSL'87, 1st Workshop on Computer Science Logic, Springer LNCS 329 (1988), 81–101

Jim Morris was a PhD student of Yuri Gurevich at the Electrical Engineering and Computer Science Department of the University of Michigan, the first PhD student working on the abstract state machine project. This is an extended abstract of Jim Morris's 1988 PhD thesis (with the same title) and the first example of the ASM semantics of a whole programming language.

78. Yuri Gurevich: On Kolmogorov machines and related issues. Originally in BEATCS 35 (June 1988), 71–82. Reprinted in: Current Trends in Theoretical Computer Science. World Scientific (1993), 225–234

One contribution of the article was to formulate the Kolmogorov-Uspensky thesis. In "To the Definition of an Algorithm" [Uspekhi Mat. Nauk 13:4 (1958), 3–28 (Russian)], Kolmogorov and Uspensky wrote that they just wanted to comprehend the notions of computable functions and algorithms, and to convince themselves that there is no way to extend the notion of computable function. In fact, they did more than that. It seems that their thesis was this:

> Every computation, performing only one restricted local action at a time, can be viewed as (not only being simulated by, but actually being) the computation of an appropriate KU machine (in the more general form).

Uspensky agreed [J. Symb. Logic 57 (1992), p. 396]. Another contribution of the paper was a popularization of the following beautiful theorem of Leonid Levin.

> Theorem. For every computable function $F(w) = x$ from binary strings to binary strings, there exists a KU algorithm A such that A conclusively inverts F and (Time of A on x) $= O$(Time of B on x) for every KU algorithm B that conclusively inverts F.

which had been virtually unknown, partially because it appeared (without a proof) in his article "Universal Search Problems" [Problems of Information Transmission 9:3 (1973), 265–266] which is hard to read.

79. Yuri Gurevich, Saharon Shelah: On the strength of the interpretation method. JSL 54:2 (1989), 305–323

The interpretation method is the main tool in proving negative results related to logical theories. We examine the strength of the interpretation method and find a serious limitation. In one of our previous papers [57], we were able to reduce true arithmetic to the monadic theory of real line. Here we show that true arithmetic cannot be interpreted in the monadic theory of the real line. The reduction of [57] is not an interpretation.

80. Yuri Gurevich, Saharon Shelah: Time polynomial in input or output. JSL 54:3 (1989), 1083–1088

There are simple algorithms with large outputs; it is misleading to measure the time complexity of such algorithms in terms of inputs only. In this connection, we introduce the class PIO of functions computable in time polynomial in the maximum of the size of input and the size of output, and some other similar classes. We observe that there is no notation system for any extension of the class of total functions computable on Turing machines in time linear in output and give a machine-independent definition of partial PIO functions.

81. Andreas Blass, Yuri Gurevich: On Matiyasevich's non-traditional approach to search problems. Information Processing Letters 32 (1989), 41–45

Yuri Matijasevich, famous for completing the solution of Hilbert's tenth problem, suggested to use differential equations inspired by real phenomena in nature to solve the satisfiability problem for boolean formulas. The initial conditions are chosen at random and it is expected that, in the case of a satisfiable formula, the process, described by differential equations, converges quickly to an equilibrium which yields a satisfying assignment. A success of the program would establish NP=R. Attracted by the approach, we discover serious complications with it.

82. Yuri Gurevich, Saharon Shelah: Nearly linear time. Symposium on Logical Foundations of Computer Science in Pereslavl-Zalessky, USSR, Springer LNCS 363 (1989) 108–118

The notion of linear time is very sensitive to the machine model. In this connection we introduce and study the class NLT of functions computable in nearly linear time $n \cdot (\log n)^{O(1)}$ on random access computers or any other "reasonable" machine model (with the standard multitape Turing machine model being "unreasonable" for that low complexity class). This gives a very robust approximation to the notion of linear time. In particular, we give a machine-independent definition of NLT and a natural problem complete for NLT.

83. Miklós Ajtai, Yuri Gurevich: Datalog vs. first-order logic. J. of Computer and System Sciences 49:3 (December 1994), 562–588 (Extended abstract in FOCS'89, 142–147)

First-order logic and Datalog are two very important paradigms for relational-database query languages. How different are they from the point of view of expressive power? What can be expressed both in first-order logic and Datalog? It is easy to see that every existential positive first-order formula is expressible by a bounded Datalog query, and the other way round. Cosmadakis suggested that there are no other properties expressible in first-order logic and in Datalog; in other words, no unbounded Datalog query is expressible in first-order logic. We prove the conjecture; that is our main theorem. It can be seen as a kind of compactness theorem for finite structures. In addition, we give counterexamples delimiting the main result.

84. Yuri Gurevich: Infinite games. Originally in BEATCS (June 1989), 93–100. Reprinted in: Current Trends in Theoretical Computer Science. World Scientific (1993), 235–244

Infinite games are widely used in mathematical logic. Recently infinite games were used in connection to concurrent computational processes that do not necessarily terminate. For example, operating system may be seen as playing a game "against" the disruptive forces of users. The classical question of the existence of winning strategies turns out to be of importance to practice. We explain a relevant part of the infinite game theory.

85. Yuri Gurevich: The challenger-solver game: Variations on the theme of P=?NP. BEATCS (October 1989), 112–121. Reprinted in: Current Trends in Theoretical Computer Science. World Scientific (1993), 245–253

The question $P \stackrel{?}{=} NP$ is the focal point of much research in theoretical computer science. But is it the right question? We find it biased toward the positive answer. It is conceivable that the negative answer is established without providing much evidence for the difficulty of NP problems in practical terms. We argue in favor of an alternative to $P \stackrel{?}{=} NP$ based on the average-case complexity.

86. Yuri Gurevich: Games people play. In: S. Mac Lane and D. Siefkes (eds.) Collected Works of J. Richard Büchi, Springer-Verlag (1990), 517–524

87. Yuri Gurevich, Saharon Shelah: Nondeterministic linear-time tasks may require substantially nonlinear deterministic time in the case of sublinear work space. Journal of the ACM 37:3 (1990), 674–687

We develop a technique to prove time-space trade-offs and exhibit natural search problems (e.g. Log-size Clique Problem) that are solvable in linear time on polylog-space (and sometimes even log-space) nondeterministic Turing machine, but no deterministic machine (in a very general sense of this term) with sequential-access read-only input tape and work space n^σ solves the problem within time $n^{1+\tau}$ if $\sigma + 2\tau < \frac{1}{2}$.

88. Yuri Gurevich: Matrix decomposition problem is complete for the average case. FOCS'90, 31st Annual Symposium on Foundations of Computer Science, IEEE Computer Society Press (1990), 802–811

The first algebraic average-case complete problem is presented. See [97] in this connection.

89. Yuri Gurevich, Lawrence S. Moss: Algebraic operational semantics and Occam. CSL'89, 3rd Workshop on Computer Science Logic, Springer LNCS 440 (1990), 176–192

We give evolving algebra semantics to the Occam programming language, generalizing in the process evolving algebras to the case of distributed concurrent computations.

Later note: this was the first example of a distributed abstract state machine.

90. Yuri Gurevich: On finite model theory. In: S. R. Buss et al. (eds.) Feasible Mathematics, (1990), 211–219

This is a little essay on finite model theory. Section 1 gives some counterexamples to classical theorems in the finite case. Section 2 gives a finite version of the classical compactness theorem. Section 3 announces two Gurevich-Shelah results. One is a new preservation theorem that implies that a first-order formula p preserved by any homomorphism from a finite structure into another finite structure is equivalent to a positive existential formula q. The other result is a lower bound result according to which a shortest q may be non-elementary longer than p.

A later note: Unfortunately, the proof of preservation theorem fell through – a unique such case in the history of the Gurevich-Shelah collaboration – and was later proved by Benjamin Rossman; see Proceedings of LICS 2005. Rossman also provided details for our lower bound proof.

91. Yuri Gurevich: On the classical decision problem. Originally in BEATCS (October 1990), 140–150. Reprinted in: Current Trends in Theoretical Computer Science. World Scientific (1993), 254–265

92. Yuri Gurevich: Evolving algebras: An introductory tutorial. Originally in BEATCS 43 (February 1991), 264–284. This slightly revised version appeared in: Current Trends in Theoretical Computer Science. World Scientific (1993), 266–292

Computation models and specification methods seem to be worlds apart. The evolving algebra project is as an attempt to bridge the gap by improving on Turing's thesis. We seek more versatile machines able to simulate arbitrary algorithms, on their natural abstraction levels, in a direct and essentially coding-free way. The evolving algebra thesis asserts that evolving algebras are such versatile machines. Here sequential evolving algebras are defined and motivated. In addition, we sketch a speculative "proof" of the sequential evolving algebra thesis: Every sequential algorithm can be lock-step simulated by an appropriate sequential evolving algebra on the natural abstraction level of the algorithm.

93. Andreas Blass, Yuri Gurevich: On the reduction theory for average-case complexity. CSL'90, 4th Workshop on Computer Science Logic, Springer LNCS 533 (1991) 17–30

A function from instances of one problem to instances of another problem is a reduction if together with any admissible algorithm for the second problem it gives an admissible algorithm for the first problem. This is an example of a descriptive definition of reductions. We slightly simplify Levin's usable definition of deterministic average-case reductions and thus make it equivalent to the appropriate descriptive definition. Then we generalize this to randomized average-case reductions.

94. Yuri Gurevich: Average case complexity. ICALP'91, International Colloquium on Automata, Languages and Programming, Madrid, Springer LNCS 510 (1991), 615–628

We motivate, justify and survey the average-case reduction theory.

95. Yuri Gurevich: Zero-one laws. Originally in BEATCS 51 (February 1991), 90–106. Reprinted in: Current Trends in Theoretical Computer Science. World Scientific (1993), 293–309

96. Andreas Blass, Yuri Gurevich: Randomizing reductions of search problems. SIAM J. on Computing 22:5 (1993), 949–975

This is the journal version of an invited talk at FST&TCS'91, 11th Conference on Foundations of Software Technology and Theoretical Computer Science, New Delhi, India; see Springer LNCS 560 (1991), 10–24.

First, we clarify the notion of a (feasible) solution for a search problem and prove its robustness. Second, we give a general and usable notion of many-one randomizing reductions of search problems and prove that it has desirable properties. All reductions of search problems to search problems in the literature on average case complexity can be viewed as such many-one randomizing reductions. This includes those reductions in the literature that use iterations and therefore do not look many-one.

97. Andreas Blass, Yuri Gurevich: Matrix transformation is complete for the average case. SIAM J. on Computing 24:1 (1995), 3–29

This is a full paper corresponding to the extended abstract [88] by the second author. We present the first algebraic problem complete for the average case under a natural probability distribution. The problem is this: Given a unimodular matrix X of integers, a set S of linear transformations of such unimodular matrices and a natural number n, decide if there is a product of at most n (not necessarily different) members of S that takes X to the identity matrix.

98. Yuri Gurevich, James K. Huggins: The semantics of the C programming language. In E. Börger et al. (eds.) CSL'92 (Computer Science Logics), Springer LNCS 702 (1993), 274–308

The method of successive refinement is used. The observation that C expressions do not contain statements gives rise to the first evolving algebra (ealgebra) which captures the command part of C; expressions are evaluated by an oracle. The second ealgebra implements the oracle under the assumptions that all the necessary declarations have been provided and user-defined functions are evaluated by another oracle. The third ealgebra handles declarations. Finally, the fourth ealgebra revises the combination of the first three by incorporating the stack discipline; it reflects all of C. (A later note: evolving algebras are now called abstract state machines.)

99. Thomas Eiter, Georg Gottlob, Yuri Gurevich: Curb your theory! A circumscriptive approach for inclusive interpretation of disjunctive information. In: R. Bajcsy, M. Kaufman (eds.) Proc. 13th Intern. Joint Conf. on AI (IJCAI'93) (1993), 634–639

We introduce, study and analyze the complexity of a new nonmonotonic technique of common sense reasoning called curbing. Like circumscription, curbing is based on model minimality, but, unlike circumscription, it treats disjunction inclusively.

100. Yuri Gurevich: Feasible functions. London Mathematical Society Newsletter 206 (June 1993), 6–7

Some computer scientists, notably Steve Cook, identify feasibility with polynomial-time computability. We argue against that point of view. Polynomial-time computations may be infeasible, and feasible computations may be not polynomial time.

101. Yuri Gurevich: Logic in computer science. In: G. Rozenberg and A. Salomaa (eds.) Current Trends in Theoretical Computer Science, World Scientific Series in Computer Science 40 (1993), 223–394

102. Yuri Gurevich: The AMAST phenomenon. Originally in BEATCS 51 (October 1993), 295–299. Reprinted in: Current Trends in Theoretical Computer Science. World Scientific (2001), 247–253

This humorous article incorporates a bit of serious criticism of algebraic and logic approaches to software problems.

103. Yuri Gurevich: Evolving algebra 1993: Lipari guide. Specification and validation methods, Oxford University Press (1995), 9–36

Computation models and specification methods seem to be worlds apart. The project on abstract state machines (a.k.a. evolving algebras) started as an attempt to bridge the gap by improving on Turing's thesis [92]. We sought more versatile machines which would be able to simulate arbitrary algorithms, on their natural abstraction levels, in a direct and essentially coding-free way. The ASM thesis asserts that ASMs are such versatile machines. The guide provided the definition of sequential and – for the first time – parallel and distributed ASMs. The denotational semantics of sequential and parallel ASMs is addressed in the Michigan guide [129].

104. Erich Grädel, Yuri Gurevich: Tailoring recursion for complexity. JSL 60:3 (September 1995), 952–969

Complexity classes are easily generalized to the case when inputs of an algorithm are finite ordered structures of a fixed vocabulary rather than strings. A logic L is said to capture (or to be tailored to) a complexity class C if a class of finite ordered structures of a fixed vocabulary belongs to C if and only if it is definable in L. Traditionally, complexity tailored logics are logics of relations. In his FOCS'83 paper, the second author showed that, on finite structures, the class of Logspace computable functions is captured by the primitive recursive calculus, and the class of PTIME computable functions is captured by the classical calculus of partially recursive functions. Here we continue that line of investigation and construct recursive calculi for various complexity classes of functions, in particular for (more challenging) nondeterministic classes, NLogspace and NPTIME.

105. Yuri Gurevich: Logic activities in Europe. ACM SIGACT NEWS 25:2 (June 1994), 11–24

This is a critical analysis of European logic activities in computer science based on a Fall 1992 European tour sponsored by the Office of Naval Research.

106. Yuri Gurevich, Raghu Mani: Group membership protocol: Specification and verification. In: Specification and Validation Methods, Oxford University Press (1995), 295–328

An interesting and useful group membership protocol of Flavio Christian involves timing constraints, and its correctness is not obvious. We construct a mathematical model of the protocol and verify the protocol (and notice that the assumptions about the environment may be somewhat weakened).

107. Egon Börger, Dean Rosenzweig, Yuri Gurevich: The bakery algorithm: Yet another specification and verification. In: Specification and Validation Methods, Oxford University Press (1995), 231–243

The so-called bakery algorithm of Lamport is an ingenious and sophisticated distributed mutual-exclusion algorithm. First we construct a mathematical model A1 which reflects the algorithm very closely. Then we construct a more abstract model A2 where the agents do not interact and the information is provided by two oracles. We check that A2 is safe and fair provided that the oracles satisfy certain conditions. Finally we check that the implementation A1 of A2 satisfies the conditions and thus A1 is safe and fair.

108. Yuri Gurevich, Neil Immerman, Saharon Shelah: McColm's conjecture. LICS 1994, Symp. on Logic in Computer Science, IEEE Computer Society Press (1994), 10–19

Gregory McColm conjectured that, over any class K of finite structures, all positive elementary inductions are bounded if every FOL + LFP formula is equivalent to a first-order formula over K. Here FOL + LFP is the extension of first-order logic with the least fixed point operator. Our main results are two model-theoretic constructions – one deterministic and one probabilistic – each of which refutes McColm's conjecture.

109. Erich Grädel, Yuri Gurevich: Metafinite model theory. Information and Computation 140:1 (1998), 26–81. Preliminary version in D. Leivant (ed.) Logic and Computational Complexity, Selected Papers, Springer LNCS 960 (1995), 313–366

Earlier, the second author criticized database theorists for admitting arbitrary structures as databases: databases are finite structures [60]. However, a closer investigation reveals that databases are not necessarily finite. For example, a query may manipulate numbers that do not even appear in the database, which shows that a numerical structure is somehow involved. It is true nevertheless that database structures are special. The phenomenon is not restricted to databases; for example, think about the natural structure to formalize the traveling salesman problem. To this end, we define metafinite structures. Typically such a structure consists of (i) a primary part, which is a finite structure, (ii) a secondary part, which is a (usually infinite) structure, e.g. arithmetic or the real line, and (iii) a set of "weight" functions from the first part into the second. Our logics do not allow quantification over the secondary part. We study definability issues and their relation to complexity. We discuss model-theoretic properties of metafinite structures, present results on descriptive complexity, and sketch some potential applications.

110. Andreas Blass, Yuri Gurevich: Evolving algebras and linear time hierarchy. In: B. Pehrson and I. Simon (eds.) IFIP 1994 World Computer Congress, Volume I: Technology and Foundations, North-Holland, Amsterdam, 383–390

A precursor of [118].

111. Yuri Gurevich, James K. Huggins: Evolving algebras and partial evaluation. In: B. Pehrson and I. Simon (eds.) IFIP 1994 World Computing Congress, Volume 1: Technology and Foundations, Elsevier, Amsterdam, 587–592

The authors present an automated (and implemented) partial evaluator for sequential evolving algebras.

112. Yuri Gurevich: Evolving algebras. In: B. Pehrson and I. Simon (eds.) IFIP 1994 World Computer Congress, Volume I: Technology and Foundations, Elsevier, Amsterdam, 423–427

The opening talk at the first workshop on evolving algebras. Sections: Introduction, The EA Thesis, Remarks, Future Work.

113. Yuri Gurevich, Saharon Shelah: On rigid structures. JSL 61:2 (June 1996), 549–562

This is related to the problem of defining linear order on finite structures. If a linear order is definable on a finite structure A, then A is rigid (which means that its only automorphism is the identity). There had been a suspicion that if K is the collection of all finite structures of a finitely axiomatizable class and if every K structure is rigid, then K permits a relatively simple uniform definition of linear order. That happens not to be the case. The main result of the paper is a probabilistic construction of finite rigid graphs. Using that construction, we exhibit a finitely axiomatizable class of finite rigid structures (called *multipedes*) such that no $L^{\omega}_{\infty,\omega}$ sentence φ with counting quantifiers defines a linear order in all the structures. Furthermore, φ does not distinguish between a sufficiently large multipede M and a multipede M' obtained from M by moving a "shoe" to another foot of the same segment.

114. Yuri Gurevich: The value, if any, of decidability. Originally in BEATCS 55 (February 1995), 129–135. Reprinted in: Current Trends in Theoretical Computer Science, World Scientific (2001), 274–280

A decidable problem can be as hard as an undecidable one for all practical purposes. So what is the value of a mere decidability result? That is the topic discussed in the paper.

115. Thomas Eiter, Georg Gottlob, Yuri Gurevich: Normal forms for second-order logic over finite structures and classification of NP optimization problems. Annals of Pure and Applied Logic 78 (1996), 111–125

We prove a new normal form for second-order formulas on finite structures and simplify the Kolaitis-Thakur hierarchy of NP optimization problems.

116. Yuri Gurevich, James K. Huggins: The railroad crossing problem: An experiment with instantaneous actions and immediate reactions. In: H. Kleine-Büning (ed.) Computer Science Logics, Selected papers from CSL'95, Springer LNCS 1092 (1996), 266–290

We give an evolving algebra (= abstract state machine) solution for the well-known railroad crossing problem, and we use the occasion to experiment with computations where agents perform instantaneous actions in continuous time and some agents fire at the moment they are enabled.

117. Yuri Gurevich, James K. Huggins: Equivalence is in the eye of the beholder. Theoretical Computer Science 179:1–2 (1997), 353–380

In a provocative paper "Processes are in the Eye of the Beholder" in the same issue of TCS (pp. 333–351), Lamport points out "the insubstantiality of processes" by proving the equivalence of two different decompositions of the same intuitive algorithm. More exactly, each of the two distributed algorithms is described by a formula in Lamport's favorite temporal logic and then the two formulas are proved equivalent. We point out that the equivalence of algorithms is itself in the eye of the beholder. In this connection, we analyze in what sense the two distributed algorithms are and are not equivalent. Our equivalence proof is direct and does not require formalizing algorithms as logic formulas.

118. Andreas Blass, Yuri Gurevich: The linear time hierarchy theorems for RAMs and abstract state machines. Springer J. of Universal Computer Science 3:4 (April 1997), 247–278

Contrary to polynomial time, linear time badly depends on the computation model. In 1992, Neil Jones designed a couple of computation models where the linear-speed-up theorem fails and linear-time computable functions form a proper hierarchy. However, the linear time of Jones's models is too restrictive. We prove linear-time hierarchy theorems for random access machines and Gurevich's abstract state machines (formerly, evolving algebras). The latter generalization is harder and more important because of the greater flexibility of the ASM model. One long-term goal of this line of research is to prove linear lower bounds for linear time problems.

119. Yuri Gurevich, Marc Spielmann: Recursive abstract state machines. Springer J. of Universal Computer Science 3:4 (April 1997), 233–246

The abstract state machine (ASM) thesis, supported by numerous applications, asserts that ASMs express algorithms on their natural abstraction levels directly and essentially coding-free. The only objection raised to date has been that ASMs are iterative in their nature, whereas many algorithms are naturally recursive. There seems to be an inherent contradiction between (i) the ASM idea of explicit and comprehensive states, and (ii) higher level recursion with its hiding of the stack.

But consider recursion more closely. When an algorithm A calls an algorithm B, a clone of B is created and this clone becomes a slave of A. This raises the idea of treating recursion as an implicitly multi-agent computation. Slave agents come and go, and the master/slave hierarchy serves as the stack.

Building upon this idea, we suggest a definition of recursive ASMs. The implicit use of distributed computing has an important side benefit: it leads naturally to concurrent recursion. In addition, we reduce recursive ASMs to distributed ASMs. If desired, one can view recursive notation as mere abbreviation.

120. Andreas Blass, Yuri Gurevich, Saharon Shelah: Choiceless polynomial time. Annals of Pure and Applied Logic 100 (1999), 141–187

The question "Is there a computation model whose machines do not distinguish between isomorphic structures and compute exactly polynomial time properties?" became a central question of finite model theory. One of us conjectured a negative answer [74]. A related question is what portion of PTIME can be naturally captured by a computation model. (Notice that we speak about computation whose inputs are arbitrary finite structures, e.g. graphs. In a special case of ordered structures, the desired computation model is that of PTIME-bounded Turing machines.) Our idea is to capture the portion of PTIME where algorithms are not allowed arbitrary choice but parallelism is allowed and, in some cases, implements choice. Our computation model is a PTIME version of abstract state machines. Our machines are able to PTIME simulate all other PTIME machines in the literature, and they are more programmer-friendly. A more difficult theorem shows that the computation model does not capture all PTIME.

121. Scott Dexter, Patrick Doyle, Yuri Gurevich: Gurevich abstract state machines and Schoenhage storage modification machines. Springer J. of Universal Computer Science 3:4 (April 1997), 279–303

We show that, in a strong sense, Schoenhage's storage modification machines are equivalent to unary basic abstract state machines without external functions. The unary restriction can be removed if the storage modification machines are equipped with a pairing function in an appropriate way.

122. Charles Wallace, Yuri Gurevich, Nandit Soparkar: A formal approach to recovery in transaction-oriented database systems. Springer J. of Universal Computer Science 3:4 (April 1997), 320–340

Failure resilience is an essential requirement for transaction-oriented database systems, yet there has been little effort to specify and verify techniques for failure recovery formally. The desire to improve performance has resulted in algorithms of considerable sophistication, understood by few and prone to errors. In this paper, we show how the formal methodology of Gurevich's Abstract State Machines can elucidate recovery and provide formal rigor to the design of a recovery algorithm. In a series of refinements, we model recovery at several levels of abstraction, verifying the correctness of each model. This initial work indicates that our approach can be applied to more advanced recovery mechanisms.

123. Yuri Gurevich: Platonism, constructivism, and computer proofs vs. proofs by hand. Originally in BEATCS 57 (October 1995), 145–166. A slightly revised version in: Current Trends in Theoretical Computer Science, World Scientific (2001), 281–302

In one of Krylov's fables, a small dog, Moska, barks at the elephant who pays no attention whatsoever to Moska. This image comes to my mind when I think of constructive mathematics versus "classical" (that is mainstream) mathematics. In this article, we put a few words into the elephant's mouth. The idea to write such an article came to me in the summer of 1995 when I came across a fascinating 1917 bet between the constructivist Hermann Weyl and George Polya, a classical mathematician. An English translation of the bet (from German) is found in the article.

Our main objection to the historical constructivism is that it has not been sufficiently constructive. The constructivists have been obsessed with computability and have not paid sufficient attention to the feasibility of algorithms. However, the constructivists' criticism of classical mathematics has a point. Instead of dismissing constructivism offhandedly, it makes sense to come up with a positive alternative, an antithesis to historical constructivism. We believe that we have found such an alternative. In fact, it is well known and very popular in computer science, namely, the principle of separating concerns.

[Added in July 2006] The additional part on computer proofs vs. proofs by hand was a result of frustration that many computer scientists would not trust informal mathematical proofs, while many mathematicians would not trust computer proofs. I seemed obvious to me that, on the large scale, proving is not only hard but also is imperfect and has an engineering character. We need informal proofs and computer proofs and more, such as stratification, experimentation.

124. Natasha Alechina, Yuri Gurevich: Syntax vs. semantics on finite structures. In: J. Mycielski et al. (eds.) Structures in Logic and Computer Science: A Selection of Essays in Honor of Andrzej Ehrenfeucht, Springer LNCS 1261 (1997), 14–33

Logic preservation theorems often have the form of a syntax/semantics correspondence. For example, the Tarski-Łoś theorem asserts that a first-order sentence is preserved by extensions if and only if it is equivalent to an existential sentence. Many of these correspondences break when one restricts attention to finite models. In such a case, one may attempt to find a new semantical characterization of the old syntactical property or a new syntactical characterization of the old semantical property. The goal of this paper is to provoke such a study. In particular, we give a simple semantical characterization of existential formulas on finite structures.

125. Anatoli Degtyarev, Yuri Gurevich, Andrei Voronkov: Herbrand's Theorem and equational reasoning: Problems and solutions. Originally in BEATCS 60 (Oct 1996), 78–95. Reprinted in: Current Trends in Theoretical Computer Science, World Scientific (2001), 303–326

The article (written in a popular form) explains that a number of different algorithmic problems related to Herbrand's theorem happen to be equivalent. Among these problems are the intuitionistic provability problem for the existential fragment of first-order logic with equality, the intuitionistic provability problem for the prenex fragment of first-order with equality, and the simultaneous rigid E-unification problem (SREU). The article explains an undecidability proof of SREU and decidability proofs for special cases. It contains an extensive bibliography on SREU.

126. Yuri Gurevich, Margus Veanes: Logic with equality: Partisan corroboration and shifted pairing. Information and Computation 152:2 (August 1999), 205–235

Herbrand's theorem plays a fundamental role in automated theorem proving methods based on tableaux. The crucial step in procedures based on such methods can be described as the corroboration (or Herbrand skeleton) problem: given a positive integer m and a quantifier-free formula, find a valid disjunction of m instantiations of the formula. In the presence of equality (which is the case in this paper), this problem was recently shown to be undecidable. The main contributions of this paper are two theorems. The Partisan Corroboration Theorem relates corroboration problems with different multiplicities. The Shifted Pairing Theorem is a finite tree-automata formalization of a technique for proving undecidability results through direct encodings of valid Turing machine computations. The theorems are used to explain and sharpen several recent undecidability results related to the corroboration problem, the simultaneous rigid E-unification problem and the prenex fragment of intuitionistic logic with equality.

127a. Anatoli Degtyarev, Yuri Gurevich, Paliath Narendran, Margus Veanes, Andrei Voronkov: The decidability of simultaneous rigid E-unification with one variable. RTA'98, 9th Conf. on Rewriting Techniques and Applications, Tsukuba, Japan, March 30 – April 1, 1998

The title problem is proved decidable and in fact EXPTIME-complete. Furthermore, the problem becomes PTIME-complete if the number of equations is bounded by any (positive) constant. It follows that the $\forall^*\exists\forall^*$ fragment of intuitionistic logic with equality is decidable, which contrasts with the undecidability of the EE fragment [126]. Notice that simultaneous rigid E-unification with two variables and only three rigid equations is undecidable [126].

127b. Anatoli Degtyarev, Yuri Gurevich, Paliath Narendran, Margus Veanes, Andrei Voronkov: Decidability and complexity of simultaneous rigid E-unification with one variable and related results. Theoretical Computer Science 243:1–2 (August 2000), 167–184

The journal version of [127a] containing also a decidability proof for the case of simultaneous rigid E-unification when each rigid equation either contains (at most) one variable or else has a ground left-hand side and the right-hand side of the form $x = y$ where x and y are variables.

128a. Yuri Gurevich, Andrei Voronkov: Monadic simultaneous rigid E-unification and related problems. ICALP'97, 24th Intern. Colloquium on Automata, Languages and Programming, Springer LNCS 1256 (1997), 154–165

We study the monadic case of a decision problem known as simultaneous rigid E-unification. We show its equivalence to an extension of word equations. We prove decidability and complexity results for special cases of this problem.

128b. Yuri Gurevich, Andrei Voronkov: Monadic simultaneous rigid E-unification. Theoretical Computer Science 222:1–2 (1999), 133–152

The journal version of [128a].

129. Yuri Gurevich: May 1997 draft of the ASM Guide. Tech Report CSE-TR-336-97, EECS Dept, University of Michigan, 1997

The draft improves upon the ASM syntax (and appears here because it is used by the ASM community and it is not going to be published).

130. Yuri Gurevich, Alexander Rabinovich: Definability and undefinability with real order at the background. JSL 65:2 (2000), 946–958

Let R be the real order, that is the set of real numbers together with the standard order of reals. Let I be the set of integer numbers, let Y range over subsets of I, let $P(I, X)$ be a monadic second-order formula about R, and let F be the collection of all subsets X of I such that $P(I, X)$ holds in R. Even though F is a collection of subsets of I, its definition may involve quantification over reals and over sets of reals. In that sense, F is defined with the background of real order. Is that background essential or not? Maybe there is a monadic second-order formula $Q(X)$ about I that defines F (so that F is the collection of all subsets X of I such that $Q(X)$ holds in I). We prove that this is indeed the case, for any monadic second-order formula $P(I, X)$. The claim remains true if the set I of integers is replaced above with any closed subset of R. The claim fails for some open subsets.

131. Yuri Gurevich: From invariants to canonization. Originally in BEATCS 63 (October 1997). Reprinted in: Current Trends in Theoretical Computer Science, World Scientific (2001), 327–331

We show that every polynomial-time full-invariant algorithm for graphs gives rise to a polynomial-time canonization algorithm for graphs.

132. Andreas Blass, Yuri Gurevich, Vladik Kreinovich, Luc Longpré: A variation on the Zero-One Law. Information Processing Letters 67 (1998), 29–30

Given a decision problem P and a probability distribution over binary strings, do this: for each n, draw independently an instance $x(n)$ of P of length n. What is the probability that there is a polynomial time algorithm that solves all instances $x(n)$? The answer is: zero or one.

133. Erich Grädel, Yuri Gurevich, Colin Hirsch: The complexity of query reliability. PODS'98, 1998 ACM Symposium on Principles of Database Systems

We study the reliability of queries on databases with uncertain information. It turns out that $FP^{\#P}$ is the typical complexity class and that many results generalize to metafinite databases which allow one to use common SQL aggregate functions.

134. Thomas Eiter, Georg Gottlob, Yuri Gurevich: Existential second-order logic over strings. Journal of the ACM 47:1 (January 2000), 77–131

We study existential second-order logic over finite strings. For every prefix class C, we determine the complexity of the model checking problem restricted to C. In particular, we prove that, in the case of the Ackermann class, for every formula φ, there is a finite automaton A that solves the model checking problem for φ.

135. Andreas Blass, Yuri Gurevich: The logic of choice. JSL 65:3 (September 2000), 1264–1310

We study extensions of first-order logic with the choice construct (choose x : $\varphi(x)$). We prove some results about Hilbert's epsilon operator, but in the main part of the paper we consider the case when all choices are independent.

136. Yuri Gurevich: The sequential ASM thesis. Originally in BEATCS 67 (February 1999), 98–124. Reprinted in: Current Trends in Theoretical Computer Science, World Scientific (2001), 363–392

The thesis is that every sequential algorithm, on any level of abstraction, can be viewed as a sequential abstract state machine. (Abstract state machines, ASMs, used to be called evolving algebras.) The sequential ASM thesis and its extensions inspired diverse applications of ASMs. The early applications were driven, at least partially, by the desire to test the thesis. Different programming languages were the obvious challenges. (A programming language L can be viewed as an algorithm that runs a given L program on given data.) From there, applications of (not necessarily sequential) ASMs spread into many directions. So far, the accumulated experimental evidence seems to support the sequential thesis. There is also a speculative philosophical justification of the thesis. It was barely sketched in the literature, but it was discussed at much greater length in numerous lectures of mine. Here I attempt to write down some of those explanations. This article does not presuppose any familiarity with ASMs.

A later note: [141] is a much revised and polished journal version.

137. Giuseppe Del Castillo, Yuri Gurevich, Karl Stroetmann: Typed abstract state machines. Unfinished manuscript (1998)

This manuscript was never published. The work, done sporadically in 1996–98, was driven by the enthusiasm of Karl Stroetmann of Siemens. Eventually he was reassigned away from ASM applications, and the work stopped. The item wasn't removed from the list because some of its explorations may be useful. (An additional minor reason was to avoid changing the numbers of the subsequent items.)

138. Yuri Gurevich, Dean Rosenzweig: Partially ordered runs: A case study. In: Abstract State Machines: Theory and Applications, Springer LNCS 1912 (2000), 131–150

We look at some sources of insecurity and difficulty in reasoning about partially ordered runs of distributed abstract state machines, and propose some techniques to facilitate such reasoning. As a case study, we prove in detail correctness and deadlock–freedom for general partially ordered runs of distributed ASM models of Lamport's Bakery Algorithm.

139. Andreas Blass, Yuri Gurevich, Jan Van den Bussche: Abstract state machines and computationally complete query languages. Information and Computation 174:1 (2002), 20–36. An earlier version in: Abstract State Machines: Theory and Applications, Springer LNCS 1912 (2000), 22–33

Abstract state machines (ASMs) form a relatively new computation model holding the promise that they can simulate any computational system in lockstep. In particular, an instance of the ASM model has recently been introduced for computing queries to relational databases [120]. This model, to which we refer as the BGS model, provides a powerful query language in which all computable queries can be expressed. In this paper, we show that when one is only interested in polynomial-time computations, BGS is strictly more powerful than both QL and WHILE_NEW, two well-known computationally complete query languages. We then show that when a language such as WHILE_NEW is extended with a duplicate elimination mechanism, polynomial-time simulations between the language and BGS become possible.

140. Yuri Gurevich, Wolfram Schulte, Charles Wallace: Investigating Java concurrency using abstract state machines. In: Abstract State Machines: Theory and Applications, Springer LNCS 1912 (2000), 151–176

We present a mathematically precise, platform-independent model of Java concurrency using the Abstract State Machine method. We cover all aspects of Java threads and synchronization, gradually adding details to the model in a series of steps. We motivate and explain each concurrency feature, and point out subtleties, inconsistencies and ambiguities in the official, informal Java specification.

141. Yuri Gurevich: Sequential abstract state machines capture sequential algorithms. ACM TOCL 1:1 (July 2000), 77–111

What are sequential algorithms exactly? Our claim, known as the sequential ASM thesis, has been that, as far as behavior is concerned, sequential algorithms are exactly sequential abstract state machines: For every sequential algorithm A, there is a sequential abstract state machine B that is behaviorally identical to A. In particular, B simulates A step for step. In this paper we prove the sequential ASM thesis, so that it becomes a theorem. But how can one possibly prove a thesis? Here is what we do. We formulate three postulates satisfied by all sequential algorithms (and, in particular, by sequential abstract state machines). This leads to the following definition: a sequential algorithm is any object that satisfies the three postulates. At this point the thesis becomes a precise statement. And we prove the statement.

This is a non-dialog version of the dialog [136]. An intermediate version was published in MSR-TR-99-65

141a. Yuri Gurevich: Sequential abstract state machines capture sequential algorithms. Russian translation of [141], by P.G. Emelyanov. In: Marchuk A.G. (ed.) Formal Methods and Models of Informatics, System Informatics 9 (2004), 7–50, Siberian Branch of the Russian Academy of Sciences

142. Andreas Blass, Yuri Gurevich: The underlying logic of Hoare logic. Originally in BEATCS 70 (February 2000), 82–110. Reprinted in: Current Trends in Theoretical Computer Science, World Scientific (2001), 409–436

Formulas of Hoare logic are asserted programs $\varphi\ P\ \psi$ where P is a program and φ, ψ are assertions. The language of programs varies; in the 1980 survey by Krzysztof Apt, one finds the language of while programs and various extensions of it. But the assertions are traditionally expressed in first-order logic (or extensions of it). In that sense, first-order logic is the underlying logic of Hoare logic. We question the tradition and demonstrate, on the simple example of while programs, that alternative assertion logics have some advantages. For some natural assertion logics, the expressivity hypothesis in Cook's completeness theorem is automatically satisfied.

143. Andreas Blass, Yuri Gurevich: Background, reserve, and Gandy machines. In: P. Clote and H. Schwichtenberg (eds.) CSL'2000, Springer LNCS 1862 (2000), 1–17

Algorithms often need to increase their working space, and it may be convenient to pretend that the additional space was really there all along but was not previously used. In particular, abstract state machines have, by definition [103], an infinite reserve. Although the reserve is a naked set, it is often desirable to have some external structure over it. For example, in [120] every state was required to include all finite sets of its atoms, all finite sets of these, etc. In this connection, we define the notion of a background class of structures. Such a class specifies the constructions (like finite sets or lists) available as "background" for algorithms.

The importation of reserve elements must be non-deterministic, since an algorithm has no way to distinguish one reserve element from another. But this sort of non-determinism is much more benign than general non-determinism. We capture this intuition with the notion of inessential non-determinism. Alternatively, one could insist on specifying a particular one of the available reserve elements to be imported. This is the approach used in [Robin Gandy, "Church's thesis and principles for mechanisms". In: J. Barwise et al. (eds.) The Kleene Symposium, North-Holland, 1980, 123–148]. The price of this insistence is that the specification cannot be algorithmic. We show how to turn a Gandy-style deterministic, non-algorithmic process into a non-deterministic algorithm of the sort described above, and we prove that Gandy's notion of "structural" for his processes corresponds to our notion of "inessential non-determinism."

144. Andreas Blass, Yuri Gurevich: Choiceless polynomial time computation and the Zero-One Law. In: P. Clote and H. Schwichtenberg (eds.) CSL'2000, Springer LNCS 1862 (2000), 18–40

This paper is a sequel to [120], a commentary on [Saharon Shelah (#634) "Choiceless polynomial time logic: inability to express", same proceedings], and an abridged version of [149] that contains complete proofs of all the results presented here. The BGS model of computation was defined in [120] with the intention of modeling computation with arbitrary finite relational structures as inputs, with essentially arbitrary data structures, with parallelism, but without arbitrary choices. It was shown that choiceless polynomial time, the complexity class defined by BGS programs subject to a polynomial time bound, does not contain the parity problem. Subsequently, Shelah proved a zero-one law for choiceless-polynomial-time properties. A crucial difference from the earlier results is this: Almost all finite structures have no non-trivial automorphisms, so symmetry considerations cannot be applied to them. Shelah's proof therefore depends on a more subtle concept of partial symmetry.

After struggling for a while with Shelah's proof, we worked out a presentation which we hope will be helpful for others interested in Shelah's ideas. We also added some related results, indicating the need for certain aspects of the proof and clarifying some of the concepts involved in it. Unfortunately, this material is not yet fully written up. The part already written, however, exceeds the space available to us in the present volume. We therefore present here an abridged version of that paper and promise to make the complete version available soon.

145. Mike Barnett, Egon Börger, Yuri Gurevich, Wolfram Schulte, Margus Veanes: Using abstract state machines at Microsoft: A case study. In: P. Clote and H. Schwichtenberg (eds.) CSL'2000, Springer LNCS 1862 (2000), 367–379

Our goal is to provide a rigorous method, clear notation and convenient tool support for high-level system design and analysis. For this purpose we use abstract state machines (ASMs). Here we describe a particular case study: modeling a debugger of a stack based runtime environment. The study provides evidence for ASMs being a suitable tool for building *executable* models of software systems on various abstraction levels, with precise refinement relationships connecting the models. High level ASM models of proposed or existing programs can be used throughout the software development cycle. In particular, ASMs can be used to model inter-component behavior on any desired level of detail. This allows one to specify application programming interfaces more precisely than it is done currently.

145.5. Colin Campbell, Yuri Gurevich: Table ASMs. In: Formal Methods and Tools for Computer Science, Eurocast 2001, eds. R. Moreno-Diaz and A. Quesada-Arencibia, Universidad de Las Palmas de Gran Canaria, Canary Islands, Spain (February 2001), 286–290

Ideally, a good specification becomes the basis for implementing, testing and documenting the system it defines. In practice, producing a good specification is hard. Formal methods have been shown to be helpful in strengthening the meaning of specifications, but despite their power, few development teams have successfully incorporated them into their software processes. This experience indicates that producing a usable formal method is also hard.

This paper is the story of how a particular theoretical result, namely the normal forms of Abstract State Machines, motivated a genuinely usable form of specification that we call ASM Tables. We offer it for two reasons. The first is that the result is interesting in and of itself and – it is to be hoped – useful to the reader. The second is that our result serves as a case study of a more general principle, namely, that in bringing rigorous methods into everyday practice, one should not follow the example of Procrustes: we find that it is indeed better to adapt the bed to the person than the other way round. We also offer a demonstration that an extremely restricted syntactical form can still contain sufficient expressive power to describe all sequential machines.

146. Andreas Blass, Yuri Gurevich: Inadequacy of computable loop invariants. ACM TOCL 2:1 (January 2001), 1–11

Hoare logic is a widely recommended verification tool. There is, however, a problem of finding easily-checkable loop invariants; it is known that decidable assertions do not suffice to verify WHILE programs, even when the pre- and post-conditions are decidable. We show here a stronger result: decidable invariants do not suffice to verify single-loop programs. We also show that this problem arises even in extremely simple contexts. Let N be the structure consisting of the set of natural numbers together with the functions $S(x) = x + 1$, $D(x) = 2x$ and function $H(x)$ that is equal to $x/2$ rounded down. There is a single-loop program P using only three variables x, y, z such that the asserted program

$$x = y = z = 0 \; \{P\} \; \mathit{false}$$

is partially correct on N but any loop invariant $I(x, y, z)$ for this asserted program is undecidable.

147. Yuri Gurevich, Alexander Rabinovich: Definability in rationals with real order in the background. Journal of Logic and Computation 12:1 (2002), 1–11

The paper deals with logically definable families of sets of rational numbers. In particular, we are interested whether the families definable over the real line with a unary predicate for the rationals are definable over the rational order alone. Let $\varphi(X, Y)$ and $\psi(Y)$ range over formulas in the first-order monadic language of order. Let Q be the set of rationals and F be the family of subsets J of Q such that $\varphi(Q, J)$ holds over the real line. The question arises whether, for every formula φ, the family F can be defined by means of a formula $\psi(Y)$ interpreted over the rational order. We answer the question negatively. The answer remains negative if the first-order logic is strengthened to weak monadic second-order logic. The answer is positive for the restricted version of monadic second-order logic where set quantifiers range over open sets. The case of full monadic second-order logic remains open.

148. Andreas Blass, Yuri Gurevich: A new zero-one law and strong extension axioms. Originally in BEATCS 72 (October 2000), 103–122. Reprinted in: Current Trends in Theoretical Computer Science, World Scientific (2004), 99–118

This article is a part of the continuing column on Logic in Computer Science. One of the previous articles in the column was devoted to the zero-one laws for a number of logics playing prominent role in finite model theory: first-order logic FO, the extension FO+LFP of first-order logic with the least fixed-point operator, and the infinitary logic where every formula uses finitely many variables [95]. Recently Shelah proved a new, powerful, and surprising zero-one law. His proof uses so-called strong extension axioms. Here we formulate Shelah's zero-one law and prove a few facts about these axioms. In the process we give a simple proof for a "large deviation" inequality à la Chernoff.

149. Andreas Blass, Yuri Gurevich: Strong extension axioms and Shelah's zero-one law for choiceless polynomial time. JSL 68:1 (2003), 65–131

This paper developed from Shelah's proof of a zero-one law for the complexity class "choiceless polynomial time," defined by Shelah and the authors. We present a detailed proof of Shelah's result for graphs, and describe the extent of its generalizability to other sorts of structures. The extension axioms, which form the basis for earlier zero-one laws (for first-order logic, fixed-point logic, and finite-variable infinitary logic) are inadequate in the case of choiceless polynomial time; they must be replaced by what we call the strong extension axioms. We present an extensive discussion of these axioms and their role both in the zero-one law and in general. ([144] is an abridged version of this paper, and [148] is a popular version of this paper.)

150. Andreas Blass, Yuri Gurevich, Saharon Shelah: On polynomial time computation over unordered structures. JSL 67:3 (2002), 1093–1125

This paper is motivated by the question whether there exists a logic capturing polynomial time computation over unordered structures. We consider several algorithmic problems near the border of the known, logically defined complexity classes contained in polynomial time. We show that fixpoint logic plus counting is stronger than might be expected, in that it can express the existence of a complete matching in a bipartite graph. We revisit the known examples that separate polynomial time from fixpoint plus counting. We show that the examples in a paper of Cai, Fürer, and Immerman, when suitably padded, are in choiceless polynomial time yet not in fixpoint plus counting. Without padding, they remain in polynomial time but appear not to be in choiceless polynomial time plus counting. Similar results hold for the multipede examples of Gurevich and Shelah, except that their final version of multipedes is, in a sense, already suitably padded. Finally, we describe another plausible candidate, involving determinants, for the task of separating polynomial time from choiceless polynomial time plus counting.

150a. Andreas Blass, Yuri Gurevich: A quick update on the open problems in Article [150] (December 2005).

151. Yuri Gurevich: Logician in the Land of OS: Abstract state machines at Microsoft. LICS 2001, IEEE Symp. on Logic in Computer Science, IEEE Computer Society (2001), 129–136

Analysis of foundational problems like "What is computation?" leads to a sketch of the paradigm of abstract state machines (ASMs). This is followed by a brief discussion on ASMs applications. Then we present some theoretical problems that bridge between the traditional LICS themes and abstract state machines.

152. Anuj Dawar, Yuri Gurevich: Fixed point logics. The Bulletin of Symbolic Logic 8:1 (2002), 65–88

Fixed-point logics are extensions of first-order predicate logic with fixed point operators. A number of such logics arose in finite model theory but they are of interest to much larger audience, e.g. AI, and there is no reason why they should be restricted to finite models. We review results established in finite model theory, and consider the expressive power of fixed-point logics on infinite structures.

153. Uwe Glässer, Yuri Gurevich, Margus Veanes: Universal plug and play machine models. MSR-TR-2001-59

Recently, Microsoft took a lead in the development of a standard for peer-to-peer network connectivity of various intelligent appliances, wireless devices and PCs. It is called the Universal Plug and Play Device Architecture (UPnP). We construct a high-level Abstract State Machine (ASM) model for UPnP. The model is based on the ASM paradigm for distributed systems with real-time constraints and is executable in principle. For practical execution, we use AsmL, the Abstract state machine Language, developed at Microsoft Research and integrated with Visual Studio and COM. This gives us an AsmL model, a refined version of the ASM model. The third part of this project is a graphical user interface by means of which the runs of the AsmL model are controlled and inspected at various levels of detail as required for simulation and conformance testing, for example.

154. Wolfgang Grieskamp, Yuri Gurevich, Wolfram Schulte, Margus Veanes: Generating finite state machines from abstract state machines. ISSTA 2002, International Symposium on Software Testing and Analysis, ACM Software Engineering Notes 27:4 (2002), 112–122

We give an algorithm that derives a finite state machine (FSM) from a given abstract state machine (ASM) specification. This allows us to integrate ASM specs with the existing tools for test-case generation from FSMs. ASM specs are executable, but have typically too many, often infinitely many, states. We group ASM states into finitely many hyperstates, which are the nodes of the FSM. The links of the FSM are induced by the ASM state transitions.

155. Yuri Gurevich, Wolfram Schulte, Margus Veanes: Toward industrial strength abstract state machines. MSR-TR-2001-98

A powerful practical ASM language, called AsmL, is being developed in Microsoft Research by the group on Foundations of Software Engineering. AsmL extends the language of original ASMs in a number of directions. We describe some of these extensions.

156. Yuri Gurevich, Nikolai Tillmann: Partial updates: Exploration. Springer J. of Universal Computer Science 7:11 (2001), 918–952

The partial update problem for parallel abstract state machines has manifested itself in the cases of counters, sets and maps. We propose a solution of the problem that lends itself to an efficient implementation and covers the three cases mentioned above. There are other cases of the problem that require a more general framework.

157-1. Andreas Blass, Yuri Gurevich: Abstract state machines capture parallel algorithms. ACM TOCL 4:4 (October 2003), 578–651

We give an axiomatic description of parallel, synchronous algorithms. Our main result is that every such algorithm can be simulated, step for step, by an abstract state machine with a background that provides for multisets. See also [157-2].

157-2. Andreas Blass, Yuri Gurevich: Abstract state machines capture parallel algorithms: Correction and extension. ACM TOCL 9:3 (June 2008), Article 19

We consider parallel algorithms working in sequential global time, for example circuits or parallel random access machines (PRAMs). Parallel abstract state machines (parallel ASMs) are such parallel algorithms, and the parallel ASM thesis asserts that every parallel algorithm is behaviorally equivalent to a parallel ASM. In an earlier paper [157-1], we axiomatized parallel algorithms, proved the ASM thesis and proved that every parallel ASM satisfies the axioms. It turned out that we were too timid in formulating the axioms; they did not allow a parallel algorithm to create components on the fly. This restriction did not hinder us from proving that the usual parallel models, like circuits or PRAMs or even alternating Turing machines, satisfy the postulates. But it resulted in an error in our attempt to prove that parallel ASMs always satisfy the postulates. To correct the error, we liberalize our axioms and allow on-the-fly creation of new parallel components. We believe that the improved axioms accurately express what parallel algorithms ought to be. We prove the parallel thesis for the new, corrected notion of parallel algorithms, and we check that parallel ASMs satisfy the new axioms.

158. Andreas Blass, Yuri Gurevich: Algorithms vs. machines. Originally in BEATCS 77 (June 2002), 96–118. Reprinted in: Current Trends in Theoretical Computer Science, World Scientific (2004), 215–236

In a recent paper, the logician Yiannis Moschovakis argues that no state machine describes mergesort on its natural level of abstraction. We do just that. Our state machine is a recursive ASM.

159. Uwe Glässer, Yuri Gurevich, Margus Veanes: Abstract communication model for distributed systems. IEEE Transactions on Software Engineering 30:7 (July 2004), 458–472

In some distributed and mobile communication models, a message disappears in one place and miraculously appears in another. In reality, of course, there are no miracles. A message goes from one network to another; it can be lost or corrupted in the process. Here we present a realistic but high-level communication model where abstract communicators represent various nets and subnets. The model was originally developed in the process of specifying a particular network architecture, namely the Universal Plug and Play architecture. But it is general. Our contention is that every message-based distributed system, properly abstracted, gives rise to a specialization of our abstract communication model. The purpose of the abstract communication model is not to design a new kind of network; rather it is to discover the common part of all message-based communication networks. The generality of the model has been confirmed by its successful reuse for very different distributed architectures. The model is based on distributed abstract state machines. It is implemented in the specification language AsmL and is being used for testing distributed systems.

160. Andreas Blass, Yuri Gurevich: Pairwise testing. Originally in BEATCS 78 (October 2002), 100–132. Reprinted in: Current Trends in Theoretical Computer Science, World Scientific (2004), 237–266

We discuss the following problem, which arises in software testing. Given some independent parameters (of a program to be tested), each having a certain finite set of possible values, we intend to test the program by running it several times. For each test, we give the parameters some (intelligently chosen) values. We want to ensure that for each pair of distinct parameters, every pair of possible values is used in at least one of the tests. And we want to do this with as few tests as possible.

161. Yuri Gurevich, Nikolai Tillmann: Partial updates. Theoretical Computer Science 336:2–3 (26 May 2005), 311–342. (A preliminary version in: Abstract State Machines 2003, Springer LNCS 2589 (2003), 57–86)

A datastructure instance, e.g. a set or file or record, may be modified independently by different parts of a computer system. The modifications may be nested. Such hierarchies of modifications need to be efficiently checked for consistency and integrated. This is the problem of partial updates in a nutshell. In our first paper on the subject [156], we developed an algebraic framework which allowed us to solve the partial update problem for some useful datastructures including counters, sets and maps. These solutions are used for the efficient implementation of concurrent data modifications in the specification language AsmL. The two main contributions of this paper are (i) a more general algebraic framework for partial updates and (ii) a solution of the partial update problem for sequences and labeled ordered trees.

162. Yuri Gurevich, Saharon Shelah: Spectra of monadic second-order formulas with one unary function. LICS 2003, 18th Annual IEEE Symp. on Logic in Computer Science, IEEE Computer Society (2003), 291–300

We prove that the spectrum of any monadic second-order formula F with one unary function symbol (and no other function symbols) is eventually periodic, so that there exist natural numbers $p > 0$ (a period) and t (a p-threshold) such that if F has a model of cardinality $n > t$ then it has a model of cardinality $n + p$.

(In the web version, some additional proof details are provided because some readers asked for them.)

163. Mike Barnett, Wolfgang Grieskamp, Yuri Gurevich, Wolfram Schulte, Nikolai Tillmann, Margus Veanes: Scenario-oriented modeling in AsmL and its instrumentation for testing. In: 2nd International Workshop on Scenarios and State Machines: Models, Algorithms, and Tools, (2003) 8–14, held at ICSE 2003, International Conference on Software Engineering 2003

We present an approach for modeling use cases and scenarios in the Abstract state machine Language and discuss how to use such models for validation and verification purposes.

164. Andreas Blass, Yuri Gurevich: Algorithms: A quest for absolute definitions. Originally in BEATCS 81 (October 2003), 195–225. Reprinted in: Current Trends in Theoretical Computer Science, World Scientific (2004), 283–311. Reprinted in: A. Olszewski et al. (eds.) Church's Thesis After 70 Years, Ontos Verlag (2006), 24–57

What is an algorithm? The interest in this foundational problem is not only theoretical; applications include specification, validation and verification of software and hardware systems. We describe the quest to understand and define the notion of algorithm. We start with the Church-Turing thesis and contrast Church's and Turing's approaches, and we finish with some recent investigations.

165. Yuri Gurevich: Abstract state machines: An overview of the project. In: D. Seipel and J. M. Turull-Torres (eds.) Foundations of Information and Knowledge Systems, Springer LNCS 2942 (2004), 6–13

We quickly survey the ASM project, from its foundational roots to industrial applications.

166. Andreas Blass, Yuri Gurevich: Ordinary interactive small-step algorithms, I. ACM TOCL 7:2 (April 2006), 363–419. A preliminary version was published as MSR-TR-2004-16

This is the first in a series of papers extending the Abstract State Machine Thesis – that arbitrary algorithms are behaviorally equivalent to abstract state machines – to algorithms that can interact with their environments during a step, rather than only between steps. In the present paper, we describe, by means of suitable postulates, those interactive algorithms that (1) proceed in discrete, global steps, (2) perform only a bounded amount of work in each step, (3) use only such information from the environment as can be regarded as answers to queries, and (4) never complete a step until all queries from that step have been answered. We indicate how a great many sorts of interaction meet these requirements. We also discuss in detail the structure of queries and replies and the appropriate definition of equivalence of algorithms. Finally, motivated by our considerations concerning queries, we discuss a generalization of first-order logic in which the arguments of function and relation symbols are not merely tuples of elements but orbits of such tuples under groups of permutations of the argument places.

167. Yuri Gurevich: Intra-step interaction. In: W. Zimmerman and B. Thalheim (eds.) Abstract State Machines 2004, Springer LNCS 3052 (2004), 1–5

For a while it seemed possible to pretend that all interaction between an algorithm and its environment occurs inter-step, but not anymore. Andreas Blass, Benjamin Rossman and the speaker are extending the Small-Step Characterization Theorem (that asserts the validity of the sequential version of the ASM thesis) and the Wide-Step Characterization Theorem (that asserts the validity of the parallel version of the ASM thesis) to intra-step interacting algorithms.

A later comment: This was my first talk on intra-step interactive algorithms. The intended audience was the ASM community. [174] is a later talk on this topic, and it is addressed to a general computer science audience.

168. Yuri Gurevich, Rostislav Yavorskiy: Observations on the decidability of transitions. In: W. Zimmerman and B. Thalheim (eds.) Abstract State Machines 2004, Springer LNCS 3052 (2004), 161–168

Consider a multiple-agent transition system such that, for some basic types $T_1, \ldots, T_n$, the state of any agent can be represented as an element of the Cartesian product $T_1 \times \cdots \times T_n$. The system evolves by means of global steps. During such a step, new agents may be created and some existing agents may be updated or removed, but the total number of created, updated and removed agents is uniformly bounded. We show that, under appropriate conditions, there is an algorithm for deciding assume-guarantee properties of one-step computations. The result can be used for automatic invariant verification as well as for finite state approximation of the system in the context of test-case generation from AsmL specifications.

169. Yuri Gurevich, Benjamin Rossman, Wolfram Schulte: Semantic essence of AsmL. Theoretical Computer Science 343:3 (17 October 2005), 370–412 Originally published as MSR-TR-2004-27

The Abstract state machine Language, AsmL, is a novel executable specification language based on the theory of Abstract State Machines. AsmL is object-oriented, provides high-level mathematical data-structures, and is built around the notion of synchronous updates and finite choice. AsmL is fully integrated into the .NET framework and Microsoft development tools. In this paper, we explain the design rationale of AsmL and provide static and dynamic semantics for a kernel of the language.

169a. Yuri Gurevich, Benjamin Rossman, Wolfram Schulte: Semantic essence of AsmL: Extended abstract. In: F. S. de Boer et al. (eds.) FMCO 2003, Formal Methods of Components and Objects, Springer LNCS 3188 (2004), 240–259

This is an extended abstract of article [169].

170. Andreas Blass, Yuri Gurevich: Ordinary interactive small-step algorithms, II. ACM TOCL 8:3 (July 2007), article 15. A preliminary version was published as a part of MSR-TR-2004-88

This is the second in a series of three papers extending the proof of the Abstract State Machine Thesis – that arbitrary algorithms are behaviorally equivalent to abstract state machines – to algorithms that can interact with their environments during a step rather than only between steps. The first paper is [166]. As in that paper, we are concerned here with ordinary, small-step, interactive algorithms. This means that the algorithms (1) proceed in discrete, global steps, (2) perform only a bounded amount of work in each step, (3) use only such information from the environment as can be regarded as answers to queries, and (4) never complete a step until all queries from that step have been answered. After reviewing the previous paper's formal description of such algorithms and the definition of behavioral equivalence, we define ordinary, interactive, small-step abstract state machines (ASM's). Except for very minor modifications, these are the machines commonly used in the ASM literature. We define their semantics in the framework of ordinary algorithms, and we show that they satisfy the postulates for these algorithms. This material lays the groundwork for the final paper in the series, in which we shall prove the Abstract State Machine Thesis for ordinary, interactive, small-step algorithms: All such algorithms are equivalent to ASMs.

171. Andreas Blass, Yuri Gurevich: Ordinary interactive small-step algorithms, III. ACM TOCL 8:3 (July 2007), article 16. A preliminary version was published as a part of MSR-TR-2004-88

This is the third in a series of three papers extending the proof of the Abstract State Machine Thesis – that arbitrary algorithms are behaviorally equivalent to abstract state machines – to algorithms that can interact with their environments during a step rather than only between steps. The first two papers are [166] and [170]. As in those papers, we are concerned here with ordinary, small-step, interactive algorithms. After reviewing the previous papers' definitions of such algorithms, of behavioral equivalence, and of abstract state machines (ASMs), we prove the main result: Every ordinary, interactive, small-step algorithm is behaviorally equivalent to an ASM. We also discuss some possible variations of and additions to the ASM semantics.

172. Andreas Blass, Yuri Gurevich: Why sets? BEATCS 84 (October 2004). Revised and published as MSR-TR-2006-138; then reprinted in: A. Avron et al. (eds.) Pillars of Computer Science: Essays Dedicated to Boris (Boaz) Trakhtenbrot on the Occasion of His 85th Birthday, Springer LNCS 4800 (2008), 179–198

Sets play a key role in foundations of mathematics. Why? To what extent is it an accident of history? Imagine that you have a chance to talk to mathematicians from a far away planet. Would their mathematics be set-based? What are the alternatives to the set-theoretic foundation of mathematics? Besides, set theory seems to play a significant role in computer science, in particular in database theory and formal methods. Is there a good justification for that? We discuss these and related issues.

173. Andreas Blass, Yuri Gurevich, Lev Nachmanson, Margus Veanes: Play to test. MSR-TR-2005-04. FATES 2005, 5th International Workshop on Formal Approaches to Testing of Software, Edinburgh (July 2005)

Testing tasks can be viewed (and organized!) as games against nature. We introduce and study reachability games. Such games are ubiquitous. A single industrial test suite may involve many instances of a reachability game. Hence the importance of optimal or near optimal strategies for reachability games. We find out when exactly optimal strategies exist for a given reachability game, and how to construct them.

174. Yuri Gurevich: Interactive algorithms 2005 with added appendix. In: D. Goldin et al. (eds.) Interactive Computation: The New Paradigm, Springer-Verlag (2006), 165–182. Originally in: J. Jedrzejowicz and A. Szepietowski (eds.) Proceedings of MFCS 2005 Math Foundations of Computer Science (2005), Gdansk, Poland, Springer LNCS 3618 (2005), 26–38 (without the appendix)

A sequential algorithm just follows its instructions and thus cannot make a nondeterministic choice all by itself, but it can be instructed to solicit outside help to make a choice. Similarly, an object-oriented program cannot create a new object all by itself; a create-a-new-object command solicits outside help. These are but two examples of intra-step interaction of an algorithm with its environment. Here we motivate and survey recent work on interactive algorithms within the Behavioral Computation Theory project.

175. Yuri Gurevich, Paul Schupp: Membership problem for modular group. SIAM Journal on Computing 37:2 (2007), 425–459.

The modular group plays an important role in many branches of mathematics. We show that the membership problem for the modular group is polynomial time in the worst case. We also show that the membership problem for a free group remains polynomial time when elements are written in a normal form with exponents.

176. Andreas Blass, Yuri Gurevich, Dean Rosenzweig, Benjamin Rossman: Interactive small-step algorithms I: Axiomatization. Logical Methods in Computer Science 3:4 (2007), paper 3. A preliminary version appeared as MSR-TR-2006-170

In earlier work, the Abstract State Machine Thesis – that arbitrary algorithms are behaviorally equivalent to abstract state machines – was established for several classes of algorithms, including ordinary, interactive, small-step algorithms. This was accomplished on the basis of axiomatizations of these classes of algorithms. Here we extend the axiomatization and, in a companion paper, the proof, to cover interactive small-step algorithms that are not necessarily ordinary. This means that the algorithms (1) can complete a step without necessarily waiting for replies to all queries from that step and (2) can use not only the environment's replies but also the order in which the replies were received.

This is essentially part one of MSR-TR-2005-113. [182] is essentially the remainder of the technical report.

177. Yuri Gurevich, Tanya Yavorskaya: On bounded exploration and bounded nondeterminism. MSR-TR-2006-07

This report consists of two separate parts, essentially two oversized footnotes to [141]. In Chapter I, Yuri Gurevich and Tatiana Yavorskaya present and study a more abstract version of the bounded exploration postulate. In Chapter II, Tatiana Yavorskaya gives a complete form of the characterization, sketched in [141], of bounded-choice sequential algorithms.

178. Andreas Blass, Yuri Gurevich: Program termination, and well partial orderings. ACM TOCL 9:3 (July 2008)

The following known observation may be useful in establishing program termination: if a transitive relation R is covered by finitely many well-founded relations $U_1, \ldots, U_n$ then R is well-founded. A question arises how to bound the ordinal height $|R|$ of the relation R in terms of the ordinals $\alpha_i = |U_i|$. We introduce the notion of the stature $\|P\|$ of a well partial ordering P and show that $|R|$ less than or equal to the stature of the direct product $\alpha_1 \times \cdots \times \alpha_n$ and that this bound is tight. The notion of stature is of considerable independent interest. We define $\|P\|$ as the ordinal height of the forest of nonempty bad sequences of P, but it has many other natural and equivalent definitions. In particular, $\|P\|$ is the supremum, and in fact the maximum, of the lengths of linearizations of P. And the stature of the direct product $\alpha_1 \times \cdots \times \alpha_n$ is equal to the natural product of these ordinals.

179. Yuri Gurevich, Margus Veanes, Charles Wallace: Can abstract state machines be useful in language theory? Theoretical Computer Science 376 (2007) 17–29. Extended Abstract in DLT 2006, Developments in Language Theory, Springer LNCS 4036 (2006), 14–19

The abstract state machine (ASM) is a modern computation model. ASMs and ASM based tools are used in academia and industry, albeit in a modest scale. They allow one to give high-level operational semantics to computer artifacts and to write executable specifications of software and hardware at the desired abstraction level. In connection with the 2006 conference on Developments in Language Theory, we point out several ways that we believe abstract state machines can be useful to the DLT community.

180. Andreas Blass, Yuri Gurevich: A note on nested words. MSR-TR-2006-139

For every regular language of nested words, the underlying strings form a context-free language, and every context-free language can be obtained in this way. Nested words and nested-word automata are generalized to motley words and motley-word automata. Every motley-word automation is equivalent to a deterministic one. For every regular language of motley words, the underlying strings form a finite intersection of context-free languages, and every finite intersection of context-free languages can be obtained in this way.

181. Yuri Gurevich: ASMs in the classroom: Personal experience. In: D. Bjørner and M. C. Henson (eds.) Logics of Specification Languages, Springer (2008), 599–602

We share our experience of using abstract state machines for teaching computation theory at the University of Michigan.

182. Andreas Blass, Yuri Gurevich, Dean Rosenzweig, Benjamin Rossman: Interactive small-step algorithms II: Abstract state machines and the Characterization Theorem. Logical Methods in Computer Science 3:4 (2007), paper 4. A preliminary version appeared as MSR-TR-2006-171

In earlier work, the Abstract State Machine Thesis – that arbitrary algorithms are behaviorally equivalent to abstract state machines – was established for several classes of algorithms, including ordinary, interactive, small-step algorithms. This was accomplished on the basis of axiomatizations of these classes of algorithms. In a companion paper [176] the axiomatization was extended to cover interactive small-step algorithms that are not necessarily ordinary. This means that the algorithms (1) can complete a step without necessarily waiting for replies to all queries from that step and (2) can use not only the environment's replies but also the order in which the replies were received. In order to prove the thesis for algorithms of this generality, we extend here the definition of abstract state machines to incorporate explicit attention to the relative timing of replies and to the possible absence of replies. We prove the characterization theorem for extended ASMs with respect to general algorithms as axiomatized in [176].

183. Dan Teodosiu, Nikolaj Bjørner, Yuri Gurevich, Mark Manasse, Joe Porkka: Optimizing file replication over limited-bandwidth networks using remote differential compression. MSR-TR-2006-157

Remote Differential Compression (RDC) protocols can efficiently update files over a limited-bandwidth network when two sites have roughly similar files; no site needs to know the content of another's files a priori. We present a heuristic approach to identify and transfer the file differences that is based on finding similar files, subdividing the files into chunks, and comparing chunk signatures. Our work significantly improves upon previous protocols such as LBFS and RSYNC in three ways. Firstly, we present a novel algorithm to efficiently find the client files that are the most similar to a given server file. Our algorithm requires 96 bits of metadata per file, independent of file size, and thus allows us to keep the metadata in memory and eliminate the need for expensive disk seeks. Secondly, we show that RDC can be applied recursively to signatures to reduce the transfer cost for large files. Thirdly, we describe new ways to subdivide files into chunks that identify file differences more accurately. We have implemented our approach in DFSR, a state-based multimaster file replication service shipping as part of Windows Server 2003 R2. Our experimental results show that similarity detection produces results comparable to LBFS while incurring a much smaller overhead for maintaining the metadata. Recursive signature transfer further increases replication efficiency by up to several orders of magnitude.

184. Martin Grohe, Yuri Gurevich, Dirk Leinders, Nicole Schweikardt, Jerzy Tyszkiewicz, Jan Van den Bussche: Database query processing using finite cursor machines. Theory of Computing Systems 44:4 (April 2009), 533–560. An earlier version appeared in: ICDT 2007, International Conference on Database Theory, Springer LNCS 4353 (2007), 284–298

We introduce a new abstract model of database query processing, finite cursor machines, that incorporates certain data streaming aspects. The model describes quite faithfully what happens in so-called "one-pass" and "two-pass query processing". Technically, the model is described in the framework of abstract state machines. Our main results are upper and lower bounds for processing relational algebra queries in this model, specifically, queries of the semijoin fragment of the relational algebra.

185. Andreas Blass, Yuri Gurevich: Zero-one laws: Thesauri and parametric conditions. BEATCS 91 (February 2007), 125–144. Reprinted in: A. Gupta et al. (eds.) Logic at the Crossroads: An Interdisciplinary View, Allied Publishers Pvt. Ltd., New Delhi (2007), 187–206

The zero-one law for first-order properties of finite structures and its proof via extension axioms were first obtained in the context of arbitrary finite structures for a fixed finite vocabulary. But it was soon observed that the result and the proof continue to work for structures subject to certain restrictions. Examples include undirected graphs, tournaments, and pure simplicial complexes. We discuss two ways of formalizing these extensions, Oberschelp's parametric conditions (Springer Lecture Notes in Mathematics 969, 1982) and our thesauri of [149]. We show that, if we restrict thesauri by requiring their probability distributions to be uniform, then they and parametric conditions are equivalent. Nevertheless, some situations admit more natural descriptions in terms of thesauri, and the thesaurus point of view suggests some possible extensions of the theory.

186. Andreas Blass, Yuri Gurevich: Background of computation. BEATCS, 92 (June 2007)

In a computational process, certain entities (for example, sets or arrays) and operations on them may be automatically available, for example by being provided by the programming language. We define background classes to formalize this idea, and we study some of their basic properties. The present notion of background class is more general than the one we introduced in an earlier paper [143], and it thereby corrects one of the examples in that paper. The greater generality requires a non-trivial notion of equivalence of background classes, which we explain and use. Roughly speaking, a background class assigns to each set (of atoms) a structure (for example, of sets or arrays or combinations of these and similar entities), and it assigns to each embedding of one set of atoms into another a standard embedding between the associated background structures. We discuss several, frequently useful, properties that background classes may have, for example that each element of a background structure depends (in some sense) on only finitely many atoms, or that there are explicit operations by which all elements of background structures can be produced from atoms.

187. Robert H. Gilman, Yuri Gurevich, Alexei Miasnikov: A geometric zero-one law. JSL 74:3 (September 2009)

Each relational structure X has an associated Gaifman graph, which endows X with the properties of a graph. If x is an element of X, let $B_n(x)$ be the ball of radius n around x. Suppose that X is infinite, connected and of bounded degree. A first-order sentence s in the language of X is almost surely true (resp. a.s. false) for finite substructures of X if for every x in X, the fraction of substructures of $B_n(x)$ satisfying s approaches 1 (resp. 0) as n approaches infinity. Suppose further that, for every finite substructure, X has a disjoint isomorphic substructure. Then every s is a.s. true or a.s. false for finite substructures of X. This is one form of the geometric zero-one law. We formulate it also in a form that does not mention the ambient infinite structure. In addition, we investigate various questions related to the geometric zero-one law.

188. Nachum Dershowitz, Yuri Gurevich: A natural axiomatization of computability and proof of Church's Thesis. Bulletin of Symbolic Logic 14:3 (September 2008), 299–350. An earlier version was published as MSR-TR-2007-85

Church's Thesis asserts that the only numeric functions that can be calculated by effective means are the recursive ones, which are the same, extensionally, as the Turing-computable numeric functions. The Abstract State Machine Theorem states that every classical algorithm is behaviorally equivalent to an abstract state machine. This theorem presupposes three natural postulates about algorithmic computation. Here, we show that augmenting those postulates with an additional requirement regarding basic operations gives a natural axiomatization of computability and a proof of Church's Thesis, as Gödel and others suggested may be possible. In a similar way, but with a different set of basic operations, one can prove Turing's Thesis, characterizing the effective string functions, and – in particular – the effectively-computable functions on string representations of numbers.

188a. Yuri Gurevich: Proving Church's Thesis. CSR 2007, Computer Science – Theory and Applications, 2nd International Symposium on Computer Science in Russia, Springer LNCS 4649 (2007), 1–3

This is an extended abstract of the opening talk of CSR 2007. It is based on [188].

189. Yuri Gurevich, Dirk Leinders, Jan Van den Bussche: A theory of stream queries. DBPL 2007, 11th International Symposium on Database Programming Languages, Springer LNCS 4797 (2007), 153–168

Data streams are modeled as infinite or finite sequences of data elements coming from an arbitrary but fixed universe. The universe can have various built-in functions and predicates. Stream queries are modeled as functions from streams to streams. Both timed and untimed settings are considered. Issues investigated include abstract definitions of computability of stream queries; the connection between abstract computability, continuity, monotonicity, and non-blocking operators; and bounded memory computability of stream queries using abstract state machines (ASMs).

190. Nikolaj Bjørner, Andreas Blass, Yuri Gurevich: Content-dependent chunking for differential compression, the local maximum approach. Journal of Computer and System Sciences 76:3–4 (May–June 2010), 154–203. Originally published as MSR-TR-2007-109

When a file is to be transmitted from a sender to a recipient and when the latter already has a file somewhat similar to it, remote differential compression seeks to determine the similarities interactively so as to transmit only the part of the new file not already in the recipient's old file. Content-dependent chunking means that the sender and recipient chop their files into chunks, with the cutpoints determined by some internal features of the files, so that when segments of the two files agree (possibly in different locations within the files), the cutpoints in such segments tend to be in corresponding locations, and so the chunks agree. By exchanging hash values of the chunks, the sender and recipient can determine which chunks of the new file are absent from the old one and thus need to be transmitted.

We propose two new algorithms for content-dependent chunking, and we compare their behavior, on random files, with each other and with previously used algorithms. One of our algorithms, the local maximum chunking method, has been implemented and found to work better in practice than previously used algorithms.

Theoretical comparisons between the various algorithms can be based on several criteria, most of which seek to formalize the idea that chunks should be neither too small (so that hashing and sending hash values become inefficient) nor too large (so that agreements of entire chunks become unlikely). We propose a new criterion, called the slack of a chunking method, which seeks to measure how much of an interval of agreement between two files is wasted because it lies in chunks that don't agree.

Finally, we show how to efficiently find the cutpoints for local maximum chunking.

191. Yuri Gurevich, Itay Neeman: DKAL: Distributed-Knowledge Authorization Language. MSR-TR-2008-09. First appeared as MSR-TR-2007-116

DKAL is an expressive declarative authorization language based on existential fixed-point logic. It is considerably more expressive than existing languages in the literature, and yet feasible. Our query algorithm is within the same bounds of computational complexity as, e.g., that of SecPAL. DKAL's distinguishing features include

- explicit handling of knowledge and information,
- targeted communication that is beneficial with respect to confidentiality, security, and liability protection,
- the flexible use and nesting of functions, which in particular allows principals to quote (to other principals) whatever has been said to them,
- flexible built-in rules for expressing and delegating trust,
- information order that contributes to succinctness.

191a. Yuri Gurevich, Itay Neeman: DKAL: Distributed-Knowledge Authorization Language. CSF 2008, 21st IEEE Computer Security Foundations Symposium, 149–162

This is an extended abstract of [191]. DKAL is a new declarative authorization language for distributed systems. It is based on existential fixed-point logic and is considerably more expressive than existing authorization languages in the literature. Yet its query algorithm is within the same bounds of computational complexity as, e.g., that of SecPAL. DKAL's communication is targeted, which is beneficial for security and for liability protection. DKAL enables flexible use of functions; in particular, principals can quote (to other principals) whatever has been said to them. DKAL strengthens the trust delegation mechanism of SecPAL. A novel information order contributes to succinctness. DKAL introduces a semantic safety condition that guarantees the termination of the query algorithm.

192. Andreas Blass, Nachum Dershowitz, Yuri Gurevich: When are two algorithms the same? Bulletin of Symbolic Logic 15:2 (2009), 145–168. An earlier version was published as MSR-TR-2008-20

People usually regard algorithms as more abstract than the programs that implement them. The natural way to formalize this idea is that algorithms are equivalence classes of programs with respect to a suitable equivalence relation. We argue that no such equivalence relation exists.

193. Andreas Blass, Yuri Gurevich: Two forms of one useful logic: Existential fixed point logic and liberal Datalog, BEATCS 95 (June 2008), 164–182

A natural liberalization of Datalog is used in the Distributed Knowledge Authorization Language (DKAL). We show that the expressive power of this liberal Datalog is that of existential fixed-point logic. The exposition is self-contained.

194. Andreas Blass, Yuri Gurevich: One useful logic that defines its own truth. MFCS 2008, 33rd International Symposium on Mathematical Foundations of Computer Science, Springer LNCS 5162 (2008), 1–15

Existential fixed point logic (EFPL) is a natural fit for some applications, and the purpose of this talk is to attract attention to EFPL. The logic is also interesting in its own right as it has attractive properties. One of those properties is rather unusual: truth of formulas can be defined (given appropriate syntactic apparatus) in the logic. We mentioned that property elsewhere, and we use this opportunity to provide the proof.

195. Nikolaj Bjørner, Andreas Blass, Yuri Gurevich, Madan Musuvathi: Modular difference logic is hard. MSR-TR-2008-140

In connection with machine arithmetic, we are interested in systems of constraints of the form $x + k \leq y + l$. Over integers, the satisfiability problem for such systems is polynomial time. The problem becomes NP complete if we restrict attention to the residues for a fixed modulus N.

196. Andreas Blass, Yuri Gurevich: Persistent queries in the behavioral theory of algorithms. ACM TOCL, to appear. An earlier version appeared as MSR-TR-2008-150

We propose a syntax and semantics for interactive abstract state machines to deal with the following situation. A query is issued during a certain step, but the step ends before any reply is received. Later, a reply arrives, and later yet the algorithm makes use of this reply. By a persistent query, we mean a query for which a late reply might be used. Syntactically, our proposal involves issuing, along with a persistent query, a location where a late reply is to be stored. Semantically, it involves only a minor modification of the existing theory of interactive small-step abstract state machines.

197. Yuri Gurevich, Arnab Roy: Operational semantics for DKAL: Application and analysis. TrustBus 2009, 6th International Conference on Trust, Privacy and Security in Digital Business, Springer LNCS 5695 (2009), 149–158

DKAL is a new authorization language based on existential fixed-point logic and more expressive than existing authorization languages in the literature. We present some lessons learned during the first practical application of DKAL and some improvements that we made to DKAL as a result. We develop operational semantics for DKAL and present some complexity results related to the operational semantics.

198. Yuri Gurevich, Itay Neeman: Infon logic: The propositional case. ACM TOCL, to appear. The TOCL version is a correction and slight extension of the version called "The infon logic" published in BEATCS 98 (June 2009), 150–178

Infons are statements viewed as containers of information (rather then representations of truth values). In the context of access control, the logic of infons is a conservative extension of logic known as constructive or intuitionistic. Distributed Knowledge Authorization Language uses additional unary connectives "p said" and "p implied" where p ranges over principals. Here we investigate infon logic and a narrow but useful primal fragment of it. In both cases, we develop the model theory and analyze the derivability problem: Does the given query follow from the given hypotheses? Our more involved technical results are on primal infon logic. We construct an algorithm for the multiple derivability problem: Which of the given queries follow from the given hypotheses? Given a bound on the quotation depth of the hypotheses, the algorithm works in linear time. We quickly discuss the significance of this result for access control.

199. Nikolaj Bjørner, Yuri Gurevich, Wolfram Schulte, Margus Veanes: Symbolic bounded model checking of abstract state machines. International Journal of Software and Informatics 3:2–3 (June/September 2009), 149–170

Abstract State Machines (ASMs) allow us to model system behaviors at any desired level of abstraction, including levels with rich data types, such as sets or sequences. The availability of high-level data types allows us to represent state elements abstractly and faithfully at the same time. AsmL is a rich ASM-based specification and programming language. In this paper we look at symbolic analysis of model programs written in AsmL with a background T of linear arithmetic, sets, tuples, and maps. We first provide a rigorous account of the update semantics of AsmL in terms of background T, and we formulate the problem of bounded path exploration of model programs, or the problem of Bounded Model Program Checking (BMPC), as a satisfiability modulo T problem. Then we investigate the boundaries of decidable and undecidable cases for BMPC. In a general setting, BMPC is shown to be highly undecidable (Σ^1_1-complete); restricted to finite sets, the problem remains RE-hard (Σ^0_1-hard). On the other hand, BMPC is shown to be decidable for a class of basic model programs that are common in practice. We apply Satisfiability Modulo Theories (SMT) tools to BMPC. The recent SMT advances allow us to directly analyze specifications using sets and maps with specialized decision procedures for expressive fragments of these theories. Our approach is extensible; background theories need in fact only be partially solved by the SMT solver; we use simulation of ASMs to support additional theories that are beyond the scope of available decision procedures.

200. Yuri Gurevich, Itay Neeman: DKAL 2 – A simplified and improved authorization language. MSR-TR-2009-11

Knowledge and information are central notions in DKAL, a logic based authorization language for decentralized systems, the most expressive among such languages in the literature. Pieces of information are called infons. Here we present DKAL 2, a surprisingly simpler version of the language that expresses new important scenarios (in addition to the old ones) and that is built around a natural logic of infons. Trust became definable, and its properties, postulated earlier as DKAL house rules, are now *proved.* In fact, none of the house rules postulated earlier is now needed. We identify also a most practical fragment of DKAL where the query derivation problem is solved in *linear time.*

201. Andreas Blass, Nachum Dershowitz, Yuri Gurevich: Exact exploration and hanging algorithms. CSL 2010, 19th EACSL Annual Conference on Computer Science Logic (August 2010), to appear

Recent analysis of sequential algorithms resulted in their axiomatization and in a representation theorem stating that, for any sequential algorithm, there is an abstract state machine (ASM) with the same states, initial states and state transitions. That analysis, however, abstracted from details of intra-step computation, and the ASM, produced in the proof of the representation theorem, may and often does explore parts of the state unexplored by the algorithm. We refine the analysis, the axiomatization and the representation theorem. Emulating a step of the given algorithm, the ASM, produced in the proof of the new representation theorem, explores exactly the part of the state explored by the algorithm. That frugality pays off when state exploration is costly. The algorithm may be a high-level specification, and a simple function call on the abstraction level of the algorithm may hide expensive interaction with the environment. Furthermore, the original analysis presumed that state functions are total. Now we allow state functions, including equality, to be partial so that a function call may cause the algorithm as well as the ASM to hang. Since the emulating ASM does not make any superfluous function calls, it hangs only if the algorithm does.

202. Andreas Blass, Yuri Gurevich, Efim Hudis: The Tower-of-Babel problem, and security assessment sharing. MSR-TR-2010-57. BEATCS 101 (June 2010), to appear

The Tower-of-Babel problem is rather general: How to enable a collaboration among experts speaking different languages? A computer security version of the Tower-of-Babel problem is rather important. A recent Microsoft solution for that security problem, called Security Assessment Sharing, is based on this idea: A tiny common language goes a long way. We construct simple mathematical models showing that the idea is sound.

Database Theory, Yuri, and Me

Jan Van den Bussche

Hasselt University and transnational University of Limburg, Diepenbeek, Belgium

Abstract. Yuri Gurevich made many varied and deep contributions to logic for computer science. Logic provides also the theoretical foundation of database systems. Hence, it is almost unavoidable that Gurevich made some great contributions to database theory. We discuss some of these contributions, and, along the way, present some personal anecdotes connected to Yuri and the author. We also describe the honorary doctorate awarded to Gurevich by Hasselt University (then called Limburgs Universitair Centrum) in 1998.

Dedicated to Yuri Gurevich, the "man with a plan", on his 70th birthday.

1 Database Theory

The theory of database systems is a very broad field of theoretical computer science, concerned with the theoretical design and analysis of all data management aspects of computer science. One can get a good idea of the current research in this field by looking at the proceedings of the two main conferences in the area: the International Conference on Database Theory, and the ACM Symposium on Principles of Database Systems. As data management research in general follows the rapid changes in computing and software technology, database theory can appear quite trendy to the outsider. Nevertheless there are also timeless topics such as the theory of database queries, to which Yuri Gurevich has made a number of fundamental contributions.

An in-depth treatment of database theory until the early 1990s can be found in the book of Abiteboul, Hull and Vianu [1]; Yuri Gurevich appears nine times in the bibliography.

A *relational database schema* is a finite relational vocabulary, i.e., a finite set of relation names with associated arities. Instead of numbering the columns of a relation with numbers, as usual in mathematical logic, in database theory it is also customary to name the columns with *attributes*. In that case the arity is replaced by a finite set of attributes (called a *relation scheme*). A *database instance* over some schema is a finite relational structure over that schema, i.e., an assignment of a concrete, finite, relation content to each of the relation names. So, if R is a relation name of arity k and D is a database, then $D(R)$ is a finite subset of U^k, where U is some universe of data elements. The idea is that the contents of a database can be updated frequently, hence the term "instance". We will often drop this term, however, and simply talk about a "database".

A. Blass, N. Dershowitz, and W. Reisig (Eds.): Gurevich Festschrift, LNCS 6300, pp. 49–60, 2010.

For example, consider a relation name Hobby of arity 3; we associate to it the relation scheme {name, hobby, location}. The intention is that an instance will store tuples (n, h, l) where n is the name of a person who has a hobby h and he performs this hobby in location l. Then a concrete instance will have a relation Hobby containing tuples like (john, birdwatching, lake) and (mary, violin, town hall).

2 Typed Template Dependencies

Usually we do not want to allow completely arbitrary relation contents in our database instances; we will typically expect certain integrity constraints to be satisfied, depending on the application. The early years of database theory focused on integrity constraints expressible in first-order logic, called "dependencies", and studied the implication problem for classes of dependencies, i.e., fragments of first-order logic. Not surprisingly, this new field of theoretical computer science attracted experts on logical decision problems, such as Gurevich. Gurevich and Lewis [27,28] settled the undecidability of the implication problem for the class of "typed template dependencies".

For an example of a template dependency, consider, for example, the following hypothetical (and admittedly rather weird) integrity constraint on the Hobby relation: if two people each perform their own hobby in some common location, they also perform there each other's hobby; in short, common location implies common hobby. We can write such a dependency in the following syntax:

$$\text{Hobby}(n_2, h_1, l) \leftarrow \text{Hobby}(n_1, h_1, l), \text{Hobby}(n_2, h_2, l).$$

The semantics is that of implication, where all variables are assumed universally quantified.

The implication problem for template dependencies then is: given a finite set Σ of template dependencies and another template dependency σ, decide whether each database instance satisfying all dependencies of Σ also satisfies σ. The typed version, which is easier but still shown undecidable by Gurevich and Lewis, restricts attention to relations with pairwise disjoint columns.

Interestingly, Gurevich and Lewis were working on this result concurrently with Moshe Vardi [36,37]. To obtain his sharpest results, Vardi could apply work by Gurevich and by Lewis on the word problem for semigroups [22,29].

3 Database Queries

One of the main purposes of a database is to query it. For many good reasons into which we cannot go here, the answer of a query to a relational database takes again the form of a relation. For example, on our Hobby relation, we could pose the query "list all pairs (n, h) such that n performs hobby h in some location where nobody else performs any hobby in that location".

So, in the most general terms, one could define a query simply as a mapping from database instances to relations. But to get a good theory, we need to impose some criteria on this mapping. First, it is convenient that all answer relations of a same query have the same arity. Second, the basic theory restricts attention to answer relations containing only values that already appear in the input database. We call such queries "domain-preserving". Third, a domain-preserving query should be "logical" in the sense of Tarski [34], i.e., it should commute with permutations of data elements.[1] This captures the intuition that the query need not distinguish between isomorphic databases; all the information required to answer the query should already be present in the database [3]. Thus, formally, a query of arity k over some database schema $\mathcal{S}$ is a domain-preserving mapping from database instances over $\mathcal{S}$ to k-ary relations on U, such that for every permutation ρ of U, and for every database instance D over $\mathcal{S}$, we have $Q(\rho(D)) = \rho(Q(D))$.

For example, if the database schema consists of a single unary relation name, so that an instance is just a naked set, a function that picks an arbitrary element out of each instance is not a query, because it is not logical. The intuition is that the database does not provide any information that would substantiate favoring one of the elements above the others.

4 The QPTIME Problem

The definition of query as recalled above was formulated by Chandra and Harel [11,12]. This notion of query comes very naturally to a logician; indeed, Gurevich independently introduced the very same notion under the name "global relation" or "global predicate" in his two seminal papers on finite model theory [23,24]. These papers also widely publicized one of the most fundamental open problems in database theory, the QPTIME problem [13,14,30,31,35]: is there a reasonable programming language in which only, and all, queries can be expressed that are computable in polynomial time? Gurevich's conjecture is that the answer is negative. The QPTIME problem has been actively investigated since its inception, as can be learned from two surveys, one by Kolaitis from 1995 [32] and one by Grohe from 2008 [19]. We will get back to it in Section 8. The problem also nicely illustrate how database theory lies at the basis of the areas of finite model theory and descriptive complexity which grew afterwards.

5 Datalog vs. First-Order Logic

In the 1980s, much attention was devoted to the query language Datalog. One of the toughest nuts in this research was cracked by Ajtai and Gurevich [4,5].

A Datalog program is a set of implications, called rules, that are much like the template dependencies we have seen above. But an essential difference is that the relation name in the head of a Datalog rule is not from the database schema; it is a so-called "intensional" relation name. Thus a Datalog program defines

[1] We mean here a permutation not just of the data elements that appear in some input database, but of the entire global universe of possible data elements.

a number of new relations from the existing ones in the database instance; the semantics is that we take the smallest expanded database instance that satisfies the rules. For example, on our Hobby relation, consider the following program:

$$T(x, y) \leftarrow \text{Hobby}(x, h, l_1), \text{Hobby}(y, h, l_2)$$
$$T(x, y) \leftarrow T(x, z), T(z, y)$$

This program computes, in relation T, the transitive closure of the binary relation that relates a person x to a person y if they have some common hobby.

Since the transitive closure is not first-order definable [3,17], the above Datalog program is not equivalent to a first-order formula. Neither is it "bounded": there is no fixed constant so that, on any database instance, the rules have to be fired only so many times until we reach a fixpoint. As a matter of fact, a non-first-order Datalog program cannot be bounded, as bounded Datalog programs are obviously first-order; an equivalent first-order formula can by obtained by unfolding the recursive rules a constant number of times.

The converse is much less obvious, however: every first-order Datalog program must in fact be bounded, and this is the above-mentioned celebrated result by Ajtai and Gurevich.

6 Metafinite Structures

In database theory, a relational database is considered to be a finite relational structure, and also in practice, relational database instances are finite. But still there is a gap between theory and practice in this manner, as in practice, relational databases do contain interpreted elements from an infinite structure, such as numbers, and queries expressed in the database language SQL can perform computations on these numbers. In a very elegant paper, Grädel and Gurevich [18] proposed the theory of metafinite structures as a way to close the gap. That theory later also inspired the development of the theory of constraint databases [33].

7 Honorary Doctorate

All the work described up to know happened before I had ever personally met Yuri. That would happen in May 1996, on the occasion of an AMS Benelux meeting at the University of Antwerp, Belgium, where I was working as a postdoc at the time. I had noticed that the famous Yuri Gurevich was scheduled to give an invited talk at the logic session, and since I had some ideas related to the QPTIME problem, I approached him and asked if he would be interested in having dinner in Antwerp together and talk mathematics. To my most pleasant surprise he readily accepted. It was my first personal encounter with Yuri and we spent an agreeable evening. He patiently listened to my ideas and made suggestions. Being a native from Antwerp I could give him a flash tour of the city and also knew a typical restaurant, things he could certainly appreciate.

Shortly afterwards, I would join the faculty of what was then known as the Limburgs Universitair Centrum in Diepenbeek, Belgium; nowadays it is called Hasselt University. The university was just preparing for its 25th anniversary in the year 1998, and there was an internal call for nominations for honorary doctorates to be awarded during the Dies Natalis ceremony. Given his fame in finite model theory and database theory, and given the pleasant experience I had had with Yuri, I nominated him before the Faculty of Sciences. Obviously he was such a strong candidate that my nomination was enthusiastically accepted. Thus in May 1998, Yuri received the honorary doctorate and became a friend of Hasselt University, and of me personally as well.

Appendix A contains a transcript, translated into English, of the nomination speech I gave (in Dutch) for the honorary degree. In the course of preparing that speech, I collected information from many people around Yuri. These people gave me so much information that much of it could not be used for the short and formal speech. On the occasion of Yuri's 60th birthday, however, Egon Börger organised a special session at the CSL 2000 conference in Fischbachau, Germany, and in a speech given there I could use that material, consisting mainly of anecdotes. Appendix B contains a transcript of that speech.

8 Choiceless Polynomial Time

Upon getting to know him better, I started to collaborate with Yuri on a scientific level. I remember a meeting on finite model theory in Oberwolfach in 1998, just a few months before the honorary doctorate ceremony, where he gave a talk on Choiceless Polynomial Time [8], a very expressive database query language in which only polynomial-time queries can be expressed. The language is nice because it borrows its high expressivity from set theory in a natural way, and also because it is based on Gurevich's Abstract State Machines (ASM [25]) . It is nice to see how Yuri's work on ASMs, originally disjoint from the QPTIME problem, is applied to that problem.

Yuri wondered about the precise relationship between choiceless polynomial time and the work that was going on in database theory, e.g., by Abiteboul and Vianu on "generic machines" [2]. We collaborated on that question and that led to our joint paper (with Andreas Blass) on ASMs and computationally complete query languages [6,7]. In short, it turns out that choiceless polynomial time is the same as the polynomial-time fragment of a natural complete query language based on first-order logic, object creation [10], and iteration. We also showed that the "non-flat" character of choiceless polynomial time, be it through arbitrarily deeply nested sets, or through object creation, is essential to its high expressive power.

We note that extensions of choiceless polynomial time are still being actively researched in connection with the QPTIME problem [9,16,15].

A small personal recollection I have on this joint research is that, when working on the proof, I visited Yuri in Paris, where he liked to spend his summers. We worked at the stuffy apartment where Yuri rented a room, and in the evening,

Yuri felt like going to the movies and asked me to suggest a good movie. I remembered my father telling me the week before that he had been impressed by the then-running movie “Seven years in Tibet” (with Brad Pitt). So I suggested we go to that movie, and indeed, the movie impressed Yuri and me greatly.

9 Finite Cursor Machines, Stream Queries

The most recent chapter in my interactions with Yuri was a great visit I made to him at Microsoft Research, Redmond, for a week in October 2006, together with my student Dirk Leinders. It was impressive to visit Microsoft Research and to see Yuri thrive in these surroundings. It was also interesting to visit Zoe and Yuri’s beautiful house. At the time I was working on querying streaming data and had an idea for an ASM-based model for computing such queries. Yuri was interested and we had fruitful brainstorm sessions on the model, which came to be called “Finite Cursor Machines”. This research led to two nice papers on stream queries [20,21,26]. I know that Yuri is still interested in modeling computation on data streams.

10 Conclusion

My life has been enriched in many ways through my encounters with such a great person as Yuri Gurevich. Yuri, I wish you much happiness on the occasion of your 70th birthday!

References

1. Abiteboul, S., Hull, R., Vianu, V.: Foundations of Databases. Addison-Wesley, Reading (1995)
2. Abiteboul, S., Vianu, V.: Generic computation and its complexity. In: Proceedings 23rd ACM Symposium on the Theory of Computing, pp. 209–219 (1991)
3. Aho, A., Ullman, J.: Universality of data retrieval languages. In: Conference Record, 6th ACM Symposium on Principles of Programming Languages, pp. 110–120 (1979)
4. Ajtai, M., Gurevich, Y.: Datalog vs first-order logic. In: Proceedings 30th IEEE Symposium on Foundations of Computer Science, pp. 142–147 (1989)
5. Ajtai, M., Gurevich, Y.: Datalog vs first-order logic. J. Comput. Syst. Sci. 49(3), 562–588 (1994)
6. Blass, A., Gurevich, Y., Van den Bussche, J.: Abstract state machines and computationally complete query languages (extended abstract). In: Gurevich, Y., Kutter, P.W., Odersky, M., Thiele, L. (eds.) ASM 2000. LNCS, vol. 1912, pp. 22–33. Springer, Heidelberg (2000)
7. Blass, A., Gurevich, Y., Van den Bussche, J.: Abstract state machines and computationally complete query languages. Information and Computation 174(1), 20–36 (2002)
8. Blass, A., Gurevich, Y., Shelah, S.: Choiceless polynomial time. Annals of Pure and Applied Logic 100, 141–187 (1999)

9. Blass, A., Gurevich, Y., Shelah, S.: On polynomial time computation over unordered structures. Journal of Symbolic Logic 67(3), 1093–1125 (2002)
10. Van den Bussche, J., Van Gucht, D., Andries, M., Gyssens, M.: On the completeness of object-creating database transformation languages. J. ACM 44(2), 272–319 (1997)
11. Chandra, A., Harel, D.: Computable queries for relational data bases. In: Proceedings 11th ACM Symposium in Theory of Computing, pp. 309–318 (1979)
12. Chandra, A., Harel, D.: Computable queries for relational data bases. J. Comput. Syst. Sci. 21(2), 156–178 (1980)
13. Chandra, A., Harel, D.: Structure and complexity of relational queries. In: Proceedings 21st IEEE Symposium on Foundations of Computer Science, pp. 333–347 (1980)
14. Chandra, A., Harel, D.: Structure and complexity of relational queries. J. Comput. Syst. Sci. 25, 99–128 (1982)
15. Dawar, A.: On the descriptive complexity of linear algebra. In: Hodges, W., de Queiroz, R. (eds.) Logic, Language, Information and Computation. LNCS (LNAI), vol. 5110, pp. 17–25. Springer, Heidelberg (2008)
16. Dawar, A., Richerby, D., Rossman, B.: Choiceless polynomial time, counting and the Cai-Fürer-Immerman graphs. Annals of Pure and Applied Logic 152(1-3), 31–50 (2008)
17. Gaifman, H., Vardi, M.: A simple proof that connectivity is not first-order definable. Bulletin of the EATCS 26, 43–45 (1985)
18. Grädel, E., Gurevich, Y.: Metafinite model theory. Information and Computation 140(1), 26–81 (1998)
19. Grohe, M.: The quest for a logic capturing PTIME. In: Proceedings 23rd Annual IEEE Symposium on Logic in Computer Science, pp. 267–271 (2008)
20. Grohe, M., Gurevich, Y., Leinders, D., Schweikardt, N., Tyszkiewicz, J., Van den Bussche, J.: Database query processing using finite cursor machines. In: Schwentick, T., Suciu, D. (eds.) ICDT 2007. LNCS, vol. 4353, pp. 284–298. Springer, Heidelberg (2006)
21. Grohe, M., Gurevich, Y., Leinders, D., Schweikardt, N., Tyszkiewicz, J., Van den Bussche, J.: Database query processing using finite cursor machines. Theory of Computing Systems 44(4), 533–560 (2009)
22. Gurevich, Y.: The word problem for some classes of semigroups (russian). Algebra and Logic 5(2), 25–35 (1966)
23. Gurevich, Y.: Toward logic tailored for computational complexity. In: Richter, M., et al. (eds.) Computation and Proof Theory. Lecture Notes in Mathematics, vol. 1104, pp. 175–216. Springer, Heidelberg (1998)
24. Gurevich, Y.: Logic and the challenge of computer science. In: Börger, E. (ed.) Current Trends in Theoretical Computer Science, pp. 1–57. Computer Science Press, Rockville (1988)
25. Gurevich, Y.: Evolving algebra 1993: Lipari guide. In: Börger, E. (ed.) Specification and Validation Methods, pp. 9–36. Oxford University Press, Oxford (1995)
26. Gurevich, Y., Leinders, D., Van den Bussche, J.: A theory of stream queries. In: Arenas, M., Schwartzbach, M.I. (eds.) DBPL 2007. LNCS, vol. 4797, pp. 153–168. Springer, Heidelberg (2007)
27. Gurevich, Y., Lewis, H.: The inference problem for template dependencies. In: Proceedings 1st ACM Symposium on Principles of Database Systems, pp. 221–229 (1982)
28. Gurevich, Y., Lewis, H.: The inference problem for template dependencies. Information and Control 55(1-3), 69–79 (1982)

29. Gurevich, Y., Lewis, H.: The word problem for cancellation semigroups with zero. Journal of Symbolic Logic 49(1), 184–191 (1984)
30. Immerman, N.: Relational queries computable in polynomial time. In: Proceedings 14th ACM Symposium on Theory of Computing, pp. 147–152 (1982)
31. Immerman, N.: Relational queries computable in polynomial time. Information and Control 68, 86–104 (1986)
32. Kolaitis, P.: Languages for polynomial-time queries: An ongoing quest. In: Gottlob, G., Vardi, M. (eds.) ICDT 1995. LNCS, vol. 893, pp. 38–39. Springer, Heidelberg (1995)
33. Kuper, G., Libkin, L., Paredaens, J. (eds.): Constraint Databases. Springer, Heidelberg (2000)
34. Tarski, A.: What are logical notions? In: Corcoran, J. (ed.) History and Philosophy of Logic, vol. 7, pp. 143–154 (1986)
35. Vardi, M.: The complexity of relational query languages. In: Proceedings 14th ACM Symposium on the Theory of Computing, pp. 137–146 (1982)
36. Vardi, M.: The implication and finite implication problems for typed template dependencies. In: Proceedings 1st ACM Symposium on Principles of Database Systems, pp. 230–238 (1982)
37. Vardi, M.: The implication and finite implication problems for typed template dependencies. J. Comput. Syst. Sci. 28, 3–28 (1984)

A Honorary Doctorate Nomination Speech

This speech was given by me (originally in Dutch) at the Dies Natalis ceremony of Hasselt University (then called Limburgs Universitair Centrum), on 28 May 1998, to nominate Yuri Gurevich for a honorary degree. Note the speech was directed to a general audience.

> Dear Guests:
>
> Dear Professor Gurevich:
>
> It is my honor and my pleasure to give a brief exposition of your life, your work, and your achievements.
>
> In these days, the field of information technology (IT) receives plenty of attention. It is hard to imagine our present-day information society without IT: in our daily lives we live and work thanks to products and services that would have been either unaffordable, or simply impossible, were it not for IT. Computer science, as an academic discipline, profits from this success, but at the same time runs some risk because of it. Indeed, the danger is that those less familiar with computer science, view IT as an obvious technology that we just use when and were we need it. This purely technological view of computer science is too limited. Computer science is just as well an exact science, a relatively young one at that, and still in full growth, that investigates the possibilities and limitations of one of the hardest tasks for us humans: the design and programming of computer systems, in a correct and efficient manner. Logical and abstract reasoning are essential skills in this endeavor.
>
> Now if there is one who is a champion in logical reasoning, it is our honored guest, professor Yuri Gurevich. Yuri was born in 1940 in Russia, and studied mathematics at Ural university. Already at the age of 24 he earned his doctorate, and four years later the Soviet state doctorate, which allowed him access

to a professor position at the highest level. Thus he found himself at the age of 29 as head of the mathematics division of the national institute for economics in Sverdlovsk. Russian colleagues have told me that such a steep career was almost unthinkable in the communist Russia of the 1960s, also because Gurevich had refused to become a member of the party.

But it was indeed impossible to ignore the scientific results he had obtained during his doctorate in mathematical logic. As a young graduate Yuri Gurevich had been directed towards a subdiscipline of mathematics, called the theory of ordered abelian groups. Fortunately I do not have to explain this theory in order to show the depth of the results that Gurevich obtained. Normally one expects of a mathematician that he or she finds some answers to specific mathematical questions posed within the discipline in which he or she is active. Gurevich, however, developed an automated procedure—think of a computer program—by which *every* question about the theory of ordered abelian groups could be answered! One might say that he replaced an entire community of mathematicians by a single computer program. At that time, as well as in the present time, it was highly unusual that an entire subdiscipline of mathematics is solved in such manner.

In the early 1970s it becomes increasingly harder for Yuri Gurevich to struggle against the discrimination against Jews in the anti-Semitic climate in Russia in those years. When he hears that the KGB has a file against him, he plans to emigrate. Unfortunately this happens under difficult circumstances, so that his scientific activities are suspended for two years. Traveling via the Republic of Georgia, from where it was easier to emigrate, he finally settles in 1974 in Israel, where he becomes a professor at the Ben-Gurion University. Yuri amazed everyone by expressing himself in Hebrew at departmental meetings already after a few months of arriving.

During his Israeli period, Yuri Gurevich develops into an absolute eminence, a world leader in research in logic. We cannot go further into the deep investigations he made, nor into the long and productive collaboration he developed with that other phenomenal logician, Saharon Shelah. The clear computer science aspect of his earlier work is less present during this period, although some of the fundamental results that he obtains here will find unexpected applications later in the area of automated verification of computer systems. The latter is not so accidental: having unexpected applications is one of the hallmarks of pure fundamental research.

Along the way, however, the computer scientist in Yuri Gurevich resurfaces. Computer science in the late 1970s was in full growth as a new, young academic discipline, and Gurevich saw the importance of a solid logical foundation for this new science. In a new orientation of his career, he accepts in 1982 an offer from the University of Michigan as professor of computer science. Since then professor Gurevich, as a leadership figure, serves as an important bridge between logic and computer science. Partly through his influence, these two disciplines have become strongly interweaved. Of the many common topics where the two disciplines interact, and where Gurevich played a decisive role, we mention finite model theory, where a logical foundation is being developed for computerised databases, an important topic in our own theoretical computer science group here at LUC; the complexity of computation, which he gave logical foundations; and software engineering, for which he designed a new logical formalism in which computer systems can be specified in a natural and correct manner.

To conclude I want to say a few words about the man Yuri Gurevich. With all his obvious talents he remains a modest—I would even say, somewhat shy—person. A good friend of his said it as follows: "Yuri loves science more than himself in it." That is all the more a reason to put him in the spotlights today.

Dear professor Gurevich: for your great achievements in mathematical logic, and for your continued contributions to the promotion and development of logical methods for the benefit of computer science, the Faculty of Sciences proposes you as a honorary doctor at our university.

B Fischbachau Speech

This speech was given by me after the dinner at the end of the symposium in honour of Yuri Gurevich's 60th birthday, held in the charming Bavarian village of Fischbachau, on 24 August 2000, co-located with the CSL conference. I thank the many people who contributed the material for this speech. Their names are mentioned explicitly.

Dear computer science logicians, and, of course, dear Yuri:

My name is Jan Van den Bussche, and during the academic year 97–98, I had the pleasure of promoting an honorary doctorate for Yuri at my university, the University of Limburg in Belgium. Because at that time I did not know Yuri personally very well yet—happily for me this has changed by now—I solicited comments from various friends and colleagues of Yuri. Within two days I received many warm responses. Unfortunately, because my laudation speech had to be rather formal, I could not directly use many of the nice things said about Yuri, and I have always found it a great pity that they remained buried in my files. Therefore I am so glad with this occasion to bring a few anecdotes about Yuri out into the open.

But first I would like to tell how Yuri and I met for the first time. In April 95, then at the university of Antwerp in Belgium, I was thinking about some issues related to one of the many nice papers Yuri wrote with Saharon Shelah. I sent him an email with a technical question, and received a very friendly reply within half an hour, giving me his intuition behind my question, telling me that for the moment he was busy with other things, but that he would be happy to discuss the matter more deeply in Antwerp, where he happened to be going soon for an AMS meeting. So we planned to meet in his hotel lobby some late afternoon, which was a bit exciting because, although I knew his face, he had never even heard about me before that email. But immediately we were both at ease, making a long walk in the old city center, me showing him various sights. I remember that I was a bit afraid of overdoing it: maybe we were walking too long and he was tired of it, but being too nice to say it. Little did I know that Yuri himself is infamous for taking people on walks they don't know when and where they will end! Taking a walk is also one of Yuri's famous ways of dealing with difficult mathematical problems. At dinner we were talking about mathematics, and he was really giving my questions genuine thought. All in all, I was really struck by his friendliness, his generosity, his accessibility. These very qualities of Yuri have been mentioned to me independently by many people.

One of these people is Joe Weisburd, who was a student of Yuri in the old Russian days back in the sixties. Joe also told me the following:

> Yuri always looked very undaunted, very powerful and very confident, which was unusual at those times, unless you were a ranked Communist. And Yuri Shlomovich—wasn't. You also didn't dare to look powerful and confident if you were Jewish. And Yuri Shlomovich—was, even blatantly so in a city where public Jewish life was completely suppressed. His patronymic, Shlomovich, was a challenge, indicating that he wasn't adjusting to sounding more Russian, as was pretty popular then. The Biblical names of his newly born daughters was another challenge. And Yuri was absolutely unprecedented in the Jewish folk song, that he suggested to sing together at the banquet after an annual mathematical Winter School of our Math department.

Yuri became a full professor at the age of 29, which again was unheard of, let alone him being Jewish. Nevertheless, it became increasingly clear that he had to emigrate. Cunningly, he discovered that for some peculiar reason, it was easier to get permission to emigrate out of the Republic of Georgia, so he requested and finally received permission to transfer there, and eventually was able to emigrate. These were severe times; they even had to go on a hunger strike. Vladik Kreinovich, who met Yuri at a seminar in St. Petersburg during this period, told me the following:

> Yuri seemed to be undisturbed by the complexity of the outside life. He radiated strength and optimism, and described very interesting results which he clearly managed to produce lately, during the extremely severe period of his life. His demeanor looked absolutely fantastic.

Once free, Yuri devoted considerable energy to help other Soviet Jews with their emigration. This happened in Israel, and also later in the United States. Vladik, himself an emigrant, continues:

> Together with a local rabbi, Yuri formed a committee which became one of the grassroots activities that finally managed to convince the American public opinion that the Soviet Jews needed help, in a period when the political climate was on the left side. History books have been written about that period, and Yuri is not described there as one of the prominent and visible leaders of the American support campaign. However, this is only because he preferred not to be in the spotlights.

Arriving in Israel, Yuri's talent for languages was very useful. Baruch Cahlon, who shared an office with Yuri in Beer-Sheva, recounts:

> Yuri hardly spoke any Hebrew, but he had to communicate with all of us in this orient-modern language, which he seemed to love, but was yet unable to utter. My wife, who is a Hebrew teacher, used to tell me how Yuri would try to answer the phone when she would call. Believe me, it was funny. It didn't take long, however, and one day, during a department meeting, Yuri stood up and spoke fluent, almost flawless Hebrew. We were totally astonished. Just couldn't believe it. At the time, of course, we had not known about his love for languages. His interest in these disciplines far surpasses that of an average mathematician. But then, of course, Yuri isn't that either!

In a similar vein, Jim Huggins, one of Yuri's later students in Michigan, told me that Yuri would ask him questions about English such as "what is the difference in pronunciation of the words 'morning' and 'mourning'?", and they would try to find English equivalents of Russian proverbs. During a number of summers spent in Paris—Yuri likes the French way of life very much—he learnt to speak more than a mouthful of French as well. And last April, when we were together at a nice restaurant in Ascona, Yuri had fun translating the Italian menu to us!

In Israel, Yuri also gave new meanings to religious symbols. Saharon Shelah told me that Yuri once turned up at a logic seminar wearing a kipah, this is the Jewish religious cap. When asked about it, he replied "why, it is very convenient: it covers exactly my bald part!" Yuri, I am sorry, but I guess that by now even an extra large won't do anymore!

I finally must mention the constant source of support in Yuri's life provided by his wife, Zoe. Not for nothing, Yuri himself describes the period before he met Zoe as his "protozoan" period! By the way, Zoe is also a mathematician by education, and she is a hell of a computer programmer.

Dear Yuri, although you are now a professor emeritus, you are definitely not yet retired, and thanks to you we will continue to see a lot of exciting things coming out of Microsoft. I congratulate you with your sixtiest birthday, and am looking forward to the next sixty years!

Tracking Evidence

Sergei Artemov*

CUNY Graduate Center, 365 Fifth Ave., New York, NY 10016, USA
SArtemov@gc.cuny.edu

For Yuri, on the occasion of his seventieth birthday.

Abstract. In this case study we describe an approach to a general logical framework for tracking evidence within epistemic contexts. We consider as basic an example which features two justifications for a true statement, one which is correct and one which is not. We formalize this example in a system of Justification Logic with two knowers: the object agent and the observer, and we show that whereas the object agent does not logically distinguish between factive and non-factive justifications, such distinctions can be attained at the observer level by analyzing the structure of evidence terms. Basic logic properties of the corresponding two-agent Justification Logic system have been established, which include Kripke-Fitting completeness. We also argue that a similar evidence-tracking approach can be applied to analyzing paraconsistent systems.

Keywords: justification, epistemic logic, evidence.

1 Introduction

In this paper, commencing from seminal works [14,21], the following analysis of basic epistemic notions was adopted: for a given agent,

$$F \text{ is known} \quad \sim \quad F \text{ holds in all epistemically possible situations.} \tag{1}$$

The notion of justification, an essential component of epistemic studies, was introduced into the mathematical models of knowledge within the framework of Justification Logic in [1,2,3,5,6,8,13,16,18,19,22] and other papers; a comprehensive account of this approach is given in [4]. At the foundational level, Justification Logic furnishes a new, evidence-based semantics for the logic of knowledge, according to which

$$F \text{ is known} \quad \sim \quad F \text{ has an adequate justification.} \tag{2}$$

Within Justification Logic, we can reason about justifications, simple and compound, and track different pieces of evidence pertaining to the same fact.

In this paper we develop a sufficiently general mechanism of evidence tracking which is crucial for distinguishing between factive and nonfactive justifications. Some preliminary observations leading to this mechanism have been discussed in [4].

* This work was supported by NSF grant 0830450.

A. Blass, N. Dershowitz, and W. Reisig (Eds.): Gurevich Festschrift, LNCS 6300, pp. 61–74, 2010.

1.1 Basics of Justification Logic

Evidence (justification) terms are built from justification variables $x, y, z, \ldots$ and evidence constants $a, b, c, \ldots$ by means of the operations *application* '$\cdot$,' *sum* '$+$,' and *evidence verifier* '!.' The list of operations is flexible: more elaborate justification logic systems also use additional operations on justifications such as negative verifier '?.' On the other hand, it makes sense to consider subsets of operations such as $\{\cdot, +\}$ or even $\{\cdot\}$ (cf. [4]). However, these features do not alter the main results of this paper and, for the sake of convenience, we choose to work with the set of operations $\{\cdot, +, !\}$.

Formulas of Justification Logic are built from logical atomic propositions by means of the usual classical logical connectives $\wedge, \vee, \neg, \ldots$ with an additional formation rule: if t is an evidence term and F a formula, then $t{:}F$ is a formula. Using p to denote any sentence letter and t for an evidence term, we define the formulas by the grammar

$$F = p \mid F \wedge F \mid F \vee F \mid F \rightarrow F \mid \neg F \mid t{:}F.$$

The basic introspective justification logic considered in this paper is called $\mathsf{J4}_0$. It contains the following postulates:

1. *Classical propositional axioms and rule Modus Ponens,*
2. *Application Axiom* $s{:}(F \rightarrow G) \rightarrow (t{:}F \rightarrow (s \cdot t){:}G)$,
3. *Monotonicity Axiom* $s{:}F \rightarrow (s + t){:}F,\ s{:}F \rightarrow (t + s){:}F$.
4. *Introspection Axiom* $t{:}F \rightarrow\ !t{:}(t{:}F)$.

Constants denote justifications of assumptions. To postulate that an axiom A is justified, one has to assume

$$c : A$$

for some evidence constant c. Furthermore, if, in addition, we want to postulate that this new principle $c : A$ is also justified, we can use Introspection and conclude $!c : (c : A)$, etc.

A *Constant Specification CS* for a given logic $\mathcal{L}$ is a set of formulas of form $c : A$ where A's are axioms of $\mathcal{L}$ and c's are evidence constants. We distinguish the following types of constant specifications:

- *axiomatically appropriate*: for each axiom A there is a constant c such that $c : A \in CS$;
- *total* (called *TCS*): for any axiom A and constant c, $c : A \in CS$.

For a given Constant Specification *CS*,

$$\mathsf{J4}_{CS} = \mathsf{J4}_0 + CS\ , \qquad\qquad \mathsf{J4} = \mathsf{J4}_0 + TCS.$$

An alternative description of $\mathsf{J4}$ is given by

$$\mathsf{J4} = \mathsf{J4}_0 + \mathrm{R4}$$

where R4 is the *Axiom Internalization Rule*:

For any axiom A and constant c, infer $c : A$.

Finite *CS*'s constitute a representative class of constant specifications: any derivation in J4 may be regarded as a derivation in $\mathsf{J4}_{CS}$ for some finite constant specification *CS*.

The Deduction Theorem holds in $\mathsf{J4}_{CS}$ for each constant specification *CS*. The following Internalization property is characteristic for Justification Logic systems.

Theorem 1 (cf. [4]). *For an axiomatically appropriate constant specification CS,* $\mathsf{J4}_{CS}$ *enjoys Internalization:*

$$\text{If} \vdash F, \text{ then} \vdash p{:}F \text{ for some justification term } p.$$

1.2 Epistemic Semantics

A Kripke-Fitting model [13] $\mathcal{M} = (W, R, \mathcal{E}, \Vdash)$ is a Kripke model $(W, R, \Vdash)$ with transitive accessibility relation R (for J4-style systems), augmented by an *admissible evidence function* $\mathcal{E}$ which for any evidence term t and formula F, specifies the set of possible worlds where t is considered admissible evidence for F, $\mathcal{E}(t, F) \subseteq W$. The admissible evidence function $\mathcal{E}$ must satisfy the closure conditions with respect to operations $\cdot, +, !$ as follows:

- *Application:* $\mathcal{E}(s, F \to G) \cap \mathcal{E}(t, F) \subseteq \mathcal{E}(s \cdot t, G)$,
- *Sum:* $\mathcal{E}(s, F) \cup \mathcal{E}(t, F) \subseteq \mathcal{E}(s+t, F)$,
- *Verifier:* $\mathcal{E}(t, F) \subseteq \mathcal{E}(!t, t{:}F)$.

In addition, $\mathcal{E}$ should be monotone with respect to R, i.e.,

$$u \in \mathcal{E}(t, F) \text{ and } uRv \text{ yield } v \in \mathcal{E}(t, F).$$

We say that $\mathcal{E}(t, F)$ *holds at a given world* u if $u \in \mathcal{E}(t, F)$.

Given $\mathcal{M} = (W, R, \mathcal{E}, \Vdash)$, the forcing relation $\Vdash$ on all formulas is defined as follows: for $u \in W$,

1. $\Vdash$ respects Boolean connectives at each world;
2. $u \Vdash t{:}F$ iff $v \Vdash F$ for every $v \in W$ with uRv (the usual Kripke condition) and $u \in \mathcal{E}(t, F)$.

According to this definition, the admissible evidence function $\mathcal{E}$ may be regarded as a Fagin-Halpern-style awareness function [12], but equipped with the structure of justifications.

A model $\mathcal{M} = (W, R, \mathcal{E}, \Vdash)$ *respects a Constant Specification CS* if $\mathcal{E}(c, A) = W$ for all formulas $c{:}A$ from *CS*.

Theorem 2 (cf. [4]). *For any Constant Specification CS,* $\mathsf{J4}_{CS}$ *is sound and complete for the corresponding class of Kripke-Fitting models respecting CS.*

The information about Kripke structure in Kripke-Fitting models can be completely encoded by the admissible evidence function; this feature is captured by *Mkrtychev models*, which are Kripke-Fitting models with a single world. Naturally, the condition of monotonicity of the evidence function $\mathcal{E}$ with respect to the accessibility relation R becomes void in Mkrtychev models.

Theorem 3. *For any Constant Specification CS,* $\mathsf{J4}_{CS}$ *is sound and complete for the class of Mkrtychev models respecting CS.*

Mkrtychev models play an important theoretical role in establishing decidability and complexity bounds in Justification Logic [4,8,15,16,17,18,19]. Kripke-Fitting models take into account both epistemic Kripke structure and evidence structure and can be useful as natural models of epistemic scenarios.

Corollary 1 (cf. [4]). *For any constant specification CS,* $\mathsf{J4}_{CS}$ *is consistent and has a model.*

1.3 Correspondence between Modal and Justification Logics

The natural modal epistemic counterpart of the evidence assertion $t{:}F$ is $\Box F$ read as

$$\textit{for some } x,\ x{:}F.$$

This observation leads to the notion of *forgetful projection* which replaces each occurrence of $t{:}F$ by $\Box F$ and hence converts a Justification Logic sentence S to a corresponding Modal Logic sentence S^o. Obviously, different Justification Logic sentences may have the same forgetful projection, hence S^o loses certain information that was contained in S. However, it is easily observed that the forgetful projection always maps valid formulas of Justification Logic (e.g., axioms of J4) to valid formulas of a corresponding Epistemic Logic (which in our case is K4). The converse also holds: any valid formula of Epistemic Logic is a forgetful projection of some valid formula of Justification Logic. This follows from Correspondence Theorem 4. We assume that forgetful projection is naturally extended from sentences to logics.

Theorem 4 (Correspondence Theorem, cf. [4]). $\mathsf{J4}^o = \mathsf{K4}$.

This correspondence holds for other pairs of Justification and Modal systems, cf. [4]. Within the core of the Correspondence Theorem is the Realizaton Theorem.

Theorem 5 (Realization Theorem). *There is an algorithm which, for each modal formula F derivable in* K4*, assigns evidence terms to each occurrence of modality in F in such a way that the resulting formula F^r is derivable in* J4*. Moreover, the realization assigns evidence variables to the negative occurrences of modality in F, thus respecting the existential reading of epistemic modality.*

The Correspondence Theorem shows that modal logic K4 has an exact Justification Logic counterpart J4. Note that the Realization Theorem is not at all trivial. Known realization algorithms which recover evidence terms in modal theorems use cut-free derivations in the corresponding modal logics [1,2,7,8].

2 Russell's Example: Induced Factivity

In this paper we offer a Justification Logic technique of handling different justifications for the same fact, e.g., when some of the justifications are factive and some are not. We will formalize and analyze Russell's well-known example from [20].

> If a man believes that the late Prime Minister's last name began with a 'B,' he believes what is true, since the late Prime Minister was Sir Henry Campbell Bannerman[1]. But if he believes that Mr. Balfour was the late Prime Minister[2], he will still believe that the late Prime Minister's last name began with a 'B,' yet this belief, though true, would not be thought to constitute knowledge.

Here we have to deal with two justifications for a true statement, one which is correct and one which is not. Let B be a sentence (propositional atom), w be a designated evidence variable for the wrong reason for B and r a designated evidence variable for the right (hence factive) reason for B. Then, Russell's example prompts the following set of assumptions[3]:

$$\mathcal{R} = \{w{:}B,\ r{:}B,\ r{:}B \rightarrow B\}.$$

Somewhat counter to our intuition, we can logically deduce factivity of w from $\mathcal{R}$:

1. $r{:}B$ - *an assumption*;
2. $r{:}B \rightarrow B$ - *an assumption*;
3. B - *from 1 and 2, by Modus Ponens*;
4. $B \rightarrow (w{:}B \rightarrow B)$ - *a propositional axiom*;
5. $w{:}B \rightarrow B$ - *from 3 and 4, by Modus Ponens.*

The question is, how can we distinguish the 'real' factivity of $r:B$ from the 'induced factivity' of $w{:}B$ when the agent can deduce both sentences $r{:}B \rightarrow B$ and $w{:}B \rightarrow B$? The intuitive answer lies in the fact that the derivation $w{:}B \rightarrow B$ is based on the factivity of r for B. Some sort of evidence-tracking mechanism is needed here to formalize this argument.

---

[1] Which was true back in 1912. There is a linguistical problem with this example. The correct spelling of this person's last name is Campbell-Bannerman; strictly speaking, this name begins with a 'C.'

[2] Which was false in 1912.

[3] Here we ignore a possible objection that the justifications 'the late Prime Minister was Sir Henry Campbell Bannerman' and 'Mr. Balfour was the late Prime Minister' are mutually exclusive since there could be only one Prime Minister at a time. If the reader is not comfortable with this, we suggest a slight modification of Russell's example in which 'Prime Minister' is replaced by 'member of the Cabinet.' The compatibility concern then disappears since justifications 'X was the member of the late Cabinet' and 'Y was the member of the late Cabinet' with different X and Y are not necessarily incompatible.

3 Two-Agent Setting: Observer and Object Agent

Let us call 'a man' from Russell's example the *object agent*. Tracking object agent reasoning does not appear to be sufficient since the object agent can easily derive the factivity of w for B from the bare assumption $s{:}B$ for any s. Indeed,

1. $s{:}B$ - *an assumption*;
2 $B \to (w{:}B \to B)$ - *a propositional axiom*;
3. $c{:}[B \to (w{:}B \to B)]$ - *constant specification of 2*;
4. $s{:}B \to (c \cdot s){:}[w{:}B \to B]$ - *from 3, by application axiom and Modus Ponens*;
5. $(c \cdot s){:}[w{:}B \to B]$ *from 1 and 4, by Modus Ponens.*

It takes an outside observer to make such a distinction. More precisely, tracking the reasoning of the object agent, an outside observer can detect the induced character of the factivity of $w{:}B$. Note that we need Justification (vs. Modal) Logic at both levels: the object agent, since he faces different justifications of the same fact, and the observer, since we need to track the observer's evidence.

So, we consider a setting with the object agent and the observer possessing a justification system obeying $\mathsf{J4}$ and hence not necessarily factive. The fact that the latter is the observer is reflected by the condition that all assumptions of the object agent and assumptions $\mathcal{R}$ are known to the observer.

Here is the formal definition of system $\mathsf{J4}(\mathsf{J4})$. The language contains two disjoint sets of evidence terms built from variables $x, y, z, \ldots$ and constants $a, b, c, \ldots$ for the observer; variables $u, v, w, \ldots$ and constants $k, l, m, \ldots$ for the object agent. We will not be using different symbols for similar operations 'application' and 'sum' for the observer and the object agent, and hope this will not lead to ambiguity. However, for better readability, we will be using different notation for evidence assertions:

$$[\![s]\!]F \quad \sim \quad s \textit{ is a justification of } F \textit{ for the observer,}$$

$$t{:}F \quad \sim \quad t \textit{ is a justification of } F \textit{ for the object agent.}$$

Using p to denote any sentence letter, s for an evidence term of the observer, and t for an evidence term of the object agent, we define the formulas of $\mathsf{J4}(\mathsf{J4})$ by the grammar

$$F = p \mid F \wedge F \mid F \vee F \mid F \to F \mid \neg F \mid t{:}F \mid [\![s]\!]F.$$

The list of postulates of $\mathsf{J4}(\mathsf{J4})$ contains the following principles:

1. *Classical propositional axioms and rule Modus Ponens*
2. *Axioms of* $\mathsf{J4}$ *(including Total Constant Specification) for the Object Agent:*
   - [A1] *Application* $t_1{:}(F \to G) \to (t_2{:}F \to (t_! \cdot t_2){:}G)$,
   - [A2] *Monotonicity* $t_1{:}F \to (t_1 + t_2){:}F$, $t_2{:}F \to (t_1 + t_2){:}F$,
   - [A3] *Verification* $t{:}F \to\, !t{:}t{:}F$,
   - [A4] *Total Constant Specification for* 1, A1–A3: TCS_A which is

$$\{k{:}F \mid k \text{ is an object agent evidence constant and } F \text{ is from 1, A1–A3}\},$$

3. *Similar axioms of* J4 *(including Total Constant Specification) for the Observer*
   - [O1] *Application* $[\![s_1]\!](F \rightarrow G) \rightarrow ([\![s_2]\!]F \rightarrow [\![s_1 \cdot s_2]\!]G)$,
   - [O2] *Monotonicity* $[\![s_1]\!]F \rightarrow [\![s_1 + s_2]\!]F$, $[\![s_2]\!]F \rightarrow [\![s_1 + s_2]\!]F$,
   - [O3] *Verification* $[\![s]\!]F \rightarrow [\![!s]\!][\![s]\!]F$,
   - [O4] *Total Constant Specification for* 1, O1–O3: TCS_O which is

$$\{[\![a]\!]F \mid a \text{ is an observer evidence constant and } F \text{ is from 1, O1–O3}\},$$

4. *Total Constant Specification for the observer of the object agent postulates*: TCS_{OA} which is

$$\{[\![b]\!]F \mid b \text{ is an observer evidence constant and } F \text{ is from 1, A1–A4}\}.$$

System J4(J4) provides a general setup with the object agent and the observer, both of reasoning type J4, but does not yet reflect the specific structure of Russell's Prime Minister example.

4 Some Model Theory

A Kripke-Fitting model for J4(J4) is

$$\mathcal{M} = (W, R_A, R_O, \mathcal{E}_A, \mathcal{E}_O, \Vdash)$$

such that
$(W, R_A, \mathcal{E}_A, \Vdash)$ is a J4-model which respects TCS_A;
$(W, R_O, \mathcal{E}_O, \Vdash)$ is a J4-model which respects TCS_O and TCS_{OA}.
Soundness of J4(J4) with respect to these models is straightforward and follows from the soundness of J4.

Theorem 6. J4(J4) *is complete with respect to the class of* J4(J4)*-models.*

Proof. By the standard maximal consistent set construction. Let W be the set of all maximal consistent sets of J4(J4)-formulas;

$$\Gamma R_A \Delta \ \text{ iff } \ \{F \mid t{:}F \in \Gamma \text{ for some } t\} \subseteq \Delta;$$

$$\Gamma R_O \Delta \ \text{ iff } \ \{F \mid [\![s]\!]F \in \Gamma \text{ for some } s\} \subseteq \Delta;$$

$$\Gamma \in \mathcal{E}_A(t, F) \ \text{ iff } \ t{:}F \in \Gamma;$$

$$\Gamma \in \mathcal{E}_O(s, F) \ \text{ iff } \ [\![s]\!]F \in \Gamma;$$

$$\Gamma \Vdash p \ \text{ iff } \ p \in \Gamma.$$

First, we notice that R_A and R_O transitive. Second, we check the closure conditions as well as the monotonicity for $\mathcal{E}_A$ and $\mathcal{E}_O$. These are all rather standard checkups, performed in the same way as the completeness proof for J4 (cf. [4]). Finally, we have to check that evidence functions $\mathcal{E}_A$and $\mathcal{E}_O$ respect the corresponding constant specifications of J4(J4). This is secured by the definition of the evidence functions, since

$$TCS_A \cup TCS_O \cup TCS_{OA} \subseteq \Gamma$$

for every maximal consistent set Γ.

Lemma 1 (Truth Lemma). *For each formula F and world $\Gamma \in W$,*

$$\Gamma \Vdash F \quad \textit{iff} \quad F \in \Gamma.$$

Proof. The proof is also rather standard and proceeds by induction on F. The base case holds by the definition of the forcing relation $\Vdash$; Boolean connectives are straightforward. Let F be $t{:}G$. If $t{:}G \in \Gamma$, then $\Gamma \in \mathcal{E}_A(t,G)$; moreover, by the definition of R_A, $G \in \Delta$ for each Δ such that $\Gamma R_A \Delta$. By the Induction Hypothesis, $\Delta \Vdash G$, therefore, $\Gamma \Vdash t{:}G$. If $t{:}G \notin \Gamma$, then $\Gamma \notin \mathcal{E}_A(t,G)$ and $\Gamma \nVdash t{:}G$.

The Induction step in case $F = [\![s]\!]G$ is considered in a similar way. □

Corollary 2. *TCS_A, TCS_O, and TCS_{OA} hold at each node.*

Indeed, $TCS_A \cup TCS_O \cup TCS_{OA} \subseteq \Gamma$ since Γ contains all postulates of $\mathsf{J4}(\mathsf{J4})$. By the Truth Lemma, $\Gamma \Vdash TCS_A \cup TCS_O \cup TCS_{OA}$.

To complete the proof of Theorem 6, consider F which is not derivable in $\mathsf{J4}(\mathsf{J4})$. The set $\{\neg F\}$ is therefore consistent. By the standard Henkin construction, $\{\neg F\}$ can be extended to a maximal consistent set Γ. Since $F \notin \Gamma$, by the Truth Lemma, $\Gamma \nVdash F$. □

5 Distinguishing Induced Factivity

Russell's Prime Minister example can be formalized over $\mathsf{J4}(\mathsf{J4})$ by the set of assumptions $\mathcal{R}$ and $\mathcal{IR}$: the latter stands for 'Internalized Russell'

$$\mathcal{IR} = \{[\![x]\!]r{:}B,\ [\![y]\!](r{:}B \to B),\ [\![z]\!]w{:}B\}.$$

Here B, r, and w are as in $\mathcal{R}$, and x, y, z are designated proof variables for the observer.

First, we check that the observer knows the factivity of w for B, e.g., that

$$\mathsf{J4}(\mathsf{J4}) + \mathcal{R} + \mathcal{IR} \vdash [\![s]\!](w{:}B \to B)$$

for some proof term s. Here is the derivation, which is merely the internalization of the corresponding derivation from Sect. 2:

1. $[\![x]\!]r{:}B$ - *an assumption*;
2. $[\![y]\!](r{:}B \to B)$ - *an assumption*;
3. $[\![y \cdot x]\!]B$ - *from 1 and 2, by application*;
4. $[\![a]\!][B \to (w{:}B \to B)]$ - *by TCS_O for a propositional axiom*;
5. $[\![a \cdot (y \cdot x)]\!](w{:}B \to B)$ - *from 3 and 4, by application.*

Finally, let us establish that the observer cannot conclude $w{:}B \to B$ other than by using the factivity of r. In our formal setting, this amounts to proving the following theorem.

Theorem 7. *If*

$$\mathsf{J4}(\mathsf{J4}) + \mathcal{R} + \mathcal{IR} \vdash [\![\tilde{s}]\!](w{:}B \to B)\ ,$$

then term $\tilde{s}$ contains both proof variables x and y.

Proof. Following [15], we axiomatize the *reflected fragment* of $\mathsf{J4(J4)} + \mathcal{R} + \mathcal{IR}$ consisting of all formulas $[\![s]\!]F$ derivable in $\mathsf{J4(J4)} + \mathcal{R} + \mathcal{IR}$.

The principal tool here is the so-called $*$-calculus (cf. [15,19]).

Calculus

$$*[\mathsf{J4(J4)} + \mathcal{R} + \mathcal{IR}]$$

has *axioms* $TCS_O \cup TCS_{OA} \cup \mathcal{IR}$ and rules of inference

Application: given $[\![s_1]\!](F \to G)$ and $[\![s_2]\!]F$, derive $[\![s_1 \cdot s_2]\!]G$;
Sum: given $[\![s_1]\!]F$, derive $[\![s_1 + s_2]\!]F$ or $[\![s_2 + s_1]\!]F$;
Proof Checker: given $[\![s]\!]F$, derive $[\![!s]\!][\![s]\!]F$.

The following Lemma connects $\mathsf{J4(J4)} + \mathcal{R} + \mathcal{IR}$ and its reflected fragment $*[\mathsf{J4(J4)} + \mathcal{R} + \mathcal{IR}]$.

Lemma 2. *For any formula* $[\![s']\!]F$,

$$\mathsf{J4(J4)} + \mathcal{R} + \mathcal{IR} \vdash [\![s']\!]F \quad \textit{iff} \quad *[\mathsf{J4(J4)} + \mathcal{R} + \mathcal{IR}] \vdash [\![s']\!]F.$$

Proof. It is obvious that if

$$*[\mathsf{J4(J4)} + \mathcal{R} + \mathcal{IR}] \vdash [\![s']\!]F \ ,$$

then

$$\mathsf{J4(J4)} + \mathcal{R} + \mathcal{IR} \vdash [\![s']\!]F.$$

Indeed, all axioms of $*[\mathsf{J4(J4)} + \mathcal{R} + \mathcal{IR}]$ are provable in $\mathsf{J4(J4)} + \mathcal{R} + \mathcal{IR}$; the rules of the former correspond to axioms of the latter.

In order to establish the converse, let us suppose that

$$*[\mathsf{J4(J4)} + \mathcal{R} + \mathcal{IR}] \nvdash [\![s']\!]F.$$

We build a singleton $\mathsf{J4(J4)}$-model $\mathcal{M} = (W, R_A, R_O, \mathcal{E}_A, \mathcal{E}_O, \Vdash)$ in which $\mathcal{R} \cup \mathcal{IR}$ holds but $[\![s']\!]F$ does not: this will be sufficient to conclude that

$$\mathsf{J4(J4)} + \mathcal{R} + \mathcal{IR} \nvdash [\![s']\!]F.$$

- $W = \{1\}$;
- $R_A = \emptyset$, $R_O = \{(1,1)\}$;
- $\mathcal{E}_A(t, G) = W$ for each t, G;
- $\mathcal{E}_O(s, G)$ holds at 1 iff $*[\mathsf{J4(J4)} + \mathcal{R} + \mathcal{IR}] \vdash [\![s]\!]G$;
- $1 \Vdash p$ for all propositional variables, including B.

Note that R_A and R_O are transitive. Let us check the closure properties of the evidence functions. $\mathcal{E}_A$ is universal and hence closed. $\mathcal{E}_O$ is closed under *application*, *sum*, and *verifier* since the calculus $*[\mathsf{J4(J4)} + \mathcal{R} + \mathcal{IR}]$ is.

Monotonicity of $\mathcal{E}_A$ and $\mathcal{E}_O$ vacuously hold since W is a singleton.

Furthermore, $TCS_{A,O,OA}$ hold in $\mathcal{M}$. To check this, we first note that since $R_A = \emptyset$, a formula $t{:}G$ holds at 1 if and only if $\mathcal{E}_A(t, G)$. Therefore, all formulas $t{:}G$ hold at 1, in particular, $1 \Vdash TCS_A$. Hence all axioms A1–A4 of $\mathsf{J4(J4)}$ hold at 1. This yields that $1 \Vdash TCS_{OA}$. Indeed, for each $[\![c]\!]A \in TCS_{OA}$, $1 \Vdash A$ (just established) and $\mathcal{E}_O(c, A)$, since $[\![c]\!]A$ is an axiom of $*[\mathsf{J4}(\mathsf{J4}) + \mathcal{R} + \mathcal{IR}]$. By the same reasons, $1 \Vdash \mathcal{R}$.

Since

$$*[\mathsf{J4}(\mathsf{J4}) + \mathcal{R} + \mathcal{IR}] \vdash TCS_O ,$$

$\mathcal{E}_O(c, A)$ holds for all $[\![c]\!]A \in TCS_O$. In addition, each such A is an axiom O1–O3, hence $1 \Vdash A$. Therefore, $1 \Vdash TCS_O$. By similar reasons, $1 \Vdash \mathcal{IR}$.

We have just established that $\mathcal{M}$ is a model for $\mathsf{J4}(\mathsf{J4}) + \mathcal{R} + \mathcal{IR}$.

We claim that $\mathcal{M} \not\Vdash [\![s']\!]F$ which follows immediately from the assumption that

$$*[\mathsf{J4}(\mathsf{J4}) + \mathcal{R} + \mathcal{IR}] \nvdash [\![s']\!]F$$

since then $\mathcal{E}_O(s', F)$ does not hold at 1. Therefore,

$$\mathsf{J4}(\mathsf{J4}) + \mathcal{R} + \mathcal{IR} \nvdash [\![s']\!]F.$$

This concludes the proof of Lemma 2. □

Lemma 3 (Subterm property of $*$-derivations). *In a tree-form derivation of a formula $[\![s]\!]F$ in $*[\mathsf{J4}(\mathsf{J4}) + \mathcal{R} + \mathcal{IR}]$, if $[\![s']\!]G$ is derived at some node, then s' is a subterm of s.*

Proof. Obvious, from the fact that all rules of $*[\mathsf{J4}(\mathsf{J4}) + \mathcal{R} + \mathcal{IR}]$ have such a subterm property. □

Lemma 4. *If $*[\mathsf{J4}(\mathsf{J4}) + \mathcal{R} + \mathcal{IR}] \vdash [\![\tilde{s}]\!](w{:}B \rightarrow B)$, then term $\tilde{s}$ contains x.*

Proof. Suppose the opposite, i.e., that $\tilde{s}$ does not contain x. Then, by the subterm property, the proof of $[\![\tilde{s}]\!](w{:}B \rightarrow B)$ in $*[\mathsf{J4}(\mathsf{J4}) + \mathcal{R} + \mathcal{IR}]$ does not use axiom $[\![x]\!]r{:}B$. Moreover, since $*[\mathsf{J4}(\mathsf{J4}) + \mathcal{R} + \mathcal{IR}]$ does not really depend on $\mathcal{R}$, $[\![\tilde{s}]\!](w{:}B \rightarrow B)$ is derivable without $\mathcal{R}$ and $[\![x]\!]r{:}B$. Since such a proof can be replicated in $\mathsf{J4}(\mathsf{J4}) + \mathcal{IR}$ without $[\![x]\!]r{:}B$, it should be the case that

$$\mathsf{J4}(\mathsf{J4}) + [\![y]\!](r{:}B \rightarrow B) + [\![z]\!]w{:}B \vdash [\![\tilde{s}]\!](w{:}B \rightarrow B).$$

To get a contradiction, it now suffices to build a $\mathsf{J4}(\mathsf{J4})$-model

$$\mathcal{M} = (W, R_A, R_O, \mathcal{E}_A, \mathcal{E}_O, \Vdash)$$

in which $[\![y]\!](r{:}B \rightarrow B)$ and $[\![z]\!]w{:}B$ hold, but $[\![\tilde{s}]\!](w{:}B \rightarrow B)$ does not. Here is the model:

- $W = \{1\}$;
- $R_A = \emptyset$, $R_O = \{(1, 1)\}$;
- $\mathcal{E}_A(r, B) = \emptyset$ and $\mathcal{E}_A(t, F) = W$ for all other pairs t, F;
- $\mathcal{E}_O(s, G) = W$ for all s, G;
- $1 \not\Vdash p$ for all propositional variables, including B.

First, we check that $\mathcal{M}$ is a J4(J4)-model. Closure and monotonicity conditions on $R_A, R_O, \mathcal{E}_A, \mathcal{E}_O$ are obviously met. We claim that $TCS_{A,O,OA}$ hold in $\mathcal{M}$. Since $R_A = \emptyset$, a formula $t{:}F$ holds at 1 if and only if $\mathcal{E}_A(t,F)$ holds at 1. Therefore, $1 \nVdash r{:}B$ and $1 \Vdash t{:}F$ for all other pairs t, and F. In particular,

$$1 \Vdash TCS_A.$$

Since $R_O = \{(1,1)\}$ and $\mathcal{E}_O(s,G) = W$, for any observer evidence term s, $1 \Vdash [\![s]\!]G$ if and only if $1 \Vdash G$. All observer axioms hold at 1 and hence

$$1 \Vdash TCS_O.$$

As we have shown, $1 \Vdash TCS_A$ and hence

$$1 \Vdash TCS_{OA}.$$

Furthermore, since $1 \Vdash r{:}B \rightarrow B$,

$$1 \Vdash [\![y]\!](r{:}B \rightarrow B)\ ,$$

and since $1 \Vdash w{:}B$,

$$1 \Vdash [\![z]\!]w{:}B.$$

Finally, since $1 \nVdash w{:}B \rightarrow B$,

$$1 \nVdash [\![\tilde{s}]\!](w{:}B \rightarrow B).$$

□

Lemma 5. *If* $*[\mathsf{J4}(\mathsf{J4}) + \mathcal{R} + \mathcal{IR}] \vdash [\![\tilde{s}]\!](w{:}B \rightarrow B)$, *then term* $\tilde{s}$ *contains* y.

Proof. Suppose the opposite, i.e., that $\tilde{s}$ does not contain y. Then, by the subterm property, the derivation of $[\![\tilde{s}]\!](w{:}B \rightarrow B)$ in $*[\mathsf{J4}(\mathsf{J4}) + \mathcal{R} + \mathcal{IR}]$ does not use axiom $[\![y]\!](r{:}B \rightarrow B)$. From this, we can find a derivation of $[\![\tilde{s}]\!](w{:}B \rightarrow B)$ in

$$\mathsf{J4}(\mathsf{J4}) + [\![x]\!]r{:}B + [\![z]\!]w{:}B.$$

To obtain a contradiction, it suffices to present a J4(J4)-model

$$\mathcal{M} = (W, R_A, R_O, \mathcal{E}_A, \mathcal{E}_O, \Vdash)$$

in which $[\![x]\!]r{:}B$ and $[\![z]\!]w{:}B$ hold, but $[\![\tilde{s}]\!](w{:}B \rightarrow B)$ does not hold. Here is this model:

- $W = \{1\}$;
- $R_A = \emptyset$, $R_O = \{(1,1)\}$;
- $\mathcal{E}_A(t,F) = W$ for all t, F;
- $\mathcal{E}_O(s,G) = W$ for all s, G;
- $1 \nVdash p$ for all propositional variables, including B.

Conditions on $R_A, R_O, \mathcal{E}_A, \mathcal{E}_O$ are obviously met. Let us check constant specifications of J4(J4). Since $R_A = \emptyset$, and $\mathcal{E}_A(t,F)$ holds at 1 for all t, F, $t{:}F$ holds at for all t, F. In particular,

$$1 \Vdash TCS_A.$$

For the same reasons, $1 \Vdash r{:}B$ and $1 \Vdash w{:}B$.

Furthermore, $1 \Vdash [\![s]\!]F$ if and only if $1 \Vdash F$, because $\mathcal{E}_O(s, F) = \{1\}$ and $R_O = \{(1, 1)\}$. Therefore,

$$1 \Vdash TCS_{OA}.$$

Since all axioms O1–O3 are true at 1,

$$1 \Vdash TCS_O.$$

Finally, since $1 \Vdash r{:}B$ and $1 \Vdash w{:}B$,

$$1 \Vdash [\![x]\!]r{:}B \text{ and } 1 \Vdash [\![z]\!]w{:}B.$$

It remains to establish that $1 \not\Vdash [\![\tilde{s}]\!](w{:}B \rightarrow B)$, for which it suffices to check that $1 \not\Vdash w{:}B \rightarrow B$, which is the case since $1 \Vdash w{:}B$ and $1 \not\Vdash B$. □

Theorem 7 now follows from Lemmas 2, 4, and 5. □

5.1 Observer's Factivity

Another natural candidate for the observer logic is the Logic of Proofs LP (cf. [2,4,13]) which is J4 augmented by the *Factivity Axiom*

$$[\![s]\!]F \rightarrow F\ ,$$

with the corresponding extension of constant specifications to include constants corresponding to this axiom. Kripke-Fitting models for LP are J4-models with a reflexive accessibility relation.

An assumption that the observer (the reader, for example) is LP-compliant is quite reasonable since, according to [2], the Logic of Proofs LP is a universal logic of mathematical reasoning for a wide range of natural formal systems (knowers)[4]. So we could therefore define a two-agent system LP(J4) and proceed with the same evidence-tracking analysis. The main result: an analogue of Theorem 7 and its proof hold for LP(J4). In particular, all models built in the proof of Theorem 7 are intentionally made reflexive with respect to the observer's accessibility relation so they fit for LP as the observer's logic.

6 Conclusions

The formalization of Russell's example given in this paper can obviously be extended to other situations with multiple justifications of the same facts. The principal technique consisting of

[4] As was shown in [9], the same LP serves as the logic of proofs for polynomially bounded agents as well.

- introducing the observer and a two-layer reasoning system;
- working in the reflected fragment of the observer's reasoning;
- formalizing dependencies of assumptions via variable occurrences in proof terms;
- reasoning in Fitting/Mkrtychev models for formally establishing independence,

is of a general character and can be useful for evidence-tracking in a general setting.

On the other hand, the whole power of J4(J4) is not needed for Russell's example, e.g., there is no use of '+' operations here. However, we have intentionally considered J4(J4) in its entirety to introduce a basic introspective system of evidence-tracking.

Verification principles for both the object agent and the observer have been used only to simplify formulations of Constant Specifications. The same evidence tracking can be done within the framework of the basic justification logic J for both the object agent and the observer. Moreover, Theorem 7 and its proof hold for a wide range of systems, e.g., J, J4, J45 for the object agent and, independently, J, JT, J4, LP, J45, JT45, etc. (cf. [4] for the definitions) for the observer.

It appears that a similar evidence-tracking approach can be applied to analyzing paraconsistent systems. For example, the set of formulas A[5]

$$A = \{p_1,\ \ p_1 \to p_2,\ \ p_2 \to p_3,\ \ \ldots,\ \ p_{n-1} \to p_n,\ \ \neg p_n\}$$

is obviously inconsistent. However, any derivation from A which does not use *all* $n+1$ assumptions of A is contradiction-free. This argument can be naturally formalized in Justification Logic.

We wish to think that this approach to evidence tracking could be also useful in distributed knowledge systems (cf. [10,11,12]).

Acknowledgements

The author is very grateful to Mel Fitting, Vladimir Krupski, Roman Kuznets, and Elena Nogina, whose advice helped with this paper. Many thanks to Karen Kletter for editing this text.

References

1. Artemov, S.: Operational modal logic. Technical Report MSI 95-29, Cornell University (1995)
2. Artemov, S.: Explicit provability and constructive semantics. Bulletin of Symbolic Logic 7(1), 1–36 (2001)

[5] Here $p_1, p_2, \ldots, p_n$ are propositional letters.

3. Artemov, S.: Justified common knowledge. Theoretical Computer Science 357(1-3), 4–22 (2006)
4. Artemov, S.: The Logic of Justification. The Review of Symbolic Logic 1(4), 477–513 (2008)
5. Artemov, S., Kuznets, R.: Logical omniscience as a computational complexity problem. In: Heifetz, A. (ed.) Theoretical Aspects of Rationality and Knowledge. Proceedings of the Twelfth Conference (TARK 2009), Stanford University, California, July 6–8, pp. 14–23. ACM, New York (2009)
6. Artemov, S., Nogina, E.: Introducing justification into epistemic logic. J. of Logic and Computation 15(6), 1059–1073 (2005)
7. Brezhnev, V.: On explicit counterparts of modal logics. Technical Report CFIS 2000-05. Cornell University (2000)
8. Brezhnev, V., Kuznets, R.: Making knowledge explicit: How hard it is. Theoretical Computer Science 357(1-3), 23–34 (2006)
9. Goris, E.: Logic of proofs for bounded arithmetic. In: Grigoriev, D., Harrison, J., Hirsch, E.A. (eds.) CSR 2006. LNCS, vol. 3967, pp. 191–201. Springer, Heidelberg (2006)
10. Gurevich, Y., Neeman, I.: DKAL: Distributed-Knowledge Authorization Language. In: 21st IEEE Computer Security Foundations Symposium (CSF 2008), pp. 149–162 (2008)
11. Gurevich, Y., Neeman, I.: The Infon Logic. Bulletin of European Association for Theoretical Computer Science 98, 150–178 (2009)
12. Fagin, R., Halpern, J., Moses, Y., Vardi, M.: Reasoning About Knowledge. MIT Press, Cambridge (1995)
13. Fitting, M.: The logic of proofs, semantically. Annals of Pure and Applied Logic 132(1), 1–25 (2005)
14. Hintikka, J.: Knowledge and Belief. Cornell University Press, Ithaca (1962)
15. Krupski, N.V.: On the complexity of the reflected logic of proofs. Theoretical Computer Science 357(1), 136–142 (2006)
16. Kuznets, R.: On the complexity of explicit modal logics. In: Clote, P.G., Schwichtenberg, H. (eds.) CSL 2000. LNCS, vol. 1862, pp. 371–383. Springer, Heidelberg (2000)
17. Kuznets, R.: Complexity Issues in Justification Logic. Ph.D. thesis. CUNY Graduate Center (2008)
18. Milnikel, R.: Derivability in certain subsystems of the Logic of Proofs is Π_2^p-complete. Annals of Pure and Applied Logic 145(3), 223–239 (2007)
19. Mkrtychev, A.: Models for the logic of proofs. In: Adian, S., Nerode, A. (eds.) LFCS 1997. LNCS, vol. 1234, pp. 266–275. Springer, Heidelberg (1997)
20. Russell, B.: The Problems of Philosophy. Williams and Norgate/Henry Holt and Company, London/New York (1912)
21. von Wright, G.H.: An essay in modal logic. North-Holland, Amsterdam (1951)
22. Yavorskaya(Sidon), T.: Multi-agent Explicit Knowledge. In: Grigoriev, D., Harrison, J., Hirsch, E.A. (eds.) CSR 2006. LNCS, vol. 3967, pp. 369–380. Springer, Heidelberg (2006)

Strict Canonical Constructive Systems*

Arnon Avron and Ori Lahav

School of Computer Science, Tel Aviv University, Israel
{aa,orilahav}@post.tau.ac.il

To Yuri, on his seventieth birthday.

Abstract. We define the notions of a canonical inference rule and a canonical constructive system in the framework of strict single-conclusion Gentzen-type systems (or, equivalently, natural deduction systems), and develop a corresponding general non-deterministic Kripke-style semantics. We show that every strict constructive canonical system induces a class of non-deterministic Kripke-style frames, for which it is strongly sound and complete. This non-deterministic semantics is used for proving a strong form of the cut-elimination theorem for such systems, and for providing a decision procedure for them. These results identify a large family of basic constructive connectives, including the standard intuitionistic connectives, together with many other independent connectives.

Keywords: sequent calculus, cut-elimination, non-classical logics, non-deterministic semantics, Kripke semantics.

1 Introduction

The standard intuitionistic connectives ($\supset$, $\wedge$, $\vee$, and $\bot$) are of great importance in theoretical computer science, especially in type theory, where they correspond to basic operations on types (via the formulas-as-types principle and Curry-Howard isomorphism). Now a natural question is: what is so special about these connectives? The standard answer is that they are all *constructive* connectives. But then what exactly is a constructive connective, and can we define other basic constructive connectives beyond the four intuitionistic ones? And what does the last question mean anyway: how do we "define" new (or old) connectives?

Concerning the last question there is a long tradition starting from [12] (see e.g. [16] for discussions and references) according to which the meaning of a connective is determined by the introduction and elimination rules which are associated with it. Here one usually has in mind natural deduction systems of an ideal type, where each connective has its own introduction and elimination rules, and these rules should meet the following conditions: in a rule for some connective this connective should be mentioned exactly once, and no other connective should be involved. The rule should also be pure in the sense of [1] (i.e. there should be no side conditions limiting its application), and its active formulas should be immediate

* This research was supported by The Israel Science Foundation (grant no. 809-06).

A. Blass, N. Dershowitz, and W. Reisig (Eds.): Gurevich Festschrift, LNCS 6300, pp. 75–94, 2010.

subformulas of its principal formula. Now an n-ary connective $\diamond$ that can be defined using such rules may be taken as constructive if in order to prove the logical validity of a sentence of the form $\diamond(\varphi_1, \ldots, \varphi_n)$, it is necessary to prove first the premises of one of its possible introduction rules (see [9]).

Unfortunately, already the handling of negation requires rules which are not ideal in the sense described above. For intuitionistic logic this problem is usually solved by not taking negation as a basic constructive connective, but defining it instead in terms of more basic connectives that can be characterized by "ideal" rules ($\neg\varphi$ is defined as $\varphi \to\perp$). In contrast, for classical logic the problem was solved by Gentzen himself by moving to what is now known as Gentzen-type systems or sequential calculi. These calculi employ single-conclusion sequents in their intuitionistic version, and multiple-conclusion sequents in their classical version. Instead of introduction and elimination rules they use left introduction rules and right introduction rules. The intuitive notions of an "ideal rule" can be adapted to such systems in a straightforward way, and it is well known that the usual classical connectives and the basic intuitionistic connectives can indeed be fully characterized by "ideal" Gentzen-type rules. Moreover: although this can be done in several ways, in all of them the cut-elimination theorem obtains. This immediately implies that the connectives of intuitionistic logic are constructive in the sense explained above, because without using cuts the only way to derive $\Rightarrow \diamond(\varphi_1, \ldots, \varphi_n)$ in single conclusion systems of this sort is to prove first the premises of one of its introduction rules (and then apply that introduction rule). Note that the only formulas that can occur in such premises are $\varphi_1, \ldots, \varphi_n$.

For the multiple-conclusion framework the above-mentioned facts about the classical connectives were considerably generalized in [6,7] by defining "multiple-conclusion canonical propositional Gentzen-type rules and systems" in precise terms. A constructive necessary and sufficient *coherence* criterion for the non-triviality of such systems was then provided, and it was shown that a system of this kind admits cut-elimination iff it is coherent. It was further proved that the semantics of such systems is provided by two-valued non-deterministic matrices (two-valued Nmatrices) – a natural generalization of the classical truth-tables. In fact, a characteristic two-valued Nmatrix was constructed for every coherent canonical propositional system. That work shows that there is a large family of what may be called semi-classical connectives (which includes all the classical connectives), each of which has both a proof-theoretical characterization in terms of a coherent set of canonical (= "ideal") rules, and a semantic characterization using two-valued Nmatrices.

In this paper we develop a similar theory for the constructive propositional framework. We define the notions of a canonical rule and a canonical system in the framework of strict single-conclusion Gentzen-type systems (or, equivalently, natural deduction systems). We prove that here too a canonical system is non-trivial iff it is coherent (where coherence is a constructive condition, defined like in the multiple-conclusion case). We develop a general non-deterministic Kripke-style semantics for such systems, and show that every constructive canonical system (i.e. coherent canonical single-conclusion system) induces a class of non-deterministic

Kripke-style frames for which it is strongly sound and complete. We use this non-deterministic semantics to show that all constructive canonical systems admit a strong form of the cut-elimination theorem. We also use it for providing decision procedures for all such systems. These results again identify a large family of basic constructive connectives, each having both a proof-theoretical characterization in terms of a coherent set of canonical rules, and a semantic characterization using non-deterministic frames. The family includes the standard intuitionistic connectives ($\supset, \wedge, \vee$, and $\perp$), as well as many other independent connectives, like the semi-implication which has been introduced and used by Gurevich and Neeman in [13].[1]

2 Strict Canonical Constructive Systems

In what follows $\mathcal{L}$ is a propositional language, $\mathcal{F}$ is its set of wffs, p, q, r denote atomic formulas, ψ, φ, θ denote arbitrary formulas (of $\mathcal{L}$), T, S denote subsets of $\mathcal{F}$, and $\Gamma, \Delta, \Sigma, \Pi$ denote finite subsets of $\mathcal{F}$. We assume that the atomic formulas of $\mathcal{L}$ are $p_1, p_2, \ldots$ (in particular: $\{p_1, p_2, \ldots, p_n\}$ are the first n atomic formulas of $\mathcal{L}$).

Definition 1. A *(Tarskian) consequence relation* for $\mathcal{L}$ is a binary relation $\vdash$ between sets of formulas of $\mathcal{L}$ and formulas of $\mathcal{L}$ that satisfies the following conditions:

strong reflexivity: if $\varphi \in T$ then $T \vdash \varphi$.
monotonicity: if $T \vdash \varphi$ and $T \subseteq T'$ then $T' \vdash \varphi$.
transitivity (cut): if $T \vdash \psi$ and $T, \psi \vdash \varphi$ then $T \vdash \varphi$.

Definition 2. A *substitution* in $\mathcal{L}$ is a function σ from the atomic formulas to the set of formulas of $\mathcal{L}$. A substitution σ is extended to formulas and sets of formulas in the obvious way.

Definition 3. A consequence relation $\vdash$ for $\mathcal{L}$ is *structural* if for every substitution σ and every T and φ, if $T \vdash \varphi$ then $\sigma(T) \vdash \sigma(\varphi)$. A consequence relation $\vdash$ is *finitary* if the following condition holds for all T and φ: if $T \vdash \varphi$ then there exists a finite $\Gamma \subseteq T$ such that $\Gamma \vdash \varphi$. A consequence relation $\vdash$ is *consistent* (or *non-trivial*) if $p_1 \not\vdash p_2$.

It is easy to see (see [7]) that there are exactly two inconsistent structural consequence relations in any given language.[2] These consequence relations are obviously trivial, so we exclude them from our definition of a *logic*:

Definition 4. A propositional *logic* is a pair $\langle \mathcal{L}, \vdash \rangle$, where $\mathcal{L}$ is a propositional language, and $\vdash$ is a consequence relation for $\mathcal{L}$ which is structural, finitary, and consistent.

[1] The results of this paper were first stated in [5] without proofs.
[2] In one $T \vdash \varphi$ for every T and φ; in the other $T \vdash \varphi$ for every nonempty T and φ.

Since a finitary consequence relation $\vdash$ is determined by the set of pairs $\langle \Gamma, \varphi \rangle$ such that $\Gamma \vdash \varphi$, it is natural to base proof systems for logics on the use of such pairs. This is exactly what is done in natural deduction systems and in (strict) single-conclusion Gentzen-type systems (both introduced in [12]). Formally, such systems manipulate objects of the following type:

Definition 5. A *sequent* is an expression of the form $\Gamma \Rightarrow \Delta$ where Γ and Δ are finite sets of formulas, and Δ is either a singleton or empty. A sequent of the form $\Gamma \Rightarrow \{\varphi\}$ is called *definite*, and we shall denote it by $\Gamma \Rightarrow \varphi$. A sequent of the form $\Gamma \Rightarrow \{\}$ is called *negative*, and we shall denote it by $\Gamma \Rightarrow$. A Horn *clause* is a sequent which consists of atomic formulas only.

Note 1. Natural deduction systems and the strict single-conclusion Gentzen-type systems investigated in this paper manipulate only definite sequents in their derivations. However, negative sequents may be used in the formulations of their rules (in the form of negative Horn clauses).

The following definitions formulate in exact terms the idea of an "ideal rule" which was described in the introduction. We first formulate these definitions in terms of Gentzen-type systems. We consider natural deduction systems in a separate subsection.

Definition 6.

1. A *strict canonical introduction rule* for a connective $\diamond$ of arity n is an expression constructed from a set of premises and a conclusion sequent, in which $\diamond$ appears in the right side. Formally, it takes the form:
$$\{\Pi_i \Rightarrow \Sigma_i\}_{1 \leq i \leq m} / \ \Rightarrow \diamond(p_1, p_2, \ldots, p_n)$$
where m can be 0, and for all $1 \leq i \leq m$, $\Pi_i \Rightarrow \Sigma_i$ is a *definite* Horn clause such that $\Pi_i \cup \Sigma_i \subseteq \{p_1, p_2, \ldots, p_n\}$.
2. A *strict canonical elimination*[3] *rule* for a connective $\diamond$ of arity n is an expression constructed from a set of premises and a conclusion sequent, in which $\diamond$ appears in the left side. Formally, it takes the form:
$$\{\Pi_i \Rightarrow \Sigma_i\}_{1 \leq i \leq m} / \ \diamond(p_1, p_2, \ldots, p_n) \Rightarrow$$
where m can be 0, and for all $1 \leq i \leq m$, $\Pi_i \Rightarrow \Sigma_i$ is a Horn clause (either definite or negative) such that $\Pi_i \cup \Sigma_i \subseteq \{p_1, p_2, \ldots, p_n\}$.
3. An *application* of the rule $\{\Pi_i \Rightarrow \Sigma_i\}_{1 \leq i \leq m} / \ \Rightarrow \diamond(p_1, p_2, \ldots, p_n)$ is any inference step of the form:
$$\frac{\{\Gamma, \sigma(\Pi_i) \Rightarrow \sigma(\Sigma_i)\}_{1 \leq i \leq m}}{\Gamma \Rightarrow \diamond(\sigma(p_1), \ldots, \sigma(p_n))}$$
where Γ is a finite set of formulas and σ is a substitution in $\mathcal{L}$.

[3] The introduction/elimination terminology comes from the natural deduction context. For the Gentzen-type context the names "right introduction rule" and "left introduction rule" might be more appropriate, but we prefer to use a uniform terminology.

4. An *application* of the rule $\{\Pi_i \Rightarrow \Sigma_i\}_{1\leq i\leq m}/ \ \diamond(p_1, p_2, \ldots, p_n) \Rightarrow$ is any inference step of the form:

$$\frac{\{\Gamma, \sigma(\Pi_i) \Rightarrow \sigma(\Sigma_i), E_i\}_{1\leq i\leq m}}{\Gamma, \diamond(\sigma(p_1), \ldots, \sigma(p_n)) \Rightarrow \theta}$$

where Γ is a finite set of formulas, σ is a substitution in $\mathcal{L}$, θ is a formula, and for all $1 \leq i \leq m$: $E_i = \{\theta\}$ in case Σ_i is empty, and E_i is empty otherwise.

Here are some examples of well-known strict canonical rules:

Example 1 (Conjunction). The two usual rules for conjunction are:

$$\{p_1, p_2 \Rightarrow \ \} \ / \ p_1 \wedge p_2 \Rightarrow \quad \text{and} \quad \{\Rightarrow p_1 \ , \ \Rightarrow p_2\} \ / \ \Rightarrow p_1 \wedge p_2.$$

Applications of these rules have the form:

$$\frac{\Gamma, \psi, \varphi \Rightarrow \theta}{\Gamma, \psi \wedge \varphi \Rightarrow \theta} \qquad \frac{\Gamma \Rightarrow \psi \quad \Gamma \Rightarrow \varphi}{\Gamma \Rightarrow \psi \wedge \varphi}$$

The above elimination rule can easily be shown to be equivalent to the combination of the two, more usual, elimination rules for conjunction.

Example 2 (Disjunction). The two usual introduction rules for disjunction are:

$$\{\Rightarrow p_1\} \ / \ \Rightarrow p_1 \vee p_2 \quad \text{and} \quad \{\Rightarrow p_2\} \ / \ \Rightarrow p_1 \vee p_2.$$

Applications of these rules have then the form:

$$\frac{\Gamma \Rightarrow \psi}{\Gamma \Rightarrow \psi \vee \varphi} \qquad \frac{\Gamma \Rightarrow \varphi}{\Gamma \Rightarrow \psi \vee \varphi}$$

The usual elimination rule for disjunction is:

$$\{p_1 \Rightarrow \ , \ p_2 \Rightarrow\} \ / \ p_1 \vee p_2 \Rightarrow .$$

Its applications have the form:

$$\frac{\Gamma, \psi \Rightarrow \theta \quad \Gamma, \varphi \Rightarrow \theta}{\Gamma, \psi \vee \varphi \Rightarrow \theta}$$

Example 3 (Implication). The two usual rules for implication are:

$$\{\Rightarrow p_1 \ , \ p_2 \Rightarrow\} \ / \ p_1 \supset p_2 \Rightarrow \quad \text{and} \quad \{p_1 \Rightarrow p_2\} \ / \ \Rightarrow p_1 \supset p_2.$$

Applications of these rules have the form:

$$\frac{\Gamma \Rightarrow \psi \quad \Gamma, \varphi \Rightarrow \theta}{\Gamma, \psi \supset \varphi \Rightarrow \theta} \qquad \frac{\Gamma, \psi \Rightarrow \varphi}{\Gamma \Rightarrow \psi \supset \varphi}$$

Example 4 (Absurdity). In intuitionistic logic there is no introduction rule for the absurdity constant $\bot$, and there is exactly one elimination rule for it: $\{\} \ / \ \bot \Rightarrow$. Applications of this rule provide new *axioms*: $\Gamma, \bot \Rightarrow \varphi$.

Example 5 (Semi-implication). In [13] a "semi-implication" $\leadsto$ was introduced using the following two rules:

$$\{\Rightarrow p_1 \ , \ p_2 \Rightarrow\} \ / \ p_1 \leadsto p_2 \Rightarrow \quad \text{and} \quad \{\Rightarrow p_2\} \ / \ \Rightarrow p_1 \leadsto p_2.$$

Applications of these rules have the form:

$$\frac{\Gamma \Rightarrow \psi \quad \Gamma, \varphi \Rightarrow \theta}{\Gamma, \psi \leadsto \varphi \Rightarrow \theta} \qquad \frac{\Gamma \Rightarrow \varphi}{\Gamma \Rightarrow \psi \leadsto \varphi}$$

Now we define the notion of *a strict canonical Gentzen-type system.*

Definition 7. A single-conclusion Gentzen-type system is called *a strict canonical Gentzen-type system* if the following hold:

- Its axioms are all sequents of the form $\varphi \Rightarrow \varphi$.
- Cut (from $\Gamma \Rightarrow \varphi$ and $\Delta, \varphi \Rightarrow \psi$ infer $\Gamma, \Delta \Rightarrow \psi$) and weakening (from $\Gamma \Rightarrow \psi$ infer $\Gamma, \Delta \Rightarrow \psi$) are among its rules.
- Each of its other rules is either a strict canonical introduction rule or a strict canonical elimination rule.

Definition 8. Let **G** be a strict canonical Gentzen-type system.

1. A derivation of a sequent s from a set of sequents $\mathcal{S}$ in **G** is a sequence of sequents, such that each sequent in it is either an axiom, belongs to $\mathcal{S}$, or follows from previous sequents by a canonical rule of **G**. If such a derivation exists, we denote $\mathcal{S} \vdash^{seq}_{\mathbf{G}} s$.
2. The consequence relation $\vdash_{\mathbf{G}}$ between *formulas* which is induced by **G** is defined by: $T \vdash_{\mathbf{G}} \varphi$ iff there exists a finite $\Gamma \subseteq T$ such that $\vdash^{seq}_{\mathbf{G}} \Gamma \Rightarrow \varphi$.

Proposition 1. *$T \vdash_{\mathbf{G}} \varphi$ iff $\{\Rightarrow \psi \mid \psi \in T\} \vdash^{seq}_{\mathbf{G}} \Rightarrow \varphi$.*

Proposition 2. *If **G** is strict canonical then $\vdash_{\mathbf{G}}$ is a structural and finitary consequence relation.*

The last proposition does not guarantee that every strict canonical system induces a *logic* (see Definition 4). For this the system should satisfy one more condition:

Definition 9. A set $\mathcal{R}$ of strict canonical rules for an n-ary connective $\diamond$ is called *coherent* if $S_1 \cup S_2$ is classically inconsistent (and so the empty clause can be derived from it using cuts) whenever $\mathcal{R}$ contains both $S_1/\diamond(p_1, p_2, \ldots, p_n) \Rightarrow$ and $S_2/ \ \Rightarrow \diamond(p_1, p_2, \ldots, p_n)$.

Example 6. All the sets of rules for the connectives $\wedge, \vee, \supset, \bot$, and $\leadsto$ which were introduced in the examples above are coherent. For example, for the two rules for conjunction we have $S_1 = \{p_1, p_2 \Rightarrow \}$, $S_2 = \{ \Rightarrow p_1 \ , \ \Rightarrow p_2\}$, and $S_1 \cup S_2$ is the classically inconsistent set $\{p_1, p_2 \Rightarrow \ , \ \Rightarrow p_1 \ , \ \Rightarrow p_2\}$ (from which the empty sequent can be derived using two cuts).

Example 7. In [15] Prior introduced a "connective" T (which he called "Tonk") with the following rules: $\{p_1 \Rightarrow \} \ / \ p_1 T p_2 \Rightarrow$ and $\{ \Rightarrow p_2\} \ / \ \Rightarrow p_1 T p_2$. Prior then used "Tonk" to infer everything from everything (trying to show by this that a set of rules might not define any connective). Now the union of the sets of premises of these two rules is $\{p_1 \Rightarrow \ , \ \Rightarrow p_2\}$, and this is a classically consistent set of clauses. It follows that Prior's set of rules for Tonk is incoherent.

Definition 10. A strict canonical single-conclusion Gentzen-type system **G** is called *coherent* if every primitive connective of the language of **G** has a coherent set of rules in **G**.

Theorem 1. *Let* **G** *be a strict canonical Gentzen-type system.* $\langle \mathcal{L}, \vdash_{\mathbf{G}} \rangle$ *is a logic (i.e.* $\vdash_{\mathbf{G}}$ *is structural, finitary and consistent) iff* **G** *is coherent.*

Proof. Proposition 2 ensures that $\vdash_{\mathbf{G}}$ is a structural and finitary consequence relation.

That the coherence of **G** implies the consistency of the *multiple* conclusion consequence relation which is naturally induced by **G** was shown in [6,7]. That consequence relation extends $\vdash_{\mathbf{G}}$, and therefore also the latter is consistent.

For the converse, assume that **G** is incoherent. This means that **G** includes two rules $S_1 / \diamond(p_1, \ldots, p_n) \Rightarrow$ and $S_2 / \ \Rightarrow \diamond(p_1, \ldots, p_n)$, such that the set of clauses $S_1 \cup S_2$ is classically satisfiable. Let v be an assignment in $\{t, f\}$ that satisfies all the clauses in $S_1 \cup S_2$. Define a substitution σ by:

$$\sigma(p) = \begin{cases} p_{n+1} & v(p) = f \\ p & v(p) = t \end{cases}$$

Let $\Pi \Rightarrow q \in S_1 \cup S_2$. Then $\vdash_G^{seq} p_1, \ldots, p_n, \sigma(\Pi) \Rightarrow \sigma(q)$. This is trivial in case $v(q) = t$, since in this case $\sigma(q) = q \in \{p_1, \ldots, p_n\}$. On the other hand, if $v(q) = f$ then $v(p) = f$ for some $p \in \Pi$ (since v satisfies the clause $\Pi \Rightarrow q$). Therefore in this case $\sigma(p) = \sigma(q) = p_{n+1}$, and so again $p_1, \ldots, p_n, \sigma(\Pi) \Rightarrow \sigma(q)$ is trivially derived from an axiom. We can similarly prove that $\vdash_{\mathbf{G}}^{seq} p_1, \ldots, p_n, \sigma(\Pi) \Rightarrow p_{n+1}$ in case $\Pi \Rightarrow \ \in S_1 \cup S_2$. Now by applying the rules $S_1 / \diamond(p_1, \ldots, p_n) \Rightarrow$ and $S_2 / \ \Rightarrow \diamond(p_1, \ldots, p_n)$ to these provable sequents we get proofs in **G** of the sequent $p_1, \ldots, p_n \Rightarrow \diamond(\sigma(p_1), \ldots, \sigma(p_n))$ and of $p_1, \ldots, p_n, \diamond(\sigma(p_1), \ldots, \sigma(p_n)) \Rightarrow p_{n+1}$. That $\vdash_{\mathbf{G}}^{seq} p_1, \ldots, p_n \Rightarrow p_{n+1}$ then follows using a cut. This easily entails that $p_1 \vdash_{\mathbf{G}} p_2$, and hence $\vdash_{\mathbf{G}}$ is not consistent. □

The last theorem implies that coherence is a minimal demand from any acceptable strict canonical Gentzen-type system **G**. It follows that not every set of such rules is legitimate for defining constructive connectives – only coherent ones do (and this is what is wrong with "Tonk"). Accordingly we define:

Definition 11. A *strict canonical constructive system* is a coherent strict canonical single-conclusion Gentzen-type system.

The following definition will be needed in the sequel:

Definition 12. Let $\mathcal{S}$ be a set of sequents.

1. A cut is called an $\mathcal{S}$-cut if the cut formula occurs in $\mathcal{S}$.
2. We say that there exists in a system $\mathbf{G}$ an $\mathcal{S}$-(cut-free) proof of a sequent s from a set of sequents $\mathcal{S}$ iff there exists a proof of s from $\mathcal{S}$ in $\mathbf{G}$ where all cuts are $\mathcal{S}$-cuts.
3. ([2]) A system $\mathbf{G}$ admits strong cut-elimination iff whenever $\mathcal{S} \vdash^{seq}_{\mathbf{G}} s$, there exists an $\mathcal{S}$-(cut-free) proof of s from $\mathcal{S}$.[4]

2.1 Natural Deduction Version

We formulated the definitions above in terms of Gentzen-type systems. However, we could have formulated them instead in terms of natural deduction systems. The definition of canonical rules in this context is exactly as above. An application of an introduction rule is also defined exactly as above, while an application of an elimination rule of the form $\{\Pi_i \Rightarrow \Sigma_i\}_{1\leq i\leq m}/\ \diamond(p_1, p_2, \ldots, p_n) \Rightarrow$ is, in the context of natural deduction, any inference step of the form:

$$\frac{\{\Gamma, \sigma(\Pi_i) \Rightarrow \sigma(\Sigma_i), E_i\}_{1\leq i\leq m} \quad \Gamma \Rightarrow \diamond(\sigma(p_1), \ldots, \sigma(p_n))}{\Gamma \Rightarrow \theta}$$

where Γ, σ, θ and E_i are as above: Γ is a finite set of formulas, σ is a substitution in $\mathcal{L}$, θ is a formula, and for all $1 \leq i \leq m$: $E_i = \{\theta\}$ in case Σ_i is empty, and E_i is empty otherwise. We present some examples of the natural deduction version of well-known strict canonical rules. Translating our other notions and results to natural deduction systems is easy.

Example 8 (Conjunction). Applications of the rule $\{p_1, p_2 \Rightarrow\ \}\ /\ p_1 \wedge p_2 \Rightarrow$ have here the form:

$$\frac{\Gamma, \psi, \varphi \Rightarrow \theta \quad \Gamma \Rightarrow \psi \wedge \varphi}{\Gamma \Rightarrow \theta}$$

Example 9 (Disjunction). Applications of the rule $\{p_1 \Rightarrow\ ,\ \ p_2 \Rightarrow\}\ /\ p_1 \vee p_2 \Rightarrow$ have here the form:

$$\frac{\Gamma, \psi \Rightarrow \theta \quad \Gamma, \varphi \Rightarrow \theta \quad \Gamma \Rightarrow \psi \vee \varphi}{\Gamma \Rightarrow \theta}$$

Example 10 (Implication). Applications of the rule $\{\Rightarrow p_1\ ,\ \ p_2 \Rightarrow\}\ /\ p_1 \supset p_2 \Rightarrow$ have here the form:

[4] By cut-elimination we mean here just the existence of proofs without (certain forms of) cuts, rather than an algorithm to transform a given proof to a cut-free one (for the assumption-free case the term "cut-admissibility" is sometimes used).

$$\frac{\Gamma \Rightarrow \psi \quad \Gamma, \varphi \Rightarrow \theta \quad \Gamma \Rightarrow \psi \supset \varphi}{\Gamma \Rightarrow \theta}$$

This form of the rule is obviously equivalent to the more usual one (from $\Gamma \Rightarrow \psi$ and $\Gamma \Rightarrow \psi \supset \varphi$ infer $\Gamma \Rightarrow \varphi$).

Example 11 (Absurdity). In natural-deduction systems applications of the rule $\{\} \ / \ \bot\Rightarrow$ for the absurdity constant allow us to infer $\Gamma \Rightarrow \varphi$ from $\Gamma \Rightarrow\bot$.

3 Semantics for Strict Canonical Constructive Systems

The most useful semantics for propositional intuitionistic logic (the paradigmatic constructive logic) is that of Kripke frames. In this section we generalize this semantics to arbitrary strict canonical constructive systems. For this we should introduce *non-deterministic* Kripke frames.[5]

Definition 13. A generalized $\mathcal{L}$-*frame* is a triple $\mathcal{W} = \langle W, \leq, v\rangle$ such that:

1. $\langle W, \leq\rangle$ is a nonempty partially ordered set.
2. v is a function from $\mathcal{F}$ to the set of persistent functions from W into $\{t, f\}$ A function $h : W \to \{t, f\}$ is *persistent* if $h(a) = t$ implies that $h(b) = t$ for every $b \in W$ such that $a \leq b$.

Notation: We shall usually write $v(a, \varphi)$ instead of $v(\varphi)(a)$.

Definition 14. A generalized $\mathcal{L}$-frame $\langle W, \leq, v\rangle$ is a *model* of a formula φ if $v(\varphi) = \lambda a \in W.t$ (i.e. $v(a, \varphi) = t$ for every $a \in W$). It is a model of a theory T if it is a model of every $\varphi \in T$.

Definition 15. Let $\mathcal{W} = \langle W, \leq, v\rangle$ be a generalized $\mathcal{L}$-frame, and let $a \in W$.

1. A sequent $\Gamma \Rightarrow \varphi$ is *locally true* in a if either $v(a, \psi) = f$ for some $\psi \in \Gamma$, or $v(a, \varphi) = t$.
2. A sequent $\Gamma \Rightarrow \varphi$ is *true* in a if it is locally true in every $b \geq a$.
3. A sequent $\Gamma \Rightarrow$ is *(locally) true* in a if $v(a, \psi) = f$ for some $\psi \in \Gamma$.
4. The generalized $\mathcal{L}$-frame $\mathcal{W}$ is a *model* of a sequent s (either of the form $\Gamma \Rightarrow \varphi$ or $\Gamma \Rightarrow$) if s is true in every $a \in W$ (iff s is locally true in every $a \in W$). It is a model of a set of sequents $\mathcal{S}$ if it is a model of every $s \in \mathcal{S}$.

Note 2. The generalized $\mathcal{L}$-frame $\mathcal{W}$ is a model of a formula φ iff it is a model of the sequent $\Rightarrow \varphi$.

Definition 16. Let $\langle W, \leq, v\rangle$ be a generalized $\mathcal{L}$-frame. A substitution σ in $\mathcal{L}$ satisfies a Horn clause $\Pi \Rightarrow \Sigma$ in $a \in W$ if $\sigma(\Pi) \Rightarrow \sigma(\Sigma)$ is true in a.

---

[5] Another type of non-deterministic (intuitionistic) Kripke frames, based on 3-valued and 4-valued non-deterministic matrices, was used in [3,4]. Non-deterministic modal Kripke frames were recently used in [11].

Note 3. Because of the persistence condition, a definite Horn clause of the form $\Rightarrow q$ is satisfied in a by σ iff $v(a, \sigma(q)) = t$.

Definition 17. Let $\mathcal{W} = \langle W, \leq, v \rangle$ be a generalized $\mathcal{L}$-frame, and let $\diamond$ be an n-ary connective of $\mathcal{L}$.

1. The frame $\mathcal{W}$ *respects* an introduction rule r for $\diamond$ if $v(a, \diamond(\psi_1, \ldots, \psi_n)) = t$ whenever all the premises of r are satisfied in a by a substitution σ such that $\sigma(p_i) = \psi_i$ for $1 \leq i \leq n$ (The values of $\sigma(q)$ for $q \notin \{p_1, \ldots, p_n\}$ are immaterial here).
2. The frame $\mathcal{W}$ *respects* an elimination rule r for $\diamond$ if $v(a, \diamond(\psi_1, \ldots, \psi_n)) = f$ whenever all the premises of r are satisfied in a by a substitution σ such that $\sigma(p_i) = \psi_i$ $(1 \leq i \leq n)$.
3. Let **G** be a strict canonical Gentzen-type system for $\mathcal{L}$. The generalized $\mathcal{L}$-frame $\mathcal{W}$ is **G**-legal if it respects all the rules of **G**.

Example 12. By definition, a generalized $\mathcal{L}$-frame $\mathcal{W} = \langle W, \leq, v \rangle$ respects the rule $(\supset\Rightarrow)$ iff for every $a \in W$, $v(a, \varphi \supset \psi) = f$ whenever $v(b, \varphi) = t$ for every $b \geq a$ and $v(a, \psi) = f$. Because of the persistence condition, this is equivalent to $v(a, \varphi \supset \psi) = f$ whenever $v(a, \varphi) = t$ and $v(a, \psi) = f$. Again by the persistence condition, $v(a, \varphi \supset \psi) = f$ iff $v(b, \varphi \supset \psi) = f$ for some $b \geq a$. Hence, we get: $v(a, \varphi \supset \psi) = f$ whenever there exists $b \geq a$ such that $v(b, \varphi) = t$ and $v(b, \psi) = f$. The frame $\mathcal{W}$ respects $(\Rightarrow\supset)$ iff for every $a \in W$, $v(a, \varphi \supset \psi) = t$ whenever for every $b \geq a$, either $v(b, \varphi) = f$ or $v(b, \psi) = t$. Hence the two rules together impose exactly the well-known Kripke semantics for intuitionistic implication ([14]).

Example 13. A generalized $\mathcal{L}$-frame $\mathcal{W} = \langle W, \leq, v \rangle$ respects the rule $(\rightsquigarrow\Rightarrow)$ under the same conditions under which it respects $(\supset\Rightarrow)$. The frame $\mathcal{W}$ respects $(\Rightarrow\rightsquigarrow)$ iff for every $a \in W$, $v(a, \varphi \rightsquigarrow \psi) = t$ whenever $v(a, \psi) = t$ (recall that this is equivalent to $v(b, \psi) = t$ for every $b \geq a$). Note that in this case the two rules for $\rightsquigarrow$ do not always determine the value assigned to $\varphi \rightsquigarrow \psi$: if $v(a, \psi) = f$, and there is no $b \geq a$ such that $v(b, \varphi) = t$ and $v(b, \psi) = f$, then $v(a, \varphi \rightsquigarrow \psi)$ is free to be either t or f. So the semantics of this connective is non-deterministic.

Example 14. A generalized $\mathcal{L}$-frame $\mathcal{W} = \langle W, \leq, v \rangle$ respects the rule $(T \Rightarrow)$ (see Example 7) if $v(a, \varphi T \psi) = f$ whenever $v(a, \varphi) = f$. It respects $(\Rightarrow T)$ if $v(a, \varphi T \psi) = t$ whenever $v(a, \psi) = t$. The two constraints contradict each other in case both $v(a, \varphi) = f$ and $v(a, \psi) = t$. This is a semantic explanation why Prior's "connective" T ("Tonk") is meaningless.

Definition 18. Let **G** be a strict canonical constructive system.

1. We denote $\mathcal{S} \models^{seq}_{\mathbf{G}} s$ (where $\mathcal{S}$ is a set of sequents and s is a sequent) iff every **G**-legal model of $\mathcal{S}$ is also a model of s.
2. The semantic consequence relation $\models_{\mathbf{G}}$ between *formulas* which is induced by **G** is defined by: $T \models_{\mathbf{G}} \varphi$ if every **G**-legal model of T is also a model of φ.

Again we have:

Proposition 3. *$T \models_{\mathbf{G}} \varphi$ iff $\{\Rightarrow \psi \mid \psi \in T\} \models^{seq}_{\mathbf{G}} \Rightarrow \varphi$.*

4 Soundness, Completeness, Cut-Elimination

In this section we show that the two logics induced by a strict canonical constructive system $\mathbf{G}$ ($\vdash_{\mathbf{G}}$ and $\models_{\mathbf{G}}$) are identical. Half of this identity is given in the following theorem:

Theorem 2. *Every strict canonical constructive system* $\mathbf{G}$ *is strongly sound with respect to the semantics of* $\mathbf{G}$*-legal generalized frames. In other words:*

1. *If* $T \vdash_{\mathbf{G}} \varphi$ *then* $T \models_{\mathbf{G}} \varphi$.
2. *If* $\mathcal{S} \vdash^{seq}_{\mathbf{G}} s$ *then* $\mathcal{S} \models^{seq}_{\mathbf{G}} s$.

Proof. We prove the second part first. Assume that $\mathcal{S} \vdash^{seq}_{\mathbf{G}} s$, and $\mathcal{W} = \langle W, \leq, v\rangle$ is a $\mathbf{G}$-legal model of $\mathcal{S}$. We show that s is locally true in every $a \subset W$. Since the axioms of G and the premises of $\mathcal{S}$ trivially have this property, and the cut and weakening rules obviously preserve it, it suffices to show that the property of being locally true is preserved also by applications of the logical rules of $\mathbf{G}$.

- Suppose $\Gamma \Rightarrow \diamond(\psi_1, \ldots, \psi_n)$ is derived from $\{\Gamma, \sigma(\Pi_i) \Rightarrow \sigma(q_i)\}_{1\leq i\leq m}$ using the introduction rule $r = \{\Pi_i \Rightarrow \Sigma_i\}_{1\leq i\leq m}/ \ \Rightarrow \diamond(p_1, p_2, \ldots, p_n)$ (σ is a substitution such that $\sigma(p_j) = \psi_j$ for $1 \leq j \leq n$). Assume that all the premises of this application have the required property. We show that so does its conclusion. Let $a \in W$. If $v(a, \psi) = f$ for some $\psi \in \Gamma$, then obviously $\Gamma \Rightarrow \diamond(\psi_1, \ldots, \psi_n)$ is locally true in a. Assume otherwise. Then the persistence condition implies that $v(b, \psi) = t$ for every $\psi \in \Gamma$ and $b \geq a$. Hence our assumption concerning $\{\Gamma, \sigma(\Pi_i) \Rightarrow \sigma(q_i)\}_{1\leq i\leq m}$ entails that for every $b \geq a$ and $1 \leq i \leq m$, either $v(b, \psi) = f$ for some $\psi \in \sigma(\Pi_i)$, or $v(b, \sigma(q_i)) = t$. It follows that for $1 \leq i \leq m$, $\Pi_i \Rightarrow q_i$ is satisfied in a by σ. Since $\mathcal{W}$ respects r, it follows that $v(a, \diamond(\psi_1, \ldots, \psi_n)) = t$, as required.
- Now we deal with the elimination rules of $\mathbf{G}$. Suppose $\Gamma, \diamond(\psi_1, \ldots, \psi_n) \Rightarrow \theta$ is derived from $\{\Gamma, \sigma(\Pi_i) \Rightarrow \sigma(\Sigma_i)\}_{1\leq i\leq m_1}$ and $\{\Gamma, \sigma(\Pi_i) \Rightarrow \theta\}_{m_1+1\leq i\leq m}$, using the elimination rule $r = \{\Pi_i \Rightarrow \Sigma_i\}_{1\leq i\leq m}/ \ \diamond(p_1, p_2, \ldots, p_n) \Rightarrow$ (where Σ_i is empty for $m_1 + 1 \leq i \leq m$, and σ is a substitution such that $\sigma(p_j) = \psi_j$ for $1 \leq j \leq n$). Assume that all the premises of this application have the required property. Let $a \in W$. If $v(a, \psi) = f$ for some $\psi \in \Gamma$ or $v(a, \theta) = t$, then we are done. Assume otherwise. Then $v(a, \theta) = f$, and (by the persistence condition) $v(b, \psi) = t$ for every $\psi \in \Gamma$ and $b \geq a$. Hence our assumption concerning $\{\Gamma, \sigma(\Pi_i) \Rightarrow \sigma(\Sigma_i)\}_{1\leq i\leq m_1}$ entails that for every $b \geq a$ and $1 \leq i \leq m_1$, either $v(b, \psi) = f$ for some $\psi \in \sigma(\Pi_i)$, or $v(b, \sigma(\Sigma_i)) = t$. This immediately implies that every definite premise of the rule is satisfied in a by σ. Since $v(a, \theta) = f$, our assumption concerning $\{\Gamma, \sigma(\Pi_i) \Rightarrow \theta\}_{m_1+1\leq i\leq m}$ entails that for every $m_1 + 1 \leq i \leq m$, $v(a, \psi) = f$ for some $\psi \in \sigma(\Pi_i)$. Hence the negative premises of the rule are also satisfied in a by σ. Since $\mathcal{W}$ respects r, it follows that $v(a, \diamond(\psi_1, \ldots, \psi_n)) = f$, as required.

The first part follows from the second by Propositions 1 and 3. □

For the converse, we first prove the following key result.

Theorem 3. *Let* **G** *be a strict canonical constructive system in* $\mathcal{L}$*, and let* $\mathcal{S} \cup \{s\}$ *be a set of sequents in* $\mathcal{L}$*. Then either there is an* $\mathcal{S}$*-(cut-free) proof of* s *from* $\mathcal{S}$*, or there is a* **G***-legal model of* $\mathcal{S}$ *which is not a model of* s*.*

Proof. Assume that $s = \Gamma_0 \Rightarrow \varphi_0$ does not have an $\mathcal{S}$-(cut-free) proof in **G**. Let $\mathcal{F}'$ be the set of subformulas of $\mathcal{S} \cup \{s\}$. Given a formula $\varphi \in \mathcal{F}'$, call a theory $\mathcal{T} \subseteq \mathcal{F}'$ *φ-maximal* if there is no finite $\Gamma \subseteq \mathcal{T}$ such that $\Gamma \Rightarrow \varphi$ has an $\mathcal{S}$-(cut-free) proof from $\mathcal{S}$, but every proper extension $\mathcal{T}' \subseteq \mathcal{F}'$ of $\mathcal{T}$ contains such a finite subset Γ. Obviously, if $\Gamma \subseteq \mathcal{F}'$, $\varphi \in \mathcal{F}'$ and $\Gamma \Rightarrow \varphi$ has no $\mathcal{S}$-(cut-free) proof from $\mathcal{S}$, then Γ can be extended to a theory $\mathcal{T} \subseteq \mathcal{F}'$ which is φ-maximal. In particular: Γ_0 can be extended to a φ_0-maximal theory $\mathcal{T}_0$.

Now let $\mathcal{W} = \langle W, \subseteq, v \rangle$, where:

- W is the set of all extensions of $\mathcal{T}_0$ in $\mathcal{F}'$ which are φ-maximal for some $\varphi \in \mathcal{F}'$.
- v is defined inductively as follows. For atomic formulas:

$$v(\mathcal{T}, p) = \begin{cases} t & p \in \mathcal{T} \\ f & p \notin \mathcal{T} \end{cases}$$

  Suppose $v(\mathcal{T}, \psi_i)$ has been defined for all $\mathcal{T} \in W$ and $1 \leq i \leq n$. We let $v(\mathcal{T}, \diamond(\psi_1, \ldots, \psi_n)) = t$ iff at least one of the following holds:
  1. There exists an introduction rule for $\diamond$ whose set of premises is satisfied in $\mathcal{T}$ by a substitution σ such that $\sigma(p_i) = \psi_i$ $(1 \leq i \leq n)$.
  2. $\diamond(\psi_1, \ldots, \psi_n) \in \mathcal{T}$ and there does not exist $\mathcal{T}' \in W$, $\mathcal{T} \subseteq \mathcal{T}'$, and an elimination rule for $\diamond$ whose set of premises is satisfied in $\mathcal{T}'$ by a substitution σ such that $\sigma(p_i) = \psi_i$ $(1 \leq i \leq n)$.[6]

First we prove that $\mathcal{W}$ is a generalized $\mathcal{L}$-frame:

- W is not empty because $\mathcal{T}_0 \in W$.
- We prove by structural induction that v is persistent:
  For atomic formulas v is trivially persistent since the order is $\subseteq$.
  Assume that v is persistent for $\psi_1, \ldots, \psi_n$. We prove its persistence for $\diamond(\psi_1, \ldots, \psi_n)$. So assume that $v(\mathcal{T}, \diamond(\psi_1, \ldots, \psi_n)) = t$ and $\mathcal{T} \subseteq \mathcal{T}^*$. By v's definition there are two possibilities:
  1. There exists an introduction rule for $\diamond$ whose set of premises is satisfied in $\mathcal{T}$ by a substitution σ such that $\sigma(p_i) = \psi_i$ $(1 \leq i \leq n)$. In such a case, the premises are all definite Horn clauses. Hence by definition, σ satisfies the same rule's premises also in $\mathcal{T}^*$, and so $v(\mathcal{T}^*, \diamond(\psi_1, \ldots, \psi_n)) = t$.

[6] This inductive definition isn't absolutely formal, since satisfaction by a substitution is defined for a generalized $\mathcal{L}$-frame, which we are in the middle of constructing, but the intention should be clear.

2. $\diamond(\psi_1, \ldots, \psi_n) \in \mathcal{T}$ and there does not exist $\mathcal{T}' \in W$, $\mathcal{T} \subseteq \mathcal{T}'$, and an elimination rule for $\diamond$ whose set of premises is satisfied in $\mathcal{T}'$ by a substitution σ such that $\sigma(p_i) = \psi_i$ $(1 \leq i \leq n)$. Then $\diamond(\psi_1, \ldots, \psi_n) \in \mathcal{T}^*$ (since $\mathcal{T} \subseteq \mathcal{T}^*$), and there surely does not exist $\mathcal{T}' \in W$, $\mathcal{T}^* \subseteq \mathcal{T}'$, and an elimination rule for $\diamond$ whose set of premises is satisfied in $\mathcal{T}'$ by a substitution σ such that $\sigma(p_i) = \psi_i$ $(1 \leq i \leq n)$ (otherwise the same would hold for $\mathcal{T}$). It follows that $v(\mathcal{T}^*, \diamond(\psi_1, \ldots, \psi_n)) = t$ in this case too.

Next we prove that $\mathcal{W}$ is **G**-legal:

1. The introduction rules are directly respected by the first condition in v's definition.
2. Let r be an elimination rule for $\diamond$, and suppose all its premises are satisfied in some $\mathcal{T} \in W$ by a substitution σ such that $\sigma(p_i) = \psi_i$. Then neither of the conditions under which $v(\mathcal{T}, \diamond(\psi_1, \ldots, \psi_n)) = t$ can hold:
   (a) The second condition explicitly excludes the option that all the premises are satisfied (in any $\mathcal{T}' \in W$, $\mathcal{T} \subseteq \mathcal{T}'$, so also in $\mathcal{T}$ itself).
   (b) The first condition cannot be met because of **G**'s coherence, which does not allow the two sets of premises (of an introduction rule and an elimination rule) to be satisfied together. To see this, assume for contradiction that S_1 is the set of premises of an elimination rule for $\diamond$, S_2 is the set of premises of an introduction rule for $\diamond$, and there exists $\mathcal{T} \in W$ in which both sets of premises are satisfied by a substitution σ such that $\sigma(p_i) = \psi_i$ $(1 \leq i \leq n)$. Let u be an assignment in $\{t, f\}$ in which $u(p_i) = v(\mathcal{T}, \psi_i)$. Since σ satisfies in $\mathcal{T}$ both sets of premises, u classically satisfies S_1 and S_2. But, **G** is coherent, i.e. $S_1 \cup S_2$ is classically inconsistent. A contradiction.

   It follows that $v(\mathcal{T}, \diamond(\psi_1, \ldots, \psi_n)) = f$, as required.

It remains to prove that $\mathcal{W}$ is a model of $\mathcal{S}$ but not of s. For this we first prove that the following hold for every $\mathcal{T} \in W$ and every formula $\psi \in \mathcal{F}'$:

(a) If $\psi \in \mathcal{T}$ then $v(\mathcal{T}, \psi) = t$.
(b) If $\mathcal{T}$ is ψ-maximal then $v(\mathcal{T}, \psi) = f$.

We prove **(a)** and **(b)** together by a simultaneous induction on the complexity of ψ. For atomic formulas they easily follow from v's definition, and the fact that $p \Rightarrow p$ is an axiom. For the induction step, assume that **(a)** and **(b)** hold for $\psi_1, \ldots, \psi_n \in \mathcal{F}'$. We prove them for $\diamond(\psi_1, \ldots, \psi_n) \in \mathcal{F}'$.

- Assume that $\diamond(\psi_1, \ldots, \psi_n) \in \mathcal{T}$, but $v(\mathcal{T}, \diamond(\psi_1, \ldots, \psi_n)) = f$. By v's definition, since $\diamond(\psi_1, \ldots, \psi_n) \in \mathcal{T}$ there should exist $\mathcal{T}' \in W$, $\mathcal{T} \subseteq \mathcal{T}'$, and an elimination rule for $\diamond$, r, whose set of premises is satisfied in $\mathcal{T}'$ by a substitution σ such that $\sigma(p_i) = \psi_i$ $(1 \leq i \leq n)$. Let $\{\Pi_i \Rightarrow\}_{1 \leq i \leq m_1}$ be the negative premises of r, and $\{\Pi_i \Rightarrow q_i\}_{m_1+1 \leq i \leq m}$ – the definite ones. Since σ satisfies in $\mathcal{T}'$ every sequent in $\{\Pi_i \Rightarrow\}_{1 \leq i \leq m_1}$, then for all $1 \leq i \leq m_1$ there exists $\psi_{j_i} \in \sigma(\Pi_i)$ such that $v(\mathcal{T}', \psi_{j_i}) = f$. By the induction hypothesis this implies that for all $1 \leq i \leq m_1$, there exists $\psi_{j_i} \in \sigma(\Pi_i)$ such that $\psi_{j_i} \notin \mathcal{T}'$.

Let φ be the formula for which $\mathcal{T}'$ is maximal. Then for all $1 \leq i \leq m_1$ there is a finite $\Delta_i \subseteq \mathcal{T}'$ such that $\Delta_i, \psi_{j_i} \Rightarrow \varphi$ has an $\mathcal{S}$-(cut-free) proof from $\mathcal{S}$, and so $\Delta_i, \sigma(\Pi_i) \Rightarrow \varphi$ has such a proof. This in turn implies that there must exist $m_1 + 1 \leq i_0 \leq m$ such that $\Gamma, \sigma(\Pi_{i_0}) \Rightarrow \sigma(q_{i_0})$ has no $\mathcal{S}$-(cut-free) proof from $\mathcal{S}$ for any finite $\Gamma \subseteq \mathcal{T}'$. Indeed, if such a proof exists for every $m_1 + 1 \leq i \leq m$, we would use the m_1 proofs of $\Delta_i, \sigma(\Pi_i) \Rightarrow \varphi$ for $1 \leq i \leq m_1$, the $m - m_1$ proofs for $\Gamma'_i, \sigma(\Pi_i) \Rightarrow \sigma(q_i)$ for $m_1 + 1 \leq i \leq m$, some trivial weakenings, and the elimination rule r to get an $\mathcal{S}$-(cut-free) proof from $\mathcal{S}$ of the sequent $\cup_{i=1}^{i=m_1} \Delta_i, \cup_{i=m_1+1}^{i=m} \Gamma_i, \diamond(\psi_1, \ldots, \psi_n) \Rightarrow \varphi$. Since $\diamond(\psi_1, \ldots, \psi_n) \in \mathcal{T}$, this would contradict $\mathcal{T}'$'s φ-maximality. Using this i_0, we extend $\mathcal{T}' \cup \sigma(\Pi_{i_0})$ to a theory $\mathcal{T}''$ which is $\sigma(q_{i_0})$-maximal. By the induction hypothesis $v(\mathcal{T}'', \psi) = t$ for all $\psi \in \sigma(\Pi_{i_0})$ and $v(\mathcal{T}'', \sigma(q_{i_0})) = f$. Since $\mathcal{T}' \subseteq \mathcal{T}''$, this contradicts the fact that σ satisfies $\Pi_{i_0} \Rightarrow q_{i_0}$ in $\mathcal{T}'$.

- Assume that $\mathcal{T}$ is $\diamond(\psi_1, \ldots, \psi_n)$-maximal, but $v(\mathcal{T}, \diamond(\psi_1, \ldots, \psi_n)) = t$. Obviously, $\diamond(\psi_1, \ldots, \psi_n) \notin \mathcal{T}$ (because $\diamond(\psi_1, \ldots, \psi_n) \Rightarrow \diamond(\psi_1, \ldots, \psi_n)$ is an axiom). Hence by v's definition there exists an introduction rule for $\diamond$, r, whose set of premises is satisfied in $\mathcal{T}$ by a substitution σ such that $\sigma(p_i) = \psi_i$ $(1 \leq i \leq n)$. Let $\{\Pi_i \Rightarrow q_i\}_{1 \leq i \leq m}$ be the premises of r. As in the previous case, there must exist $1 \leq i_0 \leq m$ such that $\Gamma, \sigma(\Pi_{i_0}) \Rightarrow \sigma(q_{i_0})$ has no $\mathcal{S}$-(cut-free) proof from $\mathcal{S}$ for any finite $\Gamma \subseteq \mathcal{T}$ (if such a proof exists for all $1 \leq i \leq m$ with finite $\Gamma_i \subseteq \mathcal{T}$ than we could have an $\mathcal{S}$-(cut-free) proof from $\mathcal{S}$ of $\cup_{i=1}^{i=m} \Gamma_i \Rightarrow \diamond(\psi_1, \ldots, \psi_n)$ using the m proofs of $\Gamma_i, \sigma(\Pi_i) \Rightarrow \sigma(q_i)$, some weakenings, and r). Using this i_0, we extend $\mathcal{T} \cup \sigma(\Pi_{i_0})$ to a theory $\mathcal{T}'$ which is $\sigma(q_{i_0})$-maximal. By the induction hypothesis, $v(\mathcal{T}', \psi) = t$ for all $\psi \in \sigma(\Pi_{i_0})$ and $v(\mathcal{T}', \sigma(q_{i_0})) = f$. Since $\mathcal{T} \subseteq \mathcal{T}'$, this contradicts the fact that σ satisfies $\Pi_{i_0} \Rightarrow q_{i_0}$ in $\mathcal{T}$.

Next we note that **(b)** can be strengthened as follows:

(c) If $\psi \in \mathcal{F}'$, $\mathcal{T} \in W$ and there is no finite $\Gamma \subseteq \mathcal{T}$ such that $\Gamma \Rightarrow \psi$ has an $\mathcal{S}$-(cut-free) proof from $\mathcal{S}$, then $v(\mathcal{T}, \psi) = f$.

Indeed, under these conditions $\mathcal{T}$ can be extended to a ψ-maximal theory $\mathcal{T}'$. Now $\mathcal{T}' \in W$, $\mathcal{T} \subseteq \mathcal{T}'$, and by **(b)**, $v(\mathcal{T}', \psi) = f$. Hence also $v(\mathcal{T}, \psi) = f$.

Now **(a)** and **(b)** together imply that $v(\mathcal{T}_0, \psi) = t$ for every $\psi \in \Gamma_0 \subseteq \mathcal{T}_0$, and $v(\mathcal{T}_0, \varphi_0) = f$. Hence $\mathcal{W}$ is not a model of s. We end the proof by showing that $\mathcal{W}$ is a model of $\mathcal{S}$. So let $\psi_1, \ldots, \psi_n \Rightarrow \theta \in \mathcal{S}$ and let $\mathcal{T} \in W$, where $\mathcal{T}$ is φ-maximal. Assume by way of contradiction that $v(\mathcal{T}, \psi_i) = t$ for $1 \leq i \leq n$, while $v(\mathcal{T}, \theta) = f$. By **(c)**, for every $1 \leq i \leq n$ there is a finite $\Gamma_i \subseteq \mathcal{T}$ such that $\Gamma_i \Rightarrow \psi_i$ has an $\mathcal{S}$-(cut-free) proof from $\mathcal{S}$. On the other hand $v(\mathcal{T}, \theta) = f$ implies (by **(a)**) that $\theta \notin \mathcal{T}$. Since $\mathcal{T}$ is φ-maximal, it follows that there is a finite $\Sigma \subseteq \mathcal{T}$ such that $\Sigma, \theta \Rightarrow \varphi$ has an $\mathcal{S}$-(cut-free) proof from $\mathcal{S}$. Now from $\Gamma_i \Rightarrow \psi_i$ $(1 \leq i \leq n)$, $\Sigma, \theta \Rightarrow \varphi$, and $\psi_1, \ldots, \psi_n \Rightarrow \theta$ one can infer $\Gamma_1, \ldots, \Gamma_n, \Sigma \Rightarrow \varphi$ by $n+1$ $\mathcal{S}$-cuts (on $\psi_1, \ldots, \psi_n$ and θ). It follows that the last sequent has an $\mathcal{S}$-(cut-free) proof from $\mathcal{S}$. Since $\Gamma_1, \ldots, \Gamma_n, \Sigma \subseteq \mathcal{T}$, this contradicts the φ-maximality of $\mathcal{T}$. $\square$

Theorem 4 (Soundness and Completeness). *Every strict canonical constructive system* $\mathbf{G}$ *is strongly sound and complete with respect to the semantics of* $\mathbf{G}$*-legal generalized frames. In other words:*

1. $\mathcal{T} \vdash_{\mathbf{G}} \varphi$ *iff* $\mathcal{T} \models_{\mathbf{G}} \varphi$.
2. $\mathcal{S} \vdash_{\mathbf{G}}^{seq} s$ *iff* $\mathcal{S} \models_{\mathbf{G}}^{seq} s$.

Proof. Immediate from Theorems 3 and 2, and Propositions 1, 3. □

Corollary 1. *If* $\mathbf{G}$ *is a strict canonical constructive system in* $\mathcal{L}$ *then* $\langle \mathcal{L}, \models_{\mathbf{G}} \rangle$ *is a logic.*

Corollary 2 (Compactness). *Let* $\mathbf{G}$ *be a strict canonical constructive system.*

1. *If* $\mathcal{S} \models_{\mathbf{G}}^{seq} s$ *then there exists a finite* $\mathcal{S}' \subseteq \mathcal{S}$ *such that* $\mathcal{S}' \models_{\mathbf{G}}^{seq} s$.
2. $\models_{\mathbf{G}}$ *is finitary.*

Theorem 5 (General Strong Cut Elimination Theorem).

1. *Every strict canonical constructive system* $\mathbf{G}$ *admits strong cut-elimination (see Definition 12).*
2. *in a strict canonical constructive system* $\mathbf{G}$ *iff it has a cut-free proof there.*

Proof. The first part follows from Theorems 4 and 3. The second part is a special case of the first, where the set $\mathcal{S}$ of premises is empty. □

Corollary 3. *The four following conditions are equivalent for a strict canonical single-conclusion Gentzen-type system* $\mathbf{G}$*:*

1. $\langle \mathcal{L}, \vdash_{\mathbf{G}} \rangle$ *is a logic (by Proposition 2, this means that* $\vdash_{\mathbf{G}}$ *is consistent).*
2. $\mathbf{G}$ *is coherent.*
3. $\mathbf{G}$ *admits strong cut-elimination.*
4. $\mathbf{G}$ *admits cut-elimination.*

Proof. Condition 1 implies condition 2 by Theorem 1. Condition 2 implies condition 3 by Theorem 5. Condition 3 trivially implies condition 4. Finally, without using cuts there is no way to derive $p_1 \Rightarrow p_2$ in a strict canonical Gentzen-type system. Hence condition 4 implies condition 1. □

5 Analycity and Decidability

In general, in order for a denotational semantics of a propositional logic to be useful and effective, it should be *analytic*. This means that to determine whether a formula φ follows from a theory $\mathcal{T}$, it suffices to consider *partial* valuations, defined on the set of all subformulas of the formulas in $\mathcal{T} \cup \{\varphi\}$. Now we show that the semantics of $\mathbf{G}$-legal frames is analytic in this sense.

Definition 19. Let $\mathbf{G}$ be a strict canonical constructive system for $\mathcal{L}$. A $\mathbf{G}$-legal *semiframe* is a triple $\mathcal{W}' = \langle W, \leq, v' \rangle$ such that:

1. $\langle W, \leq \rangle$ is a nonempty partially ordered set.
2. v' is a partial function from the set of formulas of $\mathcal{L}$ into the set of persistent functions from W into $\{t, f\}$ such that:
   - $\mathcal{F}'$, the domain of v', is closed under subformulas.
   - v' respects the rules of **G** on $\mathcal{F}'$ (e.g. if r is an introduction rule for an n-ary connective $\diamond$, and $\diamond(\psi_1, \dots, \psi_n) \in \mathcal{F}'$, then $v(a, \diamond(\psi_1, \dots, \psi_n)) = t$ whenever all the premises of r are satisfied in a by a substitution σ such that $\sigma(p_i) = \psi_i$ $(1 \leq i \leq n)$).

Theorem 6. *Let* **G** *be a strict canonical constructive system for* $\mathcal{L}$*. Then the semantics of* **G***-legal frames is* analytic *in the following sense:*
If $\mathcal{W}' = \langle W, \leq, v' \rangle$ *is a* **G***-legal* semiframe, *then* v' *can be extended to a function* v *so that* $\mathcal{W} = \langle W, \leq, v \rangle$ *is a* **G***-legal frame.*

Proof. Let $\mathcal{W}' = \langle W, \leq, v' \rangle$ be a **G**-legal semiframe. We recursively extend v' to a total function v. For atomic p we let $v(p) = v'(p)$ if $v'(p)$ is defined, and $v(p) = \lambda a \in W.t$ (say) otherwise. For $\varphi = \diamond(\psi_1, \dots, \psi_n)$ we let $v(\varphi) = v'(\varphi)$ whenever $v'(\varphi)$ is defined, and otherwise we define $v(\varphi, a) = f$ iff there exists an elimination rule r with $\diamond(p_1, \dots, p_n) \Rightarrow$ as its conclusion, and an element $b \geq a$ of W, such that all premises of r are satisfied in b (with respect to $\langle W, \leq, v \rangle$) by a substitution σ such that $\sigma(p_j) = \psi_j$ $(1 \leq j \leq n)$. Note that the satisfaction of the premises of r by σ in elements of W depends only on the values assigned by v to $\psi_1, \dots, \psi_n$, so the recursion works, and v is well defined. From the definition of v and the assumption that $\mathcal{W}'$ is a **G**-legal semiframe, it immediately follows that v is an extension of v', that $v(\varphi)$ is a persistent function for every φ (so $\mathcal{W} = \langle W, \leq, v \rangle$ is a generalized $\mathcal{L}$-frame), and that $\mathcal{W}$ respects all the elimination rules of **G**. Hence it only remains to prove that it respects also the introduction rules of **G**. Let $r = \{\Pi_i \Rightarrow q_i\}_{1 \leq i \leq m} / \Rightarrow \diamond(p_1, p_2, \dots, p_n)$ be such a rule, and assume that for every $1 \leq i \leq m$, $\sigma(\Pi_i) \Rightarrow \sigma(q_i)$ is true in a with respect to $\langle W, \leq, v \rangle$. We should show that $v(a, \diamond(\psi_1, \dots, \psi_n)) = t$.

If $v'(a, \diamond(\psi_1, \dots, \psi_n))$ is defined, then since its domain is closed under subformulas, for every $1 \leq i \leq n$ and every $b \in W$ $v'(b, \psi_i)$ is defined. In this case, our construction ensures that for every $1 \leq i \leq n$ and every $b \in W$ we have $v'(b, \psi_i) = v(b, \psi_i)$. Therefore, since for every $1 \leq i \leq m$, $\sigma(\Pi_i) \Rightarrow \sigma(q_i)$ is locally true in every $b \geq a$ with respect to $\langle W, \leq, v \rangle$, it is also locally true with respect to $\langle W, \leq, v' \rangle$. Since v' respects r, $v'(a, \diamond(\psi_1, \dots, \psi_n)) = t$, so $v(a, \diamond(\psi_1, \dots, \psi_n)) = t$ as well, as required.

Now, assume $v'(a, \diamond(\psi_1, \dots, \psi_n))$ is not defined, and assume by way of contradiction that $v(a, \diamond(\psi_1, \dots, \psi_n)) = f$. So, there exists $b \geq a$ and an elimination rule $\{\Delta_j \Rightarrow \Sigma_j\}_{1 \leq j \leq k} / \diamond(p_1, p_2, \dots, p_n) \Rightarrow$ such that $\sigma(\Delta_j) \Rightarrow \sigma(\Sigma_j)$ is locally true in b for $1 \leq j \leq k$. Since $b \geq a$, our assumption about a implies that $\sigma(\Pi_i) \Rightarrow \sigma(q_i)$ is locally true in b for $1 \leq i \leq m$. It follows that by defining $u(p) = v(b, \sigma(p))$ we get a valuation u in $\{t, f\}$ which satisfies all the clauses in the union of $\{\Pi_i \Rightarrow q_i \mid 1 \leq i \leq m\}$ and $\{\Delta_j \Rightarrow \Sigma_j \mid 1 \leq j \leq k\}$. This contradicts the coherence of **G**. □

The following two theorems are now easy consequences of Theorem 6 and the soundness and completeness theorems of the previous section:[7]

Theorem 7. *Let* **G** *be a strict canonical constructive system. Then* **G** *is strongly decidable: Given a finite set* $\mathcal{S}$ *of sequents, and a sequent* s*, it is decidable whether* $\mathcal{S} \vdash^{seq}_{\mathbf{G}} s$ *or not. In particular: it is decidable whether* $\Gamma \vdash_{\mathbf{G}} \varphi$*, where* φ *is formula and* Γ *is a finite set of formulas.*

Proof. Let $\mathcal{F}'$ be the set of subformulas of the formulas in $\mathcal{S} \cup \{s\}$. From Theorem 6 and the proof of Theorem 3 it easily follows that in order to decide whether $\mathcal{S} \vdash^{seq}_{\mathbf{G}} s$ it suffices to check all triples of the form $\langle W, \subseteq, v' \rangle$ where $W \subseteq 2^{\mathcal{F}'}$ and $v' : \mathcal{F}' \to (W \to \{t, f\})$, and see if any of them is a **G**-legal semiframe which is a model of $\mathcal{S}$ but not a model of s. □

Theorem 8. *Let* $\mathbf{G_1}$ *be a strict canonical constructive system in a language* $\mathcal{L}_1$*, and let* $\mathbf{G_2}$ *be a strict canonical constructive system in a language* $\mathcal{L}_2$*. Assume that* $\mathcal{L}_2$ *is an extension of* $\mathcal{L}_1$ *by some set of connectives, and that* $\mathbf{G_2}$ *is obtained from* $\mathbf{G_1}$ *by adding to the latter strict canonical rules for connectives in* $\mathcal{L}_2 - \mathcal{L}_1$*. Then* $\mathbf{G_2}$ *is a conservative extension of* $\mathbf{G_1}$ *(i.e. if all formulas in* $\mathcal{T} \cup \{\varphi\}$ *are in* $\mathcal{L}_1$ *then* $\mathcal{T} \vdash_{\mathbf{G_1}} \varphi$ *iff* $\mathcal{T} \vdash_{\mathbf{G_2}} \varphi$*).*

Proof. Suppose that $\mathcal{T} \not\vdash_{\mathbf{G_1}} \varphi$. Then there is $\mathbf{G_1}$-legal model $\mathcal{W}$ of $\mathcal{T}$ which is not a model of φ. Since the set of formulas of $\mathcal{L}_1$ is a subset of the set of formulas of $\mathcal{L}_2$ which is closed under subformulas, Theorem 6 implies that $\mathcal{W}$ can be extended to a $\mathbf{G_2}$-legal model of $\mathcal{T}$ which is not a model of φ. Hence $\mathcal{T} \not\vdash_{\mathbf{G_2}} \varphi$. □

Note 4. Prior's "connective" Tonk ([15]) has made it clear that not every combination of "ideal" introduction and elimination rules can be used for defining a connective. Some constraints should be imposed on the set of rules. Such a constraint was indeed suggested by Belnap in his famous [8]: the rules for a connective $\diamond$ should be *conservative*, in the sense that if $\mathcal{T} \vdash \varphi$ is derivable using them, and $\diamond$ does not occur in $\mathcal{T} \cup \varphi$, then $\mathcal{T} \vdash \varphi$ can also be derived without using the rules for $\diamond$. This solution to the problem has two problematic aspects:

1. Belnap did not provide any effective necessary and sufficient criterion for checking whether a given set of rules is conservative in the above sense. Without such criterion every connective defined by inference rules (without an independent denotational semantics) is suspected of being a Tonk-like connective, and should not be used until a proof is given that it is "innocent".
2. Belnap formulated the condition of conservativity only with respect to the basic deduction framework, in which no connectives are assumed. But nothing in what he wrote excludes the possibility of a system G having two connectives, each of them "defined" by a set of rules which is conservative

[7] The two theorems can also be proved directly from the cut-elimination theorem for strict canonical constructive systems.

over the basic system B, while G itself is not conservative over B. If this happens then it will follow from Belnap's thesis that each of the two connectives is well-defined and meaningful, but they cannot exist together. Such a situation is almost as paradoxical as that described by Prior.

Now the first of these two objections is met, of course, by our coherence criterion for strict canonical systems, since coherence of a finite set of strict canonical rules can effectively be checked. The second is met by Theorem 8. That theorem shows that a very strong form of Belnap's conservativity criterion is valid for strict canonical constructive systems, and so what a set of strict canonical rules defines in such systems is independent of the system in which it is included.

6 Related Works and Further Research

There have been several works in the past on conditions for cut-elimination. Except for [7], the closest to the present one is [10]. The range of systems dealt with there is in fact broader than ours, since it deals with various types of structural rules, while in this paper we assume the standard structural rules of minimal logic. On the other hand the results and characterization given in [10] are less satisfactory than those given here. First, in the framework of [10] any connective has essentially infinitely many introduction (and elimination) rules, while our framework makes it possible to convert these infinite sets of rules into a finite set. Second, our coherence criterion (for non-triviality and cut-elimination) is simple and constructive. In contrast, its counterpart in [10] (called *reductivity*) is not constructive. Third, our notion of *strong cut-elimination* simply limits the set of possible cut formulas used in a derivation of a sequent from other sequents to those that occur in the premises of that derivation. In contrast, *reductive cut-elimination*, its counterpart in [10], imposes conditions on *applications* of the cut rule in proofs which involve examining the whole proofs of the two premises of that application. Finally, both works use non-deterministic semantic frameworks (in [10] this is only implicit!). However, while we use the concrete framework of intuitionistic-like Kripke frames, variants of the significantly more abstract and complicated phase semantics are used in [10]. This leads to the following crucial difference: our semantics leads to decision procedures for all the systems we consider. This does not seem to be the case for the semantics used in [10].

It should be noted that unlike the present work, [10] treats *non-strict* systems (as is done in most presentations of intuitionistic logic as well as in Gentzen's original one), i.e. single-conclusion sequential systems which allow the use of negative sequents in derivations. In addition to being widely used, the non-strict framework makes it possible to define negation as a basic connective. It is natural to try to extend our methods and results to the non-strict framework. However, as the next observations show, doing it is not a straightforward matter:

- Consider a non-strict canonical Gentzen-type system **G** containing only the following rules for an unary connective, denoted by $\circ$:

$$\{p_1 \Rightarrow\} \;/\; \circ\, p_1 \Rightarrow \quad \text{and} \quad \{p_1 \Rightarrow\} \;/\; \Rightarrow \circ p_1$$

Applications of these rules have the following form (where E is either empty or a singleton):

$$\frac{\Gamma, \varphi \Rightarrow E}{\Gamma, \circ\varphi \Rightarrow E} \qquad \frac{\Gamma, \varphi \Rightarrow}{\Gamma \Rightarrow \circ\varphi}$$

Obviously, **G** is not coherent. However, in **G** there is no way to derive a negative sequent from no assumptions (this is proved by simple induction). Hence, the introduction rule for $\circ$ can never be used in proofs without assumptions. For this trivial reason, **G** is consistent. Hence, in this framework coherence is no longer equivalent to consistency.

- For the same reason, **G** from the previous example admits cut-elimination but does not admit strong cut-elimination. Hence, strong cut-elimination and cut-elimination are also no longer equivalent.
- Consider the well-known rules for intuitionistic negation:

$$\frac{\Gamma \Rightarrow \varphi}{\Gamma, \neg\varphi \Rightarrow} \qquad \frac{\Gamma, \varphi \Rightarrow}{\Gamma, \Rightarrow \neg\varphi}$$

If we naively extend our semantic definition to apply to this kind of rule, we will obtain that in every legal frame $v(a, \neg\varphi) = t$ iff $v(a, \varphi) = f$ (since a negative sequent is true iff it is locally true). This is not the well-known Kripke-style semantics for negation. Moreover, $\varphi \vee \neg\varphi$, which is obviously not provable in intuitionistic logic, is true in this semantics. Hence some changes must be done in our semantic framework if we wish to *directly* handle negation (and other negation-like connectives) in an adequate way.

References

1. Avron, A.: Simple Consequence Relations. Information and Computation 92, 105–139 (1991)
2. Avron, A.: Gentzen-Type Systems, Resolution and Tableaux. Journal of Automated Reasoning 10, 265–281 (1993)
3. Avron, A.: Nondeterministic View on Nonclassical Negations. Studia Logica 80, 159–194 (2005)
4. Avron, A.: Non-deterministic Semantics for Families of Paraconsistent Logics. In: Beziau, J.-Y., Carnielli, W., Gabbay, D.M. (eds.) Handbook of Paraconsistency. Studies in Logic, vol. 9, pp. 285–320. College Publications (2007)
5. Avron, A., Lahav, O.: Canonical constructive systems. In: Giese, M., Waaler, A. (eds.) TABLEAUX 2009. LNCS, vol. 5607, pp. 62–76. Springer, Heidelberg (2009)
6. Avron, A., Lev, I.: Canonical Propositional Gentzen-Type Systems. In: Goré, R.P., Leitsch, A., Nipkow, T. (eds.) IJCAR 2001. LNCS (LNAI), vol. 2083, pp. 529–544. Springer, Heidelberg (2001)
7. Avron, A., Lev, I.: Non-deterministic Multiple-valued Structures. Journal of Logic and Computation 15, 24–261 (2005)
8. Belnap, N.D.: Tonk, Plonk and Plink. Analysis 22, 130–134 (1962)
9. Bowen, K.A.: An extension of the intuitionistic propositional calculus. Indagationes Mathematicae 33, 287–294 (1971)

10. Ciabattoni, A., Terui, K.: Towards a Semantic Characterization of Cut-Elimination. Studia Logica 82, 95–119 (2006)
11. Fernandez, D.: Non-deterministic Semantics for Dynamic Topological Logic. Annals of Pure and Applied Logic 157, 110–121 (2009)
12. Gentzen, G.: Investigations into Logical Deduction. In: Szabo, M.E. (ed.) The Collected Works of Gerhard Gentzen, pp. 68–131. North Holland, Amsterdam (1969)
13. Gurevich, Y., Neeman, I.: The Logic of Infons, Microsoft Research Tech. Report MSR-TR-2009-10 (January 2009)
14. Kripke, S.: Semantical Analysis of Intuitionistic Logic I. In: Crossly, J., Dummett, M. (eds.) Formal Systems and Recursive Functions, pp. 92–129. North-Holland, Amsterdam (1965)
15. Prior, A.N.: The Runabout Inference Ticket. Analysis 21, 38–39 (1960)
16. Sundholm, G.: Proof theory and Meaning. In: Gabbay, D.M., Guenthner, F. (eds.) Handbook of Philosophical Logic, vol. 9, pp. 165–198 (2002)

Decidable Expansions of Labelled Linear Orderings

Alexis Bès[1] and Alexander Rabinovich[2]

[1] University of Paris-Est Créteil, LACL
bes@u-pec.fr
[2] Tel-Aviv University, The Blavatnik School of Computer Science
rabinoa@post.tau.ac.il

Dedicated to Yuri Gurevich on the occasion of his seventieth birthday

Abstract. Let $M = (A, <, \overline{P})$ where $(A, <)$ is a linear ordering and $\overline{P}$ denotes a finite sequence of monadic predicates on A. We show that if A contains an interval of order type ω or $-\omega$, and the monadic second-order theory of M is decidable, then there exists a non-trivial expansion M' of M by a monadic predicate such that the monadic second-order theory of M' is still decidable.

Keywords: monadic second-order logic, decidability, definability, linear orderings.

1 Introduction

In this paper we address definability and decidability issues for monadic second order (shortly: MSO) theories of labelled linear orderings. Elgot and Rabin ask in [9] whether there exist maximal decidable structures, i.e., structures M with a decidable first-order (shortly: FO) theory and such that the FO theory of any expansion of M by a non-definable predicate is undecidable. This question is still open. Let us mention some partial results:

- Soprunov proved in [28] that every structure in which a regular ordering is interpretable is not maximal. A partial ordering $(B, <)$ is said to be regular if for every $a \in B$ there exist distinct elements $b_1, b_2 \in B$ such that $b_1 < a$, $b_2 < a$, and no element $c \in B$ satisfies both $c < b_1$ and $c < b_2$. As a corollary he also proved that there is no maximal decidable countable structure if we replace FO by weak MSO logic.
- In [2], Bès and Cégielski consider a weakening of the Elgot-Rabin question, namely the question of whether all structures M whose FO theory is decidable can be expanded by some constant in such a way that the resulting structure still has a decidable theory. They answer this question negatively by proving that there exists a structure M with a decidable MSO theory and such that any expansion of M by a constant has an undecidable FO theory.

A. Blass, N. Dershowitz, and W. Reisig (Eds.): Gurevich Festschrift, LNCS 6300, pp. 95–107, 2010.

- The paper [1] gives a sufficient condition in terms of the Gaifman graph of M which ensures that M is not maximal. The condition is the following: for every natural number r and every finite set X of elements of the base set $|M|$ of M there exists an element $x \in |M|$ such that the Gaifman distance between x and every element of X is greater than r.

We investigate the Elgot-Rabin problem for the class of labelled linear orderings, i.e., infinite structures $M = (A; <, P_1, \ldots, P_n)$ where $<$ is a linear ordering over A and the P_i's denote unary predicates. This class is interesting with respect to the above results, since on one hand no regular ordering seems to be FO interpretable in such structures, and on the other hand their associated Gaifman distance is trivial, thus they do not satisfy the criterion given in [1].

In this paper we focus on MSO logic rather than FO. The main result of the paper is that for every labelled linear ordering M such that $(A, <)$ contains an interval of order type ω or $-\omega$ and the MSO theory of M is decidable, then there exists an expansion M' of M by a monadic predicate which is not MSO-definable in M, and such that the MSO theory of M' is still decidable. Hence, M is not maximal. The result holds in particular when $(A. <)$ is order-isomorphic to the order of the naturals $\omega = (\mathbb{N}, <)$, or to the order $\zeta = (\mathbb{Z}, <)$ of the integers, or to any infinite ordinal, or more generally any infinite scattered ordering (recall that an ordering is scattered if it does not contain any dense sub-ordering).

The structure of the proof is the following: we first show that the result holds for ω and ζ. For the general case, starting from M, we use some definable equivalence relation on A to cut A into intervals whose order type is either finite, or of the form $-\omega$, ω, or ζ. We then define the new predicate on each interval (using the constructions given for ω and ζ), from which we get the definition of M'. The reduction from $MSO(M')$ to $MSO(M)$ uses Shelah's composition theorem, which allows to reduce the MSO theory of an ordered sum of structures to the MSO theories of the summands.

The main reason to consider MSO logic rather than FO is that it actually simplifies the task. Nevertheless we discuss some partial results and perspectives for FO logic in the conclusion of the paper.

Let us recall some important decidability results for MSO theories of linear orderings (the case of labelled linear orderings will be discussed later for ω and ζ). In his seminal paper [4], Büchi proved that languages of $\omega-$words recognizable by automata coincide with languages definable in the MSO theory of ω, from which he deduced decidability of the theory. The result (and the automata method) was then extended to the MSO theory of any countable ordinal [5], to ω_1, and to any ordinal less than ω_2 [6]. Gurevich, Magidor and Shelah prove [13] that decidability of MSO theory of ω_2 is independent of ZFC. Let us mention results for linear orderings beyond ordinals. Using automata, Rabin [19] proved decidability of the MSO theory of the binary tree, from which he deduces decidability of the MSO theory of $\mathbb{Q}$, which in turn implies decidability of the MSO theory of the class of countable linear orderings. Shelah [26] improved model-theoretical techniques that allow him to reprove almost all known decidability results about MSO theories, as well as new decidability results for the case of linear orderings, and in particular dense orderings. He proved in particular that the MSO theory of $\mathbb{R}$ is undecidable. The frontier between decidable and undecidable cases was specified in later papers by Gurevich and Shelah [11,14,15]; we refer the reader to the survey [12].

Our result is also clearly related to the problem of building larger and larger classes of structures with a decidable MSO theory. For an overview of recent results in this area see [3,32].

2 Definitions, Notations and Useful Results

2.1 Labelled Linear Orderings

We first recall useful definitions and results about linear orderings. A good reference on the subject is Rosenstein's book [23].

A *linear ordering* J is a total ordering. We denote by ω (respectively ζ) the order type of $\mathbb{N}$ (respectively $\mathbb{Z}$). Given a linear ordering J, we denote by $-J$ the *backwards* linear ordering obtained by reversing the ordering relation.

Given two elements j, k of a linear ordering J, we denote by $[j; k]$ the interval $[\min(j, k), \max(j, k)]$. An ordering is *dense* if it contains no pair of consecutive elements. An ordering is *scattered* if it contains no dense sub-ordering.

In this paper we consider *labelled* linear orderings, i.e., linear orderings $(A, <)$ equipped with a function $f : A \to \Sigma$ where Σ is a finite nonempty set.

2.2 Logic

Let us briefly recall useful elements of monadic second-order logic, and settle some notations. For more details about MSO logic see e.g. [12,31]. Monadic second-order logic is an extension of first-order logic that allows to quantify over elements as well as subsets of the domain of the structure. Given a signature L, one can define the set of (MSO) formulas over L as well-formed formulas that can use first-order variable symbols $x, y, \ldots$ interpreted as elements of the domain of the structure, monadic second-order variable symbols $X, Y, \ldots$ interpreted as subsets of the domain, symbols from L, and a new binary predicate $x \in X$ interpreted as "x belongs to X". A sentence is a formula without free variable. As usual, we often confuse logical symbols with their interpretation. Given a signature L and an $L-$structure M with domain D, we say that a relation $R \subseteq D^m \times (2^D)^n$ is (MSO) definable in M if a nd only if there exists a formula over L, say $\varphi(x_1, \ldots, x_m, X_1, \ldots, X_n)$, which is true in M if and only if $(x_1, \ldots, x_m, X_1, \ldots, X_n)$ is interpreted by an $(m+n)-$tuple of R. Given a structure M we denote by $MSO(M)$ (respectively $FO(M)$) the monadic second-order (respectively first-order) theory of M. We say that M is maximal if $MSO(M)$ is decidable and $MSO(M')$ is undecidable for every expansion M' of M by a predicate which is not definable in M.

We can identify labelled linear orderings with structures of the form $M = (A, <, P_1, \ldots, P_n)$ where $<$ is a binary relation interpreted as a linear ordering over A, and the P_i's denote unary predicates. We use the notation $\overline{P}$ as a shortcut for the n-tuple $(P_1, \ldots P_n)$. The structure M can be seen as a word indexed by A and over the alphabet $\Sigma_n = \{0,1\}^n$; this word will be denoted by $w(M)$. For every interval I of A we denote by M_I the sub-structure of M with domain I.

2.3 Composition Theorems

In this paper we rely heavily on composition methods, which allow to compute the theory of a sum of structures from the ones of its summands. For an overview of the subject see [3,12,16,30]. In this section we recall useful definitions and results. For the whole section we consider signatures of the form $L = \{<, P_1, \ldots, P_n\}$ where the P_i's denote unary predicate names, and deal only with L−structures where $<$ is interpreted as a linear ordering – that is, with labelled linear orderings. Given a formula φ over L, the quantifier depth of φ is denoted by $qd(\varphi)$. The k−type of an L−structure M, which is denoted by $T^k(M)$, is the set of sentences φ such that $M \models \varphi$ and $qd(\varphi) \leq k$. Given two structures M and M', the relation $T^k(M) = T^k(M')$ is an equivalence relation with finitely many classes. Let us list some fundamental and well-known properties of k-types. The proofs of these facts can be found in several sources, see e.g. [26,31].

Proposition 1. *1. For every k there are only finitely k-types over a finite signature L*

2. *For each k-type t there is a sentence φ_t (called "characteristic sentence") which defines t, i.e., such that $M \models \varphi_t$ iff $T^k(M) = t$. For every k, a finite list of characteristic sentences for all the possible k-types can be computed. (We take the characteristic sentences as the canonical representations of k-types. Thus, for example, transforming a type into another type means to transform sentences.)*
3. *Each sentence φ is equivalent to a (finite) disjunction of characteristic sentences; moreover, this disjunction can be computed from φ.*

As a simple consequence, note that the MSO theory of a structure M is decidable if and only if the function $k \mapsto T^k(M)$ is recursive.

The sum of structures corresponds to concatenation; let us recall a general definition.

Definition 2. *Consider an index structure $Ind = (I, <^I)$ where $<^I$ is a linear ordering. Consider a signature $L = \{<, P_1, \ldots, P_n\}$, where P_i are unary predicate names, and a family $(M_i)_{i \in I}$ of L-structures $M_i = (A_i; <^i, {P_1}^i, \ldots, P_n^i)$ with disjoint domains and such that the interpretation $<^i$ of $<$ in each M_i is a linear ordering. We define the* ordered sum *of the family $(M_i)_{i \in I}$ as the L-structure $M = (A; <^M, {P_1}^M, \ldots, P_n^M)$ where*

- *A equals the union of the A_i's*
- *$x <^M y$ holds if and only if ($x \in A_i$ and $y \in A_j$ for some $i <^I j$), or ($x, y \in A_i$ and $x <^i y$)*
- *for every $x \in A$ and every $k \in \{1, \ldots, n\}$, $P_k^M(x)$ holds if and only if $M_j \models P_k^j(x)$ where j is such that $x \in A_j$.*

If the domains of the M_i are not disjoint, replace them with isomorphic chains that have disjoint domains, and proceed as before.

We shall use the notation $M = \sum_{i \in I} M_i$ for the ordered sum of the family $(M_i)_{i \in I}$. If $I = \{1, 2\}$ has two element, we denote $\sum_{i \in I} M_i$ by $M_1 + M_2$.

We need the following composition theorem on ordered sums:

Theorem 3.
(a) The k-types of labelled linear orderings M_0, M_1 determine the k-type of the ordered sum $M_0 + M_1$, which moreover can be computed from the k-types of M_0 and M_1.
(b) If the labelled linear orderings $M_0, M_1, \ldots$ all have the same k-type, then this k-type determines the k-type of $\Sigma_{i \in \mathbb{N}} M_i$, which moreover can be computed from the k-type of M_0.

Part (a) of the theorem justifies the notation $s + t$ for the k-type of a linear ordering which is the sum of two linear orderings of k-types s and t, respectively. Similarly, we write $t \times \omega$ for the k-type of a sum $\Sigma_{i \in \mathbb{N}} M_i$ where all M_i are of k-type t.

3 The Case of $\mathbb{N}$

In this section we prove that there is no maximal structure of the form $(\mathbb{N}, <, \overline{P})$ with respect to MSO logic. The proof is based upon results from [20] . Let us first briefly review results related to the decidability of the MSO theory of expansions of $(\mathbb{N}, <)$. Büchi [4] proved decidability of $MSO(\mathbb{N}, <)$ using automata. On the other hand it is known that $MSO(\mathbb{N}, +)$, and even $MSO(\mathbb{N}, <, x \mapsto 2x)$, are undecidable [22]. Elgot and Rabin study in [9] the MSO theory of structures of the form $(\mathbb{N}, <, P)$, where P is some unary predicate. They give a sufficient condition on P which ensures decidability of the MSO theory of $(\mathbb{N}, <, P)$. In particular the condition holds when P denotes the set of factorials, or the set of powers of any fixed integer. The frontier between decidability and undecidability of related theories was explored in numerous later papers [7,10,25,24,21,20,27,29]. Let us also mention that [25] proves the existence of unary predicates P and Q such that both $MSO(\mathbb{N}, <, P)$ and $MSO(\mathbb{N}, <, Q)$ are decidable while $MSO(\mathbb{N}, <, P, Q)$ is undecidable.

Most decidability proofs for $MSO(\mathbb{N}, <, P)$ are related somehow to the possibility of cutting $\mathbb{N}$ into segments whose $k-$type is ultimately constant, from which one can compute the $k-$type of the whole structure (using Theorem 3). This connection was specified in [20] (see also [21]) using the notion of homogeneous sets.

Definition 4 (k-homogeneous set). *Let $k \geq 0$. A set $H = \{h_0 < h_1 < \ldots\} \subseteq \mathbb{N}$ is called* k-homogeneous *for $M = (\mathbb{N}, <, \overline{P})$, if all sub-structures $M_{[h_i, h_j)}$ for $i < j$ (and hence all sub-structures $M_{[h_i, h_{i+1})}$ for $i \geq 0$) have the same k-type.*

This notion can be refined as follows.

Definition 5 (uniformly homogeneous set). *A set $H = \{h_0 < h_1 < \ldots\} \subseteq \mathbb{N}$ is called* uniformly homogeneous *for $M = (\mathbb{N}, <, \overline{P})$ if for each k the set $H_k = \{h_k < h_{k+1} < \ldots\}$ is k-homogeneous.*

The following result [20] settles a tight connection between $MSO(\mathbb{N}, <, \overline{P})$ and uniformly homogeneous sets.

Theorem 6. *For every $M = (\mathbb{N}, <, \overline{P})$, the MSO theory of M is decidable if and only if (the sets $\overline{P}$ are recursive and there exists a recursive uniformly homogeneous set for M).*

One can use this theorem to show that no structure $M = (\mathbb{N}, <, \overline{P})$ is maximal. Let us give the main ideas. Starting from M such that $MSO(M)$ is decidable, Theorem 6 implies the existence of a recursive uniformly homogeneous set $H = \{h_0 < h_1 < \ldots\}$ for M.

Let M' be an expansion of M by a monadic predicate P_{n+1} defined as $P_{n+1} = \{h_{n!} \mid n \in \mathbb{N}\}$.

By definition of H, the structures $M_{[h_{k!}, h_{(k+j)!}[}$ have the same $k-$type for all $j, k \geq 0$. If we combine this with the fact that successive elements of P_{n+1} are far away from each other, we can prove that P_{n+1} is not definable in M. For all $i, k \geq 0$ let us define the interval $I(i,k) = [h_{(k+i)!}, h_{(k+i+1)!}[$. In order to prove that $MSO(M')$ is decidable, we exploit the fact that all structures $M_{I(i,k)}$ have the same $k-$type for all $i, k \geq 0$, and that only the first element of each interval $I(i,k)$ belongs to P_{n+1}. This allows to compute easily the $k-$type of structures $M'_{I(i,k)}$ from the one of $M_{I(i,k)}$, and then the $k-$type of the whole structure M'. This provides a reduction from $MSO(M')$ to $MSO(M)$.

The above construction, which we described for a fixed structure M, can actually be defined uniformly in M. This leads to the following result.

Proposition 7. *There exists a function E and two recursive function g_1, g_2 such that E maps every structure $M = (\mathbb{N}, <, \overline{P})$ to an expansion M' of M by a predicate P_{n+1} such that*

1. *P_{n+1} is not definable in M;*
2. *g_1 computes $T^k(M')$ from k and $T^{g_2(k)}(M)$.*

Hence $MSO(M')$ is recursive in $MSO(M)$. In particular, if $MSO(M)$ is decidable, then $MSO(M')$ is decidable.

Let us discuss item (2). In the proof of the general result (see Sect. 5), we start from a labelled linear ordering $M = (A, <, \overline{P})$ with a decidable MSO theory and try to expand it while keeping decidability. In some case the (decidable) expansion M' of M will be defined by applying the above construction to infinitely many intervals of A of order type ω. In order to get a reduction from $MSO(M')$ to $MSO(M)$, we need that the reduction algorithm for such intervals is uniform, which is what item (2) expresses.

4 The Case of $\mathbb{Z}$

Decidability of the MSO theory of structures $M = (\mathbb{Z}, <, \overline{P})$ was studied in particular by Compton [8], Semënov [25,24], and Perrin and Schupp [18] (see also [17, chapter 9]). These works put in evidence the link between decidability of $MSO(M)$ and computability of occurrences and repetitions of finite factors in the word $w(M)$. Let us state some notations and definitions. A set X of finite words over a finite alphabet Σ is said to be regular if it is recognizable by some finite automaton. Given a $\mathbb{Z}$−word w and a finite word u, both over the alphabet Σ, we say that u occurs in w if $w = w_1 u w_2$ for some words w_1 and w_2. We say that w is *recurrent* if for every regular language X of finite words over Σ, either no element of X occurs in w, or in every prefix and every suffix of w there is an occurrence of some element of X. In particular in a recurrent word w, every finite word u either has no occurrence in w, or occurs infinitely often on both sides of w. We say that w is *rich* if every finite word occurs infinitely often on both sides of w. Given $M = (\mathbb{Z}, <, \overline{P})$, we say that M is *recurrent* if $w(M)$ is.

We have the following result.

Theorem 8. *([25,18]) Given $M = (\mathbb{Z}, <, P_1, \ldots, P_n)$,*

1. *If M is not recurrent, then every $c \in \mathbb{Z}$ is definable in M.*
2. *If M is recurrent, then no element is definable in M, and $MSO(M)$ is computable relative to an oracle which, given any regular language X of finite words over $\Sigma_n = \{0,1\}^n$, tells whether some element of X occurs in $w(M)$.*

Let $c \in \mathbb{Z}$, and let M_1 be defined as $M = M_{]-\infty,c[}$ and M_2 be defined as $M_{[c,\infty[}$. Then $M = M_1 + M_2$.

Let M_1' be the expansion of M_1 by the empty predicate P_{n+1} and let M_2' be obtained by apply the construction of Proposition 7 to M_2. Let $M' = M_1' + M_2'$.

Note that the above construction of M' from M depends on c. We denote by E_c the function described above that maps every $M = (\mathbb{Z}, <, P_1, \ldots, P_n)$ to its expansion M' by P_{n+1}.

It is easy to show that P_{n+1} is not definable in M, hence M' is a non-trivial expansion of M.

We claim that if M is not recurrent, then MSO(M') is recursive in MSO(M). Indeed, in this case, by Theorem 8, c is definable in M. Hence, M_1 and M_2 can be interpreted in M, which yields that MSO(M_1) and MSO(M_2) are recursive in MSO(M). Therefore, MSO(M_1') and MSO(M_2') are recursive in MSO(M). Finally, applying Theorem 3(a) we obtain that MSO(M') is recursive in MSO(M).

Hence, we have the following.

Proposition 9 (Expansion of non-recurrent structures). *There are two recursive function g_1, g_2 such that if $M = (\mathbb{Z}, <, P_1, \ldots, P_n)$ is not recurrent, and $c \in \mathbb{Z}$ is definable in M by a formula of quantifier depth m, then E_c maps M to an expansion M' by a predicate P_{n+1} such that*

1. *P_{n+1} is not definable in M;*
2. *g_1 computes $T^k(M')$ from k and $T^{g_2(k+m)}(M)$.*

Hence $MSO(M')$ is recursive in $MSO(M)$. In particular, if $MSO(M)$ is decidable, then $MSO(M')$ is decidable.

Remark 10. Let us discuss uniformity issues related to Proposition 7 and Proposition 9. Proposition 7 implies that there is an algorithm which reduces $MSO(M')$ to $MSO(M)$. This reduction algorithm is independent of M; it only uses an oracle for $MSO(M)$. Proposition 9 implies a weaker property. Namely, it implies that for every non-recurrent M there is an algorithm which reduces $MSO(M')$ to $MSO(M)$. However, this reduction algorithm depends on M.

Consider a recurrent structure M and let $M' = E_c(M)$ for some $c \in \mathbb{Z}$. We claim that it is possible that $MSO(M')$ is not recursive in $MSO(M)$. Indeed, we can prove that there exists a recurrent structure M over $\mathbb{Z}$ such that $MSO(M)$ is decidable, and $MSO(M_{[c',\infty[})$ is undecidable for every $c' \in Z$. Now let c' be the minimal element of P_{n+1}. Observe that c' is definable in M' and therefore, $M_{[c',\infty[}$ can be interpreted in M'. Since, $MSO(M_{[c',\infty[})$ is undecidable, we derive that $MSO(M')$ is undecidable. Hence, E_c does not preserves decidability of recurrent structures, and we need a different construction for the recurrent case.

To describe our construction for the recurrent case let us introduce first some notations.

For every word w over the alphabet $\Sigma_{n+1} = \{0,1\}^{n+1}$ which is indexed by some linear ordering $(A,<)$ we denote by $\pi(w)$ the word w' indexed by A and over the alphabet $\Sigma_n = \{0,1\}^n$, which is obtained from w by projection over the n first components of each symbol in w. The definition and notation extend to $\pi(X)$ where X is any set of words over the alphabet Σ_{n+1}. Given $M = (\mathbb{Z},<,\overline{P})$ where $\overline{P}$ is an $n-$tuple of sets, and any expansion M' of M by a predicate P_{n+1}, by definition $w(M)$ and $w(M')$ are words over Σ_n and Σ_{n+1}, respectively, and we have $\pi(w(M')) = w(M)$.

Lemma 11. *If $M = (\mathbb{Z},<,\overline{P})$ is recurrent, then there is an expansion M' of M by a predicate P_{n+1} which has the following property:*

(*) *for every $u \in \Sigma_n^*$, if u occurs infinitely often on both sides of $w(M)$, then the same holds in $w(M')$ for every word $u' \in \Sigma_{n+1}^*$ such that $\pi(u') = u$.*

The proof of Lemma 11 is similar to the proof of Proposition 2.8 in [1], which roughly shows how to deal with the case when $w(M)$ is rich.

Now $w(M')$ has a finite factor in some regular language $X' \subseteq \Sigma_{n+1}^*$ iff $w(M)$ has a finite factor in $\pi(X') \subseteq \Sigma_n^*$. The set $\pi(X')$ is regular, and a sentence which defines $\pi(X')$ is computable from a sentence that defines X', thus we obtain, by Theorem 8(2), that if $MSO(M)$ is decidable then $MSO(M')$ is decidable.

One can show that if M' is any expansion of M which has property (*), then P_{n+1} is not definable in M. This implies that no recurrent structure is maximal.

From a more detailed analysis of the proof of Theorem 8(2) we can derive the following proposition.

Proposition 12 (Expansion of recurrent structures). *There are two recursive function g_1, g_2 such that if $M = (\mathbb{Z}, <, \overline{P})$ is recurrent and M' is an expansion of M which has property (*), then*

1. *P_{n+1} is not definable in M;*
2. *g_1 computes $T^k(M')$ from k and $T^{g_2(k)}(M)$.*

Hence $MSO(M')$ is recursive in $MSO(M)$. In particular, if $MSO(M)$ is decidable, then $MSO(M')$ is decidable.

Remark 13. Proposition 12 implies that there is an algorithm which reduces $MSO(M')$ to $MSO(M)$. This reduction algorithm (like the algorithm from Proposition 7) is independent of M; it only uses an oracle for $MSO(M)$.

Proposition 9, Lemma 11 and Proposition 12 imply the following corollary.

Corollary 14. *Let $M = (\mathbb{Z}, <, \overline{P})$. There exists an expansion M' of M by some unary predicate P_{n+1} such that P_{n+1} is not definable in M, and $MSO(M')$ is recursive in $MSO(M)$. In particular, if $MSO(M)$ is decidable, then $MSO(M')$ is decidable.*

5 Main Result

The next theorem is our main result.

Theorem 15. *Let $M = (A, <, P_1, \dots, P_n)$ where $(A, <)$ contains an interval of type ω or $-\omega$. There exists an expansion M' of M by a relation P_{n+1} such that P_{n+1} is not definable in M, and $MSO(M')$ is recursive in $MSO(M)$. In particular, if $MSO(M)$ is decidable, then $MSO(M')$ is decidable.*

As an immediate consequence we obtain the following corollary.

Corollary 16. *Let $M = (A, <, P_1, \dots, P_n)$ where $(A, <)$ is an infinite scattered linear ordering. There exists an expansion M' of M by some unary predicate P_{n+1} not definable in M such that $MSO(M')$ is recursive in $MSO(M)$.*

We present a sketch of proof for Theorem 15. Let $M = (A, <, \overline{P})$ where $(A, <)$ contains an interval of type ω or $-\omega$.

Consider the binary relation defined on A by $x \approx y$ iff $[x, y]$ is finite. The relation $\approx$ is a *condensation*, i.e., an equivalence relation such that every equivalence class is an interval of A. Moreover the relation $\approx$ is definable in M. If A_i and A_2 are $\approx$-equivalence classes, we say that A_1 precedes A_2 if all elements of A_1 are less than all elements of A_2. Let I be the linear order of the $\approx$-equivalence classes for $(A, <)$. Then $M = \sum_{i \in I} M_{A_i}$ where the A_i's correspond to equivalence classes of $\approx$. Using the definition of $\approx$ and our assumption on A, it is easy to check that the A_i's are either finite, or of order type ω, or $-\omega$, or ζ, and that not all A_i's are finite.

We define the interpretation of the new predicate P_{n+1} in every interval A_i. The definition proceeds as follows:

1. if some A_i has order type ω or $-\omega$, then we apply to each substructure M_{A_i} of order type ω the construction given in Proposition 7, and add no element of P_{n+1} elsewhere. If there is no A_i of order type ω, we proceed in a similar way with each substructure M_{A_i} of order type $-\omega$, but using the dual of Proposition 7 for $-\omega$.
2. if no A_i has order type ω or $-\omega$, then at least one $\approx$ −equivalence class A_i has order type ζ. We consider two subcases:
   (a) if all $\approx$ −equivalence classes A_i with order type ζ are such that $w(M_{A_i})$ is recurrent, then we apply to each substructure M_{A_i} of order type ζ the construction given in Proposition 12. For other $\approx$ −equivalence classes A_i we set $P_{n+1} \cap A_i = \varnothing$.
   (b) otherwise there exist $\approx$ −equivalence classes A_i with order type ζ and such that $w(M_{A_i})$ is not recurrent. Let $\varphi(x)$ be a formula with minimal quantifier depth such that $\varphi(x)$ defines an element in some M_{A_i} where A_i has order type ζ. For every M_{A_i} such that A_i has order type ζ and $\varphi(x)$ defines an element c_i in M_{A_i}, we apply the construction E_{c_i} from Proposition 9 to M_{A_i}. For other $\approx$ −equivalence classes A_i we set $P_{n+1} \cap A_i = \varnothing$.

The fact that the set P_{n+1} is not definable in M follows rather easily from the construction, which ensures that there exists some A_i such that the restriction of P_{n+1} to A_i is not definable in the substructure M_{A_i}.

Let M' be the expansion of M by the predicate P_{n+1}. In order to prove that $MSO(M')$ is recursive in $MSO(M)$, we use Shelah's composition method [26, Theorem 2.4] (see also [12,30]) which allows to reduce the MSO theory of a sum of structures to the MSO theories of the components and the MSO theory of the index structure.

Theorem 17 (Composition Theorem [26]). *There exists a recursive function f and an algorithm which, given $k, l \in \mathbb{N}$, computes the k-type of any sum $M = \sum_{i \in I} M_i$ of labelled linear orderings over a signature $\{<, P_1, \ldots, P_l\}$ from the $f(k,l)$-type of the structure $(I, <, Q_1, \ldots, Q_p)$ where*

$$Q_j = \{i \in I : T^k(M_i) = \tau_j\} \quad j = 1, \ldots, p$$

and $\tau_1, \ldots, \tau_p$ is the list of all formally possible k-types for the signature L.

Let us explain the reduction from $MSO(M')$ to $MSO(M)$. We can apply Theorem 17 to $M' = \sum_{i \in I} M'_{A_i}$, which allows to show that for every k, the k−type of M' can be computed from $f(k, n+1)$−type of the structure $N' = (I, <, Q'_1, \ldots, Q'_p)$ where the Q'_i's correspond to the k−types of structures M'_{A_i} over the signature $\{<, P_1, \ldots, P_{n+1}\}$. Using the definition of P_{n+1} and Propositions 7, 9 and 12, one can prove that the k−type of M'_{A_i} can be computed form the $g(k)$−type of M_{A_i} for some recursive function g (note that g depends on M, namely whether we used case 1, 2(a) or 2(b) to construct M'). This allows to prove that N' is interpretable in the structure $N = (I, <, Q_1, \ldots, Q_q)$ where the Q_i's correspond to the $g(k)$−types of structures M_{A_i} over the signature

$\{<, P_1, \ldots, P_n\}$. It follows that $MSO(N')$ is recursive in $MSO(N)$. Now using the fact that the equivalence relation $\approx$ is definable in M, we can prove that N is interpretable in M, thus $MSO(N)$ is recursive in $MSO(M)$.

Remark 18. Let us discuss uniformity issues related to Theorem 15.

- The choice to expand "uniformly" all $\approx$ –equivalence classes is crucial for the reduction from $MSO(M')$ to $MSO(M)$. For example, if some A_i has order type ω and we choose to expand only A_i then $MSO(M')$ might become undecidable. This is the case for the structure M considered in [2] (Definition 2.4), which has decidable MSO theory, and is such that the MSO theory of any expansion of M by a constant is undecidable. For this structure all A_i's have order type ω. If we consider the structure M' obtained from M by an expansion of only one A_i, then P_{n+1} has a least element, which is definable in M', thus MSO(M') is undecidable.
- The definition of P_{n+1} in case (2) depends on whether all components A_i with order type ζ are such that $w(M_{A_i})$ is recurrent, which is not a MSO definable property. Thus that the reduction algorithm from $MSO(M')$ to $MSO(M)$ depends on M.

6 Further Results and Open Questions

Let us mention some possible extensions and related open questions.

First of all, most of our results can be easily extended to the case when the signature contains infinitely many unary predicates.

Our results can be extended to the Weak MSO logic. In the case M is countable this follows from Soprunov result [28]. However, our construction works for labelled orderings of arbitrary cardinality.

An interesting issue is to prove uniform versions of our results in the sense of items (2) in Propositions 7 and 12. A first step would be to generalize Proposition 12 to all structures $(\mathbb{Z}, <, \overline{P})$.

One can also ask whether the results of the present paper hold for FO logic. Let us emphasize some difficulties which arise when one tries to adapt the main arguments. A FO version of Theorem 6 (about the recursive homogeneous set) was already proven in [21]. Moreover, using ideas from [25] one can also give a characterization of structures $M = (\mathbb{Z}, <, \overline{P})$ with a decidable FO theory, in terms of occurrences and repetitions of finite words in $w(M)$. This allows to give a FO version of our non-maximality results for labelled orders over ω or ζ. However for the general case of $(A, <, \overline{P})$, two problems arise: (1) the constructions for $\mathbb{N}$ and $\mathbb{Z}$ cannot be applied directly since they are not uniform, and (2) the equivalence relation $\approx$ used in the proof of Theorem 15 to cut A into small intervals is not FO definable. We currently investigate these issues.

Finally, we also study the case of labelled linear orderings $(A, <, \overline{P})$ which do not contain intervals of order types ω or $-\omega$. In this case the construction presented in Sect. 5 does not work since the restriction of P_{n+1} to each A_i will be empty, i.e., our new relation is actually empty. In a forthcoming paper we show that it is possible to overcome this issue for the countable orders, and prove that no infinite countable structure $(A, <, \overline{P})$ is maximal.

Acknowledgment. This research was facilitated by the ESF project AutoMathA. The second author was partially supported by ESF project Games and EPSRC grant.

References

1. Bès, A., Cégielski, P.: Weakly maximal decidable structures. RAIRO-Theor. Inf. Appl. 42(1), 137–145 (2008)
2. Bès, A., Cégielski, P.: Nonmaximal decidable structures. Journal of Mathematical Sciences 158, 615–622 (2009)
3. Blumensath, A., Colcombet, T., Löding, C.: Logical theories and compatible operations. In: Flum, J., Grädel, E., Wilke, T. (eds.) Logic and automata: History and Perspectives, pp. 72–106. Amsterdam University Press (2007)
4. Büchi, J.R.: On a decision method in the restricted second-order arithmetic. In: Proc. Int. Congress Logic, Methodology and Philosophy of science, Berkeley 1960, pp. 1–11. Stanford University Press, Stanford (1962)
5. Büchi, J.R.: Transfinite automata recursions and weak second order theory of ordinals. In: Proc. Int. Congress Logic, Methodology, and Philosophy of Science, Jerusalem 1964, pp. 2–23. Holland (1965)
6. Büchi, J.R., Zaiontz, C.: Deterministic automata and the monadic theory of ordinals ω_2. Z. Math. Logik Grundlagen Math. 29, 313–336 (1983)
7. Carton, O., Thomas, W.: The monadic theory of morphic infinite words and generalizations. Inform. Comput. 176, 51–76 (2002)
8. Compton, K.J.: On rich words. In: Lothaire, M. (ed.) Combinatorics on words. Progress and perspectives, Proc. Int. Meet., Waterloo/Can. 1982. Encyclopedia of Mathematics, vol. 17, pp. 39–61. Addison, Reading (1983)
9. Elgot, C.C., Rabin, M.O.: Decidability and undecidability of extensions of second (first) order theory of (generalized) successor. J. Symb. Log. 31(2), 169–181 (1966)
10. Fratani, S.: The theory of successor extended with several predicates (2009) (preprint)
11. Gurevich, Y.: Modest theory of short chains.i. J. Symb. Log. 44(4), 481–490 (1979)
12. Gurevich, Y.: Monadic second-order theories. In: Barwise, J., Feferman, S. (eds.) Model-Theoretic Logics, Perspectives in Mathematical Logic, pp. 479–506. Springer, Heidelberg (1985)
13. Gurevich, Y., Magidor, M., Shelah, S.: The monadic theory of ω_2. J. Symb. Log. 48(2), 387–398 (1983)
14. Gurevich, Y., Shelah, S.: Modest theory of short chains. ii. J. Symb. Log. 44(4), 491–502 (1979)
15. Gurevich, Y., Shelah, S.: Interpreting second-order logic in the monadic theory of order. J. Symb. Log. 48(3), 816–828 (1983)
16. Makowsky, J.A.: Algorithmic uses of the Feferman-Vaught theorem. Annals of Pure and Applied Logic 126(1-3), 159–213 (2004)
17. Perrin, D., Pin, J.-E.: Infinite Words. Pure and Applied Mathematics, vol. 141. Elsevier, Amsterdam (2004), ISBN 0-12-532111-2
18. Perrin, D., Schupp, P.E.: Automata on the integers, recurrence distinguishability, and the equivalence and decidability of monadic theories. In: Symposium on Logic in Computer Science (LICS 1986), Washington, D.C., USA, June 1986, pp. 301–305. IEEE Computer Society Press, Los Alamitos (1986)

19. Rabin, M.O.: Decidability of second-order theories and automata on infinite trees. Transactions of the American Mathematical Society 141, 1–35 (1969)
20. Rabinovich, A.: On decidability of monadic logic of order over the naturals extended by monadic predicates. Inf. Comput. 205(6), 870–889 (2007)
21. Rabinovich, A., Thomas, W.: Decidable theories of the ordering of natural numbers with unary predicates. In: Ésik, Z. (ed.) CSL 2006. LNCS, vol. 4207, pp. 562–574. Springer, Heidelberg (2006)
22. Robinson, R.M.: Restricted set-theoretical definitions in arithmetic. Proc. Am. Math. Soc. 9, 238–242 (1958)
23. Rosenstein, J.G.: Linear ordering. Academic Press, New York (1982)
24. Semenov, A.L.: Decidability of monadic theories. In: Chytil, M.P., Koubek, V. (eds.) MFCS 1984. LNCS, vol. 176, pp. 162–175. Springer, Heidelberg (1984)
25. Semenov, A.L.: Logical theories of one-place functions on the set of natural numbers. Mathematics of the USSR - Izvestia 22, 587–618 (1984)
26. Shelah, S.: The monadic theory of order. Annals of Mathematics 102, 379–419 (1975)
27. Siefkes, D.: Decidable extensions of monadic second order successor arithmetic. In: Automatentheorie und Formale Sprachen, Tagung, Math. Forschungsinst, Oberwolfach (1969); Bibliograph. Inst., Mannheim, pp. 441–472 (1970)
28. Soprunov, S.: Decidable expansions of structures. Vopr. Kibern. 134, 175–179 (1988) (in Russian)
29. Thomas, W.: A note on undecidable extensions of monadic second order successor arithmetic. Arch. Math. Logik Grundlagenforsch. 17, 43–44 (1975)
30. Thomas, W.: Ehrenfeucht games, the composition method, and the monadic theory of ordinal words. In: Mycielski, J., Rozenberg, G., Salomaa, A. (eds.) Structures in Logic and Computer Science, A Selection of Essays in Honor of A. Ehrenfeucht. LNCS, vol. 1261, pp. 118–143. Springer, Heidelberg (1997)
31. Thomas, W.: Languages, automata, and logic. In: Rozenberg, G., Salomaa, A. (eds.) Handbook of Formal Languages, vol. III, pp. 389–455. Springer, Heidelberg (1997)
32. Thomas, W.: Model transformations in decidability proofs for monadic theories. In: Kaminski, M., Martini, S. (eds.) CSL 2008. LNCS, vol. 5213, pp. 23–31. Springer, Heidelberg (2008)

Existential Fixed-Point Logic, Universal Quantifiers, and Topoi

Andreas Blass*

Mathematics Department, University of Michigan, Ann Arbor, MI 48109, U.S.A.

For Yuri Gurevich on the occasion of his seventieth birthday

Abstract. When one views (multi-sorted) existential fixed-point logic (EFPL) as a database query language, it is natural to extend it by allowing universal quantification over certain sorts. These would be the sorts for which one has the "closed-world" information that all entities of that sort in the real world are represented in the database. We investigate the circumstances under which this extension of EFPL retains various pleasant properties. We pay particular attention to the pleasant property of preservation by the inverse-image parts of geometric morphisms of topoi, because, as we show, this preservation property implies many of the other pleasant properties of EFPL.

Keywords: Fixed point logic, quantifier, topos, geometric morphism, polynomial time, Hoare logic, finite model.

In a many-sorted version of existential fixed-point logic (EFPL), it may be intuitively reasonable to adjoin universal quantification over certain sorts, and this adjunction may preserve some of the desirable features of EFPL. The goal of this paper is to explain the main ingredients of the preceding sentence: What is EFPL? What is the intuition behind it? How does that intuition suggest a (limited) use of universal quantification? What are the desirable properties of EFPL? And under what circumstances will (some of) these properties survive when universal quantification over some sorts is admitted?

Some of these questions are already answered in a paper that I wrote with Yuri Gurevich long ago [6], but for the sake of completeness I'll include here a review of some of this information. Other preliminary material, also reviewed here, comes from [5] and [4]. The new material in this paper concerns the connection between topos-theoretic properties of EFPL, its more traditional properties, and universal quantification.

1 Introduction to EFPL

Existential fixed-point logic can be roughly described as the result of modifying first-order logic by

* The author is partially supported by NSF grant DMS-0653696 and by a grant from Microsoft Corporation.

A. Blass, N. Dershowitz, and W. Reisig (Eds.): Gurevich Festschrift, LNCS 6300, pp. 108–134, 2010.

- removing the universal quantifier, and
- adding the least-fixed-point construction for definable, monotone operators.

Neither of these modifications should be taken literally, because of the following two problems, both involving negation.

First, if one merely removes the universal quantifier from any of the usual formulations of first-order logic, nothing has really changed, because $\forall x$ can be simulated as $\neg\exists x\neg$. To genuinely remove $\forall$, one must also restrict the combination of $\exists$ and $\neg$. A standard solution to this problem is to allow negation only of atomic formulas. (I'm assuming here that the propositional connectives are $\neg$, $\wedge$ and $\vee$; if implication were also allowed, then it would also need restrictions, since the antecedent of an implication is, in effect, negated.)

Second, the least-fixed-point construction requires monotonicity of the operator that is iterated. So the syntax of EFPL should restrict the use of fixed-points to the monotone case. But monotonicity is not a syntactic condition. The standard solution to this problem is to enforce monotonicity by imposing the syntactic condition of positivity; the predicate(s) representing the inductively defined relation(s) should occur only positively in the definition of the operator being iterated.

Clearly, both problems – the need to prevent surreptitious reintroduction of $\forall$ and the need to ensure monotonicity of inductive definitions – would immediately disappear if we simply banished negation from the logic. Unfortunately, negation is sometimes needed. For example, the motivation for Yuri and me to study EFPL in [6] was its use as a natural assertion language for Hoare logic, and this use definitely involved negation of the guards of `if`...`then`...`else` commands. So, instead of banning negation altogether, we should impose appropriate controls on its use.

It turned out, in the context of [6], that a clean way to impose the controls is to separate, in each vocabulary, those predicate symbols that can be negated and those that cannot. The former would include the guards of conditional statements; the latter would include the inductively defined predicates in least-fixed-point constructions. We referred to the two kinds of predicates as *negatable* and *positive*, respectively. Although this partition of each vocabulary was introduced for technical reasons, it was the beginning of the train of thought that led to the idea of rehabilitating $\forall$ in a limited way, the main idea of the present paper.

But before following that train of thought, let us briefly review the definitions of the syntax and semantics of EFPL, and let us take the opportunity to make the easy generalization to a multi-sorted context.

Definition 1. A *vocabulary* consists of:

- a set of *sort symbols*;
- a set of *predicate symbols*, each with a given finite sequence of sort symbols as its arity;
- a set of *function symbols*, each with a given finite sequence of sort symbols as its input arity and an additional sort symbol as its output sort;
- a specification, for each predicate symbol and for each equality symbol,[1] whether it is *positive* or *negatable*.

[1] The logic includes an equality symbol for each sort, but it is convenient to write all these equality symbols as = as long as no confusion arises.

The language also has variables (infinitely many of each sort), equality, negation, conjunction, disjunction, existential quantification, and the least-fixed-point operator which we write as "`Let` $\cdots \leftarrow \cdots$ `then` $\cdots$." *Terms* and *atomic formulas* are defined as usual in multi-sorted first-order logic, with equality for any sort allowed between any two terms of that sort. The definition of formulas is given by recursion, for all vocabularies simultaneously, as follows (under traditional conventions for omitting parentheses).

Definition 2.
- Atomic formulas of vocabulary Υ are Υ-formulas.
- If φ is an atomic Υ-formula and its predicate (or equality) symbol is negatable, then $\neg\varphi$ is an Υ-formula.
- If φ and ψ are Υ-formulas, then so are $\varphi \wedge \psi$ and $\varphi \vee \psi$.
- If φ is an Υ-formula then so is $\exists x\, \varphi$ for any variable x.
- If $\delta_1, \dots, \delta_k$ and φ are $\Upsilon \cup \{P_1, \dots, P_k\}$-formulas, where the P_i's are distinct positive predicate symbols not in Υ, and if, for each i, $\mathbf{x}_i$ is a sequence of distinct variables whose sorts match the arity of P_i, then

  $$\texttt{Let } P_1(\mathbf{x}_1) \leftarrow \delta_1, \dots, P_k(\mathbf{x}_k) \leftarrow \delta_k \texttt{ then } \varphi$$

  is an Υ-formula.

Free and bound variables are defined as usual in first-order logic with the added convention that, in `Let` $P_1(\mathbf{x}_1) \leftarrow \delta_1, \dots, P_k(\mathbf{x}_k) \leftarrow \delta_k$ `then` φ, the occurrences of the variables of $\mathbf{x}_i$ in $P(\mathbf{x}_i)$ and in δ_i are bound. Because the P_i's, though in the vocabulary of the δ_i's and φ, are not in the vocabulary of `Let` $P_1(\mathbf{x}_1) \leftarrow \delta_1, \dots, P_k(\mathbf{x}_k) \leftarrow \delta_k$ `then` φ, it is reasonable to regard them as bound second-order variables in the latter formula.

The semantics of EFPL is defined as in multi-sorted first-order logic (where we write A_s for the base set of sort s in structure $\mathfrak{A}$) with the following additional clause for `Let` $P_1(\mathbf{x}_1) \leftarrow \delta_1, \dots, P_k(\mathbf{x}_k) \leftarrow \delta_k$ `then` φ. Suppose we are given an Υ-structure $\mathfrak{A}$ and values in its base sets for all the free variables of `Let` $P_1(\mathbf{x}_1) \leftarrow \delta_1, \dots, P_k(\mathbf{x}_k) \leftarrow \delta_k$ `then` φ, i.e., for all the variables that either occur free in φ or occur free in some δ_i but are not in the corresponding $\mathbf{x}_i$. Then the formulas δ_i collectively define an operator on k-tuples of relations of the arities of the P_i's as follows. Given such a k-tuple of relations, use them as the interpretations of the P_i's to expand $\mathfrak{A}$ to an $\Upsilon \cup \{P_1, \dots, P_k\}$-structure; interpret each δ_i in that structure (with the given, fixed values of the variables other than $\mathbf{x}_i$) to obtain new relations of the same arities; the tuple of these is the output of the operator. Now form the least fixed point[2] of that operator. Use the k components of that

[2] The use of least fixed points here presupposes that the operator is monotone; that can be proved by induction on formulas, simultaneously with the definition of semantics. Alternatively, one can define the semantics using the inflationary fixed point construction, and then afterward prove monotonicity and infer that the inflationary fixed point is in fact the least fixed point. This alternative was used in [6]. Yet another equivalent alternative is to use as a definition the second-order formulation in the next paragraph.

fixed point as the interpretations of the P_i's to produce another $\Upsilon \cup \{P_1, \ldots, P_k\}$-structure, and interpret φ in it.

Once one verifies that truth values depend monotonically on the interpretations of positive predicates, it is easy to see that the semantics of $\texttt{Let}\ P_1(\mathbf{x}_1) \leftarrow \delta_1, \ldots, P_k(\mathbf{x}_k) \leftarrow \delta_k\ \texttt{then}\ \varphi$ is the same as that of the second-order formula

$$(\forall P_1) \cdots (\forall P_k) \left[\left(\bigwedge_{i=1}^{k} (\forall \mathbf{x}_i) \, (\delta_i \implies P(\mathbf{x}_i)) \right) \implies \varphi \right] .$$

Remark 3. The requirement that the P_i's be positive predicates means not only that they occur positively in the δ_i's, to ensure monotonicity of the inductive operator, but also that they occur positively in φ. Is this additional requirement a penalty for building positivity into vocabularies rather than treating it locally, in each formula, as is traditional? No, the additional requirement is needed. If it were waived, we could use the formula $\texttt{Let}\ P(x) \leftarrow \exists y \, \neg Q(x,y)\ \texttt{then}\ \neg P(x)$ (with negatable Q) to express $\forall y \, Q(x,y)$, an unwanted universal quantifier.

2 Good Behavior of EFPL

Existential fixed-point logic has many pleasant properties, including the following, which we first list briefly and then comment on more extensively.

1. EFPL captures polynomial time (PTime) on structures with successor.
2. EFPL works well in Hoare logic of asserted programs.
3. When a formula is satisfied by some elements of a structure, only a finite part of the structure is involved.
4. The iterations that produce the fixed-points for the $\texttt{Let} \cdots \leftarrow \cdots \texttt{then} \cdots$ construction terminate in at most ω steps.
5. Satisfaction of EFPL formulas is preserved along homomorphisms.
6. Denotations of EFPL formulas are preserved by the inverse-image parts of geometric morphisms of topoi.

2.1 Capturing PTime

A famous theorem of Immerman [11] and Vardi [13] asserts that first-order logic plus the least-fixed-point operator (FO+LFP) captures polynomial time on ordered structures. In more detail, consider structures whose base set has the form $\{1, 2, \ldots, n\}$ and whose vocabulary includes a symbol $<$ interpreted as the standard ordering of natural numbers. Such structures can easily be coded as strings over a finite (2-element, if desired) alphabet, so it makes sense to talk about a collection $\mathcal{C}$ of such structures being computable in polynomial time; there should be a PTime Turing machine accepting exactly (the strings that encode) the structures in $\mathcal{C}$. The Immerman-Vardi theorem says that $\mathcal{C}$ is PTime decidable if and only if it is the collection of models of some sentence in first-order-plus-least-fixed-point logic.

It is shown in [6] that the same is true for EFPL provided the collection of structures is modified in the following way. Instead of having a symbol for the ordering, the structures should have a symbol for the immediate successor function S. This modification would make no difference in the case considered by Immerman and Vardi, because S is definable from $<$ in first-order logic and $<$ is definable from S using the least-fixed-point operator. In EFPL, only the second of these definitions is available. Since the notion of immediate successor is needed in describing the operation of Turing machines, we must assume that S is available.

It is not difficult to extend the Immerman-Vardi theorem and its EFPL analog to the case of multi-sorted structures, provided one has appropriate successor functions on all of the sorts.

We record for future use the trivial observation that these theorems imply that PTime is also captured, on structures with successor, by any logic intermediate between EFPL and FO+LFP and indeed by stronger logics as long as these admit PTime model-checking.

2.2 Hoare Logic and Expressivity

We use [1] as our standard reference for Hoare logic. This logic deals with programs Π that operate in a first-order structure by modifying the values of some variables; thus a state of the computation is given by a tuple, listing the values of the variables of Π. Hoare logic also involves an assertion logic, traditionally taken to be first-order logic, though Yuri and I have argued in [6,7] that EFPL is a better choice. (Indeed, this was our original motivation for investigating EFPL.) The central syntactic construct in Hoare logic is the *asserted program*, $\varphi\Pi\psi$ where φ and ψ are formulas in the assertion logic and Π is a program. The semantic interpretation of $\varphi\Pi\psi$ is that, if Π is started in a state satisfying φ and if the computation terminates, then ψ holds in the final state. (Strictly speaking, this is the *partial correctness interpretation*; there is also a *total correctness interpretation* in which $\varphi\Pi\psi$ means that, if Π is started in a state satisfying φ, then it will terminate and ψ will be true in the final state.) In [6] and [7], the programming language is taken to be the `while`-language with parameterless procedure calls as in [1, Sect. 3].

Cook [10] proved a completeness theorem for Hoare logic (given all valid implications between formulas of the assertion language) provided a certain expressivity hypothesis is satisfied. That expressivity hypothesis requires that the assertion logic be able to express the *strongest postcondition* for any assertion φ and any program Π; that is, there should be a formula ψ that holds in exactly those states that result from a terminating computation of Π whose initial state satisfies φ, i.e., exactly where needed to make $\varphi\Pi\psi$ true. (An alternative version of expressivity requires *weakest preconditions.*) When the assertion logic is first-order logic, the expressivity hypothesis may or may not be satisfied, depending on the class of structures under consideration. But, as was proved in [6,7], when one instead uses EFPL as the assertion logic (and takes as guards in conditional statements and `while`-statements the quantifier-free formulas that involve no positive predicate symbols), then the expressivity hypothesis needed for Cook's theorem is automatically satisfied. It is in this sense that EFPL works well – in particular works better than first-order logic – with Hoare logic.

2.3 Finite Determination

Theorem 7 of [6] says that truth of EFPL formulas is finitely determined in the following sense. Suppose an EFPL formula φ is satisfied in a structure $\mathfrak{A}$ by certain values for its free variables. Then there is a finite subset F of (the union of the base sets of) $\mathfrak{A}$ such that φ is also satisfied, by the same values of its free variables, in any structure $\mathfrak{B}$ that matches $\mathfrak{A}$ on F, i.e., any $\mathfrak{B}$ with $\mathfrak{B} \upharpoonright F = \mathfrak{A} \upharpoonright F$.

In order for structures $\mathfrak{A}$ to have finite restrictions $\mathfrak{A} \upharpoonright F$, it is in general necessary to allow function symbols to be interpreted in $\mathfrak{A} \upharpoonright F$ as partial functions, rather than total ones. Alternatively, one can replace n-place function symbols by $(n+1)$-place relation symbols. (These modifications of the notion of structure or of the vocabulary will be useful in the topos-theoretic considerations below.)

In the finite determination result cited from [6], the requirement that $\mathfrak{B} \upharpoonright F = \mathfrak{A} \upharpoonright F$ can be weakened. As it stands, it says that each atomic formula (with suitable interpretation of function symbols as above) true in either of $\mathfrak{B} \upharpoonright F$ and $\mathfrak{A} \upharpoonright F$ is also true in the other. Although this requirement is certainly needed for those atomic formulas whose predicate symbol is negatable, only half of it is needed when the predicate symbol is positive. In the latter case, it suffices to require that, if the atomic formula is true in $\mathfrak{A} \upharpoonright F$, then it is also true in $\mathfrak{B} \upharpoonright F$. The reason is that EFPL truth is preserved when the interpretations of positive predicate symbols are enlarged.

Another way to view finite determination is that, whenever an EFPL formula is satisfied in $\mathfrak{A}$ by certain values for its variables, then this fact is a consequence of finitely many atomic and negated atomic formulas (with negation used only when the predicate symbol is negatable) about those values and possibly finitely many additional elements of $\mathfrak{A}$.

2.4 No Transfinite Iteration

An easy consequence of finite determination is that the potentially transfinite iterations leading to the least fixed-points in the semantics of $\texttt{Let}\ P_1(\mathbf{x}_1) \leftarrow \delta_1, \ldots, P_k(\mathbf{x}_k) \leftarrow \delta_k\ \texttt{then}\ \varphi$ are not really transfinite; they end at stage ω if not sooner. The reason is that, if a tuple of elements were to enter the interpretation of one of the P_i's in the step from ω to $\omega + 1$, then that fact would result, by finite determination, from information in a finite part of the structure. On that finite part, the ωth stage of the induction coincides with one of the earlier stages. Then that tuple would already have entered P_i at the next stage, long before stage ω.

On an intuitive level, both finite determination and the absence of transfinite iterations support the idea that EFPL is closely related to computation. At least, it has the sort of finiteness properties that one would expect from actual computation.

2.5 Homomorphisms

A homomorphism from one Υ-structure $\mathfrak{A}$ to another $\mathfrak{B}$ consists of functions h_s, one for each sort symbol s,

- mapping the base sets of $\mathfrak{A}$ to the corresponding ones of $\mathfrak{B}$, i.e.,

$$h_s : A_s \longrightarrow B_s ,$$

- commuting with the interpretations of function symbols, i.e.,

$$h_s(F_{\mathfrak{A}}(a_1, \ldots, a_n)) = F_{\mathfrak{B}}(h_{s_1}(a_1), \ldots, h_{s_n}(a_n))$$

  where F has arity $s_1, \ldots, s_n \longrightarrow s$,
- preserving the interpretations of positive predicate symbols in the forward direction, i.e.,

$$P_{\mathfrak{A}}(a_1, \ldots, a_n)) \implies P_{\mathfrak{B}}(h_{s_1}(a_1), \ldots, h_{s_n}(a_n)) ,$$

  and
- preserving the interpretations of negatable predicate symbols in both directions, i.e.,

$$P_{\mathfrak{A}}(a_1, \ldots, a_n) \iff P_{\mathfrak{B}}(h_{s_1}(a_1), \ldots, h_{s_n}(a_n)) .$$

That is, a homomorphism must preserve truth of atomic formulas and their negations insofar as the negations are permitted, i.e., insofar as the predicate symbols involved are negatable.

Theorem 4 of [6] shows that homomorphisms preserve truth of EFPL formulas. (The set-up there was one-sorted, but the same proof works in the many-sorted case.)

2.6 Topoi and Geometric Morphisms

A topos is a category so similar to the category of sets that higher-order logic can be naturally interpreted in it. (The simplest formal definition merely requires that the category have finite products, equalizers, and power objects, but many more constructions are obtainable as consequences of these, for example, all finite limits and colimits as well as exponentiation [i.e., an object X^Y of functions from Y to X]. See [12] for details, or see [3], where the treatment is more oriented to logic rather than category theory.) Some caution is needed, however, because the logical principles guaranteed to be valid in topos interpretations are those of intuitionistic logic, not classical logic.

In more detail, one can define, for any vocabulary Υ, the notion of an Υ-structure $\mathfrak{A}$ in a topos $\mathcal{E}$. This consists of

- for each sort symbol s, an object A_s (to play the role of the base set of sort s),

- for each predicate symbol P of arity $(s_1, \dots, s_n)$, a subobject $P_{\mathfrak{A}}$ of $A_{s_1} \times \cdots \times A_{s_n}$,
- for each function symbol F of input arity $(s_1, \dots, s_n)$ and output arity s, a morphism $F_{\mathfrak{A}} : A_{s_1} \times \cdots \times A_{s_n} \longrightarrow A_s$,

subject to the requirement that, for negatable predicate symbols P, the interpretation $P_{\mathfrak{A}}$ must be a complemented subobject of $A_{s_1} \times \cdots \times A_{s_n}$.

Remark 4. In intuitionistic higher-order logic, any subobject X of any object A has a *negation*, $\neg X$, the largest subobject of A disjoint from X (where "disjoint" means that the pullback is the initial object). But the union of X and $\neg X$ need not be all of A in general. It is all of A if and only if X is *complemented* in A, i.e., some subobject Y of A satisfies $X \cap Y = \varnothing$ and $X \cup Y = A$. In view of this standard terminology in intuitionistic logic, the terminology "complemented" might have been better than "negatable" in [6], but we do not propose to change it now.

Among the constructions available in topoi are all those needed to produce interpretations of terms and formulas of second-order logic over the vocabulary Υ, once an Υ-structure $\mathfrak{A}$ is given. Specifically, a term t of sort s with (first-order) variables among $x_1, \dots, x_n$ of sorts $s_1, \dots, s_n$ is interpreted as a morphism

$$t^{\mathfrak{A}} : A_{s_1} \times \cdots \times A_{s_n} \longrightarrow A_s \ ,$$

and a formula φ with free variables among $x_1, \dots, x_n$ of sorts $s_1, \dots, s_n$ is interpreted by a subobject

$$\varphi^{\mathfrak{A}} \subseteq A_{s_1} \times \cdots \times A_{s_n} \ .$$

(The interpretation can be extended to higher-order logic, including free variables of higher types, but we shall not need this extension.) Higher-order intuitionistic logic[3] is sound for such interpretations. That is, if a formula φ is a consequence in this logic of some other formulas ψ_i (all with free variables among $\mathbf{x}$ of sorts $\mathbf{s}$), then

$$\bigcap_i \psi_i^{\mathfrak{A}} \subseteq \varphi^{\mathfrak{A}}$$

(as subobjects of $\prod A_{\mathbf{s}}$).

Recall, from the end of Sect. 1, that the least-fixed-point constructor can be expressed in second-order logic. All the other constructors (equality, connectives, and $\exists$) of EFPL are among those of second-order (indeed of first-order) logic. Therefore, EFPL Υ-formulas have interpretations, as above, in Υ-structures in arbitrary topoi.

These interpretations behave better than those of general second-order formulas, in that they are preserved by the inverse-image parts of geometric morphisms

[3] The phrase "higher-order intuitionistic logic" is a misnomer as the notion of power set is probably not acceptable to intuitionists. The phrase is a convenient shorthand for "type theory, with axioms of extensionality and comprehension, based on intuitionistic logic."

of topoi. To explain what that means, we first discuss the two sorts of morphisms commonly used in connection with topoi.

Since topoi are defined as categories with a certain amount of structure (finite limits and power objects), it is natural to define homomorphisms of topoi to be functors that preserve this structure. Such homomorphisms are called *logical morphisms* because they preserve the interpretations of all formulas of higher-order logic. That is, if $f : \mathcal{E} \longrightarrow \mathcal{F}$ is a logical morphism and if $\mathfrak{A}$ is an Υ-structure in $\mathcal{E}$, then one obtains an Υ-structure $f(\mathfrak{A})$ in $\mathcal{F}$ by applying f to all the ingredients of $\mathfrak{A}$ (base sets A_s, interpretations $P_{\mathfrak{A}}$ of predicate symbols, and interpretations $F_{\mathfrak{A}}$ of function symbols), and one has

$$f(\varphi^{\mathfrak{A}}) = \varphi^{f(\mathfrak{A})}$$

for all formulas φ of higher-order logic.

A different notion of morphism, however, was natural in the earlier, more geometric theory of topoi developed by Grothendieck (see [2]). Grothendieck had observed that much of the algebraic topology of a topological space X can be expressed in terms of the category of sheaves over X and he defined topoi as generalized sheaf-categories. Further, he defined morphisms between topoi so that, in particular, the morphisms from the topos of sheaves on X to the topos of sheaves on Y correspond (as long as X and Y are somewhat reasonable spaces) to continuous functions from X to Y. Nowadays, topoi in Grothendieck's sense are called *Grothendieck topoi*; they are a proper subclass of the topoi defined above, often called *elementary topoi.* Morphisms in Grothendieck's sense are called *geometric morphisms* because of their origin in topological considerations. It turns out that geometric morphisms can be defined not only between Grothendieck topoi but between arbitrary topoi.

Definition 5. A *geometric morphism* from a topos $\mathcal{E}$ to another topos $\mathcal{F}$ is a pair of functors $f_* : \mathcal{E} \longrightarrow \mathcal{F}$ and $f^* : \mathcal{F} \longrightarrow \mathcal{E}$ such that f_* is right adjoint to f^* and f^* preserves finite limits (equivalently, preserves finite products and equalizers). f_* is called the *direct-image* part of the geometric morphism, and f^* is called the *inverse-image* part.

Unlike logical morphisms, the constituents f_* and f^* of geometric morphisms do not in general preserve the interpretations of higher-order (or even first-order) formulas. Nevertheless, the inverse image parts f^* of geometric morphisms have some good properties with respect to logic. They preserve the interpretation of existential positive first-order formulas. (In Grothendieck topoi, the same remains true if one allows infinite disjunctions; in general elementary topoi infinite disjunctions cannot be interpreted because the category may lack the infinite unions of subobjects that one needs.) Under our convention that negatable predicate symbols must be interpreted by complemented subobjects, f^* will preserve the interpretations of all existential first-order formulas. (The point here is that f^* need not preserve negations of general subobjects, but in the case of complemented subobjects f^* will preserve the complement.) Better yet, this preservation property remains correct when the least-fixed-point operator is added to the logic.

Proposition 6 ([5]). *If* $(f_*, f^*) : \mathcal{E} \longrightarrow \mathcal{F}$ *is a geometric morphism of topoi, if* $\mathfrak{A}$ *is an* Υ*-structure in* $\mathcal{F}$*, and if* φ *is an EFPL* Υ*-formula, then*

$$f^*(\varphi^{\mathfrak{A}}) = \varphi^{f^*(\mathfrak{A})} .$$

3 Implications between Good Behaviors

At the beginning of Sect. 2, we listed six pleasant properties of EFPL. The last four of these are restrictions on what can be asserted by EFPL formulas. The fact that they hold for EFPL trivially implies that they hold for all weaker logics, i.e., all logics whose formulas are all semantically equivalent to EFPL formulas. (The first two properties in our list, capturing PTime and working well with Hoare logic, are different in this respect; weaker logics would not in general share them. But of course the easier half of the first property, the availability of a PTime model-checking algorithm for each formula, would persist when the logic is weakened.)

The four listed restrictions on EFPL formulas are not independent of each other. We already pointed out that finite determination (property (3)) easily implies that the least-fixed-point recursions end in at most ω steps (property (4)). In fact, it turns out that preservation by geometric morphisms (property (6)) implies all three of the others. The goal of the present section is to establish these implications.

In view of the easy implication from (3) to (4), what must be shown is that (6) implies both (3) and (5). That is, any logic that is preserved by the inverse-image parts of geometric morphisms of topoi is automatically preserved forward along homomorphisms and automatically enjoys the finite determination property. We shall sketch some of the topos-theoretic background needed in these proofs, but we refer the reader to [12] for a careful and detailed treatment.

A word is in order here about the phrase "any logic" in the preceding paragraph. We shall use it to refer only to logics that are included in higher-order logic, because only for such logics is there a standard interpretation in Υ-structures in topoi. Our proofs, however, will be sufficiently abstract and general to show that, if some more exotic logic were equipped with an interpretation in Υ-structures in arbitrary topoi, then the implications we prove would hold for that logic as well.

3.1 Geometric Preservation Implies Homomorphism Preservation

Suppose φ is a formula (of higher-order logic) whose interpretation in Υ-structures is preserved by the inverse-image parts of geometric morphisms of topoi. Suppose further that φ holds of certain elements $\mathbf{a}$ (as values for the free variables $\mathbf{x}$) in a certain Υ-structure $\mathfrak{A}$. (Here, as in property (3) of Sect. 2, $\mathfrak{A}$ is an ordinary, set-based Υ-structure, not one in some arbitrary topos.) Finally, suppose $h : \mathfrak{A} \longrightarrow \mathfrak{B}$ is a homomorphism of Υ-structures. Our goal in this subsection is to prove that φ holds of the elements $h(\mathbf{a})$ in the Υ-structure $\mathfrak{B}$. In other words, we seek to prove that $h(\varphi^{\mathfrak{A}}) \subseteq \varphi^{\mathfrak{B}}$, where we write simply h for the function on tuples that acts as h on all components.

For this purpose, we consider the *Sierpiński topos*, the functor category $\mathcal{S} = \mathbf{Sets}^{\mathbf{2}}$, where $\mathbf{2}$ is the category with two objects and one non-identity morphism. This means that an object in $\mathcal{S}$ amounts to a diagram $A \xrightarrow{f} B$ consisting of a single function between two sets. A morphism in $\mathcal{S}$ from one such object $A \xrightarrow{f} B$ to another $A' \xrightarrow{f'} B'$ is a commutative diagram

$$\begin{array}{ccc} A & \xrightarrow{f} & B \\ \downarrow & & \downarrow \\ A' & \xrightarrow{f'} & B' . \end{array}$$

To deal with Υ-structures in $\mathcal{S}$, one must understand the category-theoretic notions of product, subobject, and complementation used in the definition. Fortunately, these turn out to be quite easy. The product of several objects of the form $A \xrightarrow{f} B$ consists of the product of the A's, the product of the B's, and the function induced componentwise by the f's. A subobject of $A \xrightarrow{f} B$ is (up to isomorphism) $X \xrightarrow{f \restriction X} Y$ where $X \subseteq A$ and $f(X) \subseteq Y \subseteq B$. This subobject is complemented if and only if $f(A - X) \cap Y = \varnothing$, in which case the complement is $A - X \xrightarrow{f \restriction (A-X)} B - Y$. (From now on I'll omit the restriction notation over the arrows and just write f.) In general, even if $X \xrightarrow{f} Y$ is not complemented in $A \xrightarrow{f} B$, its negation is $f^{-1}(B - Y) \xrightarrow{f} B - Y$.

With these observations in place, it is easy to check that an Υ-structure in $\mathcal{S}$ amounts to two Υ-structures (in the ordinary set-based sense) and a homomorphism between them. In particular, for a negated predicate symbol, the requirement that its interpretation in a topos-based structure be complemented corresponds exactly to the requirement that a homomorphism preserve the predicate not only in the forward direction but also backward.

The Sierpiński topos has two geometric morphisms from the topos $\mathbf{Set}$ of sets. Their inverse-image parts (which are the parts of interest for us) are the domain functor D^* and the codomain functor C^*, sending $A \longrightarrow B$ to A and to B, respectively. The actions on morphisms just extract the left and right vertical arrows from the commutative square above. (The direct-image parts D_* and C_* send any set X to the unique map $X \longrightarrow 1$ and the identity map $X \longrightarrow X$, respectively, and act in the obvious way on morphisms.)

With these preliminaries, we are ready to return to the situation where a formula φ is preserved by the inverse-image parts of geometric morphisms and we want to prove that it is also preserved by homomorphisms $h : \mathfrak{A} \longrightarrow \mathfrak{B}$ between (ordinary, set-based) Υ-structures. The given $\mathfrak{A} \xrightarrow{h} \mathfrak{B}$ is an Υ-structure in $\mathcal{S}$; call it $\mathfrak{H}$ for short. To further simplify notation, let $\mathbf{s}$ be the list of sorts of the variables in φ for which $\mathbf{a}$ are the values, and write $A_{\mathbf{s}}$ for the product of the interpretations of these sorts in $\mathfrak{A}$; similarly for $B_{\mathbf{s}}$. Then the interpretation of φ in $\mathcal{H}$ is a subobject $\varphi^{\mathfrak{H}} = (X \xrightarrow{h} Y)$ of $A_{\mathbf{s}} \xrightarrow{h} B_{\mathbf{s}}$. It remains only to combine the following three facts:

– $h(X) \subseteq Y$, because $X \xrightarrow{h} Y$ is an object of $\mathcal{S}$.
– $X = D^*(\varphi^{\mathfrak{H}}) = \varphi^{D^*(\mathfrak{H})} = \varphi^{\mathfrak{A}}$ because D^* preserves the interpretation of φ.
– $Y = C^*(\varphi^{\mathfrak{H}}) = \varphi^{C^*(\mathfrak{H})} = \varphi^{\mathfrak{B}}$ because C^* preserves the interpretation of φ.

Thus, we get that $h(\varphi^{\mathfrak{A}}) \subseteq \varphi^{\mathfrak{B}}$, as required.

Remark 7. With a little additional work, essentially the same argument shows that the interpretation of φ is also preserved forward along homomorphisms between Υ-structures in arbitrary topoi. The additional work arises because we have used the fact that, in **Sets**, all subobjects are complemented. In general, if $\mathcal{S}$ is the Sierpiński topos over some other topos $\mathcal{E}$, a subobject $X \xrightarrow{f} Y$ of $A \xrightarrow{f} B$ is complemented if and only if X is complemented in A, Y is complemented in B, and f maps the complement of X into the complement of Y (and of course maps X into Y, since $X \xrightarrow{f} Y$ is an object in $\mathcal{S}$).

3.2 Geometric Preservation Implies Finite Determination

In this subsection, we shall show that the finite determination property of EFPL (property (3) in Sect. 2) is a consequence of preservation by inverse-image parts of geometric morphisms. This result was stated at the end of [4] but the proof was not published.

The proof involves part of the theory of classifying topoi, and we begin by sketching that part. For more details, see [12].

Recall that, in the discussion of finite determination in Sect. 2, as well as in [6], it was convenient to either allow function symbols to be interpreted by partial functions or replace function symbols by predicate symbols (of higher arity). The purpose was to ensure the existence of finite substructures. Accordingly, we shall also assume in the present subsection that Υ is a purely relational vocabulary, i.e., that there are no function symbols (except possibly 0-ary ones, i.e., constant symbols). At the end, we shall also comment on the alternative approach that uses partial functions.

For simplicity, we also assume that the vocabulary Υ is finite – only finitely many sorts, predicate symbols, and constant symbols.

We need the notion of classifying topos for Υ-structures. Were it not for the distinction between positive and negatable predicate symbols, this notion would be part of the standard theory developed in [12], and it could be summarized as follows. Temporarily pretend that all the predicate symbols in Υ are positive and can therefore be interpreted, in models in topoi, by arbitrary subobjects, not necessarily complemented. Let $\mathcal{M}$ be the category of finite Υ-structures (in the ordinary, set-based sense) and homomorphisms. To avoid irrelevant set-theoretic issues, include in $\mathcal{M}$ only those structures whose base sets consist of natural numbers; of course every finite Υ-structure is isomorphic to one of these. Let $\mathcal{C} = \mathbf{Sets}^{\mathcal{M}}$ be the category of functors from $\mathcal{M}$ to **Sets**. Like any functor category of the form $\mathbf{Sets}^{\mathcal{A}}$ for any small category $\mathcal{A}$, this is a topos.

Among its objects are the "base sets" functors, $G_s : \mathcal{M} \longrightarrow \mathbf{Sets}$, one for each sort s in Υ; G_s sends any Υ-structure $\mathfrak{A} \in \mathcal{M}$ to its base set A_s for sort s (and acts in the obvious way on homomorphisms). These functors G_s constitute the interpretations of the sorts in an Υ-structure $\mathfrak{G}$ in the topos $\mathcal{C}$. The interpretation $P_{\mathfrak{G}}$ of a predicate symbol P is the functor sending each Υ-structure $\mathfrak{A} \in \mathcal{M}$ to its interpretation $P_{\mathfrak{A}}$ of P. Thus, $\mathfrak{G}$ can be described as a "tautological" model in $\mathcal{C}$; each of its constituents (interpretations of sorts or of predicates) is a functor $\mathcal{M} \longrightarrow \mathbf{Sets}$ whose value at any structure $\mathfrak{A} \in \mathcal{M}$ is the corresponding constituent of $\mathfrak{A}$.

This $\mathfrak{G}$ has the following universal property among Υ-structures. Any Υ-structure $\mathfrak{A}$ in any Grothendieck topos $\mathcal{E}$ is $f^*(\mathfrak{G})$ for a unique (up to natural isomorphism) geometric morphism $f : \mathcal{E} \longrightarrow \mathcal{C}$. This property[4] is expressed by saying that $\mathcal{C}$ is the *classifying topos* for Υ-structures and $\mathfrak{G}$ is the *generic* or *universal* Υ-structure. See [12, Chapter 6] for details, with considerable simplifications because our Υ has no function symbols and because we are classifying (for now) arbitrary Υ-structures, not only the models of some geometric theory.

The preceding discussion of the classifying topos for Υ-structures was based on our temporary assumption that all predicate symbols are positive. We now consider the changes needed when some of the symbols are negatable and must therefore be interpreted as complemented subobjects. Only in very special cases will the $\mathfrak{G}$ above satisfy this requirement.

The simplest way to view the situation with a negatable predicate symbol P is that there is, in effect, another predicate symbol P', of the same arity, representing the negation of P. We then regard both P and P' as positive symbols, but we impose the logical axioms

$$\forall \mathbf{x}\, \big(P(\mathbf{x}) \wedge P'(\mathbf{x}) \implies \texttt{false}\big)$$

and

$$\forall \mathbf{x}\, \big(\texttt{true} \implies P(\mathbf{x}) \vee P'(\mathbf{x})\big)$$

which require the interpretations of P and P' to be complementary.

Fortunately, there are standard methods for constructing classifying topoi and generic models for systems of axioms of these forms – universally quantified implications between existential positive formulas. The job is easier for the first form, the axiom saying P and P' are disjoint, because this is a universal Horn sentence. The classifying topos for models of such axioms is, as before, a functor category $\mathcal{C} = \mathbf{Sets}^{\mathcal{M}}$ where $\mathcal{M}$ is now the category of finite models (again in the ordinary, set-based sense) of the universal Horn theory in question. So in our situation, the structures in $\mathcal{M}$ would be Υ-structures equipped with interpretations of the additional symbols P' for all negatable P in Υ, and subject to the requirement that the interpretations of P and P' be disjoint. The generic model is, as before, the tautologous model whose constituent functors (interpreting the sorts and predicates) take models in $\mathcal{M}$ to their constituents. For proofs of these assertions, see [8, Sect. 1].

[4] Strictly speaking, the property should be stated in a stronger form taking morphisms into account: For any topos $\mathcal{E}$, the category of geometric morphisms $f : \mathcal{E} \longrightarrow \mathcal{C}$ and natural transformations (of their inverse-image parts) is equivalent to the category of Υ-structures in $\mathcal{E}$, one direction of the equivalence being evaluation of f^* at $\mathfrak{G}$.

The second form of axioms, saying that every tuple (of the appropriate sorts) satisfies one of P and P', is not a Horn sentence, because of the disjunction in the consequent of the implication. In this situation, the standard technique for constructing classifying topoi would not produce a functor category as above but rather a subcategory of sheaves. For this discussion, we must therefore presuppose some information about sheaf topoi; at the end, it will turn out that we can, after all, use a functor category, but justifying this assertion requires some discussion of sheaves. The reader unfamiliar with sheaves could either consult [12] for the necessary information or skip the following discussion of sheaves, rejoining us at Proposition 8.

Each of the axioms

$$\forall \mathbf{x} \left(\mathtt{true} \implies P(\mathbf{x}) \lor P'(\mathbf{x})\right)$$

under consideration (one for each negatable predicate symbol P in Υ) contributes a sieve to a Grothendieck topology J on $\mathcal{M}^{\mathrm{op}}$, the dual category of $\mathcal{M}$. To describe the sieve associated to the axiom for P and P', let $\langle \mathbf{x} \rangle$ be the Υ-structure with one element $\dot{x}_i$ for each variable[5] x_i in the list $\mathbf{x}$, plus other elements serving as the values of any constant symbols in Υ, and with all predicates interpreted as empty. (Note that, although the interpretations of P and P' won't be complementary, they will satisfy the universal Horn sentence saying they're disjoint. So this structure is in $\mathcal{M}$.) Let $\langle \mathbf{x} : P(\mathbf{x}) \rangle$ be defined similarly except that P holds of exactly one tuple, namely $\dot{\mathbf{x}}$. Define $\langle \mathbf{x} : P'(\mathbf{x}) \rangle$ analogously. The two obvious homomorphisms (given, as functions, by the identity map)

$$\langle \mathbf{x} \rangle \longrightarrow \langle \mathbf{x} : P(\mathbf{x}) \rangle \quad \text{and} \quad \langle \mathbf{x} \rangle \longrightarrow \langle \mathbf{x} : P'(\mathbf{x}) \rangle$$

generate a sieve S_P on $\langle \mathbf{x} \rangle$ in $\mathcal{M}^{\mathrm{op}}$. Let J be the Grothendieck topology generated by these sieves S_P for all negatable predicate symbols P of Υ. By the general theory of classifying topoi (as in [12]), the category of J-sheaves serves as the classifying topos for Υ-structures in the original sense, with negatable predicates interpreted as complemented subobjects. The generic Υ-structure is the associated sheaf of the $\mathfrak{G} \in \mathbf{Sets}^{\mathcal{M}}$ that served as the generic model for just the universal Horn axioms.

It turns out that this description of the classifying topos for Υ-structures can be simplified. For this purpose, let us first observe that, according to the definition of "Grothendieck topology," J contains various additional sieves, beyond the generating sieves S_P.

In the first place, because Grothendieck topologies are closed under pullbacks (and pullbacks in $\mathcal{M}^{\mathrm{op}}$ are, of course, pushouts in $\mathcal{M}$), we have the following additional coverings. Suppose $\mathfrak{A}$ is an object of $\mathcal{M}$, so it interprets P and P' as disjoint but perhaps not complementary subobjects of the appropriate $A_{\mathbf{s}}$, and suppose that there really is a tuple $\mathbf{a} \in A_{\mathbf{s}}$ satisfying neither P nor P'. Let $\mathfrak{A}^+$

[5] It would do no harm to omit the dots and take the variables themselves as elements of the structure. We use the dot only to avoid any possible confusion between the syntactic role of variables and the semantic role of elements of a structure.

be exactly like $\mathfrak{A}$ except that this one tuple $\mathbf{a}$ now satisfies P; similarly, let $\mathfrak{A}^-$ be exactly like $\mathfrak{A}$ except that $\mathbf{a}$ now satisfies P'. Then the pair of homomorphisms given by identity maps

$$\mathfrak{A} \longrightarrow \mathfrak{A}^+ \quad \text{and} \quad \mathfrak{A} \longrightarrow \mathfrak{A}^-$$

generates a sieve on $\mathfrak{A}$ that is the pullback (in $\mathcal{M}^{\text{op}}$) of S_P along the homomorphism $\langle \mathbf{x} \rangle \longrightarrow \mathfrak{A}$ that sends $\dot{\mathbf{x}}$ to $\mathbf{a}$. So this pair covers $\mathfrak{A}$.

This argument can be iterated, i.e., it can be applied to other tuples that satisfy neither P nor P' in $\mathfrak{A}$ (and therefore also in $\mathfrak{A}^{\pm}$) as well as to other negatable predicate symbols. Because, in a Grothendieck topology, covers of covers are covers (and because both Υ and $\mathfrak{A}$ are finite), we find that $\mathfrak{A}$ is covered (in the topology J on $\mathcal{M}^{\text{op}}$) by homomorphisms (in $\mathcal{M}$) from $\mathfrak{A}$ to objects of $\mathcal{M}$ in which, for every negatable predicate symbol P and for every tuple $\mathbf{a}$ of the appropriate arity, either $P(\mathbf{a})$ or $P'(\mathbf{a})$ holds. In other words, every object is covered by homomorphisms to genuine Υ-structures.

In this situation, Grothendieck's "Lemme de comparaison" [2, III.4.1] applies and tells us that the topos of J-sheaves over $\mathcal{M}^{\text{op}}$ is equivalent to the topos of sheaves on the full subcategory $\mathcal{M}^{*\text{op}}$ of genuine finite Υ-structures, with the topology induced by J. Furthermore, it is easy to see that this induced topology is trivial; any object is covered only by the sieve of all morphisms to it. Thus, the category of sheaves reduces to the category of presheaves on $\mathcal{M}^{*\text{op}}$, i.e., the functor category $\mathbf{Sets}^{\mathcal{M}^*}$.

Notice also that, among objects in $\mathcal{M}^*$, the homomorphisms as defined in $\mathcal{M}$ are, in fact, homomorphisms of Υ-structures. That is, they preserve negatable predicates P not only forward but also backward. This is simply because they preserve P' forward.

We summarize the result of this sheaf discussion, adding some easily checked information about the generic model.

Proposition 8. *The classifying topos for Υ-structures is the functor category* $\mathbf{Sets}^{\mathcal{M}^*}$ *where $\mathcal{M}^*$ is the category of finite Υ-structures and homomorphisms. The generic Υ-structure is the one whose constituents (interpretations of sorts and predicates) are the functors that send each finite Υ-structure $\mathfrak{A}$ to its corresponding constituents in* **Sets**.

With this description of the classifying topos for Υ-structures and the generic structure, we are ready to prove the main result of this subsection.

Theorem 9. *Let $\varphi(\mathbf{x})$ be a formula of higher-order logic whose interpretations in Υ-structures in Grothendieck topoi are preserved by the inverse-image parts of geometric morphisms. Let Φ be the set of all those Υ-formulas $\alpha(\mathbf{x}, \mathbf{y})$ such that*

- *$\alpha(\mathbf{x}, \mathbf{y})$ is a conjunction of atomic formulas and negations of atomic formulas (where the list $\mathbf{y}$ of variables can be different for different α's) and*
- *$\alpha(\mathbf{x}, \mathbf{y})$ implies $\varphi(\mathbf{x})$ in all ordinary, set-based Υ-structures.*

Then the sentence

$$\forall \mathbf{x} \left(\varphi(\mathbf{x}) \iff \bigvee_{\alpha(\mathbf{x},\mathbf{y})\in\Phi} \exists \mathbf{y}\, \alpha(\mathbf{x},\mathbf{y}) \right) \tag{1}$$

is valid in all Υ-structures in all Grothendieck topoi. In particular, it is valid in all ordinary, set-based Υ-structures.

The conclusion (1) of this theorem asserts that, whenever $\varphi(\mathbf{x})$ is satisfied by some elements of an Υ-structure, it is "because" those elements and finitely many others satisfy some quantifier-free information $\alpha(\mathbf{x},\mathbf{y})$ that guarantees the truth of $\varphi(\mathbf{x})$. That is, we have finite determination as described in Sect. 2. And the theorem says that this will happen for any $\varphi(\mathbf{x})$ that is preserved by the inverse-image parts of geometric morphisms.

In connection with the definition of Φ, note that α is required to be an Υ-formula, so negation will be applied only to atomic formulas whose predicate symbol is negatable.

Proof. The proof proceeds in three phases, each establishing the equivalence (1) in certain circumstances.

In phase 1, we observe that (1) holds in all finite (ordinary, set-based) Υ-structures. The right-to-left implication is immediate from the definition of Φ. As for the left-to-right implication, consider any finite Υ-structure $\mathfrak{A}$ and any tuple $\mathbf{a}$ of elements in it such that $\mathfrak{A}$ satisfies $\varphi(\mathbf{a})$. Let $\mathbf{b}$ be a list (without repetitions) of all the elements of the base sets of $\mathfrak{A}$. Note that this is a finite list because, in phase 1, we are dealing with a finite structure $\mathfrak{A}$. Let $\alpha(\mathbf{x},\mathbf{y})$ be the conjunction of all the atomic formulas and negated atomic Υ-formulas that are satisfied in $\mathfrak{A}$ by the tuple $(\mathbf{a},\mathbf{b})$. There are only finitely many conjuncts here, because Υ has no function symbols and only finitely many predicate and constant symbols.

I claim that $\alpha(\mathbf{x},\mathbf{y}) \in \Phi$. Once this claim is proved, we will know that the elements $\mathbf{a}$ satisfy in $\mathfrak{A}$ the disjunct $\exists \mathbf{y}\, \alpha(\mathbf{x},\mathbf{y})$ on the right side of (1), so the proof for phase 1 will be complete.

To verify the claim, suppose our $\alpha(\mathbf{x},\mathbf{y})$ is satisfied by some tuple $(\mathbf{a}',\mathbf{b}')$ in some Υ-structure $\mathfrak{B}$. It is easy to check that, by sending each element in the list $\mathbf{b}$ (i.e., each element of any of the base sets of $\mathfrak{A}$) to the coresponding element in $\mathbf{b}'$, we define a homomorphism $h : \mathfrak{A} \longrightarrow \mathfrak{B}$ that satisfies $h(\mathbf{a}) = \mathbf{a}'$. Indeed, the fact that $(\mathbf{a}',\mathbf{b}')$ satisfies the equations in $\alpha(\mathbf{x},\mathbf{y})$ ensures that $h(\mathbf{a}) = \mathbf{a}'$, while satisfaction of the other conjuncts in $\alpha(\mathbf{x},\mathbf{y})$ is exactly what is needed to ensure that h is a homomorphism. Having already shown, in Subsection 3.1, that φ must be preserved by homomorphisms, we know that $\mathfrak{B}$ satisfies $\varphi(\mathbf{a}')$. This completes the verification of the claim and thus phase 1 of the proof.

In phase 2, we establish that the equivalence (1) is valid in the generic Υ-structure $\mathfrak{G}$ in the classifying topos $\mathbf{Sets}^{\mathcal{M}^*}$. Note that, because (1) has no free variables, its interpretation is a subobject of the empty product 1 (i.e., the interpretation is a truth value); we shall show that this interpretation is all of

1 (i.e., the truth-value `true`). For this purpose, we shall apply the assumption that $\varphi(\mathbf{x})$ is preserved by the inverse-image functors of geometric morphisms.

The relevant inverse-image functors for this phase of the proof are the evaluation functors $E^*_{\mathfrak{A}}$, one for each object $\mathfrak{A}$ of $\mathcal{M}^*$. These functors from $\mathbf{Sets}^{\mathcal{M}^*}$ to $\mathbf{Sets}$ are defined by

$$E^*_{\mathfrak{A}}(X) = X(\mathfrak{A})$$

for all objects X of $\mathbf{Sets}^{\mathcal{M}^*}$, i.e., all functors $X : \mathcal{M}^* \longrightarrow \mathbf{Sets}$. It is well known [12] and easy to check that $E^*_{\mathfrak{A}}$ is the inverse-image part of a geometric morphism; its right adjoint is the functor $E_{\mathfrak{A}*} : \mathbf{Sets} \longrightarrow \mathbf{Sets}^{\mathcal{M}^*}$ defined by

$$\begin{aligned} &E_{\mathfrak{A}*}(S)(\mathfrak{B}) = \\ &\qquad = \text{set of functions to } S \text{ from the set of homomorphisms } \mathfrak{B} \longrightarrow \mathfrak{A}\,. \end{aligned}$$

Applying the definition of $E^*_{\mathfrak{A}}$ with the (tautological) constituents of the generic $\mathfrak{G}$ in the role of X, we find that

$$E^*_{\mathfrak{A}}(\mathfrak{G}) = \mathfrak{A}$$

for all objects $\mathfrak{A}$ of $\mathcal{M}^*$, i.e., for all finite Υ-structures $\mathfrak{A}$. We know that, like the inverse-image part of any geometric morphism, $E^*_{\mathfrak{A}}$ preserves the interpretation of $\varphi(\mathbf{x})$. It also preserves the interpretation of all the α's that occur in (1). Indeed, interpretations of atomic formulas and their complements are preserved, according to the definition of how inverse-image functors act on Υ-structures, and conjunctions are preserved because inverse-image functors preserve finite limits. Furthermore, the existential quantifiers and the (in general infinite) disjunction in (1) are preserved because inverse-image functors, having right adjoints, preserve all colimits, and so preserve images and joins (even infinite joins) of subobjects.

Consider the interpretations in $\mathfrak{G}$ (in the topos $\mathbf{Sets}^{\mathcal{M}^*}$) of the two sides of the biconditional in (1), $\varphi(\mathbf{x})$ and the big disjunction. On the left we have $\varphi(\mathbf{x})_{\mathfrak{G}}$ and on the right we have another subobject D of the relevant product $G_{\mathbf{s}}$. For any finite (ordinary, set-based) Υ-structure $\mathfrak{A}$, the functor $E^*_{\mathfrak{A}}$ sends these two subobjects to $\varphi_{\mathfrak{A}}$ and the interpretation in $\mathfrak{A}$ of the right side of the biconditional in (1). We have shown in phase 1 that these two subobjects are the same. Since this holds for every object $\mathfrak{A}$ in $\mathcal{M}^*$, we have that $\varphi(\mathbf{x})_{\mathfrak{G}}$ and D are two subobjects of $G_{\mathbf{s}}$ in the functor category $\mathbf{Sets}^{\mathcal{M}^*}$ that have the same values at every $\mathfrak{A}$ in $\mathcal{M}^*$. That makes them the same functor and thereby shows that (1) holds (i.e., has interpretation 1) in $\mathfrak{G}$. This completes phase 2 of the proof.

Finally, in phase 3, we prove the full conclusion of the theorem. Let $\mathfrak{A}$ be an arbitrary Υ-structure in an arbitrary Grothendieck topos $\mathcal{E}$. Because $\mathfrak{G}$ is the generic Υ-structure, there is a geometric morphism $f : \mathcal{E} \longrightarrow \mathbf{Sets}^{\mathcal{M}^*}$ such that $\mathfrak{A} = f^*(\mathfrak{G})$. As in phase 2, we have that f^* preserves the interpretations of both sides of the biconditional in (1). Since the two sides have, according to phase 2, the same interpretation in $\mathfrak{G}$ (in $\mathbf{Sets}^{\mathcal{M}^*}$), it follows immediately that they have the same interpretation in $\mathfrak{A}$ (in $\mathcal{E}$). That completes the proof of the theorem. □

Remark 10. As indicated in Sect. 2, prohibiting function symbols (except constants) is not the only way to ensure the existence of enough finite structures as needed for finite determination. An alternative is to allow function symbols but to modify the definition of Υ-structure so that function symbols can be interpreted as partial functions.

To obtain a classifying topos for Υ-structures in this sense, it is convenient to temporarily regard the (partial) function symbols as relation symbols subject to logical axioms of the form

$$\forall \mathbf{x} \, \forall y \, \forall z \left((F(\mathbf{x}, y) \wedge F(\mathbf{x}, z)) \implies x = y \right) .$$

Since these axioms are universal Horn sentences, the classifying topos is, according to a result from [8] already used above, the topos of functors from the category $\mathcal{M}$ of models (in the new sense, with partial functions) to **Sets**.

With this classifying topos, we can proceed as above to incorporate the requirement that negatable predicate symbols be interpreted by complemented objects, and the rest of the results of this subsection work as before.

4 Negation and Universal Quantification – Closed Worlds

In this section we consider EFPL as a database query language and try to elucidate its intuitive meaning.

For computational considerations, the most important properties of EFPL are probably its capturing PTime (property (1) in Sect. 2) and its good behavior with Hoare logic (property (2)). But we shall be concerned here with more general properties of the queries expressible in EFPL, properties best seen by thinking about preservation by homomorphisms (property (5) in Sect. 2).

Homomorphism preservation means that, if a database (= structure) produces a positive answer to an EFPL query, then it will continue to do so if new elements are added and also if the relations that interpret the positive predicate symbols are increased, but not (in general) if the interpretations of negatable predicates are modified.

One way to view this situation is to regard the database as a possibly incomplete description of some reality. The elements of the database are (or represent) some elements of the real world, but the real world may well have additional elements of which the database is entirely ignorant. If a predicate P holds of some elements $\mathbf{a}$ in the database, this means that the corresponding relationship obtains between these elements in the real world, but again it might happen that P holds of some elements $\mathbf{a}$ in the real world without the database knowing that fact – even if it is aware of the elements $\mathbf{a}$ themselves. So adding new elements to the database or adding new tuples to the relations that interpret predicate symbols can bring the database closer to reality. From this point of view, it is reasonable that such improvements of the database should not invalidate any positive answers that it has already given.

What is the role of negatable predicates in this picture? Formally, their interpretations should not be increased by adding tuples when the components of those tuples were already present; such a change risks invalidating earlier positive answers to EFPL queries. Intuitively, this means that the database's information about these negatable predicates is complete, at least insofar as the elements present in the database are concerned. In other words, we have a closed-world assumption for these predicates: If the tuple $\mathbf{a}$ is available in the database but doesn't satisfy P there, then this means that $\mathbf{a}$ doesn't satisfy P in reality. (Contrast this with the situation for positive predicates, where $P(\mathbf{a})$ could fail in the database while it holds in reality, if the database simply lacked this bit of information.)

Thus, our distinction between negatable and positive predicate symbols formalizes the distinction between predicates to which such a closed-world assumption applies and others to which it does not apply.

Given this idea, it is natural to also consider another sort of closed-world assumption, one which says that the database is aware of all the elements of a certain sort; no additional elements can be added. (We formulated EFPL in a multi-sorted framework in order to be able to impose this closed-world assumption on only some sorts rather than on the entire database.) Such a closed-world assumption is not formalized in EFPL; homomorphisms can lead to new elements in any sorts. A closed-world assumption for a sort s should be reflected formally in a requirement that homomorphisms be surjective on the base sets of sort s. This restriction on the allowed homomorphisms would be reflected in a liberalization of the language; with fewer homomorphisms, we can expect them to preserve more formulas.

In fact, there is a very familiar way to extend the language so as to retain preservation properties for surjective homomorphisms but not for others: Allow universal quantification. This leads to the following proposal for extending EFPL.

A vocabulary should say which (if any) of its sorts are *closed*; the others are then called *open*. The syntax is extended by allowing universal quantification of variables of closed sorts. The semantics is the obvious one, familiar from first-order logic, in the case of set-based structures. The semantics in structures in topoi is perhaps not obvious but it is well-known. As indicated earlier, higher-order intuitionistic logic is interpreted in topoi [3,12], and that certainly includes first-order universal quantification.

Of course, it is easy to propose a new logic, especially such a slight variation of a known one. But does this extension preserve any of the nice properties of EFPL? That is the topic of the next section.

5 Geometric Preservation

In this section, we consider the pleasant properties of EFPL listed in Sect. 2 and analyze what happens to them when we introduce universal quantification over some sorts, the closed ones.

Of course there is no problem when the pleasant property is one that says the logic is rich enough for some purpose; we have only made it richer. Thus, for example, we certainly retain the "richness" half of capturing PTime; every Ptime computable property of structures with successor was expressible in EFPL and therefore is expressible in the extension by $\forall$ on closed sorts. The other half of capturing PTime, namely the availability of PTime model-checking for each formula, could be lost by enlarging the logic, but it is not lost in the present enlargement. The reason is that, even with $\forall$ adjoined, our logic is still a fragment of first-order-plus-least-fixed-point logic, which captures PTime by the Immerman-Vardi theorem.

The good behavior of EFPL in relation to Hoare logic is also quite safe under the present extension. Inspection of the relevant arguments in [6,7] shows that they depend only on the availability of the least-fixed-point construction, existential quantification, and some connectives, not on the unavailability of other things like universal quantification.

The remaining four properties of EFPL listed in Sect. 2 can, however, be lost when we introduce $\forall$ on closed sorts. In one case, preservation by homomorphisms, this loss is intentional. We introduced $\forall$ in order to match a more restrictive notion of homomorphism, surjective on the closed sorts. With this modified notion of homomorphism, this preservation property revives.

Finite determination is lost even for the simplest case, the formula $\forall x\, P(x)$, if the base set of the sort of x is infinite. The fact that all infinitely many elements of this base set satisfy P is obviously not a consequence of information about any finitely many of them. To revive finite determination, we would have to require that closed sorts be interpreted by finite base sets.

If universal quantification is allowed over infinite sets then the iterations leading to least fixed points can continue for any ordinal number of steps. An example was given at the end of [6] where, in an arbitrary wellordering, the elements are added in order, one at a time in the iteration.

Finally, we consider preservation by inverse-image parts of geometric morphisms of topoi. Here again, preservation can fail in general but will hold if the interpretations of the closed sorts satisfy an appropriate restriction, related to finiteness but considerably weaker.

Remark 11. How can it be weaker? We saw in Subsection 3.2 that geometric preservation implies finite determination, and a moment ago we saw that finite determination trivially fails unless closed sorts are finite. Therefore mustn't geometric preservation also fail unless the closed sorts are finite?

The fallacy in this argument is that the proof in Subsection 3.2 used geometric preservation for topoi like the classifying topos $\mathbf{Sets}^{\mathcal{M}^*}$ when proving finite determination in other (arbitrary) topoi. It is entirely possible that a weaker condition than finiteness, applied to the generic model in the classifying topos, may imply finiteness elsewhere, for example in **Sets**.

In order to discuss the conditions under which universal quantification over a set (the interpretation A_s of a sort s in an Υ-structure $\mathfrak{A}$) is preserved by the inverse-image parts of geometric morphisms, we must first recall the topos-theoretic interpretation of universal quantification. Consider a formula $\varphi(\mathbf{x})$ of the form $\forall y\, \psi(\mathbf{x}, y)$. Let s be the sort of the universally quantified variable y, and let

$\mathbf{r}$ be a list of the sorts of the other variables $\mathbf{x}$. So the interpretation, in an Υ-structure $\mathfrak{A}$ in a topos $\mathcal{E}$, of $\varphi(\mathbf{x})$ is a certain subobject X of $A_{\mathbf{r}}$, and similarly the interpretation of $\psi(\mathbf{x}, y)$ is a subobject Y of $A_{\mathbf{r}} \times A_s$. The relationship between these two subobjects is that X is the largest subobject of $A_{\mathbf{r}}$ whose inverse image (i.e., pullback) $p^{-1}(X)$ along the projection $p : A_{\mathbf{r}} \times A_s \longrightarrow A_{\mathbf{r}}$ is included in Y. The question we intend to answer here is under what circumstances will this relationship between X and Y be preserved by the inverse-image part f^* of an arbitrary geometric morphism $f : \mathcal{F} \longrightarrow \mathcal{E}$.

In fact, we shall answer the question in a more general context, where universal quantification is not necessarily along a product projection p but along an arbitrary morphism in $\mathcal{E}$. Specifically, suppose that $p : L \longrightarrow M$ is a morphism in $\mathcal{E}$, that Y is a subobject of L, and that X is the largest subobject of M whose pullback $p^{-1}(X)$ is included in Y. Under what circumstances is this situation preserved by the inverse-image parts f^* of arbitrary geometric morphisms $f : \mathcal{F} \longrightarrow \mathcal{E}$? That is, when can we guarantee that $f^*(X)$ is the largest subobject of $f^*(M)$ whose pullback along $f^*(p)$ is included in the subobject $f^*(Y)$ of $f^*(L)$? (To see that the general question subsumes the question in the previous paragraph, one implicitly uses that f^* preserves finite products.)

A considerable part of the answer follows easily from the fact that f^* preserves finite limits and arbitrary colimits. This fact implies that $f^*(X)$ and $f^*(Y)$ are subobjects of $f^*(M)$ and $f^*(L)$, respectively, that the pullback of $f^*(X)$ along $f^*(p)$ is f^* of the pullback of X along p, and that this is a subobject of $f^*(Y)$. The only real question is whether $f^*(X)$ is the largest subobject of $f^*(M)$ with this pullback property. That is, given a subobject Z of $f^*(M)$ in $\mathcal{F}$, whose pullback along $f^*(p)$ is included in $f^*(Y)$, can we conclude that Z is included in $f^*(X)$?

The problem can be simplified by observing that the notion of "subobject of M whose pullback along p is included in Y" can be described by an internal geometric theory (in propositional logic) in $\mathcal{E}$ and thus has a classifying topos $\mathcal{H}$ over $\mathcal{E}$. This means that there is a geometric morphism $u : \mathcal{H} \longrightarrow \mathcal{E}$ and there is a subobject G of $u^*(M)$ in $\mathcal{H}$ such that

- the pullback of G along $u^*(p)$ is a subobject of $u^*(Y)$, i.e., $u^*(p)^{-1}(G) \subseteq u^*(Y)$, and
- whenever $f : \mathcal{F} \longrightarrow \mathcal{E}$ is a geometric morphism and $Z \subseteq f^*(M)$ in $\mathcal{F}$ satisfies $f^*(p)^{-1}(Z) \subseteq f^*(Y)$, then there is a geometric morphism $g : \mathcal{F} \longrightarrow \mathcal{H}$ (unique up to natural isomorphism) such that $u \circ g = f$ and $g^*(G) = Z$.

It follows that our question, about arbitrary $f : \mathcal{F} \longrightarrow \mathcal{E}$, reduces to the same question about the single geometric morphism $u : \mathcal{H} \longrightarrow \mathcal{E}$. Indeed, if we can infer from the information we have about G that $G \subseteq u^*(X)$, then we can also infer, for any Z as above, that

$$Z = g^*(G) \subseteq g^*(u^*(X)) = (u \circ g)^*(X) = f^*(X) .$$

Our problem is thus reduced to finding the conditions on L, M, and p that ensure $G \subseteq u^*(X)$. To solve this problem, we need to take a closer look at the construction of the classifying topos $\mathcal{H}$ and the generic object G. We shall use a standard construction of classifying topoi, as in [12], with $\mathcal{E}$ as the base topos. That is, we shall work in the internal logic of $\mathcal{E}$, just as one would ordinarily work in the "real world" of **Sets**.

Working in $\mathcal{E}$ requires some caution, because the internal logic of a topos is intuitionistic. In our situation, one manifestation of intuitionistic logic will be that we must be careful with the concept of finiteness. There are various equivalent ways to define "finite" in ordinary set theory, but the proofs of their equivalence use classical logic (and in some cases even the axiom of choice). So the definitions are inequivalent intuitionistically. It turns out that the right definition for our purposes, i.e., the definition that makes the usual construction of classifying topoi work, is what is usually called K-finiteness, but we shall call it finiteness because we have no need for any other version of finiteness. According to this definition, a subset F of a set S is *finite* if it belongs to every family $\mathcal{X}$ of subsets of S such that

- $\varnothing \in \mathcal{X}$ and
- $X \cup \{s\} \in \mathcal{X}$ for all $X \in \mathcal{X}$ and all $s \in S$.

In more anthropomorphic terms, F is finite if it can be obtained by starting with the empty set and repeatedly adjoining single elements of S.

Until further notice, the following discussion takes place in the internal logic of $\mathcal{E}$. Here L and M are sets, Y and X are subsets, and p is a function.

To begin the study of the classifying topos $\mathcal{H}$, we first write, as a geometric theory, what it should classify, namely subsets of M whose pre-image along p is included in Y. Subsets of M amount to models of the theory consisting of propositional variables $m^{\bullet}$ for all $m \in M$; the truth value assigned by a model to $m^{\bullet}$ tells to what extent m is in the corresponding subset of M. The requirement that the pre-image of this subset be included in Y amounts to a geometric theory that the corresponding model must satisfy, namely the theory whose axioms are, for each $l \in L$,

$$p(l)^{\bullet} \implies \bigvee\{\mathtt{true} : l \in Y\}\ .$$

The (rather peculiar-looking) disjunction on the right is a disjunction of at most one formula, namely the formula `true`; this formula is present if and only if $l \in Y$. (If the logic were classical, there would be one disjunct, `true`, when $l \in Y$ and none, so that the disjunction is `false`, when $l \notin Y$. Intuitionistically, though, those two cases need not be exhaustive.) This disjunction has, regardless of the truth values assigned to the propositional variables, the same truth value as the statement $l \in Y$. We have written it as a disjunction to fit the general format of geometric theories and thus to enable us to apply the standard method for building classifying topoi of geometric theories.

That general method produces a topos of sheaves as follows. Begin with the partially ordered set $\mathrm{Fin}(M)$ of finite subsets of M, ordered by reverse inclusion, and make it into a category, still called $\mathrm{Fin}(M)$, in the usual way: Objects are the elements of $\mathrm{Fin}(M)$ and there is a single morphism $c \longrightarrow d$ if and only if $c \supseteq d$. (This is the dual of the category of finitely presented models and homomorphisms for the geometric theory consisting of all our propositional variables $m^\cdot$ but none of our axioms.) Each element l of L determines a sieve S_l on the object $\{p(l)\}$ as follows. The sieve contains every morphism into $p(x)$ if and only if $l \in Y$. (Classically, the sieve would be the trivial sieve of all morphisms into $\{p(l)\}$ or the empty sieve, according to whether $l \in Y$ or $l \notin Y$; but again, intuitionistically, we do not know that these alternatives are exhaustive.) Note the following unusual property of the sieves S_l: if S_l contains some morphism into $\{p(l)\}$, then it contains all morphisms into $\{p(l)\}$. (It is tempting to say that S_l contains all or none of the morphisms into $p(l)$, but this formulation presupposes classical logic and is intuitionistically too strong.)

Let J be the smallest Grothendieck topology on $\mathrm{Fin}(M)$ that contains all these sieves S_l. Then the classifying topos $\mathcal{H}$ is the topos of J-sheaves on $\mathrm{Fin}(M)$. We shall need the following explicit description of the topology J, or at least of the sieves that cover a singleton $\{m\} \in \mathrm{Fin}(M)$. Of course there are the sieves S_l described above, for all $l \in p^{-1}\{m\}$, and all sieves that are supersets of these. But there are more, because of the closure conditions on Grothendieck topologies. Closure under pullbacks doesn't yield any new covers for singletons (though it does yield covers for larger finite subsets of M). But new covers of $\{m\}$ do arise from closure under iteration, i.e., from the requirement that, if S covers $\{m\}$ and if T is a sieve on $\{m\}$ whose pullback along every $d \longrightarrow \{m\}$ in S covers d, then T covers $\{m\}$. Starting with the sieves S_l and repeatedly using this iteration closure, we find that J contains, for each $m \in M$, all the sieves described in the following definition.

Definition 12. Let $m \in M$. By induction on natural numbers n, we define sieves on $\{m\}$ of *rank* $\leq n$ as follows. The only sieve of rank ≤ 0 is the sieve of all morphisms into $\{m\}$, i.e., the sieve generated by the identity map of $\{m\}$. A sieve S on $\{m\}$ has rank $\leq n+1$ if there is a sieve T of rank $\leq n$ and there is $l \in p^{-1}\{m\}$ such that if $l \in Y$ then $T \subseteq S$.

It may be helpful to write out explicitly the first two steps of this induction. S has rank ≤ 1 if and only if

$$\exists l \in p^{-1}\{m\} \left(l \in Y \implies 1_{\{m\}} \in S\right) .$$

That is, S includes a sieve of the form S_l. S has rank ≤ 2 if and only if

$$\exists l_1 \in p^{-1}\{m\} \Big(l_1 \in Y \implies (\exists l_2 \in p^{-1}\{m\} \, (l_2 \in Y \implies 1_{\{m\}} \in S))\Big) .$$

Note that, if S has any rank n, then, just as in the case of the original S_l's, if S contains some morphism into $\{m\}$, then it contains all such morphisms.

It is routine to verify that the J-covering sieves of $\{m\}$ are just those that have some rank.

With this description of J, we can begin to characterize the circumstances under which the generic subset-of-M-with-preimage-in-Y G in $\mathcal{H}$ is included in $u^*(X)$. The requirement is that, for each $m \in M$, the truth value of $m \in G$ is below the truth value of $m \in u^*(X)$. These truth values are the J-closures of the corresponding truth values in the presheaf topos $\mathbf{Sets}^{\mathrm{Fin}(M)^{\mathrm{op}}}$. Since J-closure is an idempotent and monotone operation, this is the same as requiring the truth value of $m \in G$ in the presheaf topos to be below the J-closure of the truth value, in the presheaf topos, of $m \in u^*(X)$.

In the presheaf topos, the truth value of $m \in G$ is the sieve (on the terminal object $\varnothing$) generated by the object $\{m\}$ (or, more precisely, generated by the morphism $\{m\} \longrightarrow \varnothing$). The truth value of $m \in u^*(X)$ is the sieve T (again on $\varnothing$) that contains each morphism to $\varnothing$ if and only if $m \in X$. So the requirement that we want to analyze is that the morphism $\{m\} \longrightarrow \varnothing$ is in the J-closure of T. That is equivalent to requiring $1_{\{m\}}$ to be in the pullback to $\{m\}$ of the J-closure of T, and the latter is the J-closure of the pullback of T. The pullback of T to $\{m\}$ contains each morphism with codomain $\{m\}$ if and only if $m \in X$. Using this information and the description above of J-covering sieves, we can express the requirement that we want to analyze as the disjunction of an infinite sequence of statements ρ_n defined as follows.

For fixed $m \in M$ let ρ_0 be the statement $m \in X$ and let ρ_{n+1} be the statement

$$\exists l \in p^{-1}\{m\}\,(l \in Y \implies \rho_n)\,.$$

Again, it seems useful to explicitly exhibit ρ_1

$$\exists l_1 \in p^{-1}\{m\}\,(l_1 \in Y \implies m \in X)$$

and ρ_2

$$\exists l_1 \in p^{-1}\{m\}\,\big(l_1 \in Y \implies \exists l_2 \in p^{-1}\{m\}\,(l_2 \in Y \implies m \in X)\big)\,.$$

Summarizing the preceding discussion, we have:

Theorem 13. *For any Grothendieck topos $\mathcal{E}$ and any morphism $p : L \longrightarrow M$ in $\mathcal{E}$, the following two statements are equivalent.*

- *The inverse-image parts of all geometric morphisms into $\mathcal{E}$ preserve universal quantification along p.*
- *For each $Y \subseteq L$, one of the statements ρ_n above, with X interpreted as the universal quantification of Y along p, is valid in $\mathcal{E}$.*

In particular, we have the following corollary for the case where p is a projection from a product to one of its factors, for example the $A_{\mathrm{r}} \times A_s \longrightarrow A_{\mathrm{r}}$ that started this discussion.

Corollary 14. *For any Grothendieck topos $\mathcal{E}$ and any object A in it, the following are equivalent.*

- *Universal quantification along projections $M \times A \longrightarrow M$ is preserved by the inverse image parts of all geometric morphisms $\mathcal{F} \longrightarrow \mathcal{E}$.*
- *For each subobject $Y \subseteq A$, the following statement is valid in $\mathcal{E}$ for some n:*

$$\exists l_1 \in A\Big(l_1 \in Y \implies \exists l_2 \in A\,(l_2 \in Y \implies \dots$$
$$\dots \implies \exists l_n \in A\,(l_n \in Y \implies \forall x \in A\, x \in Y))\Big)\,. \quad (2)$$

Several comments are in order about this result. First, the formula (2) for $n = 1$ is logically valid in classical logic. The proof is to instantiate l_1 as an element of $A - Y$ if one exists, and as an arbitrary element of A if $Y = A$. (This uses that, in classical logic, one takes domains of discourse, like A, to be nonempty. In an empty A, it is not the $n = 1$ case of (2) but the $n = 0$ case that is valid.) We conclude that, if $\mathcal{E}$ satisfies classical logic, i.e., if it is a Boolean topos, then universal quantification is geometrically preserved. This is, of course, no news, because in classical logic one can express universal quantification using existential quantification and complementation, both of which are geometrically preserved. (Negation is not in general geometrically preserved, but when a complement exists, that will be preserved.)

It may be worth noting that validity of the $n = 1$ case of (2) (for inhabited A) embodies the full strength of classical logic, i.e., it implies the law of the excluded middle. To see this, let φ be an arbitrary statement, and consider the set A that contains 0 (definitely) and contains 1 if and only if φ. So A is inhabited (by 0). Let Y be the subset of A that contains 0 if and only if φ (and definitely does not contain 1). Then $\forall x \in A\, x \in Y$ is false. (It implies $0 \in Y$, hence φ, hence (as now $1 \in A$) $1 \in Y$, and hence a contradiction.) So the $n = 1$ case of (2) implies that $\exists l \in A\, \neg l \in Y$. By definition of A, such an l must be 0 or 1. If it is 0 then the definition of Y gives $\neg\varphi$. If it is 1, then the definition of A gives φ. So in both cases, we have $\varphi \vee \neg\varphi$.

It is tempting to rewrite the nested quantifications and implications in (2) as a single quantification over all n of the l_i's at once, i.e.,

$$\exists l_1, \ldots, l_n \in A\left(\left(\bigwedge_{i=1}^{n} l_i \in Y\right) \implies \forall x \in A\, x \in Y\right).$$

Unfortunately, this simplification works only in classical logic. The point is that, in the correct, nested formulation (2), l_2 need only exist (in A) to the extent that $l_1 \in Y$, whereas in the proposed simplification l_2 must exist outright. Classically, this doesn't matter; if we have a good value for l_2 when $l_1 \in Y$, then we can simply give l_2 the same value as l_1 when $l_1 \notin Y$. But intutionistically we don't know that $l_1 \in Y \vee l_1 \notin Y$, so this is not an adequate specification of a value for l_2. To see the problem in a concrete case, consider the same $A = \{0\} \cup \{1 : \varphi\}$ and $Y = \{0 : \varphi\}$ as above. Notice that the $n = 2$ case of (2) is satisfied. (Take $l_1 = 0$ and, if $l_1 \in Y$, i.e., if φ, then take $l_2 = 1$, which is legal because when φ then $1 \in A$.) But the proposed simplification can hold (for any n) only if the logic is classical. (Proof: Each l_i has to be 0 or 1. If at least one of them is 1, then $1 \in A$ and so φ holds. If all of them are 0, then the implications say that the consequent $(\forall x \in A)\, x \in Y$, which is false, follows from (n repetitions of) $0 \in Y$, i.e., from φ. So we get $\varphi \vee \neg\varphi$.)

Let us revisit Subsection 3.1, where we showed that geometric preservation implies preservation along homomorphisms, and let us try to apply the same argument in the new context where universal quantification is allowed over certain sorts, the closed sorts. The argument used geometric preservation in the case of geometric morphisms to the Sierpiński topos $\mathcal{S}$, so it applies in the new context to exactly those homomorphisms $h : \mathfrak{A} \longrightarrow \mathfrak{B}$ that satisfy the following requirement: For each closed sort s, the s-component $h_s : A_s \longrightarrow B_s$, considered as an object A of $\mathcal{S}$, satisfies the conditions in Corollary 14. So we analyze the conditions (2) in that corollary. A subobject Y of $h_s : A_s \longrightarrow B_s$ in $\mathcal{S}$ is given by subsets $Y_0 \subseteq A_s$ and $Y_1 \subseteq B_s$ such that $h_s(Y_0) \subseteq Y_1$. Working in **Sets**, where classical logic is available, we can analyze the situation by considering two cases.

Suppose first that $h_s : A_s \longrightarrow B_s$ is not surjective. Then let us take $Y_0 = A_s$ and take $Y_1 = h_s(A_s)$, a proper subset of B_s. Then $\forall x \in A\, x \in Y$ has, in $\mathcal{S}$, the truth value `false`. Indeed, this truth value would be a subobject $X = (X_0 \longrightarrow X_1)$ of the terminal object $1 \longrightarrow 1$ whose preimage in $A = (A_s \longrightarrow B_s)$ is included in $Y = (Y_0 \longrightarrow Y_1)$. Since Y_1 is not all of B_s, this forces X_1 to be 0, and then the existence of a map $X_0 \longrightarrow X_1$ forces X_0 to be 0 also. With this information, we can proceed to calculate the truth values of the formulas in (2); these are subobjects $P \longrightarrow Q$ of $1 \longrightarrow 1$, and we concentrate on the P components; are they 0 or 1? If, for some n, the P-component were 1, then there would be a witness $l_1 \in A_s$. It is automatically in Y_0 since $Y_0 = A_s$. So there would be a witness $l_2 \in A_s$. Continuing in the same way, we would get $l_3, \dots, l_n$ and finally, since $l_n \in Y_0$, the conclusion that $\forall x \in A\, x \in Y$, which we have already seen is false. This shows that $P = 0$, and therefore none of the formulas (2) are valid. Therefore, universal quantification is not geometrically preserved, and the argument from Subsection 3.1 does not apply to such an h.

There remains the case that $h : A_s \longrightarrow B_s$ is surjective. We shall show that, in this case, one of the instances $n = 0$ and $n = 1$ of (2) is valid in $A = (A_s \longrightarrow B_s)$. If A_s is empty, then (by surjectivity) so is B_s and so are both components of Y. Then $\forall x \in A\, x \in Y$ is vacuously true in $\mathcal{S}$, so the $n = 0$ instance of (2) is valid. Assume from now on that A_s and therefore also B_s are nonempty; we shall verify the $n = 1$ instance of (2). So let an arbitrary subobject $Y = (Y_0 \longrightarrow Y_1)$ of A be given. There are three subcases to consider.

First, suppose Y_0 is all of A_s. Then, as h is surjective, Y_1 is all of B_s, and so $\forall x \in A\, x \in Y$ is true in $\mathcal{S}$. We have the $n = 0$ instance of (2), but we can also get the $n = 1$ instance by instantiating l_1 with an arbitrary element of A_s (nonempty by assumption) and its h-image in B_s.

Second, suppose Y_1 is not all of B_s. Let $b \in B_s - Y_1$, and let $a \in A_s$ be such that $h(a) = b$ (possible as h is surjective). Instantiate l_1 by a in the first component and b in the second to get $l_1 \in Y$ false (in both components), so that the $n = 1$ instance is satisfied.

Finally, suppose Y_0 is not all of A_s but Y_1 is all of B_s. Let $a \in A_s - Y_0$, and instantiate l_1 as a in the first component and $h(a)$ in the second. Then $l_1 \in Y$ is false in the first component but true in the second. Those are exactly the truth values of $\forall x \in A\, x \in Y$, so again the $n = 1$ instance is verified.

The conclusion of this discussion is that our criterion for geometric preservation in Corollary 14 and the argument in Subsection 3.1 combine to give preservation under homomorphisms in exactly those cases where such preservation is wanted, namely when the homomorphisms are surjective on all the closed sorts.

References

1. Apt, K.: Ten Years of Hoare's Logic: A Survey – Part I. ACM Trans. Prog. Lang. and Systems 3, 431–483 (1981)
2. Artin, M., Grothendieck, A., Verdier, J.-L.: Théorie des Topos et Cohomologie Étale des Schémas. In: Séminaire de Géométrie Algébrique du Bois Marie 1963–64 (SGA) 4, vol. 1, Lecture Notes in Mathematics, vol. 269. Springer, Heidelberg (1972)
3. Bell, J.: Toposes and Local Set Theories. Oxford Logic Guides, vol. 14. Oxford University Press, Oxford (1988)
4. Blass, A.: Topoi and Computation. Bull. European Assoc. Theoret. Comp. Sci. 36, 57–65 (1988)
5. Blass, A.: Geometric Invariance of Existential Fixed-Point Logic. In: Gray, J., Scedrov, A. (eds.) Categories in Computer Science and Logic. Contemp. Math., vol. 92, pp. 9–22. Amer. Math. Soc., Providence (1989)
6. Blass, A., Gurevich, Y.: Existential Fixed-Point Logic. In: Börger, E. (ed.) Computation Theory and Logic. LNCS, vol. 270, pp. 20–36. Springer, Heidelberg (1987)
7. Blass, A., Gurevich, Y.: The Underlying Logic of Hoare Logic. Bull. European Assoc. Theoret. Comp. Sci. 70, 82–110 (2000); Reprinted in Paun, G., Rozenberg, G., Salomaa, A.: Current Trends in Theoretical Computer Science: Entering the 21st Century, pp. 409–436. World Scientific, Singapore (2001)
8. Blass, A., Ščedrov, A. (later simplified to Scedrov): Classifying Topoi and Finite Forcing. J. Pure Appl. Algebra 28, 111–140 (1983)
9. Chandra, A., Harel, D.: Horn Clause Queries and Generalizations. J. Logic Programming 2, 1–15 (1985)
10. Cook, S.: Soundness and Completeness of an Axiom System for Program Verification. SIAM J. Computing 7, 70–90 (1978)
11. Immerman, N.: Relational Queries Computable in Polynomial Time. Information and Control 68, 86–104 (1986); Preliminary version in 14th ACM Symp. on Theory of Computation (STOC), pp. 147–152 (1982)
12. Johnstone, P.: Topos Theory. London Mathematical Society Monographs, vol. 10. Academic Press, London (1977)
13. Vardi, M.: Complexity of Relational Query Languages. In: 14th ACM Symp. on Theory of Computation (STOC), pp. 137–146 (1982)

Three Paths to Effectiveness

Udi Boker[1,⋆] and Nachum Dershowitz[2,⋆⋆]

[1] School of Engineering and Computer Science, Hebrew University, Jerusalem 91904, Israel
udiboker@cs.huji.ac.il

[2] School of Computer Science, Tel Aviv University, Ramat Aviv 69978, Israel
nachum.dershowitz@cs.tau.ac.il

For Yuri, profound thinker, esteemed expositor, and treasured friend.

Abstract. Over the past two decades, Gurevich and his colleagues have developed axiomatic foundations for the notion of *algorithm*, be it classical, interactive, or parallel, and formalized them in a new framework of *abstract state machines*. Recently, this approach was extended to suggest axiomatic foundations for the notion of *effective computation* over arbitrary countable domains. This was accomplished in three different ways, leading to three, seemingly disparate, notions of effectiveness. We show that, though having taken different routes, they all actually lead to precisely the same concept. With this concept of effectiveness, we establish that there is – up to isomorphism – exactly one maximal effective model across all countable domains.

Keywords: ASM, effectiveness, recursive functions, Turing machines, computability, constructiveness.

1 Introduction

Church's Thesis asserts that the recursive functions are the only numeric functions that can be effectively computed. Similarly, Turing's Thesis stakes the claim that any function on strings that can be mechanically computed can be computed, in particular, by a Turing machine. For models of computation that operate over arbitrary data structures, however, these two standard notions of what constitutes effectiveness may not be directly applicable; as Richard Montague asserts [9, pp. 430–431]:

> Now Turing's notion of computability applies directly only to functions on and to the set of natural numbers. Even its extension to functions defined on (and with values in) another denumerable set S cannot be accomplished in a completely unobjectionable way. One would be inclined to choose a one-to-one correspondence between S and the set of natural numbers, and to call a function f on S computable if the function of

⋆ Supported in part by a Lady Davis postdoctoral fellowship.

⋆⋆ Supported in part by the Israel Science Foundation (grant no. 250/05).

A. Blass, N. Dershowitz, and W. Reisig (Eds.): Gurevich Festschrift, LNCS 6300, pp. 135–146, 2010.

> natural numbers induced by f under this correspondence is computable in Turing's sense. But the notion so obtained depends on what correspondence between S and the set of natural numbers is chosen; the sets of computable functions on S correlated with two such correspondences will in general differ. The natural procedure is to restrict consideration to those correspondences which are in some sense 'effective', and hence to characterize a computable function on S as a function f such that, for some effective correspondence between S and the set of natural numbers, the function induced by f under this correspondence is computable in Turing's sense. But the notion of effectiveness remains to be analyzed, and would indeed seem to coincide with computability.

One may ask, for example: What are the computable functions over the algebraic numbers? Does one obtain different sets of computable functions depending on which representation ("correspondence") one chooses for them?

Before we can answer such questions, we need a most-general notion of algorithm. Sequential algorithms – that is, deterministic algorithms without unbounded parallelism or (intra-step) interaction with the outside world – have been analyzed and formalized by Gurevich in [6]. There it was proved that any algorithm satisfying three natural formal postulates (given below) can be emulated, step by step, by a program in a very general model of computation, called "abstract state machines" (ASMs). This formalization was recently extended in [1] to handle partial functions. But an algorithm, or abstract state machine program, need not yield an effective function. Gaussian elimination, for example, is a perfectly well-defined algorithm over the real numbers, even though the reals cannot all be effectively represented and manipulated.

We adopt the necessary point of view that effectiveness is a notion applicable to collections of functions, rather than to single functions (cf. [10]). A single function over an arbitrary domain cannot be classified as effective or ineffective [9,14], since its effectiveness depends on the context. A detailed discussion of this issue can be found in [3].

To capture what it is that makes a sequential algorithm mechanically computable, three different generic formalizations of effectiveness have recently been suggested:

- In [3], the authors base their notion of effectivity on finite constructibility. Initial data are inductively defined to be effective if they only contain a Herbrand universe, in addition to finite data and functions that can be shown constructible in the same way.
- In [5], Dershowitz and Gurevich require an injective mapping between the arbitrary domain and the natural numbers. Initial data are effective if they are tracked – under that representation – by recursive functions, as in the traditional definition of "computable" algebras [15].
- In [12], Reisig bases effectiveness on the natural congruence relation between vocabulary terms that arises in the theory of ASMs. Initial data are effective if the induced congruence between terms is Turing-computable.

Properly extending these approaches to handle partial functions, and to refer to a set of algorithms, it turns out that these three notions are essentially one and the same.

2 Algorithms

We work within the abstract-state-machine framework of [6], modified to make terminal states explicit and to allow partial operation to "hang", as in [1]. We begin by recalling Gurevich's Sequential Postulates, formalizing the following intuitions: (I) we are dealing with discrete deterministic state-transition systems; (II) the information in states suffices to determine future transitions and may be captured by logical structures that respect isomorphisms; and (III) transitions are governed by the values of a finite and input-independent set of (variable-free) terms. See [5] for historical support for these postulates.

Postulate I (Sequential Time). *An algorithm determines the following:*

1. *A nonempty set* $\mathcal{S}$ *of* states *and a nonempty subset* $\mathcal{S}_0 \subseteq \mathcal{S}$ *of* initial states.
2. *A partial* next-state *transition function* $\tau : \mathcal{S} \rightharpoonup \mathcal{S}$.

A *terminal state* is one for which no transition is defined. Let $\mathcal{O} \subseteq \mathcal{S}$ denote the (possibly empty) set of terminal states. We write $x \leadsto_\tau x'$ when $x' = \tau(x)$. A *computation* is a finite or infinite chain $x_0 \leadsto_\tau x_1 \leadsto_\tau \cdots$ of states.

Since transitions are functions, the states of an algorithm must contain all the information necessary to determine the future of a computation, a full "instantaneous description" of all relevant aspects of the computation's current status.

(It may appear that a recursive function is not a state-transition system, but in fact the definition of a recursive function comes together with a computation rule for evaluating it. As Rogers [13, p. 7] writes, for instance, "We obtain the computation uniquely by working from the inside out and from left to right".)

Logical structures are ideal for capturing all the salient information stored in a state. All structures in this paper are over first-order finite vocabularies, have countably many elements in their domains (base sets), and interpret symbols as partial operations. All relations are viewed as truth-valued functions, so we refer to structures as (partial) algebras (with partial functions). We assume that structures include Boolean truth values, standard Boolean operations, and that vocabularies include symbols for these.

Postulate II (Abstract State). *The states* $\mathcal{S}$ *of an algorithm are partial algebras over a finite vocabulary* $\mathcal{F}$*, such that the following hold:*

1. *If* $x \in \mathcal{S}$ *is a state of the algorithm, then any algebra* y *isomorphic to* x *is also a state in* $\mathcal{S}$*, and* y *is initial or terminal if* x *is initial or terminal, respectively.*
2. *Transitions* τ *preserve the domain; that is,* Dom $\tau(x) =$ Dom x *for every non-terminal state* $x \in \mathcal{S} \setminus \mathcal{O}$.

3. *Transitions respect isomorphisms, so, if* $\zeta : x \cong y$ *is an isomorphism of non-terminal states* $x, y \in \mathcal{S} \setminus \mathcal{O}$*, then* $\zeta : \tau(x) \cong \tau(y)$.

Such states are "abstract", because the isomorphism requirement means that transitions do not depend in any essential way on the specific representation of the domain embodied in a given state.

Since a state x is an algebra, it interprets function symbols in $\mathcal{F}$, assigning a value $c \in \text{Dom}\ x$ to the "location" $f(a_1, \ldots, a_k)$ in x for every k-ary symbol $f \in \mathcal{F}$ and values $a_1, \ldots, a_k$ in Dom x. For location $\ell = f(a_1, \ldots, a_k)$, we write $[\![\ell]\!]_x$ for the value $f^x(a_1, \ldots, a_k)$ that x assigns to ℓ. Similarly, for term t, $[\![t]\!]_x$ is its value under the interpretations given to all the symbols in t by x. If the interpretation of any subterm is undefined, then so is the whole term. We use $\bot$ to denote an undefined value for a location or term. All terms in this paper are ground terms, that is, terms without variables.

We shall assume that all elements of the domain are accessible via terms in initial states (or else the superfluous elements may be removed with no ill effect). But note that a transition may cause accessible elements to become inaccessible, as explained in [12].

It is convenient to view each state as a collection of the graphs of its operations, given in the form of a set of location-value pairs, each written conventionally as $f(\bar{a}) \mapsto c$, for $\bar{a} \in \text{Dom}\ x$, $c \in \text{Dom}\ x$. Define the *update set* $\Delta(x)$ of state x as the changed location-value pairs, $\tau(x) \setminus x$. When x is a terminal state and $\tau(x)$ is undefined, we will indicate that by setting $\Delta(x) = \bot$.

The transition function of an algorithm must be describable in a finite fashion, so its description can only refer to finitely many locations in the state by means of finitely many terms over its vocabulary.

Postulate III (Effective Transition). *An algorithm with states* $\mathcal{S}$ *over vocabulary* $\mathcal{F}$ *determines a finite set* T *of* critical terms *over* $\mathcal{F}$*, such that states that agree on the values of the terms in* T *also share the same update sets. That is,*

$$\textit{if } x =_T y \textit{ then } \Delta(x) = \Delta(y) ,$$

for any two states $x, y \in \mathcal{S}$.

Here, $x =_T y$, for a set of terms T, means that $[\![t]\!]_x = [\![t]\!]_y$ for all $t \in T$.

Whenever we refer to an "algorithm" below, we mean an object satisfying the above three postulates, what we like to call a "classical algorithm".

Definition 1. *An algorithm* A *with states* $\mathcal{S}$ computes *a partial function* $f : D^k \rightharpoonup D$ *if there is a subset* $\mathcal{I}$ *of its initial states, with locations for input values, such that running the algorithm yields the correct output values of* f*. Specificially:*

1. *The domain of each state in* $\mathcal{I}$ *is* D.
2. *There are* k *distinct locations* $\ell_1, \ldots, \ell_k$ *such that all possible input values are covered. That is,* $\{\langle [\![\ell_1]\!]_x, \ldots, [\![\ell_k]\!]_x \rangle : x \in \mathcal{I}\} = D^k$.
3. *All states in* $\mathcal{I}$ *agree on the values of all locations other than* $\ell_1, \ldots, \ell_k$.

4. *There is a term t (in the vocabulary of the algorithm) such that for all $a_0, \ldots, a_k \in D$, if $f(a_1, \ldots, a_k) = c$, then there is some initial state $x_0 \in \mathcal{I}$, with $[\![\ell_j]\!]_{x_0} = a_j$ $(j = 1, \ldots, k)$, initiating a terminating computation $x_0 \rightsquigarrow_\tau \cdots \rightsquigarrow_\tau x_n$, where $x_n \in \mathcal{O}$ and such that $[\![t]\!]_{x_n} = c$.*
5. *Whenever $f(a_1, \ldots, a_k)$ is $\bot$, there is an initial state $x_0 \in \mathcal{I}$, with $[\![\ell_j]\!]_{x_0} = a_j$ $(j = 1, \ldots, k)$, initiating an infinite computation $x_0 \rightsquigarrow_\tau x_1 \rightsquigarrow_\tau \cdots$.*

A (finite or infinite) set of algorithms, all with the same domain, will be called a *model (of computation)*.

3 Effective Models

We turn now to examine the three different approaches to understanding effectiveness. Informally, they each add a postulate along the following lines:

Postulate IV (Effective Initial State). *The initial states $\mathcal{S}_0$ of an* effective *algorithm are finitely representable.*

3.1 Distinguishing Models

Every state x induces a congruence on all terms, under which terms are congruent whenever the state assigns them the same value:

$$s \simeq_x t \;\Leftrightarrow\; [\![s]\!]_x = [\![t]\!]_x \,.$$

Isomorphic states clearly induce the same congruence.

We call a state "distinguishing" if its induced congruence is semi-decidable in the standard sense. That is, a state is distinguishing if there is a Turing machine (or a similar device, as a partial recursive function operating on strings) that can act as "state manager", receiving two terms as input and returning true whenever both terms are defined and congruent, false when both are defined but not congruent, and diverging otherwise. This is the effectiveness notion explored in [12] (which, however, considers only a single state and total functions).

Definition 2 (Distinguishing Model).

- *A state is* distinguishing *if its induced congruence is semi-decidable.*
- *An algorithm is* distinguishing *if all its initial states are.*
- *A model is* distinguishing *if every congruence induced by a finite set of initial states (across different algorithms) is semi-decidable.*

3.2 Computable Models

We say that an algebra $\mathcal{A}$ over (a possibly infinite) vocabulary $\mathcal{F}$ with domain D *simulates* an algebra $\mathcal{B}$ over (a possibly infinite) vocabulary $\mathcal{G}$ with domain

E if there exists an injective "encoding" $\rho : E \rightharpoonup D$ such that for every partial function $f : E^k \rightharpoonup E$ of $\mathcal{B}$ there is a partial function $\widehat{f} : D^k \rightharpoonup D$ of $\mathcal{A}$, such that $f(x)$ is defined exactly when $(\widehat{f} \circ \rho)(x)$ is defined, for every $x \in E$, and that $f(x) = (\rho^{-1} \circ \widehat{f} \circ \rho)(x)$ whenever $f(x)$ is defined. In that case, we say that $\widehat{f}$ *tracks* f under ρ.

Definition 3 (Computable Model).

- *A state is* computable *if it is simulated by the partial recursive functions.*
- *An algorithm is* computable *if all its initial states are.*
- *A model is* computable *if all its algorithms are, via the same encoding.*

This is a standard notion of "computable algebra" [8,11,7,15], adopted by [5] (which, however, considers only a single algorithm and total functions).

The choice of the partial recursive functions as the starting point for defining effective algorithms over arbitrary domains is natural, considering the Church-Turing Thesis. The main question that one may raise is whether the allowance of an injective representation between the arbitrary domain and the natural numbers is sensible. We show, in the following lemma, that as long as all domain elements are reachable by ground terms, the required injective representation implies the existence of a bijection between the domain and the natural numbers. Hence, the initial functions of a computable algorithm are isomorphic to some partial-recursive functions, which makes their effectiveness hard to dispute.

Lemma 1. *Let M be a computable model over domain D. Then there is a bijection $\pi : D \leftrightarrow \mathbb{N}$ such that, for each partial function $f : D^k \rightharpoonup D$ of each initial state of each algorithm in M, there is a partial recursive function $\tilde{f} : \mathbb{N}^k \rightharpoonup \mathbb{N}$ that tracks f under π.*

Proof. Let ρ be the injective representation from D to $\mathbb{N}$, via which all initial functions of M are partial recursive. For each initial function f, we denote its partial recursive counterpart by $\widehat{f}$. That is, $f = \rho^{-1} \circ \widehat{f} \circ \rho$.

Consider one specific algorithm A of M with vocabulary $\mathcal{F}$. By definition, all elements of D are reachable in each initial state by terms of $\mathcal{F}$. Therefore, all elements of Im $\rho \subseteq \mathbb{N}$ are reachable by terms of $\mathcal{F}$, as interpreted by the partial-recursive tracking fiunctions.

Let $\{c_j\}_j$ be a computable enumeration of all terms over $\mathcal{F}$. One can construct a computable enumeration of all the $\mathcal{F}$-terms that are *defined* in the tracking interpretations by interleaving the computations of the $\{c_j\}_j$ terms in the standard (Cantor's) zigzag fashion (a computation step for the first term, followed by one for the second term and two for the first, then one for the third, two for the second and three for the first, etc.). Accordingly, an enumeration $\{d_j\}_j$ can be set up, assigning a unique $\mathcal{F}$-term for every number in Im ρ, by enumerating the defined $\mathcal{F}$-terms, as above, and ignoring those terms that evaluate to a number already obtained.

In this fashion, we can define a recursive bijection η : Im $\rho \leftrightarrow \mathbb{N}$, letting $\eta(n)$ be the unique j such that $[\![d_j]\!] = n$. Note that η^{-1} is also recursive, as $\eta^{-1}(m) = [\![d_m]\!]$.

Let $\pi : D \leftrightarrow \mathbb{N}$ be the bijection $\pi = \eta \circ \rho$, and, for each function $\widehat{f} \in \widehat{\mathcal{F}}$, define the partial recursive function $\tilde{f}$ to be $\eta \circ \widehat{f} \circ \eta^{-1}$. Then, for each function $f : D^k \rightharpoonup D$ of each initial state of each algorithm in M,

$$f = \rho^{-1} \circ \widehat{f} \circ \rho = \rho^{-1} \circ \eta^{-1} \circ \eta \circ \widehat{f} \circ \eta^{-1} \circ \eta \circ \rho = \pi^{-1} \circ \eta \circ \widehat{f} \circ \eta^{-1} \circ \pi = \pi^{-1} \circ \tilde{f} \circ \pi .$$

□

3.3 Constructive Models

Let x be an algebra over vocabulary $\mathcal{F}$, with domain D. A finite vocabulary $\mathcal{C} \subseteq \mathcal{F}$ *constructs* D if x assigns each value in D to exactly one term over C.

Definition 4 (Constructive Model).

- *A state is* constructive *if it includes constructors for its domain, plus operations that are almost everywhere "blank", meaning that all but finitely-many locations have the same default value (say* undef*).*
- *An algorithm is* constructive *if its initial states are.*
- *A model is* constructive *if all its algorithms are, via the same constructors.*

Moreover, constructive algorithms can be bootstrapped: Any state over vocabulary $\mathcal{F}$ with domain D is constructive if $\mathcal{F}$ can be extended to $\mathcal{C} \uplus \mathcal{G}$ so that $\mathcal{C}$ constructs D and every $g \in \mathcal{G}$ has a constructive algorithm over $\mathcal{C}$ that computes it.

This is the approach advocated in [3] (which, however, considers only total functions).

As expected:

Theorem 1 ([3, Thm. 3]). *The partial recursive functions form a constructive model.*

Conversely:

Theorem 2 (cf. [3, Thm. 4]). *Every constructive model can be simulated by the partial recursive functions via a bijective encoding.*

Though this theorem in [3] does not speak of a bijective encoding, its proof therein in point of fact uses a bijective encoding. That proof refers only to total functions; however, partial functions can be handled similarly. Also, one needs to show, inductively, that constructive initial functions used in bootstrapped constructive algorithms are tracked by partial recursive functions.

4 Equivalence of Definitions of Effectiveness

The following equivalence is demonstrated in the remainder of this section.

Theorem 3. *A model of computation is computable if and only if it is constructive if and only if it is distinguishing.*

Returning to the example of algebraic numbers, this means that one obtains the same set of effectively-computable (partial) functions over the domain of algebraic numbers, regardless of which definition of effectiveness one adopts.

In particular, the effective partial functions over algebraic numbers – obtained in any of these ways – are isomorphic to the partial recursive functions over the natural numbers, and, by results in [2], no representation can yield additional functions.

4.1 Computable and Distinguishing

Theorem 3a. *Computable models are distinguishing.*

Proof. Given a computable model M over domain D, there is, by definition, an injective representation $\rho : D \rightharpoonup \mathbb{N}$ such that, for every function f in an initial state of one of the algorithms in M, there is a partial recursive function $\widehat{f}$ over $\mathbb{N}$ that tracks f under ρ. Since every finite subset M' of algorithms in M can have only finitely many initial functions, it follows that a single partial recursive function can check for equality of the values of two terms in the initial states of any such M' by using partial recursive implementations of the $\widehat{f}$'s to compute the numerical counterparts of the terms and – if and when that computation terminates – testing equality of the resultant numerals. Hence, M is distinguishing. □

Theorem 3b. *Distinguishing models are computable.*

Proof. Let M be a distinguishing model over domain D and consider some specific algorithm $A \in M$ with vocabulary $\mathcal{F}$. Let $\{t_i\}_i$ be any recursive enumeration of terms over $\mathcal{F}$, and let $\simeq$ be the partial recursive function that semi-decides the congruence relation of terms in A. One can define a recursive enumeration $\{c_j\}_j$ of all $\mathcal{F}$-terms that are defined in M by interleaving the computations of $t_i \simeq t_i$ for all terms t_i, in the standard zigzag fashion. A term t_i is added to the enumeration once the corresponding congruence computation ends.

Define the injective representation $\rho : D \rightharpoonup \mathbb{N}$ by

$$\rho(x) := \min_j \{ [\![c_j]\!] = x \} \ .$$

For any initial function f of A, define the partial recursive function

$$\widehat{f}(n) := \min_i c_i \simeq f(c_n) \ ,$$

where $f(c_n)$ is the term obtained by enclosing the term c_n with the symbol f. These numerical $\widehat{f}$ track their original counterparts f over D, as follows:

$$\begin{aligned}
\widehat{f}(\rho(x)) &= \widehat{f}(\min_j \{ [\![c_j]\!] = x \}) \\
&= \min_i \{ c_i \simeq f(c_k) \} \text{ where } k = \min_j \{ [\![c_j]\!] = x \} \\
&= \min_i \{ [\![c_i]\!] = [\![f(c_k)]\!] \} \\
&= \min_i \{ [\![c_i]\!] = f([\![c_k]\!]) \} \\
&= \min_i \{ [\![c_i]\!] = f(x) \} \\
&= \rho(f(x)) \ .
\end{aligned}$$

Similarly for operators of other arities.

It is left to show how the specific injective representation from D to $\mathbb{N}$, which was defined according to one algorithm of M, suits all other algorithms of M. Consider any algorithm $B \in M$ with vocabulary $\mathcal{F}'$ and let $\{c'_j\}_j$ be some recursive enumeration of all defined terms over $\mathcal{F}'$. By the definition of distinguishing, there is a partial recursive function semi-deciding the equivalence of any two terms of the algorithms A and B. This allows one to translate between the term enumerations of A and B and also have partial recursive functions that track the initial functions of B. For any initial function g of B, define the partial recursive function

$$\widehat{g}(n) := \min_i \left\{ c_i \simeq g\left(c'_{\min_\ell \{c'_\ell \simeq c_n\}} \right) \right\} .$$

These numerical $\widehat{g}$ track their original counterparts g in B, as follows:

$$\begin{aligned}
\widehat{g}(\rho(x)) &= \widehat{g}(\min_j\{[\![c_j]\!] = x\}) \\
&= \min_i \{c_i \simeq g(c'_k)\} \text{ where } k = \min\nolimits_\ell \left\{c'_\ell \simeq c_{\min_j\{[\![c_j]\!]=x\}}\right\} \\
&\qquad\qquad\qquad\qquad\qquad\quad = \min\nolimits_\ell \left\{[\![c'_\ell]\!] = [\![c_{\min_j\{[\![c_j]\!]=x\}}]\!]\right\} \\
&\qquad\qquad\qquad\qquad\qquad\quad = \min\nolimits_\ell \{[\![c'_\ell]\!] = x\} \\
&= \min_i \{[\![c_i]\!] = [\![g(c'_k)]\!]\} \\
&= \min_i \{[\![c_i]\!] = g([\![c'_k]\!])\} \\
&= \min_i \{[\![c_i]\!] = g(x)\} \\
&= \rho(g(x)) .
\end{aligned}$$

Similarly for operators of other arities.

It follows that M is computable under the auspices of ρ. □

4.2 Computable and Constructive

Computability is based on the recursiveness of the initial functions, under an injective representation of the arbitrary domain D as natural numbers $\mathbb{N}$. Furthermore, the requirement that all domain elements are reachable by terms implies that there is also a bijective mapping from D to $\mathbb{N}$ via which the initial functions are partial recursive.

Theorem 3c. *Constructive models are computable.*

Proof. Let M be a constructive model over domain D and let M' consist of algorithms for all the constructive functions and all the almost-everywhere-blank functions in M's initial states. By Theorem 2, the set of functions computed by M' is simulatable by the partial recursive functions via some representation $\rho : D \rightharpoonup \mathbb{N}$. Hence, all initial functions of M are tracked by partial recursive functions, making M computable. □

Theorem 3d. *Computable models are constructive.*

Proof. For any computable model M over domain D, there is, by Lemma 1, a bijection $\pi : D \leftrightarrow \mathbb{N}$ such that every function f in the initial states of M's algorithms is tracked under π by some partial recursive function g. By Theorem 1, there is a constructive model that computes all the partial recursive functions $\mathbb{P}$ over $\mathbb{N}$. Since algorithms (according to Postulate II) are closed under isomorphism, so are constructive models. Hence, there is a constructive model P over $\pi^{-1}(\mathbb{N})$, with some set of constructors, that computes all functions $\pi \circ g \circ \pi^{-1}$ that are tracked by functions $g \in \mathbb{P}$, and – in particular – computes all initial functions of M. Since all M's initial functions are constructive, M is constructive. □

Theorem 3 is the conjunction of Theorems 3a–3d.

5 Conclusions

Thanks to Theorem 3, it seems reasonable to just speak of "effectiveness", without distinguishing between the three equivalent notions discussed in the previous sections. Having shown that three prima facie distinct definitions of effectiveness over arbitrary domains comprise exactly the same functions strengthens the impression that the essence of the underlying notion of computability has in fact been captured.

Fixing the concept of an effective model of computation, the question naturally arises as to whether there are "maximal" effective models, and if so, whether they are really different or basically one and the same. Formally, we consider an effective computational model M (consisting of a set of functions) over domain D to be *maximal* if adding any function $f \notin M$ over D to M gives an ineffective model $M \cup \{f\}$. It turns out that there is exactly one effective model (regardless of which of the three definitions one prefers), up to isomorphism.

Theorem 4. *The set of partial recursive functions is the unique maximal effective model, up to isomorphism, over any countable domain.*

Proof. We first note that the partial recursive functions are a maximal effective model. Their effectiveness was established in Theorem 1. As for their maximality, the partial recursive functions are "interpretation-complete", in the sense that they cannot simulate a more inclusive model, as shown in [2,4]. By Theorem 2, they can simulate every effective model, leading to the conclusion that there is no effective model more inclusive than the partial recursive functions.

Next, we show that the partial recursive functions are the unique maximal effective model, up to isomorphism. Consider some maximal effective model M over domain D. By Theorem 2, the partial recursive functions can simulate M via a bijection π. Since effectiveness is closed under isomorphism, it follows that there is an effective model M' over D isomorphic to the partial recursive functions via π^{-1}. Hence, M' contains M, and by the maximality of M we get that $M' = M$. Therefore, M is isomorphic to the partial recursive functions, as claimed. □

The Church-Turing Thesis, properly interpreted for arbitrary countable domains (see [3]), asserts that the partial recursive functions (or Turing machines) constitute the most inclusive effective model, up to isomorphism. However, this claim only speaks about the extensional power of an effective computational model, not about its internal mechanism. Turing, in his seminal work [16], justified the thesis by arguing that every "purely mechanical" human computation can be represented by a Turing machine whose steps more or less correspond to the manual computation. Indeed, Turing's argument convinced most people about the validity of the thesis, which had not been the case with Church's original thesis regarding the effectiveness of the recursive functions (let alone Church's earlier thoughts regarding the lambda calculus). Notwithstanding its wide acceptance, neither the Church-Turing Thesis nor Turing's arguments purport to characterize the internal behavior of an effective computational model over an arbitrary domain.

On the other hand, Gurevich's abstract state machines are the most general descriptive form for (sequential) algorithms known. As such, they can express the precise step-by-step behavior of arbitrary algorithms operating over arbitrary structures, whether for effective computations or for hypothetical ones. The work in [3,5,12] specializes this model by considering effective computations. The additional effectiveness axiom proposed in [3] and adopted in our Definition 4 of constructive models does not rely on the definition of Turing machines or of the partial recursive functions, thereby providing a complete, generic, stand-alone axiomatization of effective *computation* over any countable domain.

References

1. Blass, A., Dershowitz, N., Gurevich, Y.: Exact exploration. Technical Report MSR-TR-2009-99, Microsoft Research, Redmond, WA (2010), `http://research.microsoft.com/pubs/101597/Partial.pdf`; A short version to appear as Algorithms in a world without full equality. In: The Proceedings of the 19th EACSL Annual Conference on Computer Science Logic, Brno, Czech Republic. LNCS. Springer, Heidelberg (August 2010)
2. Boker, U., Dershowitz, N.: Comparing computational power. Logic Journal of the IGPL 14, 633–648 (2006)
3. Boker, U., Dershowitz, N.: The Church-Turing thesis over arbitrary domains. In: Avron, A., Dershowitz, N., Rabinovich, A. (eds.) Pillars of Computer Science. LNCS, vol. 4800, pp. 199–229. Springer, Heidelberg (2008)
4. Boker, U., Dershowitz, N.: The influence of domain interpretations on computational models. Journal of Applied Mathematics and Computation 215, 1323–1339 (2009)
5. Dershowitz, N., Gurevich, Y.: A natural axiomatization of computability and proof of Church's Thesis. Bulletin of Symbolic Logic 14, 299–350 (2008)
6. Gurevich, Y.: Sequential abstract state machines capture sequential algorithms. ACM Transactions on Computational Logic 1, 77–111 (2000)
7. Lambert Jr., W.M.: A notion of effectiveness in arbitrary structures. The Journal of Symbolic Logic 33, 577–602 (1968)
8. Mal'tsev, A.: Constructive algebras I. Russian Mathematical Surveys 16, 77–129 (1961)

9. Montague, R.: Towards a general theory of computability. Synthese 12, 429–438 (1960)
10. Myhill, J.: Some philosophical implications of mathematical logic. Three classes of ideas. The Review of Metaphysics 6, 165–198 (1952)
11. Rabin, M.O.: Computable algebra, general theory and theory of computable fields. Transactions of the American Mathematical Society 95, 341–360 (1960)
12. Reisig, W.: The computable kernel of abstract state machines. Theoretical Computer Science 409, 126–136 (2008)
13. Rogers Jr., H.: Theory of Recursive Functions and Effective Computability. McGraw-Hill, New York (1966)
14. Shapiro, S.: Acceptable notation. Notre Dame Journal of Formal Logic 23, 14–20 (1982)
15. Stoltenberg-Hansen, V., Tucker, J.V.: 4. In: Effective Algebra. Handbook of Logic in Computer Science, vol. 4, pp. 357–526. Oxford University Press, Oxford (1995)
16. Turing, A.M.: On computable numbers, with an application to the Entscheidungsproblem. Proceedings of the London Mathematical Society 42, 230–265 (1936-1937); Corrections in vol. 43, pp. 544–546 (1937), Reprinted in Davis M. (ed.): The Undecidable. Raven Press, Hewlett (1965), `http://www.abelard.org/turpap2/tp2-ie.asp`

The Quest for a Tight Translation of Büchi to co-Büchi Automata

Udi Boker* and Orna Kupferman

School of Computer Science and Engineering, Hebrew University, Israel

Abstract. The Büchi acceptance condition specifies a set α of states, and a run is accepting if it visits α infinitely often. The co-Büchi acceptance condition is dual, thus a run is accepting if it visits α only finitely often. Nondeterministic Büchi automata over words (NBWs) are strictly more expressive than nondeterministic co-Büchi automata over words (NCWs). The problem of the blow-up involved in the translation (when possible) of an NBW to an NCW has been open for several decades. Until recently, the best known upper bound was $2^{O(n \log n)}$ and the best lower bound was n. We describe the quest to the tight $2^{\Theta(n)}$ bound.

Keywords: Büchi automata, co-Büchi automata, nondeterminism, automata translation.

1 Introduction

Finite *automata on infinite objects* were first introduced in the 60's, and were the key to the solution of several fundamental decision problems in mathematics and logic [5,15,20]. Today, automata on infinite objects are used for specification verification, and synthesis of nonterminating systems. The automata-theoretic approach to verification views questions about systems and their specifications as questions about languages, and reduces them to automata-theoretic problems like containment and emptiness [13,26]. Recent industrial-strength property-specification languages such as Sugar, ForSpec, and the recent standard PSL 1.01 include regular expressions and/or automata, making specification and verification tools that are based on automata even more essential and popular [1].

Early automata-based algorithms aimed at showing decidability. The application of automata theory in practice has led to extensive research on the complexity of problems and constructions involving automata [6,19,22,24,25,27]. For many problems and constructions, our community was able to come up with satisfactory solutions, in the sense that the upper bound (the complexity of the best algorithm or the blow-up in the best known construction) coincides with the lower bound (the complexity class in which the problem is hard, or the blow-up that is known to be unavoidable). For some problems and constructions, however, the gap between the upper bound and the lower bound is significant. This situation is especially frustrating, as it implies that not only something is

* Supported in part by a Lady Davis postdoctoral fellowship.

A. Blass, N. Dershowitz, and W. Reisig (Eds.): Gurevich Festschrift, LNCS 6300, pp. 147–164, 2010.

missing in our understanding of automata on infinite objects, but also that we may be using algorithms that can be significantly improved.

One such fundamental and longstanding open problem is the translation, when possible, of a nondeterministic Büchi word automaton (NBW) to an equivalent nondeterministic co-Büchi word automaton (NCW).[1] NCWs are less expressive than NBWs. For example, the language $\{w : w \text{ has infinitely many } a\text{'s}\}$ over the alphabet $\{a, b\}$ cannot be recognized by an NCW. In fact, NCWs are not more expressive than deterministic co-Büchi automata (DCWs).[2] Hence, since deterministic Büchi automata (DBWs) are dual to DCWs, a language can be recognized by an NCW iff its complement can be recognized by a DBW.

The best translation of an NBW to an NCW (when possible) that was known until recently goes as follows. Consider an NBW $\mathcal{A}$ that has an equivalent NCW. First, co-determinize $\mathcal{A}$ and obtain a deterministic Rabin automaton (DRW) $\tilde{\mathcal{A}}$ for the complement language. By [8], DRWs are *Büchi type*. That is, if a DRW has an equivalent DBW, then the DRW has an equivalent DBW on the same structure. Since $\mathcal{A}$ can be recognized by an NCW, its complement $\tilde{\mathcal{A}}$ has an equivalent DBW, so there is a DBW $\tilde{\mathcal{B}}$ that complements $\mathcal{A}$ and has the same structure as $\tilde{\mathcal{A}}$. By viewing $\tilde{\mathcal{B}}$ as a DCW, one gets a deterministic co-Büchi automaton (DCW) equivalent to $\mathcal{A}$. The co-determinization step involves a super exponential blow-up in the number of states [22]: starting with an NBW with n states, we end up with a DCW with $2^{O(n \log n)}$ states. Beyond the super exponential blow-up, the state space that results from Safra's determinization and co-determinization constructions is awfully complex and is not amenable to optimizations and a symbolic implementation. Also, going through a deterministic automaton requires the introduction of acceptance conditions that are more complex than the Büchi and co-Büchi acceptance conditions. Piterman's construction [18] simplifies the situation only slightly; while the Rabin condition can be replaced by parity, the complication and the $2^{O(n \log n)}$ complexity of Safra's construction are still there. Note that eventhough the problem is of translating an NBW to an NCW, and thus it does not require us to end up in a deterministic automaton, the above procedure actually does result in a DCW. Thus, it is not known how to take advantage of the allowed nondeterminism, and how to keep the translation within the convenient scope of the Büchi and the co-Büchi acceptance conditions.

The $2^{O(n \log n)}$ upper bound is particularly annoying, as no non-trivial lower bound was known. In fact, there was even no counterexample to the conjecture that NBWs are *co-Büchi type*. That is, to a conjecture that if an NBW has an equivalent NCW, then the NBW has an equivalent NCW on the same structure.

---

[1] In *Büchi* automata, some of the states are designated as accepting states, and a run is accepting iff it visits states from the accepting set infinitely often [5]. Dually, in *co-Büchi* automata, a run is accepting iff it visits the set of accepting states only finitely often.

[2] When applied to universal Büchi automata, the translation in [16], of alternating Büchi automata into NBW, results in DBW. By dualizing it, one gets a translation of NCW to DCW.

The main challenge in proving a non-trivial lower bound for the translation of NBW to NCW is the expressiveness superiority of NBW with respect to NCW. Indeed, a family of languages that is a candidate for proving a lower bound for this translation has to strike a delicate balance: the languages have to somehow take advantage of the Büchi acceptance condition, and still be recognizable by a co-Büchi automaton.[3] In particular, it is not clear how to use the main feature of the Büchi condition, namely its ability to easily track infinitely many occurrences of an event, as a co-Büchi automaton cannot recognize languages that are based on such a tracking.

Beyond the theoretical challenge in tightening the gaps, and the fact they are related to other gaps in our knowledge [9], the translation of NBW to NCW has immediate important applications in formal methods. The premier example in this class is of symbolic LTL model checking. Evaluating specifications in AFMC can be done with linearly many symbolic steps. In contrast, direct LTL model checking reduces to a search for bad-cycles, whose symbolic implementation involves nested fixed-points, and is typically [4] quadratic [21]. It is shown in [12] that given an LTL formula ψ, there is an alternation-free μ-calculus (AFMC) formula equivalent to $\forall\psi$ iff ψ can be recognized by a DBW. Alternatively, an NCW for $\neg\psi$ can be linearly translated to an AFMC formula equivalent to $\exists\neg\psi$, which can be negated to a formula equivalent to $\forall\psi$. Thus, an improvement of the translation of NBW to NCW would immediately imply an improvement of the translation of LTL to AFMC.

We describe the quest to a $2^{\Theta(n)}$ tight bound for the translation. In the upper-bound front, we describe the construction in [3], which translates an NBW $\mathcal{B}$ to an NCW $\mathcal{C}$ whose underlying structure is the product of $\mathcal{B}$ with its subset construction. Thus, given an NBW $\mathcal{B}$ with n states, the translation yields an equivalent NCW with $n2^n$ states, and it has a simple symbolic implementation [17]. In the lower-bound front, we first describe the counterexample given in [11] to the NCW-typeness of NBW. We then describe the "circumventing counting" idea, according to which the ability of NBWs to easily track infinitely many occurrences of an event makes them more succinct than NCWs. The idea is to consider a family of languages $L_1, L_2, L_3, \ldots$ in which an NCW for L_k has to count to some bound that depends on k, whereas an NBW can count instead to infinity. In the first application of the idea, the NBW for the language L_k checks that an event P occurs infinitely often. The language L_k is still NCW-recognizable as other components of L_k make it possible to check instead that P has at least k occurrences. An NCW for L_k can then count occurrences of P, but it needs $O(k)$ more states for this [2]. In order to achieve a super-linear

[3] A general technique for proving lower bounds on the size of automata on infinite words is suggested in [28]. The technique is based on *full automata*, in which a word accepted by the automaton induces a language. The fact NCWs are less expressive than NBWs is a killer for the technique, as full automata cannot be translated to NCWs.

[4] Better algorithms have been suggested [7,21], but it turns out that algorithms based on nested fixed-points perform better in practice.

succinctness, we enhance the idea as follows. The NBW for the language L_k still checks that an event P occurs infinitely often. Now, however, in order for L_k to be NCW-recognizable, other components of L_k make it possible to check instead that P repeats at least once in every interval of some bounded length $f(k)$. Thus, while the NBW can detect infinitely many occurrences of P with 2 states, the NCW has to devote $O(f(k))$ states for the counting. We first use ideas from number theory in order to make $f(k)$ quadratic in k, and then use binary encoding in order to make $f(k)$ exponential in k [3].

2 Preliminaries

Given an alphabet Σ, a *word* over Σ is a (possibly infinite) sequence $w = w_1 \cdot w_2 \cdots$ of letters in Σ. For two words, x and y, we use $x \preceq y$ to indicate that x is a prefix of y and $x \prec y$ to indicate that x is a strict prefix of y. An *automaton* is a tuple $\mathcal{A} = \langle \Sigma, Q, \delta, Q_0, \alpha \rangle$, where Σ is the input alphabet, Q is a finite set of states, $\delta : Q \times \Sigma \to 2^Q$ is a transition function, $Q_0 \subseteq Q$ is a set of initial states, and $\alpha \subseteq Q$ is an acceptance condition. We define several acceptance conditions below. Intuitively, $\delta(q, \sigma)$ is the set of states that $\mathcal{A}$ may move into when it is in the state q and it reads the letter σ. The automaton $\mathcal{A}$ may have several initial states and the transition function may specify many possible transitions for each state and letter, and hence we say that $\mathcal{A}$ is *nondeterministic*. In the case where $|Q_0| = 1$ and for every $q \in Q$ and $\sigma \in \Sigma$, we have that $|\delta(q, \sigma)| \leq 1$, we say that $\mathcal{A}$ is *deterministic*. The transition function extends to sets of states and to finite words in the expected way, thus $\delta(S, x)$ is the set of states that $\mathcal{A}$ may move into when it is in a state in S and it reads the finite word x. Formally, $\delta(S, \epsilon) = S$ and $\delta(S, w \cdot \sigma) = \bigcup_{q \in \delta(S,w)} \delta(q, \sigma)$. We abbreviate $\delta(Q_0, x)$ by $\delta(x)$, thus $\delta(x)$ is the set of states that $\mathcal{A}$ may visit after reading x. For an automaton $\mathcal{A}$ and a state a of $\mathcal{A}$, we denote by $\mathcal{A}^a$ the automaton that is identical to $\mathcal{A}$, except for having $\{a\}$ as its set of initial states.

A run $r = r_0, r_1, \cdots$ of $\mathcal{A}$ on $w = w_1 \cdot w_2 \cdots \in \Sigma^\omega$ is an infinite sequence of states such that $r_0 \in Q_0$, and for every $i \geq 0$, we have that $r_{i+1} \in \delta(r_i, w_{i+1})$. Note that while a deterministic automaton has at most a single run on an input word, a nondeterministic automaton may have several runs on an input word. We sometimes refer to r as a word in Q^ω or as a function from the set of prefixes of w to the states of $\mathcal{A}$. Accordingly, we use $r(x)$ to denote the state that r visits after reading the prefix x.

Acceptance is defined with respect to the set $inf(r)$ of states that the run r visits infinitely often. Formally, $inf(r) = \{q \in Q \mid \text{ for infinitely many } i \in \mathbb{N}, \text{ we have } r_i = q\}$. As Q is finite, it is guaranteed that $inf(r) \neq \emptyset$. The run r is *accepting* iff the set $inf(r)$ satisfies the acceptance condition α. We consider here the *Büchi* and the *co-Büchi* acceptance conditions. A set $S \subseteq Q$ satisfies a Büchi acceptance condition $\alpha \subseteq Q$ iff $S \cap \alpha \neq \emptyset$; whereas S satisfies a *co-Büchi* acceptance condition $\alpha \subseteq Q$ iff $S \subseteq \alpha$. Note that the definition of co-Büchi we use is less standard than the $S \cap \alpha = \emptyset$ definition; clearly, $S \subseteq \alpha$ iff $S \cap (Q \setminus \alpha) = \emptyset$, thus the definition is equivalent. We chose to go with the $S \subseteq \alpha$ variant as it better conveys the intuition that, as with the Büchi condition, a visit in α is a "good event". An automaton accepts a word iff it has an accepting run on it. The language of an automaton $\mathcal{A}$, denoted $L(\mathcal{A})$, is the set of words that $\mathcal{A}$ accepts. We also say that $\mathcal{A}$ *recognizes* the language $L(\mathcal{A})$. For two automata $\mathcal{A}$ and $\mathcal{A}'$, we say that $\mathcal{A}$ and $\mathcal{A}'$ are *equivalent* if $L(\mathcal{A}) = L(\mathcal{A}')$.

We denote the different classes of automata by three letter acronyms in $\{\text{D}, \text{N}\} \times \{\text{B}, \text{C}\} \times \{\text{W}\}$. The first letter stands for the branching mode of the automaton (deterministic or nondeterministic); the second letter stands for the acceptance-condition type (Büchi, or co-Büchi); the third letter indicates that the automaton runs on words. We say that a language L is in a class γ if L is γ*-recognizable*, that is, L can be recognized by an automaton in the class γ.

Different classes of automata have different expressive power. In particular, while NBWs recognize all ω-regular language [15], DBWs are strictly less expressive than NBWs, and so are DCWs [14]. In fact, a language L is in DBW iff its complement is in DCW. Indeed, by viewing a DBW as a DCW, we get an automaton for the complementing language, and vice versa. The expressiveness superiority of the nondeterministic model over the deterministic one does not apply to the co-Büchi acceptance condition. There, every NCW has an equivalent DCW [16].

3 Upper Bound

In this section we present the upper-bound proof from [3] for the translation of NBW to NCW (when possible).[5] The proof is constructive: given an NBW $\mathcal{B}$ with k states whose language is NCW-recognizable, we construct an equivalent NCW $\mathcal{C}$ with at most $k2^k$ states. The underlying structure of $\mathcal{C}$ is very simple: it runs $\mathcal{B}$ in parallel to its subset construction. We refer to the construction as the *augmented subset construction*, and we describe the rationale behind it below.

Consider an NBW $\mathcal{B}$ with set $\alpha_{\mathcal{B}}$ of accepting states. The subset construction of $\mathcal{B}$ maintains, in each state, all the possible states that $\mathcal{B}$ can be at. Thus, the subset construction gives us full information about $\mathcal{B}$'s *potential* to visit $\alpha_{\mathcal{B}}$ in the future. However, the subset construction loses information about the past. In particular, we cannot know whether fulfilling $\mathcal{B}$'s potential requires us to give up past visits in $\alpha_{\mathcal{B}}$. For that reason, the subset construction is adequate for determinizing automata on finite words, but not good enough for determinizing ω-automata. A naive try to determinize $\mathcal{B}$ could be to build its subset construction and define the acceptance set as all the states for which $\mathcal{B}$ has the potential to be in $\alpha_{\mathcal{B}}$. The problem is that a word might infinitely often gain this potential via different runs. Were we only able to guarantee that the run of the subset construction follows a single run of the original automaton, we would have ensured a correct construction. Well, this is exactly what the augmented subset construction does!

[5] For readers who skipped the preliminaries, let us mention that we work here with a less standard definition of the co-Büchi condition, where a run r satisfies a co-Büchi condition α iff $inf(r) \subseteq \alpha$.

Once the above intuition is understood, there is still a question of how to define the acceptance condition on top of the augmented subset construction. Since we target for an NCW, we cannot check for infiniteness. However, the premise that the NBW is in DCW guarantees that a word is accepted iff there is a run of the augmented subset construction on it that remains in "potentially good states" from some position. We explain and formalize this property below.

We start with a property relating states of a DCW (in fact, any deterministic automaton) that are reachable via words that lead to the same state in the subset construction of an equivalent NBW.

Lemma 1. *Consider an NBW $\mathcal{B}$ with a transition function $\delta_{\mathcal{B}}$ and a DCW $\mathcal{D}$ with a transition function $\delta_{\mathcal{D}}$ such that $L(\mathcal{B}) = L(\mathcal{D})$. Let d_0 and d_1 be states of $\mathcal{D}$ such that there are two finite words x_0 and x_1 such that $\delta_{\mathcal{D}}(x_0) = d_0$, $\delta_{\mathcal{D}}(x_1) = d_1$, and $\delta_{\mathcal{B}}(x_0) = \delta_{\mathcal{B}}(x_1)$. Then, $L(\mathcal{D}^{d_0}) = L(\mathcal{D}^{d_1})$.*

For automata on finite words, if two states of the automaton have the same language, they can be merged without changing the language of the automaton. While this is not the case for automata on infinite words, the lemma below enables us to do take advantage of such states.

Lemma 2. *Consider a DCW $\mathcal{D} = \langle \Sigma, D, \delta, D_0, \alpha \rangle$. Let d_0 and d_1 be states in D such that $L(\mathcal{D}^{d_0}) = L(\mathcal{D}^{d_1})$. For all finite words u and v, if $\delta(d_0, u) = d_0$ and $\delta(d_1, v) = d_1$ then for all words $w \in (u+v)^*$ and states $d' \in \delta(d_0, w) \cup \delta(d_1, w)$, we have $L(\mathcal{D}^{d'}) = L(\mathcal{D}^{d_0})$.*

Our next observation is the key to the definition of the acceptance condition in the augmented subset construction. Intuitively, it shows that if an NCW language L is indifferent to a prefix in $(u+v)^*$, and L contains the language $(v^* \cdot u^+)^\omega$, then L must also contain the word v^ω.

Lemma 3. *Consider a co-Büchi recognizable language L. For all finite words u and v, if for every finite word $x \in (u+v)^*$ and infinite word w we have that $w \in L$ iff $x \cdot w \in L$, and $(v^* \cdot u^+)^\omega \subseteq L$, then $v^\omega \in L$.*

By considering the language of a specific state of the DCW, Lemma 3 implies the following.

Corollary 1. *Let $\mathcal{D} = \langle \Sigma, D, \delta, D_0, \alpha \rangle$ be a DCW. Consider a state $d \in D$. For all nonempty finite words v and u, if for all words $w \in (v+u)^*$ and states $d' \in \delta(d, w)$, we have $L(\mathcal{D}^{d'}) = L(\mathcal{D}^d)$, and $(v^* \cdot u^+)^\omega \subseteq L(\mathcal{D}^d)$, then $v^\omega \in L(\mathcal{D}^d)$.*

We can now present the construction together with its acceptance condition.

Theorem 1 ([3]). *For every NBW $\mathcal{B}$ with k states that is co-Büchi recognizable there is an equivalent NCW $\mathcal{C}$ with at most $k2^k$ states.*

Proof. Let $\mathcal{B} = \langle \Sigma, B, \delta_{\mathcal{B}}, B_0, \alpha_{\mathcal{B}} \rangle$. We define the NCW $\mathcal{C} = \langle \Sigma, C, \delta_{\mathcal{C}}, C_0, \alpha_{\mathcal{C}} \rangle$ on top of the product of $\mathcal{B}$ with its subset construction. Formally, we have the following.

- $C = B \times 2^B$. That is, the states of $\mathcal{C}$ are all the pairs $\langle b, E\rangle$ where $b \in B$ and $E \subseteq B$.
- For all $\langle b, E\rangle \in C$ and $\sigma \in \Sigma$, we have $\delta_{\mathcal{C}}(\langle b, E\rangle, \sigma) = \delta_{\mathcal{B}}(b, \sigma) \times \{\delta_{\mathcal{B}}(E, \sigma)\}$. That is, $\mathcal{C}$ nondeterministically follows $\mathcal{B}$ on its B-components and deterministically follows the subset construction of $\mathcal{B}$ on its 2^B-component.
- $C_0 = B_0 \times \{B_0\}$.
- A state is a member of $\alpha_{\mathcal{C}}$ if it is reachable from itself along a path whose projection on B visits $\alpha_{\mathcal{B}}$. Formally, $\langle b, E\rangle \in \alpha_{\mathcal{C}}$ if there is a state $\langle b', E'\rangle \in \alpha_{\mathcal{B}} \times 2^B$ and finite words y_1 and y_2 such that $\langle b', E'\rangle \in \delta_{\mathcal{C}}(\langle b, E\rangle, y_1)$ and $\langle b, E\rangle \in \delta_{\mathcal{C}}(\langle b', E'\rangle, y_2)$. We refer to $y_1 \cdot y_2$ as the witness for $\langle b, E\rangle$. Note that all the states in $\alpha_{\mathcal{B}} \times 2^B$ are members of $\alpha_{\mathcal{C}}$ with an empty witness.

We prove the equivalence of $\mathcal{B}$ and $\mathcal{C}$. Note that the 2^B-component of C proceeds in a deterministic manner. Therefore, each run r of $\mathcal{B}$ induces a single run of $\mathcal{C}$ (the run in which the B-component follows r). Likewise, each run r' of $\mathcal{C}$ induces a single run of $\mathcal{B}$, obtained by projecting r' on its B-component.

We first prove that $L(\mathcal{B}) \subseteq L(\mathcal{C})$. Consider a word $w \in L(\mathcal{B})$. Let r be an accepting run of $\mathcal{B}$ on w. We prove that the run r' induced by r is accepting. Consider a state $\langle b, E\rangle \in \mathit{inf}(r')$. We prove that $\langle b, E\rangle \in \alpha_C$. Since $\langle b, E\rangle \in \mathit{inf}(r')$, then $b \in \mathit{inf}(r)$. Thus, there are three prefixes x, $x \cdot y_1$, and $x \cdot y_1 \cdot y_2$ of w such that $r'(x) = r'(x \cdot y_1 \cdot y_2) = \langle b, E\rangle$ and $r'(x \cdot y_1) \in \alpha_{\mathcal{B}} \times 2^B$. Therefore, $y_1 \cdot y_2$ witnesses that $\langle b, E\rangle$ is in $\alpha_{\mathcal{C}}$. Hence, $\mathit{inf}(r) \subseteq \alpha_C$, and we are done.

We now prove that $L(\mathcal{C}) \subseteq L(\mathcal{B})$. Consider a word $w \in L(\mathcal{C})$. Let r' be an accepting run of $\mathcal{C}$ on w, let $\langle b, E\rangle$ be a state in $\mathit{inf}(r')$, and let x be a prefix of w such that $r'(x) = \langle b, E\rangle$. Since r' is accepting, $\mathit{inf}(r') \subseteq \alpha_C$, so $\langle b, E\rangle \in \alpha_C$. Let z be a witness for the membership of $\langle b, E\rangle$ in α_C. By the definition of a witness, $\delta_{\mathcal{B}}(E, z) = E$ and there is a run of $\mathcal{B}^b$ on z that visits $\alpha_{\mathcal{B}}$ and goes back to b. If $z = \epsilon$, then $b \in \alpha_B$, the run of $\mathcal{B}$ induced by r' is accepting, and we are done. Otherwise, $x \cdot z^\omega \in L(\mathcal{B})$, and we proceed as follows.

Recall that the language of $\mathcal{B}$ is NCW-recognizable. Let $\mathcal{D}=\langle \Sigma, D, \delta_D, D_0, \alpha_{\mathcal{D}}\rangle$ be a DCW equivalent to $\mathcal{B}$. Since $L(\mathcal{B}) = L(\mathcal{D})$ and $x \cdot z^\omega \in L(\mathcal{B})$, it follows that the run ρ of $\mathcal{D}$ on $x \cdot z^\omega$ is accepting. Since D is finite, there are two indices i_1 and i_2 such that $i_1 < i_2$, $\rho(x \cdot z^{i_1}) = \rho(x \cdot z^{i_2})$, and for all prefixes y of $x \cdot z^\omega$ such that $x \cdot z^{i_1} \preceq y$, we have $\rho(y) \in \alpha_{\mathcal{D}}$. Let $d_1 = \rho(x \cdot z^{i_1})$.

Consider the run η of $\mathcal{D}$ on w. Since r' visits $\langle b, E\rangle$ infinitely often and D is finite, there must be a state $d_0 \in D$ and infinitely many prefixes $p_1, p_2, \ldots$ of w such that for all $i \geq 1$, we have $r'(p_i) = \langle b, E\rangle$ and $\eta(p_i) = d_0$.

We claim that the states d_0 and d_1 satisfy the conditions of Lemma 1 with x_0 being p_1 and x_1 being $x \cdot z^{i_1}$. Indeed, $\delta_{\mathcal{D}}(p_1) = d_0$, $\delta_{\mathcal{D}}(x \cdot z^{i_1}) = d_1$, and $\delta_{\mathcal{B}}(p_1) = \delta_{\mathcal{B}}(x \cdot z^{i_1}) = E$. For the latter equivalence, recall that $\delta_{\mathcal{B}}(x) = E$ and $\delta_{\mathcal{B}}(E, z) = E$. Hence, by Lemma 1, we have $L(\mathcal{D}^{d_0}) = L(\mathcal{D}^{d_1})$.

Recall the sequence of prefixes $p_1, p_2, \ldots$. For all $i \geq 1$, let $p_{i+1} = p_i \cdot t_i$. We now claim that for all $i \geq 1$, the state d_0 satisfies the conditions of Corollary 1 with u being $z^{i_2 - i_1}$ and v being t_i. The first condition is satisfied by Lemma 2. For the second condition, consider a word $w' \in (v^* \cdot u^+)^\omega$. We prove that $w' \in L(\mathcal{D}^{d_0})$. Recall that there is a run of $\mathcal{B}^b$ on v that goes back to b and there is a run of $\mathcal{B}^b$

on u that visits $\alpha_{\mathcal{B}}$ and goes back to b. Recall also that for the word p_1, we have that $r'(p_1) = \langle b, E\rangle$ and $\eta(p_1) = d_0$. Hence, $p_1 \cdot w' \in L(\mathcal{B})$. Since $L(\mathcal{B}) = L(\mathcal{D})$, we have that $p_1 \cdot w' \in L(\mathcal{B})$. Therefore, $w' \in L(\mathcal{D}^{d_0})$.

Thus, by Corollary 1, for all $i \geq 1$ we have that $t_i^{\omega} \in L(\mathcal{D}^{d_0})$. Since $\delta_{\mathcal{D}}(d_0, t_i) = d_0$, it follows that all the states that $\mathcal{D}$ visits when it reads t_i from d_0 are in $\alpha_{\mathcal{D}}$. Note that $w = p_1 \cdot t_1 \cdot t_2 \cdots$. Hence, since $\delta_{\mathcal{D}}(p_1) = d_0$, the run of $\mathcal{D}$ on w is accepting, thus $w \in L(\mathcal{D})$. Since $L(\mathcal{D}) = L(\mathcal{B})$, it follows that $w \in L(\mathcal{B})$, and we are done. □

4 Lower Bound

In this section we describe the "circumventing counting" idea and how it has led to a matching lower bound. In the deterministic setting, DBWs are co-Büchi type. Thus, if a DBW $\mathcal{A}$ is DCW-recognizable, then there is a DCW equivalent to $\mathcal{A}$ that agrees with $\mathcal{A}$ on its structure (that is, one only has to modify the acceptance condition). The conjecture that NBW are also co-Büchi type was refuted only in [11]:

Theorem 2 ([11]). *NBWs are not co-Büchi type.*

Proof. Consider the NBW $\mathcal{A}$ described in Fig. 1. The NBW recognizes the language $a^* \cdot b \cdot (a+b)^*$ (at least one b). This language is in NCW, yet it is easy to see that there is no NCW recognizing L on the same structure. □

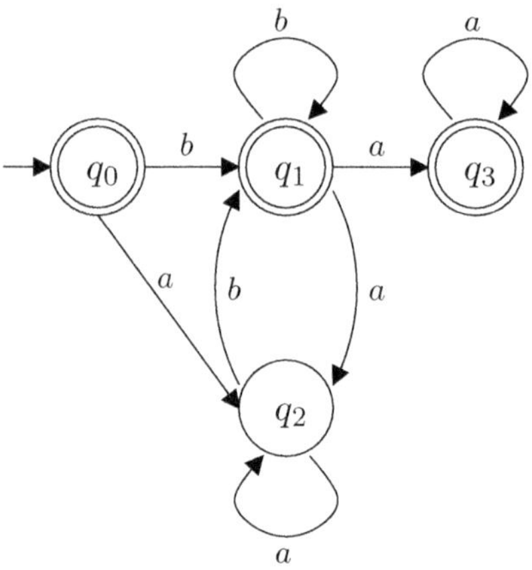

Fig. 1. NBWs for $a^* \cdot b \cdot (a+b)^*$

The result in [11] shows that there are NBWs that are NCW-recognizable and yet an NCW for them requires a structure that is different from the one of the given NBW. It does not show, however, that the NCW needs to have more states. In particular, the language of the NBW in Fig. 1 can be recognized by an NCW with two states.

4.1 A Linear Lower Bound

The first non-trivial lower bound for the NBW to NCW translation was described in [2]. It is based on an idea to which we refer as the "circumventing counting" idea. We describe the idea along with the family of languages used in [2].

Let $\Sigma = \{a, b\}$. For every $k \geq 1$, we define a language L_k as follows:

$$L_k = \{w \in \Sigma^\omega \mid \text{both } a \text{ and } b \text{ appear at least } k \text{ times in } w\}.$$

Since an automaton recognizing L_k must accept every word in which there are at least k a's and k b's, regardless of how the letters are ordered, it may appear as if the automaton must have two k-counters operating in parallel, which requires $O(k^2)$ states. This would indeed be the case if a and b had not been the only letters in Σ, of if the automaton had been deterministic or on finite words. However, since we are interested in nondeterministic automata on infinite words, and a and b are the only letters in Σ, we can do much better. Since Σ contains only the letters a and b, one of these letters must appear infinitely often in every word in Σ^ω. Hence, $w \in L_k$ iff w has at least k b's and infinitely many a's, or at least k a's and infinitely many b's. An NBW can guess which of the two cases above holds, and proceed to validate its guess (if w has infinitely many a's as well as b's, both guesses would succeed). The validation of each of these guesses requires only one k-counter, and a gadget with two states for verifying that there are infinitely many occurrences of the guessed letter. Implementing this idea results in the NBW with $2k + 1$ states appearing in Fig. 2.

The reason we were able to come up with a small NBW for L_k is that NBWs can abstract precise counting by "counting to infinity" with two states. The fact that NCWs do not share this ability [14] is what ultimately allows us to prove that NBW are more succinct than NCW. As it turns out, however, even an NCW for L_k can do much better than maintaining two k-counters with $O(k^2)$ states. To see how, note that a word w is in L_k iff w has at least k b's *after* the first k a's (this characterizes words in L_k with infinitely many b's), or a finite number of b's that is not smaller than k (this characterizes words in L_k with finitely many

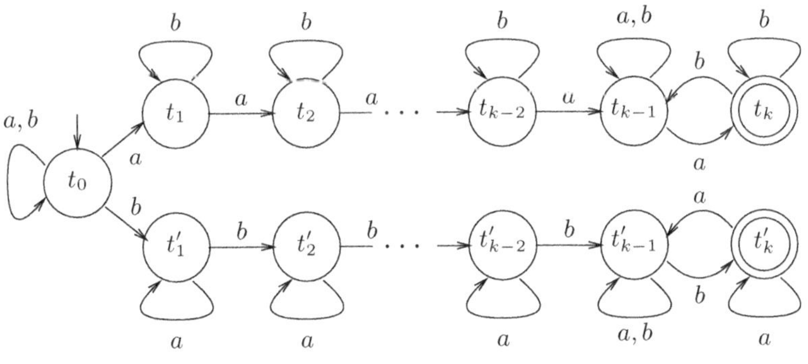

Fig. 2. An NBW for L_k with $2k + 1$ states

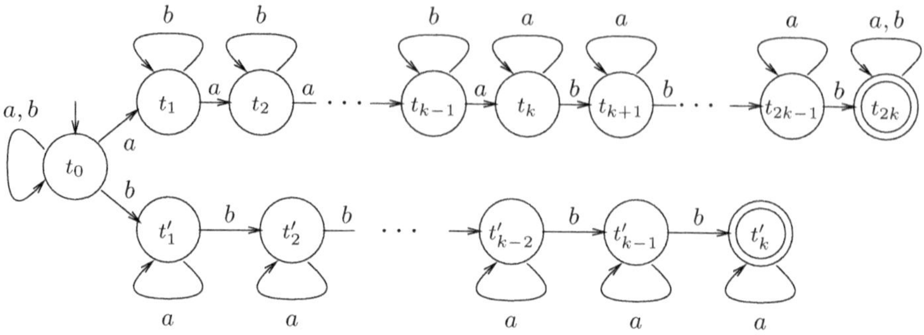

Fig. 3. An NCW for L_k with $3k + 1$ states

b's). Obviously the roles of a and b can also be reversed. Implementing this idea results in the NCW with $3k + 1$ states described in Fig. 3. As detailed in [2], up to one state this is indeed the best one can do. Thus, the family of languages $L_1, L_2, \ldots$ implies that translating an NBW with $2k + 1$ states may result in an NCW with at least $3k$ states, hence the non-trivial, but still linear, lower bound.

4.2 A Quadratic Lower Bound

In this section we enhance the circumventing-counting idea, by letting the NCW count distances between occurrences rather than number of occurrences. We demonstrate how the enhancement can lead to a quadratic lower bound. We first need some results from number theory. For $k \in \mathbb{N}$, let $S_k \subseteq \mathbb{N}$ be the set of all positive integers that can be written as $ik+j(k+1)$, for $i, j \in \mathbb{N}$. Thus, $S_k = \{ik+j(k+1) : i, j \in \mathbb{N}, i+j > 0\}$. For example, $S_4 = \{4, 5, 8, 9, 10, 12, 13, 14, 15, \ldots\}$. The following is a well-known property of S_k, which we prove below for the sake of completeness. (The set S_k is sometimes defined in the literature without the $i + j > 0$ requirement, adding the number 0 to the set.)

Theorem 3. *For every $k \in \mathbb{N}$, the number $k^2 - k - 1$ is the maximal number not in S_k. That is, $k^2 - k - 1 \notin S_k$, while for every $t \geq k^2 - k$, we have that $t \in S_k$.*

Proof. By definition, $S_k = \{ik + j(k + 1) : i, j \in \mathbb{N}, i + j > 0\}$. Equivalently, $S_k = \{(i + j)k + j : i, j \in \mathbb{N}, i + j > 0\}$. Thus, $t \in S_k$ iff $t > 0$ and there are $i', j' \in \mathbb{N}$ such that $i' \geq j'$ and $t = i'k + j'$. We first claim that $k^2 - k - 1 \notin S_k$. Let i' and j' be such that $i'k + j' = k^2 - k - 1$. We claim that $i' < j'$, implying that $k^2 - k - 1 \notin S_k$. To see this, note that $(k - 2)k + (k - 1) = k^2 - k - 1$, and that it is impossible to increase $i' = (k - 2)$ and decrease $j' = (k - 1)$, keeping the same total number. On the other hand, $(k - 1)k + 0 = k^2 - k$, thus for every $t \geq k^2 - k$, there are $i' \geq k - 1$ and $j' \leq k - 1$ such that $t = i'k + j'$. Such i' and j' witness that $t \in S_k$. □

For $k \in \mathbb{N}$, we refer to $k^2 - k - 1$ as the *threshold* of k and denote it by $th(k)$. That is, $th(k) = k^2 - k - 1$.[6]

We can now define the family of languages $L_1, L_2, \ldots$ with which we are going to prove the lower bound. Let $\Sigma = \{a, b\}$. For $k \geq 1$, let $L'_k = \{(\epsilon + \Sigma^* \cdot a) \cdot b^i \cdot a \cdot \Sigma^\omega : i \in S_k\}$. Then, $L_k = L'_k \cup (b^* \cdot a)^\omega$. Thus, $w \in L_k$ iff w starts with $b^i \cdot a$ or has a subword of the form $a \cdot b^i \cdot a$, for $i \in S_k$, or w has infinitely many a's.

Our alert readers are probably bothered by the fact the $(b^* \cdot a)^\omega$ component of L_k is not NCW-recognizable. To see why L_k is still NCW-recognizable, consider a word w with infinitely many a's. Thus, the word is of the form $b^{i_1} \cdot a \cdot b^{i_2} \cdot a \cdot b^{i_3} \cdot a \cdots$, for non-negative integers $i_1, i_2, i_3, \ldots$. If for some $j \geq 1$, we have $i_j \in S_k$, then w is in L'_k. Otherwise, for all $j \geq 1$, we have $i_j \notin S_k$. Hence, by Theorem 3, for all $j \geq 1$, we have $i_k \leq th(k)$. Accordingly, $L_k = L'_k \cup (b^{\leq th(k)} \cdot a)^\omega$. Thus, the "infinitely many a's" disjunct can be replaced by one on which the distance between two successive a's is bounded. As we are going to prove formally, while this implies that L_k can be recognized by an NCW, it forces the NCW to count to $th(k)$, and is the key to the quadratic lower bound.

Theorem 4. *For every $k \geq 1$, the language L_k can be recognized by an NBW with $k+3$ states.*

Proof. We prove that the NBW $\mathcal{B}_k$, appearing in Fig. 4, recognizes L_k.

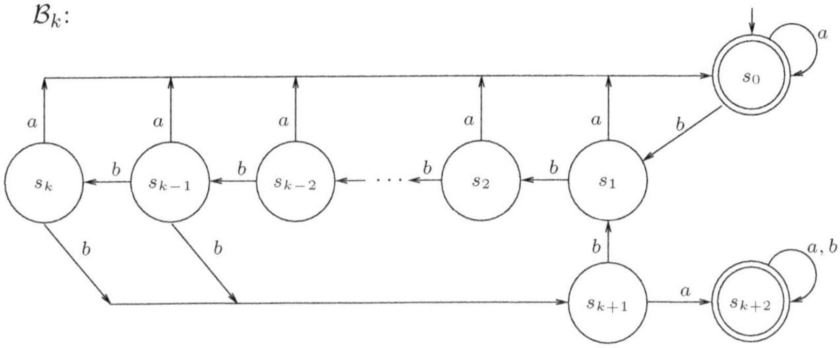

Fig. 4. The NBW $\mathcal{B}_k$ recognizing L_k

Assume first that $w \in L_k$. Then, w either have infinitely many a's, or starts with $b^r \cdot a$ or has a subword of the form $a \cdot b^r \cdot a$, for $r \in S_k$. In the first case, w is accepted by $\mathcal{B}_k$, since the automaton's transition function is total and an a-transition always goes to an accepting state. Now, assume that w has a subword of the form $a \cdot b^r \cdot a$, starting at a position t, for $r \in S_k$. Then, as argued above,

[6] In general, for a finite set of positive integers $\{n_1, n_2, \ldots, n_l\}$, we have that all integers above $\max^2\{n_1, n_2, \ldots, n_l\}$ can be written as linear combinations of the n_i's iff the greater common divisor of the n_i's is 1. For our purpose, it is sufficient to restrict attention to linear combinations of two subsequent integers.

a run of $\mathcal{B}_k$ on w will either visit s_0 or s_{k+2} at position $t+1$. If it visits s_{k+2} it is obviously an accepting run. If it visits s_0, then at position $t+1+r$ it can visit s_{k+1} if there are natural numbers i and j such that $i+j>0$ and $r=ik+j(k+1)$, which is the case by the assumption that $r \in S_k$. Thus, the run can visit s_{k+2} at position $t+r+2$, making it an accepting run. Hence, $w \in L_k$. The case in which w starts with $b^r \cdot a$, for $r \in S_k$, is handled analogously.

As for the other direction, assume that there is an accepting run of $\mathcal{B}_k$ on w. Then, the run either visits infinitely often s_0 or s_{k+2}. Since s_0 has only a-in-transitions, it follows that in the first case w has infinitely many a's, thus belonging to L_k. For the second case, note that a run visits the state s_{k+1}, only if there are natural numbers i and j such that $i+j>0$ and $\mathcal{B}_k$ has read the subword $b^{ik+j(k+1)}$ since its last visit to the state s_0. Since a run visits s_0 only at initialization or after an a is read, it follows that a word visits s_{k+2} only if it starts with $b^i \cdot a$ or has a subword of the form $a \cdot b^i \cdot a$, for $i \in S_k$. Hence, $w \in L_k$. □

Next, we show that while L_k can be recognized by an NCW, every NCW recognizing L_k cannot take advantage of its non-determinism. Formally, we present a DCW (Fig. 5) for L_k that has k^2-k+2 states, and prove that an NCW recognizing L_k needs at least that many states. For simplicity, we show that the NCW must count up to $th(k)$, resulting with at least k^2-k states, and do not consider the two additional states of the DCW.

Theorem 5. *For every $k \geq 1$, the language L_k can be recognized by a DCW with k^2-k+2 states, and cannot be recognized by an NCW with fewer than k^2-k states.*

Proof. Consider the DCW $\mathcal{D}_k$, appearing in Fig. 5. In the figure, a state s_i has an a-transition to the state $s_{th(k)+2}$ if and only if $i \in S_k$. We leave to the reader the easy task of verifying that that $L(D_k)=L_k$.

We now turn to prove the lower bound. Assume by way of contradiction that there is an NCW $\mathcal{C}_k$ with at most k^2-k-1 states that recognizes L_k. The word $w=(b^{(k^2-k-1)} \cdot a)^\omega$ belongs to L_k since it has infinitely many a's. Thus, there is an accepting run r of $\mathcal{C}_k$ on w. Let t be a position such that $r_{t'} \in \alpha$

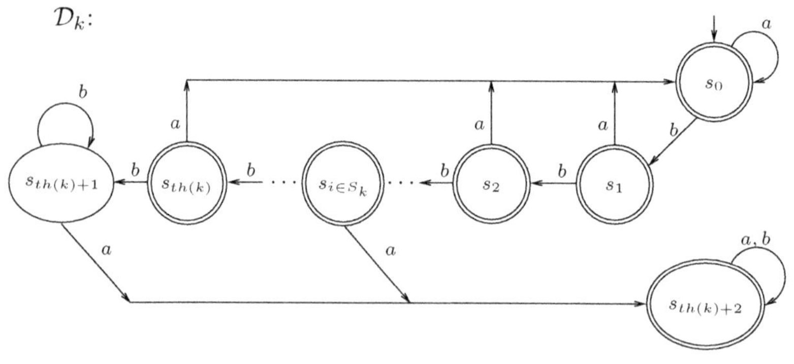

Fig. 5. The DCW $\mathcal{D}_k$ recognizing L_k

for all $t' \geq t$. Let $t' \geq t$ be the first position in which a occurs in w after t. Then, between positions $t'+1$ and $t'+k^2-k$, the run r is in α, making only b-transitions. Since $\mathcal{C}_k$ has at most k^2-k-1 states, it follows that there are positions t_1 and t_2, with $t' < t_1 < t_2 \leq t'+k^2-k$ such that $r_{t_1} = r_{t_2}$. Consider now the word $w' = w_1 \cdot w_2 \cdots w_{t_1} \cdot b^\omega$. On the one hand, w' is accepted by $\mathcal{C}_k$ via the run $r' = r_0, r_1, \ldots, r_{t_1}, (r_{t_1+1}, r_{t_1+2}, \ldots, r_{t_2})^\omega$. On the other hand, w' has only finitely many a's, and by Theorem 3, it has no i consequent b's followed by an a, such that $i \in S_k$. Indeed, all the subwords of the form $a \cdot b^i \cdot a$ in w' have $i = k^2-k-1$. Hence, $w' \notin L_k$, which leads to a contradiction. □

4.3 An Exponential Lower Bound

We now push the circumventing-counting idea to its limit, and use it in order to describe a family of languages $L_2, L_3, \ldots$ such that for all $k \geq 2$, an NBW for L_k has $O(k)$ states whereas an NCW for L_k requires at least $k2^k$ states. As in the quadratic lower bound, the NCW has to bound the distance between occurrences of an event. Now, however, the distance is exponential in k.

Let $\Sigma = \{0, 1, \$, \#\}$. The language L_k is going to be the union of a language L'_k with the language $(\Sigma^* \cdot \#)^\omega$. Before we define L'_k formally, we describe the intuition behind it. Note that the $(\Sigma^* \cdot \#)^\omega$ component of L_k is not NCW-recognizable. Thus, one task of L'_k is to neutralize the non NCW-recognizability of this component. We do this by letting L'_k contain all the words in $(\Sigma^* \cdot \#)^\omega$ that have a subword $(0+1+\$)^h$, for $h > th(k)$, for some threshold $th(k)$. As with the quadratic lower bound, this would make it possible to replace the $(\Sigma^* \cdot \#)^\omega$ component by $(\Sigma^{\leq th(k)} \cdot \#)^\omega$, which is NCW-realizable. The second task of L'_k would be to accomplish the first task with an exponential threshold.

The language L'_k is going to fulfill its second task as follows. Consider a word in Σ^ω and a subword $u \in (0+1+\$)^*$ of it. The subword u is of the form $v_0\$v_1\$v_2\$v_3\cdots$, for $v_i \in (0+1)^*$. Thus, u can be viewed as an attempt to encode a binary k-bit cyclic counter in which two adjacent values are separated by \$. For example, when $k=3$, a successful attempt might be 100\$101\$110\$111\$000. Each subword in $(0+1+\$)^*$ of length $(k+1)2^k$ must reach the value 1^k or contain an error (in its attempt to encode a counter). There are two types of errors. One type is a "syntax error", namely a value v_i of length different from k. The second type is an "improper-increase error", namely a subword $v_i \cdot \$ \cdot v_{i+1} \in (0+1)^k \cdot \$ \cdot (0+1)^k$ such that v_{i+1} is not the successor of v_i in a correct binary encoding of a cyclic k-bit counter. The language L'_k consists of all words that contain the value 1^k or an error, eventually followed by #.

We now define L'_k formally. For $v, v' \in (0+1)^*$, we use $not\_succ_k(v, v')$ to indicate that v and v' are in $(0+1)^k$ but v' is not the successor of v in the binary encoding of a k-bit counter. For example, $not\_succ_3(101, 111)$ holds, but $not\_succ_3(101, 110)$ does not hold. We define the following languages over Σ.

- $S_k = \{\$ \cdot (0+1)^m \cdot \$: m < k\} \cup \{(0+1)^m : m > k\}$,
- $I_k = \{v \cdot \$ \cdot v' \ : \ not\_succ_k(v, v')\}$, and
- $L'_k = \Sigma^* \cdot (S_k \cup I_k \cup \{1^k\}) \cdot \Sigma^* \cdot \# \cdot \Sigma^\omega$.

Finally, we define $L_k = L'_k \cup (\Sigma^* \cdot \#)^\omega$. For example, taking $k = 3$, we have that 010\$011#110\$111#$\cdots$ is in L_3 since it is in L'_3 with a 111 subword, the word 010\$\$011#$\cdots$ is in L_3 since it is in L'_3 by a syntax error, the word \$010\$010\$#$\cdots$ is in L_3 since it is in L'_3 by an improper-increase error, the word $(010\$011\#)^\omega$ is in L_3 since it has infinitely many #'s, and the word $010\$011\#000\$001\$010\#1^\omega$ is not in L_3, as it has only finitely many #'s, it does not contain an error, and while it does contain the subword 111, it does not contain a subword 111 that is eventually followed by #.

Lemma 4. *For every $k \geq 1$, the language L'_k can be recognized by an NBW with $O(k)$ states and by an NCW with $O(k)$ states.*

Proof. We show that there is an NFW with $O(k)$ states recognizing $S_k \cup I_k \cup \{1^k\}$. Completing the NFW to an NBW or an NCW for L'_k is straightforward. It is easy to construct NFWs with $O(k)$ states for S_k and for $\{1^k\}$. An NFW with $O(k)$ states for I_k is fairly standard too (see, for example, [10]). The idea is that if v' is the successor of v in a binary k-bit cyclic counter, then v' can be obtained from v by flipping the bits of the $0 \cdot 1^*$ suffix of v, and leaving all other bits unchanged (the only case in which v does not have a suffix in $0 \cdot 1^*$ is when $v \in 1^*$, in which case all bits are flipped). For example, the successor of 1001 is obtained by flipping the bits of the suffix 01, which results in 1010. Accordingly, there is an improper-increase error in $v \cdot \$ \cdot v'$ if there is at least one bit of v that does not respect the above rule. An NFW can guess the location of this bit and reveals the error by checking the bit located $k+1$ bits after it, along with the bits read in the suffix of v that starts in this bit. □

An immediate corollary of Lemma 4 is that L_k can be recognized by an NBW with $O(k)$ states. Next, we show that while L_k is NCW-recognizable, an NCW for it must be exponentially larger.

Lemma 5. *For every $k \geq 2$, the language L_k is NCW-recognizable, and every NCW recognizing L_k must have at least $k2^k$ states.*

Proof. We first prove that L_k is NCW-recognizable. Let $th(k) = (k+1)2^k$. Consider the language $B_k = (\Sigma^{\leq th(k)} \cdot \#)^\omega$. It is easy to see that B_k is NCW-recognizable. We prove that $L_k = L'_k \cup B_k$. Since, by Lemma 4, the language L'_k is NCW-recognizable, it would follow that L_k is NCW-recognizable. Clearly, $B_k \subseteq (\Sigma^* \cdot \#)^\omega$. Thus, $L'_k \cup B_k \subseteq L_k$, and we have to prove that $L_k \subseteq L'_k \cup B_k$. For that, we prove that $(\Sigma^* \cdot \#)^\omega \subseteq L'_k \cup B_k$. Consider a word $w \in (\Sigma^* \cdot \#)^\omega$. If $w \in B_k$, then we are done. Otherwise, w contains a subword $u \in (0+1+\$)^h$, for $h > th(k)$. Thus, either u does not properly encode a k-bit cyclic counter (that is, it contains a syntactic or an improper-increase error) or u has the subword 1^k. Hence, $u \in \Sigma^* \cdot (S_k \cup I_k \cup \{1^k\}) \cdot \Sigma^*$. Since $w \in (\Sigma^* \cdot \#)^\omega$, it has infinitely many occurrences of #'s. In particular, there is an occurrence of # after the subword u. Thus, $w \in L'_k$, and we are done.

We now turn to prove the lower bound. Assume by way of contradiction that there is an NCW $\mathcal{C}_k$ with acceptance set α and at most $k2^k - 1$ states that recognizes L_k. Consider the word $w = (00\cdots0\$00\cdots01\$\cdots\$11\cdots10\#)^\omega$, in which the distance between two consequent #'s is $d = (k+1)(2^k-1)$. Note that for all $k \geq 2$, we have that $d > k2^k$. The word w has infinitely many #'s and it therefore belongs to L_k. Thus, there is an accepting run r of $\mathcal{C}_k$ on w. Let t be a position such that $r_{t'} \in \alpha$ for all $t' \geq t$. Let $t_0 \geq t$ be the first position after t such that $w_{t_0} = \#$. Since $\mathcal{C}_k$ has at most $k2^k - 1$ states, there are two positions t_1 and t_2, with $t_0 < t_1 < t_2 \leq t_0 + k2^k$, such that $r_{t_1} = r_{t_2}$.

Consider the word $w' = w_1 \cdot w_2 \cdots w_{t_1} \cdot (w_{t_1+1} \cdots w_{t_2})^\omega$. The NCW $\mathcal{C}_k$ accepts w' with a run r' that pumps r between the positions t_1 and t_2. Formally, $r' = r_0, r_1, \ldots, r_{t_1}, (r_{t_1+1}, \ldots, r_{t_2})^\omega$. Note that since $r_{t'} \in \alpha$ for all $t' \geq t$, the run r' is indeed accepting. We would get to a contradiction by proving that $w' \notin L_k$.

Since $t_2 \leq t_0 + k2^k$ and $k2^k < d$, we have that $w_{t_1+1} \cdots w_{t_2}$ has no occurrence of #, thus w' has no occurrences of # after position t_0. Recall that $L_k = L'_k \cup (\Sigma^* \cdot \#)^\omega$. By the above, $w' \notin (\Sigma^* \cdot \#)^\omega$. Furthermore, since $L'_k = \Sigma^* \cdot (S_k \cup I_k \cup \{1^k\}) \cdot \Sigma^* \cdot \# \cdot \Sigma^\omega$, the fact w' has no occurrences of # after position t_0 implies that the only chance of w' to be in L_k is to have a prefix of $w_1 \cdots w_{t_0}$ in $\Sigma^* \cdot (S_k \cup I_k \cup \{1^k\}) \cdot \Sigma^* \cdot \#$. Such a prefix, however, does not exist. Indeed, all the subwords in $(0+1+\$)^*$ of $w_1 \cdots w_{t_0}$ do not contain errors in their encoding of a k-bit counter, nor they reach the value 1^k. It follows that $w \notin L_k$, and we are done. □

Lemmas 4 and 5 imply the desired exponential lower bound:

Theorem 6 ([3]). *There is a family of languages $L_2, L_3, \ldots$, over an alphabet of size 4, such that for every $k \geq 2$, the language L_k is NCW-recognizable, it can be recognized by an NBW with $O(k)$ states, and every NCW that recognizes it has at least $k2^k$ states.*

Combining the above lower bound with the upper bound in Theorem 1, we can conclude with the following.[7]

Theorem 7 ([3]). *The asymptotically tight bound for the state blow up in the translation, when possible, of an NBW to an equivalent NCW is $2^{\Theta(n)}$.*

5 Discussion

It is well known that nondeterministic automata are exponentially more succinct than deterministic ones. The succinctness is robust and it applies to all known classes of automata on finite or infinite objects. Restricting attention to nondeterministic automata makes the issue of succinctness more challenging, as now all classes of automata may guess the future, and the question is whether certain acceptance conditions can use this feature better than others. For example,

[7] Note that the lower and upper bounds are only asymptotically tight, leaving a gap in the constants. This is because the NBW that recognizes L_k requires $O(k)$ states and not strictly k states.

translating a nondeterministic Rabin word automaton with n states and index k to an NBW, results in an automaton with $O(nk)$ states, whereas translating a nondeterministic Streett automaton with n states and index k, results in an NBW with $O(n2^k)$ states. The difference between the blow-ups in the case of Rabin and Streett can be explained by viewing the acceptance condition as imposing additional nondeterminism. Indeed, simulating a Rabin automaton, an NBW has to guess not only the accepting run, but also the pair $\langle G, B\rangle$ that is going to be satisfied and the position after which no states in B are visited (hence the $O(k)$ factor in the blow up), whereas in a Streett automaton, the simulating NBW has to guess, for each pair, the way in which it is going to be satisfied (hence the $O(2^k)$ factor). This intuition is supported by matching lower bounds [23]. Starting with a Büchi automaton, no such additional nondeterminism hides in the acceptance condition, so one would not expect Büchi automata to be more succinct than other nondeterministic automata. The challenge grows with an acceptance condition like co-Büchi, whose expressive power is strictly weaker, and thus not all languages are candidates for proving succinctness. The exponential lower-bound described in Sect. 4 shows that the Büchi condition is still exponentially more succinct than its dual co-Büchi condition. The explanation to this succinctness is the ability of the Büchi condition to easily specify the fact that some event P should repeat infinitely often. Languages that involve such a specification may still be NCW-recognizable, as other components of the language force the distance between successive occurrences of P to be bounded by some fixed threshold 2^k. While an NBW for the language does not have to count to the threshold and can be of size $O(k)$, an NCW for the language has to count, which requires it to have at least 2^k states.

Co-Büchi automata can be determinized with a $2^{O(n)}$ blow up [16]. A $2^{O(n \log n)}$ translation of NBW to DCW was known, but a matching lower bound was known only for the translation of NBW to deterministic Rabin or Streett automata. As detailed in [3], the improved $2^{O(n)}$ translation of NBW to NCW described in Sect. 3, also suggests an improved $2^{O(n)}$ translation of NBW to DCW. Indeed, since the exponential component of the constructed NCW is deterministic, then applying the break-point subset construction of [16] on it does not involve an additional exponential blow-up. In particular, this implies a Safraless and symbolic translation of LTL formulas to DBW, when possible. Furthermore, by [4], one cannot expect to do better than the breakpoint construction, making the translation of NBW to DCW [3] optimal. In addition, as detailed in [3], the translation described in Sect. 3 has a one-sided error. Thus, when applied to an NBW that is not NCW-recognizable, the constructed NCW contains the language of the NBW. Accordingly, translating LTL formulas that are not DBW-recognizable to a DBW, one gets a DBW that under-approximates the specification. For many applications, and in particular synthesis, one can work with such an under-approximating automaton, and need not worry about the specification being DBW-recognizable.

References

1. Accellera: Accellera organization inc. (2006), http://www.accellera.org
2. Aminof, B., Kupferman, O., Lev, O.: On the relative succinctness of nondeterministic Büchi and co-Büchi word automata. In: Cervesato, I., Veith, H., Voronkov, A. (eds.) LPAR 2008. LNCS (LNAI), vol. 5330, pp. 183–197. Springer, Heidelberg (2008)
3. Boker, U., Kupferman, O.: Co-ing Büchi made tight and useful. In: Proc. 24th IEEE Symp. on Logic in Computer Science (2009)
4. Boker, U., Kupferman, O., Rosenberg, A.: Alternation removal in Büchi automata. In: Proc. 37th Int. Colloq. on Automata, Languages, and Programming (2010)
5. Büchi, J.R.: On a decision method in restricted second order arithmetic. In: Proc. Int. Congress on Logic, Method, and Philosophy of Science, pp. 1–12. Stanford University Press, Stanford (1962)
6. Emerson, E., Jutla, C.: The complexity of tree automata and logics of programs. In: Proc. 29th IEEE Symp. on Foundations of Computer Science, pp. 328–337 (1988)
7. Gentilini, R., Piazza, C., Policriti, A.: Computing strongly connected components in a linear number of symbolic steps. In: 14th ACM-SIAM Symp. on Discrete Algorithms, pp. 573–582 (2003)
8. Krishnan, S., Puri, A., Brayton, R.: Deterministic ω-automata vis-a-vis deterministic Büchi automata. In: Du, D.-Z., Zhang, X.-S. (eds.) ISAAC 1994. LNCS, vol. 834, pp. 378–386. Springer, Heidelberg (1994)
9. Kupferman, O.: Tightening the exchange rate beteen automata. In: Duparc, J., Henzinger, T.A. (eds.) CSL 2007. LNCS, vol. 4646, pp. 7–22. Springer, Heidelberg (2007)
10. Kupferman, O., Lustig, Y., Vardi, M.: On locally checkable properties. In: Hermann, M., Voronkov, A. (eds.) LPAR 2006. LNCS (LNAI), vol. 4246, pp. 302–316. Springer, Heidelberg (2006)
11. Kupferman, O., Morgenstern, G., Murano, A.: Typeness for ω-regular automata. International Journal on the Foundations of Computer Science 17, 869–884 (2006)
12. Kupferman, O., Vardi, M.: From linear time to branching time. ACM Transactions on Computational Logic 6, 273–294 (2005)
13. Kurshan, R.: Computer Aided Verification of Coordinating Processes. Princeton Univ. Press, Princeton (1994)
14. Landweber, L.: Decision problems for ω–automata. Mathematical Systems Theory 3, 376–384 (1969)
15. McNaughton, R.: Testing and generating infinite sequences by a finite automaton. Information and Control 9, 521–530 (1966)
16. Miyano, S., Hayashi, T.: Alternating finite automata on ω-words. Theoretical Com puter Science 32, 321–330 (1984)
17. Morgenstern, A., Schneider, K.: From LTL to symbolically represented deterministic automata. In: Logozzo, F., Peled, D.A., Zuck, L.D. (eds.) VMCAI 2008. LNCS, vol. 4905, pp. 279–293. Springer, Heidelberg (2008)
18. Piterman, N.: From nondeterministic Büchi and Streett automata to deterministic parity automata. In: Proc. 21st IEEE Symp. on Logic in Computer Science, pp. 255–264. IEEE press, Los Alamitos (2006)
19. Pnueli, A., Rosner, R.: On the synthesis of a reactive module. In: Proc. 16th ACM Symp. on Principles of Programming Languages, pp. 179–190 (1989)
20. Rabin, M.: Decidability of second order theories and automata on infinite trees. Transaction of the AMS 141, 1–35 (1969)

21. Ravi, K., Bloem, R., Somenzi, F.: A comparative study of symbolic algorithms for the computation of fair cycles. In: Johnson, S.D., Hunt Jr., W.A. (eds.) FMCAD 2000. LNCS, vol. 1954, pp. 143–160. Springer, Heidelberg (2000)
22. Safra, S.: On the complexity of ω-automata. In: Proc. 29th IEEE Symp. on Foundations of Computer Science, pp. 319–327 (1988)
23. Safra, S., Vardi, M.: On ω-automata and temporal logic. In: Proc. 21st ACM Symp. on Theory of Computing, pp. 127–137 (1989)
24. Street, R., Emerson, E.: An elementary decision procedure for the μ-calculus. In: Paredaens, J. (ed.) ICALP 1984. LNCS, vol. 172, pp. 465–472. Springer, Heidelberg (1984)
25. Vardi, M., Wolper, P.: Automata-theoretic techniques for modal logics of programs. Journal of Computer and Systems Science 32, 182–221 (1986)
26. Vardi, M., Wolper, P.: Reasoning about infinite computations. Information and Computation 115, 1–37 (1994)
27. Wolper, P., Vardi, M., Sistla, A.: Reasoning about infinite computation paths. In: Proc. 24th IEEE Symp. on Foundations of Computer Science, pp. 185–194 (1983)
28. Yan, Q.: Lower bounds for complementation of ω-automata via the full automata technique. In: Bugliesi, M., Preneel, B., Sassone, V., Wegener, I. (eds.) ICALP 2006. LNCS, vol. 4052, pp. 589–600. Springer, Heidelberg (2006)

Normalization of Some Extended Abstract State Machines

Patrick Cégielski[1] and Irène Guessarian[2,⋆]

[1] Université Paris Est–Créteil, IUT, LACL EA 4219, Route forestière Hurtault, F-77300 Fontainebleau, France
cegielski@u-pec.fr

[2] LIAFA, UMR 7089 and Université Paris 6, 2 Place Jussieu, 75254 Paris Cedex 5, France
ig@liafa.jussieu.fr

For the 70th birthday of Yuri Gurevich

Abstract. We compare the control structures of Abstract State Machines (in short ASM) defined by Yuri Gurevich and AsmL, a language implementing it. AsmL is not an algorithmically complete language, as opposed to ASM, but it is closer to usual programming languages allowing `until` and `while` iterations and sequential composition. We here give a formal definition of AsmL, its semantics, and we construct, for each AsmL program Π, a normal form (which is an ASM program) Π_n computing the same function as Π. The number of comparisons and updates during the execution of the normal form is at most three times the number of comparisons and updates in the original program.

Keywords: Abstract State Machine, Algorithm, Semantics.

1 Introduction

Yuri GUREVICH has given a schema of languages which is not only a *Turing-complete language* (a language allowing to express at least an algorithm for each computable function), but which also allows to express all algorithms for each computable function (it is an *algorithmically complete language*); this schema of languages was first called *dynamic structures*, then *evolving algebras*, and finally *ASM* (for *Abstract State Machines*) [2]. He proposed the Gurevich's thesis (*the notion of algorithm is entirely captured by the model*) in [3]. Yuri had explained us this thesis during his stay in Paris, and a fascinating russian style after-talk discussion, between Yuri and Vladimir USPENSKY, during a conference in Fontainebleau, convinced all those attending the talk of the truth of Yuri's thesis.

There exist several partial implementations of ASMs as a programming language. These implementations are *partial* by nature for two reasons: (i) ASMs allow to program functions computable by Turing machines with oracles, but

⋆ Address correspondence to this author.

A. Blass, N. Dershowitz, and W. Reisig (Eds.): Gurevich Festschrift, LNCS 6300, pp. 165–180, 2010.

obviously not the implementations; (ii) in ASMs any (computable) first order structures may be defined, which is not true for the current implementations.

Yuri GUREVICH has directed a group at Microsoft Laboratories (Redmond) which has implemented such a programming language, called *AsmL* (for *ASM Language*), written first in C++ then in C# as a language of the .NET framework of Microsoft. To invite programmers to use its language, this group extended the control structures used in ASMs. Pure ASM control structures are represented by *normal forms* of AsmL programs.

The aim of this paper is to give a formal definition of AsmL (more precisely of the part concerning control structures; we are not interested by constructions of first-order structures here), a formal definition of normal forms in AsmL, to show how to build a normal form from an AsmL program, and to compare the cost of the normal form with the cost of the original program.

2 Definitions

2.1 Definition of ASMs

We first recall the definition of ASMs and then precise our point of view on ASMs because different definitions exist.

2.2 Syntax

Definition 1. *An* ASM vocabulary, *or* signature, *is a first-order signature* $\mathcal{L}$ *with a finite number of function symbols, two boolean constant symbols (*`true` *and* `false`*), a constant symbol (denoted by* `null` *or* `undef`*), logical connectives (*`not`, `and`, *and* `or`*), and the* equality *predicate denoted by* $=$.

Signature $\mathcal{L}$ has three sorts: `Data`, `Boolean` and `Null`. *Terms* are defined by:

- if c is a nullary function symbol (a constant), then c is a term,
- if $t_1, \ldots, t_n$ are terms and f is an n-ary function symbol then $f(t_1, \ldots, t_n)$ is a term.

The above defined terms are usually called *closed* or *ground terms*; general terms with variables are disallowed. An n-ary function symbol of sort `Boolean` will be called an n-ary *predicate*.

Definition 2. Boolean terms *(or* formulæ*) are defined inductively by:*

- *if* p *is an* n*-ary predicate and* $t_1, \ldots, t_n$ *are terms then* $p(t_1, \ldots, t_n)$ *is a formula;*
- *if* t *and* t' *are terms, then* $t = t'$ *is a formula;*
- *if* F, F' *are formulæ, then* $\neg F$, $F \wedge F'$, $F \vee F'$ *are formulæ.*

Definition 3. *Let* $\mathcal{L}$ *be an ASM signature. ASM* rules *are defined inductively as follows:*

– *An* update rule *is an expression of the form:*

$$f(t_1, \ldots, t_n) := t_0$$

where f *is a* n*-ary functional symbol and* $t_0,\ t_1, \ldots,\ t_n$ *are terms of* $\mathcal{L}$ *(Recall that constants are allowed but variables are disallowed).*

– *If* $R_1, \ldots,\ R_k$ *are rules of signature* $\mathcal{L}$*, where* $k \geq 1$*, then the following expression is also a rule of* $\mathcal{L}$*, called a* block*:*

```
par
        R_1
         ⋮
        R_k
endpar
```

– *Let* φ *be a boolean term, and let* R_1 *and* R_2 *be rules of signature* $\mathcal{L}$*. The expression:*

```
if φ then R_1
     else R_2
endif
```

is also a rule of vocabulary $\mathcal{L}$*, called an* alternative rule.

– *The* test rule *is defined by:* `if` φ `then` R_1.

In early ASMs, `par` meant *parallel*, nowadays it stands for *parenthesis*.

Definition 4. *Let* $\mathcal{L}$ *be an ASM signature. A* program *on signature* $\mathcal{L}$*, or* $\mathcal{L}$*-*program, *is a synonym for a rule of that signature.*

2.3 Semantics

Definition 5. *If* $\mathcal{L}$ *is an ASM signature, an ASM* abstract state, *or more precisely an* $\mathcal{L}$-state, *is a synonym for a first-order structure* $\mathcal{A}$ *of signature* $\mathcal{L}$ *(an* $\mathcal{L}$*-structure).*

The universe of $\mathcal{A}$ consists of the disjoint union of three sets: the *basis set* A, the Boolean set $\mathbb{B} = \{true, false\}$, and a singleton set $\{\perp\}$. The values of the Boolean constant symbols `true` and `false` and of `null` (or `undef`) in $\mathcal{A}$ will be denoted by *true*, *false*, and *null* (or $\perp$).

Definition 6. *Let* $\mathcal{L}$ *be an ASM signature and* A *a non empty set. A* set of modifications *(more precisely an* $(\mathcal{L},\ A)$-modification set*) is any finite set of triples:*

$$(f,\ \overline{a},\ a),$$

where f *is a function symbol of* $\mathcal{L}$*,* $\overline{a} = (a_1, \ldots,\ a_n)$ *is an* n*-tuple of* A *(where* n *is the arity of* f*), and* a *is an element of* A*.*

Definition 7. *Let* $\mathcal{L}$ *be an ASM signature, let* $\mathcal{A}$ *be an* $\mathcal{L}$*-state and let* Π *be an* $\mathcal{L}$*-program. Let* $\Delta_\Pi(\mathcal{A})$ *denote the set defined by structural induction on* Π *as follows:*

1. *If Π is an update rule:*
$$f(t_1, \ldots, t_n) := t_0$$
then, denoting $\overline{t_1}^{\mathcal{A}}$ by a_1 , ..., $\overline{t_n}^{\mathcal{A}}$ by a_n, and $\overline{t_0}^{\mathcal{A}}$ by a, the set $\Delta_\Pi(\mathcal{A})$ is the singleton:
$$\{(f, (a_1, \ldots, a_n), a)\}.$$
2. *If Π is a block:*
$$\texttt{par } R_1 \ldots R_n \texttt{ endpar}$$
then the set $\Delta_\Pi(\mathcal{A})$ is the union:
$$\Delta_{R_1}(\mathcal{A}) \cup \ldots \cup \Delta_{R_n}(\mathcal{A}).$$
3. *If Π is a test:*
$$\texttt{if } \varphi \texttt{ then } R$$
we first have to evaluate the expression $\overline{\varphi}^{\mathcal{A}}$. If it is false then the set $\Delta_\Pi(\mathcal{A})$ is empty, otherwise it is equal to:
$$\Delta_R(\mathcal{A}).$$
The semantics of the alternative rule is similar.

We may check that $\Delta_\Pi(\mathcal{A})$ is an ($\mathcal{L}$,A)-set of modifications.

Definition 8. *A set of modifications is* incoherent *if it contains two elements $(f, \overline{a}, a)$ and $(f, \overline{a}, b)$ with $a \neq b$. It is* coherent *otherwise.*

Definition 9. *Let $\mathcal{L}$ be an ASM signature, Π an $\mathcal{L}$-program, and $\mathcal{A}$ an $\mathcal{L}$-state.*

If $\Delta_\Pi(\mathcal{A})$ is coherent, the transform *$\tau_\Pi(\mathcal{A})$ of $\mathcal{A}$ by Π is the $\mathcal{L}$-structure $\mathcal{B}$ defined by:*

- *the base set of $\mathcal{B}$ is the base set A of $\mathcal{A}$;*
- *for any element f of $\mathcal{L}$ and any element $\overline{a} = (a_1, \ldots, a_n)$ of A^n (where n is the arity of f):*
  - *if $(f, \overline{a}, a) \in \Delta_\Pi(\mathcal{A})$ for an $a \in A$, then :*
  $$\overline{f}^{\mathcal{B}}(\overline{a}) = a\ ;$$
  - *otherwise:*
  $$\overline{f}^{\mathcal{B}}(\overline{a}) = \overline{f}^{\mathcal{A}}(\overline{a}).$$

If $\Delta_\Pi(\mathcal{A})$ is incoherent then $\tau_\Pi(\mathcal{A}) = \mathcal{A}$ (hence the state is a fixed point).

Definition 10. *Let $\mathcal{L}$ be an ASM signature, Π an $\mathcal{L}$-program, and $\mathcal{A}$ an $\mathcal{L}$-state. The* computation *is the sequence of $\mathcal{L}$-states*
$$(\mathcal{A}_n)_{n\in\mathbb{N}}$$
defined by:

- *$\mathcal{A}_0 = \mathcal{A}$ (called the* initial algebra *of the computation);*
- *$\mathcal{A}_{n+1} = \tau_\Pi(\mathcal{A}_n)$ for $n \in \mathbb{N}$.*

A computation terminates *if there exists a fixed point* $\mathcal{A}_{n+1} = \mathcal{A}_n$. *This fixed point* $\mathcal{A}_n$ *is the computation result.*

Definition 11. *The* semantics *is the partial function which transforms* $\{\Pi, \mathcal{A}\}$, *where* Π *is an ASM program and* $\mathcal{A}$ *a state, in the fixed point* $\mathcal{A}_n$ *obtained by iterating* τ_Π *starting from* $\tau_\Pi(\mathcal{A})$ *until a fixed point is reached (it will be denoted* $[\![(\Pi, \mathcal{A})]\!] = \tau_\Pi^*(\mathcal{A})$*) if such a fixed point exists, otherwise it is undefined.*

3 Definition of Extended ASMs

The analog of ASM with additional rules to take into account the supplementary control structures of AsmL is called an *extended ASM*. No paper defines formally AsmL but [1] gives a precise informal description.

3.1 Syntax

Roughly speaking, AsmL considers more control structures than ASMs. We first define AsmL rules, then explain them.

Definition 12. *Let* $\mathcal{L}$ *be an ASM signature.*

- *An ASM rule is also an* extended rule *of* $\mathcal{L}$.
- *If* $R_1, \ldots, R_k$ *are extended rules of signature* $\mathcal{L}$, *where* $k \geq 1$, *then the following expression is also an extended rule of* $\mathcal{L}$, *called a* step rule*:*

  par
          step R_1
          ⋮
          step R_k
  endpar
- *If* R *is an extended rule of signature* $\mathcal{L}$, *then the following expressions are also extended rules of* $\mathcal{L}$, *called an* iteration rules*:*
  - step until φ R
  - step while φ R

  with φ *a boolean term.*
- *If* φ *is a boolean term, and* R_1 *and* R_2 *are extended rules, then:*

  if φ then R_1
         else R_2
  endif

  is an extended rule, called an alternative rule.

3.2 Semantics

We now explain the above rules. Following the semantics given above for ASMs, the meaning of the par rule:

par
        R_1
        ⋮
        R_k
endpar

is that rules R_1, ..., R_k are running simultaneously, i.e. each rule is running (with a point of view sequential or parallel for the observer, it is immaterial) independently of the other rules; incoherences might occur for some updates; if there is at least one incoherence, the program stops, otherwise updates are applied.

The paradigm of simultaneity is unusual for programmers who are used to sequentiality. Hence sequentiality was introduced using `step` in AsmL. The meaning of the extended rule

```
par
        step R_1
          ⋮
        step R_k
endpar
```

is as follows: rule R_1 is running first; then rule R_2 is running (with values of closed terms depending of the result of running rule R_1); and so on.

Finally classical iterations appear in AsmL via the *iteration rules.*

4 Normalization

In the present section, all $\mathcal{L}$-state $\mathcal{A}$ have an infinite base set which can be assumed to contain $\mathbb{N}$.

Definition 13. *Let $\mathcal{L}$ be an ASM vocabulary. If R is an extended rule of vocabulary $\mathcal{L}$, then the expression* `step until fixpoint` R *is also an extended rule of $\mathcal{L}$, called a* running rule.

Running rules are implicit in ASMs and are made explicit in AsmL. If R is an ASM rule (not extended), then `step until fixpoint` R will have the same semantics as the ASM program Π consisting of rule R.

Definition 14. *Let $\mathcal{L}$ be an ASM vocabulary. If $R_1, \ldots, R_k$ are ASM rules of vocabulary $\mathcal{L}$ (*not extended rules*), then the extended rule:*

```
step until fixpoint
     par
          if (mode = 0) then R_1
           ⋮
          if (mode = k) then R_k
     endpar
```

is a normal form *of $\mathcal{L}$, where mode is a special nullary function (constant of an ASM language with signature $\mathcal{L}_m = \mathcal{L} \cup \{mode\}$), whose value for the initial algebra is 0.*

The ASM rule under `step until fixpoint` *is called the* core *of the normal form.*

In the sequel `par` and `endpar` will be omitted: when rules have the same indentation, they will be assumed to be in a `par` ... `endpar` block.

To give a normal form for a given program is a classical issue of theoretical computer science. For ASMs, we have an extra constraint, which is very important: when running, the normal form and the original program should execute approximatively the same number of updates and comparisons.

Example 1. Consider the following programming problem: to compute the average of an array of grades and to determine the number of grades whose value is greater than this average. A natural program in extended ASMs is given below. Note that in this program, *avg*, *i*, *n*, and *nb* are not variables, but constants: the value of *i* will be changed in the next $\mathcal{L}$-state of the computation.

```
step
      avg := 0
      i := 0
step until (i = n)
      avg := avg + grade[i]
      i := i+1
step
      avg := avg/n
      i := 0
      nb := 0
step until (i = n)
      if (grade[i] > avg) then nb := nb + 1
      i := i + 1
```

using the conventions of AsmL (a same indentation indicates a block).

The core of the normal form is:

```
if (mode = 0) then
   avg := 0
   i := 0
   mode := 1
if (mode = 1) then
   avg := avg + grade[i]
   i := i + 1
   if (i = n) then mode := 2
if (mode = 2) then
   avg := avg/n
   i := 0
   nb := 0
   mode := 3
if (mode = 3) then
   if (grade[i] > avg) then nb := nb + 1
   i := i + 1
   if (i = n) then mode := 4
```

Theorem 1. *For every extended ASM program Π, there exists a normal form ASM program Π_n such that for every $\mathcal{L}$-state $\mathcal{A}$ whose base set is infinite there exists an $\mathcal{L}_m$-state $\mathcal{A}_m$ in which Π_n computes the same function as Π, where $\mathcal{L}_m$ is the signature $\mathcal{L} \cup \{mode\}$, with mode a new constant symbol, and $\mathcal{A}_m$ is the expansion of $\mathcal{A}$ to $\mathcal{L}_m$.*

Proof. We define the core of the normal form Π_n of Π by structural induction on the form of program Π. The proof that Π_n and Π compute the same function will be given in the appendix.

We will need to label some blocks. To allow this, we supposed that the base set is infinite, we used a special nullary function symbol, *mode*, which was added to the ASM signature, i.e. $\mathcal{L}_m = \mathcal{L} \cup \{mode\}$. Without loss of generality, we supposed the set $\mathbb{N}$ of natural integers is included in the base set.

- The core of the normal form R_n of an ASM rule R is:

  `if` $(mode = 0)$ `then`
  R
  $mode := 1$

- If $R'_1, \ldots, R'_k$ are the respective cores of the extended rules $R_1, \ldots, R_k$, we have to define the core of the extended rule Π:

  `par`
  `step` R_1
  $\vdots$
  `step` R_k
  `endpar`

  In the (abnormal) case $k = 1$, the core is simply R'_1. To explain the case $k \geq 2$, it is sufficient to suppose $k = 2$. In this case the core of Π_n is:

  R''_1
  $R_2^{+fin_1}$

  with

  1. $R''_1 = R'_1$.
  2. $R_2^{+fin_1}$ is defined as follows: Let fin_1 be the greatest value used for *mode* in R'_1. Rule $R_2^{+fin_1}$ is R'_2 where fin_1 is added to each *constant* occurring on the right hand side of the "=" sign of an expression rule beginning by *mode* (a boolean expression $mode = constant$ or an update rule $mode := constant$).

- If Π is an iteration rule (`step until` φ R) and R' is the core of the normal form of R then the core Π' of the normal form Π_n of Π is

```
if  (mode = 0) then
    if ¬φ then mode := 1
    if φ then mode := fin + 1
R^{+1}
```

where R^{+1} is R' in which each *constant* occurring on the right hand side of the "=" or "≥" sign of an expression beginning by *mode* (a boolean expression *mode* = *constant*, *mode* ≥ *constant* or an update rule *mode* := *constant*) is incremented by 1 but for the greatest such value fin, which is replaced by 0.

– If Π is an iteration rule (`step while` φ R), it is treated similarly:

```
if  (mode = 0) then
    if φ then mode := 1
    if ¬φ then mode := fin + 1
R^{+1}
```

– If Π is an alternative rule,

```
if φ  then R1
      else R2
endif
```

and R'_1, R'_2 are the respective cores of the normal forms of R_1, R_2, then the core of the normal form Π_n of Π is:

```
if     (mode = 0) then
       if φ then mode := 1 else mode := fin + 1
R1^{+1}
R2^{fin+1}
```

where R_i^{+k} is R'_i in which k is added to each *constant* occurring on the right hand side of the "=" (or "≥") sign of an expression rule beginning by *mode*.

Acknowledgments. We thank the referee for his very helpful comments.

References

1. Grieskamp, W., Tillmann, N.: AsmL Standard Library, Foundations of Software Engineering – Microsoft Research (2002), `http://www.codeplex.com/AsmL//AsmLReference.doc`, `http://research.microsoft.com/en-us/downloads/3444a9cb-47ce-4624-9e14-c2c3a2309a44/default.aspx`
2. Gurevich, Y.: Reconsidering Turing's Thesis: Toward More Realistic Semantics of Programs, University of Michigan, Technical Report CRL–TR–38–84, EECS Department (1984)
3. Gurevich, Y.: A New Thesis, Abstracts, p. 317. American Mathematical Society, Providence (August 1985)

A Appendix

We prove here that an AsmL program Π and its normal form Π_n have the same semantics. To this end we first define formally the semantics of AsmL programs, in the general case when `step until fixpoint` iterations are also allowed in AsmL.

A.1 Semantics of AsmL

As for ASM programs, semantics is defined by structural induction. We will denote by $[\![(\Pi,\mathcal{A})]\!]^e$ and $\tau^e_\Pi(\mathcal{A})$ the semantics and transform associated with an extended program (the ^{e} stands for "extended") while $[\![(\Pi,\mathcal{A})]\!]$ and $\tau_\Pi(\mathcal{A})$ without the superscript ^{e} represent the semantics and transform of an ASM program. Let $\mathcal{L}$ be an ASM signature, Π an $\mathcal{L}$ extended rule, and $\mathcal{A}$ an $\mathcal{L}$-state.

- If Π is a non extended $\mathcal{L}$ rule, then $[\![(\Pi,\mathcal{A})]\!]^e = \tau^e_\Pi(\mathcal{A}) = \tau_\Pi(\mathcal{A})$.
  Notice that it is not the same as the ASM semantics of Π: in AsmL, program Π is executed once, while in ASM program Π is executed until a fixed point is reached, see Definition 11.
- If Π is a step rule, it is enough to only consider the case when there are two extended rules R_1, R_2 of signature $\mathcal{L}$; in that case the semantics of Π:
  ```
  par
          step R1
          step R2
  endpar
  ```
  is given by:
  $$[\![(\Pi,\mathcal{A})]\!]^e = [\![(R_2,[\![(R_1,\mathcal{A})]\!]^e)]\!]^e \ .$$
  We need not define the transform $\tau^e_\Pi(\mathcal{A})$ in this case.
- If Π is an iteration rule, `step until` φ R, then:
  $$\tau^e_\Pi(\mathcal{A}) = \text{if } \neg\overline{\varphi}^{\mathcal{A}} \text{ then } \tau^e_R(\mathcal{A}) \text{ else } \mathcal{A} \ .$$
  Let us define inductively $\mathcal{A}^1 = \tau_\Pi(\mathcal{A})$, $\mathcal{A}^2 = \tau_\Pi(\mathcal{A}^1)$, $\ldots$, $\mathcal{A}^l = \tau_\Pi(\mathcal{A}^{l-1})$, or in short, $\mathcal{A}^l = \tau^l_\Pi(\mathcal{A})$ for $l \in \mathbb{N}$. We have $[\![(\Pi,\mathcal{A})]\!]^e = \mathcal{A}^l$ for the least l such that $\overline{\varphi}^{\mathcal{A}^l}$ is true and $[\![(\Pi,\mathcal{A})]\!]^e = \perp$ is undefined if there is no l such that $\overline{\varphi}^{\mathcal{A}^l}$ is true.
- If Π is an iteration rule, `step while` φ R, then:
  $$\tau^e_\Pi(\mathcal{A}) = \text{if } \overline{\varphi}^{\mathcal{A}} \text{ then } \tau^e_R(\mathcal{A}) \text{ else } \mathcal{A} \ .$$
  $[\![(\Pi,\mathcal{A})]\!]^e$ is defined as in the previous case.
- If Π is an iteration rule `step until fixpoint` R, its semantics is defined as in the ASM case.

- If Π is an alternative rule,

```
if φ then R1
     else R2
endif
```

its semantics is defined by:

$$[\![(\Pi, \mathcal{A})]\!]^e = \text{if } \overline{\varphi}^{\mathcal{A}} \text{ then } [\![(R_1, \mathcal{A})]\!]^e \text{ else } [\![(R_2, \mathcal{A})]\!]^e \ .$$

A.2 Proof of Theorem 1

Proof. We now prove the general case of Theorem 1, allowing for `step until fixpoint` iterations in AsmL.

Notice first that the cores of normal forms are all of the form:

```
step until fixpoint
     par
          if (mode = 0) then R1
          ...
          if (mode = j) then Rj
          ...
          if (c = 0) ∧ (mode = i) then Ri
          if (c = 1) ∧ (Mi ≥ mode ≥ i) then
               mode := i
               c := 0
          ...
          if (mode = k) then Rk
     endpar
```

Moreover, by construction, if $i < j$ all *mode* values occurring in R_i are less than *mode* values occurring in R_j, and c can take only the values 0 or 1.

We now prove that Π and its normal form Π_n have the same semantics. With each $\mathcal{L}$–state $\mathcal{A}$ we associate a $\mathcal{L}_m$–state $\mathcal{A}_m$, which is identical to $\mathcal{A}$, except for $\mathit{mode} = 0$ (and $c = 0$ when c is needed). We prove that, for each AsmL program Π, whose normal form is Π_n,

$$[\![(\Pi, \mathcal{A})]\!]^e = [\![(\Pi_n, \mathcal{A}_m)]\!]_{|\mathcal{L}} \ . \tag{1}$$

$\mathcal{B}_{|\mathcal{L}}$ denotes the restriction of $\mathcal{L}_m$–state $\mathcal{B}$ to $\mathcal{L}$ (i.e. we forget the constants *mode* and c).

We again proceed by structural induction to prove that Π and its normal form Π_n have the same semantics, i.e. that (1) holds.

- If $\Pi = R$ is an ASM rule then $\tau_{\Pi_n}(\mathcal{A}_m)$ coincides with $\tau^e_{\Pi}(\mathcal{A})$ on $\mathcal{L}$ and is such that the value of *mode* is 1; in the ASM semantics of Π_n, the rule R can be executed only once, because it can be executed only if the value of *mode* is 0, and after being executed once, *mode* is set to 1; so $[\![(\Pi_n, \mathcal{A}_m)]\!]_{|\mathcal{L}} = \tau_{\Pi_n}(\mathcal{A}_m)_{|\mathcal{L}} = \tau_{\Pi}(\mathcal{A}) = [\![(\Pi, \mathcal{A})]\!]$ and (1) holds.

– If Π is a `step` rule:

```
par
        step R_1
        step R_2
endpar
```

the core of Π_n is:

$$R_1''$$
$$R_2^{+fin_1}$$

where R_1'' is the core R_1' of R_1, and $R_2^{+fin_1}$ is defined as follows: Let fin_1 be the greatest value used for *mode* in R_1'. Rule $R_2^{+fin_1}$ is the core R_2' of R_2 where fin_1 is added to each *constant* occurring on the right hand side of the "=" (or "$\geq$") sign of an expression rule beginning by *mode* (a boolean expression $mode = constant$, $mode \geq constant$ or an update rule $mode := constant$).

In order to prove that (1) holds, note that

- Assuming by the induction hypothesis that (1) holds for R_1 and R_2, we have that

$$[\![(R_1', \mathcal{A}_m)]\!]_{|\mathcal{L}} = \left([\![(R_1, \mathcal{A})]\!]^e\right) \tag{2}$$
$$[\![(R_2', \mathcal{B}_m)]\!]_{|\mathcal{L}} = \left([\![(R_2, \mathcal{B})]\!]^e\right) . \tag{3}$$

- By construction, in every "guard" of the form `if` $mode = constant$ `then` ...which occurs in R_1'' the value of *constant* is $\leq fin_1$, and in every "guard" of the form `if` $mode = constant$ `then` ... which occurs in $R_2^{+fin_1}$ we have $fin_1 \leq constant \leq fin_1 + fin_2$.
- For $\mathcal{B}$ an $\mathcal{L}$–state, and $k \in \mathbb{N}$, let $\mathcal{B}_m^k$ denote the $\mathcal{L}_m$–state, where all function symbols are interpreted as in $\mathcal{B}$, and where *mode* has value k. Then, if Π_n is an arbitrary in normal form, and Π_n^{+k} denotes the program where all right hand side constants in expressions involving *mode* are incremented by k, we have, for any $k, l \in \mathbb{N}$

$$\tau^l_{\Pi_n^{+k}}(\mathcal{B}_m^k)_{|\mathcal{L}} = \tau^l_{\Pi_n}(\mathcal{B}_m)_{|\mathcal{L}} \tag{4}$$
$$[\![(\Pi_n^{+k}, \mathcal{B}_m^k)]\!]_{|\mathcal{L}} = [\![(\Pi_n, \mathcal{B}_m)]\!]_{|\mathcal{L}} . \tag{5}$$

- Rules in R_1' are executed only when the value of *mode* is $< fin_1$, and when execution of R_1' is finished the value of *mode* is fin_1, hence

$$[\![(R_1', \mathcal{A}_m)]\!] = \left([\![(R_1, \mathcal{A})]\!]^e\right)_m^{fin_1} . \tag{6}$$

- Rules in $R_2^{+fin_1}$ are executed only when the value of *mode* verifies $fin_1 \leq mode < (fin_1 + fin_2)$, hence rules of R'_1 can no longer be executed; when execution of $R_2^{+fin_1}$ is finished then the value of *mode* is $fin_1 + fin_2$, and no rule of Π_n can be executed, hence the fixed point of $\tau_{\Pi_n}(\mathcal{A}_m)$ is reached. We can deduce by equation (5) that

$$[\![(R'_2, \mathcal{B}_m)]\!]_{|\mathcal{L}} = [\![(R_2^{+fin_1}, \mathcal{B}_m^{+fin_1})]\!]_{|\mathcal{L}} \ . \tag{7}$$

Finally, we have:

$$\begin{aligned}
[\![(\Pi_n, \mathcal{A}_m)]\!] &= \tau^*_{R_2^{+fin_1}}(\tau^*_{R'_1}(\mathcal{A}_m)) \\
&= \tau^*_{R_2^{+fin_1}}([\![(R'_1, \mathcal{A}_m)]\!]) \quad \text{by definition} \\
&= \tau^*_{R_2^{+fin_1}}(([\![(R_1, \mathcal{A})]\!]^e)_m^{fin_1}) \quad \text{by equation (6)} \\
&= [\![(R_2^{+fin_1}, ([\![(R_1, \mathcal{A})]\!]^e)_m^{fin_1})]\!] \quad \text{by definition .}
\end{aligned}$$

Restricting the last equation to $\mathcal{L}$, we have

$$\begin{aligned}
[\![(\Pi_n, \mathcal{A}_m)]\!]_{|\mathcal{L}} &= [\![(R_2^{+fin_1}, ([\![(R_1, \mathcal{A})]\!]^e)_m^{fin_1})]\!]_{|\mathcal{L}} \\
&= [\![(R'_2, [\![(R_1, \mathcal{A})]\!]^e_m)]\!]_{|\mathcal{L}} \quad \text{by equation (7)} \\
&= [\![(R_2, [\![(R_1, \mathcal{A})]\!]^e_m)]\!]^e \quad \text{by equation (3) .}
\end{aligned}$$

Hence (1) holds for Π.

– If Π is an iteration rule (`step until` φ R) and R' is the core of the normal form of R then the core Π' of the normal form Π_n of Π is:

```
if  (mode = 0) then
    if ¬φ then mode := 1
    if φ then mode := fin + 1
R^{+1}
```

It can be seen that there is no rule with guard " `if` *mode* $= fin + 1$ `then`...", and that always $\overline{\varphi}^{\mathcal{A}_m} = \overline{\varphi}^{\mathcal{A}}$. We prove (1).

- either eventually $\varphi^{\mathcal{B}}$ is true, for some $\mathcal{B} = \tau^k_{\Pi_n}(\mathcal{A}_m)$, with $k \in \mathbb{N}$. In this case $\overline{\varphi}^{\mathcal{B}_{|\mathcal{L}}}$ is also true and we have $\mathcal{B}_{|\mathcal{L}} = \tau^{k'}_{\Pi}(\mathcal{A})$, for some $k' \in \mathbb{N}$, $k' \leq k$; then $[\![(\Pi_n, \mathcal{A}_m)]\!] = \tau^k_{\Pi_n}(\mathcal{A}_m)$ is $\tau^{k'}_{\Pi}(\mathcal{A})$ together with value of *mode* equals to $fin + 1$, and (1) holds.
- or $\overline{\varphi}^{\mathcal{B}}$ is always false, then $[\![(\Pi_n, \mathcal{A}_m)]\!]$ is undefined (because *mode* takes infinitely often the values 0 and 1), and so is $[\![(\Pi, \mathcal{A})]\!]$ whence (1).

– the case of `step while` iterations is treated similarly.

– If Π is an iteration rule (`step until fixpoint` R), let R' be the core of the normal form of R and M the largest value of *mode* occurring in R', then the core Π' of the normal form Π_n of Π is:

$$
\begin{array}{l}
\texttt{if } (c{=}1) \wedge (M \geq mode \geq 0) \texttt{ then} \\
\quad mode := 0 \\
\quad c := 0 \\
\texttt{if } (c{=}0) \wedge (mode = 0) \texttt{ then } R^c
\end{array}
$$

where R^c is R' in which each update (other than updates on c and $mode$) of the form $f_i(t_1, \dots, t_n) := t_0^i$ has been replaced by

$$
\begin{array}{l}
\texttt{if } f_i(t_1, \dots, t_n) \neq t_0^i \texttt{ then} \\
\quad f_i(t_1, \dots, t_n) := t_0^i \\
\quad c := 1
\end{array}
$$

Let now $\mathcal{L}_m = \mathcal{L} \cup \{mode, c\}$; for $\mathcal{B}$ an $\mathcal{L}$ or $\mathcal{L}_m$ algebra, let us denote by $\mathcal{B}^{00}$ the $\mathcal{L}_m$ algebra where $mode = c = 0$. In particular, if $\mathcal{A}$ is an $\mathcal{L}$ algebra, $\mathcal{B}^{00} = \mathcal{A}_m$ denotes the associated initial $\mathcal{L}_m$ algebra.
By the induction hypothesis,

$$
[\![(R', \mathcal{A}_m)]\!]_{|\mathcal{L}} = [\![(R', \mathcal{A}_m^{00})]\!]_{|\mathcal{L}} = [\![(R, \mathcal{A})]\!]^e \,, \qquad \text{which implies}
$$

$$
\left([\![(R', \mathcal{A}_m^{00})]\!]\right)^{00} = \left([\![(R, \mathcal{A})]\!]^e\right)^{00} . \tag{8}
$$

Let $\mathcal{A}_0 = \mathcal{A}$, and for $i \geq 0$, $\mathcal{A}_{i+1} = [\![(R, \mathcal{A}_i)]\!]^e$. Let also $\mathcal{A}'_0 = \mathcal{A}^{00}$, and for $i \geq 0$, $\mathcal{A}'_{i+1} = [\![(R', \mathcal{A}'_i)]\!]^{00}$. It can be seen that equation (8) implies, for all i

$$
\mathcal{A}'_i = (\mathcal{A}_i)^{00} . \tag{9}
$$

Now, on the one hand, $[\![(R, \mathcal{A})]\!]^e$ is equal to the first $\mathcal{A}_i$ such that $\mathcal{A}_i = \mathcal{A}_{i+1}$; by equation (9), we also have that $\mathcal{A}'_i = \mathcal{A}'_{i+1}$. On the other hand, let us compute $[\![(\Pi_n, \mathcal{A}_m)]\!]$.
If all updates other than updates on c and $mode$ are trivial, we let

$$
\theta_{\Pi_n}(\mathcal{B}) = [\![(R', \mathcal{B})]\!] .
$$

Otherwise, (there are non trivial updates and the fixed point is not reached), we let

$$
\theta_{\Pi_n}(\mathcal{B}) = [\![(R', \mathcal{B})]\!]^{00} .
$$

Then

1. for all $\mathcal{B}$ there is a k such that $\theta_{\Pi_n}(\mathcal{B}) = \tau^k_{\Pi_n}(\mathcal{B})$;
2. the computation defined by the sequence $\theta^i_{\Pi_n}(\mathcal{B})$ terminates if and only if the computation defined by the sequence $\tau^j_{\Pi_n}(\mathcal{B})$ terminates, and if it is the case, the limits are equal.

But $\theta^{i+1}_{\Pi_n}(\mathcal{A}_m) = \mathcal{A}'_{i+1}$ if non trivial updates (i.e. other than updates on c and $mode$) are performed in the course of the computation of R' on $\mathcal{A}'_i$, otherwise $\theta^{i+1}_{\Pi_n}(\mathcal{A}_m) = [\![(R', \mathcal{A}'_i)]\!]$.

- In the latter case, $c = 0$ and $mode > 0$ in $[\![(R', \mathcal{A}'_i)]\!]$, so no rule of Π_n can be executed, the fixed point $[\![(\Pi_n, \mathcal{A}_m)]\!]$ is reached and is equal to $[\![(R', \mathcal{A}'_i)]\!]$, i.e.

$$[\![(\Pi_n, \mathcal{A}_m)]\!] = [\![(R', \mathcal{A}'_i)]\!] \tag{10}$$

  moreover, $\mathcal{A}'_{i+1} = \mathcal{A}'_i$ (only trivial updates are performed in the course of the computation of R'), and by equation (9) this implies that also $\mathcal{A}_{i+1} = \mathcal{A}_i$, hence the fixed point of Π is reached and

$$[\![(R, \mathcal{A}_i)]\!]^e = [\![(R, \mathcal{A}_{i+1})]\!]^e = [\![(\Pi, \mathcal{A})]\!]^e\ . \tag{11}$$

  Putting together equations (9), (10), and (11) and restricting to $\mathcal{L}$, we have: $[\![(\Pi, \mathcal{A})]\!]^e = [\![(\Pi_n, \mathcal{A}_m)]\!]_{|\mathcal{L}}$, i.e. equation (1) holds.
- In the former case, if for any $i \in \mathbb{N}$ non trivial updates are applied in the course of the computation of R' on $\mathcal{A}'_i$, then for any i, $\mathcal{A}'_{i+1} \neq \mathcal{A}'_i$ implies that also $\mathcal{A}_{i+1} \neq \mathcal{A}_i$ (the non trivial updates concern functions other than c and *mode*) and neither Π nor Π_n computation terminates.

– If Π is an alternative rule,

```
if φ  then R1
      else R2
endif
```

and R'_1, R'_2 are the respective cores of the normal forms of R_1, R_2, then the core of the normal form Π_n of Π is:

```
if     (mode = 0) then
       if φ then mode := 1 else mode := fin + 1
R1^{+1}
R2^{fin+1}
```

where R_i^{+k} is R'_i where k is added to each *constant* occurring on the right hand side of the "=" (or "≥") sign of an expression rule beginning by *mode*. For $\mathcal{A}$ an $\mathcal{L}$–state, and $k \in \mathbb{N}$, recall that $\mathcal{A}_m^k$ denote the $\mathcal{L}_m$–state, where all function symbols are interpreted as in $\mathcal{A}$, and where *mode* has value k. The semantics of Π_n is:

$$[\![(\Pi_n, \mathcal{A}_m^0)]\!] = \text{if } \overline{\varphi}^{\mathcal{A}} \text{ then } [\![(R_1^{+1}, \mathcal{A}_m^1)]\!] \text{ else } [\![(R_2^{fin+1}, \mathcal{A}_m^{fin+1})]\!]\ .$$

Because:
$[\![(R_1^{+1}, \mathcal{A}_m^1)]\!]_{|\mathcal{L}} = [\![(R_1, \mathcal{A})]\!]^e$ and $[\![(R_2^{fin+1}, \mathcal{A}_m^{fin+1})]\!]_{|\mathcal{L}} = [\![(R_2, \mathcal{A})]\!]^e$,
it can be deduced that: $[\![(\Pi_n, \mathcal{A}_m^0)]\!]_{|\mathcal{L}} = [\![(\Pi, \mathcal{A})]\!]^e$. □

Proposition 1. *The number of comparisons and updates during the execution of the normal form is at most three times the number of of comparisons and updates during the execution of the original program, assuming no* **step until fixpoint** *occurs in the original program.*

Proof. By structural induction on the structure of the original program.

- If $\Pi = R$ is an ASM rule, then $cost(\Pi_n) = 2 + cost(\Pi)$, but $cost(\Pi) \geq 1$, hence $cost(\Pi_n) \leq 3 \times cost(\Pi)$.
- If Π is a step rule,

```
par
        step R1
        step R2
endpar
```

  then $cost(\Pi_n) = cost(R_{1n}) + cost(R_{2n})$ and $cost(\Pi) = cost(R_1) + cost(R_2)$, hence the result.
- If $\Pi =$ `step until` φ R then, on the one hand, we have (letting $cost(R[i])$ be the cost of the execution of the body of the loop `step until` φ R during the i-th execution of that loop body):

$$cost(\Pi) = 1 + \Sigma_i \ (1 + cost(R[i])) \ ,$$

  because we have at least a comparison and a comparison every time we enter in the body of the loop. On the other hand, we have :

$$\begin{aligned} cost(\Pi_n) &= 2 + \Sigma_i \ (2 + cost(R_n[i])) \\ &\leq 2 + \Sigma_i \ (2 + 3 \times cost(R[i])) \text{by the induction hypothesis} \\ &\leq 3 \times [1 + \Sigma_i \ (1 + cost(R[i]))] \\ &\leq 3 \times cost(\Pi) \ . \end{aligned}$$

- If Π is an alternative rule,

```
if φ then R1
     else R2
endif
```

  then

$$\begin{aligned} cost(\Pi) &= \max\{cost(R_1), cost(R_2)\} + 1 \\ cost(\Pi_n) &= \max\{cost(R_1^{+1}{}_n), cost(R_2^{fin+1}{}_n)\} + 1 \\ &\leq 3 \times \max\{cost(R_1), cost(R_2)\} + 1 \leq 3 \times cost(\Pi) \ . \end{aligned}$$

□

The `step until fixpoint` has been excluded, even though we chose a way to emulate it in normal form, because the cost of checking that the fixpoint is reached is non null in AsmL, but a precise semantics of that cost in AsmL should first be chosen before any attempt to compare costs.

Finding Reductions Automatically

Michael Crouch*, Neil Immerman*, and J. Eliot B. Moss*

Computer Science Dept., University of Massachusetts, Amherst
{mcc,immerman,moss}@cs.umass.edu

Abstract. We describe our progress building the program ReductionFinder, which uses off-the-shelf SAT solvers together with the Cmodels system to automatically search for reductions between decision problems described in logic.

Keywords: descriptive complexity, first-order reduction, quantifier-free reduction, SAT solver.

1 Introduction

Perhaps the most useful item in the complexity theorist's toolkit is the reduction. Confronted with decision problems $A, B, C, \ldots$, she will typically compare them with well-known problems, e.g., REACH, CVP, SAT, QSAT, which are complete for the complexity classes NL, P, NP, PSPACE, respectively. If she finds, for example, that A is reducible to CVP ($A \leq$ CVP), and that SAT $\leq B$, $C \leq$ REACH, and REACH $\leq C$, then she can conclude that A is in P, B is NP hard, and C is NL complete.

When Cook proved that SAT is NP complete, he used polynomial-time Turing reductions [4]. Shortly later, when Karp showed that many important combinatorial problems were also NP complete, he used the simpler polynomial-time many-one reductions [14].

Since that time, many researchers have observed that natural problems remain complete for natural complexity classes under surprisingly weak reductions including logspace reductions [13], one-way logspace reductions [9], projections [22], first-order projections, and even the astoundingly weak quantifier-free projections [11].

It is known that artificial non-complete problems can be constructed [15]. However, it is a matter of common experience that most natural problems are complete for natural complexity classes. This phenomenon is receiving a great deal of attention recently via the dichotomy conjecture of Feder and Vardi that all constraint satisfaction problems are either NP complete, or in P [7,20,1].

* The authors were partially supported by the National Science Foundation under grants CCF-0830174 and CCF-0541018 (first two authors) and CCF-0953761 and CCF-0540862 (third author). Any opinions, findings, conclusions, or recommendations expressed in this material are those of the authors and do not necessarily reflect the views of the NSF.

A. Blass, N. Dershowitz, and W. Reisig (Eds.): Gurevich Festschrift, LNCS 6300, pp. 181–200, 2010.

Since natural problems tend to be complete for important complexity classes via very simple reductions, we ask, *"Might we be able to automatically find reductions between given problems?"*

Of course this problem is undecidable in general. However, we have made progress building a program called ReductionFinder that automatically does just that. Given two decision problems A and B, ReductionFinder attempts to find the simplest possible reduction from A to B. Using off-the-shelf SAT solvers together with the Cmodels system[8], ReductionFinder finds many simple reductions between a wide class of problems, including several "clever" reductions that the authors had not realized existed.

The reader might wonder why we would want to find reductions automatically. In fact, we feel that an excellent automatic reduction finder would be an invaluable tool, addressing the following long-term problems:

1. There are many questions about the relations between complexity classes that we cannot answer. For example, we don't know whether P = NP, nor even whether NL = NP, whether P = PSPACE, etc. These questions are equivalent to the existence of quantifier-free projections between complete problems for the relevant classes [11]. For example, P = NP iff SAT $\leq_{\mathrm{qfp}}$ CVP. Similarly, NL = NP iff SAT $\leq_{\mathrm{qfp}}$ REACH and P = PSPACE iff QSAT $\leq_{\mathrm{qfp}}$ CVP. Having an automatic tool to find such reductions or determine that no small reductions exist may improve our understanding about these fundamental issues.
2. Another ambitious goal, well formulated by Jack Schwartz in the early 1980s, is to precisely describe a computational task in a high-level language such as SETL [21] and build a smart compiler that can automatically synthesize efficient code that correctly performs the task. A major part of this goal is to automatically recognize the complexity of problems. Given a problem, A, if we can automatically generate a reduction from A to CVP, then we can also synthesize code for A. On the other hand if we can automatically generate a reduction from SAT to A, then we know that A is NP hard, so it presumably has no perfect, efficient implementation and we should instead search for appropriate approximation algorithms.
3. Being able to automatically generate reductions will provide a valuable tool for understanding the relative complexity of problems. If we restrict our attention to linear reductions, then these give us true lower and upper bounds on the complexity of the problem in question compared to a known problem, K: if we find a linear reduction from A to K, then we can automatically generate code for A that runs in the same time as that for K, up to a constant multiple. Similarly if we find a linear reduction from K to A, then we know that there is no algorithm for A that runs significantly faster than the best algorithm for K.

It is an honor for us to have our paper appear in this Festschrift for Yuri Gurvich. Yuri has made many outstanding contributions to logic and computer science. We hope he is amused by what we feel is a surprising use of SAT solvers for automatically deriving complexity-theoretic relations between problems.

This paper is organized as follows: We start in Section §2 with background in descriptive complexity sufficient for the reader to understand all she needs to know about reductions and the logical descriptions of decision problems. In section §3 we explain our strategy for finding reductions using SAT solvers. In section §4 we sketch the implementation details. In section §5 we provide the main results of our experiments: the reductions found and the timing. We conclude in section §6 with directions for moving this research forward.

2 Reductions in Descriptive Complexity

In this section we present background and notation from descriptive complexity theory concerning the representation of decision problems and reductions between them. The reader interested in more detail is encouraged to consult the following texts: [10,5,16], where complete references and proofs of all the facts mentioned in this section may be found.

2.1 Vocabularies and Structures

In descriptive complexity, part of finite model theory, the main objects of interest are finite logical structures. A *vocabulary*

$$\tau = \langle R_1^{a_1}, \ldots, R_r^{a_r}; c_1, \ldots, c_s; f_1^{r_1}, \ldots, f_t^{r_t} \rangle$$

is a tuple of relation symbols, constant symbols, and function symbols. R_i is a relation symbol of arity a_i and f_j is a function symbol of arity $r_j > 0$. A constant symbol is just a function symbol of arity 0. For any vocabulary τ we let $\mathcal{L}(\tau)$ be the set of all grammatical first-order formulas built up from the symbols of τ using boolean connectives, $\neg, \vee, \wedge, \rightarrow$ and quantifiers, $\forall, \exists$.

A *structure* of vocabulary τ is a tuple,

$$\mathcal{A} \quad = \quad \langle |\mathcal{A}|; R_1^{\mathcal{A}}, \ldots, R_r^{\mathcal{A}}; c_1^{\mathcal{A}}, \ldots, c_s^{\mathcal{A}}; f_1^{\mathcal{A}}, \ldots, f_t^{\mathcal{A}} \rangle$$

whose universe is the nonempty set $|\mathcal{A}|$. For each relation symbol R_i of arity a_i in τ, $\mathcal{A}$ has a relation $R_i^{\mathcal{A}}$ of arity a_i defined on $|\mathcal{A}|$, i.e. $R_i^{\mathcal{A}} \subseteq |\mathcal{A}|^{a_i}$. For each function symbol $f_i \in \tau$, $f_i^{\mathcal{A}}$ is a total function from $|\mathcal{A}|^{r_i}$ to $|\mathcal{A}|$.

Let STRUC$[\tau]$ be the set of finite structures of vocabulary τ. For example, $\tau_g = \langle E^2; ; \rangle$ is the vocabulary of (directed) graphs and thus STRUC$[\tau_g]$ is the set of finite graphs.

2.2 Ordering

It is often convenient to assume that structures are ordered. An *ordered structure* $\mathcal{A}$ has universe $|\mathcal{A}| = \{0, 1, \ldots, n-1\}$ and *numeric* relation and constant symbols: $\leq, \mathrm{Suc}, \min, \max$ referring to the standard ordering, successor relation, minimum, and maximum elements, respectively (we take $\mathrm{Suc}(\max) = \min$). ReductionFinder may be asked to find a reduction on ordered or unordered structures. In the former case it may use the above numeric symbols. Unless otherwise noted, we from now on assume that all structures are ordered.

2.3 Complexity Classes and Their Descriptive Characterizations

We hope that the reader is familiar with the definitions of most of the following complexity classes:

$$\mathrm{AC}^0 \subset \mathrm{NC}^1 \subseteq \mathrm{L} \subseteq \mathrm{NL} \subseteq \mathrm{P} \subseteq \mathrm{NP} \tag{1}$$

where L = DSPACE[$\log n$], NL = NSPACE[$\log n$], P is polynomial time, and NP is nondeterministic polynomial time. AC^0 is the set of problems accepted by uniform families of polynomial-size, constant-depth circuits whose gates include unary "not" gates, together with unbounded-fan-in "and" and "or" gates. NC^1 is the set of problems accepted by uniform families of polynomial-size, $O(\log n)$-depth circuits whose gates include unary "not" gates, together with binary "and" and "or" gates.

Each complexity class from Equation 1 has a natural descriptive characterization. Complexity classes are sets of decision problems. Each formula in a logic expresses a certain decision problem. As is standard, we write $C = \mathcal{L}$ to mean that the complexity class C is equal to the set of decision problems expressed by the logical language $\mathcal{L}$. The following descriptive characterizations of complexity classes are well known:

Fact 1. FO = AC^0; NC^1 = FO(*Regular*); L = FO(DTC); NL = FO(TC); P = FO(IND); *and* NP = SO∃.

We now explain some of the details of Fact 1. For more information about this fact the reader should consult one of the texts [10,5,16].

2.4 Transitive Closure Operators

Given a binary relation on k-tuples, $\varphi(x_1, \ldots, x_k, y_1, \ldots, y_k)$, we let $\mathrm{TC}_{\overline{x},\overline{y}}(\varphi)$ express its transitive closure. If the free variables are understood then we may abbreviate this as $\mathrm{TC}(\varphi)$. Similarly, we let $\mathrm{RTC}(\varphi)$, $\mathrm{STC}(\varphi)$, and $\mathrm{RSTC}(\varphi)$ denote the reflexive transitive closure, symmetric transitive closure, and symmetric and reflexive transitive closure of φ, respectively.

We next define a deterministic version of transitive closure DTC. Given a first order relation, $\varphi(\overline{x}, \overline{y})$, define its deterministic reduct,

$$\varphi_d(\overline{x}, \overline{y}) \stackrel{\text{def}}{=} \varphi(\overline{x}, \overline{y}) \wedge (\forall \overline{z})(\neg\varphi(\overline{x}, \overline{z}) \vee (\overline{y} = \overline{z}))$$

That is, $\varphi_d(\overline{x}, \overline{y})$ is true just if $\overline{y}$ is the unique child of $\overline{x}$. Now define $\mathrm{DTC}(\varphi) \stackrel{\text{def}}{=} \mathrm{TC}(\varphi_d)$ and $\mathrm{RDTC}(\varphi) \stackrel{\text{def}}{=} \mathrm{RTC}(\varphi_d)$.

Let $\tau_{gst} = \langle E^2; s, t; \rangle$ be the vocabulary of graphs with two specified points. The problem $\mathrm{REACH} = \{G \in \mathrm{STRUC}[\tau_{gst}] \mid G \models \mathrm{RTC}(E)(s,t)\}$ consists of all finite graphs that have a path from s to t. Similarly, $\mathrm{REACH}_d = \{G \in \mathrm{STRUC}[\tau_{gst}] \mid G \models \mathrm{RDTC}(E)(s,t)\}$ is the subset of REACH such that there is a unique path from s to t and all vertices along this path have out-degree one. $\mathrm{REACH}_u = \{G \in \mathrm{STRUC}[\tau_{gst}] \mid G \models \mathrm{STC}(E)(s,t)\}$ is the set of graphs having an undirected path from s to t.

It is well known that REACH is complete for NL, and REACH_d and REACH_u are complete for L [10,19]. A simpler way to express deterministic transitive closure is to syntactically require that the out-degree of our graph is at most one by using a function symbol: denote the child of v as $f(v)$, with $f(v) = v$ if v has no outgoing edges. In this notation, a problem equivalent to REACH_d, and thus complete for L, is $\text{REACH}_f = \{G \in \text{STRUC}[\tau_{fst}] \mid G \models \text{RTC}(f)(s,t)\}$.

If $\mathcal{O}$ is an operator such as TC, let FO($\mathcal{O}$) be the closure of first-order logic using $\mathcal{O}$. Then L = FO(DTC) = FO(RDTC) = FO(STC) = FO(RSTC) and NL = FO(TC) = FO(RTC).

2.5 Inductive Definitions

It is useful to define new relations by induction. For example, we can express the transitive closure of the relation E inductively, and thus the property REACH, via the following Datalog program:

$$\begin{aligned} E^*(x,x) &\leftarrow \\ E^*(x,y) &\leftarrow E(x,y) \\ E^*(x,y) &\leftarrow E^*(x,z), E^*(z,y) \\ \text{REACH} &\leftarrow E^*(s,t) \end{aligned} \tag{2}$$

Define FO(IND) to be the closure of first-order logic using such positive inductive definitions. The Immerman-Vardi Theorem states that P = FO(IND). In this paper we will use stratified Datalog programs such as Equation 2 to express problems and then use ReductionFinder to automatically find reductions between them. Thus ReductionFinder can handle any problem in P or below. In the future we hope to handle problems in NP, but this will require us to go beyond SAT solvers to QBF solvers.

2.6 Reductions

Given a pair of problems $S \subseteq \text{STRUC}[\sigma]$ and $T \subseteq \text{STRUC}[\tau]$, a *many-one reduction* from S to T is an easy-to-compute function $f : \text{STRUC}[\sigma] \to \text{STRUC}[\tau]$ such that for all $\mathcal{A} \in \text{STRUC}[\sigma]$,

$$\mathcal{A} \in S \quad \Leftrightarrow \quad f(\mathcal{A}) \in T\,.$$

In descriptive complexity we use *first-order reductions* which are many-one reductions in which the function f is defined by a sequence of first-order formulas from $\mathcal{L}(\sigma)$, one for each symbol of τ. For example, the following is a reduction from REACH_f to REACH_u that ReductionFinder automatically found. Here $\sigma = \langle ; s, t; f^1 \rangle$ and $\tau = \langle E^2; s, t; \rangle$. The reduction, R_{fu}, is as follows:

$$\begin{aligned} E'(x,y) &\equiv y \neq t \ \wedge \ f(y) = x \\ s' &\equiv s \\ t' &\equiv t \end{aligned} \tag{3}$$

Note that the three formulas in R_{fu}'s definition (Equation 3) have no quantifiers, so R_{fu} is not only a first-order reduction, it is a *quantifier-free reduction* and we write $\text{REACH}_f \leq_{\text{qf}} \text{REACH}_u$.

More explicitly, for each structure $\mathcal{A} \in \text{STRUC}[\sigma]$, $\mathcal{B} = R_{fu}(\mathcal{A}) = \langle |\mathcal{A}|, E^{\mathcal{B}}, s^{\mathcal{B}}, t^{\mathcal{B}} \rangle$ is a structure in $\text{STRUC}[\tau]$ with universe the same as $\mathcal{A}$, and symbols given as follows:

$$\begin{aligned} E^{\mathcal{B}} &= \{\langle a, b\rangle \mid (\mathcal{A}, a/x, b/y) \models y \neq t \ \wedge \ f(y) = x\} \\ s^{\mathcal{B}} &= s^{\mathcal{A}} \\ t^{\mathcal{B}} &= t^{\mathcal{A}} \end{aligned}$$

In this paper we restrict ourselves to quantifier-free reductions. In general, a first-order reduction R has an arity which measures the blow-up of the size of the reduction. In [10] a first-order reduction of *arity* k maps a structure with universe $|\mathcal{A}|$ to a structure of universe $\{\langle a_1, \ldots a_k\rangle \mid (\mathcal{A}, a_1/x_1, \ldots, a_k/x_k) \models \varphi_0\}$, i.e., a first-order definable subset of $|\mathcal{A}|^k$. However, increasing the arity of a reduction beyond two is rather excessive – arity two already squares the size of the instance. In this paper, in order to keep our reductions as small and simple as possible, we use a triple of natural numbers, $\langle k, k_1, k_2\rangle$, to describe the universe of the image structure, namely

$$|R(\mathcal{A})| = |\mathcal{A}|^k \times \{1, \ldots, k_1\} \cup \{1, \ldots k_2\} \,. \tag{4}$$

That is in addition to raising the universe to the power k, we also multiply it by the constant k_1 and then we may add k_2 explicit constants to the universe. In this notation the above reduction R_{fu} has arity $\langle 1, 1, 0\rangle$. It will become apparent in our many examples in the sequel how these extra parameters keep the reductions simple and small.

3 Strategy

We are given a pair of problems $S \subseteq \text{STRUC}[\sigma]$ and $T \subseteq \text{STRUC}[\tau]$, both expressed in Datalog. We want to know if there is a quantifier-free reduction from S to T.

It is not hard to see that this problem is undecidable, and in fact complete for the second level of the arithmetic hierarchy. It asks whether there exists some reduction that is correct for all inputs from $\text{STRUC}[\sigma]$, with no bounds on the size of the reduction nor the input.

We first make the problem more tractable by bounding the complexity of the reduction: We choose a triple $a = \langle k, k_1, k_2\rangle$ describing the arity of the reduction and a tuple of parameters p bounding the size and complexity of the quantifier-free formulas expressing the reduction (e.g. how many clauses, the maximum size of each clause, etc.). This reduces the complexity of the problem to co-r.e. complete: it is still undecidable.

To make the problem decidable, we choose a bound, n, and ask whether there exists a reduction of arity a and parameters p that is correct for all structures

$\mathcal{A} \in \text{STRUC}_{\leq n}[\tau]$, i.e, whose universes have cardinality at most n. Given such a reduction we can hope to prove by machine or hand that it works on structures of all sizes. On the other hand, being told that no such small reduction exists, we learn that in a precise sense there is no "simple" reduction from S to T.

Now our problem is complete for Σ_2^p – the second level of the polynomial-time hierarchy. Let $\mathbf{R}_{a,p}$ be the set of quantifier-free reductions of arity at most a and with parameter values at most p. The following formula asks whether there exists a quantifier-free reduction of arity a and parameters p that correctly reduces S to T on all structures of size at most n:

$$(\exists R \in \mathbf{R}_{a,p})(\forall \mathcal{A} \in \text{STRUC}_{\leq n}[\sigma])(\mathcal{A} \in S \leftrightarrow R(\mathcal{A}) \in T) \tag{5}$$

3.1 Solving a Σ_2^p Problem via Repeated Calls to a SAT Solver

We solve the problem expressed in Equation 5 by starting with a random structure $G_0 \in \text{STRUC}_{\leq n}[\sigma]$ and asking a SAT solver to find a reduction $R \in \mathbf{R}_{a,p}$ that works correctly on G_0, i.e., $G_0 \in S \leftrightarrow R(G_0) \in T$. If there is no solution, then our original problem is unsolvable.

Otherwise, we ask a new question to the SAT solver: is there some other structure, $G_1 \in \text{STRUC}_{\leq n}[\sigma]$ on which R fails, i.e, $G_1 \in S \leftrightarrow R(G_1) \notin T$. If not, then we know that R is a candidate reduction that is correct for all structures of size at most n.

However, if the SAT solver produces an example G_1 where R fails, we go back to the beginning, but now searching for a reduction that is correct on our full set of candidate structures, $\mathcal{G} = \{G_0, G_1\}$.

In summary, our algorithm proceeds as follows, with $\mathcal{G}$ initialized to $\{G_0\}$:

1. Using a SAT solver, search for a reduction correct on $\mathcal{G}$:

$$\textbf{find } R \in \mathbf{R}_{a,p} \text{ s.t. } \bigwedge_{G \in \mathcal{G}} G \in S \leftrightarrow R(G) \in T \tag{6}$$

   If no such R exists: **return**("no such reduction")

2. Using a SAT solver, search for some structure G on which R fails:

$$\textbf{find } G \in \text{STRUC}_{\leq n}[\sigma] \text{ s.t. } G \in S \leftrightarrow R(G) \notin T \tag{7}$$

   If no such G exists: **return**(R)
   Else: $\mathcal{G} = \mathcal{G} \cup \{G\}$; go to 1

Figure 1 shows a schematic view of this algorithm.

This procedure is correct because each new structure G eliminates at least one potential reduction. In our experience, the procedure works within a tractable number of structures; "smaller" searches have often completed after 5-15 sample structures, while the largest spaces searched by the program have required 30-50 iterations.

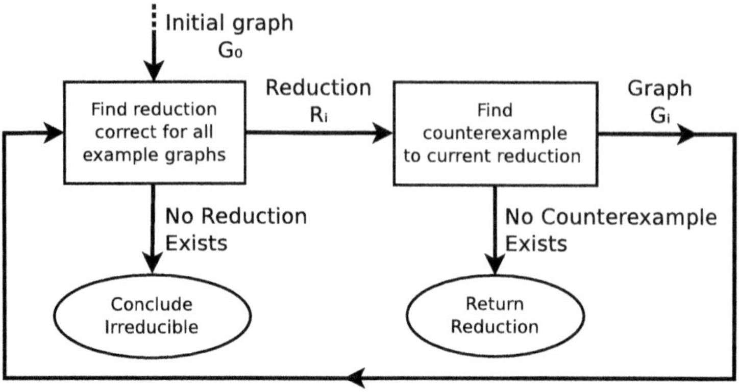

Fig. 1. A schematic view of the above algorithm

We begin searching for reductions at a very small size ($n = 3$); for search spaces without a correct reduction, even this small size is often enough to detect irreducibility. When a reduction is found at a particular size n, we examine larger structures for counterexamples; currently we look at structures of size at most $n + 2$. If a counterexample is found, we add it to $\mathcal{G}$, increment n and return to step 1.

Search time increases very rapidly as n increases. Of the 10,422 successful reductions found, 9,291 of them were found at size 3, 1076 at size 4, 38 at size 5, and 17 at sizes 6-8. See Section §5 for details of results. See Section §6 for more about the current limits of size and running time and our ideas concerning how to improve these.

4 Implementation

Figure 1 shows a schematic view of ReductionFinder's algorithm. The program is written in Scala, an object-oriented functional programming language implemented in the Java Virtual Machine[1]. ReductionFinder maintains a database of problems via a directed graph, G, whose vertices are problems. An edge (a, b) indicates that a reduction has been found from problem a to problem b, and is labelled by the parameters of a minimal such reduction that has been found so far.

When a new problem, c, is entered, ReductionFinder systematically searches for reductions to resolve the relationships between c and the problems already categorized in G.

Given a pair of problems, c, d, specified in stratified Datalog, and a search space $\mathbf{R}_{a,p}$ specifying the arity a and parameters p, ReductionFinder calls the Cmodels 3.79 answer-set system[2] to answer individual queries of the form of

[1] `http://www.scala-lang.org`

[2] `http://www.cs.utexas.edu/users/tag/cmodels.html`

Equations (6), (7). Cmodels in turn makes calls to SAT solvers. The SAT solvers we currently use are MiniSAT and zChaff [6,18].

4.1 Problem Input

Queries in ReductionFinder are input as small stratified-Datalog programs; a query on vocabulary τ has the symbols of τ available as extrinsic relations. The query is responsible for defining a single-bit intrinsic relation `satisfied`, representing the truth of the query. Input queries may use lparse rules without choice rules or disjunctive rules. When the input vocabulary contains function or constant symbols, these are translated by ReductionFinder into purely relational statements.

Equation (8) gives the ReductionFinder input for the directed-graph reachability query $\text{REACH} \subseteq \text{STRUCT}[\langle E^2; s, t\rangle]$, corresponding to the inductive definition (2). We define an intrinsic relation `reaches` to compute the transitive closure of the edge relation `E`.

```
reaches(X, X).
reaches(X, Y) :- E(X, Y).
reaches(X, Y) :- reaches(X, Z), reaches(Z, Y).
satisfied :- reaches(s, t).
```

(8)

4.2 Search Spaces

ReductionFinder restricts itself to searching for quantifier-free reductions, i.e. reductions defined by a set of quantifier-free formulas. The complexity of these quantifier-free formulas is restricted by several search parameters. The three arity numbers $\langle k, k_1, k_2\rangle$ of Section 2.6 each limit the search. The set of numeric predicates available (Section 2.2) is also a configurable parameter. The number of levels of nested function application available is a parameter.

Finally, the length of each quantifier-free formula is a parameter. Relations are defined by formulas represented in DNF; the number of disjuncts is a parameter, as is the number of conjuncts in each clause. Functions are defined as an if/else-if/else expression; the conditional of each statement is a conjunction of atomic formulas, and the resultant is a closed term. Again, the number of clauses is a parameter, as is the number of conjuncts in each clause.

The expressivity of the search space increases monotonically with most of our search parameters, inducing a natural partial ordering on search spaces. The search server respects this partial ordering, and avoids performing a search when any more-expressive space has previously been searched. The server is not restricted to increasing parameters one-at-a-time; since there are many search parameters, performing a single "large" search may be more efficient than performing many small searches. When a successful reduction is found, the server can automatically search smaller spaces to determine the smallest space containing a reduction.

4.3 The Searching Process

Once a search space and a pair of problems are fixed, ReductionFinder performs the iterative sequence of search stages described in section 3.1. Within each stage, ReductionFinder outputs a single lparse/cmodels program expressing Equations (6) or (7), and calls the Cmodels tool. The **find** statements in these equations are quantified explicitly using lparse's choice rules. The majority of the program is devoted to evaluation rules defining the structure $R(G)$ in terms of the sets of boolean variables R and G.

Figure 2 gives lparse code for a single counterexample-finding step (equation (7)). This code attempts to find a counterexample to a previously-generated reduction candidate. The specific code listed is examining reductions from REACH (Section 2.4) to its negation. The reduction candidate was $E'(x,y) \equiv (E(y,x) \wedge x = s) \vee E(x,x)$, $s' \equiv t$, $t' \equiv \text{Suc}(\text{min})$ (lines 7-9).

The counterexample is found using lparse's choice rules as existential quantifiers, directly guessing the relation `in_E` and the two constant symbols `in_s` and `in_t` (lines 12-13). Since lparse does not contain function symbols, these constants are implemented as degree-1 relations which are true at exactly one point. We specify the constraint that we cannot have `in_satisfied` == `out_satisfied` (line 16); these boolean variables will be defined later in the program, and this constraint will ensure that our graph is a counterexample to the reduction candidate.

Defining `in_satisfied` and `out_satisfied` in terms of the input and output predicates (respectively) is easy. We have already required the user to input lparse code for the input and output queries. We do some minimal processing on this code, disambiguating names and turning function symbols into relations. The user's input for directed-graph reachability, listed in Equation (8), is translated into the input query block of lines 19-22. Similarly, the output query is translated into lines 25-28.

The remainder of the lparse code exists to define the output predicates (in this case `out_E`, `out_s`, `out_t`) in terms of the input predicates and the reduction. In building the output reduction `out_E(X, Y)`, we first build up a truth table for each of the atomic formulas used; for example, line 31 states that term `e_y_x` is true at point `(X, Y)` exactly if $E(Y, X)$ in the input structure. Each position in the DNF definition is true at `(X, Y)` exactly if the atomic formula chosen for that position is true (lines 36-37). The output relation `out_E(X, Y)` is then defined via the terms in the DNF (lines 38-39). The code in lines 30-39 thus defines the output relation `out_E(X, Y)` in terms of the input relations `in_E`, `in_s`, `in_t` and the reduction candidate `reduct_E`.

Lines 41-47 similarly define the output constants `out_s` and `out_t`. Since lparse does not provide function symbols, we define these constants as unary relations `out_s(X)`, making sure that these relations are true at exactly one point. We are thus able to define the output constants in terms of the input symbols `in_s`, `in_t` and the the reduction candidate's definitions of s', t' (`reduct_s`, `reduct_t`).

The code for finding a reduction candidate (equation (6)) is very similar to the counterexample-finding code in Figure 2. We import the list $\mathcal{G}$ of counterexample

```
node(n1; n2; n3; n4).
atomic(e_x_x; e_x_y; ...; x_eq_t; y_eq_t).
closedterm(fn_s; fn_t; fn_min; fn_succ_min; fn_max).
position(pos_0_0; pos_0_1; pos_1_0).

%%% Import reduction candidate from previous stage.
reduct_E(pos_0_0, e_y_x). reduct_E(pos_0_1, x_eq_s).
reduct_E(pos_1_0, e_x_x).
reduct_s(fn_t).         reduct_t(fn_succ_min).

%%% Guess input relations E, s, t.
{ in_E(X, Y) }.
1 { in_s(X) } 1. 1 { in_t(X) } 1. % Choose exactly one s, t.

%%% A constraint on the entire program:
:- out_satisfied == in_satisfied.

%%% Translated version of input query.
in_Reaches(X, X).
in_Reaches(X, Y) :- in_E(X, Y).
in_Reaches(X, Y) :- in_Reaches(X, Z), in_Reaches(Z, Y).
in_satisfied :- in_Reaches(X, Y), in_s(X), in_t(Y).

%%% Translated version of output query.
out_Reaches(X, X).
out_Reaches(X, Y) :- out_E(X, Y).
out_Reaches(X, Y) :- out_Reaches(X, Z), out_Reaches(Z, Y).
out_satisfied :- not out_Reaches(X, Y), out_s(X), out_t(Y).

%%% Define a truth table for each atomic relation in the reduction.
true(e_y_x, X, Y) :- in_E(Y, X).
true(x_eq_s, X, Y) :- in_s(X).
true(e_x_x, X, X) :- in_E(X, X).

%%% Use these truth tables to evaluate output relations.
true(P, X, Y) :- reduct_E(P, A), true(A, X, Y),
                 position(P), atomic(A).
out_E(X, Y) :- true(pos_0_0, X, Y), true(pos_0_1, X, Y).
out_E(X, Y) :- true(pos_1_0, X, Y).

%%% Similarly, define the evaluation of each closed term.
eval_term(fn_s, X) :- in_s(X).
eval_term(fn_succ_min, n2).

%%% Define the output relations.
out_s(X) :- reduct_s(F), eval_term(F, X), closedterm(F).
out_t(X) :- reduct_t(F), eval_term(F, X), closedterm(F).
```

Fig. 2. Lparse code for a single search stage. This code implements equation (7), searching for a 4-node counterexample for a candidate reduction from REACH (Section 2.4) to its negation. Variables X, Y, Z range over nodes.

graphs, and must guess a reduction. The input query, output vocabulary, and output query are evaluated for each graph. Truth tables must be built for each relation which might appear in the reduction, and for each graph.

4.4 Timing

ReductionFinder uses the Cmodels logic programming system to solve its search problems. The Cmodels system solves answer-set programs, such as those in the lparse language, by reducing them to repeated SAT solver calls. Direct translations from answer-set programming (ASP) to SAT exist[2,12], but introduce new variables; Lifschitz and Razborov have shown that, assuming the widely-believed conjecture $P \not\subseteq NC^1/\text{poly}$, any translation from ASP must either introduce new variables or produce a program of worst-case exponential length [17].

The Cmodels system first translates the lparse program to its Clark completion [3], interpreting each rule `a :- b` as merely logical equivalence (`a` $\Leftrightarrow$ `b`). Models of this completion may fail to be answer sets if they contain *loops*, sets of variables which are true only because they assume each other. If the model found contains a loop, Cmodels adds a *loop clause* preventing this loop and continues searching, keeping the SAT solver's learned-clause database intact. A model which contains no loops is an answer set, and all answer sets can be found in this way.

The primary difficulty in finding large reductions with ReductionFinder has been computation time. The time spent finding reductions dominates over the

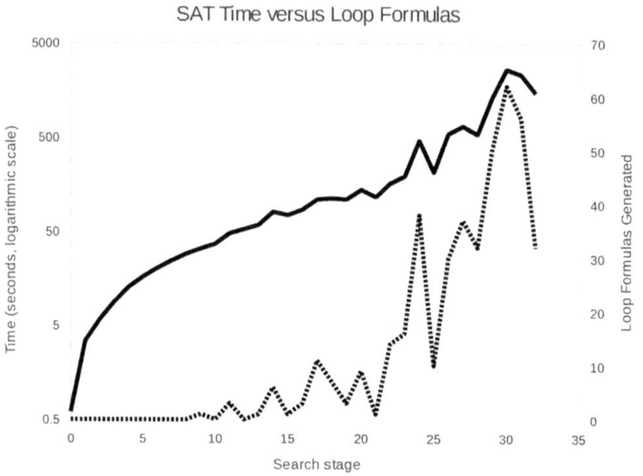

Fig. 3. Timing data for a run reducing $\neg\mathbf{RTC}[f](s,t) \leq \mathbf{RTC}[f](s,t)$ at arity 2, size 4. The solid line shows time to find each reduction candidate in seconds, on a logarithmic scale. The dotted line shows the number of loop formulas generated by Cmodels, and thus the number of SAT solver calls for each reduction candidate. This run was successful in finding a reduction.

time spent finding counterexamples; reductions must be true on each of the example graphs, and the number of lparse clauses and variables thus scales linearly with the number of example graphs. The amount of time required by Cmodels seems highly correlated with the number of loop formulas which must be generated; Figure 3 shows the time for each reduction-finding stage during a several-hour arity 2 search, versus the number of loop formulas generated in the stage. The final reduction-finding step generated an lparse program with 399,900 clauses, using 337,605 atoms.

5 Results

5.1 Size and Timing Data

We have run ReductionFinder for approximately 5 months on an 8-core 2.3 GHz Intel Xeon server with 16 GB of RAM. As of this writing, ReductionFinder has performed 331,036 searches on a database of 87 problems. Of the 7482 pairs of distinct problems, we explicitly found reductions between 2698; an additional 803 reductions could be concluded transitively. 23 pairs were manually marked as irreducible, comprising provable theorems about first-order logic plus statements that L $\lneq$ (co-)NL $\lneq$ P. From these 23, an additional 3043 pairs were transitively concluded to be irreducible. 915 pairs remained unfinished.

For many of the pairs which we reduced successfully, we found multiple successful reductions. Sometimes this occurred when we first found the reduction in a large search space, then tried smaller spaces to determine the minimal spaces containing a reduction. More interestingly, some pairs contained multiple successful reductions in distinct minimal search spaces, demonstrating trade-offs between different measures of the reduction's complexity. Some of these trade-offs were uninteresting: a reduction which simply needs "some distinguished constant" could use min, max, or c_1. Others, however, began to show non-trivial trade-offs between the formula length required and the numerics or arity available. See Equations (9), (10) for an example. Of the 12,149 correct reductions found between the 2698 explicitly-reduced pairs of problems, 5091 were in some minimal search space.

5.2 A Map of Complexity Theory

Figure 4 shows classes of queries within the ReductionFinder database. Each class contains one or more query which ReductionFinder has shown equivalent via quantifier-free reductions. An edge from class I to class J indicates that ReductionFinder has reduced $I \leq_{\mathrm{qfp}} J$. Numbers on the graph indicate the number of queries the class contains; the contents of these classes are listed in Figure 5.

ReductionFinder has placed all of the computationally-simple problems into their correct complexity classes. The trivially-true query and trivially-false query were reduced to all other queries. The class $R(s)$ contains twelve queries which lack the power of even one first-order quantifier. The classes $\exists x.R(x)$ and $\forall x.R(x)$ contain many variations of first-order quantifiers; for example, $\exists x.R(x)$

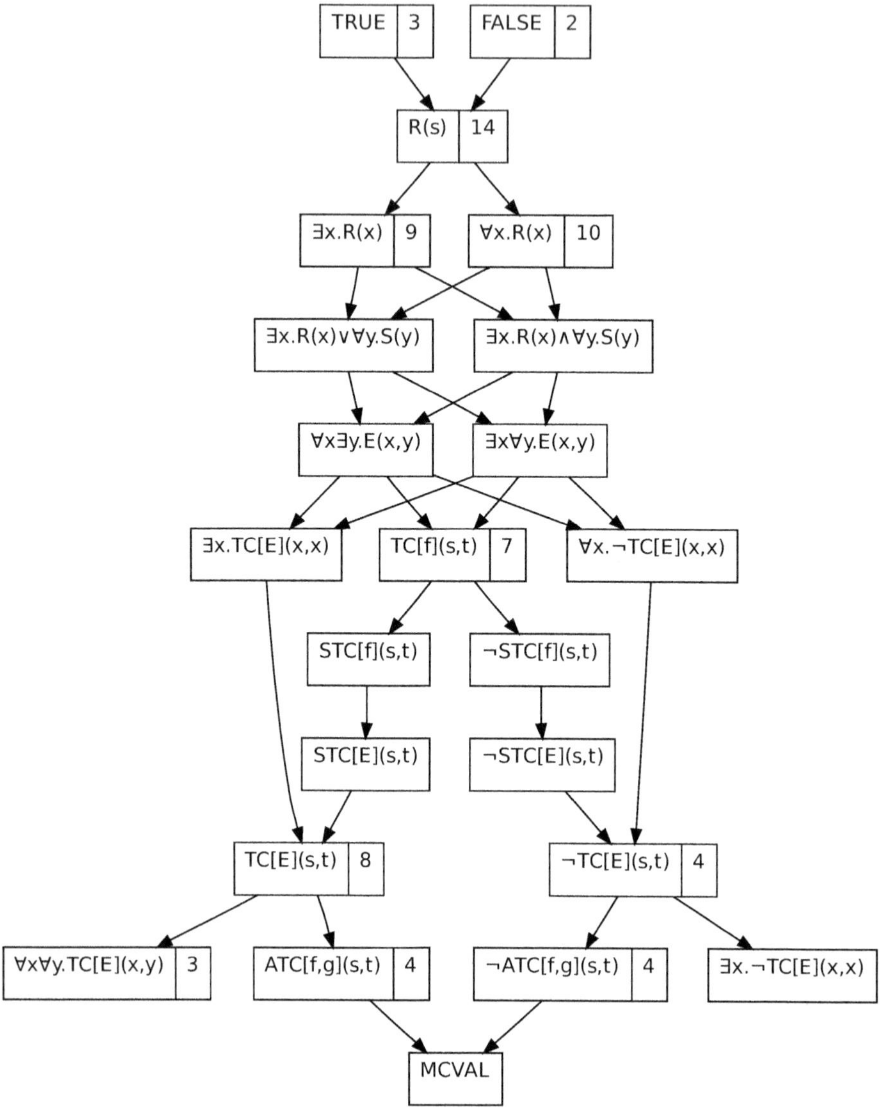

Fig. 4. A map of reductions in the query database. Nodes without numbers represent a single query. A node with number n represents n queries of the same complexity. Some queries are elided for clarity.

FALSE	$R(s) \wedge \neg R(s)$		
TRUE	$R(s) \vee \neg R(s)$	$\exists x.\mathbf{TC}[f](x,x)$	
$R(s)$	$\neg R(s)$	$R(f(s))$	
$E(s,t)$	$E(s,s)$	$E(s,t) \vee E(t,s)$	
$s = t$	$s \neq t$	$f(s) = t$	$f(s) \neq t$
$f(s) = s$	$f(s) = g(s)$	$f(s) = t \wedge f(t) = s$	$f(s) = t \vee f(t) = s$
$\exists x.R(x)$	$\exists x.R(x) \wedge sS(x)$	$\exists x.R(x) \vee S(x)$	
$\exists xy.E(x,y)$	$\exists xy.\neg E(x,y)$	$\exists xy.E(x,y) \wedge E(y,x)$	$\exists x.E(x,s)$
$\exists x.f(x) = x$	$\exists x.f(x) = s$		
$\forall x.R(x)$	$\forall x.\neg R(x)$	$\forall x.R(x) \wedge S(x)$	$\forall x.R(x) \vee S(x)$
$\forall xy.E(x,y)$	$\forall xy.\neg E(x,y)$	$\forall x.E(x,s)$	
$\forall x.f(x) = s$	$\forall x.f(x) = x$	$\forall x \neq y.\ f(x) \neq f(y)$	
$\mathbf{TC}[f](s,t)$	$\mathbf{RTC}[f](s,t)$	$\mathbf{TC}[f](s,s)$	
$\neg\mathbf{TC}[f](s,t)$	$\neg\mathbf{RTC}[f](s,t)$	$\neg\mathbf{TC}[f](s,s)$	
$(\exists y.T(y))\ \mathbf{RTC}[f](s,y)$			
$\mathbf{TC}[E](s,t)$	$\mathbf{RTC}[E](s,t)$	$\mathbf{TC}[E](s,s)$	
$\mathbf{TC}[f,g](s,t)$	$\mathbf{RTC}[f,g](s,t)$	$\mathbf{RTC}[f,g](s,s)$	
$(\exists y.T(y))\ \mathbf{RTC}[E](s,y)$		$(\exists xy.\ S(x) \wedge T(y))\ \mathbf{RTC}[E](x,y)$	
$\neg\mathbf{TC}[E](s,t)$	$\neg\mathbf{RTC}[E](s,t)$	$\neg\mathbf{TC}[E](s,s)$	$\neg\mathbf{RTC}[f,g](s,t)$
$\forall xy.\mathbf{TC}[E](x,y)$	$\forall x.\mathbf{TC}[E](x,t)$	$\forall x.\mathbf{TC}[E](x,x)$	
4 variations of **ATC**			

Fig. 5. A list of problems in the complexity classes of Figure 4. ReductionFinder has found a reduction between each pair of problems in each box. Each problem is expressed as a logical formula.

includes $\exists xy.E(x,y)$, $\exists x.f(x) = s$, $\exists x.E(s,x)$. Below this, the structure of **FO** under quantifier-free reductions is correctly represented up to two quantifier alternations.

Beyond **FO**, ReductionFinder has made significant progress in describing the complexity hierarchy. A class of 7 L-complete problems is visible at $\mathbf{TC}[f](s,t)$ (deterministic reachability), including its complement ($\neg\mathbf{TC}[f](s,t)$) and deterministic reachability with a relational target ($\exists y.T(y) \wedge \mathbf{TC}[f](s,y)$). Unfortunately, the L complete problems of cycle-finding ($\exists x.\mathbf{TC}[E](x,x)$) and its negation have not been placed in this class; nor has deterministic reachability with relations as both source and target ($\exists xy.S(x) \wedge T(y) \wedge \mathbf{TC}[E](x,y)$).

Below this level, ReductionFinder had limited success. We succeeded in reducing several problems to reachability (see Figure 5), including degree-2 reachability (reduction described in section 5.3. Not surprisingly, we did not discover a proof of the Immerman-Szelepcsényi theorem (showing co-NL $\leq$ NL by providing a reduction $\neg\mathbf{TC}[E](s,t) \leq \mathbf{TC}[E](s,t)$). We similarly did not prove Reingold's theorem [19], showing SL $\leq$ L by reducing $\mathrm{STC}[E](s,t) \leq \mathbf{TC}[f](s,t)$. These two results were historically elusive, and may require reductions above arity 2, or longer formulas than we were able to examine. Considering P-complete problems, we proved the equivalence of several variations of alternating transitive closure (**ATC**); however, we did not show the problem equivalent to its negation, or to the monotone circuit value problem (**MCVAL**).

5.3 Sample Reductions

We now list a few of the reductions that ReductionFinder has produced.

Example 1. ReductionFinder found two arity-1 reductions showing $\mathbf{RTC}[E](s,t) \leq \forall x.\mathbf{TC}[E](x,x)$. The first of these problems is simply REACH; the second states that every node of a directed graph is on some (nontrivial) cycle. The two reductions are good examples of the arity-1 reductions we have found, and also show a clear tradeoff between the formula length required to define E' and the arity parameters:

$$\begin{aligned} |R(\mathcal{A})| = &\quad \{a_1, a_2, \ldots, a_n, c_1\} \\ E'(x,y) \equiv &\quad x = t \\ &\vee\, y = s \\ &\vee\, E(x,y) \end{aligned} \tag{9}$$

The output structure $R(\mathcal{A})$ has all of the elements of the input structure $\mathcal{A}$, plus one new point c_1. The new edge relation is true wherever the old edge relation was true; in addition, all possible edges into the source and out of the target are added.

Since the new point c_1 was not part of the original edge relation, it has only one outgoing edge (to s), and only one incoming edge (to t). Therefore c_1 is on a cycle iff there is a path in the original graph from s to t. Similarly, if such a path does exist, *every* node in $R(\mathcal{A})$ is on a similar cycle. Thus the input graph satisfies $\mathbf{RTC}[E](s,t)$ iff the output satisfies $\forall x.\mathbf{TC}[E](x,x)$.

In addition to this reduction, ReductionFinder found a second arity-1 reduction. The second reduction does not use a distinguished constant element, but requires a longer formula:

$$\begin{aligned} |R(\mathcal{A})| = &\quad \{a_1, a_2, \ldots, a_n\} \\ E'(x,y) \equiv &\quad y \neq s \wedge E(x,y) \\ &\vee\, x \neq s \wedge x = y \\ &\vee\, x = t \end{aligned} \tag{10}$$

This reduction can be viewed as manipulating the graph as follows: we first remove all edges into s. We then add a self-loop on every edge except s. Finally, we add all possible edges out of t. Since the edge (t,s) is the only edge into node s, we then have that the node s is on a cycle iff there is a path from s to t. (Every other node is on a trivial cycle by construction.)

ReductionFinder has verified that neither reduction can be shortened; there is a tradeoff between the availability of the extra element c_1 and the required formula length. ReductionFinder can detect such tradeoffs, because in the partial ordering induced by our various search parameters, each of these reductions is in a minimal reduction-containing space.

Example 2. ReductionFinder successfully reduced the first-order problem $\forall x \exists y. E(x, y)$ to deterministic reachability ($\mathbf{TC}[f](s, t)$). This is a simple example of an arity-2 reduction where the successor relation is used to iteratively check all elements.

$$|R(\mathcal{A})| = \{\langle a_1, a_1 \rangle, \langle a_1, a_2 \rangle, \ldots, \langle a_n, a_n \rangle\}$$

$$\begin{aligned} f'(\langle x, y \rangle) &\equiv \begin{cases} \text{if} & E(x, y) & \text{then } \langle \text{Suc}(x), \text{Suc}(x) \rangle \\ \text{else if} & (\text{Suc}(y) \neq x) & \text{then } \langle x, \text{Suc}(y) \rangle \\ \text{else} & & \langle x, y \rangle \end{cases} \\ s' &\equiv \langle \min, \min \rangle \\ t' &\equiv \langle \min, \min \rangle \end{aligned} \tag{11}$$

Recall that each element in the output structure is a pair of elements in the input structure.

Deterministic non-reachability to deterministic reachability. Like all deterministic classes, L is closed under complement. The canonical L-complete problem is deterministic reachability. ReductionFinder was able to find a version of the canonical reduction from deterministic non-reachability to deterministic reachability, showing co-L $\leq$ L.

$$|R(\mathcal{A})| = \{\langle a_1, a_1 \rangle, \langle a_1, a_2 \rangle, \ldots, \langle a_n, a_n \rangle, c_1, c_2\}$$

$$\begin{aligned} f'(\langle x, y \rangle) &\equiv \begin{cases} \text{if} & (x = t) & \text{then } c_2 \\ \text{else if} & (y = \max) & \text{then } c_1 \\ \text{else} & & \langle f(x), \text{Suc}(y) \rangle \end{cases} \\ f'(c_i) &\equiv c_i \\ s' &\equiv \langle s, \min \rangle \\ t' &\equiv c_1 \end{aligned} \tag{12}$$

An input graph $G = \langle f; s, t \rangle$ contains no path from s to t iff the output graph $I(G) = \langle f'; s', t' \rangle$ contains a path from s to t. This arity-2 reduction walks through the original graph in the sequence $\langle s, 0 \rangle$, $\langle f(s), 1 \rangle$, $\ldots$, $\langle f^n(s), n \rangle$. If t is ever found, we move to the point c_2, representing a reject state; if t is not found after n steps, we move to the target node c_1.

Reachability to Degree-2 Reachability. Directed-graph reachability is the canonical NL-complete problem, and it is well-known that restricting ourselves to graphs with outdegree $\leq$ 2 suffices for NL-completeness. We chose to represent outdegree-2 reachability with two unary function symbols; we define $\mathbf{TC}[f, g](s, t)$

on the vocabulary $\langle ; f^1, g^1; s, t\rangle$, with the semantics that nodes can be reached through any combination of f-edges and g-edges. ReductionFinder succeeded in reducing $\mathbf{TC}[E](s,t) \le \mathbf{TC}[f,g](s,t)$ via an arity-2 reduction:[3]

$$|R(\mathcal{A})| = \{\langle a_1, a_1\rangle, \langle a_1, a_2\rangle, \ldots, \langle a_n, a_n\rangle\}$$

$$\begin{aligned} f'(\langle x,y\rangle) &\equiv \begin{cases} \text{if } E(x,y) \text{ then } \langle y,y\rangle \\ \qquad\qquad\;\; \text{else } \langle x,y\rangle \end{cases} \\ g'(\langle x,y\rangle) &\equiv \langle x, \mathrm{Suc}(y)\rangle \\ s' &\equiv \langle s,t\rangle \\ t' &\equiv \langle t,t\rangle \end{aligned} \tag{13}$$

This reduction uses the traditional technique of using successor to iterate through possible neighbors. Each node $\langle x, y\rangle$ of the output structure can be read as "we are at node x, considering y as a possible next step". If there is an edge $E(x,y)$, we nondeterministically either follow this edge (moving along f to $\langle y,y\rangle$) or move along g to the next possibility $\langle x, \mathrm{Suc}(y)\rangle$. If there is no edge $E(x,y)$, our only nontrivial movement is along g, to $\langle x, \mathrm{Suc}(y)\rangle$.

6 Conclusions and Future Directions

The ReductionFinder program successfully finds quantifier-free reductions between computational problems. The program maintains a database of known reductions between problems. Strongly connected components in this database correspond to complexity classes. When presented with a new problem, we can perform searches to automatically place the problem within the existing reduction graph.

This project has demonstrated that it is possible to find reductions between problems by using a SAT solver to search for them. Right now, ReductionFinder takes a long time to find small reductions and cannot find medium-sized reductions. We suggest some directions for future work aimed at taking automatic reduction finding to the next stage.

1. ReductionFinder searches for a small, simple reduction, R, by repeatedly calling a SAT solver as outlined in §3.1. The tasks involved are:

[3] The reduction above has undergone some syntactic simplification. ReductionFinder originally reported the reduction:

$$\begin{aligned} f'(\langle x,y\rangle) &\equiv \begin{cases} \text{if } E(x,y) \text{ then } \langle y,y\rangle \\ \qquad\qquad\;\; \text{else } \langle x, \mathrm{Suc}(x)\rangle \end{cases} \\ g'(\langle x,y\rangle) &\equiv \begin{cases} \text{if } \mathrm{Suc}(y) = x \text{ then } \langle x, \mathrm{Suc}(x)\rangle \\ \qquad\qquad\qquad\;\; \text{else } \langle x, \mathrm{Suc}(y)\rangle \end{cases} \\ s' &\equiv \langle s,t\rangle \\ t' &\equiv \langle t,t\rangle \end{aligned}$$

- Find an R that is a correct reduction on the current example graphs, $G_0, \ldots, G_k$ (Equation 6).
- Find a G_{k+1} on which the current R fails (Equation 7).

While, we would expect that such a search is exponential in the size of R, in our experience the difficulty is that the number of variables in the boolean formulas grow linearly with the number of counter-example graphs, k, and unfortunately the running time seems to increase exponentially in k. (The search for counter-example graphs in the second case does not have this problem.) Since the problem we are trying to solve is Σ_2^p – there exists a small reduction, for all small graphs – we hope to speed up our search by using strategies similar to those employed by QBF solvers. Related to this is the question of what makes a good set of counter-example graphs.

2. To show that there is a reduction from problem A to problem B, it may be that we can find a problem in the middle, M, so that reductions from A to M and M to B are simpler. We believe that finding such intermediate problems will be invaluable in searching for reductions. However, we have only found limited evidence of this so far in our work with ReductionFinder. It will be valuable to develop heuristics to find or generate appropriate intermediate problems.
3. Sufficient progress on the above two points may enable us to automatically generate linear reductions. This would have great benefits for automatic programming of optimal algorithms as discussed in Item 3 near the end of Section 1.

References

1. Allender, E., Bauland, M., Immerman, N., Schnoor, H., Vollmer, H.: The Complexity of Satisfiability Problems: Refining Schaefer's Theorem. J. Comput. Sys. Sci. 75, 245–254 (2009)
2. Ben-Eliyahu, R., Dechter, R.: Propositional semantics for disjunctive logic programs. Annals of Mathematics and Artificial Intelligence 12, 53–87 (1996)
3. Clark, K.: Negation as Failure. In: Gallaire, H., Minker, J. (eds.) Logic and Data Bases, pp. 293–322. Plenum Press, New York
4. Cook, S.: The Complexity of Theorem Proving Procedures. In: Proc. Third Annual ACM STOC Symp., pp. 151–158 (1971)
5. Ebbinghaus, H.-D., Flum, J.: Finite Model Theory, 2nd edn. Springer, Heidelberg (1999)
6. Eén, N., Sörensson, N.: An Extensible SAT-solver [extended version 1.2]. In: Giunchiglia, E., Tacchella, A. (eds.) SAT 2003. LNCS, vol. 2919, pp. 502–518. Springer, Heidelberg (2004)
7. Feder, T., Vardi, M.: The Computational Structure of Monotone Monadic SNP and Constraint Satisfaction: A Study Through Datalog and Group Theory. SAIM J. Comput. 28, 57–104 (1999)
8. Giunchiglia, E., Lierler, Y., Maratea, M.: SAT-Based Answer Set Programming. In: Proc. AAAI, pp. 61–66 (2004)
9. Hartmanis, J., Immerman, N., Mahaney, S.: One-Way Log Tape Reductions. In: IEEE Found. of Comp. Sci. Symp., pp. 65–72 (1978)

10. Immerman, N.: Descriptive Complexity. Springer Graduate Texts in Computer Science, New York (1999)
11. Immerman, N.: Languages That Capture Complexity Classes. SIAM J. Comput. 16(4), 760–778 (1987)
12. Janhunen, T.: A counter-based approach to translating normal logic programs into sets of clauses. In: Proc. ASP 2003 Workshop, pp. 166–180 (2003)
13. Jones, N.: Reducibility Among Combinatorial Problems in Log n Space. In: Proc. Seventh Annual Princeton Conf. Info. Sci. and Systems, pp. 547–551 (1973)
14. Karp, R.: Reducibility Among Combinatorial Problems. In: Miller, R.E., Thatcher, J.W. (eds.) Complexity of Computations, pp. 85–104. Plenum Press, New York (1972)
15. Ladner, R.: On the Structure of Polynomial Time Reducibility. J. Assoc. Comput. Mach. 2(1), 155–171 (1975)
16. Libkin, L.: Elements of Finite Model Theory. Springer, Heidelberg (2004)
17. Lifschitz, V., Razborov, A.A.: Why are there so many loop formulas? ACM Trans. Comput. Log. 7(2), 261–268 (2006)
18. Moskewicz, M.W., Madigan, C.F., Zhao, Y., Zhang, L., Malike, S.: Chaff: Engineering an Efficient SAT Solver. In: Design Automation Conference 2001 (2001)
19. Reingold, O.: Undirected ST-connectivity in Log-Space. In: ACM Symp. Theory of Comput., pp. 376–385 (2005)
20. Schaefer, T.: The Complexity of Satisfiability Problems. In: ACM Symp. Theory of Comput., pp. 216–226 (1978)
21. Schwartz, J.T., Dewar, R.B.K., Dubinsky, E., Schonberg, E.: Programming with Sets: an Introduction to SETL. Springer, New York (1986)
22. Valiant, L.: Reducibility By Algebraic Projections. L'Enseignement mathématique, T. XXVIII 3-4, 253–268 (1982)

On Complete Problems, Relativizations and Logics for Complexity Classes

Anuj Dawar*

University of Cambridge Computer Laboratory, Cambridge CB3 0FD, U.K.
anuj.dawar@cl.cam.ac.uk

For Yuri, on the occasion of your seventieth birthday. Thank you for always asking the most stimulating questions.

Abstract. In a paper published in 1988, Yuri Gurevich gave a precise mathematical formulation of the central question in descriptive complexity - is there a logic capturing P - and conjectured that the answer is no. In the same paper, he also conjectured that there was no logic capturing either of the complexity classes NP ∩ co-NP and RP, and presented evidence for these conjectures based on the construction of oracles with respect to which these classes do not have complete problems. The connection between the existence of complete problems and the existence of a logic capturing P was further established in (Dawar 1995). Does this imply that the question of whether there is a logic capturing P is subject to a relativization barrier? Here, we examine this question and see how the question for P differs from those for the other classes by taking a short tour through relativizations, complete problems and recursive enumerations of complexity classes.

Keywords: descriptive complexity, relativization, oracle complexity classes.

1 Introduction

One of the main drivers of research in the area of finite model theory and descriptive complexity over the last three decades has been the question of whether there is a logic that expresses exactly the polynomial-time computable properties of finite structures. In short form, we ask whether there is a logic *capturing* P. This question was first formulated by Chandra and Harel [4] but given the precise form in which it is usually cited by Yuri Gurevich [7]. In this form, the question is as follows. A logic L is a function SEN associating a recursive set of sentences to each finite vocabulary σ together with a function SAT that associates to each σ a recursive *satisfaction relation* relating finite σ-structures to sentences that is also isomorphism-invariant. That is, if $\mathbb{A}$ and $\mathbb{B}$ are isomorphic σ-structures and φ is any sentence of $\mathrm{SEN}(\sigma)$ then $(\mathbb{A}, \varphi) \in \mathrm{SAT}(\sigma)$ if, and only if, $(\mathbb{B}, \varphi) \in \mathrm{SAT}(\sigma)$. Now, a logic L *captures* P if there is a computable function that takes each sentence of L to a polynomially-clocked Turing machine that recognises the models of the sentence, and for every polynomial-time recognizable class K of structures, there is a sentence of L whose models are exactly K.

* The author carried out this work while supported by a Leverhulme Trust Study Abroad Fellowship.

A. Blass, N. Dershowitz, and W. Reisig (Eds.): Gurevich Festschrift, LNCS 6300, pp. 201–207, 2010.

Gurevich conjectured that there is no logic capturing P in this sense. He also proved in the same paper that there is such a logic if, and only if, there is a logic that captures P on graphs. Thus, the conjecture can be reformulated as the following statement.

Conjecture 1 (Gurevich). There is no recursively enumerable set S of pairs (M, p) where M is a deterministic Turing machine and p a polynomial such that:

1. for each $(M, p) \in S$, if G_1 and G_2 are isomorphic n-vertex graphs, then M accepts input G_1 in $p(n)$ steps if, and only if, it accepts G_2 in $p(n)$ steps; and
2. for any polynomial-time decidable class K of graphs, there is a pair $(M, p) \in S$ such that an n-vertex graph G is accepted by M in $p(n)$ steps if, and only if, $G \in K$.

The key difference between this and the question formulated by Chandra and Harel is that in their formulation we would require S to be a set of machines which run in polynomial-time but the polynomials need not be given explicitly. For a discussion of the relationship between the two questions see [11].

While Conjecture 1 has become well-known as Gurevich's conjecture and generated a large amount of follow-up research, in [7], Gurevich stated similar conjectures for the complexity classes NP ∩ co-NP and RP which have received somewhat less attention. To be precise, the conjecture that there is no logic for NP ∩ co-NP can be stated in the following equivalent form.

Conjecture 2 (Gurevich). There is no recursively enumerable set S of triples (M, N, p) where M and N are nondeterministic Turing machines and p a polynomial such that:

1. for each $(M, N, p) \in S$, if G_1 and G_2 are isomorphic n-vertex graphs, then M accepts input G_1 in $p(n)$ steps if, and only if, it accepts G_2 in $p(n)$ steps and the same holds for N;
2. for each $(M, N, p) \in S$ and each n-vertex graph G, M accepts G in $p(n)$ steps if, and only if, N does not accept G in $p(n)$ steps; and
3. for any class K of graphs that is in both NP and co-NP, there is a triple $(M, N, p) \in S$ such that an n-vertex graph G is accepted by M in $p(n)$ steps if, and only if, $G \in K$.

Indeed, Gurevich provided evidence in support of Conjecture 2 in the following form. He showed that if Conjecture 2 is false then there is a complete problem in NP ∩ co-NP under polynomial-time reductions. However, it is known by a result of Sipser [14] that there are oracles A such that the complexity class $\mathsf{NP}^A \cap \mathsf{co\text{-}NP}^A$ does not have complete problems with respect to polynomial-time reductions. This implies, in particular, that any refutation of Conjecture 2 would have to be *non-relativizing*.

On the other hand, it is not difficult to show that there is also an oracle A for which there is a logic capturing $\mathsf{NP}^A \cap \mathsf{co\text{-}NP}^A$ (an argument is given in Sect. 3). This means that any proof of Conjecture 2 would also have to be non-relativizing. In short, Conjecture 2 runs up against the famous *relativization barrier* in complexity theory (see [6,1]).

What about Conjecture 1? Is it also subject to the same barrier? Since a proof of the conjecture would imply that P $\neq$ NP, it is subject to all the barriers that face that question. Is a refutation of the conjecture also up against a relativization barrier? This is a question we address in this paper, which takes us through a tour of the relationship between logics capturing complexity classes, recursive enumerations and complete problems.

2 Recursive Enumerations and Complete Problems

The argument constructed by Gurevich to show that the existence of a logic capturing NP∩co-NP would imply that the class contained complete problems under polynomial-time reductions is an instance of a well-known phenomenon linking recursive enumerations of complexity classes to the existence of complete problems (see, for example, [8,9,10]). Complexity classes (in the usual sense, where they are collections of sets of strings, rather than unordered finite structures) are defined by conditions that are either *syntactic* or *semantic*. Examples of the former are the classes P and NP which can be enumerated as pairs (M, p) where M is a deterministic (or nondeterministic) machine and p is a polynomial. This enumeration then includes witnesses for all languages in P and NP respectively where we say that (M, p) is a witness for the language consisting of those strings x accepted by M in a number of steps bounded by $p(|x|)$. On the other hand, a natural set of witnesses for languages in the complexity class NP∩co-NP (as we saw in formulating Conjecture 2) might be the set of triples (M, N, p) where M and N are nondeterministic machines such that the languages they accept in p steps are complements of each other. This adds to the syntactic condition (that M and N are nondeterministic Turing machines and that p is a polynomial) the *semantic* requirement that the two machines accept complementary languages. This requirement is undecidable and the natural set of witnesses for NP ∩ co-NP is not recursively enumerable. Thus, finding a recursively enumerable set of witnesses would require a fundamentally new characterization of the class and would be a major breakthrough in complexity theory.

The situation with the class RP is analogous. This is defined as those languages L for which there exists a nondeterministic machine M and a polynomial p such that, given an input x of length n, if x is not in L, in time $p(n)$ no computation path of M will reach an accepting state while if $x \in L$, then at least half of the computation paths of M will terminate in an accepting state. It is clear that there are machines which do not accept any language in this sense because for some inputs x they will have some accepting computations but fewer than half the computations will be accepting. So, the class can be witnessed by the set of pairs (M, p) where M is a nondeterministic machine that satisfies the additional semantic (and undecidable) condition that on all inputs x, either none or at least half of all computations of M of length $p(|x|)$ will accept.

Thus, syntactic classes are those that admit recursive enumerations and from such enumerations one can construct complete problems. On the other hand, the addition of semantic conditions to the definition of a class (and generally semantic conditions, by Rice's theorem, are undecidable) makes the natural set of witnesses not recursively enumerable. Moreover, one can often exploit this to construct an oracle A with respect to which the complexity class does not have complete problems. Exactly such a construction is carried out by Sipser [14].

What then, are we to make of whether there is a logic for P? This is a syntactic class but, in considering it as a collection of sets of graphs, we have added the additional *semantic* condition of *isomorphism-invariance*. That is, we wish to restrict to those machines which, when we interpret their input as representing a graph, do not distinguish between isomorphic graphs. This condition is, again, undecidable and the natural set of witnesses is not recursively enumerable. Conjecture 1 then asserts that there is no r.e.

set of witnesses whatsoever. Could we then construct an oracle A with respect to which there is no logic capturing P?

Just as Gurevich showed that from the assumption that there is a logic for NP $\cap$ co-NP it follows that there is a complete problem in this class under polynomial-time reductions, a prospect considered unlikely, so we could also conclude from the assumption that there is a logic for P that this class contains complete problems. However, the latter prospect is not so unlikely. P certainly contains complete problems under polynomial-time reductions and indeed under much weaker reductions such as logarithmic-space or AC^0 reductions. I was able to show in [5] that the existence of a logic for P implies (and, indeed, is equivalent to) the existence in P of complete problems under reductions that are themselves *syntactically* isomorphism-invariant.

To make this precise, let us consider logical interpretations. Suppose we are given two relational signatures σ and τ and a logic L. An m-ary *L-interpretation* of τ in σ is a sequence of formulas of L in the signature σ consisting of:

- a formula $\upsilon(\mathbf{x})$;
- a formula $\eta(\mathbf{x}, \mathbf{y})$;
- for each relation symbol R in τ of arity a, a formula $\rho^R(\mathbf{x}_1, \ldots, \mathbf{x}_a)$; and
- for each constant symbol c in τ, a formula $\gamma^c(\mathbf{x})$,

where each $\mathbf{x}, \mathbf{y}$ or $\mathbf{x}_i$ is an m-tuple of free variables. We call m the *width* of the interpretation. We say that an interpretation Φ associates a τ-structure $\mathbb{B}$ to a σ-structure $\mathbb{A}$, if there is a surjective map h from the m-tuples $\{\mathbf{a} \in A^m \mid \mathbb{A} \models \upsilon[\mathbf{a}]\}$ to $\mathbb{B}$ such that:

- $h(\mathbf{a}_1) = h(\mathbf{a}_2)$ if, and only if, $\mathbb{A} \models \eta[\mathbf{a}_1, \mathbf{a}_2]$;
- $R^{\mathbb{B}}(h(\mathbf{a}_1), \ldots, h(\mathbf{a}_a))$ if, and only if, $\mathbb{A} \models \rho^R[\mathbf{a}_1, \ldots, \mathbf{a}_a]$;
- $h(\mathbf{a}) = c^{\mathbb{B}}$ if, and only if, $\mathbb{A} \models \gamma^c[\mathbf{a}]$.

Note that an interpretation Φ associates a τ-structure with $\mathbb{A}$ only if η defines an equivalence relation on A^m that is a congruence with respect to the relations defined by the formulas ρ^R and each γ^c defines a single equivalence class of this relation. In such cases however, $\mathbb{B}$ is uniquely determined up to isomorphism and we write $\mathbb{B} = \Phi(\mathbb{A})$. We are only interested in interpretations that associate a τ-structure to every $\mathbb{A}$.

We say that a class of σ-structures K_1 is *L-reducible* to a class of τ-structures K_2 if there is an L-interpretation Φ of τ in σ such that for all σ-structures $\mathbb{A}$, $\mathbb{A} \in K_1$ if, and only if, $\Phi(\mathbb{A}) \in K_2$. Finally, for a complexity class $\mathcal{C}$, a class K is $\mathcal{C}$-hard under L-reductions if every class in $\mathcal{C}$ is L-reducible to K and, as usual, it is $\mathcal{C}$-complete under L-reductions if it is $\mathcal{C}$-hard and it is in $\mathcal{C}$.

Writing FO for first-order logic, we can state the main result of [5] as:

Theorem 3. *There is a logic capturing P if, and only if, P contains a complete problem under FO-reductions.*

The construction used to prove this theorem is quite general. On the one hand, we can replace P by many other complexity classes. Indeed, in [5], the result is stated for any complexity class satisfying a *boundedness* condition and examples of such complexity classes include, besides P, L, NP and PSpace. The last two are known to have logics capturing them and so the theorem tells us that they have complete problems under

FO-reductions while for L, like for P, the existence of a logic capturing it remains an open question. On the other hand, there is also nothing special about the choice of FO in Theorem 3. The important fact is that FO-reductions are themselves syntactically defined, and so can be enumerated, and they are isomorphism-invariant. We could replace FO in the theorem by virtually any logic that has an effective syntax, is isomorphism-invariant (i.e. it is a logic in the sense defined in the introduction) and is contained in P. The important fact is that the semantic condition of isomorphism-invariance that is implied in the recursive enumeration of P is captured in the reductions themselves and we obtain the usual construction of a complete problem from the syntactic presentation of a class.

The complexity class NP ∩ co-NP does not meet the definition of a bounded class as given in [5], but with a recursive presentation in the sense of Conjecture 2, it is possible to carry through a construction analogous to the proof of Theorem 3 to obtain the following.

Theorem 4. *There is a logic capturing NP∩co-NP if, and only if, NP∩co-NP contains a complete problem under FO-reductions.*

This strengthens Theorem 1.17 of [7] by replacing polynomial-time reductions by FO-reductions and, at the same time, providing a converse.

3 Relativization Barriers

Baker, Gill and Solovay [2] proved that there is an oracle A such that the complexity classes P^A and NP^A are different and there is another oracle B so that the classes P^B and NP^B are equal. This result forms what is called the *relativization barrier* to the resolution of the question of whether or not P = NP. That is to say, it demonstrates that any resolution of the question must use methods that do not relativize to machines with oracles. In particular, methods based on diagonalization will not suffice. Since this seminal result, methods have been found that circumvent this barrier in some cases (for instance the celebrated result that IP = PSpace [13]) and other barriers have been observed to the resolution of the relationship between P and NP (see [1]). The question we address here is whether Conjectures 1 and 2 face similar barriers.

As noted in Sect. 1, Gurevich proved that if there is a logic for NP ∩ co-NP then this class contains a complete problem with respect to polynomial-time reductions. Moreover Sipser [14] (see also [8]) showed that there are oracles with respect to which this is not the case. On the other hand, it is easy to show that there is an oracle A for which there is a logic capturing NP ∩ co-NP. To be precise, take A to be a PSpace-complete problem such as satisfiability of quantified Boolean formulas. Then $\mathsf{NP}^A = \mathsf{co\text{-}NP}^A = \mathsf{PSpace}$. Moreover, it is known that there is a logic capturing PSpace (see, for example [12]) and the result follows. This implies that the question of whether there is a logic capturing NP ∩ co-NP is subject to the relativization barrier. A resolution either way would require non-relativizing methods.

How about the question of whether there is a logic for P?. Once again, it is easy to construct an oracle with respect to which there is such a logic. Indeed, taking A once again to be a PSpace-complete problem, we see that $\mathsf{P}^A = \mathsf{PSpace}$ and therefore there is a logic for P^A. Can we also construct an oracle A so that there is no logic capturing P^A? Or, equivalently, an oracle A so that P^A does not contain any complete problems with respect to FO-reductions. We show next that to construct such an oracle, we would have to separate P from NP.

Theorem 5. *If* $\mathsf{P} = \mathsf{NP}$ *then for every oracle* A*, there is a logic capturing* P^A.

Proof. A graph canonization function is a function c on strings that encode graphs with the property that for any graph G, $c(G)$ is the encoding of a graph isomorphic to G and if G and G' are isomorphic graphs then $c(G) = c(G')$. It is known that there are graph canonization functions (such as the function that given a graph G yields the lexicographically minimal string representing a graph isomorphic to G) in the polynomial hierarchy (see [3]). It follows that if $\mathsf{P} = \mathsf{NP}$ then there is a graph canonization function c that is computable in polynomial time. For any oracle machine M and polynomial p, define $C_{M,p}$ to be the machine that takes an input x, computes $c(x)$ and then simulates M for $p(n)$ steps on input $c(x)$, where n is the length of $c(x)$.

It is easy to see that the language accepted by $C_{M,p}$ is invariant under isomorphisms. Moreover, if M with oracle A and running within bounds p accepts a class of graphs K invariant under isomorphisms, then the language accepted by $C_{M,p}$ with oracle A is exactly the strings encoding graphs in K. Thus, the collection of all machines $C_{M,p}$ with oracle A is a recursive enumeration of P^A. □

Thus, it would seem that relativization is not itself a barrier to refuting Conjecture 1. Proving the conjecture, on the other hand, would separate P from NP and this is subject to all the barriers that complexity theory is familiar with.

4 Concluding Remarks

We noted that the conditions defining complexity classes come in two flavours: syntactic and semantic. Syntactic conditions are restrictions on the accepting machines (or on their resource bounding functions) that can be recognized from the form of the machines or the presentations of the functions themselves. Semantic conditions on the other hand are typically undecidable properties of the machines. Complexity classes that are defined by purely syntactic criteria such as L, P, NP and PSpace (on strings) admit recursive enumerations and from these one can construct complete problems under quite weak reductions. On the other hand, complexity classes that are defined by semantic restrictions on the witnessing machines, such as NP ∩ co-NP and RP, do not admit obvious recursive presentations or complete problems and to prove that they do would require fundamental new characterizations of these classes. Indeed establishing whether or not they have complete problems is subject to the relativization barrier in complexity theory.

The study of complexity classes on (unordered) graphs imposes a new semantic condition on machines, namely that of isomorphism invariance. However, some complexity classes (such as NP and PSpace) still admit recursive presentations even under this semantic restriction. This is because the semantic restriction can be enforced by an externally imposed pre-processing step (such as a canonization function) that does not take us out of the class. But, it remains an open question whether classes such as P or L admit recursive presentations under this restriction.

Considering NP ∩ co-NP or RP on graphs, one sees that we are dealing with two distinct semantic restrictions: one that is inherent to the definition of the class and the second arising from isomorphism invariance. This, as it were, makes it doubly unlikely that we could find logics that capture these complexity classes. Moreover, it is the first of these semantic restrictions that means that the problem of the existence of such logics runs up against the relativization barrier.

References

1. Aaronson, S., Wigderson, A.: Algebrization: a new barrier in complexity theory. In: Proc. 40th ACM Symp. on Theory of Computing, pp. 731–740 (2008)
2. Baker, T., Gill, J., Solovay, R.: Relativizations of the $P =?NP$ question. SIAM Journal on Computing 4(4), 431–442 (1975)
3. Blass, A., Gurevich, Y.: Equivalence relations, invariants, and normal forms. SIAM Journal on Computing 13(4), 682–689 (1984)
4. Chandra, A., Harel, D.: Structure and complexity of relational queries. Journal of Computer and System Sciences 25, 99–128 (1982)
5. Dawar, A.: Generalized quantifiers and logical reducibilities. Journal of Logic and Computation 5(2), 213–226 (1995)
6. Fortnow, L.: The role of relativization in complexity theory. Bulletin of the EATCS 52, 229–243 (1994)
7. Gurevich, Y.: Logic and the challenge of computer science. In: Börger, E. (ed.) Current Trends in Theoretical Computer Science, pp. 1–57. Computer Science Press, Rockville (1988)
8. Hartmanis, J., Li, M., Yesha, Y.: Containment, separation, complete sets, and immunity of complexity classes. In: Kott, L. (ed.) ICALP 1986. LNCS, vol. 226, pp. 136–145. Springer, Heidelberg (1986)
9. Kowalczyk, W.: Some connections between representability of complexity classes and the power of formal systems of reasoning. In: Proc. 11th Intl. Symp. Mathematical Foundations of Computer Science, pp. 364–369 (1984)
10. Landweber, L.H., Lipton, R.J., Robertson, E.L.: On the structure of sets in NP and other complexity classes. Theor. Comput. Sci. 15, 181–200 (1981)
11. Nash, A., Remmel, J.B., Vianu, V.: PTIME queries revisited. In: Eiter, T., Libkin, L. (eds.) ICDT 2005. LNCS, vol. 3363, pp. 274–288. Springer, Heidelberg (2004)
12. Richerby, D.: Logical characterizations of PSPACE. In: Computer Science Logic: Proc. 13th Conf. of the EACSL, pp. 370–384 (2004)
13. Shamir, A.: IP = PSpace. Journal of the ACM 39(4), 869–877 (1992)
14. Sipser, M.: On relativization and the existence of complete sets. In: Nielsen, M., Schmidt, E.M. (eds.) ICALP 1982. LNCS, vol. 140, pp. 523–531. Springer, Heidelberg (1982)

Effective Closed Subshifts in 1D Can Be Implemented in 2D*

Bruno Durand, Andrei Romashchenko**, and Alexander Shen**

LIF Marseille, CNRS and University Aix–Marseille
{bruno.durand,alexander.shen}@lif.univ-mrs.fr,
andrei.romashchenko@gmail.com

C'est avec grand plaisir que nous avons rédigé cet article pour notre ami et collègue Yuri Gurevich. En effet, l'intérêt qu'il a porté à nos travaux depuis bien longtemps nous a soutenu, ses remarques nous ont éclairés, et ses questions nous ont laissé entrevoir de nouvelles perspectives.

Не мысля гордый свет забавить...

Abstract. In this paper we use fixed point tilings to answer a question posed by Michael Hochman and show that every one-dimensional effectively closed subshift can be implemented by a local rule in two dimensions. The proof uses the fixed-point construction of an aperiodic tile set and its extensions.

Keywords: aperiodic tilings, subshifts, fixed point.

1 Introduction

Let A be a finite set (*alphabet*); its elements are called *letters*. By *A-configuration* we mean a mapping $C\colon \mathbb{Z}^2 \to A$. In geometric terms: a cell with coordinates (i, j) contains letter $C(i, j)$.

A *local rule* is defined by a positive integer M and a list of prohibited $(M \times M)$-*patterns* ($M \times M$ squares filled by letters). A configuration C *satisfies* a local rule R if none of the patterns listed in R appears in C.

Let A and B be two alphabets and let $\pi\colon A \to B$ be any mapping. Then every A-configuration can be transformed into a B-configuration (its homomorphic image) by applying π to each letter. Assume that the local rule R for A-configurations and mapping π are chosen in such a way that local rule R prohibits patterns where letters a and a' with $\pi(a) \neq \pi(a')$ are vertical neighbors. This guarantees that every A-configuration that satisfies R has an image where vertically aligned B-letters are the same. Then for each B-configuration in the image every vertical line carries one single letter of B. So we can say that π maps

* Partially supported by NAFIT ANR-09-EMER-008-01 and RFBR 09-01-00709a grants.

** On leave from IITP RAS, Moscow.

A. Blass, N. Dershowitz, and W. Reisig (Eds.): Gurevich Festschrift, LNCS 6300, pp. 208–226, 2010.

a 2-dimensional A-configuration satisfying the local rule R to a 1-dimensional B-configuration.

Thus, for every A, B, local rule R and π (with described properties) we get a subset $L(A, B, R, \pi)$ of $B^{\mathbb{Z}}$ (i.e., $L(A, B, R, \pi)$ is the set of π-images of all A-configurations that satisfy the local rule R). The following result (Theorem 1) characterizes the subsets that can be obtained in this way.

Consider a product topology in $B^{\mathbb{Z}}$. Base open sets of this topology are *intervals*. Each interval is obtained by fixing letters in finitely many places (other places may contain arbitrary letters). Each interval is therefore a finite object and we can define *effectively open* subsets of $B^{\mathbb{Z}}$ as unions of (computably) enumerable families of intervals. An *effectively closed* set is a complement of an effectively open one. A *subshift* is a subset of $B^{\mathbb{Z}}$ that is closed and invariant under left and right shifts. We are mostly interested in subshifts that are not only closed but effectively closed sets.

Theorem 1. *The subset $L(A, B, R, \pi)$ is an effectively closed subshift. For every effectively closed subshift $S \subset \mathbb{B}^Z$ one can find A, R, and π such that $S = L(A, B, R, \pi)$.*

The first part of the statement is easy. The set $L(A, B, R, \pi)$ is evidently shift invariant; it remains to show that it is effectively closed. The set of all A-configurations that satisfy R is a closed subset of a compact space and therefore is a compact space itself. The mapping of A-configurations into B-configurations is continuous. Therefore the set $L(A, B, R, \pi)$ is compact (as a continuous image of a compact set). This argument can be effectivized in a standard way. A B-string is declared *bad* if it cannot appear in the π-image of any A-configuration that satisfies R. The set of all bad strings is enumerable and $L(A, B, R, \pi)$ is the set of all bi-infinite sequences that have no bad factors.

The reverse implication is more difficult and is the main subject of this paper. It cannot be proven easily since it implies the classical result of Berger [2]: the existence of a local rule that makes all configurations aperiodic. Indeed, it is easy to construct an effectively closed subshift S that has no periodic points; if it is represented as $L(A, B, R, \pi)$, then local rule R has no periodic configurations (configurations that have two independent period vectors); indeed, those configurations have a horizontal period vector.

So it is natural to expect a proof of Theorem 1 to be obtained by modifying one of the existing constructions of an aperiodic local rule. It is indeed the case: we use the fixed-point construction described in [7]. We do not repeat this construction (assuming that the reader is familiar with that paper or has it at hand) and explain only the modifications that are needed in our case. This is done in sections 2–6; in the rest of this section we survey some other steps in the same direction.

M. Hochman [13] proved a similar result for 3D implementations of 1D subshifts (and, in general, $(k + 2)$-dimensional implementation of k-dimensional subshifts) and asked whether a stronger statement is true where 3D is replaced by 2D.

As we have mentioned, it is indeed true and can be achieved by the technique of fixed point self-similar tilings. The detailed exposition of this technique and its applications, including an answer to Hochman's question, is given in our paper [8]. Since this paper contains many other results (most boring are related to error-prone tile sets), we think that a self-contained (modulo [7]) exposition could be useful for readers that are primarily interested in this result, and provide such an exposition in the current paper.

In fact, the fixed point construction of algorithms and machines is an old and well known tool (used, e.g., for Kleene's fixed point theorem and von Neumann's self-reproducing automata) that goes back to self-referential paradoxes and Gödel's incompleteness theorem. One may only wonder why it was not used in 1960s to construct an aperiodic tile set. In a context of hierarchical constructions in the plane this technique was used by P. Gács in a much more complicated situation (see [9]); however, Gács did not bother to mention explicitly that this technique can be applied to construct aperiodic tile sets.

Fixed point tilings are not the only tool that can be used to implement subshifts. In [4] a more classical (Berger–Robinson style) construction of an aperiodic tile set is modified in several ways to implement one specific shift: the family of bi-infinite bit sequences ω such that all sufficiently long substrings x of ω have complexity greater than $\alpha|x|$ or at least ω can be cut into two pieces (left- and right-infinite) that have this property. (Here α is some constant less than 1, and $|x|$ stands for the length of x.) In fact, the construction used there is fairly general and can be applied to any enumerable set F of forbidden substrings: one may implement a shift that consists of bi-infinite sequences that have no substrings in F or at least can be cut into two parts with this property. Recently N. Aubrun and M. Sablik found a more ingenious construction that is free of this problem (splitting into two parts) and therefore provides another proof of Theorem 1 (see [1]).

The authors thank their LIF colleagues, especially E. Jeandel who pointed out that their result answers a question posed in [13].

2 The Idea of the Construction

We do not refer explicitly to our paper [7] (reproduced in the Appendix) but use the notions and constructions from that paper freely. In that paper we used local rules of special type (each letter was called a tile and has four colors at its sides; the local rule says that colors of the neighbor tile should match). In fact, any local rule can be reduced to this type by extending the alphabet; however, we do not need to worry about this since we construct a local rule and may restrict ourselves to tilings.

We superimpose two layers in our tiling. One of the layers contains B-letters; the local rule guarantees that each vertical line carries one B-letter. (Vertical neighbors should be identical.) For simplicity we assume that $B = \{0, 1\}$, so B-letters are just bits, but this is not really important for the argument.

The second layer contains an aperiodic tile set constructed in a way similar to [7]. Then some rules are used to organize the interaction between the layers;

computations in the second layer are fed with the data from the first layer and check that the first layer does not contain any forbidden string.

Indeed, the macro-tiles (at every level) in our construction contain some computation used to guarantee their behavior as building blocks for the next level. Could we run this computation in parallel with some other one that enumerates bad patterns and terminates the computation (creating a violation of the rules) if a bad pattern appears?

This idea immediately faces evident problems:

- The computation performed in macro-tiles (in [7]) was limited in time and space (and we need unlimited computations since we have infinitely many forbidden substrings and no limit on the computational resources used to enumerate them).
- Computations on high levels do not have access to bit sequence they need to check: the bits that go through these macro-tiles are "deep in the subconscious", since macro-tiles operate on the level of their sons (cells of the computation that are macro-tiles of the previous level), not individual bits.
- Even if every macro-tile checks all the bits that go through it (in some mysterious way), a "degenerate case" could happen where an infinite vertical line is not crossed by any macro-tile. Imagine a tile that is a left-most son of a father macro-tile who in its turn is the left-most son of its father and so on (see Fig. 2). They fill the right half-plane; the left half-plane is filled in a symmetric way, and the vertical dividing line between then is not crossed by any tile. Then, if each macro-tile takes care of forbidden substrings inside its zone (bits that cross this macro-tile), some substrings (that cross the dividing line) remain unchecked.

These problems are discussed in the following sections one after another; we apologize if the description of them seemed to be quite informal and vague and hope that they would become more clear when their solution is discussed.

3 Variable Zoom Factor

In our previous construction the macro-tiles of all levels were of the same size: each of them contained $N \times N$ macro-tiles of the previous level for some constant zoom factor N. Now it is not enough any more, since we need to host arbitrarily long computations in high-level macro-tiles. So we need an increasing sequence of zoom factors $N_0, N_1, N_2, \ldots$; macro-tiles of the first level are blocks of $N_0 \times N_0$ tiles; macro-tiles of the second level are blocks of $N_1 \times N_1$ macro-tiles of level 1 (and have size of $N_0N_1 \times N_0N_1$ if measured in individual tiles). In general, macro-tiles of level k are made of $N_{k-1} \times N_{k-1}$ macro-tiles of level $k-1$ and have side $N_0N_1 \ldots N_{k-1}$ measured in individual tiles.

However, all the macro-tiles (of different levels) carry the same program in their computation zone. The difference between their behavior is caused by the data: each macro-tiles "knows" its level (consciously, as a sequence of bits on its tape). Then this level k may be used to compute N_k which is then used as a modulus for coordinates in the father macro-tile. (Such a coordinate is a number between 0 and $N_k - 1$, and the addition is performed modulo N_k.)

Of course, we need to ensure that this information is correct. Two properties are required: (1) all macro-tiles of the same level have the same idea about their level, and (2) these ideas are consistent between levels (each father is one level higher than its sons). The first is easy to achieve: the level should be a part of the side macro-color and should match in neighbor tiles. (In fact an explicit check that brothers have the same idea about their levels is not really needed, since the first property follows from the second one: since all tiles on the level zero "know" their level correctly, by induction we conclude that macro-tiles of all levels have correct information about their levels.)

To achieve the second property (consistency between level information consciously known to a father and its sons) is also easy, though we need some construction. It goes as follows: each macro-tile knows its place in the father, so it knows whether the father should keep some bits of his level information in that macro-tile. If yes, the macro-tile checks that this information is correct. Each macro-tile checks only one bit of the level information, but with brothers' help they check all the bits.[1]

There is one more thing we need to take care of: the level information should fit into the tiles (and the computation needed to compute N_k knowing k should also fit into level k tile). This means that $\log k$, $\log N_k$ and the time needed to compute N_k from k should be much less than N_{k-1} (since the computation zone is some fraction of N_{k-1}). So N_k should not grow too slow (say, $N_k = \log k$ is too slow), should not grow too fast (say, $N_k = 2^{N_{k-1}}$ is too fast) and should not be too difficult to compute. However, these restriction still leave a lot of room for us: e.g., N_k can be proportional to $\sqrt{k}$, to k, to 2^k, or 2^{2^k}, or $2^{2^{2^k}}$ (any fixed height is OK). Recall that computation deals with binary encodings of k and N_k and normally is polynomial in their lengths.

In this way we are now able to embed computations of increasing sizes into the macro-tiles. Now we have to explain which data these computations would get and how the communication between levels is organized.

4 Conscious and Subconscious Bits

The problem of communication between levels can be explained using the following metaphor. Imagine you are a macro-tile; then you have some program, and process the data according to the program; you "blow up" (i.e., your interior cannot be correctly tiled) if some inconsistency in the data is found. This program makes you perform as one cell in the next-level brain (in the computation zone of the father macro-tile), but you do not worry about it: you just perform

[1] People sitting on the stadium during the football match and holding color sheets to create a slogan for their team can check the correctness of the slogan by looking at the scheme and knowing their row and seat coordinates; each person checks one pixel, but in cooperation they check the entire slogan.

the program. At the same time each cell of yourself in fact is a son macro-tile, and elementary operations of this cell (the relation between signals on its sides) are in fact performed by a lower-level computation. But this computation is your "sub-conscious", you do not have direct access to its data, though the correct functioning of the cells of your brain is guaranteed by the programs running in your sons.

Please do not took this metaphor too seriously and keep in mind that the time axis of the computations is just a vertical axis on the plane; configurations are static and do not change with time. However, it could be useful while thinking about problems of inter-level communication.

Let us decide that for each macro-tile all the bits (of the bit sequence that needs to be checked) that cross this macro-tile form its *responsibility zone*. Moreover, one of the bits of this zone may be *delegated* to the macro-tile, and in this case the macro-tile consciously knows this bit (is *responsible* for this bit). The choice of this bit depends on the vertical position of the macro-tile in its father.

More technically, recall that a macro-tile of level k is a square whose side is $L_k = N_0 \cdot N_1 \cdot \ldots \cdot N_{k-1}$, so there are L_k bits of the sequence that intersect this macro-tile. We delegate each of these bits to one of the macro-tiles it intersects. Note that every macro-tile of the next level is made of $N_k \times N_k$ macro-tiles of level k. We assume that N_k is much bigger than L_k (more about choice of N_k later); this guarantees that there are enough macro-tiles of level k (in the next level macro-tile) to serve all bits that intersect them. Let us decide that ith macro-tile of level k (from bottom to top) in a $(k+1)$-level macro-tile knows ith bit (from the left) in its zone. Since N_k is greater than L_k, we leave some unused space in each macro-tile of level $k+1$: many macro-tiles of level k are not responsible for any bit, but this does not create any problems.

This is our plan; however, we need a mechanism that ensures that the delegated bits are indeed represented correctly (are equal to the corresponding bits "on the ground", in the sequence that forms the first level of our construction). This is done in the hierarchical way: since every bit is delegated to macro-tiles

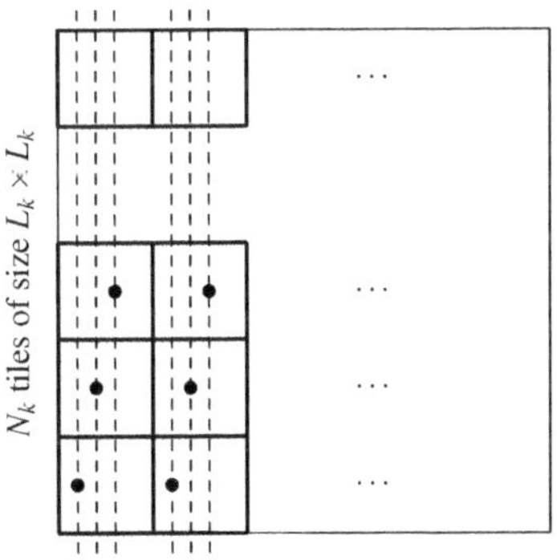

Fig. 1. Bit delegation: bits assigned to vertical lines are distributed between k-level macro-tile (according to their positions in the father macro-tile of level $k+1$)

of all levels, it is enough to ensure that the ideas about bit values are consistent between father and son.

For this hierarchical check let us agree that every macro-tile not only knows its own delegated bit (or the fact that there is no delegated bit), but also knows the bit delegated to its father (if it exists) as well as father's coordinates (in the grandfather macro-tile). This is still an acceptable amount of information (for keeping father's coordinates we need to ensure that $\log N_{k+1}$, the size of father's coordinate, is much smaller that N_{k-1}). To make this information consistent, we ensure that

- the data about the father's coordinates and bits are the same among brothers;
- if a macro-tile has the same delegated bit as its father (this fact can be checked since a macro-tile knows its coordinates in the father and father's coordinates in the grandfather), these two bits coincide;
- if a macro-tile is in the place where its father keeps its delegated bit, the actual father's information is consistent with the information about what the father should have.

So the information transfer between levels is organized as follows: the macro-tile that has the same delegated bit as its father, non-deterministically guesses this fact and distributes the information about father's coordinates and bit among the brothers. Those of the brothers who are in the correct place, check that father indeed has correct information.

On the lowest level we have direct access to the bits of the sequence, so the tile that is above the correct bit can keep its value and transmit it together with (guessed) coordinates of its father macro-tile (in the grandfather's macro-tile) to all brothers, and some brothers are in the right place and may check these values against the bits in the computation zone of the father macro-tile.

This construction makes all bits present at all levels, but this is not enough for checking: we need to check not individual bits, but bit groups (against the list of forbidden substrings). To this end we need a special arrangement described in the next section.

5 Checking Bit Groups

Here the main idea is: each macro-tile checks some substring (bit group) that is very small compared to the size of this macro-tile. However, since the size of the computation zone grows infinitely as the level increases, this does not prevent complicated checks (that may involve a long substring that appears very late in the enumeration of the forbidden patterns) from happening.

The check is performed as follows: we do some number of steps in the enumeration of forbidden patterns, and then check whether one of these patterns appears in the bit group under consideration (assigned to this macro-tile). The number of enumeration steps can be also rather small compared to the macro-tile size.

We reserve also some time and space to check that all the patterns appeared during the enumeration are not substrings of the bit group under consideration. This is not a serious time/space overhead since substring search in the given bit group can be performed rather fast, and the size of the bit group and the number of enumeration steps are chosen small enough (having in mind this overhead).

Then in the limit any violation inside some macro-tile will be discovered (and only degenerate case problem remains: substrings that are not covered entirely by any tile). The degenerate case problem is considered in the next section; it this section it remains to explain how the groups of (neighbor) bits are made available to the computation and how they are assigned to macro-tiles.

Let us consider an infinite vertical stripe of macro-tiles of level k that share the same $L_k = N_0 \cdot \ldots \cdot N_{k-1}$ columns. Together, these macro-tiles keep in their memory all L_k bits of their common zone of responsibility. Each of them perform a check for a small bit group (of length l_k, which increases extremely slowly with k and in particular is much less than N_{k-1}). We need to distribute somehow these groups among macro-tiles of this infinite stripe.

It can be done in many different ways. For example, let us agree that the starting point of the bit group checked by a macro-tile is the vertical coordinate of this macro-tile in its father (if it is not too big; recall that $N_k \gg N_0 N_1 \ldots N_{k-1}$). It remains to explain how groups of (neighbor) bits are made available to the computational zones of the corresponding macro-tiles.

We do it in the same way as for delegated bits; the difference (and simplification) is that now we may use only two levels of hierarchy since all the bits are available in the previous level (and not only in the "deep unconscious", at the ground level). We require that this group and the coordinate that determines its position are again known to all the sons of the macro-tile where the group is checked. Then the sons should ensure that (1) this information is consistent between brothers; (2) it is consistent with delegated bits where delegated bits are in the group, and (3) it is consistent with the information in the macro-tile (father of these brothers) itself. Since l_k is small, this is a small amount of information so there is no problem of its distribution between macro-tiles of the preceding level.

If a forbidden pattern belongs to a zone of responsibility of macro-tiles of arbitrarily high level, then this violation is be discovered inside a macro-tile of some level, so the tiling of the plain cannot not exist. Only the degenerate case problem remains: so far we cannot catch forbidden substrings that are not covered entirely by any macro-tile. We deal with the degenerate case problem in the next section.

6 Dealing with the Degenerate Case

The problem we need to deal with: it can happen that one vertical line is not crossed by any macro-tile of any level (see Fig. 2). In this case some substrings are not covered entirely by any macro-tile, and we do not check them. After the problem is realized, the solution is not difficult. We let every macro-tile check

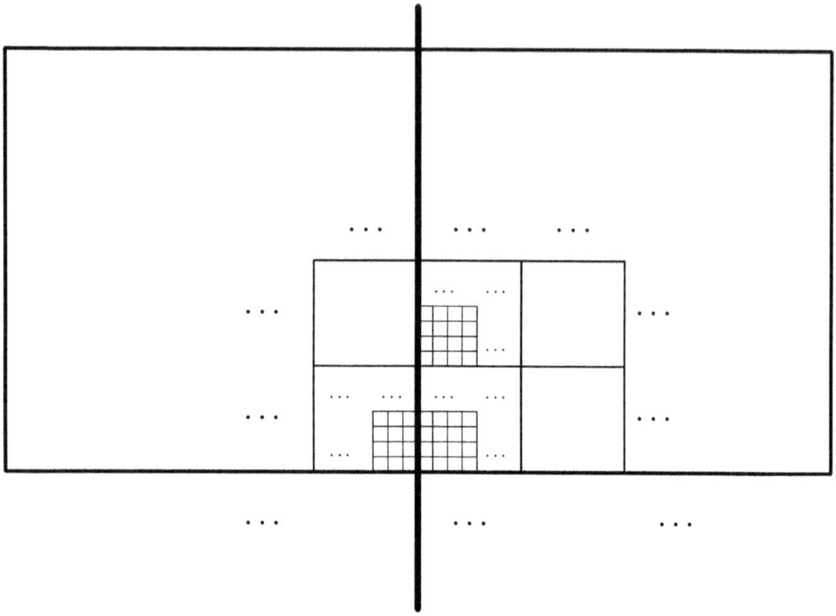

Fig. 2. Degenerate case

bit groups in its *extended responsibility zone* that is three times wider and covers not only the macro-tile itself but also its left and right neighbors.

Now a macro-tile of level k is given a small bit group which is a substring of its extended responsibility zone (the width of the extended responsibility zone is $3L_k$; it is composed of the zones of responsibility of the macro-tile itself and two its neighbors). Respectively, a macro-tile of level $(k-1)$ keeps the information about three groups of bits instead of one: for its father, left uncle, and right uncle. This information should be consistent between brothers (since they have the same father and uncles). Moreover, it should be checked across the boundary between macro-tiles: if two macro-tiles A and B are neighbors but have different fathers (B's father is A's right uncle and A's father is B's left uncle), then they should compare the information they have (about bit groups checked by fathers of A and B) and ensure it is consistent. For this we need to increase the amount of information kept in a macro-tile by a constant factor (a macro-tile keeps three bit groups instead of one, etc.), but this is still acceptable.

It is easy to see that now even in the degenerate case every substring is entirely in the extended responsibility zone of arbitrary large tiles, so all the forbidden patterns are checked everywhere.

7 Final Adjustments

We finished our argument, but we was quite vague about the exact values of parameters saying only that some quantities should be much less than others. Now we need to check again the entire construction and see that the relations between parameters that were needed at different steps could be fulfilled together.

Let us remind the parameters used at several steps of the construction: macro-tiles of level $k+1$ consist of $N_k \times N_k$ macro-tile of level k; thus, a k-level macro-tile consists of $L_k \times L_k$ tiles (of level 0), where $L_k = N_0 \cdot \ldots \cdot N_{k-1}$. Macro-tiles of level k are responsible for checking bit blocks of length l_k from their extended responsibility zone (of width $3L_k$). We have several constraints on the values of these parameters:

- $\log N_{k+1} \ll N_k$ and even $\log N_{k+2} \ll N_k$ since every macro-tile must be able to do simple arithmetic manipulations with its own coordinates in the father and with coordinates of the father in the grandfather;
- $N_k \gg L_k$ since we need enough sons of a macro-tile of level $k+1$ to keep all bits from its zone of responsibility (we use one macro-tile of level k for each bit);
- l_k and even l_{k+1} should be much less than N_{k-1} since a macro-tile of level k must contain in its computational zone the bit block of length l_k assigned to itself and three bit blocks of length l_{k+1} assigned to its father and two uncles (the left and right neighbors of the father);
- a k-level macro-tile should enumerate in its computational zone several forbidden patterns and check whether any of them is a substring of the given (assigned to this macro-tile) l_k-bits block; the number of steps in this enumeration must be small compared to the size of the macro-tile; for example, let us agree that a macro-tile of level k runs this enumeration for exactly l_k steps;
- the values N_k and l_k should be simple functions of k: we want to compute l_k in time polynomial in k, and compute N_k in time polynomial in $\log N_k$ (note that typically N_k is much greater than k, so we cannot compute or even write down its binary representation in time polynomial in k).

With all these constraints we are still quite free in the choice of parameters. For example, we may let $N_k = 2^{C2^k}$ (for some large enough constant C) and $l_k = k$.

8 Final Remarks

One may also use essentially the same construction to implement k-dimensional effectively closed subshifts using $(k+1)$-dimensional subshifts of finite type.

How far can we go further? Can we implement evert k-dimensional effectively closed subshifts by a tiling of the same dimension k? Another question (posed in [13]): let us replace a finite alphabet by a Cantor space (with the standard topology); can we represent every k-dimensional effectively closed subshifts over a Cantor space as a continuous image of the set of tilings of dimension $k+1$ (for some finite tile set)? E. Jeandel noticed that the answers to the both questions are negative (this fact is also a corollary of results from [4] and [16]).

References

1. Aubrun, N., Sablik, M.: personal communication (February 2010) (submitted for publication)
2. Berger, R.: The Undecidability of the Domino Problem. Mem. Amer. Math. Soc. 66 (1966)
3. Börger, E., Grädel, E., Gurevich, Y.: The Classical Decision Problem. Springer, Heidelberg (1987)
4. Durand, B., Levin, L., Shen, A.: Complex Tilings. J. Symbolic Logic 73(2), 593–613 (2008)
5. Durand, B., Levin, L., Shen, A.: Local Rules and Global Order, or Aperiodic Tilings. Mathematical Intelligencer 27(1), 64–68 (2005)
6. Durand, B., Romashchenko, A., Shen, A.: Fixed Point and Aperiodic Tilings. In: Ito, M., Toyama, M. (eds.) DLT 2008. LNCS, vol. 5257, pp. 276–288. Springer, Heidelberg (2008)
7. Durand, B., Romashchenko, A., Shen, A.: Fixed point theorem and aperiodic tilings (The Logic in Computer Science Column by Yuri Gurevich). Bulletin of the EATCS 97, 126–136 (2009)
8. Durand, B., Romashchenko, A., Shen, A.: Fixed-point tile sets and their applications. CoRR abs/0910.2415 (2009), `http://arxiv.org/abs/0910.2415`
9. Gács, P.: Reliable Computation with Cellular Automata. J. Comput. Syst. Sci. 32(1), 15–78 (1986)
10. Gács, P.: Reliable Cellular Automata with Self-Organization. J. Stat. Phys. 103(1/2), 45–267 (2001)
11. Grunbaum, B., Shephard, G.C.: Tilings and Patterns. W.H. Freeman & Co., New York (1986)
12. Gurevich, Y., Koryakov, I.: A remark on Berger's paper on the domino problem. Siberian Mathematical Journal 13, 319–321 (1972)
13. Hochman, M.: On the dynamic and recursive properties of multidimensional symbolic systems. Inventiones mathematicae 176, 131–167 (2009)
14. Mozes, S.: Tilings, Substitution Systems and Dynamical Systems Generated by Them. J. Analyse Math. 53, 139–186 (1989)
15. von Neumann, J.: Theory of Self-reproducing Automata. In: Burks, A. (ed.). University of Illinois Press, Urbana (1966)
16. Rumyantsev, A., Ushakov, M.: Forbidden Substrings, Kolmogorov Complexity and Almost Periodic Sequences. In: Durand, B., Thomas, W. (eds.) STACS 2006. LNCS, vol. 3884, pp. 396–407. Springer, Heidelberg (2006)
17. Wang, H.: Proving theorems by pattern recognition II. Bell System Technical Journal 40, 1–42 (1961)

A Fixed Point Theorem and Aperiodic Tilings[4]

People often say about some discovery that it appeared "ahead of time", meaning that it could be fully understood only in the context of ideas developed later. For the topic of this note, the construction of an aperiodic tile set based on the fixed-point (self-referential) approach, the situation is exactly the opposite. It should have been found in 1960s when the question about aperiodic tile sets

[4] Reproduced from [7] with permission.

was first asked: all the tools were quite standard and widely used at that time. However, the history had chosen a different path and many nice geometric *ad hoc* constructions were developed instead (by Berger, Robinson, Penrose, Ammann and many others, see [11]; a popular exposition of Robinson-style construction is given in [5]). In this note we try to correct this error and present a construction that should have been discovered first but seemed to be unnoticed for more that forty years.

A.1 The Statement: Aperiodic Tile Sets

A tile is a square with colored sides. Given a set of tiles, we want to find a tiling, i.e., to cover the plane by (translated copies of) these tiles in such a way that colors match (a common side of two neighbor tiles has the same color in both).Tiles appeared first in the context of *domino problem* posed by Hao Wang. Here is the original formulation from [17]: "Assume we are given a finite set of square plates of the same size with edges colored, each in a different manner. Suppose further there are infinitely many copies of each plate (plate type). We are not permitted to rotate or reflect a plate. The question is to find an effective procedure by which we can decide, for each given finite set of plates, whether we can cover up the whole plane (or, equivalently, an infinite quadrant thereof) with copies of the plates subject to the restriction that adjoining edges must have the same color." This question (domino problem) is closely related to the existence of aperiodic tile sets: (1) if they did not exist, domino problem would be decidable for some simple reasons (one may look in parallel for a periodic tiling or a finite region that cannot be tiled) and (2) the aperiodic tile sets are used in the proof of the undecidability of domino problem. However, in this note we concentrate on aperiodic tile sets only.

For example, if tile set consists of two tiles (one has black lower and left side and white right and top sides, the other has the opposite colors), it is easy to see that only periodic (checkerboard) tiling is possible. However, if we add some other tiles the resulting tile set may admit also non-periodic tilings (e.g., if we add all 16 possible tiles, any combination of edge colors becomes possible). It turns out that there are other tile set that have *only* aperiodic tilings.

Formally: let C be a finite set of *colors* and let $\tau \subset C^4$ be a set of *tiles*; the components of the quadruple are interpreted as upper/right/lower/left colors of a tile. Our example tile set with two tiles is represented then as

Fig. 3. Tile set that has only periodic tilings

$$\{\langle \text{white}, \text{white}, \text{black}, \text{black}\rangle, \langle \text{black}, \text{black}, \text{white}, \text{white}\rangle\}.$$

A τ-*tiling* is a mapping $\mathbb{Z}^2 \to \tau$ that satisfies matching conditions. Tiling U is called *periodic* if it has a *period*, i.e., if there exists a non-zero vector $T \in \mathbb{Z}^2$ such that $U(x+T) = U(x)$ for all x.

Now we can formulate the result (first proven by Berger [2]):

Proposition. *There exists a finite tile set τ such that τ-tilings exist but all of them are aperiodic.*

There is a useful reformulation of this result. Instead of tilings we can consider two-dimensional infinite words in some finite alphabet A (i.e., mappings of type $\mathbb{Z}^2 \to A$) and put some local constraints on them. This means that we choose some positive integer N and look at the word through a window of size $N \times N$. Local constraint then says which patterns of size $N \times N$ are allowed to appear in a window. Now we can reformulate our Proposition as follows: *there exists a local constraint that is consistent* (*some infinite words satisfy it*) *but implies aperiodicity* (*all satisfying words are aperiodic*).

It is easy to see that these two formulations are equivalent. Indeed, the color matching condition is 2×2 checkable. On the other hand, any local constraint can be expressed in terms of tiles and colors if we use $N \times N$-patterns as tiles and $(N-1) \times N$-patterns as colors; e.g., the right color of $(N \times N)$-tile is the tile except for its left column; if it matches the left color of the right neighbor, these two tiles overlap correctly.

A.2 Why Theory of Computation?

At first glance this proposition has nothing to do with theory of computation. However, the question appeared in the context of the undecidability of some logical decision problems, and, as we shall see, can be solved using theory of computations. (A rare chance to convince "normal" mathematicians that theory of computations is useful!)

The reason why theory of computation comes into play is that rules that determine the behavior of a computation device — say, a Turing machine with one-dimensional tape — can be transformed into local constraints for the space-time diagram that represents computation process. So we can try to prove the proposition as follows: consider a Turing machine with a very complicated (and therefore aperiodic) behavior and translate its rules into local constraints; then any tiling represents a time-space diagram of a computation and therefore is aperiodic.

However, this naïve approach does not work since local constraints are satisfied also at the places where no computation happens (in the regions that do not contain the head of a Turing machine) and therefore allow periodic configurations. So a more sophisticated approach is needed.

A.3 Self-similarity

The main idea of this more sophisticated approach is to construct a "self-similar" set of tiles. Informally speaking, this means that any tiling can be uniquely split by vertical and horizontal lines into $M \times M$ blocks that behave exactly like the individual tiles. Then, if we see a tiling and zoom out with scale $1 : M$, we get a tiling with the same tile set.

Let us give a formal definition. Assume that a non-empty set of tiles τ and positive integer $M > 1$ are fixed. A *macro-tile* is a square of size $M \times M$ filled with matching tiles from τ. Let ρ be a non-empty set of macro-tiles.

Definition. We say that τ *implements* ρ if any τ-tiling can be uniquely split by horizontal and vertical lines into macro-tiles from ρ.

Now we give two examples that illustrate this definition: one negative and one positive.

Negative example: Consider a set τ that consists of one tile with all white sides. Then there is only one macro-tile (of given size $M \times M$). Let ρ be a one-element set that consists of this macro-tile. Any τ-tiling (i.e., the only possible τ-tiling) can be split into ρ-macro-tiles. However, the splitting lines are not unique, so τ does *not* implements ρ.

Positive example: Let τ is a set of M^2 tiles that are indexed by pairs of integers modulo M: The colors are pairs of integers modulo M arranged as shown (Fig. 4). Then there exists only one τ-tiling (up to translations), and this tiling can be uniquely split into $M \times M$ squares whose borders have colors $(0, j)$ and $(i, 0)$. Therefore, τ implements a set ρ that consists of one macro-tile (Fig. 5).

Definition. A set of tiles τ is *self-similar* if it implements some set of macro-tiles ρ that is isomorphic to τ.

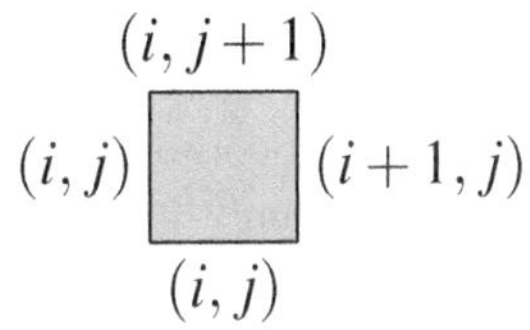

Fig. 4. Elements of τ (here i, j are integers modulo M)

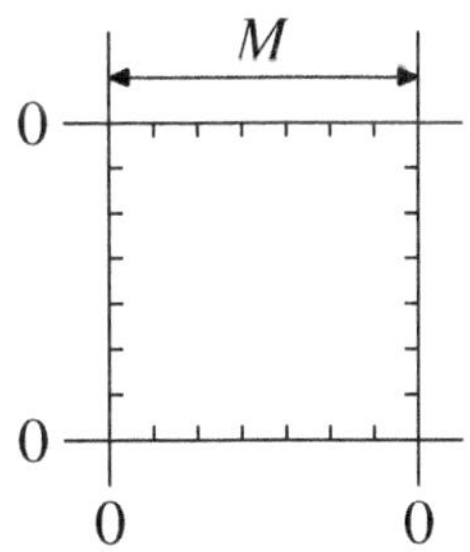

Fig. 5. The only element of ρ: border colors are pairs that contain 0

This means that there exist a 1-1-correspondence between τ and ρ such that matching pairs of τ-tiles correspond exactly to matching pairs of ρ-macro-tiles.

The following statement follows directly from the definition:

Proposition. *A self-similar tile set τ has only aperiodic tilings.*

Proof. Let T be a period of some τ-tiling U. By definition U can be uniquely split into ρ-macro-tiles. Shift by T should respect this splitting (otherwise we get a different splitting), so T is a multiple of M. Zooming the tiling and replacing each ρ-macro-tile by a corresponding τ-tile, we get a T/M-shift of a τ-tiling. For the same reason T/M should be a multiple of M, then we zoom out again etc. We conclude therefore that T is a multiple of M^k for any k, i.e., T is a zero vector. □

Note also that any self-similar set τ has at least one tiling. Indeed, by definition we can tile a $M \times M$ square (since macro-tiles exist). Replacing each τ-tile by a corresponding macro-tile, we get a τ-tiling of $M^2 \times M^2$ square, etc. In this way we can tile an arbitrarily large finite region, and then standard compactness argument (König's lemma) shows that we can tile the entire plane.

So it remains to construct a self-similar set of tiles (a set of tiles that implements itself, up to an isomorphism).

A.4 Fixed Points and Self-referential Constructions

The construction of a self-similar tile set is done in two steps. First (in Section A.5) we explain how to construct (for a given tile set σ) another tile set τ that implements σ (i.e., implements a set of macro-tiles isomorphic to σ). In this construction the tile set σ is given as a program p_σ that checks whether four bit strings (representing four side colors) appear in one σ-tile. The tile set τ then guarantees that each macro-tile encodes a computation where p_σ is applied to these four strings ("macro-colors") and accepts them.

This gives us a mapping: for every σ we have $\tau = \tau(\sigma)$ that implements σ and depends on σ. Now we need a fixed point of this mapping where $\tau(\sigma)$ is isomorphic to σ. It is done (Section A.6) by a classical self-referential trick that appeared as liar's paradox, Cantor's diagonal argument, Russell's paradox, Gödel's (first) incompleteness theorem, Tarski's theorem, undecidability of the Halting problem, Kleene's fixed point (recursion) theorem and von Neumann's construction of self-reproducing automata — in all these cases the core argument is essentially the same.

The same trick is used also in a classical programming challenge: to write a program that prints its own text. Of course, for every string s it is trivial to write a program $t(s)$ that prints s, but how do we get $t(s) = s$? It seems at first that $t(s)$ should incorporate the string s itself plus some overhead, so how $t(s)$ can be equal to s? However, this first impression is false. Imagine that our computational device is a universal Turing machine U where the program is written in a special read-only layer of the tape. (This means that the tape alphabet is a Cartesian product of two components, and one of the components is used for the program and is never changed by U.) Then the program can get access to its own text at any moment, and, in particular, can copy it to the output tape.[2] Now we explain in more details how to get a self-similar tile set according to this scheme.

A.5 Implementing a Given Tile Set

In this section we show how one can implement a given tile set σ, or, better to say, how to construct a tile set τ that implements some set of macro-tiles that is isomorphic to σ.

There are easy ways to do this. Though we cannot let $\tau = \sigma$ (recall that zoom factor M should be greater than 1), we can do essentially the same for every $M > 1$. Let us extend our "positive" example (with one macro-tile and M^2 tiles) by superimposing additional colors. Superimposing two sets of colors means the we consider the Cartesian product of color sets (so each edge carries a pair of colors). One set of colors remains the same (M^2 colors for M^2 pairs of integers modulo M). Let us describe additional (superimposed) colors. Internal edges of each macro-tile should have the same color and this color should be different for all macro-tiles, so we allocate $\#\sigma$ colors for that. This gives $\#\sigma$ macro-tiles that can be put into 1-1-correspondence with σ-tiles. It remains to provide correct border colors, and this is easy to do since each tile "knows" which σ-tile it simulates (due to the internal color). In this way we get $M^2\#\sigma$ tiles that implement the tile set σ with zoom factor M.

However, this (trivial) simulation is not really useful. Recall that our goal is to get isomorphic σ and τ, and in this implementation τ-tiles have more colors that σ-tiles (and we have more tiles, too). So we need a more creative encoding of σ-colors that makes use of the space available: a side of a macro-tile has a "macro-color" that is a sequence of M tile colors, and we can have a lot of macro-colors in this way.

So let us assume that colors in σ are k-bit strings for some k. Then the tile set is a subset $S \subset \mathbb{B}^k \times \mathbb{B}^k \times \mathbb{B}^k \times \mathbb{B}^k$, i.e., a 4-ary predicate on the set $\mathbb{B}^k$ of k-bit strings. Assume that S is presented by a program that computes Boolean value $S(x, y, z, w)$ given four k-bit strings x, y, z, w. Then we can construct a tile set τ as follows.

We start again with a set of M^2 tiles from our example and superimpose additional colors but use them in a more economical way. Assuming that $k \ll M$, we allocate k places in the middle of each side of a macro-tile and allow each of them to carry an additional color bit; then a macro-color represents a k-bit

[2] Of course, this looks like cheating: we use some very special universal machine as an interpreter of our programs, and this makes our task easy. Teachers of programming that are seasoned enough may recall the BASIC program

```
10 LIST
```

that indeed prints its own text. However, this trick can be generalized enough to show that a self-printing program exists in every language.

string. Then we need to arrange the internal colors in such a way that macro-colors (k-bit strings) x, y, z and w can appear on the four sides of a macro-tile if and only if $S(x, y, z, w)$ is true.

To achieve this goal, let us agree that the middle part (of size, say, $M/2 \times M/2$) in every $M \times M$-macro-tile is a "computation zone". Tiling rules (for superimposed colors) in this zone guarantee that it represents a time-space diagram of a computation of some (fixed) universal Turing machine. (We assume that time goes up in a vertical direction and the tape is horizontal.) It is convenient to assume that program of this machine is written on a special read-only layer of the tape (see the discussion in Section A.4).

Outside the computation zone the tiling rules guarantee that bits are transmitted from the sides to the initial configuration of a computation.

We also require that this machine should accept its input before running out of time (i.e., less than in $M/2$ steps), otherwise the tiling is impossible.

Note that in this description different parts of a macro-tile behave differently; this is OK since we start from our example where each tile "knows" its position in a macro-tile (keeps two integers modulo M). So the tiles in the "wire" zone know that they should transmit a bit, the tiles inside the computation zone know they should obey the local rules for time-space diagram of the computation, etc.

This construction uses only bounded number of additional colors since we have fixed the universal Turing machine (including its alphabet and number of states); we do not need to increase the number of colors when we increase M and k (though k should be small compared to M to leave enough space for the wires; we do not give an exact position of the wires but it is easy to see that if

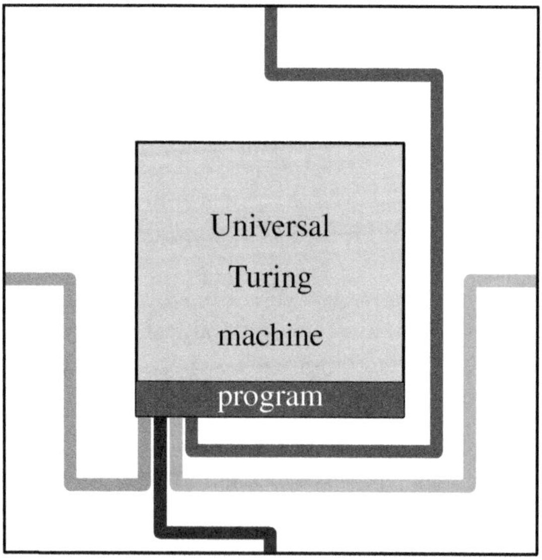

Fig. 6. k-macro-colors are transmitted to the computation zone where they are checked

k/M is small enough, there is enough space for them). So the construction uses $O(M^2)$ colors (and tiles).

A.6 A Tile Set That Implements Itself

Now we come to the crucial point in our argument: can we arrange things in such a way that the predicate S (i.e., the tile set it generates) is isomorphic to the set of tiles τ used to implement it?

Assume that $k = 2\log M + O(1)$; then macro-colors have enough space to encode the coordinates modulo M plus superimposed colors (which require $O(1)$ bits for encoding).

Note that many of the rules that define τ do not depend on σ (i.e., on the predicate S). So the program for the universal Turing machine may start by checking these rules. It should check that

- bits that represent coordinates (integers modulo M) on the four sides of a macro-tile are related in the proper way (left and lower sides have identical coordinates, on the right/upper side one of the coordinates increases modulo M);
- if the macro-tile is outside computation zone and the wires, it does not carry additional colors;
- if the macro-tile is a part of a wire, then it transmits a bit in a required direction (of course, for this we should fix the position of the wires by some formulas that are then checked by a program);
- if the macro-tile is a part of the computation zone, it should obey the local rules for the computation zone (bits of the read-only layer should propagate vertically, bits that encode the content of the tape and the head of our universal Turing machine should change as time increases according to the behavior of this machine, etc.)

This guarantees that on the next layer macro-tiles are grouped into macro-macro-tiles where bits are transmitted correctly to the computation zone of a macro-macro-tile and *some* computation of the universal Turing machine is performed in this zone. But we need more: this computation should be the same computation that is performed on the macro-tile level (fixed point!). This is also easy to achieve since in our model the text of a running program is available to it (recall the we assume that the program is written in a read-only layer): the program should check also that *if a macro-tile is in the computation zone, then the program bit it carries is correct* (program knows the x-coordinate of a macro-tile, so it can go at the corresponding place of its own tape to find out which program bit resides in this place).

This sound like some magic, but we hope that our previous example (a program for the UTM that prints its own text) makes this trick less magical (indeed, reliable and reusable magic is called technology).

A.7 So What?

We believe that our proof is rather natural. If von Neumann lived few years more and were asked about aperiodic tile sets, he would probably immediately give this argument as a solution. (He was especially well prepared to it since he used very similar self-referential tricks to construct a self-reproducing automata, see [15].) In fact this proof somehow appeared, though not very explicitly, in P. Gács' papers on cellular automata [10]; the attempts to understand these papers were our starting points.

This proof is rather flexible and can be adapted to get many results usually associated with aperiodic tilings: undecidability of domino problem (Berger [2]), recursive inseparability of periodic tile sets and inconsistent tile sets (Gurevich – Koryakov [12]), enforcing substitution rules (Mozes [14]) and others (see [3,6]). But does it give something new?

We believe that indeed there are some applications that hardly could be achieved by previous arguments. Let us conclude by mentioning two of them. First is the construction of *robust* aperiodic tile sets. We can consider tilings with holes (where no tiles are placed and therefore no matching rules are checked). A robust aperiodic tile set should have the following property: if the set of holes is "sparse enough", then tiling still should be far from any periodic pattern (say, in the sense of Besicovitch distance, i.e., the limsup of the fraction of mismatched positions in a centered square as the size of the square goes to infinity). The notion of "sparsity" should not be too restrictive here; we guarantee, for example, that a Bernoulli random set with small enough probability p (each cell belongs to a hole independently with probability p) is sparse.

While the first example (robust aperiodic tile sets) is rather technical (see [6] for details), the second is more basic. Let us split all tiles in some tile set into two classes, say, A- and B-tiles. Then we consider a fraction of A-tiles in a tiling. If a tile set is not restrictive (allows many tilings), this fraction could vary from one tiling to another. For classical aperiodic tilings this fraction is usually fixed: in a big tiled region the fraction of A-tiles is close to some limit value, usually an eigenvalue of an integer matrix (and therefore an algebraic number). The fixed-point construction allows us to get any computable number. Here is the formal statement: *for any computable real $\alpha \in [0,1]$ there exists a tile set τ divided into A- and B-tiles such that for any $\varepsilon > 0$ there exists N such that for all $n > N$ the fraction of A-tiles in any τ-tiling of $n \times n$-square is between $\alpha - \varepsilon$ and $\alpha + \varepsilon$.*

The Model Checking Problem for Prefix Classes of Second-Order Logic: A Survey*

Thomas Eiter[1], Georg Gottlob[2], and Thomas Schwentick[3]

[1] Institute of Information Systems, Vienna University of Technology, Austria
eiter@kr.tuwien.ac.at
[2] Computing Laboratory, Oxford University, United Kingdom
georg.gottlob@comlab.ox.ac.uk
[3] Fakultät für Informatik, Technische Universität Dortmund, Germany
thomas.schwentick@udo.edu

Abstract. In this paper, we survey results related to the model checking problem for second-order logic over classes of finite structures, including word structures (strings), graphs, and trees, with a focus on prefix classes, that is, where all quantifiers (both first- and second-order ones) are at the beginning of formulas. A complete picture of the prefix classes defining regular and non-regular languages over strings is known, which nearly completely coincides with the tractability frontier; some complexity issues remain to be settled, though. Over graphs and arbitrary relational structures, the tractability frontier is completely delineated for the existential second-order fragment, while it is less explored for trees. Besides surveying some of the results, we mention some open issues for research.

Keywords: Finite Model Theory, Gurevich's Classifiability Theorem, Model Checking, Monadic Second-Order Logic, Regular Languages, Second-Order Logic.

For Yuri, on the occasion of his seventieth birthday.

1 Introduction

Logicians and computer scientists have been studying for a long time the relationship between fragments of predicate logic and the solvability and complexity of decision problems that can be expressed within such fragments. Among the studied fragments, quantifier prefix classes play a predominant role. This can be explained by the syntactical simplicity of such prefix classes and by the fact that they form a natural hierarchy of increasingly complex fragments of logic that appears to be deeply related to core issues of decidability and complexity. In fact, one of the most fruitful research programs that kept logicians and computer scientists busy for decades was the exhaustive solution of Hilbert's classical Entscheidungsproblem, that is, of the problem of determining those prefix classes of first-order logic for which formula-satisfiability (resp. finite satisfiability of formulas) is decidable. A landmark reference (also sometimes called the "bible")

* Most of the material contained in this paper stems, modulo editorial adaptations, from the much longer papers [15,16,26]. This paper significantly extends the earlier report [20].

A. Blass, N. Dershowitz, and W. Reisig (Eds.): Gurevich Festschrift, LNCS 6300, pp. 227–250, 2010.

on this subject is the book by Börger, Gurevich, and Grädel [6] which gives an in depth treatment of the subject.

Quantifier prefixes emerged not only in the context of decidability theory (a common branch of recursion theory and theoretical computer science), but also in core areas of computer science such as formal language and automata theory, and later in complexity theory. In automata theory, Büchi [9,8], Elgot [18] and Trakhtenbrot [51] independently proved that a language is regular iff it can be described by a sentence of monadic second-order logic, in particular, by a sentence of monadic existential second-order logic. In complexity theory, Fagin [19] showed that a problem on finite structures is in NP iff it can be described by a sentence of existential second-order logic (ESO). These fundamental results have engendered a large number of further investigations and results on characterizing language and complexity classes by fragments of logic (see, for instance the monographs [45,42,13,32]).

While the classical research programme of determining the prefix characterizations of decidable fragments of first-order logic was successfully completed around 1984 (cf. [6]), until recently little was known on analogous problems on finite structures, in particular, on the tractability/intractability frontier of the model checking problem for prefix classes of second-order logic (SO), and in particular, of existential second-order logic (ESO) over finite structures. In the late nineties, a number of scientists, including Yuri Gurevich, Phokion Kolaitis, and the authors started to attack this new research programme in a systematic manner.

By *complexity of a prefix class* C we mean the complexity of the following model-checking problem: Given a fixed sentence Φ in C, decide for variable finite structures A whether A is a model of Φ, which we denote by $A \models \Phi$. Determining the complexity of all prefix classes is an ambitious research programme, in particular the analysis of various types of finite structures such as *strings*, that is, finite word structures with successor, *trees*, *graphs*, or arbitrary finite relational structures (corresponding to relational databases). Over strings and trees, one of the main goals of this classification is to determine the *regular* prefix classes, that is, those whose formulas express regular languages only; note that by the Büchi-Elgot-Trakhtenbrot Theorem, regular fragments over strings are (semantically) included in monadic second-order logic.

In the context of this research programme, three systematic studies were carried out recently, that shed light on the prefix classes of the existential fragment ESO (also denoted by Σ_1^1) of second-order logic:

- In [15], the ESO prefix-classes over strings are exhaustively classified. In particular, the precise frontier between regular and nonregular classes is traced out, and it is shown that every class that expresses some nonregular language also expresses some NP-complete language. There is thus a huge complexity gap in ESO: some prefix classes can express only regular languages (which are well-known to have extremely low complexity), while all others are intractable.
- In [16] this line of research was continued by systematically investigating the syntactically more complex prefix classes $\Sigma_k^1(\mathcal{Q})$ of second-order logic for each integer $k > 1$ and for each first-order quantifier prefix $\mathcal{Q}$. An exhaustive classification of the regular and nonregular prefix classes of this form is given, and complexity results for the corresponding model-checking problems are derived.

- In [26], the complexity of all ESO prefix-classes over graphs and arbitrary relational structures is analyzed, and the tractability/intractability frontier is completely delineated. Unsurprisingly, several classes that are regular over strings become NP-hard over graphs. Interestingly, the analysis shows that one of the NP-hard classes becomes polynomial for the restriction to undirected graphs without self-loops.
- The precise tractability frontier of ESO and SO prefix classes over *trees* has not yet been determined. There are some partial results, however. There are also important complexity results for MSO over trees, as well as a number of expressiveness results that show that MSO over trees is captured by regular automata and equivalent in expressive power to simpler formalisms.

In this paper, we review these results. After relevant definitions and recalling some classical results in the next section, we start with a brief survey of the results on ESO prefix-classes over strings in [15] (Sect. 3), followed by full second-order logic over strings (Sect. 4), where we also consider finer grained complexity than regularity vs. non-regularity. After that, we turn to ESO over graphs [25] (Sec. 5) and then discuss SO and MSO over trees (Sect. 6). The final Sect. 7 addresses further issues and open problems.

2 Preliminaries and Classical Results

We consider second-order logic with equality (unless explicitly stated otherwise) and without function symbols of positive arity. Predicates are denoted by capitals and individual variables by lower case letters; a bold face version of a letter denotes a tuple of corresponding symbols.

A *prefix* is any string over the alphabet $\{\exists, \forall\}$, and a *prefix set* is any language $\mathcal{Q} \subseteq \{\exists, \forall\}^*$ of prefixes. A prefix set $\mathcal{Q}$ is *trivial*, if $\mathcal{Q} = \emptyset$ or $\mathcal{Q} = \{\lambda\}$, that is, it consists of the empty prefix. In the rest of this paper, we focus on nontrivial prefix sets. We often view a prefix Q as the prefix class $\{Q\}$. A *generalized prefix* is any string over the extended prefix alphabet $\{\exists, \forall, \exists^*, \forall^*\}$. A prefix set $\mathcal{Q}$ is *standard*, if either $\mathcal{Q} = \{\exists, \forall\}^*$ or $\mathcal{Q}$ can be given by some generalized prefix.

For any prefix Q, the class $\Sigma^1_0(Q)$ is the set of all prenex first-order formulas (which may contain free variables and constants) with prefix Q, and for every $k \geq 0$, $\Sigma^1_{k+1}(Q)$ (resp., Π^1_{k+1}) is the set of all formulas $\exists \mathbf{R} \Phi$ (resp., $\forall \mathbf{R} \Phi$) where Φ is from Π^1_k (resp., Σ^1_k). For any prefix set $\mathcal{Q}$, the class $\Sigma^1_k(\mathcal{Q})$ is the union $\Sigma^1_k(\mathcal{Q}) = \bigcup_{Q \in \mathcal{Q}} \Sigma^1_k(Q)$. We write also ESO for Σ^1_1. For example, ESO($\exists^*\forall\exists^*$) is the class of all formulas $\exists \mathbf{R} \exists \mathbf{y} \forall x \exists \mathbf{z} \varphi$, where φ is quantifier-free; this is the class of ESO-prefix formulas, whose first-order part is in the well-known Ackermann class with equality.

Let $A = \{a_1, \ldots, a_m\}$ be a finite alphabet. A *string* over A is a finite first-order structure $W = \langle U, C^W_{a_1}, \ldots, C^W_{a_m}, Succ^W, min^W, max^W \rangle$, for the vocabulary $\sigma_A = \{C_{a_1}, \ldots, C_{a_m}, Succ, min, max\}$, where

- U is a nonempty finite initial segment $\{1, 2, \ldots, n\}$ of the positive integers;
- each $C^W_{a_i}$ is a unary relation over U (that is, a subset of U) for the unary predicate C_{a_i}, for $i = 1, \ldots, m$, such that the $C^W_{a_i}$ are pairwise disjoint and $\bigcup_i C^W_{a_i} = U$.

- $Succ^W$ is the usual successor relation on U and min^W and max^W are the first and the last element in U, respectively.

Observe that this representation of a string is a *successor structure* as discussed for instance in [14]. An alternative representation uses a standard linear order $<$ on U instead of the successor $Succ$. In full ESO or monadic second-order logic, $<$ is tantamount to $Succ$ since either predicate can be defined in terms of the other.

The strings W for A correspond to the nonempty finite words over A in the obvious way; in abuse of notation, we often use W in place of the corresponding word from A^* and vice versa.

A *SO sentence* Φ over the vocabulary σ_A is a second-order formula whose only free variables are the predicate variables of the signature σ_A, and in which no constant symbols except min and max occur. Such a sentence defines a language over A, denoted $\mathcal{L}(\Phi)$, given by $\mathcal{L}(\Phi) = \{W \in A^* \mid W \models \Phi\}$. We say that a language $L \subseteq A^*$ is *expressed* by Φ, if $\mathcal{L}(\Phi) = L \cap A^+$ (thus, for technical reasons, without loss of generality we disregard the empty string); L is *expressed by a set* S of sentences, if L is expressed by some $\Phi \in S$. We say that S *captures* a class C of languages, if S expresses all and only the languages in C.

Example 2.1. Let us consider some languages over the alphabet $A = \{a, b\}$, and how they can be expressed using logical sentences.

- $L_1 = \{a,b\}^* b \{a,b\}^*$: this language is expressed by the simple sentence

$$\exists x. C_b(x).$$

- $L_2 = a^* b$: this language is expressed by the sentence

$$C_b(max) \wedge \forall x. x \neq max \rightarrow C_a(x).$$

- $L_3 = (ab)^*$: using the successor predicate, we can express this language by

$$C_a(min) \wedge C_b(max) \wedge \forall x, y. Succ(x, y) \rightarrow (C_a(x) \leftrightarrow \neg C_a(y)).$$

- $L_4 = \{w \in \{a,b\}^* \mid |w| = 2n, n \geq 1\}$: we express this language by the sentence

$$\exists E \, \forall x, y. \neg E(min) \wedge E(max) \wedge Succ(x, y) \rightarrow (E(x) \leftrightarrow \neg E(y)).$$

  Note that this a monadic ESO sentence. It postulates the existence of a monadic predicate E, that is, a "coloring" of the string such that neighbored positions have different color, and the first and last position are uncolored and colored, respectively.

- $L_5 = \{a^n b^n \mid n \geq 1\}$: Expressing this language is more involved:

$$\begin{aligned} \exists R \, \forall x, x^+, y, y^-. R(min, max) \wedge [R(x, y) \rightarrow (C_a(x) \wedge C_b(y))] \wedge \\ [(Succ(x, x^+) \wedge Succ(y^-, y) \wedge R(x, y)) \rightarrow R(x^+, y^-)]. \end{aligned}$$

  Observe that this sentence is not monadic. Informally, it postulates the existence of an arc from the first to the last position of the string W, which must be an a and a b, respectively, and recursively arcs from the i-th to the $(|W| - i + 1)$-th position.

Let A be a finite alphabet. A sentence Φ over σ_A is called *regular*, if $\mathcal{L}(\Phi)$ is a regular language. A set of sentences S (in particular, any ESO-prefix class) is *regular*, if for every finite alphabet A, all sentences $\Phi \in S$ over σ_A are regular.

Büchi [9] has shown the following fundamental theorem, which was independently found by Elgot [18] and Trakhtenbrot [51]. Denote by MSO the fragment of second-order logic in which all predicate variables have arity at most one,[1] and let REG denote the class of regular languages.

Proposition 2.1 (Büchi-Elgot-Trakhtenbrot Theorem). MSO *captures* REG.

That MSO can express all regular languages is easy to see, since it is straightforward to describe runs of a finite state automaton by an existential MSO sentence. In fact, this is easily possible in monadic ESO($\forall\exists$) as well as in monadic ESO($\forall\forall$). Thus, we have the following lower expressiveness bound on ESO-prefix classes over strings.

Proposition 2.2. *Let $\mathcal{Q}$ be any prefix set. If $\mathcal{Q} \cap \{\exists, \forall\}^*\forall\{\exists, \forall\}^+ \neq \emptyset$, then* ESO($\mathcal{Q}$) *expresses all languages in* REG.

On the other hand, with non-monadic predicates allowed ESO has much higher expressivity. In particular, by Fagin's result [19], we have the following.

Proposition 2.3 (Fagin's Theorem). ESO *captures* NP.

This theorem can be sharpened to various fragments of ESO. In particular, by Leivant's results [36,14], in the presence of a successor and constants min and max, the fragment ESO($\forall^*$) captures NP; thus, ESO($\forall^*$) expresses all languages in NP.

Before proceeding with the characterization of the regular languages by nonmonadic fragments of ESO, we would like to note that many papers cover either extensions or restrictions of MSO or REG, and cite some relevant results.

Lynch [37], has studied the logic over strings obtained from existential MSO by adding addition. He proved that model checking for this logic lies in NTIME(n), that is, in nondeterministic linear time. Grandjean and Olive [29,40] obtained interesting results related to those of Lynch. They gave logical representations of the class NLIN, that is, linear time on random access machines, in terms of second-order logic with unary functions instead of relations (in their setting, also the input string is represented by a function).

Lautemann, Schwentick and Thérien [35] proved that the class CFL of context-free languages is characterized by ESO formulas of the form $\exists B\varphi$ where φ is first-order, B is a binary predicate symbol, and the range of the second-order quantifier is restricted to the class of *matchings*, that is, pairing relations without crossover. Note that this is not a purely prefix-syntactic characterization of CFL. From our results and the fact that some languages which are not context-free can be expressed in the minimal nonregular ESO-prefix classes, it follows that a syntactic characterization of CFL by means of ESO-prefix classes is impossible.

Several restricted versions of REG where studied and logically characterized by restricted versions of ESO. McNaughton and Papert [38] showed that first-order logic

[1] Observe that we assume MSO allows one to use nullary predicate variables (that is, propositional variables) along with unary predicate variables. Obviously, the Büchi-Elgot-Trakhtenbrot Theorem survives.

with a linear ordering precisely characterizes the *star-free* regular languages. This theorem was extended by Thomas [49] to ω-languages, that is, languages with infinite words. Later several hierarchies of the star-free languages were studied and logically characterized (see, for instance [49,41,42,43]). Straubing, Thérien and Thomas [46] showed that first-order logic with modular counting quantifiers characterize the regular languages whose syntactic monoids contain only solvable groups. These and many other related results can be found in the books and surveys [45,49,41,42,43].

3 Results on ESO over Strings

Combining and extending the results of Büchi et al. and Fagin, it is natural to ask: What about (nonmonadic) prefix classes ESO($\mathcal{Q}$) over finite strings? We know by Fagin's Theorem that all these classes describe languages in NP. But there is a large spectrum of languages contained in NP ranging from regular languages (at the bottom) to NP-hard languages at the top. What can be said about the languages expressed by a given prefix class ESO($\mathcal{Q}$)? Can the expressive power of these fragments be characterized? In order to clarify these issues, the following particular problems where investigated in [15]:

- Which classes ESO($\mathcal{Q}$) express only regular languages? In other terms, for which fragments ESO($\mathcal{Q}$) is it true that for any sentence $\Phi \in$ ESO($\mathcal{Q}$) the set $Mod(\Phi) = \{W \in A^* \mid W \models \Phi\}$ of all finite strings (over a given finite alphabet A) satisfying Φ constitutes a regular language? By the Büchi-Elgot-Trakthenbrot Theorem, this question is identical to the following: Which prefix classes of ESO are (semantically) included in MSO?

  Note that by Gurevich's fundamental Classifiability Theorem (cf. [6]) and by elementary closure properties of regular languages, it follows that there is *a finite number of maximal regular prefix classes* ESO($\mathcal{Q}$), and similarly, of minimal nonregular prefix classes; the latter are, moreover, standard prefix classes (cf. Sect. 2). It was the aim of [15] to determine the maximal regular prefix classes and the minimal nonregular prefix classes.
- What is the complexity of model checking (over strings) for the *nonregular* classes ESO($\mathcal{Q}$), that is, deciding whether $W \models \Phi$ for a given W (where Φ is fixed)?

  Model checking for regular classes ESO($\mathcal{Q}$) is easy: it is feasible by a finite state automaton. We also know (for instance by Fagin's Theorem) that some classes ESO($\mathcal{Q}$) allow us to express NP-complete languages. It is therefore important to know (i) which classes ESO($\mathcal{Q}$) can express NP-complete languages, and (ii) whether there are prefix classes ESO($\mathcal{Q}$) of intermediate complexity between regular and NP-complete classes.
- Which classes ESO($\mathcal{Q}$) *capture* the class REG? By the Büchi-Elgot-Trakhtenbrot Theorem, this question is equivalent to the question of which classes ESO($\mathcal{Q}$) have exactly the expressive power of MSO over strings.
- For which classes ESO($\mathcal{Q}$) is finite satisfiability decidable, that is, given a formula $\Phi \in$ ESO($\mathcal{Q}$), decide whether Φ is true on some finite string ?

Reference [15] answers all the above questions exhaustively. Some of the results are rather unexpected. In particular, a surprising dichotomy theorem is proven, which sharply classifies all ESO($\mathcal{Q}$) classes as either regular or intractable. Among the main results of [15] are the following findings.

(1) The class ESO($\exists^*\forall\exists^*$) is regular. This theorem is the technically most involved result of [15]. Since this class is nonmonadic, it was not possible to exploit any of the ideas underlying Büchi's proof for proving it regular. The main difficulty consists in the fact that relations of higher arity may connect elements of a string that are very distant from one another; it was not *a priori* clear how a finite state automaton could guess such connections and check their global consistency. To solve this problem, new combinatorial methods (related to hypergraph transversals) were developed.

Interestingly, model checking for the fragment ESO($\exists^*\forall\exists^*$) is NP-complete over *graphs*. For example, the well-known set-splitting problem can be expressed in it. Thus the fact that our input structures are monadic *strings* is essential (just as for MSO).
(2) The class ESO($\exists^*\forall\forall$) is regular. The regularity proof for this fragment is easier but also required the development of new techniques (more of logical than of combinatorial nature). Note that model checking for this class, too, is NP-complete over graphs.
(3) Any class ESO($\mathcal{Q}$) not contained in ESO($\exists^*\forall\exists^*$) $\cup$ ESO($\exists^*\forall\forall$) is not regular.

Thus ESO($\exists^*\forall\exists^*$) and ESO($\exists^*\forall\forall$) are the *maximal regular standard prefix classes*. The unique maximal (general) regular ESO-prefix class is the union of the two classes, that is, ESO($\exists^*\forall\exists^*$) $\cup$ ESO($\exists^*\forall\forall$) $=$ ESO($\exists^*\forall(\forall \cup \exists^*)$).

As shown in [15], it turns out that there are three minimal nonregular ESO-prefix classes, namely the standard prefix classes ESO($\forall\forall\forall$), ESO($\forall\forall\exists$), and ESO($\forall\exists\forall$). All these classes express nonregular languages by sentences whose list of second-order variables consists of a single binary predicate variable.

Thus, 1.-3. give a *complete characterization* of the regular ESO($\mathcal{Q}$) classes.
(4) The following dichotomy theorem is derived: Let ESO($\mathcal{Q}$) be any prefix class. Then, either ESO($\mathcal{Q}$) is regular, or ESO($\mathcal{Q}$) expresses some NP-complete language. This means that model checking for ESO($\mathcal{Q}$) is either possible by a deterministic finite automaton (and thus in constant space and linear time) or it is already NP-complete. Moreover, for all NP-complete classes ESO($\mathcal{Q}$), NP-hardness holds already for sentences whose list of second-order variables consists of a single binary predicate variable. There are no fragments of intermediate difficulty between REG and NP.
(5) The above dichotomy theorem is paralleled by the solvability of the finite satisfiability problem for ESO (and thus FO) over strings. As shown in [15], over finite strings satisfiability of a given ESO($\mathcal{Q}$) sentence is decidable iff ESO($\mathcal{Q}$) is regular.
(6) In [15], a precise characterization is given of those prefix classes of ESO which are equivalent to MSO over strings, that is of those prefix fragments that *capture* the class REG of regular languages. This provides new logical characterizations of REG. Moreover, in [15] it is established that any regular ESO-prefix class is over strings either equivalent to full MSO, or is contained in first-order logic, in fact, in FO($\exists^*\forall$).

It is further shown that ESO($\forall^*$) is the unique minimal ESO prefix class which captures NP. The proof uses results in [36,14] and well-known hierarchy theorems.

The main results of [15] are summarized in Fig. 1. In this figure, the ESO-prefix classes are divided into four regions. The upper two contain all classes that express nonregular languages, and thus also NP-complete languages. The uppermost region contains those classes which capture NP; these classes are called NP-*tailored*. The region next below,

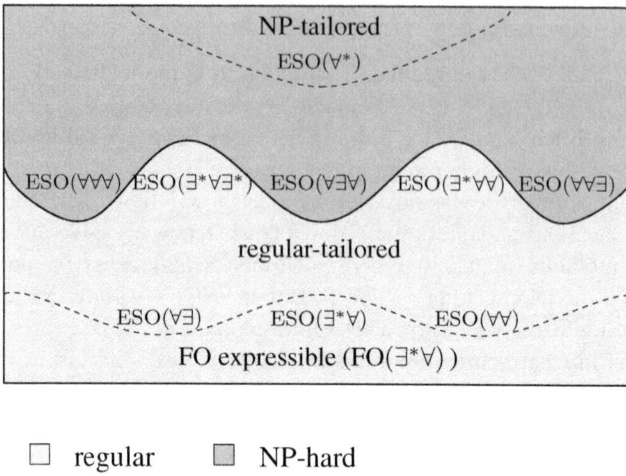

Fig. 1. Complete picture of the ESO-prefix classes on finite strings

separated by a dashed line, contains those classes which can express some NP-hard languages, but not all languages in NP. Its bottom is constituted by the minimal non-regular classes, ESO($\forall\forall\forall$), ESO($\forall\exists\forall$), and ESO($\forall\forall\exists$). The lower two regions contain all regular classes. The maximal regular standard prefix classes are ESO($\exists^*\forall\exists^*$) and ESO($\exists^*\forall\forall$). The dashed line separates the classes which capture REG(called *regular-tailored*), from those which do not; the expressive capability of the latter classes is restricted to first-order logic (in fact, to FO($\exists^*\forall$)) [15]. The minimal classes which capture REG are ESO($\forall\exists$) and ESO($\forall\forall$).

Potential Applications. Monadic second-order logic over strings is currently used in the verification of hardware, software, and distributed systems. An example of a specific tool for checking specifications based on MSO is the MONA tool developed at the BRICS research lab in Denmark [3,31].

Observe that certain interesting desired properties of systems are most naturally formulated in *nonmonadic* second-order logic. Consider, as an unpretentious example[2], the following property of a ring P of processors of different types, where two types may either be compatible or incompatible with each other. We call P *tolerant*, if for each processor p in P there exist two other distinct processors $backup_1(p) \in P$ and $backup_2(p) \in P$, both compatible to p, such that the following conditions are satisfied:

1. for each $p \in P$ and for each $i \in \{1, 2\}$, $backup_i(p)$ is not a neighbor of p;
2. for each $i, j \in \{1, 2\}$, $backup_i(backup_j(p)) \notin \{p, backup_1(p), backup_2(p)\}$.

Intuitively, we may imagine that in case p breaks down, the workload of p can be reassigned to $backup_1(p)$ or to $backup_2(p)$. Condition 1 reflects the intuition that if some processor is damaged, there is some likelihood that also its neighbors are (for instance

[2] Our goal here is merely to give the reader some intuition about a possible type of application.

in case of physical affection such as radiation), thus neighbors should not be used as backup processors. Condition 2 states that the backup processor assignment is antisymmetric and anti-triangular; this ensures, in particular, that the system remains functional, even if two processors of the same type are broken (further processors of incompatible type might be broken, provided that broken processors can be simply bypassed for communication).

Let T be a fixed set of processor types. We represent a ring of n processors numbered from 1 to n where processor i is adjacent to processor $i+1$ $(mod\, n)$ as a string of length n from T^* whose i-th position is τ if the type of the i-th processor is τ; logically, $C_\tau(i)$ is then true. The property of P being tolerant is expressed by the following second-order sentence Φ:

$$\begin{aligned}\exists R_1, R_2, \forall x \exists y_1, y_2.\ & compat(x, y_1) \wedge compat(x, y_2) \wedge \\ & R_1(x, y_1) \wedge R_2(x, y_2) \wedge \\ & \textstyle\bigwedge_{i=1,2} \bigwedge_{j=1,2} \Big(\neg R_i(y_j, x) \wedge \neg R_1(y_j, y_i) \wedge \neg R_2(y_j, y_i)\Big) \wedge \\ & x \neq y_1 \wedge x \neq y_2 \wedge y_1 \neq y_2 \wedge \\ & \neg Succ(x, y_1) \wedge \neg Succ(y_1, x) \wedge \neg Succ(x, y_2) \wedge \neg Succ(y_2, x) \wedge \\ & \Big((x = max) \rightarrow (y_1 \neq min \wedge y_2 \neq min)\Big) \wedge \\ & \Big((x = min) \rightarrow (y_1 \neq max \wedge y_2 \neq max)\Big),\end{aligned}$$

where $compat(x, y)$ is the abbreviation for the formal statement that processor x is compatible to processor y (which can be encoded as a simple boolean formula over C_τ atoms).

Φ is the natural second-order formulation of the tolerance property of a ring of processors. This formula is in the fragment ESO($\exists^*\forall\exists^*$); hence, by our results, we can immediately classify tolerance as a regular property, that is, a property that can be checked by a finite automaton.

In a similar way, one can exhibit examples of ESO($\exists^*\forall\forall$) formulas that naturally express interesting properties whose regularity is not completely obvious *a priori*. We thus hope that our results may find applications in the field of computer-aided verification.

4 Results on Full SO over Strings

In this section, we turn to an extension of the results in Sect. 3 from ESO to full second-order logic over strings, considered in [16], with a focus on the SO prefix classes which are regular and non-regular, respectively; in particular, how adding second-order variables affects regular fragments. We then also discuss the finer grained complexity of model checking, with a focus on the intractability frontier.

4.1 Regular vs. Non-regular Prefix Classes

The maximal standard Σ^1_k prefix classes which are regular and the minimal standard Σ^1_k prefix classes which are non-regular are summarized in Fig. 2.

Thus, no $\Sigma^1_k(Q)$ fragment, where $k \geq 3$ and Q contains at least two variables, is regular, and for $k = 2$, only two such fragments ($Q = \exists\forall$ or $Q = \forall\exists$) are regular.

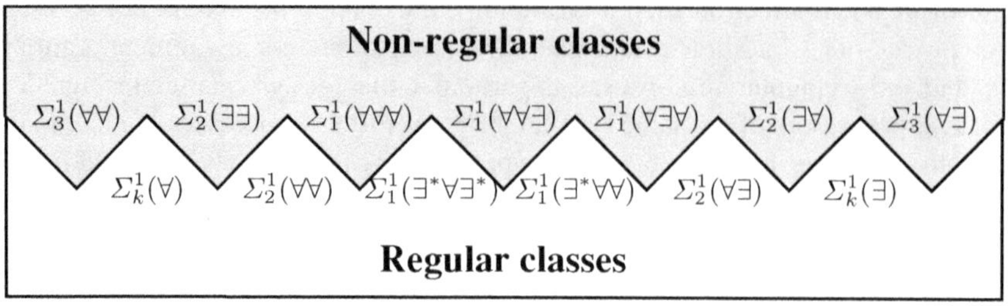

Fig. 2. Maximal regular and minimal non-regular SO prefix classes on strings

$$
\begin{array}{ccccc}
\Sigma^1_2(\exists\forall) & & & & \Sigma^1_2(\exists\exists) \\
& \supseteq & & \subseteq & \\
& \Sigma^1_2(\forall\forall) & = & \Sigma^1_2(\forall\exists) &
\end{array}
$$

Fig. 3. Semantic inclusion relations between $\Sigma^1_2(Q)$ classes over strings, $|Q| = 2$

Note that Grädel and Rosen have shown [28] that $\Sigma^1_1(\mathrm{FO}^2)$, that is, existential second-order logic with two first-order variables, is over strings regular. By the results in [16], $\Sigma^1_k(\mathrm{FO}^2)$, for $k \geq 2$, is non-regular (in fact, intractable).

Figure 3 shows inclusion relationships between the classes $\Sigma^1_2(Q)$ where Q contains two quantifiers. Similar relationships hold for $\Sigma^1_k(Q)$ classes. Furthermore, as shown in [16], we have that $\Sigma^1_2(\forall\exists) = \Sigma^1_1(\bigwedge \forall\exists)$ and $\Sigma^1_3(\exists\forall) = \Sigma^1_2(\bigvee \exists\forall)$, where $\Sigma^1_k(\bigvee \mathcal{Q})$ (resp., $\Sigma^1_k(\bigwedge \mathcal{Q})$) denotes the class of Σ^1_k sentences where the first-order part is a finite disjunction (resp., conjunction) of prefix formulas with quantifier in $\mathcal{Q}$.

We now look in more detail into these results.

Regular Fragments. Let us first consider possible generalizations of regular ESO prefix classes. It is easy to see that every ESO sentence in which only a single first-order variable occurs is equivalent to a monadic second-order sentence. This can be shown by eliminating predicate variables of arity > 1 through introducing new monadic predicate variables. The result generalizes to sentences in full SO with the same first-order part, which can be shown using the same technique. Thus, the class $\Sigma^1_k(\{\exists, \forall\})$ is regular for every $k \geq 0$.

Similarly, it can be shown that every ESO sentence in which only two first-order variables occur is equivalent to a MSO sentence [28]. However, as follows from the results below, this does not generalize beyond ESO. Nonetheless, there are still higher-rank Σ^1_k fragments with two first-order variables, in particular Σ^1_2 prefix classes, which are regular. Here, similar elimination techniques as in the case of a single first-order variable are applicable.

The following proposition, which generalizes the result for $k = 1$ from [33], says that we can eliminate second-order quantifiers easily in some cases.

Proposition 4.1 ([16]). *Every formula in $\Sigma_k^1(\exists^j)$, where $k \geq 1$ is odd, is equivalent to some formula in $\Sigma_{k-1}^1(\exists^j)$, and every formula in $\Sigma_k^1(\forall^j)$, where $k \geq 2$ is even, is equivalent to some formula in $\Sigma_{k-1}^1(\forall^j)$.*

Based on this and a generalization of the proof of Theorem 9.1 in [15], one obtains:

Theorem 4.1 ([16]). *Over strings, $\Sigma_2^1(\forall\forall) = \mathrm{MSO}$.*

For the extension of the FO prefix $\exists^j$ (resp., $\forall^j$) in Proposition 4.1 with a single universal (resp., existential) quantifier, a similar yet slightly weaker result holds. Let $\Sigma_k^1(\bigvee \mathcal{Q})$ (resp., $\Sigma_k^1(\bigwedge \mathcal{Q})$) denote the class of Σ_k^1 sentences where the first-order part is a finite disjunction (resp., conjunction) of prefix formulas with quantifier in $\mathcal{Q}$.

Proposition 4.2 ([16]). *Every formula in $\Sigma_k^1(\exists^j\forall)$, where $k \geq 1$ is odd and $j \geq 0$, is equivalent to some formula in $\Sigma_{k-1}^1(\bigvee \exists^j\forall)$, and every formula in $\Sigma_k^1(\forall^j\exists)$, where $k \geq 2$ is even and $j \geq 1$, is equivalent to some formula in $\Sigma_{k-1}^1(\bigwedge \forall^j\exists)$.*

From this and the regularity of ESO($\forall\exists$) over strings, one can easily derive that $\Sigma_2^1(\forall\exists)$ over strings is regular.

Theorem 4.2 ([16]). *Over strings, $\Sigma_2^1(\forall\exists) = \mathrm{MSO}$.*

Non-regular Fragments. We consider first Σ_2^1 and then Σ_k^1 with $k > 2$.

$\Sigma_2^1(Q)$ where $|Q| \leq 2$. While for the FO prefixes $Q = \forall\exists$ and $Q = \forall\forall$, regularity of ESO(Q) generalizes to Σ_2^1, this is not the case for $Q = \exists\forall$ and $Q = \exists\exists$.

Theorem 4.3 ([16]). *$\Sigma_2^1(\exists\forall)$ is nonregular.*

Indeed, [16] gave an example of a non-regular language defined by a $\Sigma_2^1(\exists\forall)$ sentence. Let $A = \{a, b\}$ and consider the following sentence:

$$\Phi = \exists R \forall X \exists x \forall y.[C_a(y) \vee (C_a(x) \wedge (X(y) \leftrightarrow R(x, y))].$$

Informally, this sentence is true for a string W, just if the number of b's in W (denoted $\#b(W)$) is at most logarithmic in the number of a's in W (denoted $\#a(W)$). More formally, $L(\Phi) = \{w \in \{a, b\}^* \mid \#b(W) \leq \log \#a(W)\}$; by well-known properties of regular languages, this language is not regular.

For the class $\Sigma_2^1(\exists\exists)$, the proof of non-regularity in [16] is more involved. It uses the following lemma, which shows how to emulate universal FO quantifiers using universal SO quantifiers and existential FO quantifiers over strings.

Lemma 4.1 ([16]). *Over strings, every universal first-order formula $\forall \mathbf{x}\varphi(\mathbf{x})$ which contains no predicates of arity > 2 is equivalent to some $\Pi_1^1(\exists\exists)$ formula.*

Proof. As we use techniques from the proof of this lemma later, we recall the sketch from [16].

The idea is to emulate the universal quantifier $\forall x_i$ for every x_i from $\mathbf{x} = x_1, \ldots, x_k$ using a universally quantified variable S_i ranging over singletons, and express "x_i" by "$\exists x_i.S_i(X_i)$." Then, $\forall \mathbf{x}\varphi(\mathbf{x})$ is equivalent to $\forall \mathbf{S} \exists \mathbf{x} \bigwedge_{i=1}^{k} S_i(x_k) \wedge \varphi(\mathbf{x})$.

We can eliminate all existential variables $\mathbf{x}$ but two in this formula as follows. Rewrite the quantifier-free part into a CNF $\bigwedge_{i=1}^{\ell} \delta_i(\mathbf{x})$, where each $\delta_i(\mathbf{x})$ is a disjunction of literals. Denote by $\delta_i^{j,j'}(x_j, x_{j'})$ the clause obtained from $\delta_i(\mathbf{x})$ by removing every literal which contains some variable from $\mathbf{x}$ different from x_j and $x_{j'}$. Since no predicate in φ has arity > 2, formula $\forall \mathbf{x}\varphi(\mathbf{x})$ is equivalent to the formula

$$\forall \mathbf{S} \exists \mathbf{x} \bigwedge_{i=1}^{\ell} \left(\bigvee_{j \neq j'} \exists x \exists y \delta_i^{j,j'}(x, y) \right).$$

The conjunction $\bigwedge_{i=1}^{\ell}$ can be simulated by using universally quantified Boolean variables $Z_1, \ldots, Z_\ell$ and a control formula β which states that exactly one out of $Z_1, \ldots,$ Z_n is true. By pulling existential quantifiers, we thus obtain

$$\forall \mathbf{S} \exists x \exists y \gamma,$$

where

$$\gamma = \beta \rightarrow \bigwedge_{i=1}^{\ell} \left(Z_i \rightarrow \bigvee_{j \neq j'} \delta_i^{j,j'}(x, y) \right).$$

Thus, it remains to express the variables S_i ranging over singletons. For this, we use a technique to express S_i as the difference $X_{i,1} \setminus X_{i,2}$ of two monadic predicates $X_{i,1}$ and $X_{i,2}$ which describe initial segments of the string. Fortunately, the fact that $X_{i,1}$ and $X_{i,2}$ are *not* initial segments or their difference is *not* a singleton can be expressed by a first-order formula $\exists x \exists y \psi_i(x, y)$, where ψ_i is quantifier-free, by using the successor predicate. Thus, we obtain

$$\forall \mathbf{X}_1 \mathbf{X}_2 \left(\bigvee_{i=1}^{k} \exists x \exists y \psi_i(x, y) \right) \vee \exists x \exists y \gamma^*,$$

where γ^* results from γ by replacing each S_i with $X_{i,1}$ and $X_{i,2}$. By pulling existential quantifiers, we obtain a $\Pi_1^1(\exists\exists)$ formula, as desired. □

Given that $\Sigma_1^1(\forall\forall\forall)$ contains NP-complete languages (Fig. 1), it thus follows:

Theorem 4.4 ([16]). *$\Sigma_2^1(\exists\exists)$ is nonregular.*

Therefore, the inclusions in Fig. 3 are both strict.

$\Sigma_2^1(Q)$ where $|Q| > 2$. By the results of the previous subsection and Sect. 3, we can derive that no $\Sigma_2^1(Q)$ prefix class where Q contains more than two variables is regular.

Indeed, Theorem 4.3 implies this for every prefix Q which contains $\exists$ followed by $\forall$, and Theorem 4.4 implies this for every prefix Q which contains at least two existential quantifiers. For the remaining minimal prefixes $Q \in \{\forall\forall\forall, \forall\forall\exists\}$, non-regularity of $\Sigma_2^1(Q)$ follows from the results summarized in Fig. 1. Thus,

Theorem 4.5 ([16]). *$\Sigma_2^1(Q)$ is nonregular for every prefix Q such that $|Q| > 2$.*

Table 1. Complexity of model checking for $\Sigma^1_k(Q)$ prefix classes, $k \leq 3$ ($\eta = \{\forall, \exists\}$)

$Q \in$	in REG ($\subseteq$ P)	NP-complete	Σ^p_2-complete	Σ^p_3-complete
$\Sigma^1_1(Q)$	$\exists^*\forall(\exists^* \cup \forall)$	$\eta^*\forall\eta^*(\forall\eta \cup \exists\eta^*\forall)\eta^*$	—	—
$\Sigma^1_2(Q)$	$\forall\eta$	$\forall\forall\forall^*\exists$	$\eta^*\exists\eta^*\exists\eta^*$	—
$\Sigma^1_3(Q)$	η	—	$\exists\exists(\exists^* \cup \forall)$	$\eta^*\forall\eta^*\forall\eta^*$

$\Sigma^1_k(Q)$ where $k > 2$. Let us now consider the higher fragments of SO over strings. The question is whether any of the regular two-variable prefixes $Q \in \{\forall\forall, \forall\exists\}$ for $\Sigma^1_2(Q)$ survives. However, as we shall see this is not the case.

Since $\Pi^1_2(\forall\forall)$ is contained in $\Sigma^1_3(\forall\forall)$, it follows from Theorem 4.4 that $\Sigma^1_3(\forall\forall)$ is nonregular. For the remaining class $\Sigma^1_3(\forall\exists)$, one can use a result that an existential FO quantifier, followed by another existential FO quantifier, can be emulated using an existential SO and a FO universal quantifier. This leads to the following result.

Theorem 4.6 ([16]). *Over strings, $\Sigma^1_k(\exists\exists) \subseteq \Sigma^1_k(\forall\exists)$ for every odd k, and $\Sigma^1_k(\forall\forall) \subseteq \Sigma^1_k(\exists\forall)$ for every even $k \geq 2$.*

Thus, combined with Theorem 4.4, this shows that $\Sigma^1_3(\forall\exists)$ is nonregular.

In fact, the emulation of an existential FO quantifier as above via an existential SO and an universal FO quantifier is feasible under fairly general conditions; this leads to the following result.

Theorem 4.7 ([16]). *Let $P_1 \in \{\forall\}^*$ and $P_2 \in \{\exists, \forall\}^*\forall\{\exists, \forall\}^*$ be first-order prefixes. Then,*

- *for any odd $k \geq 1$, $\Sigma^1_k(P_1\forall P_2) \subseteq \Sigma^1_{k+1}(P_1\exists P_2)$ and $\Pi^1_k(P_1\forall P_2) \subseteq \Pi^1_k(P_1\exists P_2)$,*
- *for any even $k \geq 2$, $\Sigma^1_k(P_1\forall P_2) \subseteq \Sigma^1_k(P_1\exists P_2)$, $\Pi^1_k(P_1\forall P_2) \subseteq \Pi^1_{k+1}(P_1\exists P_2)$.*

Thus, for example we obtain $\Sigma^1_2(\forall\forall\forall) \subseteq \Sigma^1_2(\exists\forall\forall)$, and by repeated application $\Sigma^1_2(\forall\forall\forall) \subseteq \Sigma^1_2(\exists\exists\forall)$.

4.2 Complexity

Generalizing Fagin's Theorem, Stockmeyer [44] showed that full SO captures the polynomial hierarchy (PH). Second-order variables turn out to be quite powerful. In fact, already two first-order variables, a single binary predicate variable, and further monadic predicate variables are sufficient to express languages that are complete for the levels of PH.

The results in [15] and [16] imply that deciding whether $W \models \Phi$ for a fixed formula Φ and a given string W is intractable for all prefix classes $\Sigma^1_k(Q)$ which are (syntactically) not included in the maximal regular prefix classes shown in Fig. 2. Table 1 shows prefix classes up to $k = 3$ that are C-complete for prominent complexity classes; the precise complexity of some classes (e.g., $\Sigma^1_2(\forall^*\exists\forall^*)$ and its analogue in Σ^1_3) is open.

Indeed, by Fig. 1, $\Sigma_2^1(\forall\forall\forall)$ is intractable; hence by Theorem 4.7 also $\Sigma_2^1(\exists\forall\forall)$ is intractability, Furthermore, the proof of Theorem 4.4 via Lemma 4.1 and Fig. 1 not only establishes that $\Sigma_2^1(\exists\exists)$ is non-regular, but in fact NP-hard.[3] As $\Pi_2^1(\forall\forall)$ is contained in $\Pi_3^1(\forall\forall)$ resp. $\Sigma_3^1(\forall\exists)$ (cf. Theorem 4.7), also the latter prefix classes are intractable.

The complexity of SO over strings increases with the number of SO quantifier alternations. Let us consider $\Sigma_2^1(\exists\exists)$ more closely.

Theorem 4.8. *Model checking for over strings is Σ_2^p-complete for each $\Sigma_2^1(Q)$ where $Q \in \eta^*\exists\eta^*\exists\eta^*$ and $\eta = \{\forall, \exists\}$.*

Proof. (Sketch) Clearly the problem is in Σ_2^p. The Σ_2^p-hardness for $Q = \exists\exists$ can be shown by encoding quantified Boolean formulas (QBFs) of the form

$$\exists p_1 \cdots \exists p_n \forall q_1 \cdots \forall q_m \neg\varphi, \tag{1}$$

where φ is a propositional CNF over the atom $p_1, \ldots, p_n, q_1, \ldots, q_m$, to model checking for $\Sigma_2^1(\exists\exists)$.

This is achieved by generalizing the SAT encoding in [15], which we first recall. Instances $F = C^1 \wedge \cdots \wedge C^m$ of SAT, where the C^i are clauses on propositional variables $p_1, \ldots, p_n$, to strings $enc(F)$ over the alphabet $A = \{\texttt{0}, \texttt{1}, \texttt{+}, \texttt{-}, \texttt{[}, \texttt{]}, \texttt{(}, \texttt{)}\}$ as follows. The variables p_i, $1 \leq i \leq n$ are encoded by binary strings of length $\lceil \log n \rceil$. Each string encoding p_i is enclosed by parentheses '(',')'. The polarity of a literal $p_i/\neg p_i$ is represented by the letter '+' or '-', respectively, which immediately follows the closing parenthesis ')' of the encoding of p_i. A clause is encoded as a sequence of literals which is enclosed in square brackets '[',']'. Without loss of generality, F is not void and each C^i contains at least one literal.

For example, $F = (p \vee q \vee \neg r) \wedge (\neg p \vee \neg q \vee r)$ is encoded by the following string:

$$enc(\mathcal{C}) = \texttt{[(00)+(01)+(10)-][(00)-(01)-(10)+]}\ .$$

Here, p, q, r are encoded by the binary strings 00, 01, 10, respectively. Clearly, $enc(F)$ is obtainable from any standard representation of F in logspace.

The formulas

$$eqcol(x, y) = \bigvee_{\ell \in A} (C_\ell(x) \wedge C_\ell(y)), \tag{2}$$

$$varenc(x) = C_{\texttt{(}}(x) \vee C_{\texttt{0}}(x) \vee C_{\texttt{1}}(x) \vee C_{\texttt{)}}(x) \tag{3}$$

state that the string has at positions x and y the same letter from A and that x is a letter of a variable encoding, respectively.

Then, let Φ be the following $\Sigma_1^1(\forall\forall\forall)$ sentence:

$$\Phi = \exists V \exists G \exists R \exists R' \forall x \forall y \forall z. \varphi(x, y, z),$$

[3] With a similar argument as in the proof of Lemma 4.1, one can show that over strings, every universal first-order formula $\forall \mathbf{x}\varphi(\mathbf{x})$ without predicates of arity >2 is equivalent to some $\Pi_1^1(\bigvee\exists\forall)$ formula. Hence, the SAT encoding in [15] is expressible in $\Sigma_2^1(\bigvee\exists\forall)$.

where G and V are unary, R' and R are binary, and $\varphi(x, y, z)$ is the conjunction of the following quantifier-free formulas φ_G, φ_V, φ_R, and $\varphi_{R'}$:

$$\varphi_G = \varphi_{G,1} \wedge \varphi_{G,2} \wedge \varphi_{G,3},$$

where

$$\varphi_{G,1} = \Big(C_{[}(x) \rightarrow \neg G(x)\Big) \wedge \Big(C_{]}(x) \rightarrow G(x)\Big),$$
$$\varphi_{G,2} = \Big(Succ(x,y) \wedge \neg C_{[}(y) \wedge \neg C_{)}(y)\Big) \rightarrow \Big(G(y) \leftrightarrow G(x)\Big),$$
$$\varphi_{G,3} = \Big(C_{)}(y) \wedge Succ(x,y) \wedge Succ(y,z)\Big) \rightarrow \Big(G(y) \leftrightarrow \Big[G(x) \vee (V(y) \wedge C_{+}(z)) \vee (\neg V(y) \wedge C_{-}(z))\Big]\Big);$$

next,

$$\varphi_V = \Big(C_{)}(x) \wedge C_{)}(y) \wedge R(x,y)\Big) \rightarrow \Big(V(x) \leftrightarrow V(y)\Big),$$
$$\varphi_R = \Big[R(x,y) \rightarrow (eqcol(x,y) \wedge varenc(x))\Big] \wedge \Big[(C_{(}(x) \wedge C_{(}(y)) \rightarrow R(x,y)\Big] \wedge \Big[\Big(\neg C_{(}(x) \wedge Succ(z,x)\Big) \rightarrow \Big(R(x,y) \leftrightarrow (R'(z,y) \wedge eqcol(x,y))\Big)\Big], \quad (4)$$

and

$$\varphi_{R'} = Succ(z,y) \rightarrow \Big[R'(x,y) \leftrightarrow \Big(R(x,z) \wedge \neg C_{)}(z)\Big)\Big].$$

As shown in [15], $enc(F) \models \Phi$ iff F is satisfiable. Back now to our QBF (1), we can choose an encoding where we simply encode the clauses in φ as described and mark in the string the occurrences of the variables q_i with an additional predicate. Furthermore, we represent truth assignments to the q_i's by a monadic variable V'. The sentence Φ for F is rewritten to

$$\Psi = \exists R \exists R' \exists V \forall V' [\alpha_1 \wedge (\alpha_2 \vee \alpha_3)],$$

where α_1 is a universal first-order formula which defines proper R, R' and V using φ_R, $\varphi_{R'}$, and φ_V; α_2 is a $\exists\exists$-prenex first-order formula which states that V' assigns two different occurrences of some universally quantified atom q_i different truth values; and α_3 states that the assignment to $p_1, \ldots, p_n, q_1, \ldots, q_m$ given by V and V' violates φ. The latter can be easily checked by a finite state automaton, and thus is expressible as a monadic $\Pi^1_1(\exists\exists)$ sentence. As Ψ contains no predicate of arity > 2, by applying the techniques of Lemma 4.1 we can rewrite Ψ to an equivalent $\Sigma^1_2(\exists\exists)$ sentence. □

Other fragments of Σ^1_2 have lower complexity. For instance,

Theorem 4.9. *Model checking for $\Sigma^1_2(\forall^*\exists)$ over strings is* NP*-complete, and* NP*-hard for each $\Sigma^1_2(Q)$ where $Q \in \forall\forall\forall^*\exists$.*

Proof. NP-hardness is inherited from the NP-completeness of $\Sigma^1_1(\forall\forall\exists)$. Membership in NP follows from Proposition 4.2: any $\Sigma^1_2(\forall^*\exists)$ sentence Φ is equivalent to some $\Sigma^1_1(\bigwedge\forall^*\exists)$ sentence Φ', for which model checking is in NP. □

On the other hand, by generalizing the QBF encoding, we can easily encode evaluating Σ_k^p-complete QBFs into $\Sigma_k^1(\exists\exists)$ for even $k > 2$ and Π_k^p-complete QBFs into $\Pi_k^1(\exists\exists)$ for odd $k > 1$ (by adding further leading quantifiers).

5 Results on ESO over Graphs

In this section, we briefly describe the main results of [26], where the computational complexity of ESO-prefix classes of is investigated and completely characterized in three contexts: over (1) directed graphs, (2) undirected graphs with self-loops, and (3) undirected graphs without self-loops.

A main theorem of [26] is that a *dichotomy* holds in these contexts, that is to say, each prefix class of ESO either contains sentences that can express NP-complete problems or each of its sentences expresses a polynomial-time solvable problem. Although the boundary of the dichotomy coincides for 1. and 2. (which we refer to as *general graphs* from now on), it changes if one moves to 3. The key difference is that a certain prefix class, based on the well-known *Ackermann class*, contains sentences that can express NP-complete problems over general graphs, but becomes tractable over undirected graphs without self-loops. Moreover, establishing the dichotomy in case 3. turned out to be technically challenging, and required the use of sophisticated machinery from graph theory and combinatorics, including results about graphs of bounded tree-width and Ramsey's Theorem.

In [26], a special notation for ESO-prefix classes was used in order to describe the results with the tightest possible precision involving both the number of SO quantifiers and their arities.[4] Expressions in this notation are built according to the following rules:

- E (resp., E_i) denotes the existential quantification over a single predicate of arbitrary arity (arity $\leq i$).
- a (resp., e) denotes the universal (existential) quantification of a single first-order variable.
- If η is a quantification pattern, then η^* denotes all patterns obtained by repeating η zero or more times.

An expression $\mathcal{E}$ in the special notation consists of a string of ESO quantification patterns (E-patterns) followed by a string of first-order quantification patterns (a or e patterns); such an expression represents the class of all prenex ESO-formulas whose quantifier prefix corresponds to a (not-necessarily contiguous) substring of $\mathcal{E}$.

For example, E_1^*eaa denotes the class of formulas $\exists P_1 \cdots \exists P_r \exists x \forall y \forall z \varphi$, where each P_i is monadic, x, y, and z are first-order variables, and φ is quantifier-free.

A prefix class C is NP-*hard* on a class $\mathcal{K}$ of relational structures, if some sentence in C expresses an NP-hard property on $\mathcal{K}$, and C is *polynomial-time* (PTIME) on $\mathcal{K}$, if for each sentence $\Phi \in C$, model checking is polynomial. Furthermore, C is called *first-order* (FO), if every $\Phi \in C$ is equivalent to a first-order formula.

The first result of [26] completely characterizes the computational complexity of ESO-prefix classes on general graphs. In fact, the same characterization holds on the

[4] For ESO over strings [15], the same level of precision was reached with simpler notation.

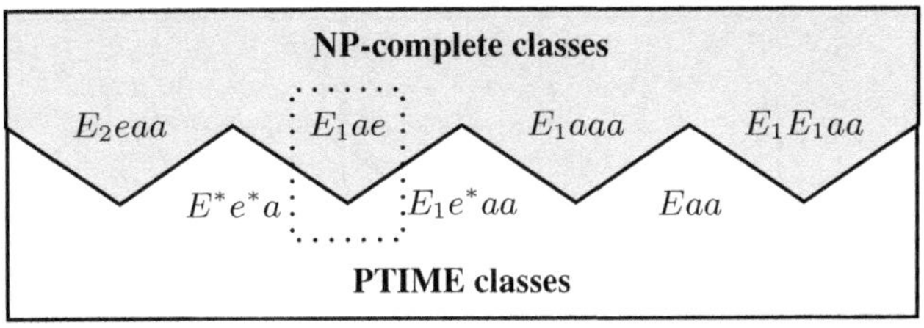

Fig. 4. ESO on arbitrary structures, directed graphs and undirected graphs with self-loops

collection of all finite structures over any relational vocabulary that contains a relation symbol of arity ≥ 2. This characterization is obtained by showing (assuming P $\neq$ NP) that there are four *minimal* NP-hard and three *maximal* PTIME prefix classes, and that these seven classes combine to give complete information about all other prefix classes. This means that every other prefix either contains one of the minimal NP-hard prefix classes as a substring (and, hence, is NP-hard) or is a substring of a maximal PTIME prefix class (and, hence, is in PTIME). Figure 4 depicts the characterization of the NP-hard and PTIME prefix classes of ESO on general graphs.

As seen in Fig. 4, the four minimal NP-hard classes are E_2eaa, E_1ae, E_1aaa, and E_1E_1aa, while the three maximal PTIME classes are E^*e^*a, E_1e^*aa, and Eaa. The NP-hardness results are established by showing that each of the four minimal prefix classes contains ESO-sentences expressing NP-complete problems. For example, a SAT encoding on general graphs can be expressed by an E_1ae sentence. Note that the first-order prefix class ae played a key role in the study of the classical decision problem for fragments of first-order logic (see [6]). As regards the maximal PTIME classes, E^*e^*a is actually FO, while the model checking problem for fixed sentences in E_1e^*aa and Eaa is reducible to 2SAT and, thus, is in PTIME (in fact, in NL).

The second result of [26] completely characterizes the computational complexity of prefix classes of ESO on undirected graphs without self-loops. As mentioned earlier, it was shown that a dichotomy still holds, but its boundary changes. The key difference is that E^*ae turns out to be PTIME on undirected graphs without self-loops, while its subclass E_1ae is NP-hard on general graphs. It can be seen that interesting properties of graphs are expressible by E^*ae-sentences. Specifically, for each integer $m > 0$, there is a E^*ae-sentence expressing that a connected graph contains a cycle whose length is divisible by m. This was shown to be decidable in polynomial time by Thomassen [50]. The class E^*ae constitutes a maximal PTIME class, because all four extensions of E_1ae by any single first-order quantifier are NP-hard on undirected graphs without self-loops [26]. The other minimal NP-hard prefixes on general graphs remain NP-hard also on undirected graphs without self-loops. Consequently, over such graphs, there are seven minimal NP-hard and four maximal PTIME prefix classes that determine the computational complexity of all other ESO-prefix classes (see Fig. 5).

Technically, the most difficult result of [26] is the proof that E^*ae is PTIME on undirected graphs without self-loops. First, using syntactic methods, it is shown that each E^*ae-sentence is equivalent to some E_1^*ae-sentence. After this, it is shown that

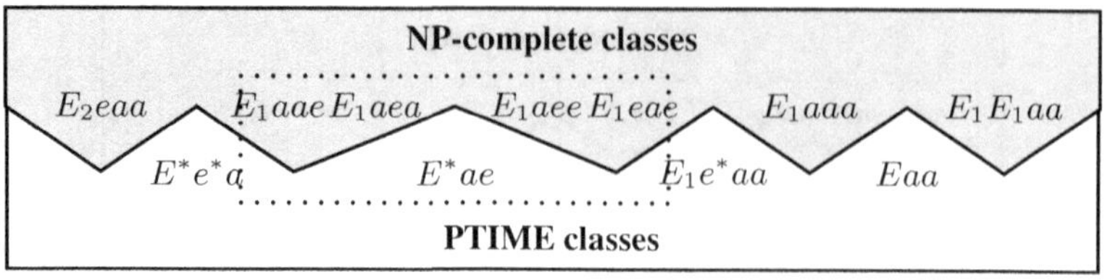

Fig. 5. ESO on undirected graphs without self-loops. The dotted boxes in Figs. 4 and 5 indicate the difference between the two cases.

for each E_1^*ae-sentence the model-checking problem over undirected graphs without self-loops is is equivalent to a natural coloring problem called the *saturation problem*. This problem asks whether there is a particular mapping from a given undirected graph without self-loops to a fixed, directed *pattern graph* P which is extracted from the E_1^*ae-formula under consideration. Depending on the labelings of cycles in P, two cases of the saturation problem are distinguished, namely *pure pattern graphs* and *mixed pattern graphs*. For each case, a polynomial-time algorithm is designed. In simplified terms and focussed on the case of connected graphs, the one for pure pattern graphs has three main ingredients. First, adapting results by Thomassen [50] and using a new graph coloring method, it is shown that if a E_1^*ae-sentence Φ gives rise to a pure pattern graph, then a fixed integer k can be found such that every undirected graph without self-loops that have tree-width bigger than k satisfies Φ. Second, Courcelle's Theorem [11] (see also Sect. 4) is used by which model-checking for MSO sentences is polynomial on graphs of bounded tree-width. Third, Bodlaender's result [5] is used that, for each fixed k, there is a polynomial-time algorithm to check whether a given graph has tree-width at most k.

The polynomial-time algorithm for mixed pattern graphs has a similar architecture, but requires the development of substantial additional technical machinery, including a generalization of the concept of graphs of bounded tree-width. The results of [26] can be summarized in the following theorem.

Theorem 5.1. *Figures 4 and 5 provide a complete classification of the complexity of all ESO prefix classes on graphs.*

6 SO over Trees

Trees are fundamental data structures widely used in computer science and mathematical linguistics. The importance of studying logical languages over trees has increased dramatically with the advent of the World Wide Web. In fact, the Web can be considered as the world's largest data and information repository, and most information on the Web is *semi-structured*, that is, presented in tree-shaped form, for example formatted in HTML or in XML [1]. Web pages and Web documents can thus be considered as finite labeled trees. Special query languages such as XPath [52] have been developed for querying XML documents. The core fragments of these languages contain constructs that are not first-order expressible, but can be defined in second-order logic. Similarly, most relevant *data extraction* tasks for selecting relevant data from a HTML Web site and for annotating the data and transforming it into a highly structured format can be expressed in MSO [21,22,23]. It is thus not astonishing that there has been a renewed interest in understanding complexity and expressiveness issues related to SO over finite trees.

There are various possible formal definitions of trees. Usually a finite tree is defined over a universe of nodes referred to as dom. We use the monadic predicates $root(.)$ and $leaf(.)$ to say that a node is the root or a leaf, respectively.

One mostly considers labeled trees, that is, trees whose nodes are labeled by letters from a finite alphabet Σ. Each label e can be represented by a monadic "color" predicate $label_e$, such that for each node $a \in dom$, $label_i(a)$ is true iff i is labeled with the letter i. (Note that these label predicates have the same role as the C_i predicates we used for strings.)

An important distinction is the one between *ranked* and *unranked trees*. A *ranked tree* is one in which each node has a number of successors, also called children bounded by K. In this case, the successors can be represented via binary functions $child_k$, $k \leq K$, where $child_i(a, b)$ means that node b is the i-th child of node a.

More formally, a finite ranked tree is thus defined as a finite relational structure

$$t_{rk} = \langle dom,\ root,\ leaf,\ (child_k)_{k \leq K},\ (label_a)_{a \in \Sigma} \rangle.$$

where, as explained, "dom" is the set of nodes in the tree, "$root$", "$leaf$", and the "$label_a$" relations are unary, and the "$child_k$" relations are binary.

In an *unranked tree*, the number of successors of a node may be unbounded. This means that we cannot directly encode successors by a fixed number of child predicates. Rather, we use the predicate $first\_child$ and $next\_sibling$ such that $first\_child(a, b)$ is true whenever node b is the first (leftmost) child of a, and $next\_sibling(a, b)$ if b is the nearest sibling to the right of a. In addition, $last\_sibling(a)$ is used to indicate that a is the last of the siblings of a node. Note that the predicates $root$, $leaf$, and $last\_sibling$ can be logically defined from $first\_child$ and $next\_sibling$. However, these definitions would require extra quantifiers and negation, so we prefer to keep these simple predicates in the signature. The signature τ_{ur} for unranked trees thus looks as follows:

$$t_{ur} = \langle dom,\ root,\ leaf,\ (label_a)_{a \in \Sigma},\ first\_child,\ next\_sibling,\ last\_sibling \rangle.$$

Unfortunately, to date, no complete complexity characterization for SO prefix classes over trees is known. Even for ESO prefix classes over trees we do not have a precise characterization. For ESO over (ranked or unranked) trees we know the following.

- All NP-hard classes for the string case remain NP-hard in the tree case. In particular, model checking for the classes ESO($\forall\forall\forall$), ESO($\forall\forall\exists$), and ESO($\forall\exists\forall$) remain NP-complete.
- Model checking for ESO($\exists^*\forall\forall$) was shown in [15] to be feasible in polynomial time *over arbitrary structures* and thus, in particular over trees. However the status of the (in)famous class ESO($\exists^*\forall\exists^*$) over trees is actually obscure. The combinatorial proof arguments used in [15] to prove tractability of model checking in the

string case do not directly carry over to trees. It is currently unclear whether they can be adapted to cover the tree case.
- For (non-monadic) full SO prefix classes, there is even less clarity. Of course, all SO prefix classes known to be NP-hard (with respect to model checking) over strings mentioned in Sect. 4 are trivially also NP-hard over trees, and all tractable ESO prefix classes over graphs mentioned in Sect. 5 are also tractable over trees.

In the rest of this section, let us concentrate on monadic second-order logic (MSO).

The Büchi-Elgot-Trakhtenbrot Theorem that MSO over strings captures the regular languages (see Sect. 2) carries over to ranked and unranked tree structures. The *regular tree languages* (for ranked as well as for unranked alphabets) are precisely those tree languages recognizable by a number of natural forms of finite automata [7]. For space reasons, we cannot discuss details of tree automata here, but refer the interested reader to standard compendia such as [10,49], as well as to [39].

The following is a classical result for ranked trees [47,12], which has been shown in [39] to hold for unranked trees as well.

Proposition 6.1. *A tree language is regular iff it is definable in MSO.*

Given that tree automata can be run in polynomial time over trees, the above result immediately yields the following well-known corollary:

Corollary 6.1. *Model checking for MSO formulas over (ranked or unranked) trees is feasible in linear time.*

This result was generalized by Courcelle to tree-like structures, more specifically, structures of *bounded treewidth*, a concept originally defined in [11]:

Theorem 6.1 ([11]). *Model checking for MSO formulas over structures of bounded treewidth is feasible in polynomial time (in fact, in linear time).*

MSO has been used as a language for computer-aided verification (CAV); model checkers for CAV such as MONA [17] are mainly based on MSO.

Other applications are, as we already mentioned, XML querying and Web data extraction [21,24,22]. However, for various reasons discussed in [22], MSO is not well-suited as a query language and query evaluation does not properly scale in the size of the query. Consequently, in [22] a different language was considered, viz. *monadic Datalog*, which restricts the well-known Datalog language to programs where all intensional predicates are either Boolean or monadic. For monadic Datalog, the following could be shown:

Theorem 6.2 ([22]). *Over (ranked or unranked) trees, Monadic Datalog is exactly as expressive as MSO.*

Moreover, for the complexity of evaluating monadic Datalog programs over ranked or unranked trees is as follows:

Theorem 6.3 ([22]). *Evaluating a monadic Datalog program P over a tree T is feasible in time $O(\|P\| \times \|T\|)$.*

The above result shows that, over trees, model checking for monadic datalog, unlike for MSO, scales linearly in the size of the datalog program P, and not only in the size of the structure T. Note that this does by no means conflict with the fact that, over trees, monadic Datalog is exactly as expressive as MSO. In fact, translating an MSO formula into an equivalent monadic Datalog program may come with a huge (exponential) blow-up in size. However, as noted in [22], it appears that most practical queries and data extraction tasks can be easily formulated by small monadic datalog programs, and hence, the worst-case blow-up is not really relevant in practice. The commercial Web data extraction system Lixto [4,23], which is successfully used for many industrial applications in various domains, is mainly based on monadic Datalog.

Theorem 6.3 was recently generalized to the setting of structures of bounded treewidth:

Theorem 6.4 ([27]). *Evaluating a monadic Datalog program P over a structures T of bounded treewidth is feasible in time $O(\|P\| \times \|T\|)$.*

7 Further Work and Conclusion

The main aim of this paper was to summarize results on determining the complexity of prefix classes over finite structures, and in particular on the prefix classes $\Sigma^1_k(\mathcal{Q})$ which over strings express regular languages only. Many of the prefix classes analyzed so far represented mathematical challenges and required novel solution methods. Some of them could be solved with automata-theoretic techniques, others with techniques of purely combinatorial nature, and yet others required graph-theoretic arguments.

While the exact "regularity frontier" for the $\Sigma^1_k(\mathcal{Q})$ classes has been charted (see Fig. 2), their tractability frontier with respect to model checking misses a single class, viz. the nonregular class $\Sigma^1_2(\exists\forall)$. If model checking is NP-hard for it, then the tractability frontier coincides with the regularity frontier (just as for ESO, cf. [15]). If, on the other hand, model checking for $\Sigma^1_2(\exists\forall)$ is tractable, then the picture is slightly different.

Moreover, it would be interesting to refine our analysis of the Σ^1_k fragments over strings by studying the second-order quantification patterns taking the number of the second-order variables and their arities into account, as done for ESO over graphs in [25] (cf. Sect. 5) and for the classical decision problem in the book by Börger, Gurevich, and Grädel [6].

We conclude this paper pointing out some interesting (and in our opinion important) open issues.

- While the work on word structures concentrated so far on strings with a successor relation $Succ$, one should also consider the cases where an additional predefined linear order $<$ is available on the word structures or the successor relation $Succ$ is replaced by such a linear order. While for full ESO or MSO, $Succ$ and $<$ are freely interchangeable, this is not the case for many of the limited ESO-prefix classes. Preliminary results suggest that most of the results in this paper carry over to the $<$ case. Other variants would be structures with function symbols [2], additional relations like set comparison operators [30], or position arithmetic (for instance $M(i,j) \equiv 0 \leq i \leq j < n \wedge i{+}j = n{-}1$ in [34]).

- Delineate the tractability/intractability frontier for all SO prefix classes over graphs, and settle the complexity characterization. Over strings, settle the complexity of the open fragments (including $\Sigma_2^1(\exists\forall)$, $\Sigma_2^1(\exists^*\forall\exists^*)$, etc).
- Study SO prefix classes over further interesting classes of structures (for instance planar graphs).
- The scope of [15,16] are *finite* strings. However, infinite strings or ω-words are another important area of research. In particular, Büchi has shown that an analogue of his theorem (Proposition 2.1) also holds for ω-words [8]. For an overview of this and many other important results on ω-words, we refer the reader to excellent survey of Thomas [48]. In this context, it would be interesting to see which of the results established so far survive for ω-words. For some results, for instance regularity of ESO($\exists^*\forall\forall$), this is obviously the case as no finiteness assumption on the input word structures was made in the proof. For determining the regularity or nonregularity of some classes such as ESO($\exists^*\forall\exists^*$), further research is needed.

References

1. Abiteboul, S., Buneman, P., Suciu, D.: Data on the Web: From Relations to Semistructured Data and XML. Morgan Kaufmann, San Francisco (1999)
2. Barbanchon, R., Grandjean, E.: The minimal logically-defined NP-complete problem. In: Diekert, V., Habib, M. (eds.) STACS 2004. LNCS, vol. 2996, pp. 338–349. Springer, Heidelberg (2004)
3. Basin, D., Klarlund, N.: Hardware verification using monadic second-order logic. In: Wolper, P. (ed.) CAV 1995. LNCS, vol. 939, pp. 31–41. Springer, Heidelberg (1995)
4. Baumgartner, R., Flesca, S., Gottlob, G.: Visual web information extraction with Lixto. In: VLDB, pp. 119–128 (2001)
5. Bodlaender, H.L.: A Linear-Time Algorithm for Finding Tree-Decompositions of Small Treewidth. SIAM Journal on Computing 25, 1305–1317 (1996)
6. Börger, E., Grädel, E., Gurevich, Y.: The Classical Decision Problem. Springer, Heidelberg (1997)
7. Brüggemann-Klein, A., Murata, M., Wood, D.: Regular tree and regular hedge languages over non-ranked alphabets: Version 1, April 3, 2001. Technical Report HKUST-TCSC-2001-05, Hong Kong University of Science and Technology, Hong Kong SAR, China (2001)
8. Büchi, J.R.: On a Decision Method in Restriced Second-Order Arithmetic. In: Nagel, E., et al. (eds.) Proc. International Congress on Logic, Methodology and Philosophy of Science, pp. 1–11. Stanford University Press, Stanford (1960)
9. Büchi, J.R.: Weak second-order arithmetic and finite automata. Zeitschrift für mathematische Logik und Grundlagen der Mathematik 6, 66–92 (1960)
10. Comon, H., Dauchet, M., Gilleron, R., Jacquemard, F., Lugiez, D., Löding, C., Tison, S., Tommasi, M.: Tree Automata Techniques and Applications (Web book) (2008), `http://tata.gforge.inria.fr/` (viewed September 25, 2009)
11. Courcelle, B.: The Monadic Second-Order Logic of Graphs I: Recognizable Sets of Finite Graphs. Information and Computation 85, 12–75 (1990)
12. Doner, J.: Tree acceptors and some of their applications. Journal of Computer and System Sciences 4, 406–451 (1970)
13. Ebbinghaus, H.D., Flum, J.: Finite Model Theory. In: Perspectives in Mathematical Logic. Springer, Heidelberg (1995)

14. Eiter, T., Gottlob, G., Gurevich, Y.: Normal Forms for Second-Order Logic over Finite Structures, and Classification of NP Optimization Problems. Annals of Pure and Applied Logic 78, 111–125 (1996)
15. Eiter, T., Gottlob, G., Gurevich, Y.: Existential Second-Order Logic over Strings. Journal of the ACM 47, 77–131 (2000)
16. Eiter, T., Gottlob, G., Schwentick, T.: Second-Order Logic over Strings: Regular and Non-Regular Fragments. In: Kuich, W., Rozenberg, G., Salomaa, A. (eds.) DLT 2001. LNCS, vol. 2295, pp. 37–56. Springer, Heidelberg (2002)
17. Elgaard, J., Klarlund, N., Møller, A.: MONA 1.x: New techniques for WS1S and WS2S. In: Y. Vardi, M. (ed.) CAV 1998. LNCS, vol. 1427, pp. 516–520. Springer, Heidelberg (1998)
18. Elgot, C.C.: Decision problems of finite automata design and related arithmetics. Transactions of the American Mathematical Society 98, 21–51 (1961)
19. Fagin, R.: Generalized First-Order Spectra and Polynomial-Time Recognizable Sets. In: Karp, R.M. (ed.) Complexity of Computation, pp. 43–74. AMS (1974)
20. Gottlob, G.: Second-order logic over finite structures - report on a research programme. In: Basin, D., Rusinowitch, M. (eds.) IJCAR 2004. LNCS (LNAI), vol. 3097, pp. 229–243. Springer, Heidelberg (2004)
21. Gottlob, G., Koch, C.: Monadic queries over tree-structured data. In: LICS, pp. 189–202 (2002)
22. Gottlob, G., Koch, C.: Monadic datalog and the expressive power of languages for web information extraction. Journal of the ACM 51, 74–113 (2004)
23. Gottlob, G., Koch, C., Baumgartner, R., Herzog, M., Flesca, S.: The Lixto data extraction project - back and forth between theory and practice. In: PODS, pp. 1–12 (2004)
24. Gottlob, G., Koch, C., Pichler, R.: Efficient algorithms for processing XPath queries. ACM Trans. Database Syst. 30, 444–491 (2005)
25. Gottlob, G., Kolaitis, P., Schwentick, T.: Existential second-order logic over graphs: Charting the tractability frontier. In: 41st Annual Symposium on Foundations of Computer Science (FOCS 2000), Redondo Beach, California, USA, November 12-14, pp. 664–674. IEEE Computer Society Press, Los Alamitos (2000)
26. Gottlob, G., Kolaitis, P., Schwentick, T.: Existential second-order logic over graphs: Charting the tractability frontier. Journal of the ACM 51, 312–362 (2004)
27. Gottlob, G., Pichler, R., Wei, F.: Monadic datalog over finite structures with bounded treewidth. In: PODS, pp. 165–174 (2007)
28. Grädel, E., Rosen, E.: Two-Variable Descriptions of Regularity. In: Proceedings 14th Annual Symposium on Logic in Computer Science (LICS 1999), Trento, Italy, July 2-5, pp. 14–23. IEEE Computer Science Press, Los Alamitos (1999)
29. Grandjean, E.: Universal quantifiers and time complexity of random access machines. Mathematical Systems Theory 13, 171–187 (1985)
30. Hachaïchi, Y.: Fragments of monadic second-order logics over word structures. Electr. Notes Theor. Comput. Sci. 123, 111–123 (2005)
31. Henriksen, J., Jensen, J., Jørgensen, M., Klarlund, N., Paige, B., Rauhe, T., Sandholm, A.: Mona: Monadic second-order logic in practice. In: Brinksma, E., Steffen, B., Cleaveland, W.R., Larsen, K.G., Margaria, T. (eds.) TACAS 1995. LNCS, vol. 1019, pp. 89–110. Springer, Heidelberg (1995)
32. Immerman, N.: Descriptive Complexity. Springer, Heidelberg (1999)
33. Kolaitis, P., Papadimitriou, C.: Some Computational Aspects of Circumscription. Journal of the ACM 37, 1–15 (1990)
34. Langholm, T., Bezem, M.: A descriptive characterisation of even linear languages. Grammars 6, 169–181 (2003)
35. Lautemann, C., Schwentick, T., Thérien, D.: Logics for context-free languages. In: Proc. 1994 Annual Conference of the EACSL, pp. 205–216 (1995)

36. Leivant, D.: Descriptive Characterizations of Computational Complexity. Journal of Computer and System Sciences 39, 51–83 (1989)
37. Lynch, J.F.: The quantifier structure of sentences that characterize nondeterministic time complexity. Computational Complexity 2, 40–66 (1992)
38. McNaughton, R., Papert, S.: Counter-Free Automata. MIT Press, Cambridge (1971)
39. Neven, F., Schwentick, T.: Query automata on finite trees. Theoretical Computer Science 275, 633–674 (2002)
40. Olive, F.: A Conjunctive Logical Characterization of Nondeterministic Linear Time. In: Nielsen, M. (ed.) CSL 1997. LNCS, vol. 1414. Springer, Heidelberg (1997) (to appear)
41. Pin, J.E.: Varieties of Formal Languages. North Oxford/Plenum, London/New York (1986)
42. Pin, J.E.: Logic On Words. Bulletin of the EATCS 54, 145–165 (1994)
43. Pin, J.E.: Semigroups and Automata on Words. Annals of Mathematics and Artificial Intelligence 16, 343–384 (1996)
44. Stockmeyer, L.J.: The Polynomial-Time Hierarchy. Theoretical Computer Science 3, 1–22 (1977)
45. Straubing, H.: Finite Automata, Formal Logic, and Circuit Complexity. Birkhäuser, Basel (1994)
46. Straubing, H., Thérien, D., Thomas, W.: Regular Languages Defined with Generalized Quantifiers. Information and Computation 118, 289–301 (1995)
47. Thatcher, J., Wright, J.: Generalized finite automata theory with an application to a decision problem of second-order logic. Mathematical Systems Theory 2, 57–81 (1968)
48. Thomas, W.: Automata on Infinite Objects. In: van Leeuwen, J. (ed.) Handbook of Theoretical Computer Science, vol. B. Elsevier Science Publishers B.V, North-Holland (1990)
49. Thomas, W.: Languages, Automata, and Logic. In: Rozenberg, G., Salomaa, A. (eds.) Handbook of Formal Language Theory, vol. III, pp. 389–455. Springer, Heidelberg (1996)
50. Thomassen, C.: On the Presence of Disjoint Subgraphs of a Specified Type. Journal of Graph Theory 12, 101–111 (1988)
51. Trakhtenbrot, B.: Finite Automata and the Logic of Monadic Predicates. Dokl. Akad. Nauk SSSR 140, 326–329 (1961)
52. World Wide Web Consortium: XPath recomendation (1999), `http://www.w3c.org/TR/xpath/` (viewed September 25, 2009)

A Logic for PTIME and a Parameterized Halting Problem

Yijia Chen[1] and Jörg Flum[2]

[1] Shanghai Jiaotong University, China
yijia.chen@cs.sjtu.edu.cn
[2] Albert-Ludwigs-Universität Freiburg, Germany
joerg.flum@math.uni-freiburg.de

For Yuri, on the occasion of his seventieth birthday.

Abstract. In [9] Yuri Gurevich addresses the question whether there is a logic that captures polynomial time. He conjectures that there is no such logic. He considers a logic, we denote it by $L_{\leq}$, that allows to express precisely the polynomial time properties of structures; however, apparently, there is no algorithm "that given an $L_{\leq}$-sentence φ produces a polynomial time Turing machine that recognizes the class of models of φ." In [12] Nash, Remmel, and Vianu have raised the question whether one can *prove* that there is no such algorithm. They give a reformulation of this question in terms of a parameterized halting problem p-$\text{Acc}_{\leq}$ for nondeterministic Turing machines. We analyze the precise relationship between $L_{\leq}$ and p-$\text{Acc}_{\leq}$. Moreover, we show that p-$\text{Acc}_{\leq}$ is not fixed-parameter tractable if "P $\neq$ NP holds for all time constructible and increasing functions." A slightly stronger complexity theoretic hypothesis implies that $L_{\leq}$ does not capture polynomial time. Furthermore, we analyze the complexity of various variants of p-$\text{Acc}_{\leq}$ and address the construction problem associated with p-$\text{Acc}_{\leq}$.

Keywords: logics for PTIME, halting problems, parameterized complexity.

1 Introduction

The existence of a logic capturing polynomial time remains the central problem in descriptive complexity. A proof that such a logic does not exist would yield that P $\neq$ NP. The problem was addressed by Yuri Gurevich in various papers (e.g. [8,9]; recent articles on this question are [7,12]). The question originated in database theory: In a fundamental paper [2] on the complexity and expressiveness of query languages, Chandra and Harel considered a related question, namely they asked for a recursive enumeration of the class of all queries computable in polynomial time.

By a result due to Immerman [10] and Vardi [13] least fixed-point logic LFP captures polynomial time on *ordered* structures. However the property of an *arbitrary* structure of having a universe of even cardinality is not expressible in

A. Blass, N. Dershowitz, and W. Reisig (Eds.): Gurevich Festschrift, LNCS 6300, pp. 251–276, 2010.

LFP. There are artificial logics capturing polynomial time on arbitrary structures, but they do not fulfill a natural requirement on logics in this context:

> There is an algorithm that decides whether $\mathcal{A}$ is a model of φ for all structures $\mathcal{A}$ and sentences φ of the logic and that does this for fixed φ in time polynomial in the size $\|\mathcal{A}\|$ of $\mathcal{A}$. (1)

In [9] Gurevich conjectures that there is no logic that captures polynomial time. The conjecture is false if one waives the effectivity condition (1) in the definition of a logic capturing polynomial time. This is shown in [9, Sect. 7, CLAIM 2] by considering a logic related to LFP and defined by Andreas Blass and Yuri Gurevich. We denote this logic by $L_\leq$ (cf. Sect. 3 for the definition of $L_\leq$). As $L_\leq$ satisfies all the other requirements on a logic capturing polynomial time, Gurevich implicitly conjectures that $L_\leq$ does not satisfy the condition (1).

In [12] Nash, Remmel, and Vianu have raised the question whether one can *prove* that there is no algorithm as required by (1) for the logic $L_\leq$. Moreover, they show that (1) can be equivalently formulated as a statement concerning the complexity of a halting problem for nondeterministic Turing machines. This reformulation is best expressed in the terminology of parameterized complexity. We consider the *parameterized acceptance problem* p-ACC$_\leq$ *for nondeterministic Turing machines*:

> p-ACC$_\leq$
> *Instance:* A nondeterministic Turing machine $\mathbb{M}$ and $n \in \mathbb{N}$ in unary.
> *Parameter:* $\|\mathbb{M}\|$, the size of $\mathbb{M}$.
> *Question:* Does $\mathbb{M}$ accept the empty input tape in at most n steps?

Then

$$L_\leq \text{ satisfies (1)} \qquad \text{if and only if} \qquad p\text{-ACC}_\leq \in \text{XP}.^1 \tag{2}$$

In this paper we mainly deal with two questions:

(a) What does "p-ACC$_\leq$ is fixed-parameter tractable" mean for the logic $L_\leq$?
(b) What is the complexity of p-ACC$_\leq$?

While we can answer question (a), we are only able to relate the statements "p-ACC$_\leq \in$ XP" and "p-ACC$_\leq \in$ FPT" with other open problems of complexity theory.

More precisely, the content of the different sections is the following. It is known that the time bound for the model-checking problem for LFP, that is, for the evaluation of a sentence φ of LFP in a structure $\mathcal{A}$, contains a factor $\|\mathcal{A}\|^{O(|\varphi|)}$; an analysis of the corresponding algorithm shows that (at least for LFP-sentences

[1] Depending on the exact formulation of (1), we need the class XP of uniform or the class XP of nonuniform parameterized complexity; see Sect. 2 for the definitions and Theorem 1 (3), (4) for the precise statement.

in normal form) a factor of the form $\|\mathcal{A}\|^{O(\text{width}(\varphi))}$ suffices, where width(φ), the width of φ, essentially is the maximum number of free variables in a subformula of φ. The main result of Sect. 4 shows that the existence of a bound of this type for the model-checking problem of the logic $L_{\leq}$ is equivalent to p-ACC$_{\leq} \in$ FPT.

Let P[TC] $\neq$ NP[TC] mean that for all *time constructible* and increasing functions h the class of problems decidable in *deterministic polynomial time* in h and the class of problems decidable in *nondeterministic polynomial time* in h are distinct, that is, DTIME($h^{O(1)}$) $\neq$ NTIME($h^{O(1)}$). In Sect. 6 we show that P[TC] $\neq$ NP[TC] implies that p-ACC$_{\leq} \notin$ FPT. Furthermore a stronger hypothesis where DTIME($h^{O(1)}$) $\neq$ NTIME($h^{O(1)}$) is replaced by NTIME($h^{O(1)}$) $\not\subseteq$ DTIME($h^{O(\log h)}$) implies that p-ACC$_{\leq} \notin$ XP (and thus by (2), it implies that $L_{\leq}$ does not capture polynomial time). In [4] we related these hypotheses to other statements of complexity theory; in particular, we saw that P[TC] $\neq$ NP[TC] holds if there is a P-*bi-immune* problem in NP.

We also study some variants of p-ACC$_{\leq}$. First we deal with p-ACC$_{=}$, the problem obtained from p-ACC$_{\leq}$ by asking for an accepting run of *exactly* n steps. We show that p-ACC$_{=}$ is related to a logic $L_{=}$ as p-ACC$_{\leq}$ is to the logic $L_{\leq}$. In Sect. 5 we improve a result of [1] by showing that p-ACC$_{=} \in$ FPT if and only if E = NE (that is, DTIME($2^{O(n)}$) = NTIME($2^{O(n)}$)). Furthermore, in Sect. 7 we introduce a halting problem for deterministic Turing machines, the "deterministic version" of p-ACC$_{\leq}$, and show that it is an example of a problem *nonuniformly* fixed-parameter tractable but provably not contained in *uniform* XP, to the best of our knowledge, the first natural such example.

Finally, in Sect. 8, we consider the construction problem associated with p-ACC$_{\leq}$ and show that it is not fpt Turing reducible to p-ACC$_{\leq}$ in case p-ACC$_{\leq} \notin$ XP.

A conference version of this paper appeared as [3].

2 Preliminaries

In this section we review some of the basic concepts of parameterized complexity and of logics and their complexity. We refer to [5] for notions not defined here.

2.1 Parameterized Complexity

We identify problems with subsets Q of $\{0,1\}^*$. Clearly, as done mostly, we present concrete problems in a verbal, hence uncodified form. We denote by P the class of problems Q such that $x \in Q$ is solvable in polynomial time. All Turing machines have $\{0,1\}$ as their alphabet.

We view *parameterized problems* as pairs (Q,κ) consisting of a problem $Q \subseteq \{0,1\}^*$ and a *parameterization* $\kappa : \{0,1\}^* \to \mathbb{N}$, which is required to be polynomial time computable. We will present parameterized problems in the form as we did for p-ACC$_{\leq}$ in the Introduction.

Recall that a parameterized problem (Q,κ) is *fixed-parameter tractable* (and in the class FPT) if $x \in Q$ is solvable by an *fpt-algorithm*, that is, by an algorithm running in time $f(\kappa(x)) \cdot |x|^{O(1)}$ for some computable $f : \mathbb{N} \to \mathbb{N}$. The parameterized problem (Q, κ) is in the class XP if $x \in Q$ is solvable in time $O(|x|^{f(\kappa(x))})$ for some computable $f : \mathbb{N} \to \mathbb{N}$.

Besides these classes of the usual (strongly uniform) parameterized complexity theory we need their *uniform* versions $\mathrm{FPT}_{\mathrm{uni}}$ and $\mathrm{XP}_{\mathrm{uni}}$ and their *nonuniform* versions $\mathrm{FPT}_{\mathrm{nu}}$ and $\mathrm{XP}_{\mathrm{nu}}$. For example, $(Q, \kappa) \in \mathrm{FPT}_{\mathrm{uni}}$ if there is an algorithm solving $x \in Q$ in time $f(\kappa(x)) \cdot |x|^{O(1)}$ for some *arbitrary* $f : \mathbb{N} \to \mathbb{N}$; and $(Q, \kappa) \in \mathrm{FPT}_{\mathrm{nu}}$ if there is a constant $c \in \mathbb{N}$, an arbitrary function $f : \mathbb{N} \to \mathbb{N}$, and for every $k \in \mathbb{N}$ an algorithm solving the (classical) problem

$$(Q, \kappa)_k := \{x \in Q \mid \kappa(x) = k\}$$

in time $f(k) \cdot |x|^c$. The problem $(Q, \kappa)_k$ is called the kth *slice* of (Q, κ).

We write $(Q, \kappa) \leq^{\mathrm{fpt}} (Q', \kappa')$ if there is an fpt reduction from (Q, κ) to (Q', κ') (this concept refers to the strongly uniform parameterized complexity theory).

2.2 Logic

A *vocabulary* τ is a finite set of relation symbols. Each relation symbol has an *arity*. A *structure* $\mathcal{A}$ of vocabulary τ, or τ*-structure* (or, simply structure), consists of a nonempty set A called the *universe*, and an interpretation $R^{\mathcal{A}} \subseteq A^r$ of each r-ary relation symbol $R \in \tau$. We say that $\mathcal{A}$ is finite, if A is a finite set. *All structures in this paper are assumed to be finite.*

For a structure $\mathcal{A}$ we denote by $\|\mathcal{A}\|$ the size of $\mathcal{A}$, that is, the length of a reasonable encoding of $\mathcal{A}$ as string in $\{0, 1\}^*$ (e.g., cf. [5] for details). If necessary, we can assume that the universe of a finite structure is $[m] := \{1, \ldots, m\}$ for some natural number $m \geq 1$, as all the properties of structures we consider are invariant under isomorphisms; in particular, it suffices that from the encoding of $\mathcal{A}$ we can recover $\mathcal{A}$ up to isomorphism. The reader will easily convince himself that we can assume that there is a computable function lgth such that for every vocabulary τ and $m \geq 1$ (we just collect the properties of lgth we use in Sect. 4):

(a) $\|\mathcal{A}\| = \mathrm{lgth}(\tau, m)$ for every τ-structure $\mathcal{A}$ with universe of cardinality m (that is, for fixed τ and m, the encoding of each τ-structure with universe of m elements has length equal to $\mathrm{lgth}(\tau, m)$);
(b) $\mathrm{lgth}(\tau, m) \geq \max\{2, m\}$;
(c) for fixed τ, the function $m \mapsto \mathrm{lgth}(\tau, m)$ is time constructible and $\mathrm{lgth}(\tau, m)$ is polynomial in m;
(d) $\mathrm{lgth}(\tau, m) < \mathrm{lgth}(\tau', m')$ for all τ, τ' with $\tau \subseteq \tau'$ and m, m' with $m < m'$;
(e) $\mathrm{lgth}(\tau, m) = O(\log |\tau| \cdot |\tau| \cdot m)$ for every τ containing only unary relation symbols;
(f) $\mathrm{lgth}(\tau \cup \{R\}, m) = O(\mathrm{lgth}(\tau, m) + m^2)$ for every binary relation symbol R not in τ.

Sometimes, for a structure $\mathcal{A}$ we denote by $<_A$ the ordering on A given by the encoding of $\mathcal{A}$.

We assume familiarity with *first-order logic* FO and its extension *least fixed-point logic* LFP. We denote by FO[τ] and LFP[τ] the set of sentences of vocabulary τ of FO and of LFP, respectively. It is known that LFP captures polynomial time on the class of ordered structures.

As we will introduce further semantics for the formulas of least fixed-point logic, we write $\mathcal{A} \models_{\text{LFP}} \varphi$ if the structure $\mathcal{A}$ is a model of the LFP-sentence φ. An algorithm based on the inductive definition of the satisfaction relation for LFP shows (see [14]):

Proposition 1. *The model-checking problem $\mathcal{A} \models_{\text{LFP}} \varphi$ for structures $\mathcal{A}$ and* LFP*-sentences φ can be solved in time*

$$|\varphi| \cdot \|\mathcal{A}\|^{O(|\varphi|)}.$$

It is known that every LFP-sentence is equivalent to an LFP-sentence in normal form, where an LFP-sentence φ is in *normal form* if it has the form

$$\varphi = \exists y[\text{LFP}_{x_1 \dots x_\ell, X} \psi(x_1 \dots x_\ell, X)]\, yy \dots y, \tag{3}$$

where ψ is a first-order formula, X is an ℓ-ary relation variable, and $x_1, \dots, x_\ell$ are the (first-order) variables free in ψ. If in addition every subformula of φ has at most ℓ free first-order variables, then the problem $\mathcal{A} \models_{\text{LFP}} \varphi$ can be solved in time $O(|\varphi| \cdot \|\mathcal{A}\|^{2\ell} \cdot \ell)$. To state the corresponding result for arbitrary LFP-sentences we introduce the width and the depth of LFP-formulas.

Let $\varphi(x_1, \dots, x_r, Y_1, \dots, Y_s)$ be an LFP-formula and let the pairwise distinct $x_1, \dots, x_r$ be the first-order variables free in φ and the pairwise distinct $Y_1, \dots, Y_s$ be the second-order variables free in φ. The *variable-weight of* φ is

$$r + \sum_{i \in [s]} \text{ar}(Y_i),$$

where $\text{ar}(Y_i)$ is the arity of Y_i. The *width* of φ, denoted by $\text{width}(\varphi)$, is the maximum of the variable-weights of the subformulas of φ. By $\text{depth}(\varphi)$, the *depth* of φ, we denote the maximum nesting depth of LFP-operators in φ.

Proposition 2. *The model-checking problem $\mathcal{A} \models_{\text{LFP}} \varphi$ for structures $\mathcal{A}$ and* LFP*-sentences φ can be solved in time*

$$|\varphi| \cdot \|\mathcal{A}\|^{O\left((1+\text{depth}(\varphi)) \cdot \text{width}(\varphi)\right)}.$$

Proof. For the reader's convenience we present a proof. Let $\varphi(\bar{x}, \bar{Y})$ with $\bar{x} = x_1, \dots, x_r$ and $\bar{Y} = Y_1, \dots, Y_s$ be an LFP-formula and let $\mathcal{A}$ be a structure. For interpretations $\bar{R}$ of $\bar{Y}$, we set

$$\varphi(\mathcal{A}, \bar{R}) := \{\bar{b} \in A^r \mid \mathcal{A} \models_{\text{LFP}} \varphi(\bar{b}, \bar{R})\}.$$

By induction on φ we show how the set $\varphi(\mathcal{A}, \bar{R})$ can be evaluated in the time bound of the claim of our proposition.

If φ does not start with an LFP-operator, we can write φ as a "first-order combination" of formulas $\varphi_1, \dots, \varphi_s$, which are atomic formulas or formulas starting with an LFP-operator. Let $\bar{x}_i$ be the sequence of first-order variables free in φ_i. By induction hypothesis we know that we can compute the sets $\varphi_i(\mathcal{A}, \bar{R}) \subseteq A^{|\bar{x}_i|}$

in the desired time. Along the first-order combination leading from $\varphi_1, \dots, \varphi_s$ to φ, we compute $\psi(\mathcal{A}, \bar{R})$ for all intermediate subformulas ψ. This can be done in the desired time (e.g., see [5, Theorem 4.24]). If φ starts with an LFP-operator, say

$$\varphi(\bar{x}, \bar{Y}) = [\mathrm{LFP}_{\bar{u},X}\, \psi(\bar{u}, \bar{y}, \bar{Y}, X)]\, \bar{z}$$

(thus, $\{\bar{x}\} = \{\bar{y}, \bar{z}\}$), then for every fixed interpretation $\bar{b}$ of $\bar{y}$, we compute the sets

$$\psi(\mathcal{A}, \bar{b}, \bar{R}, \emptyset),\ \ \psi(\mathcal{A}, \bar{b}, \bar{R}, F_\psi(\emptyset)),\ \ \psi(\mathcal{A}, \bar{b}, \bar{R}, F_\psi(F_\psi(\emptyset))), \dots; \tag{4}$$

here for $S \subseteq A^{|\bar{u}|}$

$$F_\psi(S) := \psi(\mathcal{A}, \bar{b}, \bar{R}, S) = \{\bar{a} \in A^{|\bar{u}|} \mid \mathcal{A} \models_{\mathrm{LFP}} \psi(\bar{a}, \bar{b}, \bar{R}, S)\}.$$

As the sequence in (4) is monotone, it reaches a fixed-point in at most $|A|^{|\bar{u}|}$ steps; hence we have to apply at most $|A|^{|\bar{u}|} \cdot |A|^{|\bar{u}|}$ times the evaluation procedure to ψ. As $\mathrm{depth}(\psi) = \mathrm{depth}(\varphi) - 1$, this factor $|A|^{2|\bar{u}|}$ is taken care by our definition of $\mathrm{width}(\varphi)$; note that when applying the procedure to ψ we have an interpretation of all variables of ψ not in $\bar{u}$; of course, for φ we have an additional factor $|A|^{|\bar{y}|}$. □

2.3 Logics Capturing Polynomial Time

For our purposes a *logic* L consists of

- a set $L[\tau]$ of strings, the set of *L-sentences of vocabulary* τ, for every vocabulary τ;
- an algorithm that for every vocabulary τ and every string ξ decides whether $\xi \in L[\tau]$ (in particular, $L[\tau]$ is decidable for every τ);
- a *satisfaction relation* $\models_L$; if $(\mathcal{A}, \varphi) \in {\models_L}$, then, for some τ, we have that $\mathcal{A}$ is a τ-structure and $\varphi \in L[\tau]$; furthermore for each $\varphi \in L[\tau]$ the class of structures $\mathcal{A}$ with $\mathcal{A} \models_L \varphi$ is closed under isomorphisms.

We say that $\mathcal{A}$ is a *model* of φ if $\mathcal{A} \models_L \varphi$ (that is, if $(\mathcal{A}, \varphi) \in {\models_L}$). We set

$$\mathrm{Mod}_L(\varphi) := \{\mathcal{A} \mid \mathcal{A} \models_L \varphi\}$$

and say that φ *axiomatizes* the class $\mathrm{Mod}_L(\varphi)$.

We partly take over the following terminology from [12].

Definition 1. *Let L be a logic.*

(a) L is a logic for P *if for all vocabularies τ and all classes C (of encodings) of τ-structures closed under isomorphisms we have*

$$C \in \mathrm{P} \iff C = \mathrm{Mod}_L(\varphi) \textit{ for some } \varphi \in L[\tau].$$

(b) L is a P-bounded logic for P *if (a) holds and if there is an algorithm $\mathbb{A}$ deciding $\models_L$ (that is, for every structure $\mathcal{A}$ and L-sentence φ the algorithm $\mathbb{A}$ decides whether $\mathcal{A} \models_L \varphi$) and if moreover, for every fixed φ the algorithm $\mathbb{A}$ runs in time polynomial in $\|\mathcal{A}\|$.*

Hence, if L *is a* P*-bounded logic for* P*, then for every* L*-sentence* φ *the algorithm* $\mathbb{A}$ *witnesses that* $\mathrm{Mod}_L(\varphi) \in \mathrm{P}$. *However, we do not necessarily know ahead of time the bounding polynomial.*

(c) L is an effectively P-bounded logic for P *if* L *is a* P*-bounded logic for* P *and if in addition to the algorithm* $\mathbb{A}$ *as in (b) there is a computable function that assigns to every* L*-sentence* φ *a polynomial* $q \in \mathbb{N}[X]$ *such that* $\mathbb{A}$ *decides whether* $\mathcal{A} \models_L \varphi$ *in* $\leq q(\|\mathcal{A}\|)$ *steps.*

3 Order-Invariant Variants of LFP

In this section we introduce the variants of least fixed-point logic relevant to our paper.

For a vocabulary τ let $\tau_< := \tau \cup \{<\}$, where $<$ is a binary relation symbol not in τ. For every class of τ-structures C in P closed under isomorphisms the class of $\tau_<$-structures

$$C_< := \{(\mathcal{A}, <^A) \mid \mathcal{A} \in C \text{ and } <^A \text{ an ordering of } A\} \tag{5}$$

is in P, too; hence, as the logic LFP captures polynomial time on the class of ordered structures, there is an LFP$[\tau_<]$-sentence axiomatizing $C_<$. However, we are interested in a sentence axiomatizing the class C.

In order to obtain a logic that captures polynomial time on all structures one has considered variants of LFP obtained by restricting to order-invariant sentences or by modifying the semantics such that all sentences are order-invariant. In this section we recall the corresponding logics. We start by introducing the respective notions of invariance.

Definition 2. *(a) A pair* $(\varphi, \mathcal{A})$ *is in the relation* INV *if*

- *for some vocabulary* τ *we have that* $\mathcal{A}$ *is a* τ*-structure and* $\varphi \in \mathrm{LFP}[\tau_<]$*;*
- *(*φ is order-*invariant* in $\mathcal{A}$*) for all orderings* $<_1$ *and* $<_2$ *on* A *we have*

$$(\mathcal{A}, <_1) \models_{\mathrm{LFP}} \varphi \iff (\mathcal{A}, <_2) \models_{\mathrm{LFP}} \varphi.$$

(b) An LFP$[\tau_<]$*-sentence* φ *is* order-invariant *if* $(\varphi, \mathcal{A}) \in$ INV *for all* τ*-structures* $\mathcal{A}$*.*

(c) For an LFP$[\tau_<]$*-sentence* φ *and* $m \in \mathbb{N}$ *we write*

$$(\varphi, m) \in \mathrm{INV}$$

if $(\varphi, \mathcal{A}) \in$ INV *for all* τ*-structures* $\mathcal{A}$ *with* $|A| = m$.

(d) For an LFP$[\tau_<]$*-sentence* φ *and* $m \in \mathbb{N}$ *we write*

$$(\varphi, \leq m) \in \mathrm{INV}$$

if $(\varphi, \mathcal{A}) \in$ INV *for all* τ*-structures* $\mathcal{A}$ *with* $|A| \leq m$.

Note that every LFP$[\tau_<]$-sentence axiomatizing a class of the form $C_<$ (see (5)) is order-invariant.

The different degrees of invariance lead to the following different logics. For all logics L we let

$$L[\tau] := \text{LFP}[\tau_<].$$

Hence, these logics only differ in their semantics. The logic L_{inv} is the first naive attempt to get an (effectively) P-bounded logic for P. Its semantics is fixed by

$$\mathcal{A} \models_{L_{\text{inv}}} \varphi \iff \big(\varphi \text{ is order-invariant and } (\mathcal{A}, <_A) \models_{\text{LFP}} \varphi\big)$$

(recall that $<_A$ denotes the ordering on A given by the encoding of $\mathcal{A}$).

Clearly (and this remark will also apply to the logics $L_{\text{str}}, L_=$, and $L_\leq$ to be defined yet)

all properties in P are expressible in L_{inv}.

In fact, for a class $C \in$ P of τ-structures closed under isomorphisms, every LFP$[\tau_<]$-sentence axiomatizing the class $C_<$ is an $L_{\text{inv}}[\tau]$-sentence axiomatizing C.

The logic L_{inv} is a logic for P, as

$$\text{Mod}_{L_{\text{inv}}}(\varphi) = \{\mathcal{A} \mid (\mathcal{A}, <_A) \in \text{Mod}_{\text{LFP}}(\varphi)\}$$

if $\varphi \in \text{LFP}[\tau_<]$ is invariant, and $\text{Mod}_{L_{\text{inv}}}(\varphi) = \emptyset$ otherwise. However, as already remarked by Yuri Gurevich in [9], a simple application of a theorem of Trachtenbrot shows that the set of invariant LFP$[\tau_<]$-sentences is not decidable and thus $\models_{L_{\text{inv}}}$ is not decidable; hence L_{inv} is not a P-bounded logic for P.

For the logic L_{str} we require invariance in the corresponding structure:

$$\mathcal{A} \models_{L_{\text{str}}} \varphi \iff \Big((\varphi, \mathcal{A}) \in \text{INV} \text{ and } (\mathcal{A}, <_A) \models_{\text{LFP}} \varphi\Big).$$

For a binary relation symbol E, consider an $\text{FO}\big[\{E\}_<\big]$-sentence φ expressing that E is not a graph or that in the ordering $<$ there are two consecutive elements which are not related by an edge. The class $\text{Mod}_{L_{\text{str}}}(\varphi)$ is the complement of the class of graphs having a Hamiltonian path and hence it is coNP-complete (a different coNP-complete class was axiomatized by Gurevich in [9, Theorem 1.16]).

As an easy consequence we get:

Proposition 3. *The following statements are equivalent:*

- L_{str} *is a logic for* P.
- P = NP
- L_{str} *is an effectively* P*-bounded logic for* P.

Proof. Assume that P = NP. To show that L_{str} is an effectively P-bounded logic for P we consider the problem

Instance: An L_{str}-sentence φ, a structure $\mathcal{A}$ and the number $\|\mathcal{A}\|^{|\varphi|}$ in unary.
Problem: Is $(\mathcal{A}, \varphi) \in \text{INV}$?

By Proposition 1, it is in coNP and hence it is solvable in polynomial time. This yields the algorithm $\mathbb{A}$ as required by part (b) and (c) of Definition 1. □

As coNP-complete problems can be axiomatized in L_{str}, we define the logic $L_{=}$ using a stronger invariance property:

$$\mathcal{A} \models_{L_{=}} \varphi \iff \Big((\varphi, |A|) \in \textsc{Inv} \text{ and } (\mathcal{A}, <_A) \models_{\mathrm{LFP}} \varphi\Big);$$

in particular, $\mathcal{A} \models_{L_{=}} \varphi$ can only hold if φ is invariant in all structures with universe of the same cardinality as $\mathcal{A}$.

Note that for an $L_{=}$-sentence φ it is not clear whether the class of models of φ is in P. In fact, in Sect. 7 we show (recall that E := DTIME($2^{O(n)}$) and NE := NTIME($2^{O(n)}$)):

Proposition 4. *The following statements are equivalent:*

- *$L_{=}$ is a logic for* P.
- E = NE
- *$L_{=}$ is an effectively* P*-bounded logic for* P.

Finally we introduce the logic $L_{\leq}$, where invariance in all structures of the same or smaller cardinality is required:

$$\mathcal{A} \models_{L_{\leq}} \varphi \iff \Big((\varphi, \leq |A|) \in \textsc{Inv} \text{ and } (\mathcal{A}, <_A) \models_{\mathrm{LFP}} \varphi\Big).$$

If an LFP[$\tau_<$]-sentence φ is not order-invariant, then the class $\mathrm{Mod}_{L_{\leq}}(\varphi)$ only contains (up to isomorphism) finitely many structures and hence it is in P. Therefore $L_{\leq}$ (like L_{inv}) is a logic for P.

In particular, L_{inv} and $L_{\leq}$ have less expressive power than L_{str} (if P $\neq$ NP) and less than $L_{=}$ (if E $\neq$ NE). Clearly if P = NP (and hence E = NE), then all, L_{inv}, L_{str}, $L_{=}$ and $L_{\leq}$, have the same expressive power. Otherwise we have:

Proposition 5. 1. *If* P $\neq$ NP, *then there is a class axiomatizable in L_{str} but not in $L_{=}$.*
2. *If* E $\neq$ NE, *then there is a class axiomatizable in $L_{=}$ but not in L_{str}.*

Proof. To get (1) we observe that the complement of the class of graphs having a Hamiltonian path, a class axiomatizable in L_{str} as we have seen, is not axiomatizable in $L_{=}$ if P $\neq$ NP; this is shown by the following claim.

Claim 1: Let C be a class of τ-structures. Assume that $C \notin$ P. Furthermore assume that for every $m \in \mathbb{N}$ with $m \geq 2$ there is a structure $\mathcal{A}_m \in C$ such that $|A_m| = m$. Then C is not axiomatizable in $L_{=}$.

Proof of Claim 1: Assume that $C = \mathrm{Mod}_{L_{=}}(\varphi)$. For $m \geq 2$ we have $(\varphi, m) \in$ Inv, as $\mathcal{A}_m \models_{L_{=}} \varphi$. Clearly, $(\varphi, 1) \in$ Inv. Hence φ is order-invariant and thus $\mathrm{Mod}_{\mathrm{LFP}}(\varphi) = C_<$. So $C_<$ and hence C are in P, a contradiction. ⊣

A proof of part (2), based on the following claim, will be presented in Sect. 4.1.

Claim 2: Let X be a set of natural numbers in unary with $X \notin \mathrm{P}$. Assume that the class C of τ-structures has the property: For all τ-structures $\mathcal{A}$

$$\mathcal{A} \in C \iff |A| \in X.$$

Then C is not axiomatizable in L_{str}.

Proof of Claim 2: Assume that $C = \text{Mod}_{L_{\text{str}}}(\varphi)$ with $\varphi \in L_{\text{str}}[\tau]$. Clearly $(\varphi, \mathcal{A}) \in$ INV for every τ-structure $\mathcal{A}$ of the form $\mathcal{A} = (A, (\emptyset)_{P\in\tau})$ (all relation symbols $P \in \tau$ are interpreted in $\mathcal{A}$ by the empty relation of the corresponding arity). Then for every $m \geq 1$ and for the natural ordering $<$ on $[m]$:

$$\begin{aligned}\big([m], (\emptyset)_{P\in\tau}, <\big) \models_{\text{LFP}} \varphi &\iff \big([m], (\emptyset)_{P\in\tau}\big) \models_{L_{\text{str}}} \varphi \\ &\iff m \in X\end{aligned}$$

As $\big([m], (\emptyset)_{P\in\tau}, <\big) \models_{\text{LFP}} \varphi$ can be checked in time polynomial in

$$\Big\|\big([m], (\emptyset)_{P\in\tau}, <\big)\Big\|$$

and hence, polynomial in m, we see that $X \in \mathrm{P}$, a contradiction. □

This paper mainly addresses the question whether $L_\leq$ is an effectively P-bounded logic for P, a question raised in [12]. Even though not everyone is convinced that there is no effectively P-bounded logic for P, it is conjectured that $L_\leq$ is not such a logic. In [12] this question (or conjecture) is reformulated as an effectivity property for a halting problem for nondeterministic Turing machines. We analyze the relationship between the logic and the halting problem in the next section.

For later purposes we remark that as

$$(\varphi, \leq m) \in \text{INV} \iff (\neg\varphi, \leq m) \in \text{INV},$$

we get

$$(\varphi, \leq |A|) \in \text{INV} \iff (\mathcal{A} \models_{L_\leq} \varphi \text{ or } \mathcal{A} \models_{L_\leq} \neg\varphi). \tag{6}$$

4 A Halting Problem and Its Relationship to $L_\leq$

From the Introduction we recall the parameterized acceptance problem p-ACC$_\leq$ for nondeterministic Turing machines:

p-ACC$_\leq$	
Instance:	A nondeterministic Turing machine $\mathbb{M}$ and $n \in \mathbb{N}$ in unary.
Parameter:	$\|\mathbb{M}\|$.
Question:	Does $\mathbb{M}$ accept the empty input tape in at most n steps?

We shall see in this section how the complexity of this problem is related to properties of the logic $L_{\leq}$. We start with the following simple observation on the complexity of p-ACC$_{\leq}$.

Proposition 6. *The problem* p-ACC$_{\leq}$ *is in the class* $\mathrm{FPT}_{\mathrm{nu}}$.

Proof. Fix $k \in \mathbb{N}$; then there are only finitely many nondeterministic Turing machines $\mathbb{M}$ with $\|\mathbb{M}\| = k$, say, $\mathbb{M}_1, \dots, \mathbb{M}_s$. For each $i \in [s]$ let ℓ_i be the smallest natural number ℓ such that there exists an accepting run of $\mathbb{M}_i$, started with empty input tape, of length ℓ. We set $\ell_i = \infty$ if $\mathbb{M}_i$ does not accept the empty input tape. Hence the algorithm $\mathbb{A}_k$ that on any instance $(\mathbb{M}, n)$ of p-ACC$_{\leq}$ with $\|\mathbb{M}\| = k$ determines the i with $\mathbb{M} = \mathbb{M}_i$, and then accepts if and only if $\ell_i \leq n$, decides the kth slice of p-ACC$_{\leq}$. It has running time $O(\|\mathbb{M}\| + n)$; thus it witnesses that p-ACC$_{\leq} \in \mathrm{FPT}_{\mathrm{nu}}$. □

This observation can easily be generalized. We call a parameterized problem (Q, κ) *slicewise monotone* if its instances have the form (x, n), where $x \in \{0,1\}^*$ and $n \in \mathbb{N}$ is given in unary, if $\kappa(x, n) = |x|$, and finally if for all $x \in \{0,1\}^*$ and $n, n' \in \mathbb{N}$ we have

$$(x, n) \in Q \text{ and } n < n' \text{ imply } (x, n') \in Q.$$

In particular, p-ACC$_{\leq}$ is slicewise monotone and the preceding argument shows:

Lemma 1. $(Q, \kappa) \in \mathrm{FPT}_{\mathrm{nu}}$ *for slicewise monotone* (Q, κ).

Is p-ACC$_{\leq} \in \mathrm{FPT}_{\mathrm{uni}}$ or, at least, p-ACC$_{\leq} \in \mathrm{XP}_{\mathrm{uni}}$? By [12] the conjecture "$L_{\leq}$ is not a P-bounded logic for P" mentioned in the previous section is equivalent to the statement p-ACC$_{\leq} \notin \mathrm{XP}_{\mathrm{uni}}$ (and similarly, the statement "$L_{\leq}$ is not an effectively P-bounded logic for P" is equivalent to p-ACC$_{\leq} \notin \mathrm{XP}$).

However it is not even clear whether p-ACC$_{\leq} \notin \mathrm{FPT}$. Do the statements p-ACC$_{\leq} \in \mathrm{FPT}$ and p-ACC$_{\leq} \in \mathrm{FPT}_{\mathrm{uni}}$ also correspond to natural properties of the logic $L_{\leq}$? We address this problem in this section.

Proposition 2 motivates the introduction of the following notion. We say that $L_{\leq}$ is an *(effectively) depth-width* P*-bounded logic for* P if there is an algorithm $\mathbb{A}$ deciding $\models_{L_{\leq}}$ in such a way that there is a (computable) function h such that $\mathcal{A} \models_{L_{\leq}} \varphi$ can be solved in time

$$h(|\varphi|) \cdot \|\mathcal{A}\|^{O\left((1+\mathrm{depth}(\varphi)) \cdot \mathrm{width}(\varphi)\right)}.$$

By Proposition 2, the logic LFP "is an effectively depth-width P-bounded logic for P on ordered structures." Parts (1) and (2) of the following theorem are the main result of this section, (3) and (4) are already mentioned in [12].

Theorem 1. *1. $L_{\leq}$ is an effectively depth-width* P*-bounded logic for* P *if and only if* p-ACC$_{\leq} \in \mathrm{FPT}$.
2. $L_{\leq}$ is a depth-width P*-bounded logic for* P *if and only if* p-ACC$_{\leq} \in \mathrm{FPT}_{\mathrm{uni}}$.
3. $L_{\leq}$ is an effectively P*-bounded logic for* P *if and only if* p-ACC$_{\leq} \in \mathrm{XP}$.
4. $L_{\leq}$ is a P*-bounded logic for* P *if and only if* p-ACC$_{\leq} \in \mathrm{XP}_{\mathrm{uni}}$.

4.1 Proof of Theorem 1 and Some Consequences

The following observations will lead to a proof of the direction "from right to left" in the statements of Theorem 1.

For an $L_{\leq}$-sentence φ let τ_φ be the set of relation symbols distinct from $<$ that do occur in φ. For a suitable time constructible function $t : \mathbb{N} \to \mathbb{N}$ we will need a nondeterministic Turing machine $\mathbb{M}_\varphi(t)$ that, started with empty tape, operates as follows: In a first phase it writes a word of the form 1^m for some $m \geq 1$ on some tape. The second phase (the main phase) consists of at most $t(m) + 1$ steps (this can be ensured as t is time constructible). If $\mathbb{M}_\varphi(t)$ does not stop during the first $t(m)$ steps of the main phase, then it stops in the next step and rejects. During these $t(m)$ steps, $\mathbb{M}_\varphi(t)$ guesses (the encoding of) a τ_φ-structure $\mathcal{A}$ with universe $[m]$ and two orderings $<_1$ and $<_2$ on $[m]$ and checks whether $\big((\mathcal{A}, <_1) \models_{\text{LFP}} \varphi \iff (\mathcal{A}, <_2) \models_{\text{LFP}} \varphi\big)$. If this is not the case, then $\mathbb{M}_\varphi(t)$ accepts; otherwise it rejects.

The first phase takes m steps. To guess a τ_φ-structure $\mathcal{A}$ with universe $[m]$ and two orderings $<_1$ and $<_2$ requires $O(\text{lgth}(\tau_\varphi, m) + 2m^2)$ bits (see Sect. 2.2); thus for some $d_1 \in \mathbb{N}$ the machine $\mathbb{M}_\varphi(t)$ needs

$$t^1_\varphi(m) := d_1 \cdot (\text{lgth}(\tau_\varphi, m) + 2m^2)$$

steps. Finally, by Proposition 2, to check the equivalence $\big((\mathcal{A}, <_1) \models_{\text{LFP}} \varphi$ iff $(\mathcal{A}, <_2) \models_{\text{LFP}} \varphi\big)$ takes at most

$$t^2_\varphi(m) := |\varphi| \cdot \text{lgth}((\tau_\varphi)_<, m)^{d_2 \cdot (1+\text{depth}(\varphi)) \cdot \text{width}(\varphi)}$$

steps for some $d_2 \in \mathbb{N}$. By the time constructibility of the function $m \mapsto \text{lgth}((\tau_\varphi)_<, m)$ we can arrange the machine in such a way that it needs exactly the number of steps given by the upper bounds $t^1_\varphi(m)$ and $t^2_\varphi(m)$ (if it is not stopped by the time bound $t(m)$). Thus, if in the first phase of a run $\mathbb{M}_\varphi(t)$ has written the word 1^m, then $\mathbb{M}_\varphi(t)$ performs exactly

$$t_\varphi(m) := t^1_\varphi(m) + t^2_\varphi(m) \tag{7}$$

additional steps before it stops (assuming $t_\varphi(m) \leq t(m)$). Note that t_φ is increasing. Therefore we have

$$\begin{aligned}(\varphi, \leq m) \in \text{INV} \iff\ & \mathbb{M}_\varphi(t_\varphi) \text{ does not accept the empty string} \\ & \qquad\qquad \text{in } \leq m + t_\varphi(m) \text{ steps} \\ \iff\ & (\mathbb{M}_\varphi(t_\varphi), m + t_\varphi(m)) \notin p\text{-ACC}_{\leq}.\end{aligned} \tag{8}$$

We collect some facts we are going to use:

(i) There is an algorithm assigning to every $L_{\leq}$-sentence φ the machine $\mathbb{M}_\varphi(t_\varphi)$.

(ii) For every $L_{\leq}$-sentence φ and all τ_φ-structures $\mathcal{A}$:

$$\mathcal{A} \models_{L_\leq} \varphi \iff \Big((\varphi, \leq |A|) \in \text{INV} \text{ and } (\mathcal{A}, <_A) \models_{\text{LFP}} \varphi\Big)$$

(by definition of $\models_{L_\leq}$)

$$\iff \Big((\mathbb{M}_\varphi(t_\varphi), m + t_\varphi(|A|)) \notin p\text{-ACC}_{\leq} \text{ and } (\mathcal{A}, <_A) \models_{\text{LFP}} \varphi\Big)$$

(by (8)).

(iii) There is a computable function g such that for every $L_\leq$-sentence φ and all τ_φ-structures $\mathcal{A}$ we have

$$t_\varphi(|A|) \leq g(|\varphi|) \cdot \|\mathcal{A}\|^{O\big((1+\text{depth}(\varphi))\cdot \text{width}(\varphi)\big)}$$

(by the definition (7) of the function t_φ and the properties of the lgth-function mentioned in Sect. 2.2).

Now we can show the direction "from right to left" in the statements of Theorem 1. We give the proof for the claims (1) and (2); obvious modifications yield (3) and (4).

Assume p-ACC$_\leq \in$ FPT$_{\text{uni}}$ (p-ACC$_\leq \in$ FPT), that is, assume that $(\mathbb{M}, n) \in p$-ACC$_\leq$ can be solved in time

$$f(\|\mathbb{M}\|) \cdot n^e$$

for some $e \in \mathbb{N}$ and some (computable) function $f : \mathbb{N} \to \mathbb{N}$. We consider the problem $\mathcal{A} \models_{L_\leq} \varphi$, where $\mathcal{A}$ is a structure and φ an $L_\leq$-sentence. We may assume that $\mathcal{A}$ is a τ_φ-structure (if $\mathcal{A}$ contains more relations, we omit them; this can be done in $O(|\varphi| + \|\mathcal{A}\|)$ steps). Using (i), (iii), and Proposition 2 we see that there are an algorithm and a (computable) function h such that the condition

$$(\mathbb{M}_\varphi(t_\varphi), |A| + t_\varphi(|A|)) \notin p\text{-ACC}_{\leq} \quad \text{and} \quad (\mathcal{A}, <_A) \models_{\text{LFP}} \varphi$$

and hence, by (ii), the problem

$$\mathcal{A} \models_{L_\leq} \varphi$$

can be solved in time $h(|\varphi|) \cdot \|\mathcal{A}\|^{O\big((1+\text{depth}(\varphi))\cdot \text{width}(\varphi)\big)}$. □

Before we proceed with the proof of Theorem 1, it is worthwhile to extract from the previous argument information relevant for the logic $L_=$. We introduce the corresponding halting problem:

> p-ACC$_=$
> *Instance:* A nondeterministic Turing machine $\mathbb{M}$ and $n \in \mathbb{N}$ in unary.
> *Parameter:* $\|\mathbb{M}\|$.
> *Question:* Does $\mathbb{M}$ accept the empty input tape in *exactly* n steps?

Then:

Lemma 2. *If p-ACC$_=$ $\in$ FPT, then $L_=$ is an effectively depth-width* P*-bounded logic for* P.[2]

Proof. Note that the following variant of (8) holds:

$$(\varphi, m) \in \text{INV} \iff \mathbb{M}_\varphi(t_\varphi) \text{ does not accept the empty string in exactly } m + t_\varphi(m) \text{ steps}$$
$$\iff (\mathbb{M}_\varphi(t_\varphi), m + t_\varphi(m)) \notin p\text{-ACC}_=,$$

and thus,

$$\mathcal{A} \models_{L_=} \varphi \iff \Big((\mathbb{M}_\varphi(t_\varphi), |A| + t_\varphi(|A|)) \notin p\text{-ACC}_= \quad \text{and} \quad (\mathcal{A}, <_A) \models_{\text{LFP}} \varphi\Big).$$

Thus our claim can be derived in exactly the same way as the corresponding statement for $L_\leq$. □

We turn to a proof of the directions "from left to right" in Theorem 1. Let $\mathbb{M}$ be a nondeterministic Turing machine and let $m_0 := m_0(\mathbb{M})$ be the maximum of the number of states and the number of tapes. We can assume that $[k]$ is the set of states of $\mathbb{M}$ (for some $k \leq m_0$) and that 1 is its initial state. Furthermore, we may assume that every two distinct successor configurations of a given configuration of $\mathbb{M}$ have distinct states. We let $P_0, P_1, \dots, P_k$ be unary relation symbols. We shall see that for $\tau := \{P_0, \dots, P_k\}$ there is a LFP$[\tau_<]$-sentence $\varphi_\mathbb{M}$ in normal form with the following properties: For every τ-structure $\mathcal{A}$

(a) If $|A| < m_0$, then $(\mathcal{A}, <^A) \models_{\text{LFP}} \varphi_\mathbb{M}$ for all orderings $<^A$ on A.
(b) If $|A| \geq m_0$ and the subsets $P_0^A, \dots, P_k^A$ do not form a partition of A, then $(\mathcal{A}, <^A) \models_{\text{LFP}} \varphi_\mathbb{M}$ for all orderings $<^A$ on A.
(c) Let $|A| \geq m_0$ and assume that $P_0^A, \dots, P_k^A$ form a partition of A and $<^A$ is an ordering on A. Let $a_1, \dots, a_{|A|}$ be the enumeration of the elements of A according to the ordering $<^A$ and choose i_s such that $a_s \in P_{i_s}^A$ for $s \in [|A|]$.
  (i) If there is a $j \in [|A| - 1]$ such that $1, i_1, \dots, i_j$ is the sequence of states of a complete run of $\mathbb{M}$, started with empty input tape (in particular, $i_s \neq 0$ for all $s \in [j]$), then $(\mathcal{A}, <^A) \models_{\text{LFP}} \varphi_\mathbb{M}$ if and only if this run of $\mathbb{M}$ is a rejecting one.
  (ii) If for all $j \in [|A| - 1]$ the sequence $1, i_1, \dots, i_j$ does not correspond to a complete run of $\mathbb{M}$ with empty input tape, then $(\mathcal{A}, <^A) \models_{\text{LFP}} \varphi_\mathbb{M}$.

We show that for every $m \geq m_0(\mathbb{M})$

$$(\mathbb{M}, m) \in p\text{-ACC}_\leq \iff (\varphi_\mathbb{M}, \leq m) \notin \text{INV}. \tag{9}$$

[2] Along the lines of the proof the reader will easily verify the analogues for p-ACC$_=$ and $L_=$ of the directions "from right to left" of all statements of Theorem 1. However, all the others will follow from Corollary 2.

First assume that $(\mathbb{M}, m) \in p\text{-}\textsc{Acc}_{\leq}$. Then there are $j \in [m-1]$ and $i_1, \ldots, i_j \in [k]$ such that $1, i_1, \ldots, i_j$ is the sequence of states of an accepting run of $\mathbb{M}$. By (c)(i) there is a structure $\mathcal{A}$ on $[m]$ such that $(\mathcal{A}, <^A) \not\models_{\text{LFP}} \varphi_{\mathbb{M}}$ for the natural ordering $<^A$ on $[m]$ and $P_0^{\mathcal{A}} = \{m\}$. We choose an ordering $<'$ on $[m]$ such that m is the first element of $<'$ and hence, $i_1 = 0$ under $<'$. By (c)(ii) we see that $(\mathcal{A}, <') \models_{\text{LFP}} \varphi_{\mathbb{M}}$. Hence, $(\varphi_{\mathbb{M}}, \leq m) \notin \textsc{Inv}$.

Conversely, if $(\mathbb{M}, m) \notin p\text{-}\textsc{Acc}_{\leq}$, it is easy to see, using (a)–(c), that $\mathcal{A} \models_{L_{\leq}} \varphi_{\mathbb{M}}$ for every structure $\mathcal{A}$ with $|A| \leq m$; hence, $(\varphi_{\mathbb{M}}, \leq m) \in \textsc{Inv}$.

The sentence $\varphi_{\mathbb{M}}$ (in normal form and hence of depth 1) is obtained by standard techniques. We sketch its construction. Recall that $\{0, 1\}$ is the alphabet of $\mathbb{M}$. The sentence $\varphi_{\mathbb{M}}$ will make use of the binary relation variable *State* and the ternary relation variables *Head*, *Zero*, *One* with the intended meaning

$$\begin{aligned} \textit{State}\, t\, s &\text{ iff at time } t \text{ the state is } s \\ \textit{Head}\, t\, i\, j &\text{ iff at time } t \text{ the head on tape } i \text{ scans the } j\text{th cell} \\ \textit{Zero}\, t\, i\, j &\text{ iff at time } t \text{ there is a 0 on the } j\text{th cell of tape } i \\ \textit{One}\, t\, i\, j &\text{ iff at time } t \text{ there is a 1 on the } j\text{th cell of tape } i. \end{aligned}$$

To be able to apply the least fixed-point operator to a formula *positive* in the corresponding variable(s), we need also relation variables *CState*, *CHead*, *CZero* and *COne* for the *complements* of the relations just introduced. Then we can express the intended meaning of $\varphi_{\mathbb{M}}$ using a single simultaneous least fixed-point over all the relation variables we have introduced in the form

$$\exists x \exists y [\text{S-LFP}_{t\,s\,State,\ t\,i\,j\,Head,\ t\,i\,j\,Zero,\ t\,i\,j\,One,\ t\,s\,CState,\ t\,i\,j\,CHead,\ t\,i\,j\,CZero,\ t\,i\,j\,COne} \psi_{\mathbb{M}}]\, xy$$

(even though $\mathbb{M}$ is nondeterministic, in the formula $\psi_{\mathbb{M}}$ we always get the state to be chosen in the next step using the relations $P_1, \ldots, P_k$). By standard means this sentence can be converted in an LFP-sentence $\varphi_{\mathbb{M}}$ in normal form (and hence of depth 1). Of course, the sentence $\varphi_{\mathbb{M}}$ depends on the machine $\mathbb{M}$, however, as in $\varphi_{\mathbb{M}}$ we have to take care of a run of at most as many steps as the cardinality of the universe, it can be defined in such a way that its width is independent of $\mathbb{M}$; here we use the fact that we can address the ith element in the ordering $<$ by a formula of width 3.

Now we are able to finish the proof of the directions "from left to right" in Theorem 1. Again we present the argument for claims (1) and (2) of this theorem, as claims (3) and (4) are obtained by the obvious modifications. Assume that $L_{\leq}$ is an (effectively) depth-width P-bounded logic for P and choose $c \in \mathbb{N}$ and a (computable) function $h : \mathbb{N} \to \mathbb{N}$ such that the model-checking problem $\mathcal{A} \models_{L_{\leq}} \varphi$ for structures $\mathcal{A}$ and $L_{\leq}$-sentences φ can be solved in time

$$h(|\varphi|) \cdot \|\mathcal{A}\|^{c \cdot (1 + \text{depth}(\varphi)) \cdot \text{width}(\varphi)}. \tag{10}$$

We show that $p\text{-}\textsc{Acc}_{\leq} \in \text{FPT}_{\text{uni}}$ ($p\text{-}\textsc{Acc}_{\leq} \in \text{FPT}$). Let $(\mathbb{M}, m)$ be an arbitrary instance of $p\text{-}\textsc{Acc}_{\leq}$. If $m < m_0(\mathbb{M})$ ($\leq \|\mathbb{M}\|$), we check whether $(\mathbb{M}, m) \in p\text{-}\textsc{Acc}_{\leq}$

by brute force. Otherwise, we construct from $\mathbb{M}$ the sentence $\varphi_{\mathbb{M}}$. Moreover we choose the τ-structure $\mathcal{A}_m$ with $A_m = [m]$ and empty relations. Then, by the property (d) of the lgth-function (see Sect. 2.2), we have

$$\|\mathcal{A}_m\| = O(\log|\tau| \cdot |\tau| \cdot |A_m|) = O(|\varphi_{\mathbb{M}}|^2 \cdot m). \tag{11}$$

As by (6)

$$(\varphi_{\mathbb{M}}, \leq m) \in \text{INV} \iff (\mathcal{A}_m \models_{L_\leq} \varphi_{\mathbb{M}} \ \text{ or } \ \mathcal{A}_m \models_{L_\leq} \neg\varphi_{\mathbb{M}}),$$

we obtain by (9)

$$(\mathbb{M}, m) \notin p\text{-ACC}_\leq \iff (\mathcal{A}_m \models_{L_\leq} \varphi_{\mathbb{M}} \ \text{ or } \ \mathcal{A}_m \models_{L_\leq} \neg\varphi_{\mathbb{M}}). \tag{12}$$

Therefore, by (11) and (10), we see that there is a (computable) function f and a constant $e \in \mathbb{N}$ (recall that for all nondeterministic Turing machines $\mathbb{M}$ the depth of $\varphi_{\mathbb{M}}$ is one and that there is a constant bounding the width of $\varphi_{\mathbb{M}}$) such that $(\mathbb{M}, m) \in p\text{-ACC}_\leq$ can be solved in time $f(\|\mathbb{M}\|) \cdot m^e$. This finishes the proof of Theorem 1. □

Again we extract from the proof the information on $p\text{-ACC}_=$ and $L_=$ that we shall need in Sect. 5.

Lemma 3. *If $L_=$ is a logic for* P, *then $p\text{-ACC}_= \in \text{XP}_{\text{nu}}$.*

Proof. A minor change in the definition of $\varphi_{\mathbb{M}}$ in the previous proof yields an LFP-sentence $\chi_{\mathbb{M}}$ with

$$(\mathbb{M}, m) \in p\text{-ACC}_= \iff (\chi_{\mathbb{M}}, m) \notin \text{INV}$$

instead of (9), and hence

$$(\mathbb{M}, m) \notin p\text{-ACC}_= \quad \iff \quad (\mathcal{A}_m \models_{L_=} \chi_{\mathbb{M}} \ \text{ or } \ \mathcal{A}_m \models_{L_=} \neg\chi_{\mathbb{M}}) \tag{13}$$

instead of (12). Assume $L_=$ is a logic for P. Fix $k \in \mathbb{N}$ and let $\mathbb{M}_1, \ldots, \mathbb{M}_s$ be the finitely many NTMs $\mathbb{M}$ with $\|\mathbb{M}\| = k$. As $L_=$ is a logic for P, for all $i \in [s]$ there is an algorithm solving $\mathcal{A} \models_{L_=} \chi_{\mathbb{M}_i}$ in time polynomial in $\|\mathcal{A}\|$. Now (13) yields the claim $p\text{-ACC}_= \in \text{XP}_{\text{nu}}$. □

By refining the proof of the second part of Theorem 1 we obtain the following result (related to Gurevich's result [9, Theorem 1.16]):

Corollary 1. *For sufficiently large $w \in \mathbb{N}$ the problem*

Instance: A structure $\mathcal{A}$ and an $L_\leq[1]$-sentence φ of width $\leq w$.
Problem: Is $\mathcal{A} \models_{L_\leq} \varphi$?

is coNP*-complete.*

Proof. Clearly the problem is in coNP. Now let Q be any problem in coNP. We give a polynomial reduction of Q to the problem in the statement. Let $\mathbb{M}$ be a polynomial time nondeterministic Turing machine such that for all $x \in \{0,1\}^*$ we have

$$x \in Q \iff \text{no run of } \mathbb{M} \text{ accepts } x.$$

Choose $c, d \in \mathbb{N}$ such that the running time of $\mathbb{M}$ on input x is bounded by $c \cdot |x|^d$. We fix $x \in \{0,1\}^*$. Again we assume that $[k]$ is the set of states of $\mathbb{M}$ and that 1 is its starting state and set $\tau := \{P_0, \ldots, P_k\}$ with unary relation variables $P_0, \ldots, P_k$. Along the lines of the proof of the second part of Theorem 1, in time polynomial in $|x|$ one can obtain an LFP$[\tau_<]$-sentence $\varphi_{\mathbb{M},x}$ in normal form with the properties (a)–(c) (on page 264), where now in (c)(i),(ii) we consider complete runs *started with the word x in the input tape.* Again we can achieve that the width of $\varphi_{\mathbb{M},x}$ does not depend on x (once more, we use the fact that for $i \in [|x|]$ we can address the ith element in the ordering $<$ by a formula of width 3). Let $\mathcal{A}_x$ be any τ-structure with $|A_x| \geq \max\{m_0(\mathbb{M}),\ c \cdot |x|^d\}$. Then

$$x \in Q \iff \mathcal{A}_x \models_{L_\leq} \varphi_{\mathbb{M},x},$$

and hence $x \mapsto (\mathcal{A}_x, \varphi_{\mathbb{M},x})$ is the desired reduction. □

We close this section by a proof of Proposition 5 (2).

Proof of Proposition 5 (2): Let X be a set of natural numbers in binary in NE $\setminus$ E. Then $X(\text{un}) \in \text{NP} \setminus \text{P}$, where $X(\text{un})$ is the set of natural numbers of X in unary. Hence there is a nondeterministic Turing machine $\mathbb{M}$ that given $m \in \mathbb{N}$ in *unary* decides whether $m \in X(\text{un})$ in polynomial time, say, in time $c \cdot m^d$. We may assume that every run of $\mathbb{M}$ on input m has length $c \cdot m^d$. Similarly to the sentence $\varphi_{\mathbb{M}}$ in the proof of Theorem 7, we construct an LFP-sentence $\rho_{\mathbb{M}}$ expressing that

> if for some $m \in \mathbb{N}$ the universe has cardinality $c \cdot m^d$ and the relations $P_0, \ldots, P_k$ code a run of $\mathbb{M}$ with input 1^m, then it is not accepting.

Then for every $\{P_0, \ldots, P_s\}$-structure $\mathcal{A}$ we have

$$\mathcal{A} \models_{L_=} \rho_{\mathbb{M}} \iff |A| \notin \{c \cdot m^d \mid m \in X(\text{un})\}. \tag{14}$$

As the set $\{c \cdot m^d \mid m \in X(\text{un})\}$ of natural numbers in unary is not in P, we get that $\text{Mod}_{L_=}(\rho_{\mathbb{M}})$ is not axiomatizable in L_{str} by Claim 2 in the proof of Proposition 5. □

5 The Parameterized Complexity of p-Acc$_=$

Let $\mathbb{M}$ be a nondeterministic Turing machine. By suitably adding to $\mathbb{M}$ a state, which can be accessed and left nondeterministically, one obtains a machine $\mathbb{M}^*$ such that for all $n \in \mathbb{N}$ we have:

$$\mathbb{M} \text{ accepts the empty input tape in } \leq n \text{ steps}$$
$$\iff \mathbb{M}^* \text{ accepts the empty input tape in exactly } n \text{ steps.}$$

Hence p-ACC$_{\leq} \leq^{\text{fpt}} p$-ACC$_{=}$. Recall that p-ACC$_{\leq} \in \text{FPT}_{\text{nu}}$. On the other hand, p-ACC$_{=} \notin \text{FPT}_{\text{nu}}$ if E $\neq$ NE, as shown by the main result of this section:

Theorem 2. *The following statements are equivalent:*

- p-ACC$_{=} \notin$ FPT.
- p-ACC$_{=} \notin \text{XP}_{\text{nu}}$.
- E $\neq$ NE.

In [1] it is shown that p-ACC$_{=} \in$ XP implies E = NE. By Lemma 2 and Lemma 3 we get as a consequence of this theorem the following improvement of Proposition 4.

Corollary 2. *The following statements are equivalent:*

- $L_{=}$ *is a logic for* P.
- E = NE.
- $L_{=}$ *is an effectively depth-width* P*-bounded logic for* P.

We prove Theorem 2 by the following two lemmas.

Lemma 4. *If* E = NE, *then* p-ACC$_{=} \in$ FPT.

Proof. Consider the classical problem:

Instance: A nondeterministic Turing machine $\mathbb{M}$ and $n \in \mathbb{N}$ in *binary*.
Problem: Does $\mathbb{M}$ accept the empty input tape in exactly n steps?

Clearly, it is NE. By the assumption E = NE we can solve it in time

$$2^{O(\|\mathbb{M}\|+\log n)}.$$

It follows that p-ACC$_{=}$ is decidable in time

$$O(n) + 2^{O(\|\mathbb{M}\|+\log n)} = 2^{O(\|\mathbb{M}\|)} \cdot n^{O(1)},$$

and hence p-ACC$_{=} \in$ FPT. □

Lemma 5. *If* p-ACC$_{=} \in \text{XP}_{\text{nu}}$, *then* E = NE.

Proof. Assume that p-ACC$_{=} \in \text{XP}_{\text{nu}}$. Let $Q \subseteq \{0,1\}^*$ be in NE. We have to show $Q \in$ E. Without loss of generality we may assume that every $x \in Q$ starts with a "1." Let $n(x)$ be the natural number with binary representation x; then

$$n(x) \neq n(y) \text{ for } x, y \in Q \text{ with } x \neq y. \tag{15}$$

As $Q \in$ NE there is a nondeterministic Turing machine $\mathbb{M}$ and a $c \in \mathbb{N}$ such that $\mathbb{M}$ decides whether $x \in Q$ in time

$$2^{c \cdot |x|}$$

and every run of $\mathbb{M}$ on input x has length at most $2^{c \cdot |x|}$. Note that for x starting with a "1", we have $2^{c \cdot |x|} = n(x)^c$.

We define a nondeterministic Turing machine $\mathbb{M}^*$ that started with empty input tape runs as follows:

1. Guess a string $y \in \{0,1\}^*$ *2.* **if** y does not start with a "1", **then** reject *3.* simulate $\mathbb{M}$ on input y for $n(y)^c$ many steps *4.* **if** $\mathbb{M}$ rejects, **then** reject *5.* make some additional dummy steps such that so far the total running time of $\mathbb{M}^*$ is $2 \cdot n(y)^c - 1$ *6.* accept.

By (15) we have for every $x \in \{0,1\}^*$ starting with a "1":

$$x \in Q \iff \mathbb{M}^* \text{ accepts the empty input tape in exactly } 2 \cdot n(x)^c \text{ many steps.} \tag{16}$$

As p-ACC$_=$ $\in$ XP$_{\text{nu}}$, for some $d \in \mathbb{N}$ we can decide whether $\mathbb{M}^*$ accepts the empty string in exactly $2 \cdot n(x)^c$ many steps in time

$$(2 \cdot n(x)^c)^d.$$

Hence, $x \in Q$ can be decided in time $2^{O(|x|)}$. □

Proof of Theorem 2: Immediate by Lemmas 4 and 5. □

6 The Parameterized Complexity of *p*-Acc$_\leq$.

We already know that the parameterized problem p-ACC$_\leq$ is in FPT$_{\text{nu}}$; however, is it fixed-parameter tractable or at least in XP? We address these questions in this section.

Let

$$\text{P[TC]} \neq \text{NP[TC]}$$

mean that DTIME($h^{O(1)}$) $\neq$ NTIME($h^{O(1)}$) for all time constructible and increasing functions h.

The assumption P[TC] $\neq$ NP[TC] implies P $\neq$ NP, even E $\neq$ NE, as seen by taking as h the identity function and the function 2^n, respectively. At the end of this section we are going to relate P[TC] $\neq$ NP[TC] to further statements of complexity theory. The main result of this section is:

Theorem 3. *If* P[TC] $\neq$ NP[TC], *then* p-ACC$_\leq$ $\notin$ FPT.

The following idea underlies the proof of this result. Assume that p-ACC$_\leq$ $\in$ FPT. Then, in particular we have a *deterministic* algorithm deciding p-ACC$_\leq$, the (parameterized) acceptance problem for *nondeterministic* Turing machines. This yields a way (different from brute force) to translate nondeterministic algorithms into deterministic ones; a careful analysis of this translation shows that NTIME($h^{O(1)}$) $\subseteq$ DTIME($h^{O(1)}$) for a suitable time constructible and increasing function h.

For our detailed proof we need the following simple lemma:

Lemma 6. *For every computable and increasing function $f : \mathbb{N} \to \mathbb{N}$ and every $e \in \mathbb{N}$ there is a time constructible and increasing function $h : \mathbb{N} \to \mathbb{N}$ such that for all $x, y \in \mathbb{N}$*

$$f(e \cdot (x + y^2)) \leq h(x) + h(y).$$

Proof. We define $h_0 : \mathbb{N} \to \mathbb{N}$ by

$$h_0(n) := f(2e \cdot n^2)$$

and let $h : \mathbb{N} \to \mathbb{N}$ be a time constructible and increasing function with $h_0(n) \leq h(n)$ for all $n \in \mathbb{N}$. □

Proof of Theorem 3: For any nondeterministic Turing machine $\mathbb{M}$ and every $x \in \{0,1\}^*$ we let $\mathbb{M}_x$ be the nondeterministic Turing machine that, started with empty input tape, first writes x on some tape and then simulates $\mathbb{M}$ started with x. Clearly we can define $\mathbb{M}_x$ such that $\|\mathbb{M}_x\| = O(\|\mathbb{M}\| + |x| \cdot \log |x|)$. We choose $e \in \mathbb{N}$ such that

$$\|\mathbb{M}_x\| \leq e \cdot (\|\mathbb{M}\| + |x|^2). \tag{17}$$

Now by contradiction assume that p-ACC$_{\leq} \in$ FPT. Then there is an algorithm $\mathbb{A}$ that for every nondeterministic Turing machine $\mathbb{M}$ and every natural number n decides whether $\mathbb{M}$ accepts the empty input tape in $\leq n$ steps in time

$$f(\|\mathbb{M}\|) \cdot n^{O(1)}$$

for a computable and increasing function $f : \mathbb{N} \to \mathbb{N}$. For this f and the e of (17) we choose h according to Lemma 6 and show that NTIME$(h^{O(1)}) \subseteq$ DTIME$(h^{O(1)})$.

Let $Q \subseteq \{0,1\}^*$ be in NTIME$(h^{O(1)})$. We choose a nondeterministic Turing machine $\mathbb{M}$ and constants $c, d \in \mathbb{N}$ such that the machine $\mathbb{M}$ decides whether $x \in Q$ in time $c \cdot h(|x|)^d$ and every run of $\mathbb{M}$ on input x is $c \cdot h(|x|)^d$ time-bounded (recall that h is time constructible).

For $x \in \{0,1\}^*$ we have (recall the definition of $\mathbb{M}_x$)

$$\begin{aligned} x \in Q &\iff \mathbb{M}_x \text{ accepts the empty string in at most } |x| + c \cdot h(|x|)^d \text{ steps} \\ &\iff \mathbb{M}_x \text{ accepts the empty string in at most } 2c \cdot h(|x|)^d \text{ steps} \\ &\iff \mathbb{A} \text{ accepts } (\mathbb{M}_x, 2c \cdot h(|x|)^d). \end{aligned}$$

By (17) the running time of $\mathbb{A}$ on input $(\mathbb{M}_x, 2c \cdot h(|x|)^d)$ is bounded by

$$f(e \cdot (\|\mathbb{M}\| + |x|^2)) \cdot (2c \cdot h(x))^{O(1)} \leq \Big(h(\|\mathbb{M}\|) + h(|x|)\Big) \cdot h(x)^{O(1)}.$$

As $\|\mathbb{M}\|$ is a constant, this shows $Q \in$ DTIME$(h(x)^{O(1)})$. □

We can refine the previous argument to get p-ACC$_{\leq} \notin$ XP; however we need a complexity-theoretic assumption (apparently) stronger than P[TC] $\neq$ NP[TC].

Theorem 4. *Assume that*

$$\mathrm{NTIME}(h^{O(1)}) \not\subseteq \mathrm{DTIME}(h^{O(\log h)})$$

for every time constructible and increasing function h. Then p-$\mathrm{ACC}_{\leq} \notin \mathrm{XP}$.

Proof. Assume that p-$\mathrm{ACC}_{\leq} \in \mathrm{XP}$. Then there is an algorithm $\mathbb{A}$ that for every nondeterministic Turing machine $\mathbb{M}$ and every natural number n decides whether $\mathbb{M}$ accepts the empty input tape in $\leq n$ steps in time

$$O(n^{f(\|\mathbb{M}\|)})$$

for a computable and increasing function $f : \mathbb{N} \to \mathbb{N}$. For this function f and e as in (17) we choose h according to Lemma 6. We define $g : \mathbb{N} \to \mathbb{N}$ by $g(n) := 2^{h(n)}$; clearly g is time constructible and increasing, too. We show that $\mathrm{NTIME}(g^{O(1)}) \subseteq \mathrm{DTIME}(g^{O(\log g)})$.

Let $Q \subseteq \{0,1\}^*$ be an arbitrary problem in $\mathrm{NTIME}(g^{O(1)})$. Let $\mathbb{M}$, c,d, and $\mathbb{M}_x$ for $x \in \{0,1\}^*$ be defined as in the previous proof, but now g takes over the role of h. As there we get

$$x \in Q \iff \mathbb{A} \text{ accepts } (\mathbb{M}_x, 2c \cdot g(|x|)^d).$$

The running time of $\mathbb{A}$ on input $(\mathbb{M}_x, 2c \cdot g(|x|)^d)$ is bounded by

$$O\big(g(|x|)^{d \cdot f(e \cdot (\|\mathbb{M}\| + |x|^2))}\big) \leq O\Big(g(|x|)^{d \cdot \left(h(\|\mathbb{M}\|) + h(|x|)\right)}\Big)$$
$$\leq O\Big(g(|x|)^{d \cdot \left(h(\|\mathbb{M}\|) + \log g(|x|)\right)}\Big).$$

As $\|\mathbb{M}\|$ is a constant, this shows $Q \in \mathrm{DTIME}(g^{O(\log g)})$. □

By Theorem 1 we obtain from the previous result:

Corollary 3. *Assume that*

$$\mathrm{NTIME}(h^{O(1)}) \not\subseteq \mathrm{DTIME}(h^{O(\log h)})$$

for every time constructible and increasing function h. Then $L_{\leq}$ is not an effectively P*-bounded logic for* P.

6.1 Relating P[TC] ≠ NP[TC] to Other Statements

We partly report on results from [4] relating P[TC] ≠ NP[TC] and the hypothesis in Theorem 4 to further statements of complexity theory.

Let C be a classical complexity class. Recall that a problem $Q \subseteq \{0,1\}^*$ is C-*bi-immune* if both Q and the complement of Q do not have an *infinite subset* that belongs to C. It has been conjectured (cf. [11]) that

NP contains a P-bi-immune problem.

In particular, there exist P-bi-immune sets inside NP in some *relativized* worlds [6]. We showed in [4]:

Proposition 7. *The following statement (a) implies (b).*

(a) NP *contains a* P*-bi-immune problem.*
(b) $\mathrm{P}[\mathrm{TC}] \neq \mathrm{NP}[\mathrm{TC}]$.

It seems that the statement (a) is much stronger than (b). In fact as shown in [4] "not (b)" implies

> there is an infinite $I \in \mathrm{P}$ such that for all $Q \in \mathrm{NP}$ at least one of the sets $Q \cap I$ and $(\{0,1\}^* \setminus Q) \cap I$ is an infinite set in P,

while "not (a)" can be reformulated as

> for all $Q \in \mathrm{NP}$ there is an infinite $I \in \mathrm{P}$ such that at least one of the sets $Q \cap I$ or $(\{0,1\}^* \setminus Q) \cap I$ is an infinite set in P.

Furthermore it is shown in [4]:

Proposition 8. *The following statement (a) implies (b).*

(a) NP *contains an* E*-bi-immune problem.*
(b) For every time constructible and increasing function h

$$\mathrm{NTIME}(h^{O(1)}) \not\subseteq \mathrm{DTIME}(h^{O(\log h)}).$$

7 A Deterministic Variant of p-Acc$_\leq$

If in the problem p-ACC$_\leq$ we replace the nondeterministic Turing machine $\mathbb{M}$ by a deterministic Turing machine simulating all computation paths of length n of $\mathbb{M}$ with empty input tape we "arrive at" p-DTM-EXP-ACC$_\leq$:

p-DTM-EXP-ACC$_\leq$	
Instance:	A *deterministic* Turing machine $\mathbb{M}$ and $n \in \mathbb{N}$ in unary.
Parameter:	$\|\mathbb{M}\|$.
Question:	Does $\mathbb{M}$ accept the empty input tape in at most 2^n steps?

Thus, p-ACC$_\leq \leq^{\mathrm{fpt}}$ p-DTM-EXP-ACC$_\leq$. As p-DTM-EXP-ACC$_\leq$ is a slicewise monotone parameterized problem, we know that it is in $\mathrm{FPT}_{\mathrm{nu}}$ by Lemma 1. Clearly, $\mathrm{FPT}_{\mathrm{nu}} \subseteq \mathrm{XP}_{\mathrm{nu}}$ and $\mathrm{XP} \subseteq \mathrm{XP}_{\mathrm{uni}} \subseteq \mathrm{XP}_{\mathrm{nu}}$. The problem p-DTM-EXP--ACC$_\leq$ lies in $\mathrm{FPT}_{\mathrm{nu}} \setminus \mathrm{XP}_{\mathrm{uni}}$, as we show:

Theorem 5. p-DTM-EXP-ACC$_\leq \notin \mathrm{XP}_{\mathrm{uni}}$.

Proof. One easily verifies that p-DTM-EXP-ACC$_\leq$ is fpt equivalent to

> p-DTM-INP-EXP-ACC$_{\leq}$
> *Instance:* A deterministic Turing machine $\mathbb{M}$, $x \in \{0,1\}^*$, and $n \in \mathbb{N}$ in unary.
> *Parameter:* $\|\mathbb{M}\| + |x|$.
> *Question:* Does $\mathbb{M}$ accept x in $\leq 2^n$ steps?

Thus, it suffices to show that p-DTM-INP-EXP-ACC$_{\leq} \notin \mathrm{XP}_{\mathrm{uni}}$. By contradiction, assume there exists an algorithm $\mathbb{A}$ that for every instance $(\mathbb{M}, x, n)$ decides whether $(\mathbb{M}, x, n) \in p$-DTM-INP-EXP-ACC$_{\leq}$ in time

$$c \cdot n^{f(\|\mathbb{M}\|+|x|)}$$

for some (not necessarily computable) function $f : \mathbb{N} \to \mathbb{N}$ and $c \in \mathbb{N}$.

We denote by $\mathrm{enc}(\mathbb{M})$ the encoding of a Turing machine $\mathbb{M}$ by a string in $\{0,1\}^*$. We consider the following deterministic Turing machine $\mathbb{M}_0$:

> $\mathbb{M}_0(x)$
> // $x \in \{0,1\}^*$
>
> 1. **if** x is not the encoding of a deterministic Turing machine, **then** reject
> 2. determine the deterministic Turing machine $\mathbb{M}$ with $x = \mathrm{enc}(\mathbb{M})$
> 3. $m \leftarrow$ the number of steps performed by $\mathbb{M}_0$ so far
> 4. simulate at most 2^m steps of the computation of $\mathbb{A}$ on input $(\mathbb{M}, x, m+3)$
> 5. **if** the simulation does not halt, **then** $m \leftarrow m+1$ and goto 4
> 6. **if** $\mathbb{A}$ accepts $(\mathbb{M}, x, m+3)$ in at most 2^m steps **then** reject **else** accept.

We finish the proof by a diagonal argument: We set $x_0 := \mathrm{enc}(\mathbb{M}_0)$ and start $\mathbb{M}_0$ with input x_0. For sufficiently large $m \in \mathbb{N}$ we have

$$c \cdot (m+3)^{f(\|\mathbb{M}_0\|+|x_0|)} \leq 2^m.$$

Therefore eventually $\mathbb{M}_0$ with input x_0 reaches an m, we call it m_0, such that the simulation in Line 4 halts, more precisely,

$$\mathbb{A} \text{ halts on } (\mathbb{M}_0, x_0, m_0+3) \text{ in at most } 2^{m_0} \text{ steps.} \tag{18}$$

At that point the number of steps (of the run of $\mathbb{M}_0$ on input x_0) is bounded by 2^{m_0+2}. Hence

$$\mathbb{M}_0 \text{ on } x_0 \text{ halts in } \leq 2 + 2^{m_0+2} \leq 2^{m_0+3} \text{ steps.} \tag{19}$$

Thus

$$\begin{aligned}
\mathbb{M}_0 \text{ accepts } x_0 &\iff \mathbb{M}_0 \text{ accepts } x_0 \text{ in } \leq 2^{m_0+3} \text{ steps} && \text{(by (19))}\\
&\iff \mathbb{A} \text{ accepts } (\mathbb{M}_0, x_0, m_0+3) && \text{(by definition of } \mathbb{A})\\
&\iff \mathbb{A} \text{ accepts } (\mathbb{M}_0, x_0, m_0+3) \text{ in at most } 2^{m_0} \text{ steps} && \text{(by (18))}\\
&\iff \mathbb{M}_0 \text{ rejects } x_0 && \text{(by Line 6 in the definition of } \mathbb{M}_0),
\end{aligned}$$

the desired contradiction. □

8 The Construction Problem Associated with p-Acc$_\leq$

We consider the construction problem associated with p-Acc$_\leq$:

p-Constr-Acc$_\leq$	
Instance:	A nondeterministic Turing machine $\mathbb{M}$ and $n \in \mathbb{N}$ in unary.
Parameter:	$\|\mathbb{M}\|$.
Problem:	Construct an accepting run of $\leq n$ steps of $\mathbb{M}$ started with empty input tape if there is one (otherwise report that there is no such run).

Similarly as we showed that p-Acc$_\leq \in \text{FPT}_{\text{nu}}$ (cf. Proposition 6), one gets that p-Constr-Acc$_\leq$ is nonuniformly fixed-parameter tractable (it should be clear what this means).

Definition 3. *An* fpt$_{\text{uni}}$ Turing reduction *(*fpt Turing reduction*) from a parameterized construction problem (Q, κ) to a parameterized decision problem (Q', κ') is a deterministic algorithm $\mathbb{T}$ with an oracle to (Q', κ') solving the construction problem (Q, κ) and with the property that there are (computable) functions $f, g : \mathbb{N} \to \mathbb{N}$, and $c \in \mathbb{N}$ such that for every instance x of Q*

- *the run of $\mathbb{T}$ with input x has length $\leq f(\kappa(x)) \cdot |x|^c$;*
- *for every oracle query "$x' \in Q'$?" of the run of $\mathbb{A}$ with input x we have $\kappa(x') \leq g(\kappa(x))$.*

Often the construction problem has the same complexity as the corresponding decision problem, that is, the construction problem is fpt Turing reducible to the decision problem; for p-Constr-Acc$_\leq$ we can show:

Theorem 6. *1. There is an fpt$_{\text{uni}}$ Turing reduction from p-Constr-Acc$_\leq$ to p-Acc$_\leq$.*
2. If p-Acc$_\leq \notin$ XP, then there is no *fpt Turing reduction from p-Constr-Acc$_\leq$ to p-Acc$_\leq$.*

Proof. (1) On an instance $(\mathbb{M}, n)$ of p-Constr-Acc$_\leq$ the desired reduction $\mathbb{T}$ first asks the oracle query "$(\mathbb{M}, n) \in p$-Acc$_\leq$?". If the answer is no, then $\mathbb{T}$ answers accordingly. Otherwise $\mathbb{T}$, by brute force, constructs an accepting run of at most n steps of $\mathbb{M}$. We analyze the running time of $\mathbb{T}$. For $m \in \mathbb{N}$ let $\mathbb{M}_1, \ldots, \mathbb{M}_\ell$ be the finitely many nondeterministic Turing machines with $\|\mathbb{M}_i\| \leq m$ and with an accepting run started with empty input tape. Let ρ_i be such a run of $\mathbb{M}_i$ of minimum length. We set

$$f(m) := \max\{|\rho_1|, \ldots, |\rho_\ell|\}.$$

It is not hard to see that the running time of $\mathbb{T}$ on the instance $(\mathbb{M}, n)$ can be bounded by $\|\mathbb{M}\|^{O(f(\|\mathbb{M}\|))} \cdot n$.

(2) By contradiction, assume there is an fpt Turing reduction $\mathbb{T}$ from p-CONSTR-ACC$_{\leq}$ to p-ACC$_{\leq}$. We show how $\mathbb{T}$ can be turned into an algorithm witnessing p-ACC$_{\leq} \in$ XP.

According to the definition of fpt Turing reduction there are computable functions f, g and $c \in \mathbb{N}$ such that for every instance $(\mathbb{M}, n)$ of p-CONSTR-ACC$_{\leq}$, the algorithm $\mathbb{T}$ will only make queries "$(\mathbb{M}', n') \in p$-ACC$_{\leq}$?" with

$$\|\mathbb{M}'\| \leq g(\|\mathbb{M}\|) \text{ and } n' \leq f(\|\mathbb{M}\|) \cdot n^c.$$

There are at most $2^{g(\|\mathbb{M}\|)+1}$ machines $\mathbb{M}'$ with $\|\mathbb{M}'\| \leq g(\|\mathbb{M}\|)$. For each such machine $\mathbb{M}'$ the answer to queries of the form "$(\mathbb{M}', n') \in p$-ACC$_{\leq}$?" with $n' \leq f(\|\mathbb{M}\|) \cdot n^c$ is determined by every one of the following $f(\|\mathbb{M}\|) \cdot n^c + 1$ many statements: "the length of an accepting run of $\mathbb{M}'$ of minimum length is 1",..., "the length of an accepting run of $\mathbb{M}'$ of minimum length is $f(\|\mathbb{M}\|) \cdot n^c$", and "there is no accepting run of $\mathbb{M}'$ of length $\leq f(\|\mathbb{M}\|) \cdot n^c$." Therefore the table of theoretically possible answers contains at most

$$\left(f(\|\mathbb{M}\|) \cdot n^c + 1\right)^{2^{g(\|\mathbb{M}\|)+1}}$$

entries, that is $O(n^{h(\|\mathbb{M}\|)})$ many for some computable h. For each such possibility we simulate $\mathbb{T}$ by replacing the oracle queries accordingly. For those possibilities where $\mathbb{T}$ yields a purported accepting run of $\mathbb{M}$, we check whether it is really an accepting run of $\mathbb{M}$. □

An analysis of the previous proof shows that we can even rule out the existence of a Turing reduction with running time $O(|x|^{f(\kappa(x))})$ instead of $f(\kappa(x)) \cdot |x|^c$. We call such reductions xp Turing reductions.

Furthermore, Theorem 6 is a special case of a result for slicewise monotone problems: Let (Q, κ) be a slicewise monotone parameterized problem (this concept was defined just before Lemma 1) and assume that Q has a representation of the form

$$\begin{aligned} &(x, n) \in Q \\ &\quad \iff \text{there is } y \in \{0,1\}^*\text{: } \Big(|y| \leq f(|x|) \cdot n^c \text{ and } (x, n, y) \in Q_W\Big), \end{aligned} \tag{20}$$

where $f : \mathbb{N} \to \mathbb{N}$ is computable, $c \in \mathbb{N}$ and Q_W is decidable in polynomial time. A string y with the properties on the right hand side is a *witness for* (x, n).

The construction problem p-CONSTR-(Q, κ) for every instance (x, n) of Q (with parameter $|x|$) asks for a witness for (x, n) if there is one (otherwise the nonexistence should be reported). Along the lines of the previous proofs one can show:

Proposition 9. *Let* (Q, κ) *be a slicewise monotone parameterized problem with a representation as in* (20).

1. p-CONSTR-(Q, κ) *is nonuniformly fixed-parameter tractable.*
2. *There is an* fpt$_{\text{uni}}$ *Turing reduction from* p-CONSTR-(Q, κ) *to* (Q, κ).
3. *If* $(Q, \kappa) \notin$ XP, *then there is no fpt Turing reduction (even no xp Turing reduction) from* p-CONSTR-(Q, κ) *to* (Q, κ).

9 Conclusions

We have studied the relationship between the complexity of the model-checking problems of the logics $L_=$ and $L_\leq$ and the complexity of the parameterized problems p-ACC$_=$ and p-ACC$_\leq$. We have introduced the assumption P[TC] $\neq$ NP[TC] and seen that it implies that p-ACC$_\leq \notin$ FPT. A slightly stronger hypothesis shows that p-ACC$_\leq \notin$ XP and hence that the logic $L_\leq$ is not an effectively P-bounded logic for P. What are reasonable complexity theoretic assumptions that imply p-ACC$_\leq \notin$ XP$_{\text{uni}}$ and hence that the logic $L_\leq$ is not a P-bounded logic for P?

We believe that a study of the strength of the assumption P[TC] $\neq$ NP[TC] and of its consequences deserves further attention.

References

1. Aumann, Y., Dombb, Y.: Fixed structure complexity. In: Grohe, M., Niedermeier, R. (eds.) IWPEC 2008. LNCS, vol. 5018, pp. 30–42. Springer, Heidelberg (2008)
2. Chandra, A.K., Harel, D.: Structure and complexity of relational queries. Journal of Computer and System Science 25, 99–128 (1982)
3. Chen, Y., Flum, J.: A logic for PTIME and a parameterized halting problem. In: Proceedings of the 24th Annual IEEE Symposium on Logic in Computer Science (LICS 2009), pp. 397–406. IEEE Computer Society, Los Alamitos (2009)
4. Chen, Y., Flum, J.: On the complexity of Gödel's proof predicate. The Journal of Symbolic Logic 75, 239–254 (2010)
5. Flum, J., Grohe, M.: Parameterized Complexity Theory. Springer, Heidelberg (2005)
6. Gasarch, W.I., Homer, S.: Relativizations comparing NP and exponential time. Information and Control 58, 88–100 (1983)
7. Grohe, M.: The quest for a logic capturing PTIME. In: Proceedings of the Twenty-Third Annual IEEE Symposium on Logic in Computer Science (LICS 2008), pp. 267–271. IEEE Computer Society, Los Alamitos (2008)
8. Gurevich, Y.: Toward logic tailored for computational complexity. In: Computation and Proof Theory, pp. 175–216. Springer, Heidelberg (1984)
9. Gurevich, Y.: Logic and the challenge of computer science. In: Current Trends in Theoretical Computer Science, pp. 1–57. Computer Science Press, Rockville (1988)
10. Immerman, N.: Relational queries computable in polynomial time. Information and Control 68, 86–104 (1986)
11. Mayordomo, E.: Almost every set in exponential time is p-bi-immune. Theoretical Computer Science 136, 487–506 (1994)
12. Nash, A., Remmel, J.B., Vianu, V.: PTIME queries revisited. In: Eiter, T., Libkin, L. (eds.) ICDT 2005. LNCS, vol. 3363, pp. 274–288. Springer, Heidelberg (2004)
13. Vardi, M.Y.: The complexity of relational query languages (extended abstract). In: Proceedings of the Fourteenth Annual ACM Symposium on Theory of Computing (STOC 1982), pp. 137–146. ACM, New York (1982)
14. Vardi, M.Y.: On the complexity of bounded-variable queries. In: Proceedings of the Fourteenth ACM SIGACT-SIGMOD-SIGART Symposium on Principles of Database Systems (PODS 1995), pp. 266–276. ACM Press, New York (1995)

Inferring Loop Invariants Using Postconditions

Carlo Alberto Furia and Bertrand Meyer

Chair of Software Engineering, ETH Zurich, Switzerland
{caf,bertrand.meyer}@inf.ethz.ch

To Yuri Gurevich in joyful celebration of his 70th birthday, and with thanks for his many contributions to computer science, including his original leadership of the group on whose tools the present work crucially relies.

Abstract. One of the obstacles in automatic program proving is to obtain suitable *loop invariants*. The invariant of a loop is a weakened form of its postcondition (the loop's goal, also known as its contract); the present work takes advantage of this observation by using the postcondition as the basis for invariant inference, using various heuristics such as "uncoupling" which prove useful in many important algorithms. Thanks to these heuristics, the technique is able to infer invariants for a large variety of loop examples. We present the theory behind the technique, its implementation (freely available for download and currently relying on Microsoft Research's Boogie tool), and the results obtained.

Keywords: Correctness proofs, formal specifications, loop invariants, assertion inference.

1 Overview

Many of the important contributions to the advancement of program proving have been, rather than grand new concepts, specific developments and simplifications; they have removed one obstacle after another preventing the large-scale application of proof techniques to realistic programs built by ordinary programmers in ordinary projects. The work described here seeks to achieve such a practical advance by automatically generating an essential ingredient of proof techniques: *loop invariants*. The key idea is that invariant generation should use not just the text of a loop but its *postcondition*. Using this insight, the gin-pink tool can infer loop invariants for non-trivial algorithms including array partitioning (for Quicksort), sequential search, coincidence count, and many others. The tool is available for free download.[1]

1.1 Taking Advantage of Postconditions

In the standard Floyd-Hoare approach to program proving, loop invariants are arguably the biggest practical obstacle to full automation of the proof process.

[1] `http://se.inf.ethz.ch/people/furia/`

A. Blass, N. Dershowitz, and W. Reisig (Eds.): Gurevich Festschrift, LNCS 6300, pp. 277–300, 2010.

Given a routine's specification (contract), in particular its postcondition, the proof process consists of deriving intermediate properties, or *verification conditions*, at every point in the program text. Straightforward techniques yield verification conditions for basic instructions, such as assignment, and basic control structures, such as sequence and conditional. The main difficulty is the loop control structure, where the needed verification condition is a *loop invariant*, which unlike the other cases cannot be computed through simple rules; finding the appropriate loop invariant usually requires human invention.

Experience shows, however, that many programmers find it hard to come up with invariants. This raises the question of devising automatic techniques to infer invariants from the loop's text, in effect extending to loops the mechanisms that successfully compute verification conditions for other constructs. Loops, however, are intrinsically more difficult constructs than (for example) assignments and conditionals, so that in the current state of the art we can only hope for *heuristics* applicable to specific cases, rather than general algorithms guaranteed to yield a correct result in all cases.

While there has been considerable research on loop invariant generation and many interesting results (reviewed in the literature survey of Section 6), most existing approaches are constrained by a fundamental limitation: to obtain the invariant they only consider the *implementation* of a loop. In addition to raising epistemological problems explained next, such techniques can only try to discover relationships between successive loop iterations; this prevents them from discovering many important classes of invariants.

The distinctive feature of the present work is that it uses postconditions of a routine for inferring the invariants of its loops. The postcondition is a higher-level view of the routine, describing its goal, and hence allows inferring the correct invariants in many more cases. As will be explained in Section 4.1, this result follows from the observation that a loop invariant is always a weakened form of the loop's postcondition. Invariant inference as achieved in the present work then relies on implementing a number of *heuristics for mutating postconditions into candidate invariants*; Section 2 presents four such heuristics, such as *uncoupling* and *constant relaxation*, which turn out to cover many practical cases.

1.2 Inferring Assertions: The Assertion Inference Paradox

Any program-proving technique that attempts to infer specification elements (such as loop invariants) from program texts faces a serious epistemological objection, which we may call the Assertion Inference Paradox.

The Assertion Inference Paradox is a risk of vicious circle. The goal of program proving is to establish program correctness. A program is correct if its implementation satisfies its specification; for example a square root routine implements a certain algorithm, intended to reach a final state satisfying the specification that the square of the result is, within numerical tolerance, equal to the input. To talk about correctness requires having both elements, the implementation and the specification, and assessing one against the other. But if we infer the specification from the implementation, does the exercise not become vacuous?

Surely, the proof will succeed, but it will not teach us anything since it loses the fundamental property of independence between the mathematical property to be achieved and the software artifact that attempts to achieve it – the problem and the solution.

To mitigate the Assertion Inference Paradox objection, one may invoke the following arguments:

- The Paradox only arises if the goal is to prove correctness. Specification inference can have other applications, such as reverse-engineering legacy software.
- Another possible goal of inferring a specification may be to present it to a programmer, who will examine it for consistency with an intuitive understanding of its intended behavior.
- Specification inference may produce an inconsistent specification, revealing a flaw in the implementation.

For applications to program proving, however, the contradiction remains; an inferred specification not exhibiting any inconsistencies cannot provide a sound basis for a proof process.

For that reason, the present work refrains from attempting specification inference for the principal units of a software system: routines (functions, methods) and those at an even higher level of granularity (such as classes). It assumes that these routine specifications are available. Most likely they will have been written explicitly by humans, although their origin does not matter for the rest of the discussion.

What does matter is that once we have routine specifications it is desirable to infer the specifications of all lower-level constructs (elementary instructions and control structures such as conditionals and loops) automatically. At those lower levels, the methodological objection expressed by the Assertion Inference Paradox vanishes: the specifications are only useful to express the semantics of implementation constructs, not to guess the software's intent. The task then becomes: given a routine specification – typically, a precondition and postcondition – derive the proof automatically by inferring verification conditions for the constructs used in the routine and proving that the constructs satisfy these conditions. No vicious circle is created.

For basic constructs such as assignments and conditional instructions, the machinery of Floyd-Hoare logic makes this task straightforward. The principal remaining difficulty is for loops, since the approach requires exhibiting a *loop invariant*, also known as an *inductive assertion*, and proving that the loop's initialization establishes the invariant and that every execution of the body (when the exit condition is not satisfied) preserves it.

A loop invariant captures the essence of the loop. Methodologically, it is desirable that programmers devise the invariant while or before devising the loop. As noted, however, many programmers have difficulty coming up with loop invariants. This makes invariants an attractive target for automatic inference.

In the present work, then, postconditions are known and loop invariants inferred. The approach has two complementary benefits:

– It does not raise the risk of circular reasoning since the specification of every program unit is explicitly provided, not inferred.
– Having this specification of a loop's context available gives a considerable boost to loop invariant inference techniques. While there is a considerable literature on invariant inference, it is surprising that none of the references with which we are familiar use postconditions. Taking advantage of postconditions makes it possible – as described in the rest of this paper – to derive the invariants of many important and sometimes sophisticated loop algorithms that had so far eluded other techniques.

2 Illustrative Examples

This section presents the fundamental ideas behind the loop-invariant generation technique detailed in Section 5 and demonstrates them on a few examples. It uses an Eiffel-like [30] pseudocode, which facilitates the presentation thanks to the native syntax for contracts and loop invariants.

As already previewed, the core idea is to generate candidate invariants by mutating postconditions according to a few commonly recurring patterns. The patterns capture some basic ways in which loop iterations modify the program state towards achieving the postcondition. Drawing both from classic literature [19,29] and our own more recent investigations we consider the following fundamental patterns.

Constant relaxation [29,19]: replace one or more constants by variables.
Uncoupling [29]: replace two occurrences of the same variable each by a different variable.
Term dropping [19]: remove a term, usually a conjunct.
Variable aging: replace a variable by an expression that represents the value the variable had at previous iterations of the loop.

These patterns are then usually used in combination, yielding a number of mutated postconditions. Each of these candidate invariants is then tested for initiation and consecution (see Section 4.1) over any loop, and all verified invariants are retained.

The following examples show each of these patterns in action. The tool described in Sections 3 and 5 can correctly infer invariants of these (and more complex) examples.

2.1 Constant Relaxation

Consider the following routine to compute the maximum value in an array.

```
max (A: ARRAY [T]; n: INTEGER): T
   require A.length = n ≥ 1
   local i: INTEGER
   do
      from i := 0; Result := A[1];
```

```
6          until i ≥ n
7          loop
8             i := i + 1
9             if Result ≤ A[i] then Result := A[i] end
10         end
11      ensure ∀j • 1 ≤ j ∧ j ≤ n ⟹ A[j] ≤ Result
```

Lines 5–10 may modify variables i and **Result** but they do not affect the input argument n, which is therefore a constant with respect to the loop body. The *constant relaxation* technique replaces every occurrence of the constant n by a variable i. The modified postcondition $\forall\ j \bullet 1 \leq j \ \wedge \ j \leq i \Longrightarrow A[j] \leq$ **Result** is indeed an invariant of the loop: after every iteration, the value of **Result** is the maximum value of array A over range $[1..i]$.

2.2 Variable Aging

Sometimes replacing a constant by a variable in the postcondition does not yield any loop invariant because of how the loop body updates the variable. It may happen that the loop body does not need the latest value of the substituted variable until the next iteration. Consider another implementation of computing the maximum of an array, which increments the variable i *after* using it, so that only the range $[1..i-1]$ of array A has been inspected after every iteration:

```
max_v2 (A: ARRAY [T], n: INTEGER): T
   require A.length = n ≥ 1
   local i: INTEGER
   do
      from i := 1; Result := A[1];
      until i > n
      loop
         if Result ≤ A[i] then Result := A[i] end
         i := i + 1
      end
   ensure ∀ j • 1 ≤ j ∧ j ≤ n ⟹ A[j] ≤ Result
```

The *variable aging* heuristics handles these cases by introducing an expression that represents the value of the variable at the previous iteration in terms of its current value. In the case of routine *max_v2* it is straightforward that such an expression for variable i is $i-1$. The postcondition can be modified by first replacing variable n by variable i and then "aging" variable i into $i-1$. The resulting formula $\forall\ j \bullet 1 \leq j \ \wedge \ j \leq i-1 \Longrightarrow A[j] \leq$ **Result** correctly captures the semantics of the loop.

Computing the symbolic value of a variable at the "previous" iteration can be quite complex in the general case. In practice, however, a simple (e.g., flow-insensitive) approximation is often enough to get significant results. The experiments of Section 3 provide a partial evidence to support this conjecture.

2.3 Uncoupling

Consider the task (used as part of the Quicksort algorithm) of partitioning an array A of length n into two parts such that every element of the first part is less than or equal to a given *pivot* value and every element of the second part is greater than or equal to it. The following contracted routine specifies and implements this task, working on an integer variable *position*.

```
 1 partition (A: ARRAY [T]; n: INTEGER; pivot: T)
 2   require A.length = n ≥ 1
 3   local low_index, high_index: INTEGER
 4   do
 5     from low_index := 1; high_index := n
 6     until low_index = high_index
 7     loop
 8       from -- no loop initialization
 9       until low_index = high_index ∨ A[low_index] > pivot
10       loop low_index := low_index + 1 end
11       from -- no loop initialization
12       until low_index = high_index ∨ pivot> A[high_index]
13       loop high_index := high_index - 1 end
14       A.swap (A, low_index, high_index)
15     end
16     if pivot ≤ A[low_index] then
17       low_index := low_index - 1
18       high_index := low_index
19     end
20     position := low_index
21   ensure ( ∀ k • 1 ≤ k ∧ k < position + 1 ⟹ A[k] ≤ pivot )
22            ∧ ( ∀ k • position < k ∧ k ≤ n ⟹ A[k] ≥ pivot )
```

The postcondition consists of the conjunction of two formulas (lines 21 and 22). If we try to mutate it by replacing constant *position* by variable *low_index* or by variable *high_index* we obtain no valid loop invariant. This is because the two clauses of the postcondition should refer, respectively, to portion $[1..\,low\_index-1]$ and $[high\_index+1..n]$ of the array. We achieve this by first uncoupling *position*, which means replacing its first occurrence (in line 21) by variable *low_index* and its second occurrence (in line 22) by variable *high_index*. After "aging" variable *low_index* we get the formula:

$$(\forall k \bullet 1 \leq k \wedge k < low\_index \Longrightarrow A[k] \leq pivot) \\ \wedge (\forall k \bullet high\_index < k \wedge k \leq n \Longrightarrow A[k] \geq pivot) .$$

The reader can check that this indeed a loop invariant of all loops in routine *partition* and that it allows a straightforward partial correctness proof of the implementation.

2.4 Term Dropping

The last mutation pattern that we consider consists simply of removing a part of the postcondition. The formula to be modified is usually assumed to be in conjunctive normal form, that is, expressed as the conjunction of a few clauses: then term dropping amounts to removing one or more conjuncts. Going back to the example of *partition*, let us drop the first conjunct in the postcondition. The resulting formula

$$\forall\ k\ \bullet\ \textbf{Result} < k \wedge\ k \leq n\ \implies\ A[k]\ \geq pivot$$

can be further transformed through constant relaxation, so that we end up with a conjunct of the invariant previously obtained by uncoupling: $\forall\ k\ \bullet\ high\_index < k \wedge k \leq n \implies A[k] \geq pivot$. This conjunct is also by itself an invariant. In this example term dropping achieved by different means the same result as uncoupling.

3 Implementation and Experiments

Before presenting the technical details of the loop-invariant generation technique, this section describes experiments with a tool implementing the technique. The results, discussed in more detail in Section 6, demonstrate the feasibility and practical value of the overall approach.

gin-pink (Generation of INvariants by PostcondItioN weaKening) is a command-line tool implementing the loop-invariant inference technique described in Section 5. While we plan to integrate gin-pink into EVE (the Eiffel Verification Environment[2]) where it will analyze the code resulting from the translation of Eiffel into Boogie [40], its availability as a stand-alone tool makes it applicable to languages other than Eiffel provided a Boogie translator is available.

gin-pink applies the algorithm described in Section 5 to a selected procedure in a Boogie file provided by the user. After generating all mutated postconditions it invokes the Boogie tool to determine which of them is indeed an invariant: for every candidate invariant I, a new copy of the original Boogie file is generated with I declared as invariant of *all* loops in the procedure under analysis. It then repeats the following until either Boogie has verified all current declarations of I in the file or no more instances of I exist in the procedure:

1. Use Boogie to check whether the current instances of I are verified invariants, that is they satisfy initiation and consecution.
2. If any candidate fails this check, comment it out of the file.

In the end, the file contains all invariants that survive the check, as well as a number of commented out candidate invariants that could not be checked. If no invariant survives or the verified invariants are unsatisfactory, the user can still manually inspect the generated files to see if verification failed due to the limited reasoning capabilities of Boogie.

[2] `http://eve.origo.ethz.ch`

Table 1. Experiments with gin-pink

Procedure	LOC	#	LP.	M.V.	CND.	INV.	REL.	T.	SRC.
Array Partitioning (v1)	58(22)	1	(1)	2+1	38	9	3 (33%)	93	
Array Partitioning (v2)	68(40)	3	(2)	2+1	45	2	2(100%)	205	[29]
Array Stack Reversal	147(34)	2	(1)	1+2	134	4	2 (50%)	529	
Array Stack Reversal (ann.)	147(34)	2	(1)	1+2	134	6	4 (67%)	516	
Bubblesort	69(29)	2	(2)	2+1	14	2	2(100%)	65	[33]
Coincidence Count	59(29)	1	(1)	3+0	1351	1	1(100%)	4304	[27]
Dutch National Flag	77(43)	1	(1)	3+1	42	10	2 (20%)	117	[16]
Dutch National Flag (ann.)	77(43)	1	(1)	3+1	42	12	4 (33%)	122	[16]
Longest Common Sub. (ann.)	73(59)	4	(2)	2+2	508	22	2 (9%)	4842	
Majority Count	48(37)	1	(1)	3+0	23	5	2 (40%)	62	[4,32]
Max of Array (v1)	27(17)	1	(1)	2+0	13	1	1(100%)	30	
Max of Array (v2)	27(17)	1	(1)	2+0	7	1	1(100%)	16	
Plateau	53(29)	1	(1)	3+0	31	6	3 (50%)	666	[19]
Sequential Search (v1)	34(26)	1	(1)	3+0	45	9	5 (56%)	120	
Sequential Search (v2)	29(21)	1	(1)	3+0	24	6	6(100%)	58	
Shortest Path (ann.)	57(44)	1	(1)	1+4	23	2	2(100%)	53	[3]
Stack Search	196(49)	2	(1)	1+3	102	3	3(100%)	300	
Sum of Array	26(15)	1	(1)	2+0	13	1	1(100%)	44	
Topological Sort (ann.)	65(48)	1	(1)	2+4	21	3	2 (67%)	101	[31]
Welfare Crook	53(21)	1	(1)	3+0	20	2	2(100%)	586	[19]

When generating candidate invariants, gin-pink does not apply all heuristics at once but it tries them incrementally, according to user-supplied options. Typically, the user starts out with just constant relaxation and checks if some non-trivial invariant is found. If not, the analysis is refined by gradually introducing the other heuristics – and thus increasing the number of candidate invariants as well. In the examples below we briefly discuss how often and to what extent this is necessary in practice.

Examples. Table 1 summarizes the results of a number of applications of gin-pink to Boogie procedures obtained from Eiffel code. We carried out the experimental evaluation as follows. First, we collected examples from various sources [29,19,33,27,3] and we manually completed the annotations of every algorithm with full pre and postconditions as well as with any loop invariant or intermediate assertion needed in the correctness proof. Then, we coded and tried to verify the annotated programs in Boogie, supplying some background theory to support the reasoning whenever necessary. The latest Boogie technology cannot verify certain classes of properties without a sophisticated *ad hoc* background theory or without abstracting away certain aspects of the implementation under verification. For example, in our implementation of Bubblesort, Boogie had difficulties proving that the output is a permutation of the input. Correspondingly, we omitted the parts of the specification that Boogie could not prove even with a detailedly annotated program. Indeed, "completeness" (full functional correctness) should not be a primary concern, because its significance depends on properties of the prover (here Boogie), orthogonal to the task of inferring invariants. Finally, we ran gin-pink on each of the examples after commenting out all annotations except for pre and postconditions (but leaving the simple background theories in); in a few difficult cases (discussed next) we ran additional experiments with some of the annotations left in. After running the tests, we measured the relevance of every automatically inferred invariant: we call an inferred invariant *relevant* if the correctness proof needs it. Notice that our choice of omitting postcondition clauses that Boogie cannot prove does not influence relevance, which only measures the fraction of inferred invariants that are useful for proving correctness.

For each example, Table 1 reports: the name of the procedure under analysis; the length in lines of codes (the whole file including annotations and auxiliary procedures and, in parentheses, just the main procedure); the total number of loops (and the maximum number of nested loops, in parentheses); the total number of variables modified by the loops (scalar variables/array or map variables); the number of mutated postconditions (i.e., candidate invariants) generated by the tool; how many invariants it finds; the number and percentage of verified invariants that are relevant; the total run-time of gin-pink in seconds; the source (if any) of the implementation and the annotations. The experiments where performed on a PC equipped with an Intel Quad-Core 2.40 GHz CPU and 4 Gb of RAM, running Windows XP as guest operating system on a VirtualBox virtual machine hosted by Ubuntu GNU/Linux 9.04 with kernel 2.6.28.

Most of the experiments succeeded with the application of the most basic heuristics. Procedures *Coincidence Count* and *Longest Common Subsequence* are the only two cases that required a more sophisticated uncoupling strategy where two occurrences of the same constant within the same formula were modified to two different aged variables. This resulted in an explosion of the number of candidate invariants and consequently in an experiment running for over an hour.

A few programs raised another difficulty, due to Boogie's need for user-supplied loop invariants to help automated deduction. Boogie cannot verify any invariant in *Shortest Path*, *Topological Sort*, or *Longest Common Subsequence* without additional invariants obtained by means other than the application of the algorithm itself. On the other hand, the performance with programs *Array Stack Reversal* and *Dutch National Flag* improves considerably if user-supplied loop invariants are included, but fair results can be obtained even without any such annotation. Table 1 reports both experiments, with and without user-supplied annotations.

More generally, Boogie's reasoning abilities are limited by the amount of information provided in the input file in the form of axioms and functions that postulate sound inference rules for the program at hand. We tried to limit this amount as much as possible by developing the necessary theories before tackling invariant generation. In other words, the axiomatizations provided are enough for Boogie to prove functional correctness with a properly annotated program, but we did not strengthen them only to ameliorate the inference of invariants.

A richer axiomatization may have removed the need for user-supplied invariants in the programs considered.

4 Foundations

Having seen typical examples we now look at the technical choices that support the invariant inference tools. To decouple the loop-invariant generation technique from the specifics of the programming language, we adopt Boogie from Microsoft Research [26] as our concrete programming language; Section 4.2 is then devoted to a concise introduction to the features of Boogie that are essential for the remainder. Sections 4.1 and 4.3 introduce definitions of basic concepts and some notational conventions that will be used. We assume the reader is familiar with standard formal definitions of the axiomatic semantics of imperative programs.

4.1 Invariants

Proving a procedure correct amounts to verifying that:

1. Every computation terminates.
2. Every call to another procedure is issued only when the preconditions of the callee hold.
3. The postconditions hold upon termination.

It is impossible to establish these facts automatically for all programs but the most trivial ones without programmer-provided annotations. The crucial aspect is the characterization of loops, where the expressive power of universal computation lies. A standard technique to abstract the semantics of any number of iterations of a loop is by means of *loop invariants.*

Definition 1 (Inductive loop invariant). *Formula ϕ is an inductive invariant of loop*

$$\textbf{from } \textit{Init} \textbf{ until } \textit{Exit} \textbf{ loop } \textit{Body} \textbf{ end}$$

iff:

- Initiation*: ϕ holds after the execution of Init*
- Consecution*: the truth of ϕ is preserved by every execution of Body where Exit does not hold*

In the rest of the discussion, inductive invariants will be called just invariants for short. Note, however, that an invariant in the weaker sense of a property that stays true throughout the loop's execution is not necessarily an inductive invariant: in

$$\textbf{from } x := 1 \textbf{ until } \textit{False} \textbf{ loop } x := -\, x \textbf{ end}$$

the formula $x \geq -1$ will remain true throughout, but is not considered an inductive invariant because $\{x \geq -1\}\ x := -x\ \{x \geq -1\}$ is not a correct

Hoare triple. In the remainder we will deal solely with inductive loop invariants, as is customary in the program proving literature.

From a design methodology perspective, the invariant expresses a weakened form of the loop's postcondition. More precisely [31,19], the invariant is a form of the loop's postcondition that applies to a subset of the data, and satisfies the following three properties:

1. It is *strong enough* to yield the postcondition when combined with the exit condition (which states that the loop has covered the *entire* data).
2. It is *weak enough* to make it easy to write an algorithm (the loop initialization *Init*) that will satisfy the invariant on a subset (usually empty or trivial) of the data.
3. It is *weak enough* to make it easy to write an algorithm (the loop body *Body*) that, given that the invariant holds on a subset of the data that is not the entire data, extends it to cover a slightly larger subset.

"Easy", in the last two conditions, means "much easier than solving the entire original problem". The loop consists of an approximation strategy that starts with the initialization, establishing the invariant, then retains the invariant while extending the scope by successive approximations to an ever larger set of the input through repeated executions of the loop body, until it hits the exit condition, signaling that it now covers the entire data and hence satisfies the loop's postcondition. This explains that the various strategies of Section 2, such as constant relaxation and uncoupling, are heuristics for mutating the loop's postcondition into a weaker form. The present work applies the same heuristics to mutate postconditions of the *routine* that encapsulates the loop. The connection between the routine's and the loop's postcondition justifies the rationale behind using weakening heuristics as mutation heuristics to generate invariant candidates.

4.2 Boogie

Boogie, now in its second version, is both an intermediate verification language and a verification tool.

The Boogie language combines a typed logical specification language with an in-the-small imperative programming language with variables, procedures, contracts, and annotations. The type system comprises a few basic primitive types as well as type constructors such as one- and two-dimensional arrays. It supports a relatively straightforward encoding of object-oriented language constructs. Boogie is part of the Spec# programming environment; mappings have been defined for other programming languages, including Eiffel [40] and C [39]. This suggests that the results described here can be generalized to many other contexts.

The Boogie tool verifies conformance of a procedure to its specification by generating verification conditions (VC) and feeding them to an automated theorem prover (the standard one being Z3). The outcome of a verification attempt can be successful or unsuccessful. In the latter case the tool provides some feedback on what might be wrong in the procedure, in particular by pointing out what contracts or annotations it could not verify. Verification with Boogie is sound but incomplete: a verified procedure is always guaranteed to be correct, while an unsuccessful verification attempt might simply be due to limitations of the technology.

The Boogie Specification Language. The Boogie specification language is essentially a typed predicate calculus with equality and arithmetic. Correspondingly, formulas – that is, logic expressions – are built by combining atomic constants, logic variables, and program variables with relational and arithmetic operators, as well as with Boolean connectives and quantifiers. For example, the following formula (from Section 2.1) states that no element in array X within positions 1 and n is larger than v: in other words, the array has upper bound v.

$$\forall\, j : \mathbf{int} \;\bullet\; 1 \leq j \,\wedge\, j \leq n \implies X[j] \leq v\,.$$

The syntactic classes Id, $Number$ and Map represent constant and variable identifiers, numbers and mappings.

Complex formulas and expressions can be postulated in *axioms* and parameterized by means of logic *functions*. Functions are a means of introducing encapsulation and genericity for formulas and complex expressions. For example, the previous formula can be parameterized into function *is_upper* with the following signature and definition:

$$\begin{array}{l} \mathbf{function}\ \mathit{is\_upper}\,(m\text{: } \mathbf{int},\ A\text{: } \mathit{array}\ \mathbf{int},\ \mathit{low}\text{: } \mathbf{int},\ \mathit{high}\text{: } \mathbf{int}) \\ \quad \mathbf{returns}\ (\ \mathbf{bool}\ ) \\ \quad \{\ \forall\, j : \mathbf{int} \;\bullet\; \mathit{low} \leq j \,\wedge\, j \leq \mathit{high} \implies A[j] \leq m\ \}\,. \end{array}$$

Axioms constrain global constants, variables, and functions; they are useful to supply Boogie with domain knowledge to facilitate inference and guide the automated reasoning over non-trivial programs. In certain situations it might for example be helpful to introduce the property that if an array has the upper bound m over the range $[low..high]$ and the element in position $high + 1$ is smaller than m then m is also the upper bound over the range $[low..high + 1]$. The following Boogie axiom expresses this property.

$$\begin{array}{l} \mathbf{axiom}\ (\ \forall m\text{: } \mathbf{int},\ A\text{: } \mathit{array}\ \mathbf{int},\ \mathit{low}\text{: } \mathbf{int},\ \mathit{high}\text{: } \mathbf{int} \;\bullet \\ \qquad \mathit{is\_upper}\ (m,\ A,\ \mathit{low},\ \mathit{high}) \,\wedge\, A[\mathit{high}+1] < m \\ \qquad\qquad \implies \mathit{is\_upper}\ (m,\ A,\ \mathit{low},\ \mathit{high}{+}1)\ )\,. \end{array}$$

The Boogie Programming Language. A Boogie program is a collection of *procedures*. Each procedure consists of a signature, a *specification* and (optionally) an *implementation* or body. The signature gives the procedure a name and declares its formal input and output arguments. The specification is a collection of contract clauses of three types: *frame conditions*, *preconditions*, and *postconditions*.

A frame condition, introduced by the keyword **modifies**, consists of a list of global variables that can be modified by the procedure; it is useful in evaluating

Statement ::=	Assertion | Modification
|	ConditionalBranch | Loop
Annotation ::=	**assert** Formula | **assume** Formula
Modification ::=	**havoc** VariableId | VariableId := Expression
|	**call** [VariableId$^+$:=] ProcedureId (Expression*)
ConditionalBranch ::=	**if** (Formula) Statement* [**else** Statement*]
Loop ::=	**while** (Formula) Invariant* Statement*
Invariant ::=	**invariant** Formula

Fig. 1. Simplified abstract syntax of Boogie statements

the side-effects of procedure call within any context. A precondition, introduced by the keyword **requires**, is a formula that is required to hold upon procedure invocation. A postcondition, introduced by the keyword **ensures**, is a formula that is guaranteed to hold upon successful termination of the procedure. For example, procedure *max_v2*, computing the maximum value in an array A given its size n, has the following specification.

procedure *max_v2* (A: *array* **int**, n: **int**) **returns** (m: **int**)
  requires $n \geq 1$;
  ensures *is_max* (m, A, 1, n);

The implementation of a procedure consists of a declaration of local variables, followed by a sequence of (possibly labeled) program statements. Figure 1 shows a simplified syntax for Boogie statements. Statements of class *Annotation* introduce checks at any program point: an *assertion* is a formula that must hold of every execution that reaches it for the program to be correct and an *assumption* is a formula whose validity at the program point is postulated. Statements of class *Modification* affect the value of program variables, by nondeterministically drawing a value for them (**havoc**), assigning them the value of an expression (:=), or calling a procedure with actual arguments (**call**). The usual conditional **if** statement controls the execution flow. Finally, the **while** statement supports loop iteration, where any loop can be optionally annotated with a number of *Invariants* (see Section 4.1). Boogie can check whether Definition 1 holds for any user-provided loop invariant.

The implementation of procedure *max_v2* is:

```
var i: int;
i := 1;   m := A[1];
while (i≤ n)
{
    if (m≤A[i])  { m := A[i]; }
      i := i + 1;
}
```

While the full Boogie language includes more types of statement, any Boogie statement can be desugared into one of those in Figure 1. In particular, the only looping construct we consider is the structured **while**; this choice simplifies the presentation of our loop invariant inference technique and makes it closer as if it was defined directly on a mainstream high-level programming language.

Also, there is a direct correspondence between Boogie's **while** loop and Eiffel's **from** ... **until** loop, used in the examples of Section 2 and the definitions in Section 4.1.

4.3 Notational Conventions

$\mathsf{subExp}(\phi, SubType)$ denotes the set of sub-expressions of formula ϕ that are of syntactic type $SubType$. For example, $\mathsf{subExp}(is\_upper(v, X, 1, n), Map)$ denotes all mapping sub-expressions in $is\_upper(v, X, 1, n)$, that is only $X[j]$.

$\mathsf{replace}(\phi, old, new, *)$ denotes the formula obtained from ϕ by replacing every occurrence of sub-expression old by expression new. Similarly, $\mathsf{replace}(\phi, old, new, n)$ denotes the formula obtained from ϕ by replacing only the n-th occurrence of sub-expression old by expression new, where the total ordering of sub-expressions is given by a pre-order traversal of the expression parse tree. For example, $\mathsf{replace}(is\_upper(v, X, 1, n), j, h, *)$ is

$$\forall\ h : \mathbf{int} \ \bullet \ low \leq h \land h \leq high \implies A[h] \leq m\,,$$

while $\mathsf{replace}(is\_upper(v, X, 1, n), j, h, 4)$ is:

$$\forall\ j : \mathbf{int} \ \bullet \ low \leq j \land j \leq high \implies A[h] \leq m\,.$$

Given a while loop ℓ: **while** (...) { *Body* }, $\mathsf{targets}(\ell)$ denotes the set of its *targets*: variables (including mappings) that can be modified by its *Body*; this includes global variables that appear in the **modifies** clause of called procedures.

Given a **procedure** *foo*, $\mathsf{variables}(foo)$ denotes the set of all variables that are visible within *foo*, that is its locals and any global variable.

A loop ℓ' is *nested* within another loop ℓ, and we write $\ell' \prec \ell$, iff ℓ' belongs to the *Body* of ℓ. Notice that if $\ell' \prec \ell$ then $\mathsf{targets}(\ell') \subseteq \mathsf{targets}(\ell)$. Given a **procedure** *foo*, its *outer* while loops are those in its body that are not nested within any other loop.

5 Generating Loop Invariants from Postconditions

This section presents the loop invariant generation algorithm.

5.1 Main Algorithm

The pseudocode in Figure 2 describes the main algorithm. The algorithm operates on a given procedure and returns a set of formulas that are invariant of *some* loop in the procedure. Every postcondition *post* among all postconditions $\mathsf{postconditions}(a\_procedure)$ of the procedure is considered separately (line 5). This is a coarse-grained yet effective way of implementing the *term-dropping* strategy outlined in Section 2.4: the syntax of the specification language supports splitting postconditions into a number of conjuncts, each introduced by the **ensures** keyword, hence each of these conjuncts is modified in isolation. It is reasonable to assume that the splitting into **ensures** clauses performed by the

```
1 invariants ( a_procedure: PROCEDURE )
2                 : SET_OF [FORMULA]
3   do
4     Result := ∅
5     for each post in postconditions(a_procedure) do
6       for each loop in outer_loops(a_procedure) do
7         -- compute all mutations of post
8         -- according to chosen strategies
9         mutations := mutations(post, loop)
10        for each formula in mutations do
11          for each any_loop in loops(a_procedure) do
12            if is_invariant(formula, any_loop) then
13              Result := Result ∪ {formula}
```

Fig. 2. Procedure invariants

user separates logically distinct portions of the postcondition, hence it makes sense to analyze each of them separately. This assumption might fail, of course, and in such cases the algorithm can be enhanced to consider more complex combinations of portions of the postcondition. However, one should only move to this more complex analysis if the basic strategy – which is often effective – fails. This enhancement belongs to future work.

The algorithm of Figure 2 then considers every *outer* while loop (line 6). For each, it computes a set of mutations of the postcondition *post* (line 9) according to the heuristics of Section 2. It then examines each mutation to determine if it is invariant to *any* loop in the procedure under analysis (lines 10–13), and finally returns the set **Result** of mutated postconditions that are invariants to some loop. For is_invariant($formula, loop$), the check consists of verifying whether initiation and consecution hold for $formula$ with respect to *loop*, according to Definition 1. This check is non-trivial, because loop invariants of different loops within the same procedure may interact in a circular way: the validity of one of them can be established only if the validity of the others is known already and *vice versa*.

This was the case of *partition* presented in Section 2.3. Establishing consecution for the modified postcondition in the outer loop requires knowing that the same modified postcondition is invariant to each of the two internal while loops (lines 8–13) because they belong to the body of the outer while loop. At the same time, establishing initiation for the first internal loop requires that consecution holds for the outer while loop, as every new iteration of the external loop initializes the first internal loop. Section 3 discussed a solution to this problem.

5.2 Building Mutated Postconditions

Algorithm mutations(*post*, *loop*), described in Figure 3, computes a set of modified versions of postcondition formula *post* with respect to outer while loop *loop*. It first includes the unchanged postcondition among the mutations (line 4). Then,

```
1 mutations ( post: FORMULA; loop: LOOP )
2                    : SET_OF [FORMULA]
3   do
4     Result := {post}
5     all_subexpressions := subExp(post, Id) ∪
6                           subExp(post, Number) ∪
7                           subExp(post, Map)
8     for each constant in all_subexpressions \ targets(loop) do
9        for each variable in targets(loop) do
10          Result := Result ∪
11             coupled_mutations(post, constant, variable) ∪
12             uncoupled_mutations(post, constant, variable)
```

Fig. 3. Procedure mutations

it computes (lines 5–7) a list of sub-expressions of *post* made of atomic variable identifiers (syntactic class Id), numeric constants (syntactic class $Number$) and references to elements in arrays (syntactic class Map). Each such sub-expression that *loop* does not modify (i.e., it is not one of its targets) is a *constant* with respect to the loop. The algorithm then applies the *constant relaxation* heuristics of Section 2.1 by relaxing *constant* into any *variable* among the *loop*'s targets (lines 8–9). More precisely, it computes two sets of mutations for each pair $\langle constant, variable \rangle$: in one *uncoupling*, described in Section 2.3, is also applied (lines 12 and 11, respectively).

The justification for considering outer loops only is that any target of the loop is a candidate for substitution. If a loop ℓ' is nested within another loop ℓ then $\mathsf{targets}(\ell') \subseteq \mathsf{targets}(\ell)$, so considering outer while loops is a conservative approximation that does not overlook any possible substitution.

```
1 coupled_mutations
2      ( post: FORMULA; constant, variable: EXPRESSION )
3      : SET_OF [FORMULA]
4   do
5     Result := replace(post, constant, variable, *)
6     aged_variable := aging(variable, loop)
7     Result := Result ∪
8                   replace(post, constant, aged_variable, *)
```

Fig. 4. Procedure coupled_mutations

5.3 Coupled Mutations

The algorithm in Figure 4 applies the constant relaxation heuristics to postcondition *post* without uncoupling. Hence, relaxing *constant* into *variable* simply amounts to replacing every occurrence of *constant* by *variable* in *post* (line 5); i.e., $\mathsf{replace}(post, constant, variable, *)$ using the notation introduced in Section 4.3. Afterward, the algorithm applies the *aging* heuristics (introduced in

Section 2.2): it computes the "previous" value of *variable* in an execution of *loop* (line 6) and it substitutes the resulting expression for *constant* in *post* (lines 7–8).

While the implementation of function aging could be very complex we adopt the following unsophisticated approach. For every possible acyclic execution path in *loop*, we compute the symbolic value of *variable* with initial value v_0 as a symbolic expression $\epsilon(v_0)$. Then we obtain aging(*variable*, *loop*) by solving the equation $\epsilon(v_0) = variable$ for v_0, for every execution path.[3] For example, if the loop simply increments *variable* by one, then $\epsilon(v_0) = v_0 + 1$ and therefore aging(*variable*, *loop*) $= variable - 1$. Again, while the example is unsophisticated it is quite effective in practice; indeed, most of the times it is enough to consider simple increments or decrements of *variable* to get a "good enough" aged expression.

5.4 Uncoupled Mutations

The algorithm of Figure 5 is a variation of the algorithm of Figure 4 applying the *uncoupling* heuristics outlined in Section 2.3. It achieves this by considering every occurrence of *constant* in *post* separately when performing the substitution of *constant* into *variable* (line 6). Everything else is as in the non-uncoupled case; in particular, aging is applied to every candidate for substitution.

This implementation of uncoupling relaxes one occurrence of a constant at a time. In some cases it might be useful to substitute different occurrences of the same constant by different variables. This was the case of *partition* discussed in Section 2.3, where relaxing two occurrences of the same constant *position* into two different variables was needed in order to get a valid invariant. Section 2.4 showed, however, that the term-dropping heuristics would have made this double relaxation unnecessary for the procedure.

6 Discussion and Related Work

6.1 Discussion

The experiments in Section 3 provide a partial, yet significant, assessment of the practicality and effectiveness of our technique for loop invariant inference. Two important factors to evaluate any inference technique deserve comment: relevance of the inferred invariants and scalability to larger programs.

A large portion of the invariants retrieved by gin-pink are relevant – i.e., required for a functional correctness proof – and complex – i.e., involving first-order quantification over several program elements. To some extent, this is unsurprising because deriving invariants from postconditions ensures by construction that they play a central role in the correctness proof and that they are at least as complex as the postcondition.

[3] Note that aging(*variable*, *loop*) is in general a set of expressions, so the notation at lines 6–8 in Figure 4 is a shorthand.

```
uncoupled_mutations
      (post: FORMULA; constant, variable: EXPRESSION)
      : SET_OF [FORMULA]
  do
    Result := ∅; index := 1
    for each occurrence of constant in post do
      Result := Result ∪
         {replace(post, constant, variable, index)}
      aged_variable := aging(variable, loop)
      Result := Result ∪
         {replace(post, constant, aged_variable, index)}
      index := index + 1
```

Fig. 5. Procedure uncoupled_mutations

As for scalability to larger programs, the main problem is the combinatorial explosion of the candidate invariants to be checked as the number of variables that are modified by the loop increases. In properly engineered code, each routine should not be too large or call too many other routines. The empirical observations mentioned in [24, Sec. 9] seem to support this assumption, which ensures that the candidate invariants do not proliferate and hence the inference technique can scale within reasonable limits. The examples of Section 3 are not trivial in terms of length and complexity of loops and procedures, if the yardstick is well-modularized code. On the other hand, there is plenty of room for finessing the application order of the various heuristics in order to analyze the most "promising" candidates first; the Houdini approach [17] might also be useful in this context. The investigation of these aspects belongs to future work.

6.2 Limitations

Relevant invariants obtained by postcondition mutation are most of the times significant, practically useful, and complementary to a large extent to the categories that are better tackled by other methods (see next sub-section). Still, the postcondition mutation technique cannot obtain *every* relevant invariant. Failures have two main different origins: conceptual limitations and shortcomings of the currently used technology.

The first category covers invariants that are not expressible as mutations of the postcondition. This is the case, in particular, whenever an invariant refers to a local variable whose final state is not mentioned in the postcondition. For example, the postcondition of procedure *max* in Section 2.1 does not mention variable i because its final value n is not relevant for the correctness. Correspondingly, invariant $i \leq n$ – which is involved in the partial correctness proof – cannot be obtained from by mutating the postcondition. A potential solution to these conceptual limitations is two-fold: on the one hand, many of these invariants that escape postcondition mutations can be obtained reliably with other inference techniques that do not require postconditions – this is the case of invariant $i \leq n$ in procedure *max* which is retrieved automatically by Boogie. On the other hand, if we can augment postconditions with complete information about local variables, the mutation approach can have a chance to work. In the case of *max*, a dynamic technique could suggest the supplementary postcondition $i \leq n \wedge i \geq n$ which would give the sought invariant by dropping the second conjunct.

Shortcomings of the second category follow from limitations of state-of-the-art automated theorem provers, which prevent reasoning about certain interesting classes of algorithms. For the sake of illustration, consider the following idealized[4] implementation of Newton's algorithm for the square root of a real number, more precisely the variant known as the Babylonian algorithm [29]:

```
square_root (a: REAL): REAL
  require a ≥ 0
  local y: REAL
  do
    from Result := 1; y := a
    until Result = y
    loop
      Result := (Result + y)/2
      y := a / Result
    end
  ensure Result ≥ 0 ∧ Result ∗ Result = a
```

Postcondition mutation would correctly find invariant **Result** $* \; y = a$ (by term dropping and uncoupling), but Boogie cannot verify that it is an invariant because the embedded theorem prover Z3 does not handle reasoning about properties of products of numeric variables [27]. If we can verify by other means that a candidate is indeed an invariant, the postcondition mutation technique of this paper would be effective over additional classes of programs.

6.3 Related Work

The amount of research work on the automated inference of invariants is formidable and spread over more than three decades; this reflects the cardinal role that invariants play in the formal analysis and verification of programs. This section outlines a few fundamental approaches and provides some evidence that this paper's technique is complementary, in terms of kinds of invariants inferred, to previously published approaches. For more references, in particular regarding software engineering applications, see the "related work" section of [15].

Static methods. Historically, the earliest methods for invariant inference where *static* as in the pioneering work of Karr [23]. Abstract interpretation and the constraint-based approach are the two most widespread frameworks for static invariant inference (see also [6, Chap. 12]).

Abstract interpretation is, roughly, a symbolic execution of programs over abstract domains that over-approximates the semantics of loop iteration. Since

[4] The routine assumes infinite-precision reals and does not terminate.

the seminal work by Cousot and Cousot [11], the technique has been updated and extended to deal with features of modern programming languages such as object-orientation and heap memory-management (e.g., [28,9]).

Constraint-based techniques rely on sophisticated decision procedures over non-trivial mathematical domains (such as polynomials or convex polyhedra) to represent concisely the semantics of loops with respect to certain template properties.

Static methods are sound – as is the technique introduced in this paper – and often complete with respect to the class of invariants that they can infer. Soundness and completeness are achieved by leveraging the decidability of the underlying mathematical domains they represent; this implies that the extension of these techniques to new classes of properties is often limited by undecidability. In fact, state-of-the-art static techniques can mostly infer invariants in the form of "well-behaving" mathematical domains such as linear inequalities [12,10], polynomials [38,37], restricted properties of arrays [7,5,20], and linear arithmetic with uninterpreted functions [1]. Loop invariants in these forms are extremely useful but rarely sufficient to prove full functional correctness of programs. In fact, one of the main successes of abstract interpretation has been the development of sound but incomplete tools [2] that can verify the absence of simple and common programming errors such as division by zero or void dereferencing. Static techniques for invariant inference are now routinely part of modern static checkers such as ESC/Java [18], Boogie/Spec# [26], and Why/Krakatoa/Caduceus [22].

The technique of the present paper is complementary to most static techniques in terms of the kinds of invariant that it can infer, because it derives invariants directly from postconditions. In this respect "classic" static inference and our inference by means of postcondition mutation can fruitfully work together to facilitate functional verification; to some extent this happens already when complementing Boogie's built-in facilities for invariant inference with our own technique.

[34,21,8,25,24] are the approaches that, for different reasons, share more similarities with ours. To our knowledge, [34,21,8,25] are the only other works applying a static approach to derive loop invariants from annotations. [21] relies on user-provided assertions nested within loop bodies and essentially tries to check whether they hold as invariants of the loop. This does not release the burden of writing annotations nested within the code, which is quite complex as opposed to providing only contracts in the form of pre and postconditions. In practice, the method of [21] works only when the user-provided annotations are very close to the actual invariant; in fact the few examples where the technique works are quite simple and the resulting invariants are usually obtainable by other techniques that do not need annotations. [8] briefly discusses deriving the invariant of a *for* loop from its postcondition, within a framework for reasoning about programs written in a specialized programming language. [25] also leverages specifications to derive intermediate assertions, but focusing on lower-level and type-like properties of pointers. On the other hand, [34] derives candidate invariants from postconditions in a very different setting than ours, with symbolic execution and model-checking techniques.

Finally, [24] derives complex loop invariants by first encoding the loop semantics as recurring relations and then instructing a rewrite-based theorem prover to try to remove the dependency on the iterator variable(s) in the relations. It shares with our work a practical attitude that favors powerful heuristics over completeness and leverages state-of-the-art verification tools to boost the inference of additional annotations.

Dynamic methods. More recently, dynamic techniques have been applied to invariant inference. The Daikon approach of Ernst et al. [15] showed that dynamic inference is practical and sprung much derivative work (e.g., [35,13,36] and many others). In a nutshell, the Daikon approach consists in testing a large number of candidate properties against several program runs; the properties that are not violated in any of the runs are retained as "likely" invariants. This implies that the inference is not sound but only an "educated guess": dynamic invariant inference is to static inference what testing is to program proofs. Nonetheless, just like testing is quite effective and useful in practice, dynamic invariant inference is efficacious and many of the guessed invariants are indeed sound.

Our approach shares with the Daikon approach the idea of guessing a candidate invariant and testing it *a posteriori*. There is an obvious difference between our approach, which retains only invariants that can be soundly verified, and dynamic inference techniques, which rely on a finite set of tests. A deeper difference is that Daikon guesses candidate invariants almost blindly, by trying out a pre-defined set of user-provided templates (including comparisons between variables, simple inequalities, and simple list comprehensions). On the contrary, our technique assumes the availability of contracts (and postconditions in particular) and leverages it to restrict quickly the state-space of search and get to good-quality loop invariants in a short time. As it is the case for static techniques, dynamic invariant inference methods can also be usefully combined with our technique, in such a way that invariants discovered by dynamic methods boost the application of the postcondition-mutation approach.

Program construction. Classical formal methods for program construction [14,19,29,32] have first described the idea of deriving loop invariants from postconditions. Several of the heuristics that we discussed in Section 2 are indeed a rigorous and detailed rendition of some ideas informally presented in [29,19]. In addition, the focus of the seminal work on program construction is to derive systematically an implementation from a complete functional specification. In this paper the goal is instead to enrich the assertions of an already implemented program and to exploit its contracts to annotate the code with useful invariants that facilitate a functional correctness proof.

7 Conclusion and Future Work

We have shown that taking advantage of postconditions yields loop invariants through effective techniques – not as predictable as the algorithms that yield verification conditions for basic constructs such as assignments and conditionals, but sufficiently straightforward to be applied by tools, and yielding satisfactory results in many practical cases.

The method appears general enough, covering most cases in which a programmer with a strong background in Hoare logic would be able at some effort to derive the invariant, but a less experienced one would be befuddled. So it does appear to fill what may be the biggest practical obstacle to automatic program proving.

The method requires that the programmer (or a different person, the "proof engineer", complementing the programmer's work, as testers traditionally do) provide the postcondition for every routine. As has been discussed in Section 1, we feel that this is a reasonable expectation for serious development, reflected in the Design by Contract methodology. For some people, however, the very idea of asking programmers or other members of a development team to come up with contracts of any kind is unacceptable. With such an a priori assumption, the results of this paper will be of little practical value; the only hope is to rely on invariant inference techniques that require the program only (complemented, in approaches such as Daikon, by test results and a repertoire of invariant patterns).

Some of the results that the present approach yields (sometimes trivially) when it is applied manually, are not yet available through the tools used in the current implementation of gin-pink. Although undecidability results indicate that program proving will never succeed in all possible cases, it is fair to expect that many of these limitations – such as those following from Z3's current inability to handle properties of products of variables – will go away as proof technology continues to progress.

We believe that the results reported here can play a significant role in the effort to make program proving painless and even matter-of-course. So in addition to the obvious extensions – making sure the method covers all effective patterns of postcondition mutation, and taking advantage of progress in theorem prover technology – our most important task for the near future is to integrate the results of this article, as unobtrusively as possible for the practicing programmer, in the background of a verification environment for contracted object-oriented software components.

Acknowledgements. A preliminary version of this work has been presented at the IFIP TC2 WG 2.3 meeting in Lachen, Switzerland, March 2010. The authors thank the attendees for their useful comments and criticism.

References

1. Beyer, D., Henzinger, T.A., Majumdar, R., Rybalchenko, A.: Invariant synthesis for combined theories. In: Cook, B., Podelski, A. (eds.) VMCAI 2007. LNCS, vol. 4349, pp. 378–394. Springer, Heidelberg (2007)
2. Blanchet, B., Cousot, P., Cousot, R., Feret, J., Mauborgne, L., Miné, A., Monniaux, D., Rival, X.: A static analyzer for large safety-critical software. In: Proceedings of the 2003 ACM SIGPLAN Conference on Programming Language Design and Implementation (PLDI 2003), pp. 196–207. ACM, New York (2003)

3. Böhme, S., Leino, K.R.M., Wolff, B.: HOL-Boogie — an interactive prover for the Boogie program-verifier. In: Mohamed, O.A., Muñoz, C., Tahar, S. (eds.) TPHOLs 2008. LNCS, vol. 5170, pp. 150–166. Springer, Heidelberg (2008)
4. Boyer, R.S., Moore, J.S.: MJRTY: A fast majority vote algorithm. In: Automated Reasoning: Essays in Honor of Woody Bledsoe, pp. 105–118 (1991)
5. Bozga, M., Habermehl, P., Iosif, R., Konečný, F., Vojnar, T.: Automatic verification of integer array programs. In: Bouajjani, A., Maler, O. (eds.) CAV 2009. LNCS, vol. 5643, pp. 157–172. Springer, Heidelberg (2009)
6. Bradley, A.R., Manna, Z.: The Calculus of Computation. Springer, Heidelberg (2007)
7. Bradley, A.R., Manna, Z., Sipma, H.B.: What's decidable about arrays? In: Emerson, E.A., Namjoshi, K.S. (eds.) VMCAI 2006. LNCS, vol. 3855, pp. 427–442. Springer, Heidelberg (2005)
8. de Caso, G., Garbervetsky, D., Gorín, D.: Reducing the number of annotations in a verification-oriented imperative language. In: Proceedings of Automatic Program Verification (2009)
9. Chang, B.Y.E., Leino, K.R.M.: Abstract interpretation with alien expressions and heap structures. In: Cousot, R. (ed.) VMCAI 2005. LNCS, vol. 3385, pp. 147–163. Springer, Heidelberg (2005)
10. Colón, M., Sankaranarayanan, S., Sipma, H.: Linear invariant generation using non-linear constraint solving. In: Hunt Jr., W.A., Somenzi, F. (eds.) CAV 2003. LNCS, vol. 2725, pp. 420–432. Springer, Heidelberg (2003)
11. Cousot, P., Cousot, R.: Abstract interpretation: A unified lattice model for static analysis of programs by construction or approximation of fixpoints. In: Proceedings of the 4th Annual ACM Symposium on Principles of Programming Languages (POPL 1977), pp. 238–252 (1977)
12. Cousot, P., Halbwachs, N.: Automatic discovery of linear restraints among variables of a program. In: Proceedings of the 5th Annual ACM Symposium on Principles of Programming Languages (POPL 1978), pp. 84–96 (1978)
13. Csallner, C., Tillman, N., Smaragdakis, Y.: DySy: dynamic symbolic execution for invariant inference. In: Schäfer, W., Dwyer, M.B., Gruhn, V. (eds.) Proceedings of the 30th International Conference on Software Engineering (ICSE 2008), pp. 281–290. ACM, New York (2008)
14. Dijkstra, E.W.: A Discipline of Programming. Prentice-Hall, Englewood Cliffs (1976)
15. Ernst, M.D., Cockrell, J., Griswold, W.G., Notkin, D.: Dynamically discovering likely program invariants to support program evolution. IEEE Transactions of Software Engineering 27(2), 99–123 (2001)
16. Filliâtre, J.C.: The WHY verification tool (2009), version 2.18, `http://proval.lri.fr`
17. Flanagan, C., Leino, K.R.M.: Houdini, an annotation assistant for ESC/Java. In: Oliveira, J.N., Zave, P. (eds.) FME 2001. LNCS, vol. 2021, pp. 500–517. Springer, Heidelberg (2001)
18. Flanagan, C., Leino, K.R.M., Lillibridge, M., Nelson, G., Saxe, J.B., Stata, R.: Extended static checking for Java. In: Proceedings of the 2002 ACM SIGPLAN Conference on Programming Language Design and Implementation (PLDI'02). SIGPLAN Notices, vol. 37(5), pp. 234–245. ACM, New York (2002)
19. Gries, D.: The science of programming. Springer, Heidelberg (1981)
20. Henzinger, T.A., Hottelier, T., Kovács, L., Voronkov, A.: Invariant and type inference for matrices. In: Barthe, G., Hermenegildo, M. (eds.) VMCAI 2010. LNCS, vol. 5944, pp. 163–179. Springer, Heidelberg (2010)

21. Janota, M.: Assertion-based loop invariant generation. In: Proceedings of the 1st International Workshop on Invariant Generation, WING 2007 (2007)
22. Jean-Christophe Filliâtre, C.M.: The Why/Krakatoa/Caduceus platform for deductive program verification. In: Damm, W., Hermanns, H. (eds.) CAV 2007. LNCS, vol. 4590, pp. 173–177. Springer, Heidelberg (2007)
23. Karr, M.: Affine relationships among variables of a program. Acta Informatica 6, 133–151 (1976)
24. Kovács, L., Voronkov, A.: Finding loop invariants for programs over arrays using a theorem prover. In: Chechik, M., Wirsing, M. (eds.) FASE 2009. LNCS, vol. 5503, pp. 470–485. Springer, Heidelberg (2009)
25. Lahiri, S.K., Qadeer, S., Galeotti, J.P., Voung, J.W., Wies, T.: Intra-module inference. In: Bouajjani, A., Maler, O. (eds.) CAV 2009. LNCS, vol. 5643, pp. 493–508. Springer, Heidelberg (2009)
26. Leino, K.R.M.: This is Boogie 2 (June 2008), (Manuscript KRML 178), `http://research.microsoft.com/en-us/projects/boogie/`
27. Leino, K.R.M., Monahan, R.: Reasoning about comprehensions with first-order SMT solvers. In: Shin, S.Y., Ossowski, S. (eds.) Proceedings of the 2009 ACM Symposium on Applied Computing (SAC 2009), pp. 615–622. ACM Press, New York (2009)
28. Logozzo, F.: Automatic inference of class invariants. In: Steffen, B., Levi, G. (eds.) VMCAI 2004. LNCS, vol. 2937, pp. 211–222. Springer, Heidelberg (2004)
29. Meyer, B.: A basis for the constructive approach to programming. In: Lavington, S.H. (ed.) Proceedings of IFIP Congress 1980, pp. 293–298 (1980)
30. Meyer, B.: Object-oriented software construction, 2nd edn. Prentice-Hall, Englewood Cliffs (1997)
31. Meyer, B.: Touch of Class: learning to program *well* with objects and contracts. Springer, Heidelberg (2009)
32. Morgan, C.: Programming from Specifications, 2nd edn. Prentice-Hall, Englewood Cliffs (1994)
33. Parberry, I., Gasarch, W.: Problems on Algorithms (2002), `http://www.eng.ent.edu/ian/books/free/`
34. Păsăreanu, C.S., Visser, W.: Verification of Java programs using symbolic execution and invariant generation. In: Graf, S., Mounier, L. (eds.) SPIN 2004. LNCS, vol. 2989, pp. 164–181. Springer, Heidelberg (2004)
35. Perkings, J.H., Ernst, M.D.: Efficient incremental algorithms for dynamic detection of likely invariants. In: Taylor, R.N., Dwyer, M.B. (eds.) Proceedings of the 12th ACM SIGSOFT International Symposium on Foundations of Software Engineering (SIGSOFT 2004/FSE-12), pp. 23–32. ACM, New York (2004)
36. Polikarpova, N., Ciupa, I., Meyer, B.: A comparative study of programmer-written and automatically inferred contracts. In: Proceedings of the ACM/SIGSOFT International Symposium on Software Testing and Analysis (ISSTA 2009), pp. 93–104 (2009)
37. Rodríguez-Carbonell, E., Kapur, D.: Generating all polynomial invariants in simple loops. Journal of Symbolic Computation 42(4), 443–476 (2007)
38. Sankaranarayanan, S., Sipma, H., Manna, Z.: Non-linear loop invariant generation using Gröbner bases. In: Jones, N.D., Leroy, X. (eds.) Proceedings of the 31st ACM SIGPLAN-SIGACT Symposium on Principles of Programming Languages (POPL 2004), pp. 318–329. ACM, New York (2004)
39. Schulte, W., Xia, S., Smans, J., Piessens, F.: A glimpse of a verifying C compiler (extended abstract). In: C/C++ Verification Workshop (2007)
40. Tschannen, J.: Automatic verification of Eiffel programs. Master's thesis, Chair of Software Engineering, ETH Zürich (2009)

ASMs and Operational Algorithmic Completeness of Lambda Calculus

Marie Ferbus-Zanda and Serge Grigorieff

LIAFA, CNRS & Université Paris Diderot,
Case 7014 75205 Paris Cedex 13
{ferbus,seg}@liafa.jussieu.fr
http://www.liafa.jussieu.fr/~seg/

Discussions with Yuri, during his many visits in Paris or Lyon, have been a source of great inspiraton for the authors. We thank him for so generously sharing his intuitions around the many faces of the notion of algorithm.

Abstract. We show that lambda calculus is a computation model which can step by step simulate any sequential deterministic algorithm for any computable function over integers or words or any datatype. More formally, given an algorithm above a family of computable functions (taken as primitive tools, i.e., kind of oracle functions for the algorithm), for every constant K big enough, each computation step of the algorithm can be simulated by exactly K successive reductions in a natural extension of lambda calculus with constants for functions in the above considered family.

The proof is based on a fixed point technique in lambda calculus and on Gurevich sequential Thesis which allows to identify sequential deterministic algorithms with Abstract State Machines.

This extends to algorithms for partial computable functions in such a way that finite computations ending with exceptions are associated to finite reductions leading to terms with a particular very simple feature.

Keywords: ASM, lambda calculus, theory of algorithms, operational semantics.

1 Introduction

1.1 Operational versus Denotational Completeness

Since the pioneering work of Church and Kleene, going back to 1935, many computation models have been shown to compute the same class of functions, namely, using Turing Thesis, the class of all computable functions. Such classes are said to be *Turing complete* or *denotationally algorithmically complete*.

This is a result about crude input/output behaviour. What about the ways to go from the input to the output, i.e., the executions of algorithms in each of

A. Blass, N. Dershowitz, and W. Reisig (Eds.): Gurevich Festschrift, LNCS 6300, pp. 301–327, 2010.

these computation models? Do they constitute the same class? Is there a Thesis for algorithms analog to Turing Thesis for computable functions?

As can be expected, denotational completeness does not imply operational completeness. Clearly, the operational power of machines using massive parallelism cannot be matched by sequential machines. For instance, on networks of cellular automata, integer multiplication can be done in real time (cf. Atrubin, 1962 [1], see also Knuth, [21] p.394-399), whereas on Turing machines, an $\Omega(n/\log n)$ time lower bound is known. Keeping within sequential computation models, multitape Turing machines have greater operational power than one-tape Turing machines. Again, this is shown using a complexity argument: palindromes recognition can be done in linear time on two-tapes Turing machines, whereas it requires computation time $O(n^2)$ on one-tape Turing machines (Hennie, 1965 [18], see also [5,24]).

Though resource complexity theory may disprove operational algorithmic completeness, there was no formalization of a notion of operational completeness since the notion of algorithm itself had no formal mathematical modelization. Tackled by Kolmogorov in the 50's [20], the question for *sequential algorithms* has been answered by Gurevich in the 80's [11,12,13] (see [6] for a comprehensive survey of the question), with their formalization as "*evolving algebras*" (now called *"abstract state machines" or ASMs*) which has lead to *Gurevich's sequential Thesis.*

Essentially, an ASM can be viewed as a first order multi-sorted structure and a program which modifies some of its predicates and functions (called dynamic items). Such dynamic items capture the moving environment of a procedural program. The run of an ASM is the sequence of structures – also called states – obtained by iterated application of the program. The program itself includes two usual ingredients of procedural languages, namely affectation and the conditional "if... then... else... ", plus a notion of parallel block of instructions. This last notion is a key idea which is somehow a programming counterpart to the mathematical notion of system of equations.

Gurevich's sequential Thesis [12,16,17] asserts that ASMs capture the notion of sequential algorithm. Admitting this Thesis, the question of operational completeness for a sequential procedural computation model is now the comparison of its operational power with that of ASMs.

1.2 Lambda Calculus and Operational Completeness

In this paper we consider lambda calculus, a subject created by Church and Kleene in the 30's, which enjoys a very rich mathematical theory. It may seem a priori strange to look for operational completeness with such a computation model so close to an assembly language (cf. Krivine's papers since 1994, e.g., [22]). It turns out that, looking at reductions by groups (with an appropriate but constant length), and allowing one step reduction of primitive operations, lambda calculus simulates ASMs in a very tight way. Formally, our translation of ASMs in lambda calculus is as follows. Given an ASM, we prove that, for every integer K big enough (the least such K depending on the ASM), there exists a lambda term θ with the following property. Let $a_1^t, \ldots, a_p^t$ be the values (coded as lambda terms) of all dynamic items of the ASM at step t, if the run does not stop at step t then

$$\theta a_1^t \ldots a_p^t \quad \overbrace{\rightarrow \quad \cdots \quad \rightarrow}^{K \text{ reductions}} \quad \theta a_1^{t+1} \ldots a_p^{t+1} \ .$$

If the run stops at step t then the left term reduces to a term in normal form which gives the list of outputs if they are defined. Thus, representing the state of the ASM at time t by the term $\theta a_1^t \ldots a_p^t$, a group of K successive reductions gives the state at time $t+1$. In other words, K reductions faithfully simulate one step of the ASM run. Moreover, this group of reductions is that obtained by the *leftmost redex reduction strategy*, hence it is a deterministic process. Thus, *lambda calculus is operationally complete for deterministic sequential computation.*

Let us just mention that adding to lambda calculus one step reduction of primitive operations is not an unfair trick. Every algorithm has to be "above" some basic operations which are kind of oracles: the algorithm decomposes the computation in elementary steps which are considered as atomic steps though they obviously themselves require some work. In fact, such basic operations can be quite complex: when dealing with integer matrix product (as in Strassen's algorithm in time $O(n^{\log 7})$), one considers integer addition and multiplication as basic... Building algorithms on such basic operations is indeed what ASMs do with the so-called static items, cf. §2.3, Point 2.

The proof of our results uses Curry's fixed point technique in lambda calculus plus some padding arguments.

1.3 Road Map

This paper deals with two subjects which have so far not been much related: ASMs and lambda calculus. To make the paper readable to both ASM and lambda calculus communities, Sect. 2, 3 recall all needed prerequisites in these two domains (*so that most readers may skip one of these two sections*).

What is needed about ASMs is essentially their definition, but it cannot be given without a lot of preliminary notions and intuitions. Our presentation of ASMs in §2 differs in inessential ways from Gurevich's one (cf. [13,15,17,10]). Crucial in the subject (and for this paper) is Gurevich's sequential Thesis that we state in §2.2. We rely on the literature for the many arguments supporting this Thesis.

§3 recalls the basics of lambda calculus, including the representation of lists and integers and Curry fixed point combinator.

The first main theorem in §5.3 deals with the simulation in lambda calculus of sequential algorithms associated to ASMs in which all dynamic symbols are constant ones (we call them type 0 ASMs). The second main theorem in §5.4 deals with the general case.

2 ASMs

2.1 The Why and How of ASMs on a Simple Example

Euclid's Algorithm Consider Euclid's algorithm to compute the greatest common divisor (gcd) of two natural numbers. It turns out that such a simple algorithm already allows to pinpoint an operational incompleteness in usual programming languages. Denoting by $\mathtt{rem}(u,v)$ the remainder of u modulo v, this algorithm can be described as follows[1]

Given data: two natural numbers a, b
While $b \neq 0$ replace the pair (a,b) by $(b, \mathtt{rem}(a,b))$
When $b = 0$ halt: a is the wanted gcd

Observe that the the pair replacement in the above while loop involves some elementary parallelism which is the algorithmic counterpart to co-arity, i.e., the consideration of functions with range in multidimensional spaces such as the $\mathbb{N}^2 \to \mathbb{N}^2$ function $(x,y) \mapsto (y, \mathtt{rem}(x,y))$.

Euclid's Algorithm in Pascal In usual programming languages, the above simultaneous replacement is impossible: affectations are not done in parallel but sequentially. For instance, *no Pascal program implements it as it is*, one can only get a distorted version with an extra algorithmic contents involving a new variable z, cf. Figure 1.

Euclid's algorithm in Pascal
while $b > 0$ **do begin**
$z := a$;
$a := b$;
$b := rem\ (z, b)$;
end;
$gcd := a$.

Euclid's algorithm in ASM	
if $0 < b$ **then**	$a := b$ $b := rem\ (a, b)$

(In both programs, a, b are inputs and a is the output)

Fig. 1. Pascal and ASM programs for Euclid's algorithm

An ASM for Euclid's Algorithm Euclid's algorithm has a faithful formalization using an ASM. The vertical bar on the left in the ASM program (cf. Figure 1) tells that the two updates are done simultaneously and independently. Initialization gives symbols a, b the integer values of which we want to compute the gcd. The semantical part of the ASM involves the set $\mathbb{N}$ of integers to interpret all symbols. Symbols $0, <, =, rem$ have fixed interpretations in integers which are the expected ones. Symbols a, b have varying interpretations in the integers. The sequence of values taken by a, b constitutes the run of the ASM.

When the instruction gets void (i.e., when b is null) the run stops and the value of the symbol a is considered to be the output.

[1] Sometimes, one starts with a conditional swap: if $a < b$ then a, b are exchanged. But this is done in the first round of the while loop.

2.2 Gurevich Sequential Thesis

Yuri Gurevich has gathered as three Sequential Postulates (cf. [17,10]) some key features of deterministic sequential algorithms for partial computable functions (or type 1 functionals).

I *(Sequential time).* An algorithm is a deterministic state-transition system. Its transitions are partial functions.
Non deterministic transitions and even nonprocedural input/output specifications are thereby excluded from consideration.

II *(Abstract states).* States are multitructures[2], sharing the same fixed, finite vocabulary. States and initial states are closed under isomorphism. Transitions preserve the domain, and transitions and isomorphisms commute.

III *(Bounded exploration).* Transitions are determined by a fixed finite "glossary" of "critical" terms. That is, there exists some finite set of (variable-free) terms over the vocabulary of the states such that states that agree on the values of these glossary terms also agree on all next-step state changes.

Gurevich, 2000 [17], stated an operational counterpart to Church's Thesis:

Thesis. [Gurevich's sequential Thesis] *E*very sequential algorithm satisfies the Sequential Postulates I-III.

2.3 The ASM Modelization Approach

Gurevich's postulates lead to the following modelization approach (we depart in non essential ways from [10], see Remark 2.1).

1. *The base sets.* Find out the underlying families of objects involved in the given algorithm, i.e., objects which can be values for inputs, outputs or environmental parameters used during the execution of the algorithm. These families constitute the base sets of the ASM. In Euclid's algorithm, a natural base set is the set $\mathbb{N}$ of natural integers.
2. *Static items.* Find out which particular fixed objects in the base sets are considered and which functions and predicates over/between the base sets are viewed as atomic in the algorithm, i.e., are not given any modus operandi. Such objects, functions and predicates are called the primitive or static items of the ASM. They do not change value through transitions. In Euclid's algorithm, static items are the integer 0, the *rem* function and the $<$ predicate.
3. *Dynamic items.* Find out the diverse objects, functions and predicates over the base sets of the ASM which vary through transitions. Such objects, functions and predicates are called the dynamic items of the ASM. In Euclid's algorithm, these are a, b.
4. *States: from a multi-sorted partial structure to a multi-sorted partial algebra.* Collecting all the above objects, functions and predicates leads to a first-order multi-sorted structure of some logical typed language: any function

[2] In ASM theory, an ASM is, in fact, a multialgebra (cf. point 1 of Remark §2.1).

goes from some product of sorts into some sort, any predicate is a relation over some sorts. However, there is a difference with the usual logical notion of multi-sorted structure: predicates and functions may be partial. A feature which is quite natural for any theory of computability, a fortiori for any theory of algorithms.

To such a multi-sorted structure one can associate a multi-sorted algebra as follows. First, if not already there, add a sort for Booleans. Then replace predicates by their characteristic functions In this way, we get a multi-sorted structure with partial functions only, i.e. a multialgebra.

5. *Programs.* Finally, the execution of the algorithm can be viewed as a sequence of states. Going from one state to the next one amounts to applying to the state a particular program – called the ASM program – which modifies the interpretations of the sole dynamic symbols (but the universe itself and the interpretations of the static items remain unchanged). Thus, the execution of the algorithm appears as an iterated application of the ASM program. It is called the run of the ASM.

   Using the three above postulates, Gurevich [16,17] proves that quite elementary instructions – namely blocks of parallel conditional updates – suffice to get ASM programs able to simulate step by step any deterministic procedural algorithm.
6. *Inputs, initialization map and initial state.* Inputs correspond to the values of some distinguished static symbols in the initial state, i.e., we consider that all inputs are given when the algorithm starts (though questionable in general, this assumption is reasonable when dealing with algorithms to compute a function). All input symbols have arity zero for algorithms computing functions. Input symbols with non zero arity are used when dealing with algorithms for type 1 functionals.

   The initialization map associates to each dynamic symbol a term built up with static symbols. In an initial state, the value of a dynamic symbol is required to be that of the associated term given by the initialization map.
7. *Final states and outputs.* There may be several outputs, for instance if the algorithm computes a function $\mathbb{N}^k \to \mathbb{N}^\ell$ with $\ell \geq 2$.
   A state is final when, applying the ASM program to that state,
   (a) either the `Halt` instruction is executed *(Explicit halting)*,
   (b) or no update is made (i.e. all conditions in conditional blocks of updates get value *False*) *(Implicit halting)* .

   In that case, the run stops and the outputs correspond to the values of some distinguished dynamic symbols. For algorithms computing functions, all output symbols are constants (i.e. function symbols with arity zero).
8. *Exceptions.* There may be a finite run of the ASM ending in a non final state. This corresponds to exceptions in programming (for instance a division by 0) and there is no output in such cases. This happens when
   (a) either the `Fail` instruction is executed *(Explicit failing)*,
   (b) or there is a clash between two updates which are to be done simultaneously *(Implicit failing)*.

Remark 2.1. Let us describe how our presentation of ASMs (slightly) departs from [10].

1. We stick to what Gurevich says in §.2.1 of [14] (Lipari Guide, 1993): *"Actually, we are interested in multi-sorted structures with partial operations"*. Thus, we do not regroup sorts into a single universe and do not extend functions with the *undef* element.
2. We add the notion of initialization map which brings a syntactical counterpart to the semantical notion of initial state. It also rules out any question about the status of initial values of dynamic items which would not be inputs.
3. We add explicit acceptance and rejection as specific instructions in ASM programs. Of course, they can be simulated using the other ASM instructions (so, they are syntactic sugar) but it may be convenient to be able to explicitly tell there is a failure when something like a division by zero is to be done. This is what is done in many programming languages with the so-called exceptions. Observe that `Fail` has some common flavor with `undef`. However, `Fail` is relative to executions of programs whereas `undef` is relative to the universe on which the program is executed.
4. As mentioned in §2.1, considering several outputs goes along with the idea of parallel updates.

2.4 Vocabulary and States of an ASM

ASM vocabularies and ASM states correspond to algebraic signatures and algebras. The sole difference is that an ASM vocabulary comes with an extra classification of its symbols as static, dynamic, input and output carrying the intuitions described in points 2, 3, 6, 7 of §2.3.

Definition 2.2. *1. An ASM vocabulary is a finite family of sorts $s_1, \ldots, s_m$ and a finite family $\mathcal{L}$ of function symbols with specified types of the form s_i or $s_{i_1} \times \cdots \times s_{i_k} \to s_i$ (function symbols with type s_i are also called constants of type s_i). Four subfamilies of symbols are distinguished:*

$$\begin{array}{lcl} \mathcal{L}^{sta} \text{ (static symbols)} & , & \mathcal{I} \text{ (input symbols)} \\ \mathcal{L}^{dyn} \text{ (dynamic symbols)} & , & \mathcal{O} \text{ (output symbols)} \end{array}$$

such that $\mathcal{L}^{sta}, \mathcal{L}^{dyn}$ is a partition of $\mathcal{L}$ and $\mathcal{I} \subseteq \mathcal{L}^{sta}$ and $\mathcal{O} \subseteq \mathcal{L}^{dyn}$. We also require that there is a sort to represent Booleans and that $\mathcal{L}^{sta}$ contains symbols to represent the Boolean items (namely symbols `True`, `False`, *$\neg$, $\wedge$, $\vee$) and, for each sort s, a symbol $=_s$ to represent equality on sort s.*

*2. Let $\mathcal{L}$ be an ASM vocabulary with n sorts. An $\mathcal{L}$-*state *is any n-sort multialgebra $\mathcal{S}$ for the vocabulary $\mathcal{L}$. The multi-domain of $\mathcal{S}$ is denoted by $(\mathcal{U}_1, \ldots, \mathcal{U}_m)$. We require that*

i. one of the $\mathcal{U}_i$'s is `Bool` *with the expected interpretations of symbols* `True`, `False`,*s $\neg$, $\wedge$, $\vee$,*
ii. the interpretation of the symbol $=_i$ is usual equality in the interpretation $\mathcal{U}_i$ of sort s_i.

In the usual way, using variables typed by the n sorts of $\mathcal{L}$, one constructs typed $\mathcal{L}$-terms and their types. The type of a term t is of the form s_i or $s_{i_1} \times \cdots \times s_{i_k} \to s_i$ where $s_{i_1}, \ldots, s_{i_k}$ are the types of the different variables occurring in t. Ground terms are those which contain no variable. The semantics of typed terms is the usual one.

Definition 2.3. *Let $\mathcal{L}$ be an ASM vocabulary and $\mathcal{S}$ an ASM $\mathcal{L}$-state. Let t be a typed term with type $s_{i_1} \times \cdots \times s_{i_1} \to s_i$. We denote by $t_\mathcal{S}$ its interpretation in $\mathcal{S}$, which is a function $\mathcal{U}_{i_1} \times \cdots \times \mathcal{U}_{i_\ell} \to \mathcal{U}_i$. In case $\ell = 0$, i.e., no variable occurs, then $t_\mathcal{S}$ is an element of $\mathcal{U}_i$.*

It will be convenient to lift the interpretation of a term with ℓ variables to be a function with any arity k greater than ℓ.

Definition 2.4. *Let $\mathcal{L}$ be an ASM vocabulary and $\mathcal{S}$ an ASM $\mathcal{L}$-state with universe $\mathcal{U}$. Suppose $\sigma : \{1, \ldots, \ell\} \to \{1, \ldots, p\}$ is any map and $\tau : \{1, \ldots, p\} \to \{1, \ldots, m\}$ is a distribution of (indexes of) sorts. Suppose t is a typed term of type $s_{\tau(\sigma(1))} \times \cdots \times s_{\tau(\sigma(\ell))} \to s_i$. We let $t_\mathcal{S}^{\tau,\sigma}$ be the function $\mathcal{U}_{s_{\tau(1)}} \times \cdots \times \mathcal{U}_{s_{\tau(p)}} \to \mathcal{U}_i$ such that, for all $(a_1, \cdots, a_p) \in \mathcal{U}_{s_{\tau(1)}} \times \cdots \times \mathcal{U}_{s_{\tau(p)}}$,*

$$t_\mathcal{S}^{\tau,\sigma}(a_1, \cdots, a_k) = t_\mathcal{S}(a_{\sigma(1)}, \cdots, a_{\sigma(\ell)}) \ .$$

2.5 Initialization Maps

$\mathcal{L}$-terms with no variable are used to name particular elements in the universe $\mathcal{U}$ of an ASM whereas $\mathcal{L}$-terms with variables are used to name particular functions over $\mathcal{U}$.

Using the lifting process described in Definition 2.4, one can use terms containing less than k variables to name functions with arity k.

Definition 2.5. *1, Let $\mathcal{L}$ be an ASM vocabulary. An $\mathcal{L}$-*initialization map *ξ has domain family $\mathcal{L}^{(dyn)}$ of dynamic symbols and satisfies the following condition:*

> *if α is a dynamic function symbol with type $s_{\tau(1)} \times \cdots \times s_{\tau(\ell)} \to s_i$ then $\xi(\alpha)$ is a pair (σ, t) such that $\sigma : \{1, \ldots, \ell\} \to \{1, \ldots, p\}$ and t is a typed $\mathcal{L}$-term with type $s_{\tau(\sigma(1))} \times \cdots \times s_{\tau(\sigma(\ell))} \to s_i$ which is built with the sole static symbols (with $\tau : \{1, \ldots, p\} \to \{1, \ldots, m\}$).*

2. Let ξ be an $\mathcal{L}$-initialization map. An $\mathcal{L}$-state $\mathcal{S}$ is ξ-initial if, for any dynamic function symbol α, if $\xi(\alpha) = (\sigma, t)$ then the interpretation of α in $\mathcal{S}$ is $t_\mathcal{S}^{\tau,\sigma}$.

3. An $\mathcal{L}$-state is initial if it is ξ-initial for some ξ.

Remark 2.6. Of course, the values of static symbols are basic ones, they are not to be defined from anything else: either they are inputs or they are the elementary pieces upon which the ASM algorithm is built.

2.6 ASM Programs

Definition 2.7. *1. The vocabulary of ASM programs is the family of symbols*

$$\{\texttt{Skip}\ ,\ \texttt{Halt}\ ,\ \texttt{Fail}\ ,\ :=\ ,\ \Big|\ ,\ \texttt{if}\ \ldots\ \texttt{then}\ \ldots\ \texttt{else}\ \ldots\}$$

2. (ℒ-updates). *Given an ASM vocabulary* $\mathcal{L}$*, a sequence of* $k+1$ *ground typed* $\mathcal{L}$*-terms* $t_1, \ldots, t_k, u$ *(i.e. typed terms with no variable), a dynamic function symbol* α*, if* $\alpha(t_1, \ldots, t_k)$ *is a typed* $\mathcal{L}$*-term with the same type as* u *then the syntactic object* $\alpha(t_1, \ldots, t_k) := u$ *is called an* $\mathcal{L}$*-update.*

3. (ℒ-programs). *Given an ASM vocabulary* $\mathcal{L}$*, the* $\mathcal{L}$ *programs are obtained via the following clauses.*

i. (Atoms). Skip, Halt, Fail *and all* $\mathcal{L}$*-updates are* $\mathcal{L}$*-programs.*
ii. (Conditional constructor). *Given a ground typed term* C *with Boolean type and two* $\mathcal{L}$*-programs* P, Q*, the syntactic object*

$$\texttt{if}\ C\ \texttt{then}\ P\ \texttt{else}\ Q$$

is an $\mathcal{L}$*-program.*
iii. (Parallel block constructor). *Given* $n \geq 1$ *and* $\mathcal{L}$*-programs* $P_1, \ldots, P_n$*, the syntactic object (with a vertical bar on the left)*

$$\left|\begin{array}{l} P_1 \\ \vdots \\ P_n \end{array}\right.$$

is an $\mathcal{L}$*-program.*

The intuition of programs is as follows.

- Skip is the program which does nothing. Halt halts the execution in a successful mode and the outputs are the current values of the output symbols. Fail also halts the execution but tells that there is a failure, so that there is no meaningful output.
- Updates modify the interpretations of dynamic symbols, they are the basic instructions. The left member has to be of the form $\alpha(\cdots)$ with α a dynamic symbol because the interpretations of static symbols do not vary.
- The conditional constructor has the usual meaning whereas the parallel constructor is a new control structure to get *simultaneous and independent executions* of programs $P_1, \ldots, P_n$.

2.7 Action of an ℒ-Program on an ℒ-State

Active Updates and Clashes. In a program the sole instructions which have some impact are updates. They are able to modify the interpretations of dynamic symbols on the sole tuples of values which can be named by tuples of ground terms. Due to conditionals, not every update occurring in a program will really be active. it does depend on the state to which the program is applied. Which symbols on which tuples are really active and what is their action? This is the object of the next definition.

Definition 2.8 (Active updates). *Let $\mathcal{L}$ be an ASM vocabulary, P an $\mathcal{L}$-program and $\mathcal{S}$ an $\mathcal{L}$-state. Let* $\texttt{Update}(P)$ *be the family of all updates occurring in P. The subfamily* $\texttt{Active}\ (\mathcal{S}, P) \subseteq \texttt{Update}(P)$ *of so-called $(\mathcal{S}, P)$-*active updates *is defined via the following induction on P :*

$$\texttt{Active}\ (\mathcal{S}, \texttt{Skip}) = \emptyset$$

$$\texttt{Active}\ (\mathcal{S}, \alpha(t_1, \ldots, t_k) := u) = \{\alpha(t_1, \ldots, t_k) := u\}$$

$$\texttt{Active}\ (\mathcal{S},\ \texttt{if}\ C\ \texttt{then}\ Q\ \texttt{else}\ R) = \begin{cases} \texttt{Active}\ (\mathcal{S}, Q) & \text{if } C_\mathcal{S} = \texttt{True} \\ \texttt{Active}\ (\mathcal{S}, R) & \text{if } C_\mathcal{S} = \texttt{False} \\ \emptyset & \text{if } C_\mathcal{S} \notin \texttt{Bool} \end{cases}$$

$$\texttt{Active}\ (\mathcal{S}, \left| \begin{array}{l} P_1 \\ \vdots \\ P_n \end{array} \right.) = \texttt{Active}\ (\mathcal{S}, P_1) \cup \ldots \cup \texttt{Active}\ (\mathcal{S}, P_n)$$

The action of a program P on a state $\mathcal{S}$ is to be seen as the conjunction of updates in $\texttt{Active}\ (\mathcal{S}, P)$ provided these updates are compatible. Else, P clashes on $\mathcal{S}$.

Definition 2.9. *An $\mathcal{L}$-program P clashes on an $\mathcal{L}$-state $\mathcal{S}$ if there exists two active updates $\alpha(s_1, \ldots, s_k) := u$ and $\alpha(t_1, \ldots, t_k) := v$ in* $\texttt{Active}\ (\mathcal{S}, P)$ *relative to the same dynamic symbol α such that $s_{1\mathcal{S}} = t_{1\mathcal{S}}, \ldots, s_{k\mathcal{S}} = t_{k\mathcal{S}}$ but $u_\mathcal{S}$ and $v_\mathcal{S}$ are not equal (as elements of the universe).*

Remark 2.10. A priori, another case could also be considered as a clash. We illustrate it for a parallel block of two programs P, Q and the update of a dynamic constant symbol c. Suppose $c_\mathcal{S} \neq u_\mathcal{S}$ and $c := u$ is an active update in $\texttt{Active}\ (\mathcal{S}, P)$. Then P wants to modify the value of $c_\mathcal{S}$. Suppose also that there is no active update with left member c in $\texttt{Active}\ (\mathcal{S}, Q)$. Then Q does not want to touch the value of $c_\mathcal{S}$. Thus, P and Q have incompatible actions: P modifies the interpretation of c whereas Q does nothing about c. One could consider this as a clash for the parallel program $\left| \begin{array}{l} P \\ Q \end{array} \right.$. Nevertheless, *this case is not considered to be a clash.* A moment reflection shows that this is a reasonable choice. Otherwise, a parallel block would always clash except in case all programs $P_1, \ldots, P_n$ do exactly the same actions... Which would make parallel blocks useless.

Halt and Fail

Definition 2.11. *Let $\mathcal{L}$ be an ASM vocabulary, $\mathcal{S}$ be an $\mathcal{L}$-state and P an $\mathcal{L}$-program. By induction, we define the two notions: P halts (resp. fails) on $\mathcal{S}$.*

- *If P is* `Skip` *or an update then P neither halts nor fails on $\mathcal{S}$.*
- *If P is* `Halt` *(resp.* `Fail`*) then P halts and does not fail (resp. fails and does not halt) on $\mathcal{S}$.*
- $\texttt{if}\ C\ \texttt{then}\ Q\ \texttt{else}\ R$ *halts on $\mathcal{S}$ if and only if*

$$\begin{cases} \text{either } C_\mathcal{S} = \texttt{True} \text{ and } Q \text{ halts on } \mathcal{S} \\ \text{or } C_\mathcal{S} = \texttt{False} \text{ and } R \text{ halts on } \mathcal{S} \end{cases}$$

– `if` C `then` Q `else` R *fails on* $\mathcal{S}$ *if and only if*

$$\begin{cases} \textit{either } C_{\mathcal{S}} = \texttt{True} \textit{ and } Q \textit{ fails on } \mathcal{S} \\ \textit{or } C_{\mathcal{S}} = \texttt{False} \textit{ and } R \textit{ fails on } \mathcal{S} \;. \end{cases}$$

– *The parallel block of programs* $P_1, \ldots, P_n$ *halts on* $\mathcal{S}$ *if and only if some* P_i *halts on* $\mathcal{S}$ *and no* P_j *fails on* $\mathcal{U}$.
– *The parallel block of programs* $P_1, \ldots, P_n$. *fails on* $\mathcal{S}$ *if and only if some* P_i *fails on* $\mathcal{S}$.

Successor State

Definition 2.12. Let $\mathcal{L}$ be an ASM vocabulary and $\mathcal{S}$ be an $\mathcal{L}$-state.

The *successor state* $\mathcal{T} = \texttt{Succ}(\mathcal{S}, P)$ of state $\mathcal{S}$ relative to an $\mathcal{L}$-program P is defined if only if P does not clash nor fail nor halt on $\mathcal{S}$.

In that case, the successor is inductively defined via the following clauses.

1. $\mathcal{T} = \texttt{Succ}(\mathcal{S}, P)$ and $\mathcal{S}$ have the same base sets $\mathcal{U}_1, \ldots, \mathcal{U}_n$.
2. $\alpha_{\mathcal{T}} = \alpha_{\mathcal{S}}$ for any static symbol α.

3a. $\texttt{Succ}(\mathcal{S}, \texttt{Skip}) = \mathcal{S}$ (recall that `Skip` does nothing. . . .)

3b. Suppose P is an update program $\alpha(t_1, \ldots, t_k) := u$ where α is a dynamic symbol with type $s_{i_1} \times \cdots \times s_{i_k} \to s_i$ and $\boldsymbol{a} = (t_{1\mathcal{S}}, \ldots, t_{k\mathcal{S}})$. Then all dynamic symbols different from α have the same interpretation in $\mathcal{S}$ and $\mathcal{T}$ and, for every $\boldsymbol{b} \in \mathcal{U}_{i_1} \times \cdots \times \mathcal{U}_{i_k}$, we have $\alpha_{\mathcal{T}}(\boldsymbol{b}) = \begin{cases} \alpha_{\mathcal{S}}(\boldsymbol{b}) & \text{if } \boldsymbol{b} \neq \boldsymbol{a} \\ u_{\mathcal{S}} & \text{if } \boldsymbol{b} = \boldsymbol{a} \end{cases}$.

3c. Suppose P is the conditional program `if` C `then` Q `else` R. Then

$$\begin{cases} \texttt{Succ}(\mathcal{S}, P) = \texttt{Succ}(\mathcal{S}, Q) \text{ if } C_{\mathcal{S}} = \texttt{True} \\ \texttt{Succ}(\mathcal{S}, P) = \texttt{Succ}(\mathcal{S}, R) \text{ if } C_{\mathcal{S}} = \texttt{False} \end{cases}$$

(since P does not fail on $\mathcal{S}$, we know that $C_{\mathcal{S}}$ is a Boolean).

3d Suppose P is the parallel block program $\left| \begin{array}{l} P_1 \\ \vdots \\ P_n \end{array} \right.$ and P does not clash on $\mathcal{S}$.

Then $\mathcal{T} = \texttt{Succ}(\mathcal{S}, P)$ is such that, for every dynamic symbol α with type $s_{i_1} \times \cdots \times s_{i_k} \to s_i$ and every tuple $\boldsymbol{a} = (a_1, \ldots, a_k)$ in $\mathcal{U}_{i_1} \times \cdots \times \mathcal{U}_{i_k}$,
  – if there exists an update $\alpha(t_1, \ldots, t_k) := u$ in `Active` $(\mathcal{S}, P)$ such that $\boldsymbol{a} = (t_{1\mathcal{S}}, \ldots, t_{k\mathcal{S}})$ then $\alpha(\boldsymbol{a})_{\mathcal{T}}$ is the common value of all $v_{\mathcal{S}}$ for which there exists some update $\alpha(s_1, \ldots, s_k) := v$ in `Active` $(\mathcal{S}, P)$ such that $\boldsymbol{a} = (s_{1\mathcal{S}}, \ldots, s_{k\mathcal{S}})$.
  – Else $\alpha(\boldsymbol{a})_{\mathcal{T}} = \alpha(\boldsymbol{a})_{\mathcal{S}}$.

Remark 2.13. In particular, $\alpha_{\mathcal{T}}(\boldsymbol{a})$ and $\alpha_{\mathcal{S}}(\boldsymbol{a})$ have the same value in case $\boldsymbol{a} = (a_1, \ldots, a_k)$ is not the value in $\mathcal{S}$ of any k-tuple of ground terms $(t_1, \ldots, t_k)$ such that `Active` $(\mathcal{S}, P)$ contains an update of the form $\alpha(t_1, \ldots, t_k) := u$ for some ground term u.

2.8 Definition of ASMs and ASM Runs

At last, we can give the definition of ASMs and ASM runs.

Definition 2.14. *1. An ASM is a triple $(\mathcal{L}, P, (\xi, \mathcal{J}))$ (with two morphological components and one semantico-morphological component) such that:*

- *$\mathcal{L}$ is an ASM vocabulary as in Definition 2.2,*
- *P is an $\mathcal{L}$-program as in Definition 2.7,*
- *ξ is an $\mathcal{L}$-initialization map and $\mathcal{J}$ is a ξ-initial $\mathcal{L}$-state as in Definition 2.5.*

An ASM has type 0 if all its dynamic symbols have arity 0 (i.e., they are constants).

2. The run *of an ASM $(\mathcal{L}, P, (\xi, \mathcal{J}))$ is the sequence of states $(\mathcal{S}_i)_{i \in I}$ indexed by a finite or infinite initial segment I of $\mathbb{N}$ which is uniquely defined by the following conditions:*

- *$\mathcal{S}_0$ is $\mathcal{J}$.*
- *$i+1 \in I$ if and only if P does not clash nor fail nor halt on $\mathcal{S}_i$ and* `Active` *$(\mathcal{S}_i, P) \neq \emptyset$ (i.e. there is an active update[3]).*
- *If $i+1 \in I$ then $\mathcal{S}_{i+1} =$* `Succ`*$(\mathcal{S}_i, P)$.*

3. Suppose I is finite and i is the maximum element of I.
The run is successful if `Active` *$(\mathcal{S}_i, P)$ is empty or P halts on $\mathcal{S}_i$. In that case the outputs are the interpretations on $\mathcal{S}_i$ of the output symbols.*
The run fails if P clashes or fails on $\mathcal{S}_i$. In that case the run has no output.

Remark 2.15. In case `Active` $(\mathcal{S}_i, P) \neq \emptyset$ and P does not clash nor fail nor halt on $\mathcal{S}_i$ and $\mathcal{S}_i = \mathcal{S}_{i+1}$ (i.e., if the active updates do not modify $\mathcal{S}_i$) then the run is infinite: $\mathcal{S}_j = \mathcal{S}_i$ for every $j > i$.

2.9 Operational Completeness: The ASM Theorem

Let us now state the fundamental theorem of ASMs.

Theorem 2.16 (ASM Theorem, 1999 [16,17], cf. [10]). *Every process satisfying the Sequential Postulates (cf. §2.2) can be emulated by an ASM with the same vocabulary, sets of states and initial states.*

In other words, using Gurevich Sequential Thesis 2.2, every sequential algorithm can be step by step emulated by an ASM with the same values of all environment parameters. I.e., ASMs are operationally complete as concerns sequential algorithms.

The proof of the ASM Theorem also shows that ASM programs of a remarkably simple form are sufficient.

[3] Nevertheless, it is possible that $\mathcal{S}_i$ and `Succ`$(\mathcal{S}_i, P)$ coincide, cf. Remark 2.15.

Definition 2.17. *Let $\mathcal{L}$ be an ASM vocabulary. Two ASM $\mathcal{L}$-programs P, Q are equivalent if, for every $\mathcal{L}$-initialization map ξ and every ξ-initial state $\mathcal{J}$, the two ASMs $(\mathcal{L}, P, (\xi, \mathcal{J}))$ and $(\mathcal{L}, Q, (\xi, \mathcal{J}))$ have exactly the same runs.*

Theorem 2.18 (Gurevich, 1999 [16]). *Every ASM program is equivalent to a program which is a parallel block of conditional blocks of updates, halt or fail instructions, namely a program of the form:*

$$\left|\begin{array}{l} \texttt{if } C_1 \texttt{ then } \left|\begin{array}{l} I_{1,1} \\ \vdots \\ I_{1,p_1} \end{array}\right. \\ \vdots \\ \texttt{if } C_n \texttt{ then } \left|\begin{array}{l} I_{n,1} \\ \vdots \\ I_{n,p_n} \end{array}\right. \end{array}\right.$$

where the $I_{i,j}$'s are updates or `Halt` *or* `Fail` *and the interpretations of $C_1, \ldots, C_n$ in any state are Booleans such that at most one of them is* `True`.

Proof. For `Skip`, `Halt`, `Fail` consider an empty parallel block. For an update or `Halt` or `Fail` consider a block of one conditional with a tautological condition. Simple Boolean conjunctions allow to transform a conditional of two programs of the wanted form into the wanted form. The same for parallel blocks of such programs. □

3 Lambda Calculus

As much as possible, our notations are taken from Barendregt's book [3] (which is a standard reference on Λ-calculus).

3.1 Lambda Terms

Recall that the family Λ of λ-terms of the Λ-calculus is constructed from an infinite family of variables via the following rules:

1. Any variable is a λ-term.
2. *(Abstraction)* If x is a variable and M is a λ-term then $\lambda x \,.\, M$ is a λ-term.
3. *(Application)* If M, N are λ-terms then $(M\ N)$ is a λ-term.

Free and bound occurrences of a variable in a λ-term are defined as in logical formulas, considering that abstraction $\lambda x \,.\, M$ bounds x in M.

One considers λ-terms up to a renaming (called α-conversion) of their bound variables. In particular, one can always suppose that, within a λ-term, no variable has both free occurrences and bound occurrences and that any two abstractions involve distinct variables.

To simplify notations, it is usual to remove parentheses in terms, according to the following conventions:

- applications associate leftwards: in place of $(\cdots((N_1\ N_2)\ N_3)\cdots\ N_k)$ we write $N_1 N_2 N_3 \cdots N_k$,
- abstractions associate rightwards: $\lambda x_1 . (\lambda x_2 . (\cdots . (\lambda x_k . M)\cdots))$ is written $\lambda x_1 \cdots x_k . M$.

3.2 β-Reduction

Note 3.1. Symbols := are used for updates in ASMs and are also commonly used in Λ-calculus to denote by $M[x := N]$ the substitution of all occurrences of a variable x in a term M by a term N. To avoid any confusion, we shall rather denote such a substitution by $M[N/x]$.

Decorated rules of reduction in Λ-calculus		
(Id) $M \to_0 M$	$(\lambda x.M)\ N \to_1 M[N/x]$ (β)	
(App) $\dfrac{M \to_i M'}{MN \to_i M'N}$	$\dfrac{N \to_i N'}{MN \to_i MN'}$	$\dfrac{M \to_i M'}{(\lambda x.M) \to_i \lambda x.M'}$ (Abs)

Fig. 2. Reductions with decorations

The family of λ-terms is endowed with a *reducibility relation*, called β-reduction and denoted by $\to$.

Definition 3.2. *1. Let P be a λ-term. A subterm of P the form $(\lambda x.M)N$ is called a β-redex (or simply redex) of P. Going from P to the λ-term Q obtained by substituting in P this redex by $M[N/x]$ (i.e., substituting N to every free occurrence of x in M) is called a β-reduction and we write $P \to Q$.*

2. The iterations $\to_i$ of $\to$ and the reflexive and transitive closure $\twoheadrightarrow$ are defined as follows:

$$
\begin{aligned}
\to_0 &= \{(M, M) \mid M\} \\
\to_{i+1} &= \to_i \circ \to \qquad \textit{(so that } \to\ =\ \to_1\textit{)} \\
&= \{(M_0, M_i) \mid \exists M_1, \ldots, M_i \mid M_0 \to M_1 \to \cdots \to M_i \to M_{i+1}\} \\
\twoheadrightarrow &= \textstyle\bigcup_{i\in\mathbb{N}} \to_i
\end{aligned}
$$

These reduction relations are conveniently expressed via axioms and rules (cf. Figure 1): the schema of axioms (β) gives the core transformation whereas rules (App) and (Abs) insure that this can be done for subterms.

Relations $\to_i$ are of particular interest to analyse the complexity of the simulation of one ASM step in Λ-calculus. Observe that axioms and rules for $\to$ extend to $\twoheadrightarrow$.

3.3 Normal Forms

Definition 3.3. *1. A λ-term M is in normal form if it contains no redex.*

2. A λ-term M has a normal form if there exists some term N in normal form such that $M \twoheadrightarrow N$.

Remark 3.4. There are terms with no normal form. The classical example is $\Omega = \Delta\Delta$ where $\Delta = \lambda x \,.\, xx$. Indeed, Ω is a redex and reduces to itself.

In a λ-term, there can be several subterms which are redexes, so that iterating $\rightarrow$ reductions is a highly non deterministic process. Nevertheless, going to normal form is a functional process.

Theorem 3.5 (Church-Rosser [7], 1936). *The relation* $\twoheadrightarrow$ *is confluent: if* $M \twoheadrightarrow N'$ *and* $M \twoheadrightarrow N''$ *then there exists* P *such that* $N' \twoheadrightarrow P$ *and* $N'' \twoheadrightarrow P$*. In particular, there exists at most one term* N *in normal form such that* $M \twoheadrightarrow N$.

Remark 3.6. Theorem 3.5 deals with $\twoheadrightarrow$ exclusively: relation $\rightarrow_i$ is *not* confluent for any $i \geq 1$.

A second fundamental property is that going to normal form can be made a deterministic process.

Definition 3.7. *Let* R', R'' *be two occurrences of redexes in a term* P*. We say that* R' *is left to* R'' *if the first lambda in* R' *is left to the first lambda in* R'' *(all this viewed in* P*). If terms are seen as labelled ordered trees, this means that the top lambda in* R' *is smaller than that in* R'' *relative to the prefix ordering on nodes of the tree* P.

Theorem 3.8 (Curry & Feys [9], 1958). *Reducing the leftmost redex of terms not in normal form is a deterministic strategy which leads to the normal form if there is some.*

In other words, if M *has a normal form* N *then the sequence* $M = M_0 \rightarrow M_1 \rightarrow M_2 \rightarrow \cdots$ *where each reduction* $M_i \rightarrow M_{i+1}$ *reduces the leftmost redex in* M_i *(if* M_i *is not in normal form) is necessarily finite and ends with* N.

3.4 Lists in Λ-Calculus

We recall the usual representation of lists in Λ-calculus with special attention to decoration (i.e., the number of β reductions in sequences of reductions).

Proposition 3.9. *Let* $\langle u_1, \ldots, u_k \rangle = \lambda z \,.\, zu_1 \ldots u_k$ *and, for* $i = 1, \ldots, k$*, let* $\pi_i^k = \lambda x_1 \ldots x_k \,.\, x_i$*. Then* $\langle u_1, \ldots, u_k \rangle\ \pi_i^k \rightarrow_{1+k} u_i$.

Moreover, if all u_i*'s are in normal form then so is* $\langle u_1, \ldots, u_k \rangle$ *and these reductions are deterministic: there exists a unique sequence of reductions from* $\langle u_1, \ldots, u_k \rangle$ *to* u_i.

3.5 Booleans in Λ-Calculus

We recall the usual representation of Booleans in Λ-calculus.

Proposition 3.10. *Boolean elements* True, False *and usual Boolean functions can be represented by the following* λ*-terms, all in normal form:*

$$\begin{array}{|rl|}\hline \ulcorner \mathtt{True} \urcorner = \lambda xy.x & \begin{array}{rl} \mathtt{neg} = & \lambda x \,.\, x \ulcorner \mathtt{False} \urcorner \ulcorner \mathtt{True} \urcorner \\ \mathtt{and} = & \lambda xy \,.\, xy \ulcorner \mathtt{False} \urcorner \\ \mathtt{or} = & \lambda xy \,.\, x \ulcorner \mathtt{True} \urcorner\, y \\ \mathtt{implies} = & \lambda xy \,.\, xy \ulcorner \mathtt{True} \urcorner \\ \mathtt{iff} = & \lambda xy \,.\, xy(\ulcorner \neg \urcorner y) \end{array} \\ \ulcorner \mathtt{False} \urcorner = \lambda xy.y & \\ \hline \end{array}$$

For $a, b \in \{\mathtt{True}, \mathtt{False}\}$, *we have* neg $\ulcorner a \urcorner \to \ulcorner \neg a \urcorner$, and $\ulcorner a \urcorner \ulcorner b \urcorner \twoheadrightarrow \ulcorner afb \urcorner, \ldots$.

Proposition 3.11 (If Then Else). *For all terms* M, N,

$$(\lambda z \,.\, zMN) \ulcorner \mathtt{True} \urcorner \to_2 M \quad , \quad (\lambda z \,.\, zMN) \ulcorner \mathtt{False} \urcorner \to_2 N \,.$$

We shall use the following version of iterated conditional.

Proposition 3.12. *For every* $n \geq 1$ *there exists a term* $Case_n$ *such that, for all normal terms* $M_1, \ldots, M_n$ *and all* $t_1, \ldots, t_n \in \{\ulcorner \mathtt{True} \urcorner, \ulcorner \mathtt{False} \urcorner\}$,

$$Case_n\ M_1 \ldots M_n\ t_1 \ldots t_n \to_{3n} M_i$$

relative to leftmost reduction in case $t_i = \ulcorner \mathtt{True} \urcorner$ *and* $\forall j < i\ t_j = \ulcorner \mathtt{False} \urcorner$.

Proof. Let $u_i = y_i(\lambda x_{i+1} \,.\, I) \ldots (\lambda x_n \,.\, I)$, set

$$Case_n = \lambda y_1 \ldots y_n z_1 \ldots z_n \,.\, z_1 u_1 (z_2 u_2 (\ldots (z_{n-1} u_{n-1} (z_n u_n I)) \ldots))$$

and observe that, for leftmost reduction, letting $M'_i = u_i[M_i/y_i]$,

$$\begin{array}{rl} Case_n\ M_1 \ldots M_n\ t_1 \ldots t_n & \to_{2n}\ t_1 M'_1 (t_2 M'_2 (\ldots (t_{n-1} M'_{n-1} (t_n M'_n I)) \ldots)) \\ & \to_i\ M'_i \\ & \to_{n-i}\ M_i . \end{array}$$

□

3.6 Integers in Λ-Calculus

There are several common representations of integers in Λ-calculus. We shall consider a slight variant of the standard one (we choose another term for $\ulcorner 0 \urcorner$), again with special attention to decoration.

Proposition 3.13. *Let*

$$\begin{array}{|rll|}\hline \ulcorner 0 \urcorner = & \lambda z \,.\, z \ulcorner \mathtt{True} \urcorner \ulcorner \mathtt{False} \urcorner & \\ \ulcorner n+1 \urcorner = & \langle \ulcorner \mathtt{False} \urcorner, \ulcorner n \urcorner \rangle & = \lambda z \,.\, z \ulcorner \mathtt{False} \urcorner \ulcorner n \urcorner \\ Zero = & \lambda x \,.\, x \ulcorner \mathtt{True} \urcorner & \\ Succ = & \lambda z \,.\, \langle \ulcorner \mathtt{False} \urcorner, z \rangle & \\ Pred = & \lambda z \,.\, x \ulcorner \mathtt{False} \urcorner & \\ \hline \end{array}$$

The above terms are all in normal form and

$$\begin{array}{r} Zero \ulcorner 0 \urcorner \to_3 \ulcorner \mathtt{True} \urcorner \\ Zero \ulcorner n+1 \urcorner \to_3 \ulcorner \mathtt{False} \urcorner \end{array} \quad , \quad \begin{array}{r} Succ \ulcorner n \urcorner \to_3 \ulcorner n+1 \urcorner \\ Pred \ulcorner n+1 \urcorner \to_3 \ulcorner n \urcorner \\ Pred \ulcorner 0 \urcorner \to_3 \ulcorner \mathtt{False} \urcorner \end{array} .$$

Moreover, all these reductions are deterministic.

Remark 3.14. The standard definition sets $\ulcorner 0 \urcorner = \lambda x \,.\, x$. Observe that $Zero(\lambda x \,.\, x) \to_2 \ulcorner \mathtt{True} \urcorner$. The chosen variant of $\ulcorner 0 \urcorner$ is to get the same decoration (namely 3) to go from $Zero\ulcorner 0 \urcorner$ to $\ulcorner \mathtt{True} \urcorner$ and to go from $Zero\ulcorner n+1 \urcorner$ to $\ulcorner \mathtt{False} \urcorner$.

Let us recall Kleene's fundamental result.

Theorem 3.15 (Kleene, 1936). *For every partial computable function $f : \mathbb{N}^k \to \mathbb{N}$ there exists a λ-term M such that, for every tuple $(n_1, \cdots, n_k)$,*

- *$M\ulcorner n_1 \urcorner \cdots \ulcorner n_k \urcorner$ admits a normal form (i.e., is $\twoheadrightarrow$ reducible to a term in normal form) if and only if $(n_1, \cdots, n_k)$ is in the domain of f,*
- *in that case, $M\ulcorner n_1 \urcorner \cdots \ulcorner n_k \urcorner \twoheadrightarrow \ulcorner f(n_1, \cdots, n_k) \urcorner$ (and, by Theorem 3.5, this normal form is unique).*

3.7 Datatypes in Λ-Calculus

We just recalled some representations of Booleans and integers in Λ-calculus. In fact, any inductive datatype can also be represented. Using computable quotienting, this allows to also represent any datatype used in algorithms.

Though we will not extend on this topic, let us recall Scott encoding of inductive datatypes in the Λ-calculus (cf. Mogensen [23]).

1. If the inductive datatype has constructors $\psi_1, \ldots, \psi_p$ having arities $k_1, \ldots, k_p$, constructor ψ_i is represented by the term

$$\lambda x_1 \ldots x_{k_i} \alpha_1 \ldots \alpha_p \,.\, \alpha_i x_1 \ldots x_{k_i} \,.$$

In particular, if ψ_i is a generator (i.e., an arity 0 constructor) then it is represented by the projection term $\lambda \alpha_1 \ldots \alpha_p \,.\, \alpha_i$.
2. An element of the inductive datatype is a composition of the constructors and is represented by the similar composition of the associated λ-terms.

Extending the notations used for Booleans and integers, we shall also denote by $\ulcorner a \urcorner$ the λ-term representing an element a of a datatype.

Scott's representation of inductive datatypes extends to finite families of datatypes defined via mutual inductive definitions. It suffices to endow constructors with types and to restrict compositions in point 2 above to those respecting constructor types.

3.8 Lambda Calculus with Benign Constants

We consider an extension of the lambda calculus with constants to represent particular computable functions and predicates. Contrary to many $\lambda\delta$-calculi (Church $\lambda\delta$-calculus, 1941 [8], Statman, 2000 [26], Ronchi Della Rocca, 2004 [25], Barendregt & Statman, 2005 [4]), this adds no real additional power: it essentially allows for shortcuts in sequences of reductions. The reason is that axioms in Definition 3.16 do not apply to all terms but only to codes of elements in datatypes.

Definition 3.16. *Let $\mathbb{F}$ be a family of functions with any arities over some datatypes $A_1, \ldots, A_n$. The $\Lambda_{\mathbb{F}}$-calculus is defined as follows:*

- *The family of $\lambda_{\mathbb{F}}$-terms is constructed as in §3.1 from the family of variables augmented with constant symbols: one constant c_f for each $f \in \mathbb{F}$.*
- *The axioms and rules of the top table of Figure 2 are augmented with the following axioms: if $f : A_{i_1} \times \cdots \times A_{i_k} \to A_i$ is in $\mathbb{F}$ then, for all $(a_1, \cdots, a_k) \in A_{i_1} \times \cdots \times A_{i_k}$,*

$$(Ax_f) \qquad c_f \ulcorner a_1 \urcorner \cdots \ulcorner a_k \urcorner \to \ulcorner f(a_1, \cdots, a_k) \urcorner .$$

Definition 3.17. *1. We denote by $\to_\beta$ the classical β-reduction (with the contextual rules (Abs), (App)) extended to terms of $\Lambda_{\mathbb{F}}$.*
2. We denote by $\to_{\mathbb{F}}$ the reduction given by the sole (Ax_f)-axioms and the contextual rules (Abs), (App).
3. We use double decorations: $M \to_{i,j} N$ means that there is a sequence consisting of i β-reductions and j $\mathbb{F}$-reductions which goes from t to u.

The Church-Rosser property still holds.

Proposition 3.18. *The $\Lambda_{\mathbb{F}}$-calculus is confluent (cf. Theorem 3.5).*

Proof. Theorem 3.5 insures that $\twoheadrightarrow_\beta$ is confluent. It is immediate to see that any two applications of the $\mathbb{F}$ axioms can be permuted: this is because two distinct $\mathbb{F}$-redexes in a term are always disjoint subterms. Hence $\to_{\mathbb{F}}$ is confluent. Observe that $\twoheadrightarrow$ is obtained by iterating finitely many times the relation $\twoheadrightarrow_\beta \cup \to_{\mathbb{F}}$. Using Hindley-Rosen Lemma (cf. Barendregt's book [3], Proposition 3.3.5, or Hankin's book [19], Lemma 3.27), to prove that $\twoheadrightarrow$ is confluent, it suffices to prove that $\twoheadrightarrow_\beta$ and $\to_{\mathbb{F}}$ commute. One easily reduces to prove that $\to_\beta$ and $\to_{\mathbb{F}}$ commute, i.e.,

$$\exists P\ (M \to_\beta P \to_{\mathbb{F}} N) \iff \exists Q\ (M \to_{\mathbb{F}} Q \to_\beta N) .$$

Any length two such sequence of reductions involves two redexes in the term M: a β-redex $R = (\lambda x \,.\, A)B$ and a $\mathbb{F}$-redex $C = c \ulcorner a_1 \urcorner \cdots \ulcorner a_k \urcorner$. There are three cases: either R and C are disjoint subterms of M or C is a subterm of A or C is a subterm of B. Each of these cases is straightforward. □

We adapt the notion of leftmost reduction in the $\Lambda_{\mathbb{F}}$-calculus as follows.

Definition 3.19. *The leftmost reduction in $\Lambda_{\mathbb{F}}$ reduces the leftmost $\mathbb{F}$-redex if there is some else it reduces the leftmost β-redex.*

3.9 Good $\mathbb{F}$-Terms

To functions which can be obtained by composition from functions in $\mathbb{F}$ we associate canonical terms in $\Lambda_{\mathbb{F}}$ and datatypes. These canonical terms are called good $\mathbb{F}$-terms, they contain no abstraction, only constant symbols c_f, with $f \in \mathbb{F}$, and variables.

Problem 3.20. We face a small problem. Functions in $\mathbb{F}$ are to represent static functions of an ASM. Such functions are typed whereas $\Lambda_{\mathbb{F}}$ is an untyped lambda calculus. In order to respect types when dealing with composition of functions in $\mathbb{F}$, the definition of good $\mathbb{F}$-terms is done in two steps: the first step involves typed variables and the second one replaces them by untyped variables.

Definition 3.21. *1. Let $A_1, \ldots, A_n$ be the datatypes involved in functions of the family $\mathbb{F}$. Consider typed variables $x_j^{A_i}$ where $j \in \mathbb{N}$ and $i = 1, \ldots, n$. The family of pattern $\mathbb{F}$-terms, their types and semantics are defined as follows: Let $f \in \mathbb{F}$ be such that $f : A_{i_1} \times \cdots \times A_{i_k} \to A_q$.*

- *If $x_{j_1}^{A_{i_1}}, \ldots, x_{j_k}^{A_{i_k}}$ are typed variables then the term $c_f\ x_{j_1}^{A_{i_1}} \ldots x_{j_k}^{A_{i_k}}$ is a pattern $\mathbb{F}$-term with type $A_{i_1} \times \cdots \times A_{i_k} \to A_q$ and semantics $[\![c_f\ x_{j_1}^{A_{i_1}} \ldots x_{j_k}^{A_{i_k}}]\!] = f$.*
- *For $j = 1, \ldots, k$, let t_j be a pattern $\mathbb{F}$-term with datatype A_j or a typed variable $x_i^{A_j}$. Suppose the term $t = c_f\ t_1 \cdots t_k$ contains exactly the typed variables $x_j^{A_i}$ for $(i,j) \in I$ and, for $\ell = 1, \ldots, k$, the term t_ℓ contains exactly the typed variables $x_j^{A_i}$ for $(i,j) \in I_j \subseteq I$.*
  Then the term $c_f\ t_1 \cdots t_k$ is a pattern $\mathbb{F}$-term with type $\prod_{i \in I} A_i \to A_q$ and a semantics $[\![c_f\ t_1 \cdots t_k]\!]$ such that, for every tuple $(a_i)_{i \in I} \in \prod_{i \in I} A_i$,

$$[\![t]\!]((a_i)_{i \in I}) = f([\![t_1]\!]((a_i)_{i \in I_1})), \ldots, [\![t_k]\!]((a_k)_{i \in I_k}))) \ .$$

2. Good $\mathbb{F}$-terms are obtained by substituting in a pattern $\mathbb{F}$-term untyped variables to the typed variables so that two distinct typed variables are substituted by two distinct untyped variables.

The semantics of good $\mathbb{F}$-terms is best illustrated by the following example: the function h associated to the term $c_g(c_h y)x(c_g zzx)$ is the one given by equality $f(x, y, z) = g(h(y), x, g(z, z, x))$ which corresponds to Figure 3.9.

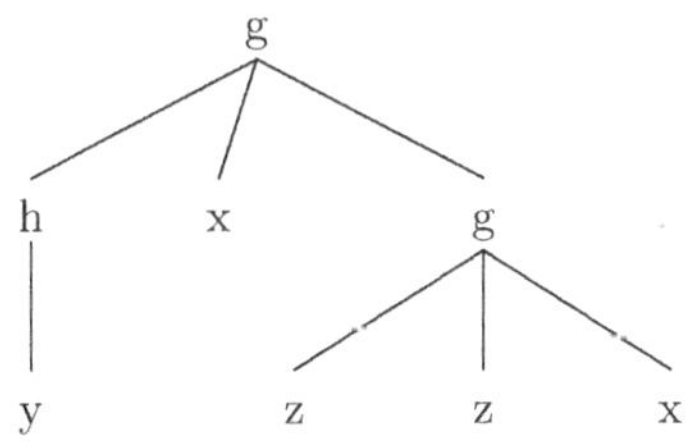

Fig. 3. Composition tree

The reason for the above definition is the following simple result about reductions of good terms obtained via substitutions. It is proved via a straightforward induction on good $\mathbb{F}$-terms and will be used in §4.3, 4.4

Proposition 3.22. *Let t be a good $\mathbb{F}$-term with k variables $y_1, \ldots, y_k$ such that $[\![t]\!] = f : A_{i_1} \times \cdots \times A_{i_k} \to A_q$. Let N be the number of nodes of the tree associated to the composition of functions in $\mathbb{F}$ giving f (cf. Figure 3.9).*

There exists $L_t = O(N)$ such that, for every $(a_1, \ldots, a_k) \in A_{i_1} \times \cdots \times A_{i_k}$,

$$t[\ulcorner a_1 \urcorner / y_1, \ldots, \ulcorner a_k \urcorner / y_k] \twoheadrightarrow_{\mathbb{F}} \ulcorner f(a_1, \ldots, a_k) \urcorner$$

and, using the leftmost reduction strategy, this sequence of reductions consists of exactly L_t $\mathbb{F}$-reductions.

4 Variations on Curry's Fixed Point

4.1 Curry's Fixed Point

Let us recall Curry's fixed point.

Definition 4.1. *The Curry operator $\varphi \mapsto \theta_\varphi$ on λ-terms is defined as follows*

$$\theta_F = (\lambda x \,.\, F(xx))(\lambda x \,.\, F(xx)) \ .$$

Theorem 4.2 (Curry's fixed point). *For every λ-term F, $\theta_F \to F\theta_F$.*

Proof. One β-reduction suffices: θ_F is of the form XX and is itself a redex (since X is an abstraction) which β-reduces to $F(XX)$, i.e., to $F\theta_F$. □

4.2 Padding Reductions

We show how to pad leftmost reduction sequences so as to get prescribed numbers of β and $\mathbb{F}$-reductions.

Lemma 4.3 (Padding lemma). *Suppose that $\mathbb{F}$ contains some function $\omega : B_1 \times \cdots \times B_\ell \to B_i$ (with $1 \leq i \leq \ell$) and some constants $\nu_1 \in B_1, \ldots, \nu_\ell \in B_\ell$.*
1. For every $K \geq 2$ and $L \geq 0$, there exists a λ-term $pad_{K,L}$ in $\Lambda_{\mathbb{F}}$ with length $O(K+L)$ such that, for any finite sequence of λ-terms $\theta, t_1, \ldots, t_k$ in $\Lambda_{\mathbb{F}}$ which contain no $\mathbb{F}$-redex,

i. $pad_{K,L}\ \theta\ t_1 \cdots t_k \ \twoheadrightarrow\ \theta\ t_1 \cdots t_k$.
ii. The leftmost derivation consists of exactly L $\mathbb{F}$-reductions followed by K β-reductions.

2. Moreover, if $K \geq 3$, one can also suppose that $pad_{K,L}$ contains no $\mathbb{F}$-redex.

Proof. 1. For the sake of simplicity, we suppose that ω has arity 1, the general case being a straightforward extension. Let $I = \lambda x \,.\, x$ and $I^\ell = I \cdots I$ (ℓ times I). Observe that $I^\ell\ s_0 \cdots s_p \ \twoheadrightarrow\ s_0 \cdots s_p$ and the leftmost derivation consists of exactly ℓ β-reductions. So it suffices to set $pad_{K,0} = I^K$ and, for $L \geq 1$,

$$pad_{K,L} = I^{K-2}\ (\lambda xy \,.\, y)\ (\overbrace{\ulcorner\omega\urcorner(\ldots(\ulcorner\omega\urcorner}^{L \text{ times}}\ulcorner\nu_1\urcorner)\ldots)) \ .$$

2. To suppress the $\mathbb{F}$-redex $\ulcorner\omega\urcorner\ulcorner\nu_1\urcorner$, modify $pad_{K,L}$ as follows:

$$pad_{K,L} = I^{K-3}\ (\lambda xy \,.\, xy)\ ((\lambda z \,.\, (\overbrace{\ulcorner\omega\urcorner(\ldots(\ulcorner\omega\urcorner}^{L \text{ times}} z)\ldots)))\ \ulcorner\nu_1\urcorner) \ .$$

□

4.3 Constant Cost Updates

We use Curry's fixed point Theorem and the above padding technique to insure constant length reductions for any given update function for tuples.

Lemma 4.4. *Let $A_1, \ldots, A_n$ be the datatypes involved in functions of the family $\mathbb{F}$. Suppose that $\mathbb{F}$ contains some function $\omega : B_1 \times \cdots \times B_\ell \to B_i$ (with $1 \leq i \leq \ell$) and some constants $\nu_1 \in B_1, \ldots, \nu_\ell \in B_\ell$. Let $\tau : \{1, \ldots, k\} \to \{1, \ldots, n\}$ be a distribution of indexes of sorts. For $j = 1, \ldots, k$, let φ_j be a good $\mathbb{F}$-term with variables x_i for $i \in I_j \subseteq \{1, \ldots, k\}$ such that $[\![\varphi_j]\!] = f_j : \prod_{i \in I_j} A_{\tau(i)} \to A_{\tau(j)}$.*

There exists constants K_{min} and L_{min} such that, for all $K \geq K_{min}$ and $L \geq L_{min}$, there exists a λ-term θ such that,

1. *Using the leftmost reduction strategy, for all $(a_1, \ldots, a_k) \in A_{\tau(i)} \times \cdots \times A_{\tau(k)}$, denoting by $\boldsymbol{a}_I$ the tuple $(a_j)_{j \in I}$,*

$$\theta \ulcorner a_1 \urcorner \cdots \ulcorner a_k \urcorner \twoheadrightarrow \theta \ulcorner f_1(\boldsymbol{a}_{I_1}) \urcorner \cdots \ulcorner f_k(\boldsymbol{a}_{I_k}) \urcorner . \tag{1}$$

2. *This sequence of reductions consists of K β-reductions and L $\mathbb{F}$-reductions.*

Proof. Let K', L' be integers to be fixed later on. Set

$$F = pad_{K',L'} \ \lambda \alpha x_1 \ldots x_k \ . \ \alpha \varphi_1 \ldots \varphi_k \qquad \theta = (\lambda z \ . \ F(zz)) \ (\lambda z \ . \ F(zz)) \ .$$

Since θ and the φ_i's have no $\mathbb{F}$-redex, we have the following leftmost reduction:

$$\begin{array}{lll} \theta \ulcorner a_1 \urcorner \cdots \ulcorner a_k \urcorner & \to_{1,0} & F \ \theta \ulcorner a_1 \urcorner \cdots \ulcorner a_k \urcorner \qquad \text{(cf. Theorem 4.2)} \\ & = & pad_{K',L'} \ (\lambda \alpha x_1 \ldots x_k \ . \ \alpha \varphi_1 \ldots \varphi_k) \ \theta \ulcorner a_1 \urcorner \cdots \ulcorner a_k \urcorner \\ & \to_{K',L'} & (\lambda \alpha x_1 \ldots x_k \ . \ \alpha \varphi_1 \ldots \varphi_k) \ \theta \ulcorner a_1 \urcorner \cdots \ulcorner a_k \urcorner \\ & & \quad \text{(apply Lemma 4.3)} \\ & \to_{k+1,0} & \theta \ \varphi_1[\ulcorner a_1 \urcorner / x_1, \ldots, \ulcorner a_k \urcorner / x_k] \\ & & \qquad \cdots \varphi_k[\ulcorner a_1 \urcorner / x_1, \ldots, \ulcorner a_k \urcorner / x_k] \\ & \to_{0,S} & \theta \ulcorner f_1(\boldsymbol{a}_{I_1}) \urcorner \cdots \ulcorner f_k(\boldsymbol{a}_{I_k}) \urcorner \\ & & \quad \text{(apply Proposition 3.22)} \end{array}$$

where $S = \sum_{j=1,\ldots,k} L_{\varphi_j}$. The total cost is $K' + k + 2$ β-reductions plus $L' + S$ $\mathbb{F}$-reductions. We conclude by setting $K' = K - (k+2)$ and $L' = L - S$. □

4.4 Constant Cost Conditional Updates

We refine Lemma 4.4 to conditional updates.

Lemma 4.5. *Let $A_1, \ldots, A_n$ be the datatypes involved in functions of the family $\mathbb{F}$. Suppose that $\mathbb{F}$ contains some function $\omega : B_1 \times \cdots \times B_\ell \to B_i$ (with $1 \leq i \leq \ell$) and some constants $\nu_1 \in B_1, \ldots, \nu_\ell \in B_\ell$. Let $\tau : \{1, \ldots, k\} \to \{1, \ldots, n\}$, $\iota_1, \ldots, \iota_q \in \{1, \ldots, n\}$ be distributions of indexes of sorts. Let $(\rho_s)_{s=1,\ldots,p+q}$,*

$(\varphi_{i,j})_{i=1,\dots,p,j=1,\dots,k}$, $(\gamma_\ell)_{i=1,\dots,q}$ *be sequences of good* $\mathbb{F}$*-terms with variables* x_i *with* i *varying in the respective sets* $I_s, I_{i,j}, J_\ell \subseteq \{1,\dots,k\}$. *Suppose that*

$$\begin{array}{rl} [\![\rho_s]\!] = r_s & : \prod_{i\in I_s} A_{\tau(i)} \to \mathtt{Bool} \;, \\ [\![\varphi_{i,j}]\!] = f_{i,j} & : \prod_{i\in I_{i,j}} A_{\tau(i)} \to A_{\tau(j)} \;, \\ [\![\gamma_\ell]\!] = g_\ell & : \prod_{i\in J_\ell} A_{\tau(i)} \to A_{\iota(\ell)} \end{array}$$

(in particular, $f_{1,j},\dots,f_{p,j}$ *all take values in* $A_{\tau(j)}$*). There exists constants* K_{min} *and* L_{min} *such that, for all* $K \geq K_{min}$ *and* $L \geq L_{min}$*, there exists a* λ*-term* θ *such that,*

1. *Using the leftmost reduction strategy, for all* $(a_1,\dots,a_k) \in A_{\tau(1)} \times \cdots \times A_{\tau(k)}$ *and* $s \in \{1,\dots,p,p+1,\dots,p+q\}$,

$$\text{If} \quad r_s(\boldsymbol{a}_{I_s}) = \mathtt{True} \;\wedge\; \forall t < s \; r_t(\boldsymbol{a}_{I_t}) = \mathtt{False} \qquad (\dagger)_s$$

$$\text{then } \theta \ulcorner a_1 \urcorner \cdots \ulcorner a_k \urcorner \twoheadrightarrow \begin{cases} \theta \ulcorner f_{s,1}(\boldsymbol{a}_{I_{s,1}}) \urcorner \cdots \ulcorner f_{s,k}(\boldsymbol{a}_{I_{s,k}}) \urcorner & \text{if } s \leq p \\ \ulcorner g_\ell(\boldsymbol{a}_{J_\ell}) \urcorner & \text{if } s = p+\ell \end{cases} .$$

2. *In all cases, this sequence of reductions consists of exactly* K β*-reductions and* L $\mathbb{F}$*-reductions.*

Proof. Let K', L' be integers to be fixed at the end of the proof. For $i = 1,\dots,p$ and $\ell = 1,\dots,q$, let

$$M_i = \alpha\varphi_{i,1} \cdots \varphi_{i,k} \qquad M_{p+\ell} = \gamma_\ell \;.$$

Using the $Case_n$ term from Proposition 3.12, set

$$\begin{aligned} H &= Case_{p+q} M_1 \dots M_p M_{p+1} \dots M_{p+q} \\ G &= \lambda\alpha x_1 \dots x_k \,.\, (H\; \rho_1(x_1 \dots x_k) \dots \rho_{p+q}(x_1 \dots x_k)) \\ F &= pad_{K',L'}\; G \\ \theta &= (\lambda z \,.\, F(zz))\; (\lambda z \,.\, F(zz)) \end{aligned}$$

The following sequence of reductions is leftmost because, as long as $pad_{K',L'}$ is not completely reduced, there is no $\mathbb{F}$-redex on its right.

$$(R_1) \quad \begin{array}{rll} \theta \ulcorner a_1 \urcorner \cdots \ulcorner a_k \urcorner \to_{1,0} & F\, \theta \ulcorner a_1 \urcorner \cdots \ulcorner a_k \urcorner & \text{(cf. Theorem 4.2)} \\ = & pad_{K',L'}\; G\, \theta \ulcorner a_1 \urcorner \cdots \ulcorner a_k \urcorner & \\ \to_{K',L'} & G\, \theta \ulcorner a_1 \urcorner \cdots \ulcorner a_k \urcorner & \end{array}$$

Let us denote by A^σ the term $A_i^\sigma = A[\theta/\alpha, \ulcorner a_1 \urcorner/x_{,1}, \dots, \ulcorner a_k \urcorner/x_k]$. The leftmost reduction sequence goes on with β-reductions as follows:

$$(R_2) \quad \begin{array}{rl} G\, \theta \ulcorner a_1 \urcorner \cdots \ulcorner a_k \urcorner = & (\lambda\alpha x_1 \dots x_k \,.\, (H\; \rho_1 \dots \rho_{p+q}))\, \theta \ulcorner a_1 \urcorner \cdots \ulcorner a_k \urcorner \\ \to_{k+1,0} & H^\sigma\; \rho_1^\sigma \cdots \rho_{p+q}^\sigma \end{array}$$

Now, using Proposition 3.22, the following leftmost reductions are $\mathbb{F}$-reductions:

$$\begin{array}{rll} \varphi_{i,j}^\sigma & \to_{0,L_{\varphi_{i,j}}} & \ulcorner f_{i,j}(\boldsymbol{a}_{I_{i,j}}) \urcorner \\ M_i^\sigma & \to_{0,\sum_{j=1}^{j=k} L_{\varphi_{i,j}}} & \theta \ulcorner f_{i,1}(\boldsymbol{a}_{I_{i,1}}) \urcorner \dots \ulcorner f_{i,k}(\boldsymbol{a}_{I_{i,k}}) \urcorner \\ M_{p+\ell}^\sigma = \gamma_{p+\ell}^\sigma & \to_{0,L_{\gamma_\ell}} & \ulcorner g_\ell(\boldsymbol{a}_{J_\ell}) \urcorner \\ \rho_s^\sigma & \to_{0,L_{\rho_s}} & \ulcorner r_s(\boldsymbol{a}_{I_s}) \urcorner \end{array}$$

Going with our main leftmost reduction sequence, letting

$$N = (\sum_{i=1}^{i=p}\sum_{j=1}^{j=k} L_{\varphi_{i,j}}) + \sum_{\ell=1}^{\ell=q} L_{\gamma_\ell} + \sum_{s=1}^{s=p+q} L_{\rho_s}$$

and s be as in condition $(\dagger)_s$ in the statement of the Lemma, we get

$$(R_3) \quad \begin{array}{ll} H^\sigma\ \rho_1^\sigma \ldots \rho_{p+q}^\sigma = & Case_{p+q} M_1^\sigma \ldots M_p^\sigma M_{p+1}^\sigma \ldots M_{p+q}^\sigma\ \rho_1^\sigma \ldots \rho_{p+q}^\sigma \\ \to_{0,N} & Case_{p+q} \\ & (\theta \ulcorner f_{1,1}(\boldsymbol{a}_{I_{1,1}})\urcorner \ldots \ulcorner f_{1,k}(\boldsymbol{a}_{I_{1,k}})\urcorner) \\ & \ldots \\ & (\theta \ulcorner f_{p,1}(\boldsymbol{a}_{I_{p,1}})\urcorner \ldots \ulcorner f_{p,k}(\boldsymbol{a}_{I_{p,k}})\urcorner) \\ & (\ulcorner g_1(\boldsymbol{a}_{J_1})\urcorner) \quad \ldots \quad (\ulcorner g_q(\boldsymbol{a}_{J_q})\urcorner) \\ & \rho_1^\sigma \quad \ldots \quad \rho_{p+q}^\sigma \\ \to_{3(p+q),0} & \begin{cases} \theta \ulcorner f_{s,1}(\boldsymbol{a}_{I_{s,1}})\urcorner \ldots \ulcorner f_{s,k}(\boldsymbol{a}_{I_{s,k}})\urcorner & \textit{if } s \leq p \\ \ulcorner g_\ell(\boldsymbol{a}_{J_\ell})\urcorner) & \textit{if } s = p + \ell \end{cases} \end{array}$$

Summing up reductions (R_1), (R_2), (R_3), we see that

$$\theta \ulcorner a_1 \urcorner \ldots \ulcorner a_k \urcorner \to_{\eta,\zeta} \begin{cases} \theta \ulcorner f_{s,1}(\boldsymbol{a}_{I_{s,1}})\urcorner \ldots \ulcorner f_{s,k}(\boldsymbol{a}_{I_{s,k}})\urcorner & \textit{if } s \leq p \\ \ulcorner g_\ell(\boldsymbol{a}_{J_\ell})\urcorner) & \textit{if } s = p + \ell \end{cases}$$

where $\eta = 1 + K' + (k+1) + 3(p+q)$ and $\zeta = L' + N$.
To conclude, set $K_{min} = k + 5 + 3(p+q)$ and $L_{min} = N$. If $K \geq K_{min}$ and $L \geq L_{min}$ it suffices to set $K' = K - (K_{min} - 3)$ and $L' = L - L_{min}$ and to observe that $K' \geq 3$ as needed in Lemma 4.3. □

5 ASMs and Lambda Calculus

All along this section, $\mathcal{S} = (\mathcal{L}, P, (\xi, \mathcal{J}))$ is some fixed ASM (cf. Definition 2.14).

5.1 Datatypes and ASM Base Sets

The definition of ASMs does not put any constraint on the base sets of the multialgebra. However, only elements which can be named are of any use, i.e. elements which are in the range of compositions of (static or dynamic) functions on the ASM at the successive steps of the run.

The following straightforward result formalizes this observation.

Proposition 5.1. *Let $(\mathcal{L}, P, (\xi, \mathcal{J}))$ be an ASM. Let $\mathcal{U}_1, \ldots, \mathcal{U}_n$ be the base sets interpreting the different sorts of this ASM. For $t \in \mathbb{N}$, let $A_1^{(t)} \subseteq \mathcal{U}_1, \ldots, A_n^{(t)} \subseteq \mathcal{U}_n$ be the sets of values of all ground good $\mathbb{F}$-terms (i.e. with no variable) in the t-th successor state $\mathcal{S}_t$ of the initial state $\mathcal{J}$ of the ASM.*

1. For any $t \in \mathbb{N}$, $A_1^{(t)} \supseteq A_1^{(t+1)}, \ldots, A_n^{(t)} \supseteq A_n^{(t+1)}$.

2. $(A_1^{(t)}, \ldots, A_n^{(t)})$ is a submultialgebra of $\mathcal{S}_t$, i.e. it is closed under all static and dynamic functions of the state $\mathcal{S}_t$.

Thus, the program really works only on the elements of the sets $(A_1^{(0)}, \ldots, A_n^{(0)})$ of the initial state which are datatypes defined via mutual inductive definitions using ξ and $\mathcal{J}$.

5.2 Tailoring Lambda Calculus for an ASM

Let $\mathbb{F}$ be the family of interpretations of all static symbols in the initial state. The adequate Lambda calculus to encode the ASM is $\Lambda_{\mathbb{F}}$.

Let us argue that this is not an unfair trick. An algorithm does decompose a task in elementary ones. But "elementary" does not mean "trivial" nor "atomic", it just means that we do not detail how they are performed: they are like oracles. There is no absolute notion of elementary task. It depends on what big task is under investigation. For an algorithm about matrix product, multiplication of integers can be seen as elementary. Thus, algorithms go with oracles.

Exactly the same assumption is done with ASMs: static and input functions are used for free.

5.3 Main Theorem for Type 0 ASMs

We first consider the case of type 0 ASMs.

Theorem 5.2. *Let $(\mathcal{L}, P, (\xi, \mathcal{J}))$ be an ASM with base sets $\mathcal{U}_1, \ldots, \mathcal{U}_n$. Let $A_1, \ldots, A_n$ be the datatypes $A_1^{(0)}, \ldots, A_n^{(0)}$ (cf. Proposition 5.1). Let $\mathbb{F}$ be the family of interpretations of all static symbols of the ASM restricted to the datatypes $A_1, \ldots, A_n$. Suppose all dynamic symbols have arity 0, i.e. all are constants symbols. Suppose these dynamic symbols are $\eta_1, \ldots, \eta_k$. and $\eta_1, \ldots, \eta_\ell$ are the output symbols.*

Let us denote by e_i^t the value of the constant η_i in the t-th successor state $\mathcal{S}_t$ of the initial state $\mathcal{J}$.

There exists K_0 such that, for every $K \geq K_0$, there exists a λ-term θ in $\Lambda_{\mathbb{F}}$ such that, for all initial values $e_1^0, \ldots, e_k^0$ of the dynamic constants and for all $t \geq 1$,

$$\begin{array}{lll} \theta \ulcorner e_1^0 \urcorner \ldots \ulcorner e_k^0 \urcorner \rightarrow_{Kt} \theta \ulcorner e_1^t \urcorner \ldots \ulcorner e_k^t \urcorner & \left\{ \begin{array}{l} \textit{if the run does not haltnor fail nor clash} \\ \textit{for steps} \leq t \end{array} \right. \\ \theta \ulcorner e_1^0 \urcorner \ldots \ulcorner e_k^0 \urcorner \rightarrow_{Ks} \langle \ulcorner 1 \urcorner, \ulcorner e_1^s \urcorner \ldots \ulcorner e_\ell^s \urcorner \rangle & \textit{if the run halts at step } s \leq t \\ \theta \ulcorner e_1^0 \urcorner \ldots \ulcorner e_k^0 \urcorner \rightarrow_{Ks} \ulcorner 2 \urcorner & \textit{if the run fails at step } s \leq t \\ \theta \ulcorner e_1^0 \urcorner \ldots \ulcorner e_k^0 \urcorner \rightarrow_{Ks} \ulcorner 3 \urcorner & \textit{if the run clashes at step } s \leq t \end{array}$$

Thus, groups of K successive reductions simulate in a simple way the successive states of the ASM, and give the output in due time when it is defined.

Proof. Use Theorem 2.18 to normalize the program P. We stick to the notations of that Theorem. Since there is no dynamic function, only dynamic constants, the ASM terms C_i and $I_{i,j}$ name the result of applying to the dynamic constants a composition of the static functions (including static constants). Thus, one can associate good $\mathbb{F}$-terms $\rho_i, \varphi_{i,j}$ to these compositions.

Observe that one can decide if the program halts or fails or clashes via some composition of functions in $\mathbb{F}$ (use the static equality function which has been assumed, cf. Definition 2.14). So enter negative answers to these decisions in the existing conditions $C_1, \ldots, C_n$. Also, add three more conditions to deal with the positive answers to these decisions. These three last conditions are associated to terms $\gamma_1, \gamma_2, \gamma_3$. Finally, apply Lemma 4.5 (with $p = n$ and $q = 3$). □

Remark 5.3. A simple count in the proof of Lemma 4.5 allows to bound K_0 as follows: $K_0 = O((\text{size of } P)^2)$.

5.4 Main Theorem for All ASMs

Let ψ be a dynamic symbol. Its initial interpretation $\psi_{\mathcal{S}_0}$ is given by a composition of the static objects (cf. Definition 2.5) hence it is available in each successor state of the initial state. In subsequent states $\mathcal{S}_t$, its interpretation $\psi_{\mathcal{S}_t}$ is different but remains almost equal to $\psi_{\mathcal{S}_0}$: the two differ only on finitely many tuples. This is so because, at each step, any dynamic symbol is modified on at most N tuples where N depends on the program. Let $\Delta\psi$ be a list of all tuples on which $\psi_{\mathcal{S}_0}$ has been modified. What can be done with ψ can also be done with $\psi_{\mathcal{S}_0}$ and $\Delta\psi$. Since $\psi_{\mathcal{S}_0}$ is available in each successor state of the initial state, we are going to encode $\Delta\psi_{\mathcal{S}_t}$ rather than $\psi_{\mathcal{S}_t}$. Now, $\Delta\psi_{\mathcal{S}_t}$ is a list and we need to access in constant time any element of the list. And we also need to manage the growth of the list.

This is not possible in constant time with the usual encodings of datatypes in Lambda calculus. So the solution is to make $\Lambda_{\mathbb{F}}$ bigger: put new constant symbols to represent lists and allow new $\mathbb{F}$-reduction axioms to get *in one step* the needed information on lists.

Now, is this fair? We think it is as regards simulation of ASMs. In ASM theory, one application of the program is done in one unit of time though it involves a lot of things to do. In particular, one can get in one unit of time all needed information about the values of static or dynamic functions on the tuples named by the ASM program. What we propose to do with the increase of $\Lambda_{\mathbb{F}}$ is just to get more power, as ASMs do on their side.

Definition 5.4. *Let $A_1, \ldots, A_n$ be the datatypes involved in functions of $\mathbb{F}$. If $\varepsilon = (i_1, \ldots, i_m, i)$ is an $(m+1)$-tuple of elements in $\{1, \ldots, n\}$, we let L_ε be the datatype of finite sequences of $(m+1)$-tuples in $A_{i_1} \times \cdots \times A_{i_m} \times A_i$.*

Let E be a family of tuples of elements of $\{1, \ldots, n\}$. The Lambda calculus $\Lambda_{\mathbb{F}}^E$ is obtained by adding to $\Lambda_{\mathbb{F}}$ families of symbols

$$(F_\varepsilon, B_\varepsilon, V_\varepsilon, Add_\varepsilon, Del_\varepsilon)_{\varepsilon \in E}$$

and the axioms associated to the following intuitions. For $\varepsilon = (i_1, \ldots, i_m, i)$,

i. Symbol F_ε is to represent the function $L_\varepsilon \to$ `Bool` such that, for $\sigma \in L_\varepsilon$, $F_\varepsilon(\sigma)$ is `True` if and only if σ is functional in its first m components. In other words, F_ε checks if any two distinct sequences in σ always differ on their first m components.

ii. Symbol B_ε is to represent the function $L_\varepsilon \times (A_{i_1} \times \cdots \times A_{i_m}) \to$ `Bool` such that, for $\sigma \in L_\varepsilon$ and $\boldsymbol{a} \in A_{i_1} \times \cdots \times A_{i_m}$, $B_\varepsilon(\sigma, \boldsymbol{a})$ is `True` if and only if $\boldsymbol{a}$ is a prefix of some $(m+1)$-tuple in the finite sequence σ.

iii. Symbol V_ε is to represent the function $L_\varepsilon \times (A_{i_1} \times \cdots \times A_{i_m}) \to A_i$ such that, for $\sigma \in L_\varepsilon$ and $\boldsymbol{a} \in A_{i_1} \times \cdots \times A_{i_m}$,
- $V_\varepsilon(\sigma, \boldsymbol{a})$ is defined if and only if $F_\varepsilon(\sigma) =$ `True` and $B_\varepsilon(\sigma, \boldsymbol{a}) =$ `True`,

- *when defined,* $V_\varepsilon(\sigma, \boldsymbol{a})$ *is the last component of the unique* $(m+1)$*-tuple in the finite sequence* σ *which extends the* m*-tuple* $\boldsymbol{a}$.

iv. *Symbol* Add_ε *is to represent the function* $L_\varepsilon \times (A_{i_1} \times \cdots \times A_{i_m} \times A_i) \to L_\varepsilon$ *such that, for* $\sigma \in L_\varepsilon$ *and* $\boldsymbol{a} \in A_{i_1} \times \cdots \times A_{i_m} \times A_i$, $Add_\varepsilon(\sigma, \boldsymbol{a})$ *is obtained by adding the tuple* $\boldsymbol{a}$ *as last element in the finite sequence* σ.

v. *Symbol* Del_ε *is to represent the function* $L_\varepsilon \times (A_{i_1} \times \cdots \times A_{i_m} \times A_i) \to L_\varepsilon$ *such that, for* $\sigma \in L_\varepsilon$ *and* $\boldsymbol{a} \in A_{i_1} \times \cdots \times A_{i_m} \times A_i$, $Del_\varepsilon(\sigma, \boldsymbol{a})$ *is obtained by deleting all occurrences of the tuple* $\boldsymbol{a}$ *in the finite sequence* σ.

Now, we can extend Theorem 5.2.

Theorem 5.5. *Let* $(\mathcal{L}, P, (\xi, \mathcal{J}))$ *be an ASM with base sets* $\mathcal{U}_1, \ldots, \mathcal{U}_n$. *Let* $A_1, \ldots, A_n$ *be the datatypes* $A_1^{(0)}, \ldots, A_n^{(0)}$ *(cf. Proposition 5.1). Let* $\mathbb{F}$ *be the family of interpretations of all static symbols of the ASM restricted to the datatypes* $A_1, \ldots, A_n$. *Let* $\eta_1, \ldots, \eta_k$ *be the dynamic symbols of the ASM. Suppose* η_i *has type* $\mathcal{U}_{\tau(i,1)} \times \cdots \times \mathcal{U}_{\tau(i,p_i)} \to \mathcal{U}_{q_i}$ *for* $i = 1, \ldots, k$.
Set $E = \{(\tau(i,1), \ldots, \tau(i,p_i), q_i) \mid i = 1, \ldots, k\}$.
The conclusion of Theorem 5.2 is still valid in the Lambda calculus $\Lambda_{\mathbb{F}}^{E}$ *with the following modification:*

> e_i^t *is the list of* $p_i + 1$*-tuples describing the differences between the interpretations of* $(\eta_i)_{\mathcal{S}_0}$ *and* $(\eta_i)_{\mathcal{S}_t}$.

References

1. Atrubin, A.J.: A One-Dimensional Real-Time Iterative Multiplier. Trans. on Electronic Computers EC 14(3), 394–399 (1965)
2. Börger, E., Stärk, R.: Abstract State Machines: A Method for High-Level System Design and Analysis. Springer, Heidelberg (2003)
3. Barendregt, H.P.: The Lambda calculus. Its syntax and semantics. North-Holland, Amsterdam (1984)
4. Barendregt, H., Statman, R.: Böhm's Theorem, Church's Delta, Numeral Systems, and Ershov Morphisms. In: Middeldorp, A., van Oostrom, V., van Raamsdonk, F., de Vrijer, R. (eds.) Processes, Terms and Cycles: Steps on the Road to Infinity. LNCS, vol. 3838, pp. 40–54. Springer, Heidelberg (2005)
5. Biedl, T., Buss, J.F., Demaine, E.D., Demaine, M.L., Hajiaghayi, M., Vinař, T.: Palindrome recognition using a multidimensional tape. Theoretical Computer Science 302(1-3), 475–480 (2003)
6. Börger, E.: The Origins and the Development of the ASM Method for High Level System Design and Analysis. Journal of Universal Computer Science 8(1), 2–74 (2002)
7. Church, A., Rosser, J.B.: Some properties of conversion. Trans. Amer. Math. Soc. 39, 472–482 (1937)
8. Church, A.: The Calculi of Lambda Conversion. Princeton University Press, Princeton (1941)
9. Curry, H., Feys, R.: Combinatory logic, vol. I. North-Holland, Amsterdam (1958)
10. Dershowitz, N., Gurevich, Y.: A natural axiomatization of computability and proof of Church's Thesis. Bulletin. of Symbolic Logic 14(3), 299–350 (2008)

11. Gurevich, Y.: Reconsidering Turing's Thesis: towards more realistic semantics of programs. Technical Report CRL-TR-38-84, EEC Department. University of Michigan (1984)
12. Gurevich, Y.: A new Thesis. Abstracts, American Math. Soc., Providence (1985)
13. Gurevich, Y.: Evolving Algebras: An Introductory Tutorial. Bulletin of the European Association for Theoretical Computer Science 43, 264–284 (1991); Reprinted in Current Trends in Theoretical Computer Science, pp. 266–269. World Scientific, Singapore (1993)
14. Gurevich, Y.: Evolving algebras 1993: Lipari guide. In: Specification and Validation Methods, pp. 9–36. Oxford University Press, Oxford (1995)
15. Gurevich, Y.: May 1997 Draft of the ASM Guide. Tech. Report CSE-TR-336-97, EECS Dept., University of Michigan (1997)
16. Gurevich, Y.: The Sequential ASM Thesis. Bulletin of the European Association for Theoretical Computer Science 67, 93–124 (1999); Reprinted in Current Trends in Theoretical Computer Science, pp. 363–392. World Scientific, Singapore (2001)
17. Gurevich, Y.: Sequential Abstract State Machines capture Sequential Algorithms. ACM Transactions on Computational Logic 1(1), 77–111 (2000)
18. Hennie, F.C.: One-tape off-line Turing machine complexity. Information and Computation 8, 553–578 (1965)
19. Hankin, C.: Lambda calculi. In: A guide for computer scientists. Graduate Texts in Computer. Oxford University Press, Oxford (1994)
20. Kolmogorov, A.N.: On the definition of algorithm. Uspekhi Mat. Nauk. 13(4), 3–28 (1958); Translations Amer. Math. Soc. 29, 217–245 (1963)
21. Knuth, D.: The Art of Computer Programming, 3rd edn., vol. 2. Addison-Wesley, Reading (1998)
22. Krivine, J.L.: A call-by-name lambda-calculus machine. Higher Order and Symbolic Computation 20, 199–207 (2007)
23. Mogensen, T.: Efficient Self-Interpretation in Lambda Calculus. J. of Functional Programming 2(3), 345–363 (1992)
24. Paul, W.: Kolmogorov complexity and lower bounds. In: Budach, L. (ed.) Second Int. Conf. on Fundamentals of Computation Theory, pp. 325–334. Akademie, Berlin (1979)
25. Ronchi Della Rocca, S., Paolini, L.: The Parametric Lambda-calculus. In: A Metamodel for Computation. Springer, Heidelberg (2004)
26. Statman, R.: Church's Lambda Delta Calculus. In: Parigot, M., Voronkov, A. (eds.) LPAR 2000. LNCS (LNAI), vol. 1955, pp. 293–307. Springer, Heidelberg (2000)

Fixed-Point Definability and Polynomial Time on Chordal Graphs and Line Graphs

Martin Grohe

Humboldt Universität zu Berlin
grohe@informatik.hu-berlin.de

For Yuri, in recognition of his inspiring work in finite model theory and elsewhere.

Abstract. The question of whether there is a logic that captures polynomial time was formulated by Yuri Gurevich in 1988. It is still wide open and regarded as one of the main open problems in finite model theory and database theory. Partial results have been obtained for specific classes of structures. In particular, it is known that fixed-point logic with counting captures polynomial time on all classes of graphs with excluded minors. The introductory part of this paper is a short survey of the state-of-the-art in the quest for a logic capturing polynomial time.

The main part of the paper is concerned with classes of graphs defined by excluding induced subgraphs. Two of the most fundamental such classes are the class of chordal graphs and the class of line graphs. We prove that capturing polynomial time on either of these classes is as hard as capturing it on the class of all graphs. In particular, this implies that fixed-point logic with counting does not capture polynomial time on these classes. Then we prove that fixed-point logic with counting does capture polynomial time on the class of all graphs that are both chordal and line graphs.

Keywords: Descriptive complexity theory, fixed-point logic.

1 The Quest for a Logic Capturing PTIME

Descriptive complexity theory started with Fagin's Theorem [25] from 1974, stating that existential second-order logic *captures* the complexity class NP. This means that a property of finite structures is decidable in nondeterministic polynomial time if and only if it is definable in existential second order logic. Similar logical characterisations where later found for most other complexity classes. For example, in 1982 Immerman [44] and independently Vardi [60] characterised the class PTIME (polynomial time) in terms of least fixed-point logic, and in 1983 Immerman [46] characterised the classes NLOGSPACE (nondeterministic logarithmic space) and LOGSPACE (logarithmic space) in terms of transitive closure logic and its deterministic variant. However, these logical characterisations of the classes PTIME, NLOGSPACE, and LOGSPACE, and all other known logical characterisations of complexity classes contained in PTIME, have a serious drawback: They

A. Blass, N. Dershowitz, and W. Reisig (Eds.): Gurevich Festschrift, LNCS 6300, pp. 328–353, 2010.

only apply to properties of *ordered structures*, that is, relational structures with one distinguished relation that is a linear order of the elements of the structure. It is still an open question whether there are logics that characterise these complexity classes on arbitrary, not necessarily ordered structures. We focus on the class PTIME from now on. In this section, which is an updated version of [32], we give a short survey of the quest for a logic capturing PTIME.

1.1 Logics Capturing PTIME

The question of whether there is a logic that characterises, or *captures*, PTIME is subtle. If phrased naively, it has a trivial, but completely uninteresting positive answer. Yuri Gurevich [37] was the first to give a precise formulation of the question. Instead of arbitrary finite structures, we restrict our attention to graphs in this paper. This is no serious restriction, because the question of whether there is a logic that captures PTIME on arbitrary structures is equivalent to the restriction of the question to graphs. We first need to define what constitutes a logic. Following Gurevich, we take a very liberal, semantically oriented approach. We identify *properties* of graphs with classes of graphs closed under isomorphism. A logic L (on graphs) consists of a computable set of *sentences* together with a semantics that associates a property $\mathcal{P}_\varphi$ of graphs with each sentence φ. We say that a graph G *satisfies* a sentence φ, and write $G \models \varphi$, if $G \in \mathcal{P}_\varphi$. We say that a property $\mathcal{P}$ of graphs is *definable* in L if there is a sentence φ such that $\mathcal{P}_\varphi = \mathcal{P}$. A logic L *captures* PTIME if the following two conditions are satisfied:

(G.1) Every property of graphs that is decidable in PTIME is definable in L.

(G.2) There is a computable function that associates with every L-sentence φ a polynomial $p(X)$ and an algorithm A such that A decides the property $\mathcal{P}_\varphi$ in time $p(n)$, where n is the number of vertices of the input graph.

While condition (G.1) is obviously necessary, condition (G.2) may seem unnecessarily complicated. The natural condition we expect to see instead is the following condition (G.2'): Every property of graphs that is definable in L is decidable in PTIME. Note that (G.2) implies (G.2'), but that the converse does not hold. However, (G.2') is too weak, as the following example illustrates:

Example 1. Let $\mathcal{P}_1, \mathcal{P}_2, \ldots$ be an arbitrary enumeration of all polynomial time decidable properties of graphs. Such an enumeration exists because there are only countably many Turing machines and hence only countably many decidable properties of graphs. Let L′ be the "logic" whose sentences are the natural numbers and whose semantics is defined by letting sentence i define property $\mathcal{P}_i$. Then L′ is a logic according to our definition, and it does satisfy (G.1) and (G.2'). But clearly, L′ is not a "logic capturing PTIME" in any interesting sense.

Let me remark that most natural logics that are candidates for capturing PTIME trivially satisfy (G.2). The difficulty is to prove that they also satisfy (G.1), that is, define all PTIME-properties.

There is a different route that leads to the same question of whether there is a logic capturing PTIME from a database-theory perspective: After Aho and

Ullman [2] had realised that SQL, the standard query language for relational databases, cannot express all database queries computable in polynomial time, Chandra and Harel [10] asked for a recursive enumeration of the class of all relational database queries computable in polynomial time. It turned out that Chandra and Harel's question is equivalent to Gurevich's question for a logic capturing PTIME, up to a minor technical detail.[1]

The question of whether there is a logic that captures PTIME is still wide open, and it is considered one of the main open problems in finite model theory and database theory. Gurevich conjectured that there is no logic capturing PTIME. This would not only imply that PTIME $\neq$ NP — remember that by Fagin's Theorem there is a logic capturing NP — but it would actually have interesting consequences for the structure of the complexity class PTIME. Dawar [15] proved a dichotomy theorem stating that, depending on the answer to the question, there are two fundamentally different possibilities: If there is a logic for PTIME, then the structure of PTIME is very simple; all PTIME-properties are variants or special cases of just one problem. If there is no logic for PTIME, then the structure of PTIME is so complicated that it eludes all attempts for a classification. The formal statement of the first possibility is that there is a complete problem for PTIME under first-order reductions. The formal statement of the second possibility is that the class of PTIME-properties is not recursively enumerable.[2]

1.2 Fixed-Point Logics

Fixed-point logics play an important role in finite-model theory, and in particular in the quest for a logic capturing PTIME. Very briefly, the fixed-point logics considered in this context are extensions of first-order logic by operators that formalise inductive definitions. We have already mentioned that *least fixed-point logic* LFP captures polynomial time on ordered structures; this result is known as the *Immerman-Vardi Theorem*. For us, it will be more convenient to work with *inflationary fixed-point logic* IFP, which was shown to have the same expressive power as LFP on finite structures by Gurevich and Shelah [39] and on infinite structures by Kreutzer [50].

IFP does not capture polynomial time on all finite structures. The most immediate reason is the inability of the logic to count. For example, there is no IFP-sentence stating that the vertex set of a graph has even cardinality; obviously, the graph property of having an even number of vertices is decidable in polynomial time. This led Immerman [45] to extending fixed-point logic by "counting

[1] In Chandra and Harel's version of the question, condition (G.2) needs to be replaced by the following condition (CH.2): There is a computable function that associates with every L-sentence φ an algorithm A such that A decides the property $\mathcal{P}_\varphi$ in polynomial time. The difference between (G.2) and (CH.2) is that in (CH.2) the polynomial bounding the running time of the algorithm A is not required to be computable from φ.

[2] The version of recursive enumerability used here is not exactly the same as the one considered by Chandra and Harel [10]; the difference is essentially the same as the difference between conditions (G.2) and (CH.2) discussed earlier.

operators". The formal definition of fixed-point logic with counting operators that we use today, *inflationary fixed-point logic with counting* IFP+C, is due to Grädel and Otto [29]. IFP+C comes surprisingly close to capturing PTIME. Even though Cai, Fürer, and Immerman [9] gave an example of a property of graphs that is decidable in PTIME, but not definable in IFP+C, it turns out that the logic does capture PTIME on many interesting classes of structures.

1.3 Capturing PTIME on Classes of Graphs

Let $\mathcal{C}$ be a class of graphs, which we assume to be closed under isomorphism. We say that a logic L captures PTIME *on* $\mathcal{C}$ if it satisfies the following two conditions:

(G.1)$_\mathcal{C}$ For every property $\mathcal{P}$ of graphs that is decidable in PTIME there is an L-sentence φ such that for all graphs $G \in \mathcal{C}$ it holds that $G \models \varphi$ if and only if $G \in \mathcal{P}$.

(G.2)$_\mathcal{C}$ There is a computable function that associates with every L-sentence φ a polynomial $p(X)$ and an algorithm A such that given a graph $G \in \mathcal{C}$, the algorithm A decides if $G \models \varphi$ in time $p(n)$, where n is the number of vertices of G.

Note that these conditions coincide with conditions (G.1) and (G.2) if $\mathcal{C}$ is the class of all graphs.

The first positive result in this direction is due to Immerman and Lander [48], who proved that IFP+C captures PTIME on the class of all trees. In 1998, I proved that IFP+C captures PTIME on the class of all planar graphs [30] and around the same time, Julian Mariño and I proved that IFP+C captures PTIME on all classes of structures of bounded tree width [34]. In [31], I proved the same result for the class of all graphs that have no complete graph on five vertices, K_5, as a minor. A *minor* of graph G is a graph H that can be obtained from a subgraph of G by contracting edges. We say that a class $\mathcal{C}$ of graphs *excludes a minor* if there is a graph H that is not a minor of any graph in $\mathcal{C}$. Very recently, I proved that IFP+C captures PTIME on all classes of graphs that exclude a minor [33].

In the last few years, maybe as a consequence of Chudnowsky, Robertson, Seymour, and Thomas's [11] proof of the strong perfect graph theorem, the focus of many graph theorists has shifted from graph classes with excluded minors to graph classes defined by excluding induced subgraphs. One of the most basic and important example of such a class is the class of *chordal graphs*. A cycle C of a graph G is *chordless* if it is an induced subgraph. A graph is *chordal* (or *triangulated*) if it has no chordless cycle of length at least four. Fig. 1(a) shows an example of a chordal graph. All chordal graphs are *perfect*, which means that the graphs themselves and all their induced subgraphs have the chromatic number equal to the clique number. Chordal graphs have a nice and simple structure; they can be decomposed into a tree of cliques. A second important example is the class of *line graphs*. The line graph of a graph G is the graph $L(G)$ whose vertices are the edges of G, with two edges being adjacent in $L(G)$ if they have a common endvertex in G. Fig. 1(b) shows an example of a line graph. The class of all line graphs is closed under taking induced subgraphs. Beineke [5] gave a characterisation of the class of line graphs (more precisely, the class of all graphs isomorphic to a line graph) by a family of nine excluded subgraphs. An extension of the class of line graphs, which has also received a lot of attention in the literature, is the class of *claw-free* graphs. A graph is claw-free if it does not have a vertex with three pairwise nonadjacent neighbours, that is, if it does not have a *claw* (displayed in Fig. 2) as an induced subgraph. It is easy to see that all line graphs are claw-free. Recently, Chudnowsky and Seymour (see [12]) developed a structure theory for claw-free graphs.

It would be tempting to use this structure theory for claw free graphs, or at least the simple treelike structure of chordal graphs, to prove that IFP+C captures PTIME on these classes in a similar way as the structure theory for classes of graphs with excluded minors is used to prove that IFP+C captures PTIME on classes with excluded minors. Unfortunately, this is only possible on the very restricted class of graphs that are both chordal and line graphs (an example of such a graph is shown in Fig. 3 on p.343). We prove the following theorem:

Theorem 2

1. IFP+C *does not capture* PTIME *on the class of chordal graphs or on the class of line graphs.*
2. IFP+C *captures* PTIME *on the class of chordal line graphs.*

Our construction to prove (1) is so simple that it will apply to any reasonable logic, which means that if a "reasonable" logic captures PTIME on the class of chordal graphs or on the class of line graphs, then it captures PTIME on the class of all graphs.

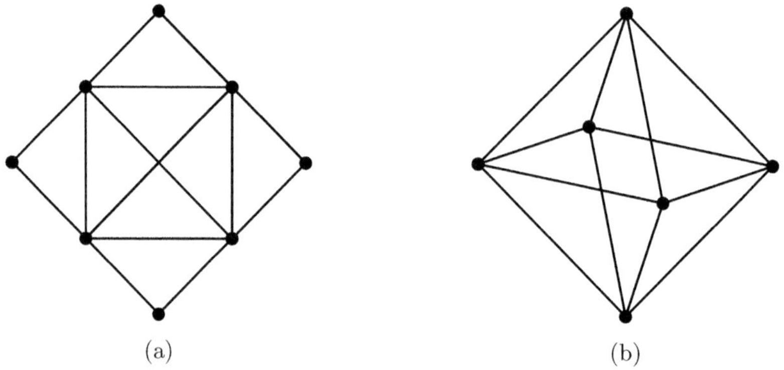

Fig. 1. (a) a chordal graph, which is not a line graph, and (b) the line graph of K_4, which is not chordal

Fig. 2. A claw

Further interesting graph classes closed under taking induced subgraphs are various classes of intersection graphs. Very recently, Laubner [51] proved that IFP+C captures PTIME on the class of all interval graphs. To conclude our discussion of classes of graphs on which IFP+C captures PTIME, let me mention a result due to Hella, Kolaitis, and Luosto [41] stating that IFP+C captures PTIME on almost all graphs (in a precise technical sense). Thus it seems that the results for specific classes of graphs are not very surprising, but it should be mentioned that almost no graphs fall in one of the natural graphs classes discussed before.

Instead of capturing all PTIME on a specific class of structures, Otto [55,56,57] studied the question of capturing all PTIME properties satisfying certain invariance conditions. Most notably, he proved that bisimulation-invariant properties are decidable in polynomial time if and only if they are definable in the *higher-dimensional μ-calculus.*

1.4 Isomorphism Testing and Canonisation

As an abstract question, the question of whether there is a logic capturing polynomial time is linked to the graph isomorphism and canonisation problems. Otto [55] was the first to systematically study the connection between canonisation and descriptive complexity theory. Specifically, if there is a polynomial time canonisation algorithm for a class $\mathcal{C}$ of graphs, then there is a logic that captures polynomial time on this class $\mathcal{C}$. This follows from the Immerman-Vardi Theorem. To explain it, let us assume that we represent graphs by their adjacency matrices. A *canonisation mapping* gets as argument some adjacency matrix representing a graph and returns a *canonical* adjacency matrix for this graph, that is, it maps *isomorphic* adjacency matrices to *equal* adjacency matrices. As an adjacency matrix for a graph is completely fixed once we specify the ordering of the rows and columns of the matrix, we may view a canonisation as a mapping associating with each graph a canonical ordered copy of the graph. Now we can apply the Immerman-Vardi Theorem to this ordered copy.

Clearly, if there is a polynomial time canonisation mapping for a class of graphs (or other structures) then there is a polynomial time isomorphism test for this class. It is open whether the converse also holds. It is also open whether the existence of a logic for polynomial time implies the existence of a polynomial time isomorphism test or canonisation mapping.

Polynomial time canonisation mappings are known for many natural classes of graphs, for example planar graphs [43,42], graphs of bounded genus [26,54], graphs of bounded eigenvalue multiplicity [3], graphs of bounded degree [4,53], and graphs of bounded tree width [8]. Hence for all theses classes there are logics capturing PTIME. However, the logics obtained through canonisation hardly qualify as natural logics. If a logic is to contribute to our understanding of the complexity class PTIME— and from my perspective this is the main reason for being interested in such a logic — we have to look for natural logics that derive their expressiveness from clearly visible basic principles like inductive definability, counting or other combinatorial operations, and maybe fundamental algebraic operations like computing the rank or the determinant of a matrix. If such a logic captures polynomial time on a class of structures, then this shows that all polynomial time properties of structures in this class are based on the principles underlying the logic. Thus even for classes for which we know that there is a logic capturing PTIME through a polynomial-time canonisation algorithm, I think it is important to find "natural" logics capturing PTIME on these classes. In particular, I view it as an important open problem to find a natural logic that captures PTIME on classes of graphs of bounded degree. It is known that IFP+C does not capture PTIME on the class of all graphs of maximum degree at most three.

Most known capturing results are proved by showing that there is a canonisation mapping that is definable in some logic. In particular, all capturing results for IFP+C mentioned above are proved this way. It was observed by Cai, Fürer, and Immerman [9] that for classes $\mathcal{C}$ of structures which admit a canonisation mapping definable in IFP+C, a simple combinatorial algorithm known as the Weisfeiler-Lehman (WL) algorithm [23,24] can be used as a polynomial time isomorphism test on $\mathcal{C}$. Thus the the WL-algorithm correctly decides isomorphism on the class of chordal line graphs and on all classes of graphs with excluded minors. A refined version of the same approach was used by Verbitsky and others [35,49,61] to obtain parallel isomorphism tests running in polylogarithmic time for planar graphs and graphs of bounded tree width.

1.5 Stronger Logics

Early on, a number of results regarding the possibility of capturing polynomial time by adding Lindström quantifiers to first-order logic or fixed-point logic were obtained. Hella [40] proved that adding finitely many Lindström quantifiers (or infinitely many of bounded arity) to fixed-point logic does not suffice to capture polynomial time (also see [17]). Dawar [14] proved that if there is a logic capturing polynomial time, then there is such a logic obtained from fixed-point logic by adding one vectorised family of Lindström quantifiers. Another family of logics that have been studied in this context consists of extensions of fixed-point logic with nondeterministic choice operators [1,18,27].

Currently, the two main candidates for logics capturing PTIME are *choiceless polynomial time with counting* CP+C and *inflationary fixed-point logic with a rank operator* IFP+R. The logic CP+C was introduced by Blass, Gurevich and Shelah [6] (also see [7,19]). The formal definition of the logic is carried out in the framework of *abstract state machines* (see, for example, [38]). Intuitively CP+C may be viewed as a version of IFP+C where quantification and fixed-point operators not only range over elements of a structure, but instead over all objects that can be described by $O(\log n)$ bits, where n is the size of the structure. This intuition can be formalised in an expansion of a structure by all hereditarily finite sets which use the elements of the structure as atoms. The logic IFP+R [16] is an extension of IFP by an operator that determines the rank of definable matrices in a structure. This may be viewed as a higher dimensional version of a counting operator. (Counting appears as a special case of diagonal $\{0, 1\}$-matrices.)

Both CP+C and IFP+R are known to be strictly more expressive than IFP+C. Indeed, both logics can express the property used by Cai, Fürer, and Immerman to separate IFP+C from PTIME. For both logics it is open whether they capture polynomial time, and it is also open whether one of them semantically contains the other.

2 Preliminaries

$\mathbb{N}_0$ and $\mathbb{N}$ denote the sets of nonnegative integers and positive integers, respectively. For $m, n \in \mathbb{N}_0$, we let $[m, n] := \{\ell \in \mathbb{N}_0 \mid m \leq \ell \leq n\}$ and $[n] := [1, n]$. We denote the power set of a set S by 2^S and the set of all k-element subsets of S by $\binom{S}{k}$.

We often denote tuples $(v_1, \ldots, v_k)$ by $\vec{v}$. If $\vec{v}$ denotes the tuple $(v_1, \ldots, v_k)$, then by $\tilde{v}$ we denote the set $\{v_1, \ldots, v_k\}$. If $\vec{v} = (v_1, \ldots, v_k)$ and $\vec{w} = (w_1, \ldots, w_\ell)$, then by $\vec{v}\vec{w}$ we denote the tuple $(v_1, \ldots, v_k, w_1, \ldots, w_\ell)$. By $|\vec{v}|$ we denote the length of a tuple $\vec{v}$, that is, $|(v_1, \ldots, v_k)| = k$.

2.1 Graphs

Graphs in this paper are always finite, nonempty, and simple, where simple means that there are no loops or parallel edges. Unless explicitly called "directed", graphs are undirected. The vertex set of a graph G is denoted by $V(G)$ and the edge set by $E(G)$. We view graphs as relational structures with $E(G)$ being a binary relation on $V(G)$. However, we often find it convenient to view edges (of undirected graphs) as 2-element subsets of $V(G)$ and use notations like $e = \{u, v\}$ and $v \in e$. Subgraphs, induced subgraphs, union, and intersection of graphs are defined in the usual way. We write $G[W]$ to denote the induced subgraph of G with vertex set $W \subseteq V(G)$, and we write $G \setminus W$ to denote $G[V(G) \setminus W]$. The set $\{w \in V(G) \mid \{v, w\} \in E(G)\}$ of *neighbours* of a node v is denoted by $N^G(v)$, or just $N(v)$ if G is clear from the context, and the *degree* of v is the cardinality of $N(v)$. The *order* of a graph, denoted by $|G|$, is the number of vertices of G. The class of all graphs is denoted by $\mathcal{G}$. A *homomorphism* from a graph G to a graph H is a mapping $h : V(G) \to V(H)$ that preserves adjacency, and an *isomorphism* is a bijective homomorphism whose inverse is also a homomorphism.

For every finite nonempty set V, we let $K[V]$ be the *complete graph* with vertex set V, and we let $K_n := K\big[[n]\big]$. A *clique* in a graph G is a set $W \subseteq V(G)$ such that $G[W]$ is a complete graph. *Paths* and *cycles* in graphs are defined in the usual way. The *length* of a path or cycle is the number of its edges. *Connectedness* and *connected* components are defined in the usual way. A set $W \subseteq V(G)$ is *connected* in a graph G if $W \neq \emptyset$ and $G[W]$ is connected. For sets $W_1, W_2 \subseteq V(G)$, a set $S \subset V(G)$ *separates* W_1 *from* W_2 if there is no path from a vertex in $W_1 \setminus S$ to vertex in $W_2 \setminus S$ in the graph $G \setminus S$.

A *forest* is an undirected acyclic graph, and a *tree* is a connected forest. It will be a useful convention to call the vertices of trees and forests *nodes*. A *rooted tree* is a triple $T = (V(T), E(T), r(T))$, where $(V(T), E(T))$ is a tree and $r(T) \in V(T)$ is a distinguished node called the *root*.

We occasionally have to deal with *directed graphs*. We allow directed graphs to have loops. We use standard graph theoretic terminology for directed graphs, without going through it in detail. Homomorphisms and isomorphisms of directed graphs preserve the direction of the edges. Paths and cycles in a directed graph are always meant to be directed; otherwise we will call them "paths or cycles of the underlying undirected graph". Note that cycles in directed graphs may have length 1 or 2. For a directed graph D and a vertex $v \in V(D)$, we let $N^D(v) := \{w \in V(D) \mid (v, w) \in E(D)\}$. *Directed acyclic graphs* will be of particular importance in this paper, and we introduce some additional terminology for them: Let D be a directed acyclic graph. A node w is a *child* of a node v, and v is a *parent* of w, if $(v, w) \in E(D)$. We let $\trianglelefteq^D$ be the reflexive transitive closure of the edge relation $E(D)$ and $\triangleleft^D$ its irreflexive version. Then $\trianglelefteq^D$ is a partial order on $V(D)$.

A *directed tree* is a directed acyclic graph T in which every node has at most one parent, and for which there is a vertex r called the *root* such that for all $t \in V(t)$ there is a path from r to t. There is an obvious one-to-one correspondence between rooted trees and directed trees: For a rooted tree T with root $r := r(T)$ we define the corresponding directed tree T' by $V(T') := V(T)$ and $E(T') := \{(t, u) \mid \{t, u\} \in E(T) \text{ and } t \text{ occurs on the path } rTu\}$. We freely jump back and forth between rooted trees and directed trees, depending on which will be more convenient. In particular, we use the terminology introduced for directed acyclic graphs (parents, children, the partial order $\trianglelefteq$, et cetera) for rooted trees.

2.2 Relational Structures

A *relational structure* A consists of a finite set $V(A)$ called the *universe* or *vertex set* of A and finitely many relations on A. The only types of structures we will use in this paper are *graphs*, viewed as structures $G = \big(V(G), E(G)\big)$ with one binary relation $E(G)$, and *ordered graphs*, viewed as structures $G = \big(V(G), E(G), \leqslant (G)\big)$ with two binary relations $E(G)$ and $\leqslant (G)$, where $\big(V(G), E(G)\big)$ is a graph and $\leqslant (G)$ is a linear order of the vertex set $V(G)$.

2.3 Logics

We assume that the reader has a basic knowledge in logic. In this section, we will informally introduce the two main logics IFP and IFP+C used in this paper. For background and a precise definition, I refer the reader to one of the textbooks [21,28,47,52]. It will be convenient to start by briefly reviewing *first-order logic* FO. Formulae of first-order logic in the language of graphs are built from atomic formulae $E(x, y)$ and $x = y$, expressing adjacency and equality of vertices, by the usual Boolean connectives and existential and universal quantifiers ranging over the vertices of a graph. First-order formulae in the language of ordered graphs may also contain atomic formulae of the form $x \leqslant y$ with the obvious meaning, and formulae in other languages may contain atomic formulae defined for these languages. We write $\varphi(x_1, \dots, x_k)$ to denote that the free variables of a formula φ are among $x_1, \dots, x_k$. For a graph G and vertices $v_1, \dots, v_k$, we write $G \models \varphi[v_1, \dots, v_k]$ to denote that G satisfies φ if x_i is interpreted by v_i, for all $i \in [k]$.

Inflationary fixed-point logic IFP is the extension of FO by a fixed-point operator with an inflationary semantics. To introduce this operator, let $\varphi(X, \vec{x})$ be a formula that, besides a k-tuple $\vec{x} = (x_1, \dots, x_k)$ of free *individual variables* ranging over the vertices of a graph, has a free k-ary *relation variable* ranging over k-ary relations on the vertex set. For every graph G we define a sequence $R_i = R_i(G, \varphi, X, \vec{x})$, for $i \in \mathbb{N}_0$, of k-ary relations on $V(G)$ as follows:

$$\begin{aligned} R_0 &:= \emptyset \\ R_{i+1} &:= R_i \cup \{\vec{v} \mid G \models \varphi[R_i, \vec{v}]\} \qquad \text{for all } i \in \mathbb{N}_0. \end{aligned}$$

Since we have $R_0 \subseteq R_1 \subseteq R_2 \subseteq \cdots \subseteq V(G)^k$ and $V(G)$ is finite, the sequence reaches a fixed-point $R_n = R_{n+1} = R_i$ for all $i \geq n$, which we denote by $R_\infty = R_\infty(G, \varphi, X, \vec{x})$. The *ifp-operator* applied to $\varphi, X, \vec{x}$ defines this fixed-point. We use the following syntax:

$$\underbrace{\text{ifp}\,(X \leftarrow \vec{x} \mid \varphi)\vec{x}'}_{=:\psi(\vec{x}')} . \tag{1}$$

Here $\vec{x}'$ is another k-tuple of individual variables, which may coincide with $\vec{x}$. The variables in the tuple $\vec{x}'$ are the free variables of the formula $\psi(\vec{x}')$, and for every graph G and every tuple $\vec{v} \in V(G)^k$ of vertices we let $G \models \psi[\vec{v}] \iff \vec{v} \in R_\infty$. These definitions can easily be extended to a situation where the formula φ contains other free variables than X and and the variables in $\tilde{x}$; these variables remain free variables of ψ. Now formulae of inflationary fixed-point logic IFP in the language of graphs are built from atomic formulae $E(x, y)$, $x = y$, and $X\vec{x}$ for relation variables X and tuples of individual variables $\vec{x}$ whose length matches the arity of X, by the usual Boolean connectives and existential and universal quantifiers ranging over the vertices of a graph, and the ifp-operator.

Example 3. The IFP-sentence

$$\mathsf{conn} := \forall x_1 \forall x_2 \text{ ifp}\left(X \leftarrow (x_1, x_2) \;\middle|\; \begin{array}{l} x_1 = x_2 \vee E(x_1, x_2) \vee \\ \exists x_3\big(X(x_1, x_3) \wedge X(x_3, x_2)\big) \end{array}\right)(x_1, x_2)$$

states that a graph is connected.

Inflationary fixed-point logic with counting, IFP+C, is the extension of IFP by counting operators that allow it to speak about cardinalities of definable sets and relations. To define IFP+C, we interpret the logic IFP over two sorted extensions of graphs (or other relational structures) by a numerical sort. For a graph G, we let $N(G)$ be the initial segment $[0, |G|]$ of the nonnegative integers. We let G^+ be the two-sorted structure $G \cup (N(G), \leq)$, where $\leq$ is the natural linear order on $N(G)$. To avoid confusion, we always assume that $V(G)$ and $N(G)$ are disjoint. We call the elements of the first sort $V(G)$ *vertices* and the elements of the second sort $N(G)$ *numbers*. Individual variables of our logic range either over the set $V(G)$ of vertices of G or over the set $N(G)$ of numbers of G. Relation variables may range over mixed relations, having certain places for vertices and certain places for numbers. Let us call the resulting logic, inflationary fixed-point logic over the two-sorted extensions of graphs, IFP^+. We may still view IFP^+ as a logic over plain graphs, because the extension G^+ is uniquely determined by G. More precisely, we say that a sentence φ of IFP^+ is satisfied by a graph G if it $G^+ \models \varphi$. *Inflationary fixed-point logic with counting* $\mathsf{IFP{+}C}$ is the extension of IFP^+ by *counting terms* formed as follows: For every formula φ and every vertex variable x we add a term $\#x\ \varphi$; the value of this term is the number of assignments to x such that φ is satisfied.

With each $\mathsf{IFP{+}C}$-sentence φ in the language of graphs we associate the graph property $\mathcal{P}_\varphi := \{G \mid G \models \varphi\}$. As the set of all $\mathsf{IFP{+}C}$-sentences is computable, we may thus view $\mathsf{IFP{+}C}$ as an abstract logic according to the definition given in Section 1.1. It is easy to see that $\mathsf{IFP{+}C}$ satisfies condition (G.2) and therefore condition $(\mathrm{G.2})_\mathcal{C}$ for every class $\mathcal{C}$ of graphs. Thus to prove that $\mathsf{IFP{+}C}$ captures PTIME on a class $\mathcal{C}$ it suffices to verify $(\mathrm{G.1})_\mathcal{C}$.

In the following examples, we use the notational convention that x and variants such as x_1, x' denote vertex variables and that y and variants denote number variables.

Example 4. The $\mathsf{IFP{+}C}$-term $\mathsf{0} := \#x\ \neg x = x$ defines the number $0 \in N(G)$. The formula

$$\mathsf{succ}(y_1, y_2) := y_1 \leq y_2 \wedge \neg y_1 = y_2 \wedge \forall y(y \leq y_1 \vee y_2 \leq y)$$

defines the successor relation associated with the linear order $\leq$. The following $\mathsf{IFP{+}C}$-formula defines the set of even numbers in $N(G)$:

$$\mathsf{even}(y) := \mathrm{ifp}\Big(Y \leftarrow y \;\Big|\; y = \mathsf{0} \vee \exists y' \exists y'' \big(Y(y') \wedge \mathsf{succ}(y', y'') \wedge \mathsf{succ}(y'', y)\big)\Big) y\ .$$

Example 5. An *Eulerian cycle* in a graph is a closed walk on which every edge occurs exactly once. A graph is *Eulerian* if it has a Eulerian cycle. It is a well-known fact that a graph is Eulerian if and only if it is connected and every vertex has even degree. Then the following $\mathsf{IFP{+}C}$-sentence defines the class of Eulerian graphs:

$$\mathsf{eulerian} := \mathsf{conn} \wedge \forall x_1\ \mathsf{even}\big(\#x_2\ E(x_1, x_2)\big)\ ,$$

where conn is the sentence from Example 3 and $\mathsf{even}(y)$ is the formula from Example 4. By standard techniques from finite model theory, it can be proved that the class of Eulerian graphs is neither definable in IFP nor in the counting extension $\mathsf{FO{+}C}$ of first-order logic.

2.4 Syntactical Interpretations

In the following, L is one of the logics $\mathsf{IFP{+}C}$, IFP, or FO, and λ, μ are relational languages such as the language $\{E\}$ of graphs or the language $\{E, \leqslant\}$ of ordered graphs. An $\mathsf{L}[\lambda]$-*formula* is an L-formula in the language λ, and similarly for μ. We need some additional notation:

- Let $\approx$ be an equivalence relation on a set U. For every $u \in U$, by $u/_\approx$ we denote the $\approx$-equivalence class of u, and we let $U/_\approx := \{u/_\approx \mid u \in U\}$ be the set of all equivalence classes. For a tuple $\vec{u} = (u_1, \dots, u_k) \in U^k$ we let $\vec{u}/_\approx := (u_1/_\approx, \dots, u_k/_\approx)$, and for a relation $R \subseteq U^k$ we let $R/_\approx := \{\vec{u}/_\approx \mid \vec{u} \in R\}$.
- Two tuples $\bar{x} = (x_1, \dots, x_k), (y_1, \dots, y_\ell)$ of individual variables have the same *type* if $k = \ell$ and for all $i \in [k]$ either both x_i and y_i range over vertices or both x_i and y_i range over numbers. For every structure G, we let $G^{\vec{x}}$ be the set of all tuples $\vec{a} \in (V(G) \cup N(G))^k$ such that for all $i \in [k]$ we have $a_i \in V(G)$ if x_i is a vertex variable and $a_i \in N(G)$ if x_i is a number variable.

Definition 6. *1. An* L-interpretation of μ in λ *is a tuple*

$$\Gamma(\vec{x}) = \Big(\gamma_{app}(\vec{x}), \gamma_V(\vec{x}, \vec{y}), \gamma_\approx(\vec{x}, \vec{y}_1, \vec{y}_2), \big(\gamma_R(\vec{x}, \vec{y}_R)\big)_{R \in \mu}\Big),$$

of $\mathsf{L}[\lambda]$*-formulae, where* $\vec{x}$, $\vec{y}$, $\vec{y}_1$, $\vec{y}_2$, *and* $\vec{y}_R$ *for* $R \in \mu$ *are tuples of individual variables such that* $\vec{y}, \vec{y}_1, \vec{y}_2$ *all have the same type, and for every* k*-ary* $R \in \mu$ *the tuple* $\vec{y}_R$ *can be written as* $\vec{y}_{R1} \dots \vec{y}_{R,k}$, *where the* $\vec{y}_{R,i}$ *have the same type as* $\vec{y}$.

In the following, let $\Gamma(\vec{x})$ *be an* L*-interpretation of* μ *in* λ. *Let* G *be a* λ*-structure and* $\vec{a} \in G^{\vec{x}}$*:*

3. $\Gamma(\vec{x})$ *is* applicable *to* $(G, \vec{a})$ *if* $G \models \gamma_{app}[\vec{a}]$.
4. If $\Gamma(\vec{x})$ *is applicable to* $(G, \vec{a})$, *we let* $\Gamma[G; \vec{a}]$ *be the* μ*-structure with vertex set*

$$V\big(\Gamma[G;\vec{a}]\big) := \big\{\vec{b} \in G^{\vec{y}} \mid G \models \gamma_V[\vec{a}, \vec{b}]\big\}/_\approx,$$

where $\approx$ *is the reflexive, symmetric, transitive closure of the binary relation* $\big\{(\vec{b}_1, \vec{b}_2) \in (G^{\vec{y}})^2 \mid G \models \gamma_\approx[\vec{a}, \vec{b}_1, \vec{b}_2]\big\}$. *Furthermore, for* k*-ary* $R \in \mu$, *we let*

$$R\big(\Gamma[G;\vec{a}]\big) := \Big\{(\vec{b}_1, \dots, \vec{b}_k) \in V\big(\Gamma[G;\vec{a}]\big) \;\Big|\; G \models \gamma_R[\vec{a}, \vec{b}_1, \dots, \vec{b}_k]\Big\}\Big/_\approx.$$

Syntactical interpretations map λ-structures to μ-structures. The crucial observation is that they also induce a reverse translation from $\mathsf{L}[\mu]$-formulae to $\mathsf{L}[\lambda]$-formulae.

Fact 7 (Lemma on Syntactical Interpretations). *Let* $\Gamma(\vec{x})$ *be an* L*-interpretation of* μ *in* λ. *Then for every* $\mathsf{L}[\mu]$*-sentence* φ *there is an* $\mathsf{L}[\lambda]$*-formula* $\varphi^{-\Gamma}(\vec{x})$ *such that the following holds for all* λ*-structures* G *and all tuples* $\vec{a} \in G^{\vec{x}}$*: If* $\Gamma(\vec{x})$ *is applicable to* $(G, \vec{a})$, *then*

$$G \models \varphi^{-\Gamma}[\vec{a}] \iff \Gamma[G;\vec{a}] \models \varphi.$$

A proof of this fact for first-order logic can be found in [22]. The proof for the other logics considered here is an easy adaptation of the one for first-order logic.

2.5 Definable Canonisation

A *canonisation mapping* for a class of $\mathcal{C}$ graphs associates with every graph $G \in \mathcal{C}$ an *ordered copy* of G, that is, an ordered graph $(H, \leq)$ such that $H \cong G$. We are interested in canonisation mappings definable in the logic IFP+C by syntactical interpretations of $\{E, \leqslant\}$ in $\{E\}$. The easiest way to define a canonisation mapping is by defining a linear order $\leq$ on the universe of a structure G and then take $(G, \leq)$ as the canonical copy. However, defining an ordered copy of a structure is not the same as defining a linear order on the universe, as the following example illustrates:

Example 8. Let $\mathcal{K}$ be the class of all complete graphs. It is easy to see that there is no IFP+C-formula $\varphi(x_1, x_2)$ such that for all $K \in \mathcal{K}$ the binary relation $\varphi[K; x_1, x_2]$ is a linear order of $V(K)$.

However, there is an FO+C-definable canonisation mapping for the class $\mathcal{K}$: Let

$$\Gamma = \big(\gamma_{app}, \gamma_V(\vec{y}), \gamma_{\approx}(y_1, y_2), \gamma_E(y_1, y_2), \gamma_{\leqslant}(y_1, y_2)\big)$$

be the numerical FO+C-interpretation of $\{E, \leqslant\}$ in $\{E\}$ defined by:

- $\gamma_{app} := \forall x\, x = x$;
- $\gamma_V(y) := 1 \leq y \wedge y \leq \mathsf{ord}$, where $\mathsf{ord} := \#x\, x = x$;
- $\gamma_{\approx}(y_1, y_2) := y_1 = y_2$;
- $\gamma_E(y_1, y_2) := \neg y_1 = y_2$;
- $\gamma_{\leqslant}(y_1, y_2) := y_1 \leq y_2$.

It is easy to see that the mapping $K \mapsto \Gamma[K]$ is a canonisation mapping for the class $\mathcal{K}$.

Our notion of *definable canonisation* slightly relaxes the requirement of defining a canonisation mapping; instead of just one ordered copy, we associate with each structure a parametrised family of polynomially many ordered copies.

Definition 9. *1. Let $\Gamma(\vec{x})$ be an* L*-interpretation of $\{E, \leqslant\}$ in $\{E\}$. Then $\Gamma(\vec{x})$* canonises *a graph G if there is at least one tuple $\vec{a} \in G^{\vec{x}}$ such that $\Gamma(\vec{x})$ is applicable to $(G, \vec{a})$, and for all tuples $\vec{a} \in G^{\vec{x}}$ such that $\Gamma(\vec{x})$ is applicable to $(G, \vec{a})$ it holds that $\Gamma[G; \vec{a}]$ is an ordered copy of G.*

2. A class $\mathcal{C}$ of graphs admits L-definable canonisation *if there is an* L*-interpretation $\Gamma(\vec{x})$ of $\{E, \leqslant\}$ in $\{E\}$ that canonises all $G \in \mathcal{C}$.*

The following well-known fact is a consequence of the Immerman-Vardi Theorem. It is used, at least implicitly, in [30,31,34,48,55]:

Fact 10. *Let $\mathcal{C}$ be a class of graphs that admits* IFP+C*-definable canonisation. Then* IFP+C *captures* PTIME *on $\mathcal{C}$.*

3 Negative Results

In this section, we prove that IFP+C does not capture PTIME on the classes of chordal graphs and line graphs. Actually, our proof yields a more general result: Any logic that captures PTIME on any of these two classes and that is "closed under first-order reductions" captures PTIME on the class of all graphs. It will be obvious what we mean by "closed under first-order reductions" from the proofs, and it is also clear that most "natural" logics will satisfy this closure condition. It follows from our constructions that if there is a logic capturing PTIME on one of the two classes, then there is a logic capturing PTIME on all graphs.

Our negative results for IFP+C are based on the following theorem:

Fact 11 (Cai, Fürer, and Immerman [9]). *There is a* PTIME*-decidable property $\mathcal{P}_{\mathrm{CFI}}$ of graphs that is not definable in* IFP+C*.*

Without loss of generality we assume that all $G \in \mathcal{P}_{\mathrm{CFI}}$ are connected and of order at least 4.

3.1 Chordal Graphs

Let us denote the class of chordal graphs by $\mathcal{CD}$.

For every graph G, we define a graph $\hat{G}$ as follows:

- $V(\hat{G}) := V(G) \cup \{v_e \mid e \in E(G)\}$, where for each $e \in E(G)$ we let v_e be a new vertex;
- $E(\hat{G}) := \binom{V(G)}{2} \cup \big\{\{v, v_e\} \bigm| v \in V(G), e \in E(G), v \in e\big\}$.

The following lemmas collect the properties of the transformation $G \mapsto \hat{G}$ that we need here. We leave the straightforward proofs to the reader.

Lemma 12. *For every graph G the graph $\hat{G}$ is chordal.*

Note that for the graphs K_2 and $I_3 := \big([3], \emptyset\big)$ it holds that $\hat{K}_2 \cong \hat{I}_3 \cong K_3$. It turns out that K_2 and I_3 are the only two nonisomorphic graphs that have isomorphic images under the mapping $G \mapsto \hat{G}$. It is easy to verify this by observing that for G with $|G| \geq 4$ and $v \in V(\hat{G})$, it holds that $v \in V(G)$ if and only if $\deg(v) \geq 3$. Let $\hat{\mathcal{G}}$ be the class of all graphs H such that $H \cong \hat{G}$ for some graph G.

Lemma 13. *The class $\hat{\mathcal{G}}$ is polynomial time decidable. Furthermore, there is a polynomial time algorithm that, given a graph $H \in \hat{\mathcal{G}}$, computes the unique (up to isomorphism) graph $G \in \mathcal{G} \setminus \{K \mid K \cong K_2\}$ with $\hat{G} \cong H$.*

Lemma 14. *There is an* FO*-interpretation $\hat{\Gamma}$ of $\{E\}$ in $\{E\}$ such that for all graphs G it holds that $\hat{\Gamma}[G] \cong \hat{G}$.*

Theorem 15. IFP+C *does not capture* PTIME *on the class $\mathcal{CD}$ of chordal graphs.*

Proof. Let $\mathcal{P}_{\mathrm{CFI}}$ be the graph property of Fact 11 that separates PTIME from IFP+C. Note that $K_2 \notin \mathcal{P}_{\mathrm{CFI}}$ by our assumption that all graphs in $\mathcal{P}_{\mathrm{CFI}}$ have order at least 4. By Lemma 13, the class $\hat{\mathcal{P}} := \{H \mid H \cong \hat{G} \text{ for some } G \in \mathcal{P}_{\mathrm{CFI}}\}$ is a polynomial time decidable subclass of $\mathcal{CD}$.

Suppose for contradiction that IFP+C captures polynomial time on $\mathcal{CD}$. Then by (G.1)$_{\mathcal{CD}}$ there is an IFP+C-sentence φ such that for all chordal graphs G it holds that $G \models \varphi \iff G \in \hat{\mathcal{P}}$. We apply the Lemma on Syntactical Interpretations to φ and the interpretation $\hat{\Gamma}$ of Lemma 14 and obtain an IFP+C-sentence $\varphi^{-\hat{\Gamma}}$ such that for all graphs G it holds that

$$G \models \varphi^{-\hat{\Gamma}} \iff \hat{G} \cong \hat{\Gamma}[G] \models \varphi \,.$$

Thus $\varphi^{-\hat{\Gamma}}$ defines $\mathcal{P}_{\mathrm{CFI}}$, which is a contradiction. □

3.2 Line Graphs

Let $\mathcal{L}$ denote the class of all line graphs, or more precisely, the class of all graphs L such that there is a graph G with $L \cong L(G)$. Observe that a triangle and a claw have the same line graph, a triangle. Whitney [62] proved that for all nonisomorphic connected graphs G, H except the claw and triangle, the line graphs of G and H are nonisomorphic. The following fact, corresponding to Lemma 13, is essentially an algorithmic version of Whitney's result:

Fact 16 (Roussopoulos [59]). *The class $\mathcal{L}$ is polynomial time decidable. Furthermore, there is a polynomial time algorithm that, given a connected graph $H \in \mathcal{L}$, computes the unique (up to isomorphism) graph $G \in \mathcal{G} \setminus \{K \mid K \cong K_3\}$ with $L(G) \cong H$.*

Lemma 17. *There is an* FO*-interpretation Λ of $\{E\}$ in $\{E\}$ such that for all graphs G it holds that $\Lambda[G] \cong L(G)$.*

Proof. We define $\Lambda := \big(\lambda_{\mathrm{app}}, \lambda_V(y_1, y_2), \lambda_{\approx}(y_1, y_2, y_1', y_2'), \lambda_E(y_1, y_2, y_1', y_2')\big)$ by:

- $\lambda_{\mathrm{app}} := \forall x\, x = x$;
- $\lambda_V(y_1, y_2) := E(y_1, y_2)$;
- $\lambda_{\approx}(y_1, y_2, y_1', y_2') := (y_1 = y_1' \wedge y_2 = y_2') \vee (y_1 = y_2' \wedge y_2 = y_1')$;
- $\lambda_E(y_1, y_2, y_1', y_2') := (y_1 = y_1' \wedge \neg y_2 = y_2') \vee (y_2 = y_2' \wedge \neg y_1 = y_1') \vee (y_1 = y_2' \wedge \neg y_2 = y_1') \vee (y_2 = y_1' \wedge \neg y_2 = y_1')$. □

Theorem 18. IFP+C *does not capture* PTIME *on the class $\mathcal{L}$ of line graphs.*

Proof. The proof is completely analogous to the proof of Theorem 15, using Fact 16 and Lemma 17 instead of Lemmas 13 and 14. □

4 Capturing Polynomial Time on Chordal Line Graphs

In this section, we shall prove that IFP+C captures PTIME on the class $\mathcal{CD} \cap \mathcal{L}$ of graphs that are both chordal and line graphs. As we will see, such graphs have a simple treelike structure. We can exploit this structure and canonise the graphs in $\mathcal{CD} \cap \mathcal{L}$ in a similar way as trees or graphs of bounded tree width.

Example 19. Fig. 3 shows an example of a chordal line graph.

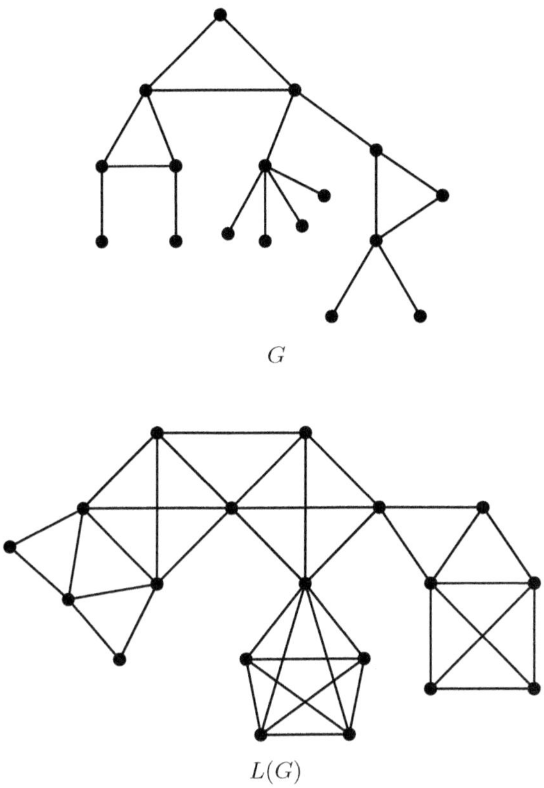

Fig. 3. A graph G and its line graph $L(G)$, which is chordal

4.1 On the Structure of Chordal Line Graphs

It is a well-known fact that chordal graphs can be decomposed into cliques arranged in a tree-like manner. To state this formally, we review tree decompositions of graphs. A *tree decomposition* of a graph G is a pair (T, β), where T is a tree and $\beta : V(T) \to 2^{V(G)}$ is a mapping such that the following two conditions are satisfied:

(T.1) For every $v \in V(G)$ the set $\{t \in V(T) \mid v \in \beta(t)\}$ is connected in T.

(T.2) For every $e \in E(G)$ there is a $t \in V(T)$ such that $e \subseteq \beta(t)$.

The sets $\beta(t)$, for $t \in V(T)$, are called the *bags* of the decomposition. It will be convenient for us to always assume the tree T in a tree decomposition to be rooted. This gives us the partial tree order $\trianglelefteq^T$. We introduce some additional notation. Let (T, β) be a tree decomposition of a graph G. For every $t \in V(T)$ we let:

$$\gamma(t) := \bigcup_{u \in V(T) \text{ with } t \trianglelefteq^T u} \beta(u) ,$$

The set $\gamma(t)$ is called the *cone* of (T, β) at t. It easy to see that for every $t \in V(T) \setminus \{r(T)\}$ with parent s the set $\beta(t) \cap \beta(s)$ separates $\gamma(t)$ from $V(G) \setminus \gamma(t)$. Furthermore, for every clique X of G there is a $t \in V(T)$ such that $X \subseteq \beta(t)$. (See Diestel's textbook [20] for proofs of these facts and background on tree decompositions.) Another useful fact is that every tree decomposition (T, β) of a graph G can be transformed into a tree decomposition (T', β') such that for all $t' \in V(T')$ there exists a $t \in V(T)$ such that $\beta'(t') = \beta(t)$, and for all $t, u \in V(T')$ with $t \neq u$ it holds that $\beta'(t) \not\subseteq \beta'(u)$.

Fact 20. *A nonempty graph G is chordal if and only if G has a tree decomposition into cliques, that is, a tree decomposition (T, β) such that for all $t \in V(T)$ the bag $\beta(t)$ is a clique of G.*

For a graph G, we let $MCL(G)$ be the set of all maximal cliques in G with respect to set inclusion. If we combine Fact 20 with the observations about tree decomposition stated before the fact, we obtain the following lemma:

Lemma 21. *Let G be a nonempty chordal graph. Then G has a tree decomposition (T, β) with the following properties:*

(i) For every $t \in V(T)$ it holds that $\beta(t) \in MCL(G)$.

(ii) For every $X \in MCL(G)$ there is exactly one $t \in V(T)$ such that $\beta(t) = X$.

We call a tree decomposition satisfying conditions (i) and (ii) a good tree decomposition *of G.*

Let us now turn to line graphs. Let $L := L(G)$ be the line graph of a graph G. For every $v \in V(G)$, let $X(v) := \{e \in E(G) \mid v \in e\} \subseteq V(L)$. Unless v is an isolated vertex, $X(v)$ is a clique in L. Furthermore, we have

$$L = \bigcup_{v \in V(G)} L[X(v)] \,.$$

Observe that for all $v, w \in V(G)$, if $e := \{v, w\} \in E(G)$ then $X(v) \cap X(w) = \{e\}$, and if $\{v, w\} \notin E(G)$ then $X(v) \cap X(w) = \emptyset$. The following proposition, which is probably well-known, characterises the line graphs that are chordal:

Proposition 22. *Let $L = L(G) \in \mathcal{L}$. Then*

$$L \in \mathcal{CD} \iff \text{all cycles in } G \text{ are triangles.}$$

Note that on the right hand side, we do not only consider chordless cycles.

Proof. For the forward direction, suppose that $L \in \mathcal{CD}$, and let $C \subseteq G$ be a cycle. Then $L[E(C)]$ is a chordless cycle in L. Hence $|C| \leq 3$, that is, C is a triangle.

For the backward direction, suppose that all cycles in G are triangles, and let $C \subseteq L$ be a chordless cycle of length k. Let $e_1, \ldots, e_k$ be the vertices of C in cyclic order. To simplify the notation, let $e_0 := e_k$. Then for all $i \in [k]$ it holds that $\{e_{i-1}, e_i\} \in E(L)$ and thus $e_{i-1} \cap e_i \neq \emptyset$. Let $v_0, v_1 \in V(G)$ such that $e_1 = \{v_0, v_1\}$, and for $i \in [2, k]$, let $v_i \in e_i \setminus e_{i-1}$. Then $v_i \neq v_j$ for all $j \in [i-2]$, and if $i < k$ even for $j \in [0, i-2]$, because the cycle C is chordless and thus $e_i \cap e_j = \emptyset$. Furthermore, $v_k = v_0$. Thus $\{v_1, \ldots, v_k\}$ is the vertex set of a cycle in G, and we have $k = 3$. □

Lemma 23. *Let $L = L(G) \in \mathcal{CD} \cap \mathcal{L}$, and let $X \in MCL(L)$ and $e = \{v, w\} \in X$. Then $X = X(v)$ or $X = X(w)$ or there is an $x \in V(G)$ such that $\{x, v\}, \{x, w\} \in E(G)$ and $X = \{e, \{x, v\}, \{x, w\}\}$.*

Proof. For all $f \in X$, either $v \in f$ or $w \in f$, because f is adjacent to e. Hence $X \subseteq X(v) \cup X(w)$. If $X \subseteq X(v)$, then $X = X(v)$ by the maximality of X. Similarly, if $X \subseteq X(w)$ then $X = X(w)$. Suppose that $X \setminus X(v) \neq \emptyset$ and $X \setminus X(w) \neq \emptyset$. Let $f \in X \setminus X(v)$ and $g \in X \setminus X(w)$. As X is a clique, we have $\{f, g\} \in E(L)$ and thus $f \cap g \neq \emptyset$. Hence there is an $x \in V(G)$ such that $f = \{x, w\}$ and $g = \{x, v\}$. Furthermore, $X = \{e, f, g\}$. To see this, let $h \in X$. Then $\{h, e\} \in E(L)$ and thus $v \in h$ or $w \in h$. Say, $v \in h$. If $w \in h$, then $h = e$. Otherwise, we have $x \in h$, because h is adjacent to g. Thus $h = g$. □

Lemma 24. *Let $L \in \mathcal{CD} \cap \mathcal{L}$, and let $X_1, X_2 \in MCL(L)$ be distinct. Then $|X_1 \cap X_2| \leq 2$.*

Proof. Let $L = L(G)$ for some graph G. Suppose for contradiction that $|X_1 \cap X_2| \geq 3$. Then $|X_1|, |X_2| \geq 4$, because X_1 and X_2 are distinct maximal cliques. By Lemma 23, it follows that there are vertices $v_1, v_2 \in V(G)$ such that $X_1 = X(v_1)$ and $X_2 = X(v_2)$, which implies $|X_1 \cap X_2| \leq 1$. This is a contradiction. □

Lemma 25. *Let $L \in \mathcal{CD} \cap \mathcal{L}$, and let $X_1, X_2, X_3 \in MCL(L)$ be pairwise distinct such that $X_1 \cap X_2 \cap X_3 \neq \emptyset$. Then there are i, j, k such that $\{i, j, k\} = [3]$ and $X_i \subseteq X_j \cup X_k$ and $|X_i| = 3$.*

Proof. Let $L = L(G)$ for some graph G. Let $e \in X_1 \cap X_2 \cap X_3$. Suppose that $e = \{v, w\} \in E(G)$. As the cliques X_1, X_2, X_3 are distinct, it follows from Lemma 23 that there is an $i \in [3]$ and an $x \in V(G)$ such that $X_i = \{e, \{x, v\}, \{x, w\}\}$. Choose such i and x.

Claim 1. For all $j \in [3] \setminus \{i\}$, either $X_j = X(v)$ or $X_j = X(w)$.

Proof. Suppose for contradiction that $X_j \neq X(v)$ and $X_j \neq X(w)$. Then by Lemma 23, there exists a $y \in V(G)$ such that $\{y, v\}, \{y, w\} \in E(G)$ and $X_j = \{e, \{y, v\}, \{y, w\}\}$. But then

$$L\big[\{y, v\}, \{v, x\}, \{x, w\}, \{w, y\}\big]$$

is a chordless cycle in L, which contradicts L being chordal. ⌟

Thus there are j, k such that $\{i, j, k\} = [3]$ and $X_j = X(v)$ and $X_k = X(w)$. Then $X_i \subseteq X_j \cup X_k$. □

Lemma 26. *Let $L \in \mathcal{CD} \cap \mathcal{L}$. Then every good tree decomposition (T, β) of L satisfies the following conditions (in addition to conditions (i) and (ii) of Lemma 21):*

(iii) For all $t \in V(T)$,
- *either $|\beta(t)| = 3$ and t has at most three neighbours in T (the* neighbours *of a node are its children and the parent),*
- *or for all distinct neighbours u, u' of t in T it holds that $\beta(u) \cap \beta(u') = \emptyset$.*

(iv) For all $t, u \in V(T)$ with $t \neq u$ it holds that $|\beta(t) \cap \beta(u)| \leq 2$.

Proof. Let (T, β) be a good tree decomposition of L. Such a decomposition exists because L is chordal. As all bags of the decomposition are maximal cliques of L, condition (iii) follows from Lemma 25 and condition (iv) follows from Lemma 24. □

4.2 Canonisation

Theorem 27. *The class $\mathcal{CD} \cap \mathcal{L}$ of all chordal line graphs admits* IFP+C*-definable canonisation.*

Corollary 28. IFP+C *captures* PTIME *on the class of all chordal line graphs.*

Proof (Proof of Theorem 27). The proof resembles the proof that classes of graphs of bounded tree width admit IFP+C-definable canonisation [34] and also the proof of Theorem 7.2 (the "Second Lifting Theorem") in [31]. Both of these proofs are generalisations of the simple proof that the class of trees admits IFP+C-definable canonisation (see, for example, [36]). We shall describe an inductive construction that associates with each chordal line graph G a canonical copy G' whose universe is an initial segment of the natural numbers. For readers with some experience in finite model theory, it will be straightforward to formalise the construction in IFP+C. We only describe the canonisation of *connected* chordal line graphs that are not complete graphs. It is easy to extend it to arbitrary chordal line graphs. For complete graphs, which are chordal line graphs, cf. Example 8.

To describe the construction, we fix a connected graph $G \in \mathcal{CD} \cap \mathcal{L}$ that is not a complete graph. Note that this implies $|G| \geq 3$. Let (T, β^T) be a good tree decomposition of G. As G is not a complete graph, we have $|T| \geq 2$. Without loss of generality we may assume that the root $r(T)$ has exactly one child in T, because every tree has at least one node of degree at most 1 and properties (i), (ii) of a good decomposition do not depend on the choice of the root. It will be convenient to view the rooted tree T as a directed graph, where the edges are directed from parents to children.

Let U be the set of all triples $(u_1, u_2, u_3) \in V(G)^3$ such that $u_3 \neq u_1, u_2$ (possibly, $u_1 = u_2$), and there is a unique $X \in MCL(G)$ such that $u_1, u_2, u_3 \in X$. For all $\vec{u} = (u_1, u_2, u_3) \in U$, let $A(\vec{u})$ be the connected component of $G \setminus \{u_1, u_2\}$ that contains u_3 (possibly, $A(\vec{u}) = G \setminus \{u_1, u_2\}$). We define mappings

$\sigma^U, \alpha^U, \gamma^U, \beta^U : U \to 2^{V(G)}$ as follows: For all $\vec{u} = (u_1, u_2, u_3) \in U$, we let $\sigma^U(\vec{u}) := \{u_1, u_2\}$ and $\alpha^U(\vec{u}) := V(A(\vec{u}))$. We let $\gamma^U(\vec{u}) := \sigma^U(\vec{u}) \cup \alpha^U(\vec{u})$, and we let $\beta^U(\vec{u})$ the unique $X \in \mathit{MCL}(G)$ with $u_1, u_2, u_3 \in X$. We define a partial order $\trianglelefteq$ on U by letting $\vec{u} \trianglelefteq \vec{v}$ if and only if $\vec{u} = \vec{v}$ or $\alpha(\vec{u}) \supset \alpha(\vec{v})$. We let F be the successor relation of $\trianglelefteq$, that is, $(\vec{u}, \vec{v}) \in F$ if $\vec{u} \triangleleft \vec{v}$ and there is no $\vec{w} \in U \setminus \{\vec{u}, \vec{v}\}$ such that $\vec{u} \triangleleft \vec{w} \triangleleft \vec{v}$. Finally, we let $D := (U, F)$. Then D is a directed acyclic graph. It is easy to verify that for all $\vec{u} \in U$ we have

$$\beta^U(\vec{u}) = \gamma^U(\vec{u}) \setminus \bigcup_{\vec{v} \in N^D(\vec{u})} \alpha^U(\vec{v}), \tag{2}$$

where $N^D(\vec{u}) = \{\vec{v} \in U \mid (\vec{u}, \vec{v}) \in F\}$.

Recall that we also have mappings $\beta^T, \gamma^T : V(T) \to 2^{V(G)}$ derived from the tree decomposition. We define a mapping $\sigma^T : V(T) \to 2^{V(G)}$ as follows:

- For a node $t \in V(T) \setminus \{r(T)\}$ with parent s, we let $\sigma^T(t) := \beta^T(t) \cap \beta^T(s)$.
- For the root $r := r(T)$, we first define a set $S \subseteq V(G)$ by letting $S := \beta^T(r) \setminus \beta^T(t)$, where t is the unique child of r. (Remember our assumption that r has exactly one child.) Then if $|S| \geq 2$, we choose distinct $v, v' \in S$ and let $\sigma^T(r) := \{v, v'\}$, and if $|S| = 1$ we let $\sigma^T(r) := S$.

Note that $\beta^T(t) \setminus \sigma^T(t) \neq \emptyset$ and $1 \leq |\sigma^T(t)| \leq 2$ for all $t \in V(T)$. For the root, this follows immediately from the definition of $\sigma^T(t)$, and for nodes $t \in V(T) \setminus \{r(T)\}$ it follows from Lemma 26. We define a mapping $\alpha^T : V(T) \to 2^{V(G)}$ by letting $\alpha^T(t) := \gamma^T(t) \setminus \sigma^T(t)$ for all $t \in V(T)$. We define a mapping $g : V(T) \to U$ by choosing, for every node $t \in V(T)$, vertices u_1, u_2 such that $\sigma^T(t) = \{u_1, u_2\}$ (possibly $u_1 = u_2$) and a vertex $u_3 \in \beta(t) \setminus \sigma(t)$ and letting $g(t) := (u_1, u_2, u_3)$. Note that $(u_1, u_2, u_3) \in U$, because $\beta^T(t)$ is the unique maximal clique in $\mathit{MCL}(G)$ that contains u_1, u_2, u_3.

Claim 1. The mapping g is a directed graph embedding of T into D. Furthermore, for all $t \in V(T)$ it holds that $\alpha^T(t) = \alpha^U(g(t))$, $\beta^T(t) = \beta^U(g(t))$, $\gamma^T(t) = \gamma^U(g(t))$, and $\sigma^T(t) = \sigma^U(g(t))$.

Proof. We leave the straightforward inductive proof to the reader. ┘

Let $\vec{u}_0 := g(r(T))$, and let U_0 be the subset of U consisting of all $\vec{u} \in U$ such that $\vec{u}_0 \trianglelefteq \vec{u}$. Let F_0 be the restriction of F to U_0 and $D_0 := (U_0, F_0)$. Note that U_0 is upward closed with respect to $\trianglelefteq$ and that $g(T) \subseteq D_0$.

Claim 2. There is a mapping $h : U_0 \to V(T)$ such that h is a directed graph homomorphism from D_0 to T and $h \circ g$ is the identity mapping on $V(T)$. Furthermore, for all $\vec{u} \in U_0$ it holds that $\alpha^U(\vec{u}) = \alpha^T(h(\vec{u}))$, $\beta^U(\vec{u}) = \beta^T(h(\vec{u}))$, $\gamma^U(\vec{u}) = \gamma^T(h(\vec{u}))$, and $\sigma^U(\vec{u}) = \sigma^T(h(\vec{u}))$.

Proof. We define h by induction on the partial order $\trianglelefteq$. The unique $\trianglelefteq$-minimal element of U_0 is $\vec{u}_0$. We let $h(\vec{u}_0) := r(T)$. Now let $\vec{v} = (v_1, v_2, v_3) \in U_0$, and suppose that $h(\vec{u})$ is defined for all $\vec{u} \in U_0$ with $\vec{u} \triangleleft \vec{v}$. Let $\vec{u} \in U_0$ such that $(\vec{u}, \vec{v}) \in F_0$, and let $s := h(\vec{u})$. By the induction hypothesis, we have $\alpha^U(\vec{u}) =$

$\alpha^T(s)$, $\beta^U(\vec{u}) = \beta^T(s)$, $\gamma^U(\vec{u}) = \gamma^T(s)$, and $\sigma^U(\vec{v}) = \sigma^T(s)$. The set $\alpha^U(\vec{v})$ is the vertex set of a connected component of $G \setminus \sigma^U(\vec{v})$ which is contained in $\alpha^U(\vec{u}) \subseteq \gamma^U(\vec{u}) = \gamma^T(s)$, and by (2) it holds that $\alpha^U(\vec{v}) \cap \beta^U(\vec{u}) = \emptyset$. Hence there is a child t of s such that $\alpha^U(\vec{v}) \subseteq \alpha^T(t)$. Let $\vec{v}' := g(t)$. If $\alpha^U(\vec{v}) \subset \alpha^T(t) = \alpha^U(\vec{v}')$, then $\vec{u} \lhd \vec{v}' \lhd \vec{v}$, which contradicts $(\vec{u}, \vec{v}) \in F$. Hence $\alpha^U(\vec{v}) = \alpha^T(t)$ and thus $\sigma^U(\vec{v}) = \sigma^T(t)$. This also implies $\gamma^U(\vec{v}) = \gamma^T(t)$ and $\beta^U(\vec{v}) = \beta^T(t)$. We let $h(\vec{v}) := t$.

To prove that h is really a homomorphism, it remains to prove that for all $\vec{u}' \in U_0$ with $(\vec{u}', \vec{v}) \in F_0$ we also have $h(\vec{u}') = s$. So let $\vec{u}' \in U_0$ with $(\vec{u}', \vec{v}) \in F_0$, and let $s' = h(\vec{u}')$. Suppose for contradiction that $s \neq s'$. If $s' \lhd^T s$ then $\alpha^U(\vec{u}') \supset \alpha^U(\vec{u})$ and thus $\vec{u}' \lhd \vec{u}$, which contradicts $(\vec{u}', \vec{v}) \in F_0$. Thus $s' \not\trianglelefteq^T s$, and similarly $s \not\trianglelefteq^T s'$. But then both $\sigma^T(s)$ and $\sigma^T(s')$ separate $\gamma^T(s)$ from $\gamma^T(s')$ in G. This contradicts $\alpha^U(\vec{v}) \subseteq \alpha^T(s) \cap \alpha^T(s') \subseteq \big(\gamma^T(s) \cap \gamma^T(s')\big) \setminus \big(\sigma^T(s) \cup \sigma^T(s')\big)$. ⌟

Thus essentially, the "treelike" decomposition (D_0, β^U) is the same as the tree decomposition (T, β^T). However, the decomposition (D_0, β^U) is IFP-definable with three parameters fixing the tuple $\vec{u}_0 = g(r(T))$.

Let us now turn to the canonisation. For every $\vec{u} \in U_0$, we let $G(\vec{u}) := G[\gamma(\vec{u})]$. Then $G = G(\vec{u}_0)$. We inductively define for every $\vec{u} = (u_1, u_2, u_3) \in U_0$ a graph $H(\vec{u})$ with the following properties:

(i) $V\big(H(\vec{u})\big) = [n_{\vec{u}}]$, where $n_{\vec{u}} := |\gamma(\vec{u})| = \big|V(G_{\vec{u}})\big|$.

(ii) There is an isomorphism $f_{\vec{u}}$ from $G(\vec{u})$ to $H(\vec{u})$ such that if $u_1 \neq u_2$ it holds that $f_{\vec{u}}(u_1) = 1$ and $f_{\vec{u}}(u_2) = 2$, and if $u_1 = u_2$ it holds that $f_{\vec{u}}(u_1) = 1$.

For the induction basis, let $\vec{u} \in U_0$ with $N^{D_0}(\vec{u}) = \emptyset$. Then $\gamma^U(\vec{u}) = \beta^U(\vec{u})$, and $G(\vec{u}) = K[\beta^U(\vec{u})]$. We let $n := n_{\vec{u}} = |\beta^U(\vec{u})|$ and $H(\vec{u}) := K_n$. Then (i) and (ii) are obviously satisfied.

For the induction step, let $\vec{u} \in U_0$ and $N^{D_0}(\vec{u}) = \{\vec{v}^1, \dots, \vec{v}^n\} \neq \emptyset$. It follows from Claim 2 that for all $i, j \in [n]$, either $\gamma(\vec{v}^i) = \gamma(\vec{v}^j)$ or $\gamma(\vec{v}^i) \cap \gamma(\vec{v}^j) = \sigma(\vec{v}^i) \cap \sigma(\vec{v}^j) \subseteq \beta(\vec{u})$. We may assume without loss of generality that there are $i_1, \dots, i_m \in [n]$ such that $i_1 < i_2 < \dots < i_m$ and for all $j, j' \in [m]$ with $j \neq j'$ we have $\gamma(\vec{v}^{i_j}) \neq \gamma(\vec{v}^{i_{j'}})$ and for all $j \in [m]$, $i \in [i_j, i_{j+1} - 1]$ we have $\gamma(\vec{v}^i) = \gamma(\vec{v}^{i_j})$. Here and in the following we let $i_{m+1} := n + 1$.

The class of all graphs whose vertex set is a subset of $\mathbb{N}$ may be ordered lexicographically; we let $H \leq_{\text{s-lex}} H'$ if either $V(H)$ is lexicographically smaller than $V(H')$, that is, the first element of the symmetric difference $V(H) \triangle V(H')$ belongs to $V(H')$, or $V(H) = V(H')$ and $E(H)$ is lexicographically smaller than $E(H')$ with respect to the lexicographical ordering of unordered pairs of natural numbers, or $H = H'$. Without loss of generality we may assume that for each $j \in [m]$ it holds that

$$H(\vec{v}^{i_j}) \leq_{\text{s-lex}} H(\vec{v}^{i_j+1}) \leq_{\text{s-lex}} H(\vec{v}^{i_j+2}) \leq_{\text{s-lex}} \cdots \leq_{\text{s-lex}} H(\vec{v}^{i_{j+1}-1})$$

and, furthermore,

$$H(\vec{v}^{i_1}) \leq_{\text{s-lex}} H(\vec{v}^{i_2}) \leq_{\text{s-lex}} \cdots \leq_{\text{s-lex}} H(\vec{v}^{i_m}) . \tag{3}$$

Note that, even though the graphs $G(\vec{v}^{i_1}), G(\vec{v}^{i_2}), \ldots, G(\vec{v}^{i_m})$ are vertex disjoint subgraphs of $G(\vec{u})$, they may be isomorphic, and hence not all of the inequalities in (3) need to be strict. For all $j \in [m]$, let $\vec{v}_j := \vec{v}^{i_j}$ and $G_j := G(\vec{v}_j)$ an $H_j := H(\vec{v}_j)$. Then $H_1 \leq_{\text{s-lex}} H_2 \leq_{\text{s-lex}} \cdots \leq_{\text{s-lex}} H_m$. Let $j_1, \ldots, j_\ell \in [m]$ such that $j_1 < j_2 < \ldots < j_\ell$ and $H_j = H_{j_i}$ for all $i \in [\ell]$, $j \in [j_i, j_{i+1} - 1]$, where $j_{\ell+1} = m+1$, and $H_{j_i} \neq H_{j_{i+1}}$ for all $i \in [\ell - 1]$. For all $i \in [\ell]$, let $J_i := H_{j_i}$. Furthermore, let $n_i := |J_i|$ and $k_i := j_{i+1} - j_i$ and $q_i := |\sigma^U(\vec{v}^{i_j})|$ and

$$q := \left| \beta^U(\vec{u}) \setminus \bigcup_{j=1}^{m} \beta^U(\vec{v}_j) \right| .$$

Case 1: For all neighbours t, t' of $h(\vec{u})$ in the undirected tree underlying T it holds that $\beta^T(t) \cap \beta^T(t') = \emptyset$.
We define $H(\vec{u})$ by first taking a complete graph K_q, then k_1 copies of J_1, then k_2 copies of J_2, et cetera, and finally k_ℓ copies of J_ℓ. The universes of all these copies are disjoint, consecutive intervals of natural numbers. Let K be the union of $[q]$ with the first q_i vertices of each of the k_i copies of J_i for all $i \in [\ell]$. Then K is the set of vertices of $H(\vec{u})$ that corresponds to the clique $\beta(\vec{u})$. We add edges among the vertices in K to turn it into a clique. It is not hard to verify that the resulting structure satisfies (i) and (ii).

Case 2: There are neighbours t, t' of $h(\vec{u})$ in the undirected tree underlying T such that $\beta^T(t) \cap \beta^T(t') \neq \emptyset$.
Then by Lemma 26(iii) we have $|\beta^U(\vec{u})| = 3$, and $h(\vec{u})$ has at most two children. Hence $m \leq 2$, and essentially this means we only have two possibilities of how to combine the parts H_1, H_2 to the graph $H(\vec{u})$; either H_1 comes first or H_2. We choose the lexicographically smaller possibility. We omit the details.

This completes our description of the construction of the graphs $H(\vec{u})$.

It remains to prove that $H(\vec{u})$ is IFP+C-definable. We first define IFP-formulae $\theta_U(\vec{x})$, $\theta_F(\vec{x}, \vec{y})$, $\theta_\alpha(\vec{x}, y)$, $\theta_\beta(\vec{x}, y)$, $\theta_\gamma(\vec{x}, y)$, $\theta_\sigma(\vec{x}, y)$ such that

$$\begin{aligned} U &= \{\vec{u} \in V(G)^3 \mid G \models \theta_U[\vec{u}]\}, \\ F &= \{(\vec{u}, \vec{v}) \in U^2 \mid G \models \theta_F[\vec{u}, \vec{v}]\}, \\ \alpha^U(\vec{u}) &= \{v \in V(G) \mid G \models \theta_\alpha[\vec{u}, v]\} \qquad \text{for all } \vec{u} \in U, \end{aligned}$$

and similarly for β, γ, σ. Then we define formulae $\theta_U^0(\vec{x}_0, \vec{x})$, $\theta_F^0(\vec{x}_0, \vec{x})$ that define D_0. We have no canonical way of checking that a tuple $\vec{u}_0$ really is the image $g(r(T))$ of the root of a good tree decomposition, but all we need is that the graph $D^0(\vec{u}_0)$ with vertex set $\{\vec{u} \in V(G)^3 \mid G \models \theta_U^0[\vec{u}_0, \vec{u}]\}$ and edge set $\{(\vec{u}, \vec{v}) \in U^2 \mid G \models \theta_F[\vec{u}_0, \vec{u}, \vec{v}]\}$ has the properties we derive from T being a good tree decomposition. In particular, if a node $\vec{u}$ has a child $\vec{v}$ with $\sigma^U(\vec{u}) \cap \sigma^U(\vec{v}) \neq \emptyset$ or children $\vec{v}_1 \neq \vec{v}_2$ with $\sigma^U(\vec{v}_1) \cap \sigma^U(\vec{v}_1) \neq \emptyset$, then $|\beta^U(\vec{u})| \leq 3$. Once we have defined D^0, it is straightforward to formalise the definition of the graphs $H(\vec{u})$

in IFP+C and define an IFP+C-interpretation $\Gamma(\vec{x}_0)$ that canonises G. We leave the (tedious) details to the reader. □

Remark 29. Implicitly, the previous proof heavily depends on the concepts introduced in [31]. In particular, the definable directed graph D together with the definable mappings σ and α constitute a *definable tree decomposition*. However, our theorem does not follow directly from Theorem 7.2 of [31].

The class $\mathcal{CD} \cap \mathcal{L}$ of chordal line graphs is fairly restricted, and there may be an easier way to prove the canonisation theorem by using Proposition 22. The proof given here has the advantage that it generalises to the class of all chordal graphs that have a good tree decomposition where the bags of the neighbours of a node intersect in a "bounded way". We omit the details.

5 Further Research

I mentioned several important open problems related to the quest for a logic capturing PTIME in the survey in Section 1. Further open problems can be found in [32]. Here, I will briefly discuss a few open problems related to classes closed under taking induced subgraphs, or equivalently, classes defined by excluding (finitely or infinitely many) induced subgraphs.

A fairly obvious, but not particularly interesting generalisation of our positive capturing result is pointed out in Remark 29. I conjecture that our theorem for chordal line graphs can be generalised to the class of chordal claw-free graphs, that is, I conjecture that the class of chordal claw-free graphs admits IFP+C-definable canonisation. Further natural classes of graphs closed under taking induced subgraphs are the classes of disk intersection graphs and unit disk intersection graphs. It is open whether IFP+C or any other logic captures PTIME on these classes. A very interesting and rich family of classes of graphs closed under taking induced subgraphs is the family of classes of graphs of bounded rank width [58], or equivalently, bounded clique width [13]. It is conceivable that IFP+C captures polynomial time on all classes of bounded rank width. To the best of my knowledge, currently it is not even known whether isomorphism testing for graphs of bounded rank width is in polynomial time.

Acknowledgements

I would like to thank Yijia Chen and Bastian Laubner for valuable comments on an earlier version of this paper.

References

1. Abiteboul, S., Vianu, V.: Non-deterministic languages to express deterministic transformations. In: Proceedings of the 9th ACM Symposium on Principles of Database Systems, pp. 218–229 (1990)

2. Aho, A., Ullman, J.: The universality of data retrieval languages. In: Proceedings of the Sixth Annual ACM Symposium on Principles of Programming Languages, pp. 110–120 (1979)
3. Babai, L., Grigoryev, D., Mount, D.: Isomorphism of graphs with bounded eigenvalue multiplicity. In: Proceedings of the 14th ACM Symposium on Theory of Computing, pp. 310–324 (1982)
4. Babai, L., Luks, E.: Canonical labeling of graphs. In: Proceedings of the 15th ACM Symposium on Theory of Computing, pp. 171–183 (1983)
5. Beineke, L.: Characterizations of derived graphs. Journal of Combinatorial Theory 9, 129–135 (1970)
6. Blass, A., Gurevich, Y., Shelah, S.: Choiceless polynomial time. Annals of Pure and Applied Logic 100, 141–187 (1999)
7. Blass, A., Gurevich, Y., Shelah, S.: On polynomial time computation over unordered structures. Journal of Symbolic Logic 67, 1093–1125 (2002)
8. Bodlaender, H.: Polynomial algorithms for graph isomorphism and chromatic index on partial k-trees. Journal of Algorithms 11, 631–643 (1990)
9. Cai, J., Fürer, M., Immerman, N.: An optimal lower bound on the number of variables for graph identification. Combinatorica 12, 389–410 (1992)
10. Chandra, A., Harel, D.: Structure and complexity of relational queries. Journal of Computer and System Sciences 25, 99–128 (1982)
11. Chudnovsky, M., Robertson, N., Seymour, P., Thomas, R.: The strong perfect graph theorem. Annals of Mathematics 164, 51–229 (2006)
12. Chudnovsky, M., Seymour, P.: The structure of claw-free graphs. In: Webb, B. (ed.) Surveys in Combinatorics. London Mathematical Society Lecture Note Series, vol. 327, pp. 153–171. Cambridge University Press, Cambridge (2005)
13. Courcelle, B., Olariu, S.: Upper bounds to the clique-width of graphs. Discrete Applied Mathematics 101, 77–114 (2000)
14. Dawar, A.: Generalized quantifiers and logical reducibilities. Journal of Logic and Computation 5, 213–226 (1995)
15. Dawar, A.: A restricted second order logic for finite structures. In: Leivant, D. (ed.) LCC 1994. LNCS, vol. 960, Springer, Heidelberg (1995)
16. Dawar, A., Grohe, M., Holm, B., Laubner, B.: Logics with rank operators. In: Proceedings of the 24th IEEE Symposium on Logic in Computer Science, pp. 113–122 (2009)
17. Dawar, A., Hella, L.: The expressive power of finitely many generalized quantifiers. In: Proceedings of the 9th IEEE Symposium on Logic in Computer Science (1994)
18. Dawar, A., Richerby, D.: A fixed-point logic with symmetric choice. In: Baaz, M., Makowsky, J.A. (eds.) CSL 2003. LNCS, vol. 2803, pp. 169–182. Springer, Heidelberg (2003)
19. Dawar, A., Richerby, D., Rossman, B.: Choiceless polynomial time, counting and the Cai-Fürer-Immerman graphs: (Extended abstract). Electronic Notes on Theoretical Compututer Science 143, 13–26 (2006)
20. Diestel, R.: Graph Theory, 3rd edn. Springer, Heidelberg (2005)
21. Ebbinghaus, H.D., Flum, J.: Finite Model Theory, 2nd edn. Springer, Heidelberg (1999)
22. Ebbinghaus, H.D., Flum, J., Thomas, W.: Mathematical Logic, 2nd edn. Springer, Heidelberg (1994)
23. Evdokimov, S., Karpinski, M., Ponomarenko, I.: On a new high dimensional Weisfeiler-Lehman algorithm. Journal of Algebraic Combinatorics 10, 29–45 (1999)
24. Evdokimov, S., Ponomarenko, I.: On highly closed cellular algebras and highly closed isomorphism. Electronic Journal of Combinatorics 6, #R18 (1999)

25. Fagin, R.: Generalized first–order spectra and polynomial–time recognizable sets. In: Karp, R.M. (ed.) Complexity of Computation. SIAM-AMS Proceedings, vol. 7, pp. 43–73 (1974)
26. Filotti, I.S., Mayer, J.N.: A polynomial-time algorithm for determining the isomorphism of graphs of fixed genus. In: Proceedings of the 12th ACM Symposium on Theory of Computing, pp. 236–243 (1980)
27. Gire, F., Hoang, H.: An extension of fixpoint logic with a symmetry-based choice construct. Information and Computation 144, 40–65 (1998)
28. Grädel, E., Kolaitis, P., Libkin, L., Marx, M., Spencer, J., Vardi, M., Venema, Y., Weinstein, S.: Finite Model Theory and Its Applications. Springer, Heidelberg (2007)
29. Grädel, E., Otto, M.: On Logics with Two Variables. Theoretical Computer Science 224, 73–113 (1999)
30. Grohe, M.: Fixed-point logics on planar graphs. In: Proceedings of the 13th IEEE Symposium on Logic in Computer Science, pp. 6–15 (1998)
31. Grohe, M.: Definable tree decompositions. In: Proceedings of the 23rd IEEE Symposium on Logic in Computer Science, pp. 406–417 (2008)
32. Grohe, M.: The quest for a logic capturing PTIME. In: Proceedings of the 23rd IEEE Symposium on Logic in Computer Science, pp. 267–271 (2008)
33. Grohe, M.: Fixed-point definability and polynomial time on graphs with excluded minors. In: Proceedings of the 25th IEEE Symposium on Logic in Computer Science (2010) (to appear)
34. Grohe, M., Mariño, J.: Definability and descriptive complexity on databases of bounded tree-width. In: Beeri, C., Bruneman, P. (eds.) ICDT 1999. LNCS, vol. 1540, pp. 70–82. Springer, Heidelberg (1998)
35. Grohe, M., Verbitsky, O.: Testing graph isomorphism in parallel by playing a game. In: Bugliesi, M., Preneel, B., Sassone, V., Wegener, I. (eds.) ICALP 2006, Part I. LNCS, vol. 4051, pp. 3–14. Springer, Heidelberg (2006)
36. Grädel, E.: Finite model theory and descriptive complexity. In: Grädel, E., Kolaitis, P.G., Libkin, L., Marx, M., Spencer, J., Vardi, M.Y., Venema, Y., Weinstein, S. (eds.) Finite Model Theory and Its Applications, pp. 125–230. Springer, Heidelberg (2007)
37. Gurevich, Y.: Logic and the challenge of computer science. In: Börger, E. (ed.) Current trends in theoretical computer science, pp. 1–57. Computer Science Press, Rockville (1988)
38. Gurevich, Y.: Sequential abstract-state machines capture sequential algorithms. ACM Transaction on Computational Logic 1, 77–111 (2000)
39. Gurevich, Y., Shelah, S.: Fixed point extensions of first–order logic. Annals of Pure and Applied Logic 32, 265–280 (1986)
40. Hella, L.: Definability hierarchies of generalized quantifiers. Annals of Pure and Applied Logic 43, 235–271 (1989)
41. Hella, L., Kolaitis, P., Luosto, K.: Almost everywhere equivalence of logics in finite model theory. Bulletin of Symbolic Logic 2, 422–443 (1996)
42. Hopcroft, J. E., Wong, J.: Linear time algorithm for isomorphism of planar graphs. In: Proceedings of the 6th ACM Symposium on Theory of Computing, pp. 172–184 (1974)
43. Hopcroft, J.E., Tarjan, R.: Isomorphism of planar graphs (working paper). In: Miller, R.E., Thatcher, J.W. (eds.) Complexity of Computer Computations. Plenum Press, New York (1972)
44. Immerman, N.: Relational queries computable in polynomial time. Information and Control 68, 86–104 (1986)

45. Immerman, N.: Expressibility as a complexity measure: results and directions. In: Proceedings of the 2nd IEEE Symposium on Structure in Complexity Theory, pp. 194–202 (1987)
46. Immerman, N.: Languages that capture complexity classes. SIAM Journal on Computing 16, 760–778 (1987)
47. Immerman, N.: Descriptive Complexity. Springer, Heidelberg (1999)
48. Immerman, N., Lander, E.: Describing graphs: A first-order approach to graph canonization. In: Selman, A. (ed.) Complexity theory retrospective, pp. 59–81. Springer, Heidelberg (1990)
49. Köbler, J., Verbitsky, O.: From invariants to canonization in parallel. In: Hirsch, E.A., Razborov, A.A., Semenov, A., Slissenko, A. (eds.) Computer Science – Theory and Applications. LNCS, vol. 5010, pp. 216–227. Springer, Heidelberg (2008)
50. Kreutzer, S.: Expressive equivalence of least and inflationary fixed-point logic. Annals of Pure and Applied Logic 130, 61–78 (2004)
51. Laubner, B.: Capturing polynomial time on interval graphs. In: Proceedings of the 25th IEEE Symposium on Logic in Computer Science (2010) (to appear)
52. Libkin, L.: Elements of Finite Model Theory. Springer, Heidelberg (2004)
53. Luks, E.: Isomorphism of graphs of bounded valance can be tested in polynomial time. Journal of Computer and System Sciences 25, 42–65 (1982)
54. Miller, G.L.: Isomorphism testing for graphs of bounded genus. In: Proceedings of the 12th ACM Symposium on Theory of Computing, pp. 225–235 (1980)
55. Otto, M.: Bounded variable logics and counting – A study in finite models. Lecture Notes in Logic, vol. 9. Springer, Heidelberg (1997)
56. Otto, M.: Canonization for two variables and puzzles on the square. Annals of Pure and Applied Logic 85, 243–282 (1997)
57. Otto, M.: Bisimulation-invariant PTIME and higher-dimensional μ-calculus. Theoretical Computer Science 224, 237–265 (1999)
58. Oum, S.I., Seymour, P.: Approximating clique-width and branch-width. Journal of Combinatorial Theory, Series B 96, 514–528 (2006)
59. Roussopoulos, N.: A $\max\{m, n\}$ algorithm for determining the graph H from its line graph G. Information Processing Letters 2, 108–112 (1973)
60. Vardi, M.: The complexity of relational query languages. In: Proceedings of the 14th ACM Symposium on Theory of Computing, pp. 137–146 (1982)
61. Verbitsky, O.: Planar graphs: Logical complexity and parallel isomorphism tests. In: Thomas, W., Weil, P. (eds.) STACS 2007. LNCS, vol. 4393, pp. 682–693. Springer, Heidelberg (2007)
62. Whitney, H.: Congruent graphs and the connectivity of graphs. American Journal of Mathematics 54, 150–168 (1932)

Ibn Sīnā on Analysis: 1. Proof Search. Or: Abstract State Machines as a Tool for History of Logic

Wilfrid Hodges

Herons Brook, Sticklepath, Okehampton EX20 2PY, England
wilfrid.hodges@btinternet.com

For Yuri Gurevich on the occasion of his seventieth birthday.

Abstract. The 11th century Arabic-Persian logician Ibn Sīnā (Avicenna) in Sect. 9.6 of his book *Qiyās* gives what appears to be a proof search algorithm for syllogisms. We confirm that it is indeed a proof search algorithm, by extracting all the essential ingredients of an Abstract State Machine from Ibn Sīnā's text. The paper also contains a translation of the passage from Ibn Sīnā's Arabic, and some notes on the text and translation.

Keywords: abstract state machine, proof search, Ibn Sīnā, Avicenna, syllogism.

1 Introduction

This paper contains a translation and commentary on Sect. 9.6 of Ibn Sīnā's major work on logic, the volume 'Syllogism' (*Qiyās*) from his encyclopedic *Šifā'*, a work written in Arabic in the 1020s. (Sect. 9.6 is the first of four sections, 9.6–9.9, on what Ibn Sīnā calls 'analysis'; hence 'Analysis: 1' in the title of this paper.) The section is itself a loose commentary on some lines in Aristotle's *Prior Analytics* i.32. It falls into two parts. In the first part Ibn Sīnā describes what he sees as the task of logical 'analysis' (*taḥlīl*). One ingredient of that task is to complete formal proofs which have a piece missing, and Ibn Sīnā gives his account of this in the second part. A special case of this problem (though not one mentioned by Ibn Sīnā himself) is to find a formal proof where everything is missing except the conclusion, and this is precisely the task of proof search. To the best of my knowledge, Ibn Sīnā's account is the first work to come anywhere near describing a proof search algorithm in formal logic.

Abstract State Machines (ASMs [6]) were introduced by Yuri Gurevich [10], in whose honour this essay is written. They give a framework for describing algorithms with complete precision at whatever level of refinement we choose. The main business of this paper is to describe Ibn Sīnā's intended algorithm. The fact that Ibn Sīnā himself is less than explicit about some details is no excuse for us to lapse into vagueness. If we want to record with decent precision what Ibn

A. Blass, N. Dershowitz, and W. Reisig (Eds.): Gurevich Festschrift, LNCS 6300, pp. 354–404, 2010.

Sīnā used or understood, and what he didn't, we need the best descriptive tools; and so I turned to ASMs. Fortunately the work is already partly done, because a famous early application of ASMs was Börger and Rosenzweig's specification of the proof search algorithm of Prolog to meet the ISO 1995 standard [5].

As far as I know, the use of ASMs below is the first application of ASMs to the history of logic, and one of the first applications of ASMs in the humanities. (A recent paper [11] calls for applications in linguistics, but these go in a rather different direction.) In practice the task of constructing an ASM was an invaluable research tool; it kept raising questions to be addressed to Ibn Sīnā's text. Remarkably often Ibn Sīnā does answer these questions in his text, though I often had to refer to other sections of the *Qiyās* for clarifications. I doubt that any other logician between Aristotle and Leibniz would have come through this test as successfully as Ibn Sīnā does.

The paper has an unusually wide spread of prerequisites. First there is the Arabic text of Ibn Sīnā and its historical background. Second there are the mathematical facts about syllogisms. Third there is the methodology of Abstract State Machines. Unfortunately papers are linear strings of text, so some prerequisites will have to wait their turn. The structure of the paper is as follows:

Sect. 1. Introduction.
Sect. 2. Historical background (on logic from Aristotle to Ibn Sīnā).
Sect. 3. *taḥṣīl* (roughly the counterpart in Ibn Sīnā of Tarski's notion of setting up a deductive theory).
Sect. 4. Mathematical prerequisites on syllogisms.
Sect. 5. Extracting the algorithm.
Sect. 6. Review.
Appendix A. Translation of *Qiyās* 9.6.
Appendix B. Notes on the text translated.
Appendix C. The ASM.

The passage translated in Appendix A (*Qiyās* 9.6) needs to be matched up with *Qiyās* Sects. 2.4 on categorical syllogisms, 9.3 on compound syllogisms, 9.4 on supplying missing premises of simple syllogisms and 9.7–9 on other aspects of analysis. I will do my best to get translations of these sections onto my website at `http://wilfridhodges.co.uk`. Meanwhile Tony Street [30] gives a useful summary of Ibn Sīnā's theory of predicative syllogisms.

I thank Egon Börger, Jamal Ouhalla, Roshdi Rashed, Gabriel Sabbagh and the referee for some valuable remarks and suggestions, and Amirouche Moktefi for advice on the Arabic translation. But I take full responsibility for errors; there are bound to be some, though I believe the use of ASMs has eliminated many of the more serious ones.

2 Historical Background

In the middle of the 4th century BC, Aristotle noticed that many arguments in mathematics, metaphysics and elsewhere have one of a small number of forms, and that any argument of any of these forms is guaranteed to be convincing. He referred to arguments of these forms as 'syllogisms', and he classified them into three 'figures'. He listed and discussed the argument forms in lectures or writings which have reached us as a book called *Prior Analytics* [3]. That book was part of the edition of Aristotle's writings which was put together by Andronicus in the first century BC. Apart from the text itself, we have virtually no evidence of what Andronicus did with his raw materials. He may have put things together in ways that Aristotle never intended.

Andronicus' edition of Aristotle came to form a collection of textbooks for the offspring of cultured parents in the Roman Empire. By the late second century AD it had become clear that some explanatory commentaries on Aristotle's text were needed, and Alexander of Aphrodisias wrote a set. His commentaries were followed by many others, mostly now lost. The two surviving Roman Empire commentaries on the parts of the *Prior Analytics* that will concern us are those of Alexander and the 6th century Alexandrian scholar John Philoponus. (See Ebbesen [8] Chapter III for an account of the intellectual climate in which commentaries on Aristotle's logic arose.)

By the middle of the 8th century the new Arab empire had started to absorb western scholarship, including Aristotle's logic. Ibn Sīnā reports that in the 990s he visited the large library of the Sultan of Bukhara (in present-day Uzbekistan), and found that it contained a catalogued collection of books of 'the ancients', which included a number of rare items, presumably in Arabic or Persian translation (Gutas [12] p. 28f). Most of this material has now gone missing, together with the Greek originals. In the 10th century Al-Fārābī wrote a lengthy Arabic commentary on the *Prior Analytics*. Ibn Sīnā probably knew this work, but today very little of it survives, and nothing that will help us in this paper.

The longest and fullest of Ibn Sīnā's writings in logic is the Logic section of his encyclopedic *Šifā'*, written in the 1020s. It takes the form of a commentary on Aristotle's logic, some of it very close to Aristotle and some of it apparently quite new. In *Šifā'* the book *Qiyās* ('Syllogism' [15]) is his commentary on the *Prior Analytics*. In *Qiyās*, Sect. 9.6 is his commentary on just twenty-three lines of the *Prior Analytics*, namely 46b40–47a22. These are the opening lines of Sect. i.32 of *Prior Analytics*.

Aristotle begins these twenty-three lines by announcing that his next task is to 'explain how we can lead deductions back into the figures stated previously' ([3] p. 50). He adds that this is a matter of analysing (*analúoimen*) arguments that 'have already been produced' ([3] p. 50) into the three syllogistic figures. The arguments could have been already produced 'in writing or in speech' ([3] p. 51). He makes it clear that analysis includes both identifying the underlying form of an argument, and also *repairing* the argument, for example adding missing premises or removing redundancies.

Alexander of Aphrodisias and Philoponus both report Aristotle's views faithfully. Alexander adds that since the whole book is called *Analytics*, this section on 'analysis' of arguments must be the heart of it. The modern commentator David Ross agrees that the use of 'analyse' in this passage is the source of the name *Analytics*, both for this book and for Aristotle's work *Posterior Analytics* on the theory of knowledge ([28] p. 400). He also calls attention to the mathematical use of *analúein* to mean working backwards from conclusions to premises. This usage agrees with the part of Aristotle's 'analysis' that consists of finding a missing premise.

Ibn Sīnā's Sect. 9.6 falls into two parts. The first part, from paragraph [9.6.1] to [9.6.5], more or less matches Aristotle's text. The second part, consisting of paragraphs [9.6.6] to [9.6.12], is completely new. It picks up Aristotle's brief remark that the syllogism being analysed may have a premise missing, and it discusses how to fill the hole. Ibn Sīnā's presentation in this part is very unusual: instead of explaining his method, he illustrates it with sixty-four examples and some comments. For example here are two of his examples for completing a syllogism whose conclusion (the 'goal') is the sentence 'Some C is an A':

> If the *ḥāṣil* [premises] are 'Every D is a C' and 'Every B is an A', and 'Every (or some) D is a B' is attached, then this makes the syllogism *ḥāṣil*. (1)
>
> If the *ḥāṣil* [premises] are 'Every D is a C' and 'Some B is an A', it can't be used. (From Problems 9, 10 in Appendix A below.)

Clearly this text needs some interpretation. We begin with the word '*ḥāṣil*', which expresses a central notion in Ibn Sīnā's methodology.

3 *taḥṣīl*

There are two notions to be brought together here. One is *taḥlīl*, which is the Arabic word that Ibn Sīnā uses to translate Aristotle's *análusis*. Ibn Sīnā regarded *Prior Analytics* i.32–46 as a manual of analysis, and he commented on these sections in Sects. 9.6 to 9.9 of his *Qiyās*. The material in Sects. 9.7–9 is not directly related to that in 9.6, but it is needed for a full picture of Ibn Sīnā's understanding of analysis.

The second notion is *taḥṣīl*, which means 'making *ḥāṣil*'. There are no easy English translations of *taḥṣīl* and *ḥāṣil*, and even if there were, we would still need to explain how the notions fit into Ibn Sīnā's view of philosophical activity.

At the most literal level, *ḥāṣil* means 'available for use', so that *taḥṣīl* means 'making available for use'. The word *ḥāṣil* occurs nine times in *Qiyās* 9.6, and its grammatical relatives many times more. A thing is *muḥaṣṣal* if it has been made available for use. Here is a remarkable example of the literal usage:

> ... some people demonstrate without any rule, like Archimedes who demonstrated mathematically, since in his time logic wasn't yet available (*lam yakun muḥaṣṣal*). (*Qiyās* [15] 15.10f.) (2)

Ibn Sīnā has his history confused – Archimedes was born a hundred years later than Aristotle. The idea that Archimedes demonstrated 'without any rule' is puzzling. Roshdi Rashed (personal communication) suggests that the point is that geometrical reasoning, of the kind that Archimedes used, is not algorithmic.

For Ibn Sīnā, one of the main tasks of a philosopher was to apply *taḥṣīl* to the ideas of earlier philosophers. He refers several times to commentators on Aristotle as *muḥaṣṣilūn*, people who make *ḥāṣil*. A typical example is in *Išārāt*:

> Nothing but this has been stated by earlier scholars (*muḥaṣṣilūn*), but in a manner overlooked by recent ones. ([20] I.9.2, p. 150 of Inati.) (3)

Likewise at *Najāt* [18] i p. 35.4 he refers to 'Alexander and a number of later *muḥaṣṣilūn*'.

For readabilty, henceforth I follow a suggestion of the referee and use an English word to stand for *ḥāṣil* where Ibn Sīnā uses it as a technical term. For the rest of this paper, including the translation in Appendix A, 'determinate' means *ḥāṣil*. It is not an exact translation; there is no exact English translation of *ḥāṣil*.

What were the commentators doing that counted as 'making determinate (*ḥāṣil*)'? The answer has to depend on exactly what they were making determinate. We find places where Ibn Sīnā describes the following things as being made determinate: (a) concepts, (b) propositions, (c) syllogisms, (d) knowledge. Usages (b) and (c) are frequent in *Qiyās* 9.6.

Usage (d) is illustrated by the following passage near the beginning of *Burhān*:

> Knowledge – whether it is obtained through reflective reasoning (*fikr*) or is determinate (*ḥāṣil*) without being obtained through reflective reasoning – is of two kinds. One of them is assent (*taṣdīq*) and the other is conceptualisation (*taṣawwur*). Knowledge in the form of assent, when it is obtained through reflective reasoning, becomes determinate to us through a syllogism. Knowledge in the form of conceptualisation, when it is obtained through reflective reasoning, becomes determinate to us through a definition. ([16] 3.10–12.) (4)

Here Ibn Sīnā sets out two independent classifications of kinds of knowledge. The first classification is into those forms of knowledge which depend on reflective thinking and those which come to us without our having to think reflectively. The second classification, which is fundamental throughout Ibn Sīnā's logic and epistemology, is between two processes that lead to knowledge. The first of these processes is conceptualisation (*taṣawwur*); it leads us to having a concept, and Ibn Sīnā counts this as a kind of knowledge. The second process is assent (*taṣdīq*), i.e. coming to recognise that a proposition is true; it leads to knowledge of the fact stated by the proposition. Although Ibn Sīnā in his first sentence uses 'determinate' only for knowledge not dependent on reflective thinking, the rest of his text shows that this is just an accident of style, and both kinds of knowledge can be determinate. In fact the passage suggests that for knowledge, being determinate and being 'obtained' amount to the same thing.

The passage gives us strong clues about usages (a) and (b), because *taṣawwur* leads to knowledge of concepts and *taṣdīq* leads to knowledge of propositions.

Take concepts first. Here Ibn Sīnā's usage slots in with a philosophical usage that had been around already for many decades. The 9th century translator of

Aristotle, Isḥāq bin Ḥunain, rendered Aristotle's 'indefinite' (*aóristos*) as *gair muḥaṣṣal*, i.e. 'not *muḥaṣṣal*' (*Peri Hermeneías* 16b14, translated at [21] p. 111). The implication is that a concept is *muḥaṣṣal* if it is well-defined. Kutsch [23] assembles a large number of references where *muḥaṣṣal* means this.

In (4) above, Ibn Sīnā is saying that a concept is made determinate or well-defined by being given a definition. He certainly regarded this as one of the main tasks of the Aristotelian commentators. At ^{c}*Ibāra* [14] 2.9f he refers to 'those commentators who are experts on definition' (*al-muḥaṣṣilūn min 'ahl ṣinā^{c}a t-taḥdīd*). Incidentally there is a close analogy here with Ernst Zermelo's notion of a 'definit' criterion for class membership in mathematics [33]. Just as Ibn Sīnā expected a commentator to define in genus-differentia form, so Skolem proposed that Zermelo's 'definit' should be read in set theory as 'first-order definable'.

We turn to propositions. Ibn Sīnā says in several places that we can't assent to a proposition until we have conceptualised its meaning, i.e. until we understand it. Thus in *Easterners*:

> So when the conceptualisation is made determinate (*ḥāṣil*) for us, assent to [the proposition] is made determinate for us [too]. But the conceptualisation comes first; so if we don't conceptualise a meaning then we don't get assent to [the proposition]. Sometimes we get the conceptualisation without assent attached to it. ([19] 9.12–14.) (5)

So there is a sense in which making a proposition determinate is like making a concept determinate; we clarify the construction of the proposition and the meanings of the words in it. But in both (4) and (5), Ibn Sīnā mentions another sense in which a proposition can become determinate, namely that we come to recognise that it is true. In (4) he says that this happens when we deduce the proposition through a syllogism. We note that for this to work, the premises of the syllogism must already be determinate in this sense.

Ibn Sīnā regarded this second kind of making propositions determinate as central to the activity of philosophical commentators. The activity consists of taking a claim made by Aristotle (for example), or on his behalf, and looking to see how much of an argument is offered to support the claim. Then one works on the argument to fill in gaps, remove irrelevances etc. etc.

> Sometimes a person is addressed with a well-crafted and definitive syllogism, or he finds such a syllogism written in a book. But then ... sometimes the pieces are jumbled out of their natural order, or a part of the syllogism is hidden, or something superfluous is added. ... If we don't have rules to guide us, on how to seek with due deliberation the syllogism that proves a given goal, [and to confirm] the soundness of the connection between a given syllogism [and its goal], so that we can analyse the syllogism into a group of premises, put them in the natural order, strip off defects and add any part that is missing, reducing the syllogism to the syllogistic figure that produces it – [if we don't have rules for all this,] then the new information that the syllogism provides will escape us. (From paragraph [9.6.1] in Appendix A.) (6)

In fact one performs exactly the 'analysis' that we saw Aristotle himself describing in *Prior Analytics* i.32ff. But while for Aristotle and Alexander this kind of analysis was one of the general tools of logic, Ibn Sīnā thought he could point to a large body of published work specifically devoted to it, namely the philosophical commentaries. (There is a hint of this view already in Philoponus [24] p. 315 l. 20, where he says that the syllogism to be analysed may come from 'the ancients'.)

To fill in the history a little, the idea of commenting on a philosopher by reducing that philosopher's arguments to syllogistic form seems to have surfaced first among the Middle Platonist commentators on Plato's dialogues in the first century AD. It may have been encouraged by a desire to show that Plato was just as good a logician as Aristotle (a view that Ibn Sīnā explicitly rejects with contempt [17] pp. 114f). For example Alcinous [1] 158.42–159.3 finds the following second-figure syllogism in Plato's *Parmenides* 137d8–138a1:

> A thing that has no parts is neither straight nor circular. A thing that has a shape is either straight or circular. Therefore a thing that has no parts has no shape. (7)

(The second premise is obviously false. In any case Plato as I read him gives 'straight' and 'circular' as typical examples of shapes, not as an exhaustive list. But Alcinous wasn't the world's greatest logician.)

Most of the surviving Roman Empire or Arabic commentaries on Aristotle, including those of Ibn Sīnā, do contain explicit reductions of particular arguments to syllogistic form. These reductions form a very small proportion of the text of the commentaries. But probably Ibn Sīnā regarded it as a criterion of the quality of a commentary that it should be straightforward to analyse the commentator's arguments into syllogistic form. The analogy with modern set theory applies here too. We don't expect set theorists to set out their arguments as first-order deductions from the Zermelo-Fraenkel axioms, but we do take it as a criterion of a sound set-theoretic argument that it should be routine to reduce the argument to this form.

By implication we have already said what it should mean to describe a syllogism as determinate. We make a syllogism determinate by analysing it into a form so that it makes its conclusion determinate. This involves putting it into one of the standard syllogistic moods, and ensuring that its premises are determinate.

There are a couple of nuts-and-bolts points about *taḥṣīl* that can be made here as well as anywhere. First, the notion of 'determinate' is relational: a thing can be determinate for me but not for you. This is explicit in both (4) and (5). As far as I'm aware, there is no notion in Ibn Sīnā of a thing being 'determinate in itself but not for us', such as we might expect in 13th or 14th century Scholastics.

And second, the set of propositions that are determinate for you is dynamic: you can add new items to the set by deducing them from things already in the set. This causes some problems of terminology. In proof search we assume we have a database T of sentences, and we search for proofs of given sentences from assumptions that are in T. In Ibn Sīnā's case the set T is the set of propositions that are *already* determinate. But it's natural for him to say that a successful proof search makes another proposition ϕ determinate, and it could look as if he is saying that ϕ is added to the database. Granted, Prolog has a function *assert* which does exactly that. But adding ϕ to T is completely different from deducing ϕ from things already in T, and it's the latter that is important for the proof search algorithm. The remedy is to distinguish strictly between those propositions that were *already* determinate and those that *become* determinate through application of the algorithm. Ibn Sīnā's choice of words doesn't always help us to make this distinction; see Problem 32 in Appendix A and the note on it in Appendix B.

4 Mathematical Prerequisites on Syllogisms

4.1 Syllogistic Sentences

In the part of *Qiyās* [15] Sect. 9.6 that relates to proof search, Ibn Sīnā discusses four kinds of sentence, namely those in (8) below. (Sometimes he varies the wording of these sentence types; see for example the note on Problem 14.) In Ibn Sīnā's logical theory the logical properties of sentences of these types are determined by their truth conditions. These truth conditions were explained in *ᶜIbāra* [14] Sect. 2.2, which comes before *Qiyās* in the *Šifā'*.

The four sentence types are as follows, together with their names and their truth conditions:

– Universally quantified affirmative, 'Every A is a B'. This counts as true if there are As, but there are no As that are not Bs, and false otherwise.
– Universally quantified negative, 'No A is a B'. This counts as true if there are no As that are Bs, and false otherwise.
– Existentially quantified affirmative, 'Some A is a B'. This counts as true if there is some A that is a B, and false otherwise.
– Existentially quantified negative, 'Some A is not a B'. This counts as true in two cases: (a) there is an A that is not a B, and (b) there are no As. Otherwise it counts as false.

(8)

The letters 'A' and 'B' are place-holders for two distinct 'terms' (*ḥadd*). Warning: terms in traditional logic are not at all the same thing as terms in modern logic. For present purposes we can think of terms in Ibn Sīnā as being the meanings of actual or possible common nouns. (There is no requirement that the nouns describe nonempty classes.)

Ibn Sīnā believed that when reasoning we manipulate terms in our minds through linguistic expressions that mean them. This allowed him to do the same in his logical theory, for example using common nouns as surrogates for their meanings. A syllogistic sentence of the form 'Every A is a B' is got by putting common nouns in place of 'A' and 'B', with the sole restriction that the two common nouns must have different meanings. By 'syllogistic sentence' we will mean a sentence of one of the four forms in (8).

A syllogistic sentence can be identified by four features. The first is the 'subject' (*mawḍū*^{c}), which is the term put for 'A'. The second is the 'predicate' (*maḥmūl*), which is the term put for 'B'. The third is the 'quantity' (*kam*), which is either 'existentially quantified' (*juz'ī*) or 'universally quantified' (*kullī*). The fourth is the 'quality' (*kaifa*), which is either 'affirmative' (*mūjib*) or 'negative' (*maslūb*). For purposes of the ASM I treat a syllogistic sentence as a 4-tuple

$$[\text{subject,predicate,quantity,quality}] \tag{9}$$

using 0 for existentially quantified and affirmative, and 1 for universally quantified and negative. (See (Def1) in Appendix C.)

The conditions for 'Every A is a B' to be true are satisfied exactly when those for 'Some A is not a B' are not satisfied. So each of these syllogistic sentences means the same as the negation of the other. We say they are 'contradictories' of each other, and we write $\bar{\phi}$ for the contradictory of ϕ. Likewise 'No A is a B' and 'Some A is a B' are contradictories.

By 'formal sentences' I mean the expressions that we get if we put uninterpreted 1-ary relation symbols (we call them 'term symbols') in place of 'A', 'B' in (8) above. The truth conditions translate at once into conditions for a formal sentence to be true in a structure. So we have a model-theoretic notion of entailment: a set T of formal sentences entails a formal sentence ψ if and only if there is no structure in which all the formal sentences in T are true but ψ is not true. Though this notion was unknown to Ibn Sīnā, it gives us some mathematics that will be helpful for understanding various things that Ibn Sīnā does.

For example it allows us to demonstrate all the cases where one formal sentence entails another. They are as follows (where we write $\Rightarrow$ for 'entails'):

Every A is a B.		Every B is an A.	
$\Downarrow$		$\Downarrow$	
Some A is a B.	$\Leftrightarrow$	Some B is an A.	
			(10)
Some A is not a B.		Some B is not an A.	
$\Uparrow$		$\Uparrow$	
No A is a B.	$\Leftrightarrow$	No B is an A.	

The top and bottom halves of this diagram are not independent. Each sentence in the bottom half is the contradictory of its counterpart in the top half. Hence the arrows in the bottom half go the opposite way to those in the top half. Ibn Sīnā recognised all the instances of these entailments as examples of 'following from'.

4.2 Inconsistent Sets

A set T of formal sentences is 'consistent' if there is a structure in which all the formal sentences in T are true, and 'inconsistent' if there is no such structure.

It is 'minimal inconsistent' if it is inconsistent but every proper subset of it is consistent.

We can characterise the minimal inconsistent sets of formal sentences as follows. First, by a 'minimal circle' we mean a set of formal sentences arranged in a circle

$$[\phi_1, \ldots, \phi_n] \ \ (n \geqslant 2) \tag{11}$$

where ϕ_1 is immediately after ϕ_n in the circle, in such a way that every term symbol appearing in the sentences occurs exactly twice, and the two occurrences are in adjacent sentences in the circle.

Theorem 1. *Every minimal inconsistent set of formal sentences can be arranged into a minimal circle.*

We say that a term symbol t in a formal sentence ϕ is either 'distributed' or 'undistributed' in ϕ as follows. If t is subject of ϕ then t is distributed in ϕ if ϕ is universally quantified, and undistributed otherwise. If t is predicate of ϕ then t is distributed in ϕ if ϕ is negative, and undistributed otherwise.

Theorem 2. *A minimal circle C is inconsistent if and only if it meets the following two conditions:*

1. *Each term occurring in sentences of C has at least one distributed occurrence.*
2. *Exactly one of the sentences in C is negative.*

These two theorems are equivalent to results in §46 of Thom [31], which Thom proves proof-theoretically. But they can be proved directly from the truth conditions in (8). Ibn Sīnā himself probably knew Theorem 1 from experience, though it's hard to see how he could have proved it. On the other hand he almost certainly didn't know Theorem 2. Any form of this result involves partitioning occurrences of terms in syllogistic sentences into the two classes that we called distributed and undistributed, and no such partition has been found in Ibn Sīnā's logical writings.

Now given an inconsistent circle as in (11), we can take out any one sentence, say ϕ_i. Then the remaining sentences entail $\overline{\phi_i}$; moreover all entailments between formal sentences, where there are no redundant sentences in the entailing set, are formed in this way. List the entailing sentences in their order in the circle:

$$[\phi_{i+1}, \ldots, \phi_n, \phi_1, \ldots, \phi_{i-1}] \tag{12}$$

Then the sequence (12) has the property that every term symbol occurs twice, in two adjacent sentences of the sequence, except for one term symbol that occurs only in the first sentence and another one that occurs only in the last sentence. We describe a sequence (12) with this property as a 'linkage' (*qarīna*, though strictly Ibn Sīnā uses the term only for such sequences of length 2). The sequence (12) and the sentence $\overline{\phi_i}$ together form a 'formal separated syllogism' whose 'premises' (*muqaddamāt*) are the sentences in (12) and whose 'conclusion' (*natīja*) is the sentence $\overline{\phi_i}$. The expression 'separated syllogism' (*qiyās mafṣūl*) is from Ibn Sīnā (*Qiyās* [15] p. 436.1), though strictly he uses it only when there are more than two premises.

So we can speak of a 'separated syllogism', meaning an entailment between syllogistic sentences, got by taking a formal separated syllogism and replacing the distinct term symbols by distinct terms. The separated syllogisms that Ibn Sīnā recognises all have the property that their premises entail their conclusion (model-theoretically); in his terminology the conclusion 'follows from' (*yalzam*) the premises. But later in this section it will take us some time to unpick the relationship between Ibn Sīnā's notion of following from and our notion of entailment.

But first we turn to the notion that the proof search algorithm is meant to deal with: separated syllogisms with a premise missing. Suppose for example that we have a separated syllogism with premises $[\phi_1, \ldots, \phi_m]$ and conclusion χ, and we remove one or more adjacent premises, say ϕ_j and ϕ_{j+1}. In the inconsistent circle the contradictory of χ belongs at the beginning or the end; we will put it at the end:

$$[\phi_1, \ldots, \phi_{j-1}, \phi_{j+2}, \ldots, \phi_m, \bar{\chi}]. \tag{13}$$

Now we can describe the gap as follows. It comes immediately after the $(j-1)$-th sentence in the sequence (13); we call the number $j-1$ the 'gap site'. If ϕ_1 and ϕ_2 had been removed, the gap would be immediately after $\bar{\chi}$, which is the $(m-1)$-th sentence in $[\phi_3, \ldots, \phi_m, \bar{\chi}]$, so the gap site would be $m-1$. Also when the linkage (13) contains at least two sentences, there is a unique term shared by the lefthand missing sentence and the one to the left of it; we call this term the 'left edge' of the gap. Likewise there is a unique term shared by the righthand missing sentence and the one to the right of it in (13); we call this term the 'right edge' of the gap.

Thus in Problem 20 Ibn Sīnā gives the following example:

> Conclusion (understood from Problem 12) 'Some C is not an A'.
> Premises 'Some D is a C' and 'No A is a B'. (14)

Putting the contradictory of the conclusion at the end gives the sequence

[Some D is a C', 'No A is a B', 'Every C is an A']. (15)

The gap site is 1, the left edge is D and the right edge is B. The definitions just given are more formal than Ibn Sīnā himself uses. But he provides several types of example with different gap sites. At Problem 7 he uses the left and right edges of the gap. In this problem he does also include an irrelevant term, and clearly he knows that it's irrelevant; perhaps he wants to encourage the student to work out that only the left and right gaps are needed at that stage in the algorithm. (See the notes on Problem 7.)

Ibn Sīnā doesn't consider the case where all the premises are missing – which is actually the case that corresponds to the proof search problem for Prolog. In this case the gap comes immediately after the contradictory of the conclusion, so the gap site is 1. But with only one sentence present, there is no way of telling which of its terms is the left edge and which is the right. We need a definite choice; I stipulate that in this case the left edge is the subject of the conclusion and the right edge is its predicate. (This is not quite arbitrary; it reconciles two messages that Ibn Sīnā sends about which end of the gap to start with when we fill it. Namely in *Qiyās* [15] Sect. 9.3 he works from the left side to the right, and when finding middles in Sect. 9.4 he starts with the subject of the conclusion.)

Identifying the gap site and the left and right edges is necessary for the algorithm, so I made it a module of the ASM. See (ASM3) in Appendix C for the module DESCRIBE. I haven't bothered to spell out the formal definition in cases like this where there is a purely book-keeping manipulation that can be specified unambiguously in English.

4.3 Simple Syllogisms

A 'simple syllogism' (*qiyās basīṭ*) is a separated syllogism with two premises. Ibn Sīnā often abbreviates this to 'syllogism'.

The term which occurs in both premises of a simple syllogism is called the 'middle' (*wasṭ*). The term which is subject of the conclusion is called 'lesser' (*asḡar*), and the premise containing it is the 'minor' (*ṣuḡrā*). The term which is predicate of the conclusion is called 'greater' (*akbar*), and the premise containing it is the 'major' (*kubrā*).

Buried in the definition of 'simple syllogism' there is the condition that the conclusion follows from the premises. But we remarked earlier that Ibn Sīnā's notion of 'follows from' is some way distant from the model-theoretic definition of entailment. In the case of two-premise arguments he doesn't recognise as syllogisms any that are not model-theoretically valid. But he puts two further requirements, as follows.

Sometimes Ibn Sīnā speaks as if a syllogism consists of just the premises. For this terminology to work, there has to be a unique way of reading off the conclusion from the premises. By Theorem 3 below, if two syllogistic premises do entail a syllogistic conclusion, then there is a strongest conclusion that they entail. But there are two cases where this strongest conclusion is not uniquely determined, namely 'Some A is a B' (which is logically equivalent to 'Some B is an A') and 'No A is a B' (which is logically equivalent to 'No B is an A'. In these two cases Ibn Sīnā resolves the question by the following condition:

Premise order condition The minor premise is listed before the major premise.

In other words, the subject of the conclusion is the term that occurs in the first premise. In fact Ibn Sīnā follows this rule uniformly for all simple syllogisms, even where there is no ambiguity to be resolved.

Thus for example at *Qiyās* [15] 114.6 he gives the mood *Cesare* in the form

Every C is a B and no A is a B, so no C is an A. (16)

while at 115.17 he cites *Camestres*:

No C is a B and every A is a B, so no C is an A. (17)

One of the very few counterexamples to this convention is in Problem 3 below, where he infers 'No C is a D' from 'Every D is a B' and 'No C is a D' in that order. This is probably an accident of his exposition; see the note on that problem.

Ibn Sīnā also imposes a further condition, which rules out what are sometimes known as 'fourth figure syllogisms':

Fourth figure condition. The middle term is not both predicate of the first premise and subject of the second premise.

At *Qiyās* 107.12 he describes arguments that violate the fourth figure condition as 'unnatural, unacceptable and unsuitable for the practice of serious study'.

With the help of Theorem 2 one can show that a model-theoretically valid simple syllogism which fails the fourth figure condition either has a premise of one of the forms 'Some A is a B', 'Every A is a B' where the syllogism would still be model-theoretically valid if we replaced this premise by 'Some B is an A'; or it has a premise of the form 'No A is a B', which can be replaced by 'No B is an A' without loss of validity. So it's always possible to bring such a syllogism into a form that Ibn Sīnā accepts, by using the implications in (10). This will involve a 'conversion' (^{c}*aks*), which swaps the order of the two terms in one premise. Ibn Sīnā notes at Problem 59 that a positive solution for the problem is impossible unless one makes a conversion in the premise. At Problems 29 and 44 he comments that conversion makes no difference to the outcome. Note also his remark about conversion at 466.3f.

Ibn Sīnā classifies the possible shapes of simple syllogisms into three figures; the first and second figures have four shapes each, called 'moods', and the third figure has six 'moods'. Ibn Sīnā expects his students to know this catalogue by heart. In fact at 466.4ff he says that a student who hasn't memorised the catalogue is not going to be able to follow the algorithm. At first sight this is puzzling, because his account of the algorithm doesn't ever seem to use the figures and moods. Closer inspection reveals one hidden reference to 'first [figure]' in Problem 3, and this reference shows what is going on. Ibn Sīnā expects his students to be able to recognise, given two syllogistic sentences ϕ_1 and ϕ_2, whether there is a syllogism with these as its premises; and where there is such a syllogism, to state its conclusion. In the style of education that he favours, the students memorise this information, and he expects them to do it in terms of figures and moods. (But don't assume that he would have used Theorem 2 if he had known it. We come back to this in Subsect. 6.2 below.)

To carry out the proof search algorithm, the student needs to be able to find, for any pair of sentences $[\phi_1, \phi_2]$ that are the premises of a syllogism, the strongest consequence of these premises. In our ASM the function *consequence* will perform this operation; we leave it to the implementer to decide how to compute the values of the function. See (Def3) in Appendix C. A sequence of sentences that are not the premises of any separated syllogism is said to be 'sterile' (^{c}*aqīm* – again we generalise Ibn Sīnā's usage from two premises to any number). When $[\phi_1, \phi_2]$ is sterile, we give $consequence(\phi_1, \phi_2)$ the formal value *sterile*.

We mentioned a theorem about strongest consequences. It says the following:

Theorem 3. *Let T be a consistent set of formal sentences and C the set of all formal sentences ψ such that T entails ψ and there is no proper subset of T that entails ψ. Then if C is not empty, C contains a sentence ψ which entails all the other sentences in C.*

Theorem 3 can be proved from Theorem 2. The sentence ψ in the conclusion of Theorem 3 is what we have been calling the 'strongest consequence' of T; it's unique up to the equivalences in (10).

For simple syllogisms, i.e. the case where T has size 2, Theorem 3 seems to have been common knowledge in Ibn Sīnā's time, and it would have been easy to prove by enumerating the possible cases. Probably the better logicians had a shrewd idea that it was true for any size of T, but I don't recall seeing it stated in the middle ages, and I doubt they could have proved it.

By cutting the inconsistent circle at a different place, Theorem 3 yields a corollary:

Corollary 1. *Let T be a consistent set of formal sentences and ψ a formal sentence. Let K be the set of all sentences χ such that $T \cup \{\chi\}$ entails ψ but there is no proper subset T' of T such that $T' \cup \{\chi\}$ entails ψ. Then if K is not empty, it contains a sentence χ which is entailed by each of the other sentences in K.*

We call the sentence χ in the conclusion of Corollary 1 the 'weakest fill'; it's unique up to the equivalences in (10).

4.4 Connected Syllogisms

Ibn Sīnā never attempts to apply any definition of 'follows from' directly to separated syllogisms with more than two premises. For simple syllogisms he understands 'follows from' in terms of how our minds manipulate ideas, and it would hardly be plausible to assume that we could hold in our minds a set of a thousand premises. Instead he maintains that a separated syllogism is shorthand for a more complex kind of syllogism, namely a tree of simple syllogisms. At *Qiyās* [15] p. 436.1 he describes such a tree as a 'connected syllogism' (*qiyās mawṣūl*). He explains at *Qiyās* [15] p. 442.8 that separated syllogisms are so-called because in them the intermediate conclusions (the conclusions of all the simple syllogisms except the one at the root of the tree) are separated from the premises (presumably he means the premises at the leaves of the tree), so that the premises are mentioned explicitly but the intermediate conclusions are left out. At *Burhān* [16] 141.15ff he comments that a connected syllogism with a thousand intermediate steps is no big deal provided we are 'mentally prepared for the drudgery'.

So part of the job of analysis is to find these intermediate conclusions. Ibn Sīnā discusses an example in detail at *Qiyās* Sect. 9.3, p. 442.8–443.13. The text is corrupt, but on one reconstruction Ibn Sīnā is discussing the separated syllogism with premises

$$\text{'Every } J \text{ is a } D\text{', 'Every } D \text{ is an } H\text{', 'Every } H \text{ is a } Z\text{', 'Every } Z \text{ is an } I\text{'} \quad (18)$$

and conclusion 'Every J is an I'. The intermediate conclusions are 'potential', he says. To find them, we start with two explicitly stated premises and draw a conclusion ϕ from them, and then we form a syllogism with ϕ as first premise and another of the explicit premises as second premise, and so on. An example would be to prove 'Every J is an H' first, and then 'Every J is a Z'. He warns us against starting with the second and third premises to deduce 'Every D is Z' – this is not 'the arrangement that we chose'. (He adds that we could have chosen a different arrangement.)

Exactly this procedure, starting from the lefthand end, appears in Problem 3. (In Arabic of course it is the righthand end. I won't say this again.) Ibn Sīnā takes a supposed separated syllogism of length 3 with the middle premise missing. He suggests a way of filling it, so that the three premises are

$$\text{'No } C \text{ is a } B\text{', 'Every } D \text{ is a } B\text{' and 'Every } A \text{ is a } D\text{'.} \quad (19)$$

He first infers 'No C is a D' from the first two premises, and then he infers the required conclusion 'No C is an A' from this and the third premise. Since this is one of the first problems, it's presumably meant as a strong clue about the procedure to be followed.

So the procedure appears in the ASM of Appendix C as module (ASM4), called SYNTHESISE. Ibn Sīnā's word for 'synthesis' is *tarkīb*, which means forming a compound; he also uses it for the compound formed. At *Qiyās* [15] p. 434.11 he explains that 'synthesising a syllogism' means forming a connected compound syllogism, which is the main thing that this module does.

Now it's clear that if $\phi_1, \ldots, \phi_5$ are formal sentences such that ϕ_1 and ϕ_2 entail ϕ_4, and ϕ_3 and ϕ_4 entail ϕ_5, then ϕ_1, ϕ_2, ϕ_3 together entail ϕ_5. But Ibn Sīnā needs more than this. His procedure is also meant to tell us when the raw materials *can't* be filled out into a syllogism. Suppose we infer ϕ_4 from ϕ_1, ϕ_2 and then find that ϕ_5 doesn't follow from ϕ_3, ϕ_4; what does this show? How do we know we couldn't have proved ϕ_5 from ϕ_1, ϕ_2, ϕ_3 by choosing ϕ_4 differently, or by starting at the righthand end? If Ibn Sīnā had tried to prove the correctness of his algorithm, he would have had to face this question.

In fact there is a positive answer, at least in terms of model-theoretic entailment. The heart of the matter is the following result.

Theorem 4. *Suppose* $[\phi_1, \ldots, \phi_n]$ *and* $[\psi_1, \ldots, \psi_m]$ *are linkages of formal sentences. Then the following are equivalent:*

(a) $[\phi_1, \ldots, \phi_n, \psi_1, \ldots, \psi_m]$ *forms an inconsistent minimal circle.*
(b) The set $\psi_1, \ldots, \psi_m$ *has a strongest consequence* θ*, and* $[\phi_1, \ldots, \phi_n, \theta]$ *is an inconsistent minimal circle.*

The theorem tells us that (so long as there are no irredundancies in the premises) we can take any segment of the premises of a separated syllogism, and shrink it down to its strongest consequence. The result will still be a separated syllogism entailing the same conclusion. At least this is true for model-theoretic entailment.

But consider for example the syllogism

Every B is a C. Every D is a B. Some D is an A. Therefore some C is an A. (20)

Model-theoretically the three premises do entail the conclusion. But if we try to build a connected syllogism, starting from the lefthand end as in Ibn Sīnā's examples, we immediately hit a problem. The first two premises violate the fourth figure condition.

A possible way around this is to start by drawing a conclusion from the second and third premises. By the premise order condition this conclusion must be 'Some B is an A', by the third-figure mood *Disamis*. So we have the intermediate syllogism

Every B is a C. Some B is an A. Therefore some C is an A. (21)

This again is a valid instance of *Disamis*.

Hence Ibn Sīnā's procedure for constructing a connected syllogism from a separated one, in the form in which it appears in the problems of his Sect. 9.6, is inadequate. In fact he hits exactly this inadequacy at Problem 33. His solution is to switch the order of the first two premises in (20). The fact that he does this, rather than plough ahead with a different justification of the syllogism, is confirmation that he expects the student to start by drawing a conclusion from the two leftmost premises. This could be because he works from left to right, or because he shrinks the sequence of premises before filling the gap. Either way, this will fit our reading of the algorithm.

Until this glitch is sorted out, some doubt remains about exactly what separated syllogisms Ibn Sīnā would accept. Perhaps closer examination of *Qiyās* [15] Sect. 9.3 will settle the point.

Theorem 4 merits a couple of further comments.

First, when we are trying to fill a gap in a sequence of premises, Theorem 4 tells us that if we can fill it at all without making any of the premises redundant, then we can fill it with a single sentence. Then Corollary 1 adds that there is a weakest single-sentence fill χ. When looking for linkages to fill the gap, we can confine ourselves to linkages that entail χ. It's not clear how far Ibn Sīnā was aware of this. For example at Problem 9 he notes that both 'Every D is a B' and 'Some D is a B' will fill the gap, but he fails to note that we have a better chance of finding a proof of 'Some D is a B' (the weakest fill) than of 'Every D is a B'. (But as always, maybe he is encouraging his better students to see this point for themselves.)

The second comment is a technical warning. Suppose $m = 2$, ψ_1 has terms A, B and ψ_2 has terms B, C. Because of Ibn Sīnā's assumptions (8) about truth when a term is empty, θ in the theorem need not be logically equivalent to $\exists B(\psi_1 \wedge \psi_2)$. (Syllogisms don't have quantifier elimination.) So passing to θ might throw away information about A and C. But we can show that the lost information is recoverable from the rest of the circle.

4.5 Other Kinds of Syllogism

In [9.6.4] and [9.6.5] of *Qiyās* 9.6, Ibn Sīnā refers briefly to some other kinds of syllogism.

Earlier in the *Qiyās* ([15] p. 106) Ibn Sīnā has distinguished between two kinds of syllogism which he calls respectively 'recombinant' (*iqtirānī*) and 'duplicative' (*istiṯnā'ī*). A recombinant syllogism has two premises, each of them built out of two parts; one of these parts is the same in both premises. The conclusion is formed by recombining the two remaining parts. Simple syllogisms as in Subsect. 4.3 above fit this description. But so do some propositional (*šarṭī*) syllogisms, for example

If p then q. If q then r. Therefore if p then r. (22)

Ibn Sīnā's view is that recombinant syllogisms are a generalisation of simple syllogisms, and that generally speaking the rules for simple syllogisms transfer to recombinant syllogisms too. (This is presumably what he has in mind at 468.7.)

Duplicative syllogisms are propositional. They have two premises. One of the two premises has two parts. The other premise consists of one of these two parts (or its contradictory), and the conclusion consists of the other part (or its contradictory). The shorter premise and the conclusion are said to be 'duplications' (i.e. of parts of the longer premise). Besides modus ponens:

If p then q. p. Therefore q. (23)

this description covers inferences like:

Not both p and q. p. Therefore not q. (24)

Ibn Sīnā regards duplicative syllogisms as incomplete in themselves; they only make sense as part of a longer argument. There seems to be no natural way of generalising his proof search procedure to them.

Ibn Sīnā classifies binary sentence connectives and the compounds that are formed using them as 'meet-like' (*muttaṣil*) or 'difference-like' (*munfaṣil*). This is a soft classification based on some supposed resemblance to meet ('and') or difference (exclusive 'or'). But he doesn't use the classification consistently, and my present impression is that he never settled on a satisfactory principle for classifying binary sentence connectives. The 'If ... then' in (23) counts as meet-like, while 'Not both' in (24) counts as difference-like; so these two syllogisms are respectively meet-like duplicative and difference-like duplicative.

There is more on Ibn Sīnā's propositional syllogisms in Shehaby [29], together with a translation of the propositional part of *Qiyās*. Shehaby translates the technical terms differently: he has 'conjunctive' for 'recombinant', 'exceptive' for 'duplicative', 'conditional' for 'propositional', 'connective' for 'meet-like' and 'separative' for 'difference-like'.

5 Extracting the Algorithm

Can we be sure that Ibn Sīnā really meant to describe an algorithm for proof search?

Ibn Sīnā himself doesn't say anything to indicate that he regards the procedure that he is teaching as comparable with the kinds of algorithm known to medieval Arabic mathematicians (see Subsect. 6.2 below). He does say in [9.6.1] that we need 'rules' (*qawānīn*, plural of *qānūn*) to guide us in analysis. But in [9.6.2] he explains this as 'rules in the form of dos and don'ts', which doesn't sound like an algorithm. At *Qiyās* [15] p. 537.3 he uses the same phrase 'dos and don'ts' for advice about how to conduct a debate.

In fact the passage that I interpret as describing an algorithm (paragraphs [9.6.6] to [9.6.11]) consists of 64 problems, with answers given and some remarks about how the answers are found. The problems are divided into four groups according to their patterns; Ibn Sīnā explains the patterns and tells the reader 'Do the remaining cases of this pattern for yourself' (463.12, 464.12, 466.2, 467.7). The problems are introduced without any explanation of what they are for, apart from the fact that they appear in a discussion of analysis. At the end of them Ibn Sīnā comments (468.4f):

> When you put the steps in this order, as I have shown you, you will reach the terms, figures and moods. (25)

Strangely the 64 problems make no mention at all of syllogistic moods, and only one mention of figures. But figures and moods are a classification that makes sense only for simple syllogisms, so Ibn Sīnā is implying here that by 'putting the steps' in the right order we will be able to reduce the raw material in the problems to simple syllogisms. Since no new kinds of 'step' are described here, it seems to follow that Ibn Sīnā means we can use steps already discussed earlier in *Qiyās* to reduce the raw material to simple syllogisms. This exactly matches the use of the module SYNTHESISE in our ASM, which rests on procedures described in *Qiyās* Sect. 9.3. So it confirms that we are on the right track.

In any event, paragraphs [9.6.6] to [9.6.11] are clearly intended to teach the reader a procedure for taking data of a certain kind and coming up with answers to certain questions about the data. The decision whether to call this procedure an algorithm is for us, not for Ibn Sīnā. Our choice should rest on three issues: (1) Is the class of data to which the procedure applies well-defined? (2) Is it clear what question or questions the procedure is meant to answer? (3) Is the procedure mechanical?

I take these issues in turn. Of course the procedure defined by the ASM in Appendix C is an algorithm, but we need to ask how much of that algorithm is already in Ibn Sīnā's text.

5.1 The Class of Input Data

The 64 problems all share a common format. They involve syllogistic sentences with letters for terms; but we are not told what the letters stand for. So there is no loss in thinking of the sentences as formal sentences, provided that we don't impose our definition of entailment on Ibn Sīnā.

Each problem begins with a syllogistic sentence called the 'goal' (*maṭlūb*), except where we have to understand that the goal is the same as in the previous problem. Then follows a sequence of one or more formal sentences. Ibn Sīnā's commonest description for this sequence is that 'you have' (*kāna* ^{c}*indak*) the sentences in it; he uses this or closely similar phrases in 38 problems. The only name that he offers for the sequence is 'thing found' (*mawjūd*, in 8 problems). This word has a variety of meanings in Ibn Sīnā's logic, and another variety in his metaphysics. But since he also says in 6 problems that 'you have found' (*wajadta*) the sequence, I assume it just means 'what has been found'. For the sake of English style I shorten this to 'datum'; I follow Ibn Sīnā's lead in using the singular even when there is more than one sentence in the sequence.

So each problem has a goal and a datum. In every case but one, the sequence consisting of the datum followed by the goal is a linkage in the sense of Subsect. 4.2 above. The one exception is Problem 33, where the linkage order would run foul of the fourth figure condition (see the note on the problem). This exception shows that Ibn Sīnā does expect the student to be able to handle a larger class of inputs than the ASM in Appendix C is designed for. But I think we can regard Problem 33 as a freak case.

Ibn Sīnā indicates at 464.12f, 465.5, 466.6, 467.7 and 467.10 that we should think of the datum as having two parts, one that has a sentence containing the subject of the goal and one with a sentence containing the predicate of the goal; except that one of these two parts may be empty. What he describes here is exactly the gap site that we calculated in Subsect. 4.2; the part of the datum linked to the subject is precisely the part before the gap indicated by the gap site.

It's curious that Ibn Sīnā explains this structure of the datum only after the 26th problem. Perhaps some text has gone missing, but I doubt it. He has a tendency to explain himself only after he has given you a chance to work out for yourself what he must have meant. My impression is that the Arabic mathematicians of his time would have regarded this as poor style.

All four kinds of formal sentence appear in Ibn Sīnā's problems, in a wide variety of combinations. So it seems that the class of properly ordered possible goal-datum pairs is well defined, except perhaps for the questions of length and of the number of gaps. To begin with length, in all Ibn Sīnā's examples the datum has length 1 or 2. Did he intend his procedure to apply only in these cases?

I believe not, for two reasons. The first is that in Problems 1 and 2 he points out that if we can't find a suitable single sentence to fill the gap, we may need to look for a pair. He doesn't say what happens next, but one reasonable way forward would be to guess (say) the first sentence of the pair and put it into the datum. Finding the second sentence would then be the original problem but with a longer datum. (And so on recursively, though he never says this.)

Strictly this is not the only way forward. As we will see, the question of looking for a pair of sentences only arises after we have discovered a weakest fill ϕ for the gap in the original datum. Then by Theorem 4 above, it suffices to continue with ϕ as new goal and an empty datum. But even this would add 0 to the possible lengths of data. I haven't followed this route, because it would imply some mechanism for feeding back the result of the calculation with ϕ as goal into the original problem.

But the case of length 0 is interesting anyway, not least because it corresponds to the Prolog proof search problem. For that reason I set up the ASM to handle data of length 0. Ibn Sīnā himself may have reckoned that he had said enough about the case of data of length 0 already in *Qiyās* [15] Sect. 9.4 'On obtaining premises, and on *taḥṣīl* of syllogisms with a given goal'.

The second reason for doubting that Ibn Sīnā intends a restriction to lengths 1 and 2 is his statement at 465.2 that he will deal with the case of 'more than two premises' in the appendices. We don't have the promised appendix; see the note on this passage. Of course he might have said in the appendix that these longer data can be handled, but only by a different procedure. I think this is unlikely, for the first reason just given.

Nevertheless there is a good reason for Ibn Sīnā to concentrate on the case of length $\leqslant 2$. If the datum has length greater than 2, it always contains two adjacent sentences that share a term. So we can reduce the length of the datum immediately, by replacing these two sentences by their strongest consequence – unless they are sterile, in which case the problem has no positive solution. We can't be sure that Ibn Sīnā intended this way of working, but it makes good sense and I have built it into the ASM.

The other possibility is that Ibn Sīnā intends his procedure to apply where the datum contains more than one gap, or perhaps even when it contains no gap at all. He does in fact discuss the case of more than one gap in paragraph [9.6.7]. His view is that it can be handled but at the cost of a more complicated procedure, which again he will describe in the appendices. The main thing we would need to do in order to extend our ASM to more than one gap would be to incorporate some further machinery to control the search; see Subsect. 6.2 below for a discussion of what would be required. Presumably Ibn Sīnā's appendix would have said something about this too. The case of no gaps is covered by the procedures of *Qiyās* Sect. 9.3, which we have incorporated into the module SYNTHESISE; so this case is at least implicitly in Ibn Sīnā's algorithm already.

In his initial remarks on analysis in [9.6.1], Ibn Sīnā says that the text to be analysed may contain 'something superfluous', and our rule will need to tell us how to 'strip off defects'. This suggests that the procedure should also eliminate redundant parts of the datum. None of the 64 problems suggests any way of doing this. Indeed it's not clear what the aim would be if Ibn Sīnā did allow this. One could always start by removing the entire datum and working from the goal alone; would this count? If not, would the aim be to throw away as little as possible of the datum? This could lead to serious complexities. So I think we can sensibly assume that the procedure is not meant to eliminate redundant parts of the datum.

5.2 What Question Is Answered?

Alongside each one of his 64 problems, Ibn Sīnā provides an answer. With trivial variations, all the answers take one of two forms. The affirmative form is: If the sentence χ is attached (*ittasal*, 27 problems) then 'it has been made determinate' (*qad ḥuṣṣil*, 17 problems). Usually Ibn Sīnā doesn't tell us what has been made determinate. But at Problems 8 and 9 he does: it's the syllogism. (See also Problem 1: 'your syllogism is in good order'.) The translation in Appendix A reflects this.

By 'attached' he clearly means 'put into the gap in the datum'. So the procedure involves an operation that does this. I was tempted to call this operation 'attach', but unfortunately this is a reserved word in the vocabulary of ASMs. Since the operation is a syntactic triviality, I made it not a module but a basic function: (Def5) in Appendix C.

The negative form of answer is: 'It can't be used' (*lam yuntafa*^{c} *bih*, 23 problems). This time we can hardly expand to 'The syllogism can't be used', because in these problems there is no syllogism. A more accurate expansion would be 'The goal and datum can't be used to generate a determinate syllogism'; but for brevity I stick with 'it' in the translation.

The problems with a negative answer are exactly those in which there is no sentence that can be put in the gap of the datum so as to yield a separated syllogism with the goal as conclusion. Also in the problems with an affirmative answer Ibn Sīnā nearly always names a sentence that can be put in the gap so as to yield the required syllogism. So this is at least one of the aims of the procedure:

> *Determine whether or not there is a sentence χ that can be put into the gap of the datum so that the datum becomes the premise sequence of a separated syllogism whose conclusion is the goal. When the answer is Yes, supply a sentence χ with this property.* (26)

I call this the 'logical task'. Note that it makes no reference at all to sentences that are already determinate.

Note also that the logical task, as stated, doesn't include classifying the resulting syllogism by means of figures and moods. We saw earlier that in fact Ibn Sīnā's procedure, if we have reconstructed it correctly, does yield enough information to convert the separated syllogism into a connected one, and then the figures and moods can be read off. So we could add a further module to the ASM which delivers the connected syllogism with its simple syllogisms labelled by figure and mood. But this would mean introducing a new datatype for connected syllogisms, and it would be just a unit bolted onto what is already in the ASM. So I take a lead from Ibn Sīnā, who mentions in [9.6.12] that the procedure will yield this information but gives no further details. I add only that one possible implementation of the ASM is in terms of diagrams written on paper, very likely as Ibn Sīnā's students would have drawn them. These diagrams would almost certainly have included the connected syllogisms.

But Ibn Sīnā also says a number of other things that only make sense if he is expecting the procedure to deliver a syllogism that is determinate in the sense we studied in Sect. 3 above. First and foremost, there is the wording that we quoted in the affirmative case: '[the syllogism] has been made determinate'. Add to this that in 10 problems he says that the premises in the datum are determinate; this is irrelevant for the logical task. In 6 of the problems with an affirmative answer, he requires that the attached sentence is 'true' or 'true for you' or 'clear' (*bayyin* – this must mean 'clearly true'). Finally there are two problems (1 and 2) where Ibn Sīnā finds a sentence χ that solves the logical task, and then adds that if the sentence is not 'clear' or true, then it doesn't solve the problem and one 'needs a middle' (i.e. has to look for a two-sentence filling for the gap).

So there is clear evidence that Ibn Sīnā also has in mind another task:

Given that the datum consists of sentences that are already determinate, discover whether or not there is a sequence of sentences $[\chi_1, \ldots, \chi_m]$ *that are already determinate, which can be put into the gap of the datum so that the datum becomes the premise sequence of a determinate separated syllogism whose conclusion is the goal. When the answer is Yes, supply a sequence* $[\chi_1, \ldots, \chi_m]$ *with this property.* (27)

I call this the '*taḥṣīl* task'. The two tasks are connected by the fact that a negative answer to the logical task implies a negative answer to the *taḥṣīl* task, but otherwise the tasks are independent.

I think it's inconceivable that Ibn Sīnā was in any way confused about the difference between the logical task and the *taḥṣīl* task. But I wouldn't put it past him to be deliberately ambiguous in hopes of catching both tasks under the same general description. There is some evidence of deliberate ambiguity. In Subsect. 5.1 we interpreted the word 'found' (*mawjūd*) as meaning datum, i.e. 'the thing you found in front of you when you were given the problem'; but it would be entirely in keeping with Ibn Sīnā's logical vocabulary if we read it as 'found to be true', i.e. determinate. Likewise the phrase 'you have' (*kāna* ^{c}*indak*) could also mean 'according to you', in other words, 'it's determinate for you that ...'.

It would also be in character for Ibn Sīnā to leave the ambiguity as a deliberate trap for idle or unintelligent students.

In sum, we have identified two tasks that the procedure is meant to perform. The logical task is well-defined apart from the uncertainty about what separated syllogisms Ibn Sīnā accepts. But at least we can rigorously check the correctness of Ibn Sīnā's own solutions of his 64 problems. The *taḥṣīl* task is well-defined apart from the same uncertainty about separated syllogisms, though it does require us to know what sentences are 'already determinate'. The set of things that are already determinate is the counterpart of the set of clauses of the Prolog program in the Prolog case. Börger and Rosenzweig [5] build this set of clauses into their ASM through a predicate *PROGRAM* and a basic operation *clause_list*. I prefer not to do that here, because it would pre-empt a question we have to discuss in a moment, namely whether Ibn Sīnā considers that the set of sentences that are already determinate can be read off mechanically.

5.3 Is the Procedure Mechanical?

Ibn Sīnā doesn't ringfence his procedure; we have some discretion to decide what counts as part of it and what involves an appeal to the environment. The real question here is whether the procedure has a purely mechanical core, and if so, what that core contains.

Most of the procedure is quite obviously mechanical. Although Ibn Sīnā refers at [9.6.12] to 'putting the steps in this order', he is a little vague about what that order is. But as far as I can see, the indeterminacies are all of the kind where it doesn't matter what order we choose, and it's routine to find a mechanical arrangement of the steps that does the required job.

There are three places where Ibn Sīnā relies on the reader to have a certain skill. The first is the computation of the strongest consequence of a non-sterile pair of premises. This is the job of the basic function *consequence* at (Def3) in Appendix C. If the worst comes to the worst, the function can be implemented by simply listing the possible cases, as in an appeal to the student's memory.

The second place is where, in the *taḥṣīl* task, Ibn Sīnā asks the student whether a certain named sentence is already determinate. I see no problem about taking this as a basic function *hasil* of the ASM, as in (Def6) in Appendix C. It doesn't necessarily follow that the sentences that are already determinate can be listed by listing all sentences and then filtering through the function *hasil*, because the set *SENTENCE* could be dynamic. More precisely there could be infinitely many sentences, or more than are listed in the set *SENTENCE* in the ASM at the outset of the computation, and the ASM may be able to add further sentences *SENTENCE* as the computation proceeds. (Since *SENTENCE* is defined in terms of *TERM*, this would involve adding new terms to *TERM* too.) This possibility doesn't arise when a single given sentence is being evaluated for being determinate.

The third place where the reader needs a skill is where Ibn Sīnā says (in Problems 1 and 2) 'it needs a middle'. The situation is that a sentence χ has been identified as the weakest fill for a certain datum, and the function *hasil* has been used to reveal that χ is not already determinate. I have to mention another glitch hidden here. It could happen that χ is not itself already determinate, but it is a consequence of a one-premise inference (as in (10)) from a premise θ that is already determinate. The algorithm should identify θ and put it in place of χ. This needs an extra piece of machinery which I haven't included in the ASM. One excuse I can offer is that putting θ in place of χ could possibly lead to a violation of the fourth figure condition, and we don't know what Ibn Sīnā thinks about this possibility.

In any case, the statement 'it needs a middle' is shorthand for:

> We need to look for a term C and sentences ϕ_1, ϕ_2 using the terms of χ and the term C, so that ϕ_1, ϕ_2 are already determinate and are the premises of a syllogism with conclusion χ and C as middle.

Ibn Sīnā discusses this situation in a number of places.

For example *Qiyās* [15] Sect. 9.4 is about this question. Ibn Sīnā advises that we start by looking at the form of χ. Thus suppose it has the form 'Every A is a B'. Then we should unpack the definition of the term A, and extract from it sentences of the form 'Every A is a C'. For each of these, we should see whether we can also prove 'Every C is a B'. If we have no success with the definition of A, Ibn Sīnā advises looking next at the properties that we can prove for A, using the principles of the relevant science.

In the cases where χ has the form 'No A is a B' or 'Some A is a B', the situation is symmetrical and we can start with either A or B. In the case of 'Some A is not a B', Ibn Sīnā's wording suggests – I can't put it stronger than that – that we start with properties that some A is known to have. So a general rule that covers all cases would be that we start by looking for determinate sentences that involve the subject term of χ. (Note that the subject term could be either the left edge or the right edge of the gap.)

Ibn Sīnā comes back to the matter at *Burhān* [16] pp. 138.22ff and 139.10ff. He claims that in mathematics most sentences have the form 'Every A is a B' (here he is agreeing with Aristotle *Posterior Analytics* A14). He suggests that when χ has this form in mathematics, if there is a middle as required, then one can be found by unpacking the definition of the subject term of χ. (This seems to me a gross oversimplification outside elementary linear algebra.) In this case it would be reasonable to say that the list of possible terms can be found mechanically from the definition of the subject term, so we would only need to include in the ASM a basic function for finding the definitions of terms. But Ibn Sīnā goes on to say that outside mathematics things are not so straightforward. We would need to consider the inherent accidents of the subject term of χ, and in the worst case even its non-inherent accidents.

He comes back again to the same question in his autobiography. He tells us that sometimes he was 'at a loss about a problem, concerning which I was unable to find the middle term in a syllogism', and so he resorted to prayer, then to alcohol and then to sleep; 'many problems became clear to me while asleep' ([12] p. 27f). Prayer, alcohol and sleep are not mechanical procedures.

All in all, I think it would be very unwise to assume that Ibn Sīnā thinks we can list in advance all the determinate sentences that involve the subject term of χ. This is a pity, because the backtracking algorithm of [5] (which Börger and Stärk display as an ASM module on page 114 of [6]) assumes that we can make this list.

At this point I am going to cheat and call on a relatively advanced kind of Abstract State Machine called an asynchronous multi-agent ASM ([6] Chapter 6). This multi-agent ASM has a family of 'agents' who each perform according to their own ASMs, at their own speeds and for the most part independently. But there can be super-global procedures that pass messages to and from the agents. The set of agents can be 'potentially dynamic', in other words there can be super-global procedures that add new agents. In ASMs one can treat the set of threads in a Java program as a dynamic set of agents; I thank Egon Börger for this example. (The term 'super-global' is to distinguish from those features of the agent ASMs that are global within these ASMs.)

In this setting, suppose an agent reaches a point where 'it needs a middle'. The agent then sends a message to the super-global agent who operates the super-global procedures; prayer, alcohol and sleep might be ways of sending this message. The super-global agent responds by listing all the possible options; but instead of sending the list to the agent, it splits the agent into a family of agents, each of whom has one of the options to work on. I see Ibn Sīnā identifying the global agent as the Active Intellect, and the agents who carry out the algorithm as possible intellects, so that

> when a connection occurs between our souls and [the Active Intellect], there are imprinted from it in them the intellected forms which are specific for this specific preparation for specific judgements. (*Išārāt* [20] II.3 *iš.* 13.) (28)

But that's an aside – the super-global agent has a precise job to do, which is encoded in the ASM as a super-global basic function.

All the agents do the same calculation for the logical task. When the logical task has delivered an affirmative answer, they switch to the *taḥṣīl* task and may have to split. So for the *taḥṣīl* task we need to clarify the notion of correctness of the ASM, as follows. *The ASM is correct for the* taḥṣīl *task if: (1) when the task has a negative answer, all (lower level) agents return a negative answer; (2) when the task has an affirmative answer, at least one agent returns an affirmative answer; and (3) every agent returning an affirmative answer also returns a sequence of sentences which is a correct fill for the gap in the datum.*

A fragment of the backtracking procedure is still needed, but for a more limited purpose, namely to find the weakest fill in a datum. Ibn Sīnā shows at Problems 3 and 7 that he expects the student to find it by listing possibilities and trying each in turn. The edges of the gap are known, and they provide the two terms of the weakest fill. So there are eight possible sentences to consider. Given this approach, it makes sense to list the possibilities in an order where ψ comes before χ whenever χ entails ψ; so when we first find a possible fill we know it is a weakest one. The function *listsentences* at (Def2) in Appendix C provides such a list.

Ibn Sīnā allows the student to use background knowledge to cut down from eight to a shorter list of possible fills; see the notes on Problems 3 and 7. I count this move as a shortcut, not as a part of the algorithm.

Are we sure that no further backtracking is needed? For example, perhaps we find a weakest fill, but then further down the line we discover that the resulting connected syllogism runs into trouble with the fourth figure condition, so that we need to backtrack and try the converse of the weakest fill instead. I believe that this problem doesn't arise, because the premise order condition fixes the order of the terms in all the intermediate sentences in the connected syllogism, independent of the order of the terms in the premises of the separated syllogism. To be sure of this we need a correctness proof; but I think this would be wasted effort until we have an answer to the question about which connected syllogisms to accept.

6 Review

We must do two things here. The first is to give an informal summary of the algorithm, and the second is to place it in the history of logic and mathematics. A more formal description of the algorithm is given in Appendix C, in the form of an asynchronous multi-agent ASM, where each agent follows its own agent ASM within the multi-agent ASM.

6.1 Summary of the Algorithm

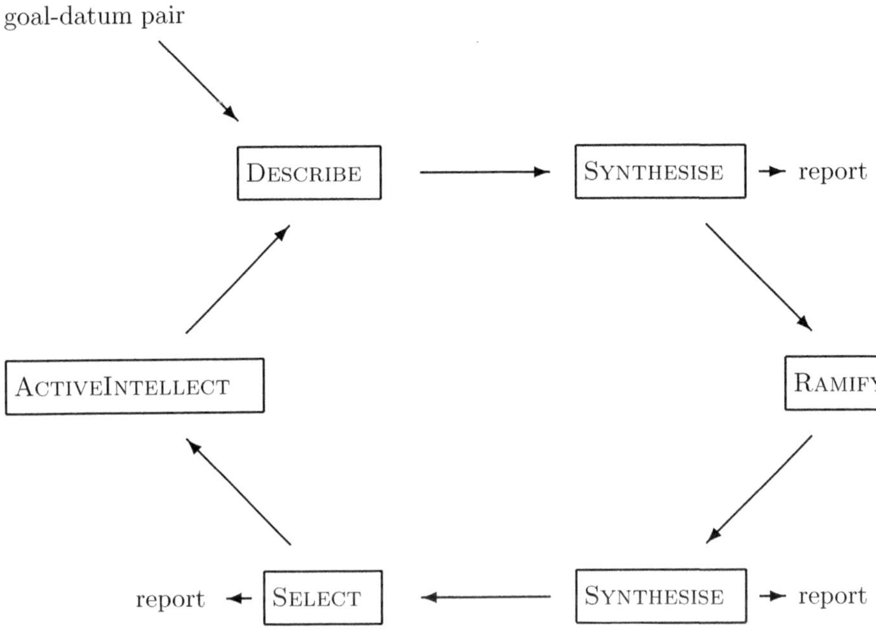

We describe what happens to a goal-datum pair as it proceeds through the diagram above.

Entering the module DESCRIBE, the goal-datum pair is measured up to discover where its gap site is and what the left and right edges of the gap are. This information is attached to it for future use.

Then it proceeds to the module SYNTHESISE, which shrinks it down. If there is a pair of adjacent sentences in the datum that have a term in common, SYNTHESISE works out the strongest consequence of these two sentences, and replaces the sentences by this strongest consequence. It does this starting with the leftmost such pair of sentences, and continues until either there are no such pairs left, or it reaches a pair that is sterile. In the latter case it reports failure and the algorithm halts.

If the shrunken goal-datum pair survives through SYNTHESISE, it passes to the module RAMIFY. This module finds the eight sentences $\phi_1, \ldots, \phi_8$ whose terms are the two edges of the gap. The sentences are listed so that if ϕ_i entails

ϕ_j but not vice versa, then $j < i$. Then the module splits the goal-datum into eight clones, and it fills the gap in the i-th clone with the sentence ϕ_i. So now there are eight goal-datum pairs, none of which has a gap.

There is a subtlety if the goal-datum pair that passes to RAMIFY has an empty datum. In this case there is always a sentence that fills the gap and entails the goal, namely the goal itself. So in this case RAMIFY makes just one new page, in which the datum is changed to the goal sentence.

After RAMIFY has done its work, the first of the resulting gap-free goal-datum pairs passes to SYNTHESISE, which shrinks down any adjacent pair of sentences in the datum with a term in common, until either the datum consists of a single sentence, or a sterile pair of sentences has come to light. If a sterile pair of sentences comes to light, the goal-datum pair is discarded and the next of the eight clones passes into SYNTHESISE for similar treatment; and so on. If none of the eight clones are left, the module reports failure.

If a goal-datum pair with a single-sentence datum survives, it passes to the module SELECT. This module checks which of three cases hold: (1) the datum equals the goal, and it is already determinate; (2) the datum equals the goal, but it is not already determinate; (3) the datum doesn't equal the goal. In case (1) the module reports success in the *taḥṣīl* task and the algorithm halts. In case (2) the module reports success in the logical task (if it hasn't already been reported), restores the gappy goal-datum pair that RAMIFY had filled, and sends this pair to the Active Intellect with a request for a determinate sentence that attaches at one side of the restored gap. The Active Intellect compiles a list of all the determinate sentences that could be used, and it makes one clone of the goal-datum pair for each such sentence ψ. The clone that goes with ψ has ψ inserted into its gap; but the gap is not completely filled, so we once again have a goal-datum pair with a gap. All these new goal-datum pairs are sent back into DESCRIBE in parallel, and so on around the cycle. In case (3) the same happens as the failure case in the previous paragraph: the goal-datum pair is discarded and the next of the eight is called for, unless none of the eight are left, in which case the module reports failure.

There are several places where a module reports success or failure. If no success has been reported yet, then the first report of success or failure is a report on the logical task, except in case (1) for SELECT. If logical failure has been reported, the algorithm halts. If logical success has been reported, a later report of failure is a report on the *taḥṣīl* task, and again the algorithm halts. If logical success has been reported, the only further report of success that makes any difference is a report of *taḥṣīl* success in case (1) for SELECT.

This is the algorithm in broad outline. We need to clarify what are the separate steps, and how the algorithm decides which step happens when – what Ibn Sīnā refers to as the 'order'. The description below is very much based on Gurevich's notion of an ASM and the use made of it by Börger and Rosenzweig in [5].

The idea of goal-datum pairs swimming around between modules is only a metaphor. A different metaphor is more realistic: the calculator (or 'agent') does each piece of calculation by writing out one or more pages that state the results of the calculation. (The pages are the 'nodes' of [5].) A step of the calculation could involve writing several pages, but only where the pages can be written simultaneously. For example when RAMIFY makes eight clones and fills them, in principle this can be done on eight pages simultaneously (though eight hands would be useful), so it counts as a single step. But when SYNTHESISE shrinks down the datum, the result of shrinking down the first pair of sentences is generally an input to the operation of shrinking the next pair. So shrinking down a single pair of sentences to their strongest consequence is a whole step. In general SYNTHESISE will process a goal-datum pair for several steps until there is no fat left on the datum; this will involve producing a succession of new pages with shorter datum sequences.

In principle the agent could go to work on any existing page at any time, using any one of the four modules DESCRIBE, SYNTHESISE, RAMIFY or SELECT. What decides which page and which module the agent will take next?

Written in a separate place, not on the pages, there are three further pieces of information stored in 'global variables'. The first is the label of the 'current page', i.e. the page now being processed. The agent reads the current page and acts according to instructions in the algorithm; these instructions refer to the contents of the current page, and to the values of the global variables. The instructions tell the agent what new pages to produce, and what changes to make to the global variables. So for example if the agent is looking at page 5, the instructions may tell the agent to change the current page variable to 6; the effect is that when page 5 has been dealt with, the agent turns next to page 6. And so on.

There are two other global variables besides 'current page'. One of them records the goal (which is fixed at the start and never changes). The other global variable stores reports of success or failure (and starts with the value 'ignorance').

The rest of the information needed for controlling the calculations consists of six records on each page, as follows. (In Appendix C these six records are called 'properties' of the page.) The first is a record of the datum on that page. (The starting page carries the datum given by the problem to be solved.) The second records the gap site for the current goal-datum pair; the record may also show that there is no gap, or that the gap site needs calculating. The third is a record of the left and right edges of the gap. The fourth is a record of the fill, i.e. the sentence that was put in the gap when RAMIFY was last used.

The fifth and sixth records on the page store information about the movement between the pages. One of them records 'previous page'; what this means is that when a page p is being read and a new page q is constructed according to the information in p, then p is recorded as 'previous page' on q. (After the algorithm has reported success, one will need to work backwards from the final page to its previous page, its previous page's previous page and so on in order to reconstruct the required connected syllogism.) The other record is called 'next'. The main function of 'next' is that when a group of pages $p_1, \ldots, p_n$ are constructed simultaneously, 'next' on page p_i (where $i < n$) indicates p_{i+1}. When the agent is reading p_i and has to discard it, the value of 'next' on p_i tells the agent which page to try next. The agent makes this happen by changing the value of the global variable 'current page' to p_{i+1}.

If we have the algorithm set up correctly, then at any stage the records on the current page p will determine uniquely which module takes care of this stage of the calculation – unless the algorithm has halted with a report of success or failure. For example if the record of the gap site on p says that there is no gap, the module that applies will be one of the two at the bottom of the diagram. If and only if the record says that the gap needs calculating, the module that applies will be DESCRIBE. If the record says that there is a gap, the module that applies will be one of SYNTHESISE, RAMIFY and ACTIVEINTELLECT.

The module ACTIVEINTELLECT, which is operated by a higher force, comes into play when and only when the record 'next' on the current page indicates prayer, alcohol or sleep – or more prosaically when it has the value 'needs a middle'.

Assuming that neither DESCRIBE nor ACTIVEINTELLECT has been called, what settles the choice between SYNTHESISE, RAMIFY and SELECT? The answer is that SYNTHESISE applies if and only if the datum on the current page has two adjacent sentences with a term in common. (The module appears twice in the flow diagram above, but it has the same job to perform in both cases.) If SYNTHESISE doesn't apply, then RAMIFY applies if there is a gap in the goal-datum pair, and SELECT applies if there isn't.

It may be helpful to note that when RAMIFY or SELECT applies to a page, then the datum on the page has been slimmed down as much as possible by SYNTHESISE. So if the goal-datum pair has a gap (as at RAMIFY), the datum consists of at most two sentences; if the pair has no gap (as at SELECT), the datum is a single sentence. I suggested earlier that this explains why Ibn Sīnā confines his 64 problems to cases where the datum has at most two sentences.

For further details refer to Appendix C.

6.2 The Place of the Algorithm in History

My remarks here will be very incomplete – this is already a long paper. I am very much indebted to Roshdi Rashed for his comments and information, though he should not be held responsible for any particular claims I make.

A 'search algorithm' is a mechanical procedure which allows its user to find a solution of a problem, or establish that there is no solution, by running systematically through a set of possible partial or total solutions. (The set is called the 'search space'.) Ibn Sīnā's algorithm, insofar as it really is an algorithm, is a search algorithm for finding solutions to the logical and *taḥṣīl* problems. It searches through partial or total compound syllogisms that extend the datum. I know of no other examples of search algorithms in the medieval Arabic literature. In modern times search algorithms go back at least to Tarry's maze-solving algorithm of 1895 ([4] p. 18ff), though the best known examples are from the second half of the 20th century.

Closely related to search algorithms are two other kinds of algorithm. A 'counting algorithm' allows its user to calculate the number of elements of a given set. A 'listing algorithm' allows its user to list without repetition all and only the elements of a given set. Search algorithms sometimes use listing algorithms to list the elements of the search space; but unless the listing is appropriate for the problem, the resulting search algorithm can be very inefficient. Sometimes a counting algorithm can be proved correct by examining a listing algorithm. For example one can show that the number of elements of the cartesian product $X \times Y$ is the product of the number of elements of X and the number of elements of Y by examining the lexicographic product of X and Y (as done for example by Ibn Mun'im in the 13th century, Katz [22]).

The earliest reported algorithms in Arabic mathematics are listing and counting algorithms in connection with strings of letters. They appear first in the *Kitāb al-ᶜAyn*, an 8th century linguistic text normally attributed to the polymath al-K̲alīl ibn Aḥmad (though the situation must be more complicated, because the book includes third-party reports of al-K̲alīl's views, cf. Versteegh [32] Chapter 2). The basic problem that al-K̲alīl addresses is how to list the words of Arabic in a dictionary. His preferred ordering is not lexicographic; rather he lists unordered sets of consonants, and for each set he makes a sublist of its permutations. He introduces this ordering with some combinatorial calculations that count numbers of permutations. In the 9th century Ibn Duraid developed these calculations, describing them as a sort of calculus (*ḥisāb*). Cf. Rashed [27] p. 18ff for this aspect of the *Kitāb al-ᶜAyn* and its later influence.

In the 9th century Muḥammad ibn Mūsā al-K̲wārizmī introduced the classical algorithm for solving quadratic equations. More than this, he demonstrated the 'cause' (*ᶜilla*) of the algorithm; in other words, he gave a mathematical demonstration that the algorithm always yields a correct answer when the coefficients of the equations are real numbers (or less anachronistically, when they can be represented as lengths). For this he converted from numbers to lengths, and then invoked geometrical arguments in the style of Euclid. Details are in [27]. In the rich tradition inspired by al-K̲wārizmī's work, the word for 'algebraic algorithm' is *bāb*, translated into Latin as *regula*.

None of the algorithms of al-K̲alīl or al-K̲wārizmī are search algorithms, and Ibn Sīnā gives no indication that he sees himself as doing anything similar to what they did. He never describes his procedure as a *ḥisāb* or a *bāb*. There is some overlap with al-K̲alīl's calculations, namely that Ibn Sīnā uses lists of possibilities. But unlike al-K̲alīl, Ibn Sīnā spends no time discussing systematic ways of listing the possibilities. Ibn Sīnā would have had to consider some kind of backtracking algorithm if he had taken more seriously the implications of the *taḥṣīl* problem; but this would have moved him into territory unknown to any medieval mathematician (as far as we know).

To adapt his algorithm to problems with more than one gap, Ibn Sīnā would have had to search systematically through the cartesian product of the sets of possible fills at the separate gaps. If he proposed to do this by listing the possibilities lexicographically – and it's hard to think of any other reasonable procedure

– then this would have brought him close to al-Ḵalīl's listing procedures, and he would very likely have given us the earliest description of a search through lexicographic listing. This makes it all the more painful that we don't have the appendix which he said would discuss problems with more than one gap. (See the note on 465.2.)

There is a major difference between Ibn Sīnā's discussion and that of al-Ḵwārizmī. Namely, Ibn Sīnā never makes any attempt to show that his algorithm is correct. (If he had done, he would certainly have given a much better algorithm.) There are several aspects to this difference. First, al-Ḵwārizmī is following every mathematician's dream: to solve a problem by reducing it to some apparently quite different problem that is easy to solve or has already been solved. The reduction of a problem in algebra to one in geometry is a beautiful example; and incidentally it runs clean counter to the aristotelian tendency to keep the various sciences in a rigid hierarchy. I doubt that Ibn Sīnā ever had this mathematical dream. In the same way as Aristotle, he writes mathematics like an intelligent outsider, not like a true addict.

The second aspect concerns how Ibn Sīnā sees the nature of logic. For Ibn Sīnā logic is not about when this follows from that; it's about *how we can see from first principles that this follows from that.* For example if we are given a linkage and a sentence, Theorem 2 gives a fast way of testing whether the linkage entails the sentence without needing to construct any simple syllogisms at all. In Ibn Sīnā's time this theorem wasn't yet known. But even if it had been, it would have established a logical fact by going outside the basic processes of deduction, and so Ibn Sīnā very probably wouldn't have used it. The fact that Ibn Sīnā uses only direct and bottom-level methods was a great help for extracting the algorithm from Ibn Sīnā's text. One knew in advance that there were no hidden tricks or changes of viewpoint or appeals to intuition. The student was expected to solve the problems by direct application of basic facts of logic, and all that Ibn Sīnā was teaching him was how to apply the steps in the right order (as he himself says at [9.6.12]).

For balance one should add that in general Ibn Sīnā was certainly prepared to use metatheorems of logic as well as theorems. In fact he despised logicians who couldn't do this. But the metatheorems that he used were ones that summed up elementary facts about syllogisms, not ones that introduced new ideas.

In one other respect Ibn Sīnā's algorithm matches the mathematics of his time. He achieves the effect of induction by reducing more complex cases to simpler ones, until he reaches ground level. We might compare Proposition 8 of Ṯābit ibn Qurra in Rashed [26] p. 337ff, and Rashed's analysis on page 159. Ṯābit computes an n-term sum by writing the terms to be summed, then below them $n-1$ terms to be summed, and so on down to a single term. This produces a two-dimensional array, and Ṯābit computes the sum of the top line from properties of the whole array. For his exposition he takes $n = 4$ as a typical case. We saw that Ibn Sīnā takes cases of length 2 or 3, but here the parallel may break down, because we found that these cases play a special role in the calculation.

Appendices

A Translation of *Qiyās 9.6*

IX.6 The analysis of syllogisms, with a mention of dos and don'ts that can be relied on and used in that [analysis].

[9.6.1] Sometimes a person is addressed with a well-crafted and
definitive syllogism, or he finds such a syllogism written in a book.
But then [sometimes] the syllogism is not simple but compound; or 460.5
it appears not as a connected whole but as scattered pieces. And
sometimes moreover the pieces are jumbled out of their natural
order, or a part of the syllogism is hidden, or something superfluous
is added. [Even] when it is simple, sometimes it is jumbled out of its
natural order, or missing a piece, or with a piece added. You already
know how this happens. If we don't have rules to guide us, on how
to seek with due deliberation the syllogism that proves a given
goal, [and to confirm] the soundness of the connection between a 460.10
given syllogism [and its goal], so that we can analyse the syllogism
into a group of premises, put them in the natural order, strip off
defects and add any part that is missing, reducing the syllogism to
the syllogistic figure that produces it – [if we don't have rules for
all this,] then the new information that the syllogism provides will
escape us. If the syllogism is sound then [so is] what it entails. If
it's faulty, one should locate the fault either in its premises or in
its construction.

[9.6.2] So we need to have rules in the form of dos and don'ts, to be 460.15
used in the analysis of a syllogism. The rules should apply, not on
the basis that the syllogism is demonstrative or dialectical or some
other kind, but on the basis that it is an absolute syllogism. Then
when you are given [the syllogism], you reach what the analysis
leads you to, and it agrees with your starting point when you fol- 461.1
lowed the route of synthesis. Thus you find the truth agreeing with
itself, however you come to it, and standing as witness to its essence.
For the truth, insofar as it is what is the case, stands witness to its
essence insofar as [its essence] is how the truth is conceptualised.
Likewise insofar as [the essence of truth] is the starting point of [the
truth], [the truth] witnesses to its essence insofar as [the truth] is
where [the essence] leads us to; and insofar as [the essence of truth]
is where [the truth] leads us to, [the truth] stands as witness to its
essence insofar as [the truth] is the starting point of [its essence].

461.4,5 [9.6.3] So when you have found a syllogism, you start by looking for
its two premises. You do this before looking for the terms, because
gathering up fewer things is easier [than gathering up many]. Also
when you start with the terms, it can be that there are more than
two ways of combining them into two premises, so that the cases you
would need to consider would ramify. The reason for that is that by
locating the terms you don't thereby locate the premises as things
composed [from the terms]. You would have to examine the case of
each term, and then examine four possible ways of combining [pairs
461.10 of terms]. So you would have to consider five items: first you would
consider the terms [themselves], and then you would consider the
four cases which arise from the ways of composing the premises
from two terms. But if you locate the two premises, it's enough
for you to consider one more thing, namely to list the terms. Thus
when you have found two premises, locating the syllogism and how
it behaves will be easy for you.

461.12 [9.6.4] Then the first step is to investigate whether each of the
premises shares one of its terms with the goal but is distinguished
from the goal by another [term]. Suppose [it does, and] one of the
two premises shares both its terms with one part of the second
premise, while another part of the second premise – not the whole
of it – shares both the terms of the goal. Then the syllogism is
duplicative, and the premise which has one part overlapping the
goal and another part overlapping the other premise is a proposi-
462.1 tional compound, while the other premise is a duplication. So look
carefully at [the sentence] which has a part overlapping the goal in
two terms: is it meet-like or difference-like? If it is meet-like then
find out whether its overlap [with the goal] is its first or second
clause, and find out whether that other [sentence] is the same [as
this part of the premise], or is its contradictory. If the premise is
difference-like, then find out whether the overlapping [clauses] are
the same or contradictories. Do the same with the other [premise],
which is the duplicating one. In this way your syllogism is analysed
462.5 into the propositional moods.

462.5 [9.6.5] If this is not the case, and for every [sentence] of the syllogism
the goal (which is proved through [the syllogism]) overlaps it in just
one term, then you know that the syllogism is recombinant. If you
have found that each of the premises overlaps the conclusion, then
look for the middle term, so that you find the figure. Then connect
the terms to the conclusion, so as to find the major and minor
[premises] and the other things that you should be looking for. If
you can't find a middle term, then the syllogism is not simple;
462.10 instead you have a compound syllogism with at least four terms.

[9.6.6] [First case: two given premises, each sharing one term with
the goal]

[Problem 1.] Suppose the goal is universally quantified affirmative, 462.10
namely 'Every C is an A', and suppose that the found premises are 'Every C is a B' and 'Every D is an A'. Then if it's clear that 'Every B is a D', your syllogism is in good order; otherwise it needs a middle.

[Problem 2.] Suppose the goal is universally quantified negative, 462.12
[namely 'No C is an A'], and suppose the found [premises] are 'Every C is a B' and 'No D is an A'. Then consider whether 'Every B is a D'. If so, then a syllogism can be composed. If not, then it needs a middle.

[Problem 3.] Suppose the found premises are 'No C is a B' and 462.15
'Every A is a D'. Then it will be no help to you in this case to find 'Every B is a D', so that the negative [premise] becomes the minor [premise of a syllogism] in the first [figure] and the remaining two premises are affirmative. So consider whether it's true for you that 'Every D is a B'. If it is, then you say 'Every D is a B' and 'No C
is a B', which entails 'No C is a D'. Then you add to it that 'Every 463.1
A is a D', so that it entails 'No C is an A'.

[Problem 4.] Suppose the found [premises] are 'No C is a B' and 463.2
'Every D is an A'. Then it can't be used.

[Problem 5.] Suppose the goal is 'Some C is an A', and you have 463.2
found [the premises] 'Some C is a D' and 'Every B is an A'. Then if 'Every D is a B' is attached, you have found [the syllogism].

[Problem 6.] If the found [premises] are 'Every D is a C' and 'Every 463.4
B is an A', then if 'Every D is a B' is attached, you have found [the syllogism].

[Problem 7.] If the determinate [premises] are 'Every C is a D' and 463.5
'Some B is an A', then if 'Every D is a B' or 'Some D is a B' is attached, it can't be used. If 'Every C is a B' or 'Some C is a B' is attached, it can't be used. Likewise if 'Some B is a C', or 'Some B is a D' is attached, it can't be used. And likewise if 'Every B is a D' is attached, it can't be used. And if 'Every B is a C' is attached, it doesn't entail to ['Some] C is an A'.

[Problem 8.] If the found determinate [premises] are 'Some D is a 463.9
C' and 'Every B is an A', and 'Every D is a B' is attached, then this makes the syllogism determinate.

[Problem 9.] If the determinate [premises] are 'Every D is a C' and 463.10
'Every B is an A', and 'Every (or some) D is a B' is attached, then this makes the syllogism determinate.

[Problem 10.] If the determinate [premises] are 'Every D is a C' 463.11
and 'Some B is an A', it can't be used.

[Problem 11.] If the determinate [premises] are 'Some D is a C' and 463.12
'Every A is a B', it can't be used.

So consider the remaining cases [with existentially quantified affirmative goal] in the same way.

463.13 [Problem 12.] Suppose that the goal is existentially quantified negative: 'Not every C is an A', and that you have found [the premises] 'Some C is a B' and 'No D is an A'. Then if [an appropriate sentence with terms] B, D is attached, then you can use it – for example 'Every B is a D'.

463.15 [Problem 13.] If you have [the premises] 'No C is a B' and 'Some D is an A', it can't be used.

463.16 [Problem 14.] Likewise if you have [the premises] 'Every C is a B' and 'Not some D is an A', [the syllogism can't be used].

464.1 [Problem 15.] If you have [the premises] 'Not every C is a B' and 'Every D is an A', then it can't be used.

464.1 [Problem 16.] If you have [the premises] 'Some B is a C' and 'No D is an A', and 'Every B is a D' is attached, you can use it.

464.3 [Problem 17.] If [the premises] are 'No B is a C' and 'Some D is an A', it can't be used.

464.3 [Problem 18.] If [the premises] are 'Every B is a C' and 'Every D is an A', it can't be used.

464.4 [Problem 19.] If you have [the premises] 'Not every B is a C' and 'Every D is an A', it can't be used.

464.5 [Problem 20.] If you have [the premises] 'Some D is a C' and 'No A is a B', and 'Every D is a B' is attached, then you can use it.

464.6 [Problem 21.] If you have [the premises] 'No C is a B', and 'Some A is a D', it can't be used.

464.7 [Problem 22.] If the determinate [premises] are 'Every C is a B', and 'Not some A is a D', it can't be used.

464.8 [Problem 23.] If the determinate [premises] are 'Not every B is a C', and 'Every A is a D', it can't be used.

464.8 [Problem 24.] If you have: 'Some C is a B' and 'No A is a D', and 'Every B is a D' is attached, you can use it.

464.10 [Problem 25.] If you have [the premises] 'No B is a C' and 'Some A is a D', then it can't be used.

464.10 [Problem 26.] If you have [the premises] 'Every B is a C' and 'Not every A is a D', it can't be used.

464.12 [9.6.7] Likewise in the other remaining cases. This is when the two premises each share a term with the goal. If the two [premises] share [a term] with each other, and they don't share with the goal at all, then don't bother to analyse it, because in this case the shortfall is too great. And likewise when only one of the two shares [a term]
464.15 with the goal, and the other doesn't share with the goal or with its companion, then [the argument] is not straightforward to analyse. In order to explain how to analyse it we would need to apply a
465.1 lengthy principle that is not expressible in a rule that one can take on board briefly. Analysis of [such an argument] is possible, but the appropriate place for this is the appendices, which will also [extend] analysis to more than two premises.

[9.6.8] [Second case.] If you have found two premises that share [a term] with each other, and one of them shares [a term] with the goal, then this shared [term] is either the subject or the predicate of the goal. Suppose it is the subject. 465.5

[Problem 27.] First suppose the conclusion is universally quanti- 465.5
fied and affirmative, thus: 'Every C is an A.' Suppose the found [premises] are 'Every C is a B' and 'Every B is a D'. Then if you have found [a premise] linking D to A, this makes [the syllogism] determinate.
[Problem 28.] Suppose the conclusion is universally quantified neg- 465.7
ative [thus: 'No C is an A'], and the found [premises] are: 'Every C is a B' and 'Every B is a D'. Then if you have found [the premise] 'No D is an A', this makes [the syllogism] determinate.
[Problem 29.] If you have found [the premises] 'Every C is a B' 465.8
and 'No B is a D', and then you found [the attachment] 'Every A is a D', this makes [the syllogism] determinate without needing a conversion.
[Problem 30.] If you have found [the premises] 'No C is a B' and 465.10
'Every B is a D', it can't be used.
[Problem 31.] If you have found [the premises] 'No C is a B' and 465.10
'Every D is a B', and then you found the premise 'Every A is a D', this makes [the syllogism] determinate.
[Problem 32.] Suppose the conclusion is existentially quantified af- 465.12
firmative [thus: 'Some C is an A']. Suppose [the premises] 'Some C is a B' and 'Every B is a D' are already determinate, and 'Every D is an A' is attached, then this makes [the syllogism] determinate.
[Problem 33.] Suppose [we have] 'Every D is a B' and 'Every B is 465.13
a C'. Then if 'Every D is an A' or 'Some D is an A' is attached, this makes [the syllogism] determinate.
[Problem 34.] Suppose [the premises] are 'Every C is a B' and 'Some 465.14
B is a D'; then this [syllogism] can't be used.
[Problem 35.] If the existentially quantified [goal] is negative [thus: 465.15
'Some C is not an A'], and you have found [the premises] 'Some C is a D' and 'Every D is a B', and 'No B is an A' is attached, this makes [the syllogism] determinate.
[Problem 36.] If you have found [the premises] 'Some C is a B' and 466.1
'No B is a D', and 'Every A is a D' is attached, this makes [the syllogism] determinate.

Work through the remaining cases of this kind for yourself, taking the compound [syllogisms] in turn.

466.3 [9.6.9] You should know that when we said: 'This makes [the syllogism] determinate', this meant determinate without having to alter [the syllogism] by making a conversion in the found [premises]. Also you should know that we are not putting ourselves to the trouble of telling you now what figure the determinate [syllogism] is [proved]
466.5 in. If you don't understand that, and didn't memorise what was said [about it earlier], you won't have been able to make any use of this [lesson].
[9.6.10] [Third case: Two premises which share one term with each other, and one of them shares a term with the predicate of the goal.]

466.6 [Problem 37.] If the shared [term] is in the predicate of the goal, and the goal is universally quantified affirmative [thus: 'Every C is an A']; and you have [the premises] 'Every D is a B' and 'Every B is an A', and 'Every C is a D' is attached, this makes [the syllogism] determinate.

466.7 [Problem 38.] If the goal is universally quantified negative [thus: 'No C is an A'], and the found [premises] are 'Every D is a B' and 'No B is an A', and 'Every C is a D' is attached, this makes [the syllogism] determinate.

466.9 [Problem 39.] If the found [premises] that you have are 'No D is a B' and 'Every A is a B', and 'Every C is a D' is attached, this makes [the syllogism] determinate.

466.10 [Problem 40.] If you have [the premises] 'Every D is a B' and 'No A is a B', and 'Every C is a D' is attached, this makes [the syllogism] determinate.

466.11 [Problem 41.] If the goal is existentially quantified affirmative [thus: 'Some C is an A'], and you have [the premises] 'Some B is a D' and 'Every D is an A', and 'Every B is a C' is attached, you can use it.

466.13 [Problem 42.] If you have: 'Some B is a D', and 'Every A is a D', it can't be used.

466.13 [Problem 43.] If you have 'Some D is a B' and 'Every B is an A', and [the attached premise] is 'Every D is a C', you can use it.

466.14 [Problem 44.] If you have 'Some D is a B' and 'Some A is a D', it can't be used, even with the order [of the terms in a premise] converted.

466.15 [Problem 45.] If your goal is existentially quantified negative [thus: 'Some C is not an A'], and you have [the premises] 'Some B is a D' and 'No D is an A', and 'Every B is a C' is attached, you can use it.

467.1 [Problem 46.] Or you have 'Every B is a D' and 'Some D is not an A' – then you can't use it.

467.2 [Problem 47.] If you have [the premises] 'Not every B is a D' and 'Every D is an A', you can't use it.

467.2 [Problem 48.] If you have 'No B is a D' and 'Some D is an A', you can't use it.

[Problem 49.] If you have 'Some D is a B' and 'No A is a B', and 467.3
'Every D is a C' is attached, you can use it.
[Problem 50.] If you have 'No D is a B' and 'Every A is a B', and 467.4
'Some C is a D' is attached, you can use it.
[Problem 51.] If you have 'Not every D is a B', and 'Some A is a 467.5
B', it can't be used.
Try out for yourself the compound [syllogisms] where the overlap 467.7
is with the predicate of the goal, in the same relation as above.
These, and similar [examples] that we handle by comparison with
them, are instances of analysis where you have two premises. 467.9

[9.6.11] [Fourth case: One premise, which shares a term with the
goal.]
[Problem 52.] In the case where you have a single premise, which 467.9
overlaps the predicate of the conclusion, and the goal is universally
quantified affirmative, namely 'Every C is an A', and you have [the
premise] 'Every D is an A', then if 'Every C is a D' is attached,
this makes [the syllogism] determinate.
[Problem 53.] If you have 'Every A is a D', it can't be used. 467.12
[Problem 54.] If the goal is universally quantified negative [thus: 467.12
'No C is an A'], and you have [the premise] 'No D is an A' or
'No A is a D', and 'Every C is a D' is attached, this makes [the
syllogism] determinate.
[Problem 55.] If you have [the premise] 'Every D is an A', then [the 467.14
syllogism] can't be made determinate.
[Problem 56.] Rather, if you have 'Every A is a D', and it's true 467.14
that 'No C is a D', this makes [the syllogism] determinate.
[Problem 57.] If the goal is existentially quantified affirmative [thus: 467.15
'Some C is an A'], and you have [the premise] 'Some D is an A',
and 'Every D is a C' is attached, you can use it.
[Problem 58.] If you have [the premise] 'Every D is an A', and 467.16
'Some C is a D' is attached, you can use it.
[Problem 59.] If you have 'Some A is a D', you can't use it at all, 467.17
unless you convert [the premise].
[Problem 60.] If the goal is existentially quantified negative [thus: 467.18
'Some C is not an A'], and you have [the premise] 'Every D is an
A', you can't use it at all.
[Problem 61.] Rather, if [the premise] is 'No D is an A', and 'Some 467.19
C is a D' is attached, you can use it. 468.1
[Problem 62.] Likewise if you have 'Some D is an A', it can't be 468.1
used.
[Problem 63.] If you have [the premise] 'Not every D is an A', and 468.2
'Every D is a C' is attached, you can use it.
[Problem 64.] If [the premise] is 'Not every A is a D', it can't be 468.3
used.

468.4 [9.6.12] When you put the steps in this order, as I have shown you, you will reach the [required] terms, figures and moods. And the terms that you encounter will be ones within the formats mentioned above as ones that can be used.

468.7 Apply exactly the same considerations to propositional compounds.

B Notes on the Text Translated

The text above is translated from the Arabic text in [15], which is a volume from the Cairo edition of the *Šifā'*, published under the overall editorship of Ibrahim Madkour.

Title

Ibn Sīnā writes '*intafa*^{c} X *bi*-Y' to express 'X can use Y'. The passive form, which occurs in the title, is '*untufi*^{c} *bi*-Y', meaning 'Y can be used'. I haven't found this meaning in the dictionaries, including Goichon [9]. But it's fairly common in Ibn Sīnā's logical writing. For example in *Burhān* [16] 13.14 one can't use (*lam yantafi*^{c} *bi-*) what a teacher says unless one thinks for oneself; 63.8 there are students who can use (*intafa*^{c} *bi-*) a compass but are still stupid; 141.13 in debate one can't use (*lā yantafi*^{c} *bi-)* a proof that requires very many middle terms; ^{c}*Ibāra* [14] 2.12f sciences are developed so that later generations can use (*yantafi*^{c} *bi-*) them. Dozy [7] comes nearest with the meaning 'trouver son compte à', which Gabriel Sabbagh kindly tells me can be translated as 'finds advantageous or useful'.

[9.6.1]

460.5f 'not connected but separated' (*gair mawṣūl bal mafṣūl*): See Subsects. 4.2 and 4.4 above for these notions.

[9.6.2]

460.17 'absolute syllogism' (*qiyās muṭlaq*): 'absolute' (*muṭlaq*) means without any restriction or condition being imposed. The kind of restriction he has in mind here is to syllogisms that are appropriate for a particular purpose. For example demonstrative syllogisms are for demonstrating that something is true by deducing it from things that are self-evident or already demonstrated, so their premises must be necessary truths. Dialectical syllogisms must have premises that are true for the most part and generally accepted. (*Qiyās* [15] p. 4, *Išārāt* [20] Method 9.)

461.2 'standing as witness to its essence' (*šāhid li-ḏātih*): This rhetorical flourish apparently comes from the translator into Arabic, Taḏârî; it is not in the Greek original. Ibn Sīnā seems to have replied with a similar blossom of rhetoric. But was Taḏârî quoting something?

[9.6.3]

461.8 'as thing composed' : i.e. rather than as segments of text.

At *Prior Analytics* i.32 47a11 Aristotle claims that it's easier to divide a thing into large parts than into small, but offers no argument in support of this. Ibn Sīnā may be right about which order is easier, but his reason doesn't convince. If you first locate the premises as segments of the text, you don't thereby locate them 'as composed from the terms'. There may be many different ways of carving a subject-predicate form out of one and the same sentence. For example if the sentence in front of you is 'This line and that line meet', should you parse it as '(This line) (meets that line)' or as '(These two lines) (meet)'? Or to take Ramsey's more philosophical example, should you parse 'Socrates is wise' as '(Socrates) (is wise)' or as '(Wisdom) (is a characteristic of Socrates)' ([25] p. 21)? The nub of the matter is that Ibn Sīnā in this section ignores the possibilities of local formalising; cf. [13].

[9.6.4]

461.13 'its terms': Ibn Sīnā is discussing propositional syllogisms here, so for example the 'terms' of the proposition '*if p then q*' are p and q, both of which are sentences and not terms in the usual sense. See Subsect. 4.5 above.

[9.6.6]

Problem 3. The two terms that occur once only in the given goal and premises are B and D, so we are looking for a sentence ϕ with terms B and D. The goal is universally quantified, so all the premises are universally quantified, and in particular ϕ is universally quantified. The goal is negative, so there is exactly one negative premise; hence the remaining two premises including ϕ must be affirmative. Thus ϕ must be either 'Every B is a D' or 'Every D is a B'. We try both in turn. If we combine 'Every B is a D' with 'No C is a B' as the premises of a simple syllogism, then since B is subject in one and predicate in the other, the syllogism is in first figure, and its minor premise is 'No C is a B' since this is the one with the middle term D as its predicate. But the only mood in first figure with two universally quantified premises and one of them negative is the second mood (*Celarent* in the Latin nomenclature), whose minor premise is affirmative. So 'Every B is a D' can't be used, and we have to try 'Every D is a B' instead. The result is the following connected compound syllogism, which meets the requirements:

$$\dfrac{\dfrac{\text{No } C \text{ is a } B. \qquad \text{Every } D \text{ is a } B.}{\text{No } C \text{ is a } D.} \qquad \text{Every } A \text{ is a } D.}{\text{No } C \text{ is an } A.} \tag{29}$$

In his discussion Ibn Sīnā seems to derive 'No C is a D' from the premises 'Every D is a B' and 'No C is a B' in that order, breaking the premise order condition. Probably the reason is in the immediately preceding text: 'Consider whether … 'Every D is a B'. If it is, then' (etc.) The premise 'Every D is a B' follows on naturally from this, and Ibn Sīnā need not be claiming that it serves as first premise. (The case is quite different from Problem 33, where the order is unexpected until we remember Ibn Sīnā's conventions.)

Problem 5. 'found': What is found? In Problems 8, 9 it's explicitly the syllogism, and there are no examples where it's explicitly the goal. So I infer the syllogism is meant here.

Problem 6. The problem is the same as Problem 9 below. The solutions are different; at Problem 9 Ibn Sīnā gives the weakest fill and one other, but here he gives only an unnecessarily strong fill. (See the end of Subsect. 4.4.) One might be tempted to change the text so as to remove the doublet. But there is another doublet: Problem 43 is Problem 41 with the letters B and D transposed. So I left the text alone.

Problem 7. Assuming the text is sound, here is a sequence of thoughts that it could represent.

First, the goal is affirmative, so there is no need to consider an added premise θ that is negative. Second, we saw in Problem 3 that θ should be taken with the first premise to yield an intermediate conclusion. The first premise is 'Every C is a D', so either C or D is a term in θ. The other term can't be A, since A already occurs twice; so it must be B. (This is clumsy: the same reasoning would eliminate C too, since it also occurs twice. Maybe Ibn Sīnā wanted his student to say 'That's clumsy' and formulate the reason.) So first we try affirmative sentences with B as predicate. There are two with D as subject, and combining with 'Every C is a D' they yield respectively 'Every C is a B' and nothing at all. 'Every C is a B' can't combine with 'Some B is an A', because it would give a first figure syllogism with existentially quantified major premise, which is impossible. There are two with C as subject, and they both give 'Some B is a D' (or conversely). This can't combine with 'Some B is an A' because both are existentially quantified. Next we try the possibilities with B as subject. There are two existentially quantified, but again these will yield an existentially quantified intermediate sentence that won't combine with 'Some B is a D'. There remain 'Every B is a D' and 'Every B is a C'. The first yields nothing with 'Every C is a D'. The second yields 'Every B is a D', which combines with 'Some B is an A' to yield 'Some D is an A', not the conclusion we want.

The last case, namely 'Every B is a C', is important because it does in fact combine with the given premises to yield the required goal; but the resulting syllogism uses only the second premise. Since

Ibn Sīnā doesn't count this as a solution of the problem, we have confirmation that the algorithm is not intended to eliminate unnecessary premises.

But there is also a problem about how to read the text at the end of the example. The Cairo edition has *lam yuntij 'ilā j*, which is meaningless. The sense has to be either that the inference using 'Some B is an A' doesn't use the premise 'Every C is a D', or that when used with all the other premises, it doesn't yield the required conclusion. The normal usage is that one entails *min* or *ᶜan* the premises and *ᶜalā* the conclusion; the preposition *'ilā* 'to' could hardly be a variant of the first two, but it could be a variant of *ᶜalā* 'onto', though I've found no other examples. Hence we can guess that *j* should be *j a*, a shorthand for the goal (since our C corresponds to Arabic *j*). Therefore I propose to emend to *lam yuntij 'ilā j a*, and I have translated accordingly. See Problem 32 for the opposite error; there is very little difference between a handwritten Arabic *a* and a short scratch on the paper.

Problem 8. 'makes the syllogism determinate' (*qad ḥuṣṣila l-qiyās*): For this translation see the notes on Problem 32.

Problem 9. See on Problem 6.

Problem 14. If we add the premise 'Every B is a D', then from 'Every C is a B' we get 'Every C is a D'. To deduce 'Not every C is an A' we need 'No D is an A'; 'Some D is not an A' wouldn't be enough. So it looks as if Ibn Sīnā here understands 'Not some D is an A' (*laisa baᶜḍu d a*) as 'Some D is not an A'. At first sight this is at odds with his treatment of *laisa kullu*, which he always interprets as 'Not every'. There is a similar discrepancy at *ᶜIbāra* [14] 67.10, where Ibn Sīnā says that *lā kull* and *lā baᶜḍ* are equivalent. The interactions of quantifiers and negation in Arabic are complicated; Jamal Ouhalla alerts me to the fact that focus can affect scope. But as far as I can see, the relevant phrases in this Problem and at the *ᶜIbāra* reference are in topic position, though there are grounds for thinking that logic blinded Ibn Sīnā to questions of topic.

Problem 22. Here 'Not some A is a D' should be interpreted as 'Some A is not a D'. If it was read as 'No A is a D' and we attached 'Every B is a D', the first two premises would yield 'Every C is a D', which combines with 'No A is a D' to yield 'No C is an A' by *Cesare*, and hence 'Not every C is an A' by (10). Compare Problem 14.

Problem 23. The Cairo text has 'Every A is a D' (*kullu a d*) rather than 'Every D is an A' (*kullu d a*). But with that reading Ibn Sīnā should say that attaching 'Every D is a B' yields the required syllogism. I've adopted the easiest correction that doesn't introduce a doublet.

[9.6.8]

465.2 'in the appendices' (*bil-lawāḥiq*): In several places in the *Šifā'* Ibn Sīnā refers to things that will appear in the appendices. But no work of this name or with exactly the required contents has been found. It has been suggested that Ibn Sīnā's two other works *Taᶜlīqāt* and *Mubāḥaṯāt* contain material that was intended for the appendices. (Gutas [12] pp. 141–144.) But the published versions of these two works contain only philosophical material, and nothing about proof search. More's the pity, because Ibn Sīnā's treatment of incomplete syllogisms with two or more gaps would have shown us more about how he handled problems of search. See Subsect. 6.2 for more on the historical context.

Problem 27. Ibn Sīnā doesn't say what premise linking D to A will work. There may be a subtle reason. This is the first example with two premises ϕ_1, ϕ_2 to the left of the gap, so the student has a choice between first combining ϕ_1 with ϕ_2 before combining the result with the test sentence; or first combining ϕ_2 with the test sentence and then bringing in ϕ_1. The first route is clearly more sensible, because the result of combining ϕ_1 with ϕ_2 will be the same for each test sentence. Ibn Sīnā forces the student to see this, by putting pressure on the student to try several test sentences. But the effect is slightly spoiled by the fact that in this particular case the answer 'Every D is an A' is obvious without any calculation.

Problem 29. Paragraph [9.6.9] below suggests that Ibn Sīnā is talking about converting a premise. But why should anybody think of converting a premise in this example? A possible explanation lies in the fact that this is the first example in this block where a premise has its terms out of the obvious order. We might expect $(C, B)(B, D), (D, A)$, but instead the last premise gives (A, D). Perhaps Ibn Sīnā had students who (apparently like Smith [3] Note to 42b5–26) assumed that switches like this don't occur. Ibn Sīnā had made the same point already at *Qiyās* 444.5, where it seems to have confused the copyists.

Problem 32. The Arabic contains two occurrences of *qad ḥuṣṣil*, but they must mean different things. In general *qad* with the past tense is a perfective marker: it indicates that the present state is the outcome of a previous action described by the verb. But previous to what? At the first occurrence here the phrase must mean previous to the problem having been posed, hence 'already determinate'. But the second occurrence describes the outcome of the algorithm, so it can't mean that; it must mean that the application of the algorithm created the present situation, hence 'this makes it determinate'.

Also the *'a* at the end of the sentence in line 465.13 should be deleted (as in one ms); cf. Problem 7.

Problem 33. This is the one problem where Ibn Sīnā gives the premises in an order that doesn't form a linkage where the goal subject points leftwards and the goal predicate points rightwards. The reason for this is explained in Subsect. 4.4 above.

The solution 'Every D is an A' is redundant since it implies 'Some D is an A', which is already a solution (it's the weakest fill). Ibn Sīnā's procedure of trying all options is likely to throw up redundancies of this kind. But maybe he expects his brighter students to note the redundancy and formulate a policy.

Problem 34. Delete the second occurrence of *wa-ba*^{c}*ḍ b d*.

466.1 'taking the compound [syllogisms] in turn' (*bi-ḥasabi l-tarākīb*): *ḥasb* means calculation (as in the modern *ḥāsib* 'computer'). But there is probably no reference to computing or algorithms here. 'Calculations' in Ibn Sīnā's time were normally assumed to be numerical. So *bi-ḥasab* here probably has its usual meaning of 'according to'.

[9.6.10]

Problem 37. Delete *māḏā* at the beginning of line 466.6. Also the *'i* in the Cairo edition is a misprint for *'in*.

Problem 44. The student might worry that these two premises violate the fourth figure condition. Strictly this is not relevant, because the connected syllogism wouldn't combine these two premises in a simple syllogism; but it may explain why Ibn Sīnā remarks that no conversion is needed.

Problem 48. In the Cairo text the first premise is 'No B is a C', violating the case assumption for [9.6.10]. Read 'No B is a D', following two manuscripts.

Problem 51. The Cairo text has 'Every A is a B' for the second found premise. There must be a slip, because in that case we get a syllogism by attaching 'Every D is a C'. But on the Cairo reading this is also the only example in this block where the second found premise is the same as in the previous example. So I have replaced 'Every A' by 'Some A'.

[9.6.11]

Problem 59. The only sentence that will complete the syllogism logically is 'Every D is a C'. The middle term is D, which is subject in 'Every D is a C' and predicate in 'Some A is a D', so the syllogism violates the fourth figure condition. Converting the premise 'Some A is a D' to 'Some D is an A' yields a third-figure syllogism in *Disamis*.

Problem 61. We have to correct 'Some C is an A' to 'Some C is a D'.

Problem 62. The Cairo text reads 'Likewise if [the premise] is 'No A is a D', and you have (^{c}*indak*) 'Some D is an A' or 'Some A is a C', it can't be used.' There are several problems with this. First, with the datum 'No A is a D' we get the goal by appending 'Some C is a D'; so the datum is presumably wrong. Second, this is the one problem where Ibn Sīnā seems to introduce the appended sentence with

^{c}*indak*; in 28 other problems ^{c}*indak* introduces the datum. Third, the sentence 'Some A is a C' is silly here, because it has the same terms as the goal. We can get a reasonable problem by deleting the first and third syllogistic sentences and the text around them, as I have done in the translation. Then Ibn Sīnā is saying correctly that the goal can't be reached from the datum 'Some D is an A'.

[9.6.12]

468.7 'propositional connectives': An example of a simple recombinant propositional syllogism, using the propositional connective 'If ... then', is

If p then q. If q then r. Therefore: If p then r. (30)

where p, q and r are declarative sentences. Ibn Sīnā brings this to a form analogous to a predicative syllogism by the device of 'replacing "if" by "whenever" ' (e.g. *Qiyās* 471.5), so that the sense becomes

Every occasion on which p is an occasion on which q. Every occasion on which q is an occasion on which r. Therefore: Every occasion on which p is an occasion on which r. (31)

This reduction, together with similar ones for some other propositional connectives, allows the proof search algorithm to be carried over routinely from predicative syllogisms to propositional ones.

C The ASM

Briefly, an ASM consists of a set of rules operated by modules; for example the module ProofSearch below has four modules, namely Describe, Synthesise, Ramify and Select. At each step in a computation, *all* the rules are applied once and simultaneously; if they clash, the machine stops. Rules normally begin with a condition, so they do nothing unless the condition is met. They can activate other rules by resetting parameters so that the conditions for the other rules are met. For example when the conditions for Synthesise are met, the rules of Synthesise have the effect of shortening the datum by 1 every time they operate. They continue to operate until there are no consecutive formulas in the datum with a term in common; at this point the condition for Synthesise fails, but that for Ramify may be met, so that the rules of Ramify take over.

The notation $X := Y$ means that the value of the parameter X becomes Y. The notation $X^\star$ means the set of nonempty finite sequences of elements of X. I hope the rest is reasonably self-explanatory.

The logical part of this ASM was implemented in Perl 5 and run on all of Ibn Sīnā's 64 problems. There were discrepancies from Ibn Sīnā's solution (as reported in the Cairo text) at Problems 23, 33, 51, 61 and 62. These are all discussed in the notes to the relevant problems in Sect. B above. Problem 33 is the only one of the 64 problems that lies outside the domain of the algorithm. The other four discrepancies are well within the range of transcription errors that one might expect from the state of the manuscript tradition. Some could possibly be misprints in the Cairo edition – I haven't consulted any manuscripts to check this.

Multi-agent ASM

Universe: *INTELLECT* (dynamic set of agents)

(ASM1) ActiveIntellect=
forall $\iota \in INTELLECT$ **if** $next(currpage).\iota = needsmiddle$ **then**
  let $k = |hasils(fill(currpage).\iota)|$
  if $k > 0$ **then**
    let $\iota_k := \iota$
    if $k > 1$ **then**
      let $\iota_1, \ldots, \iota_{k-1} = new(INTELLECT)$
      forall $1 \leqslant i \leqslant k$
        let $p = new(PAGE).\iota_i$
        $datum(p).\iota_i := insert(datum(currpage).\iota_i,$
          $i\text{-th}(hasils(fill(currpage).\iota_i)), gapsite(currpage).\iota_i)$
        $gapsite(p).\iota_i := needscalculating$
        $previous(p).\iota_i := currpage.\iota_i$
        $next(p).\iota_i := 0$
        $X(p).\iota_i := X(currpage).\iota_i$
          where $X = edges, fill$
        $currpage.\iota_i := p$

where $hasils(\phi).\iota$ is the set of all sentences that are determinate for the intellect ι and have exactly one term in common with ϕ; we assume this set is finite. In general the i-th sentence in the set will contain a term t that is not already in $TERM.\iota_i$, so the Active Intellect will need to add t to $TERM.\iota_i$; the term t is the imprinted form that we met in (28).

Signature of the agent ASMs

Universes: *TERM*, *PAGE*

Globals:

$goal \in SENTENCE$
  (input from the problem)
$currpage \in \mathbb{N}$
  (initially 0)
$report \in \{ignorance, logicalfailure, logicalsuccess, tahsilsuccess\}$
  (initially $= ignorance$)

Page properties:

$datum : PAGE \to SENTENCE^{\star}$
(initially input from the problem)
$gapsite : PAGE \to \mathbb{N} \cup \{needscalculating\}$
(initially = $needscalculating$)
$edges : PAGE \to TERM^2$
$fill : PAGE \to SENTENCE$
$previous : PAGE \to PAGE$
$next : PAGE \to PAGE \cup \{needsmiddle\}$
(initially 0)

Agent modules

(ASM2) ProofSearch = {Describe, Synthesise, Ramify, Select}

(ASM3) Describe =
if $gapsite(currpage) = needscalculating$ **then**
let $p = new(PAGE)$
take $datum(currpage)$, $goal$ and identify the gap site, the left edge and the right edge. (If no gap then the gap site is 0.)
$gapsite(p) :=$ calculated gap site
$edges(p) :=$ (left edge,right edge)
$previous(p) := currpage$
$X(p) := X(currpage)$
where $X = datum, fill, next$
$currpage := p$

(ASM4) Synthesise =
if $gapsite(currpage) \geqslant 0$ **and** $next(currpage) \geqslant 0$ **and**
$(length(datum(currpage)) > 2$ **or**
$(length(datum(currpage)) = 2$ **and** $gapsite(currpage) \neq 1))$ **then**
let $k = \begin{cases} 1 & \textbf{if } gapsite(currpage) \neq 1, \\ 2 & \textbf{otherwise} \end{cases}$
let $\ell = length(datum(currpage))$
let $\phi = consequence(k\text{-th}(datum(currpage)),$
$(k+1)\text{-th}(datum(currpage)))$
if $\phi \neq sterile$ **then**
let $\alpha = replacepair(datum(currpage), \phi, k)$
let $p = new(PAGE)$
$datum(p) := \alpha$
$previous(p) := currpage$
if $gapsite(currpage) > 1$ **then**
$gapsite(p) := gapsite(currpage) - 1$
else

```
          gapsite(p) := gapsite(currpage)
        X(p) := X(currpage)
          where X = edges, fill, next
        currpage := p
    else
        if next(currpage) > 0 then
          currpage := next(currpage)
        else
          if report = ignorance then
            report := logicalfailure
```

(ASM5) RAMIFY=

```
if gapsite(currpage) > 0 and next(currpage) ⩾ 0 and
(length(datum(currpage)) ⩽ 1 or
(length(datum(currpage)) = 2 and gapsite(currpage) = 1)) then
    if length(datum(currpage)) ⩾ 1 then
        let p_1, ..., p_8 = new(PAGE)
        forall 1 ⩽ i ⩽ 8
            let φ = listsentences(1-th(edges(currpage)),
                2-th(edges(currpage)), i)
          datum(p_i) := insert(datum(currpage), φ, gapsite(currpage))
          fill(p_i) := φ
          gapsite(p_i) := 0
          edges(p_i) := edges(currpage)
          previous(p_i) := currpage
        forall 1 ⩽ j ⩽ 7
          next(p_i) := p_{i+1}
        next(p_8) := 0
        currpage := p_1
    else
        let p = new(PAGE)
        datum(p) := insert(datum(currpage), goal, 1)
        gapsite(p) := 0
        edges(p) := edges(currpage)
        fill(p) := goal
        previous(p) := currpage
        next(p) := 0
```

(ASM6) SELECT=

```
if gapsite(currpage) = 0 and length(datum(currpage)) = 1 then
    if 1-th(datum(currpage) = goal then
        if hasil(fill) = true then
            report := tahsilsuccess
        else
            let k = least k ⩾ 1 such that
```

$gapsite(previous^k(currpage)) > 0$
let $p := new(PAGE)$
$X(p) := X(previous^k(currpage))$
where $X = datum, gapsite, edges$
$fill(p) := fill(currpage)$
$next(p) := needsmiddle$
$previous(p) := currpage$
$report := logicalsuccess$
$currpage := p$
else
if $next(currpage) > 0$ **then**
$currpage := next(currpage)$
else
if $report = ignorance$ **then**
$report := logicalfailure$

Basic functions

(Def1) $SENTENCE \subseteq TERM^2 \times \{0, 1\}^2$
$SENTENCE(s, t, i, j) \Leftrightarrow s \neq t$

(Def2) $listsentences : TERM^2 \times \{1, \ldots, 8\} \to SENTENCE$
$listsentences(s, t, 1) = (s, t, 0, 0)$
$listsentences(s, t, 2) = (s, t, 0, 0)$
$listsentences(s, t, 3) = (s, t, 0, 1)$
$listsentences(s, t, 4) = (s, t, 0, 1)$
$listsentences(s, t, 5) = (s, t, 1, 0)$
$listsentences(s, t, 6) = (s, t, 1, 1)$
$listsentences(s, t, 7) = (s, t, 1, 1)$
$listsentences(s, t, 8) = (s, t, 1, 1)$

(Def3) $consequence : SENTENCE^2 \to SENTENCE \cup \{sterile\}$
$consequence(\phi, \psi) =$
$$\begin{cases} \text{strongest consequence of } [\phi, \psi] & \textbf{if } [\phi, \psi] \text{ is not sterile,} \\ sterile & \textbf{otherwise.} \end{cases}$$

(Def4) $replacepair : SENTENCE^\star \times SENTENCE \times \mathbb{N} \to SENTENCE$
$replacepair([\phi_1, \ldots, \phi_n], \psi, i) = [\phi_1, \ldots, \phi_{i-1}, \psi, \phi_{i+2}, \ldots, \phi_n]$.

(Def5) $insert : SENTENCE^\star \times SENTENCE \times \mathbb{N} \to SENTENCE$
$insert([\phi_1, \ldots, \phi_n], \psi, i) = [\phi_1, \ldots, \phi_i, \psi, \phi_{i+1}, \ldots, \phi_n]$.

(Def6) $hasil : SENTENCE \to \{true, false\}$
a user-defined basic function

References

1. Whittaker, J. (ed.): Alcinoos: Enseignement des Doctrines de Platon. Budé, Paris (1990)
2. Mueller, I. (Trans.): Alexander of Aphrodisias: On Aristotle Prior Analytics, 1.32–46. Duckworth, London (2006)
3. Smith, R. (Trans.& ed.): Aristotle: Prior Analytics. Hackett, Indianapolis Indiana (1989)
4. Biggs, N.L., Lloyd, E.K., Wilson, R.J.: Graph Theory 1736–1936. Clarendon Press, Oxford (1976)
5. Börger, E., Rosenzweig, D.: A mathematical definition of full Prolog. Science of Computer Programming 24, 249–286 (1995)
6. Börger, E., Stärk, R.: Abstract State Machines. Springer, Berlin (2003)
7. Dozy, R.P.A.: Supplément aux Dictionnaires Arabes. Librairie du Liban, Beirut (1968)
8. Ebbesen, S.: Commentators and Commentaries on Aristotle's Sophistici Elenchi, vol. 1, The Greek Tradition. Brill, Leiden (1981)
9. Goichon, A.-M.: Lexique de la Langue Philosophique d'Ibn Sīnā. Desclée de Brouwer, Paris (1938)
10. Gurevich, Y.: Evolving algebras. A tutorial introduction. Bulletin of European Association for Theoretical Computer Science 43, 264–284 (1991)
11. Gurevich, Y., Veanes, M., Wallace, C.: Can abstract state machines be useful in language theory? Theoretical Computer Science 376, 17–29 (2007)
12. Gutas, D.: Avicenna and the Aristotelian Tradition: Introduction to Reading Avicenna's Philosophical Works. Brill, Leiden (1988)
13. Hodges, W.: Traditional logic, modern logic and natural language. Journal of Philosophical Logic 38, 589–606 (2009)
14. Madkour, I., et al. (eds.): Ibn Sīnā: Al-ᶜIbāra. Dār al-Kātib al-ᶜArabī lil-Ṭabāᶜ wal-Našr, Cairo (1970)
15. Madkour, I., et al. (eds.): Ibn Sīnā: Al-Qiyās. Našr Wizāra al-Ṯaqāfa wal-'Iršād al-Qūmī (1964) (referred to above as the Cairo edition)
16. Badawī, A. (ed.): Ibn Sīnā: Al-Burhān. Dār al-Nahḍa al-ᶜArabīyya, Cairo (1966)
17. Madkour, I., et al. (eds.): Ibn Sīnā: Al-Sufista. Našr Wizāra al-Ṯaqāfa wal-Taᶜlīm, Cairo (1956)
18. Ibn Sīnā: Al-Najāt. Jamīᶜ al-Ḥuqūq, Beirut (1992)
19. Ibn Sīnā: Manṭiq al-Mašriqiyyīn. Al-Maktaba al-Salafiyya, Cairo (1910)
20. Zare'i, M. (ed.): Ibn Sīnā: Al-Išārāt wal-Tanbiyyāt, Qum. Būstān-e Ketab-e Qom, Iran (2000); (The logical part is translated: Inati, S. C.: Ibn Sīnā, Remarks and Admonitions, Part One: Logic. Toronto: Pontifical Institute of Mediaeval Studies) (1984)
21. Jabre, F.: Al-Naṣṣ al-Kāmil li-Manṭiq Arisṭū, vol. 1. Dār al-Fikr al-Libnānī, Beirut (1999)
22. Katz, V.J.: Combinatorics and induction in medieval Hebrew and Islamic mathematics. In: Calinger, R. (ed.) Vita Mathematica: Historical Research and Integration with Teaching, pp. 99–106. Mathematical Association of America (1996)
23. Kutsch, W.: Muḥaṣṣal – Ġayr Muḥaṣṣal. Mélanges de l'Université Saint Joseph 27(8), 169–176 (1947-1948)
24. John Philoponus: Aristotelis Analytica Priora Commentaria. In: Wallies, M. (ed.). Reimer, Berlin (1905)

25. Ramsey, F.P.: Foundations: Essays in Philosophy, Logic, Mathematics and Economics. In: Mellor, D.H. (ed.). Routledge & Kegan Paul, London (1978)
26. Rashed, R.: Les Mathématiques Infinitésimales du IXe au XIe Siècle, vol. 1, Fondateurs et Commentateurs. Al-Furqān, London (1996)
27. Rashed, R.: Al-Khwārizmī, Le Commencement de l'Algèbre. Blanchard, Paris (2007)
28. Ross, W.D.: Aristotle's Prior and Posterior Analytics. Clarendon Press, Oxford (1949)
29. Shehaby, N.: The Propositional Logic of Avicenna. Reidel, Dordrecht (1973)
30. Street, T.: An outline of Avicenna's syllogistic. Archiv für Geschichte der Philosophie 84(2), 129–160 (2002)
31. Thom, P.: The Syllogism. Philosophia Verlag, Munich (1981)
32. Versteegh, K.: Landmarks in Linguistic Thought III: The Arabic Linguistic Tradition. Routledge, London (1997)
33. Zermelo, E.: Untersuchungen über die Grundlagen der Mengenlehre I. Mathematische Annalen 65, 261–281 (1908)

Abstract State Machines and the Inquiry Process

James K. Huggins[1] and Charles Wallace[2]

[1] Kettering University, Flint, MI
jhuggins@kettering.edu
[2] Michigan Technological University, Houghton, MI
wallace@mtu.edu

Abstract. Abstract State Machines have long played a valuable role as a catalyst for inquiry into software problems. In the ASM literature, however, there is a tendency to omit reflection on the *process* of ASM-based design and analysis, focusing instead on final, complete ASM *products*. As educators, we believe it is important to expose our students to a full, explicit process of inquiry, using ASMs as a vehicle to motivate active questioning. We report on our experiences in bringing ASM-based inquiry to the classroom. A course plan that combines ASMs and Problem Frames has proved effective in eliciting critical inquiry among students.

Keywords: formal methods, software requirements, education, refinement, inquiry.

1 Introduction

The idea of a mathematically rigorous semantic basis for software has long held an allure for software engineers, for a variety of reasons. Software problems and products can be captured in a precise language; sophisticated analysis techniques, including automated verification, can be brought to bear; complex systems can be synthesized from multiple partial specifications. The advent of tools for "lightweight" analysis [21] has made formal methods an option for a broad range of applications.

Another, less heralded benefit is the role formal methods can play as a catalyst for inquiry, provoking the constructive questioning that uncovers tacit assumptions and unforeseen consequences. This holds special interest for us as software engineering educators. From the narrow perspective of professional training, we must prepare our students to engage in the challenging workplace tasks of requirements elicitation and analysis – tasks that are complicated by the invisibility of software and the wide range of application domains [10]. From a wider pedagogical perspective, our primary mission is to expose our students to complex problems and to promote active, inquiry-based problem solving.

A. Blass, N. Dershowitz, and W. Reisig (Eds.): Gurevich Festschrift, LNCS 6300, pp. 405–413, 2010.

Our principal aim in this paper is to highlight the importance of ASMs as a vehicle for inquiry into software problems. We review the history of ASMs in uncovering hidden implications and unresolved ambiguities in real-world software products. We argue that in the ASM literature, the *process* of ASM-based inquiry has remained largely tacit, and that students need a more explicit exposition of the process. We report on our efforts to encourage a practice of inquiry among undergraduate software engineering students, using ASMs along with Problem Frames [22]. Finally, we provide some advice on what to emphasize in an ASM-based approach to inquiry.

2 The Practice of Problem Solving

At the heart of software engineering and computer science lies *problem solving*, a practice that is fundamentally heuristic due to epistemic, cognitive and temporal constraints on the human problem solver [25,24]. An expert problem solver typically employs heuristics tacitly, in a way that is invisible to any observer [23]. This presents a problem for the educator: how can we convey our internalized "know-how" to our students? Clearly, presenting the "final product" is not sufficient; students must be explicitly encouraged to evaluate and refine the work at hand, rather than accepting it at face value. This requires constant, systematic questioning – both internal (analyzing given information) and external (eliciting further information from other stakeholders).

In the field of technical communication, overreliance on the "final copy" has generally been acknowledged as a mistake in past educational practice. In current practice, revision is viewed as a valuable practice in itself, not simply as a disposable step toward an end product [14]. Inspired by this example, some software engineering educators are emphasizing a process of inquiry mediated by writing and revision. For example, students in Wright's software engineering course [27] use a template structure to document the design decisions made during a project. The author notes that "the processes students use to think about and organize what they know and do not know about the design problems is more important to the students' learning than the artifacts they generate." Also, case studies developed at Penn State [12] and Michigan Tech [10] contain documents produced through the lifespan of various software projects. The documents are hyperlinked to show the evolution of the project over time.

3 The Unknown as Known

In his classic treatise *How to Solve It* [25], Pólya paraphrases Pappus' explanation of Heuristic (*analyomenos*), the "art of problem solving" expounded by Euclid and others. Included in this paraphrase is the following:

> [I]n analysis we assume what is required to be done as already done (what is sought as already found, what we have to prove as true). (142)

Pólya remarks that the original text sounds as strange and counterintuitive as the English paraphrase: "is it not mere self-deception to assume that the problem we have to solve is solved?" (146). Upon reflection, however, the meaning is clear: "[i]n order to examine the condition, we have to conceive, to represent to ourselves, or to visualize geometrically the relations which the condition prescribes between [the unknown] and [the data]; how could we do so without conceiving, representing, or visualizing [the unknown] as existent?" (146). To derive the consequences of the problem condition, thereby uncovering hidden contradictions or ambiguities, we must take a provisional step of representing the unknown as known. Doing so allows us to inhabit and explore the problem space.

How best to represent the unknown depends on the nature of the problem. In geometrical problems, for instance, sketching figures is a natural and effective approach. In the context of invisible, arbitrarily complex software, the choice of representation is not as simple. Accuracy is important, yet excessive effort in building an accurate representation draws time and attention away from the main problem-solving task. Furthermore, in a problem solving session involving multiple stakeholders, complex notation may marginalize those without the necessary expertise.

Börger makes the case for ASMs as the right choice for software (and hardware, for that matter) [4]. Most relevant to us is his discussion of constructing an appropriate *ground model* – an expression of the software problem as stated by the stakeholders, and the basis for subsequent refinements. Subsequent analysis and synthesis "will remain an intellectual exercise if we do not 'know' that [the ground model] is 'correct' " (261). Börger enumerates the qualities of a good ground model, some of which are shared by formal methods in general. The positive characteristics that seem particularly applicable to ASMs are abstraction, simplicity and conciseness, to ensure acceptance and understanding by all stakeholders. Thus ASM models can be sketched in a quick, flexible way – "on the back of an envelope", to use a popular metaphor – yet with sufficient accuracy (abstraction and falsifiability) to elicit critical questions.

4 ASMs and Inquiry

Critical inquiry has been at the heart of much ASM-related work. The tone was set with Gurevich's early tutorial [15], which takes the form of a Socratic dialogue with the author's student, Quisani. The intensive and incisive questioning, familiar to anyone acquainted with the tutorial's author, reveals important theoretical and practical decisions in the design of the ASM language. The entire ASM enterprise stems from Gurevich's desire to illuminate the dark corners of the Pascal programming language [13]. Through their attempts to formulate ASM models of programming languages, researchers uncovered problems in the original descriptions of Prolog [5] and Java [7,18]. Börger *et al.* [6] acknowledge the importance of ASM ground models in supporting the productive process of asking "ignorant questions" [2], by "providing a precise ground against which questions can be formulated" (18).

In the early days of ASMs, much work focused on applying ASMs to complex real-world software problems, investigating the thesis (later proved by Gurevich [16]) that ASMs are powerful enough to capture (sequential) algorithms at any level of detail. A pattern emerges throughout this investigatory genre: while the process of constructing the ASM models invariably digs up ambiguities or contradictions buried in the original natural language, the authors typically do not reflect on the role that ASMs played in uncovering them. Usually, the ASM models are generally presented as *faits accomplis*: fully specified, with little or no discussion of alternatives. The process of envisioning, selecting and discarding alternatives, and the rationale for selecting a particular design [11], are left to the reader's imagination. Given the audience of this work – primarily academics and professionals – and the space limitations enforced on scholarly work, it is perhaps not surprising that reflection on analysis and design is omitted; such readers have an implicit understanding of these activities and are more interested in the details of the final product.

One can find occasional hints of the design process buried in the exposition of ASM papers. For instance, in an overview of their ASM models of the ARM processor [19], the authors state, "[o]ne can always tailor an ASM to a higher or a lower level of abstraction, depending on one's interests [...] Our models were chosen in order to effectively demonstrate the difference between pipelined and non-pipelined versions of the same microprocessor." (576) But this describes only the design motivation, not the process. In other papers, there are intimations of a design process in cases where peculiarities of the problem domain necessitate unusual design choices. For example, in the ASM paper on the C programming language [17], the authors justify the use of a dynamic external function to resolve operator evaluation order, which is officially underspecified (and therefore potentially dynamically determined). In the ASM describing the compilation of Prolog code to the Warren Abstract Machine [5], the authors discuss the dilemma presented by occur checks during unification: either follow the mathematical definition, or follow the example of numerous implementations and ignore the issue. In the ASM for the Kerberos authentication protocol [1], the authors must first formalize the limits of an adversary before the security of the system can be proven; several alternative adversaries are discussed before one is chosen and formalized as an ASM.

By and large, these hints at a design process are the exception rather than the rule. For newcomers to ASMs, the models given in the research literature can be difficult to use as a model for their own applications. Generally, students have little trouble understanding the ASM models as presented in the papers – activity corresponding to the "comprehension" level of Bloom's taxonomy [3]. However, the paucity of information on how to create their own ASMs presents an obstacle to the higher levels of "application", "analysis" and "synthesis". While reading ASMs may certainly grant insight into the particular problems being studied, and may convince readers of the general utility of ASMs, there are few clues as to how to develop a new ASM for oneself.

5 Encouraging Inquiry through ASMs and Problem Frames

The second author has used ASMs as a basis for inquiry into problems for several years, in a senior-level undergraduate course on software quality assurance. ASMs are used in conjunction with Problem Frames [26]. Both Problem Frames and ASMs are used in the portion of the course on requirements elicitation and analysis.

In Problem Frames, software problems are viewed as complex interactions between the target system (the *machine*) and its environment. Attention is directed to carefully *locating* and *bounding* the problem, identifying the roles of the machine and of environmental factors. This encourages precise description of the problem, in an appropriately abstract way that supposes no machine-specific knowledge. Each Problem Frame is a generic class of simple, commonly occuring problem, similar to a design pattern. Associated with a frame is a set of characteristic *concerns*: complicating issues that must be addressed in documentation and communicated to stakeholders. Problem Frames therefore provide not only a basis for inquiry into problems but also heuristics for problem solving: each frame provides rules of thumb for effective requirements elicitation and documentation.

Problem Frames and ASMs have some synergistic properties. Both allow for a clean separation of the "machine" or "ground model" from the "environment". Both gently encourage precision, but also allow things to remain abstract. With Problem Frames, students get a detailed picture of the environment, then move to details of the machine gradually with ASMs. Problem Frame "interfaces" between machine and environment translate naturally into "monitored" functions in ASMs.

The authors developed the ASM Primer [20] to be used as a means for teaching students not merely *about* ASMs, but *how* to develop an ASM. Inspired by Gurevich's conversations with his student Quisani, the authors engage in a dialogue with an undergraduate student, Questor. Over the course of the dialogue, the authors and the student discuss the development of several ASMs dealing with three classic "textbook" problems: greatest common divisor, string matching, and minimum spanning trees.

The primer attempts to show the inquiry process by explicitly discussing alternative design choices. For example, after giving a high-level ASM for string matching, the dialogue goes on to develop revisions of that high-level algorithm which yield the brute-force, Knuth-Morris-Pratt, and Boyer-Moore algorithms. The character of Questor embodies our ideal of critical thinking about the design process. Questor often asks questions about particular design choices; the response of the authors allows the reader to hear about how certain choices might be made. It is hoped that this dialogue inspires its readers to think about why they make certain choices as they develop their own ASMs.

Given the interactions in the primer as a model, students then engage in activities that exercise their own powers of critical inquiry. In one in-class exercise, for example, students are given an assortment of small electronic devices (*e.g.* digital watches, handheld video games, pocket electronic dictionaries), all obtained at low cost from various secondhand stores. Many of these "black boxes" exhibit unusual behavior, and documentation is nonexistent. In the first phase of the exercise, each team of 3–4 students "plays" with a device, exploring its behavior in reaction to human input. The team then maps the machine-environment relationships to a set of Problem Frames. Finally, the students represent the behavior they have observed in terms of an ASM. In the second phase, students present what they have discovered to the entire class, and audience members join in with their own lines of questioning. In the exercise, the Problem Frame and ASM documents serve as focus points from which team members explore possible behaviors. Furthermore, the value of these formalisms as a communication medium is made evident during the class presentations; the documents are not created purely to satisfy the instructor, but to convey information to classmates.

6 Results

Our experience shows that ASMs can be a valuable classroom tool – if introduced with care. Evaluations of the second author's software quality assurance course showed that students significantly broadened their range of inquiry techniques [9]. However, students can easily develop a cynical attitude toward ASMs as just another form of "useless" documentation. The pragmatic value of ASM-mediated inquiry must be made clear early and often.

One of the major risks is also one of the major advantages touted by ASM supporters: the "natural" character of ASM code and its similarity in form to pseudocode. This similarity may lead students to treat ASMs as a freeform "style" rather than a well-defined language. This in turn can lead to a skeptical or cynical attitude that anything is a valid ASM. The use of automated tools can mitigate this risk. The error checking functionality of these tools can help to avoid fundamental misunderstandings, and simulation and automated test generation allows for deeper investigation than possible by hand. On the other hand, with the introduction of a programming environment, we lose the spontaneous "back of the envelope" feel that is useful in an active inquiry session.

A related risk stems from the (well-placed) ASM emphasis on abstraction. To a student who has spent many hours writing long programs in earlier projects, a high-level ASM of only a few lines may seem like a worthless or even fraudulent artifact. In one humorous episode, a clever student presented his ASM program, consisting of a single update $\mathsf{CurrentState} := \mathsf{NextState}$, and praised its "high level of abstraction". Holding to the dogma that "abstraction is good", without illustrating *why* it is good, can actually have the effect of immunizing students against the concept.

Above all, formalization for the sake of formalization must be avoided. As Jackson and Wing contend [21], "[t]here can be no point embarking on the construction of a specification until it is known exactly what the specification is for; which risks it is intended to mitigate; and in which respects it will inevitably prove inadequate. [21]" While students who only encounter complete, fully developed ASMs may attain a passive, "read-only" attitude to ASMs, those who write their own without using them in a meaningful way may pick up an equally unproductive "write-only" attitude. Students must be presented with the technical details of ASMs within a teleological context that gives purpose to the whole enterprise.

7 Conclusion

In the introduction to the ASM Primer [20], we claimed a need "to provide a gentler introduction [to ASMs], focusing more on the use of the technique than on formal definitions." (1) In retrospect, it seems that our focus on "use" can be more precisely described as a contextualization of ASMs within a culture of inquiry. The formal documents that comprise the ASM literature reflect such a culture. One admires the mathematical beauty of the resulting work, and the skill of those who produce it. Yet such documents yield few clues regarding how those works were produced, or how someone could produce a similar result on their own. Problem Frames, which combine a simple problem representation technique with heuristics for problem analysis, may provide an instructive example. Along similar lines, a set of design patterns could be developed for ASMs, gently guiding design and analysis.

ASMs are a powerful tool, capable of representing a wide variety of types of algorithms at multiple levels of abstraction. But their value depends crucially upon the training given to the practitioner. One does not train a craftsman by simply showing completed works; one usually works alongside a senior craftsman, who shows the potential of the tools in the proper hands. Much of the growth of the ASM community has happened precisely because of this type of mentoring, as each generation of researchers mentors the next into maturity. But if ASMs are to become a widely-used tool, we will need to find different ways to teach about ASMs. The spirit of inquiry common to the ASM community will need to find expression in forms in addition to our collective oral history.

Acknowledgments. We wish to acknowledge our mentor, Yuri Gurevich, in whose honor this volume has been prepared. His work epitomizes the spirit of inquiry that we champion here. A highly adept Quisani himself, he has shown us by splendid example how to play the role.

We also thank Robert R. Johnson for his guidance regarding the problem solving process in technical communication.

References

1. Bella, G., Riccobene, E.: Formal Analysis of the Kerberos Authentication System. Journal of Universal Computer Science 3(12), 1337–1381 (1997)
2. Berry, D.M.: The Importance of Ignorance in Requirements Engineering. Journal of Systems and Software 28(2), 179–184 (1995)
3. Bloom, B.S.: Taxonomy of Educational Objectives. In: Handbook I: The Cognitive Domain. Longman, White Plains, New York (1956)

4. Börger, E.: Why Use Evolving Algebras for Hardware and Software Engineering? In: Bartosek, M., Staudek, J., Wiedermann, J. (eds.) SOFSEM 1995. LNCS, vol. 1012, pp. 236–271. Springer, Heidelberg (1995)
5. Börger, E., Rosenzweig, D.: The WAM – Definition and Compiler Correctness. In: Beierle, L.C., Pluemer, L. (eds.) Logic Programming: Formal Methods and Practical Applications. North-Holland Series in Computer Science and Artificial Intelligence (1994)
6. Börger, E., Stärk, R.: Abstract State Machines: A Method for High-Level System Design and Analysis. Springer, Heidelberg (2003)
7. Börger, E., Stärk, R., Schmid, J.: Java and the Java Virtual Machine: Definition, Verification, Validation. Springer, Heidelberg (2001)
8. Börger, E., Schulte, W.: Initialization Problems for Java. Software: Principles and Tools 19(4), 175–178 (2000)
9. Brady, A., Seigel, M., Vosecky, T., Wallace, C.: Addressing Communication Issues in Software Development through Case Studies. In: Conference on Software Engineering Education & Training (2007)
10. Brady, A., Seigel, M., Vosecky, T., Wallace, C.: Speaking of Software: Case Studies in Software Communication. In: Ellis, H.J.C., Demurjian, S.A., Naveda, J.F. (eds.) Software Engineering: Effective Teaching and Learning Approaches and Practices (2008)
11. Burge, J.E., Carroll, J.M., McCall, R., Mistrík, I.: Rationale-Based Software Engineering. Springer, Heidelberg (2008)
12. Carroll, J.M., Rosson, M.B.: A Case Library for Teaching Usability Engineering: Design Rationale, Development, and Classroom Experience. Journal of Educational Resources in Computing 5(1), 3 (2005)
13. DeSanto, F.: Gurevich Abstract State Machines. Communicator: EECS Department Newsletter, University of Michigan (December 1997)
14. Flower, L.: Problem Solving Strategies for Writing in College and Community. Wadsworth Publishing, Belmont (1997)
15. Gurevich, Y.: Evolving Algebras: An Attempt to Discover Semantics. In: Rozenberg, G., Salomå A. (ed.) Current Trends in Theoretical Computer Science, pp. 266–292. World Scientific, Singapore (1993)
16. Gurevich, Y.: Sequential Abstract State Machines Capture Sequential Algorithms. ACM Transactions on Computational Logic 1(1), 77–111 (2000)
17. Gurevich, Y., Huggins, J.K.: The Semantics of the C Programming Language. In: Martini, S., Börger, E., Kleine Büning, H., Jäger, G., Richter, M.M. (eds.) CSL 1992. LNCS, vol. 702, pp. 274–308. Springer, Heidelberg (1993)
18. Gurevich, Y., Schulte, W., Wallace, C.: Investigating Java Concurrency using Abstract State Machines. In: Gurevich, Y., Kutter, P.W., Odersky, M., Thiele, L. (eds.) ASM 2000. LNCS, vol. 1912. Springer, Heidelberg (2000)
19. Huggins, J.K., van Campenhout, D.: Specification and Verification of Pipelining in the ARM2 RISC Microprocessor. ACM Transactions on Design Automation of Electronic Systems 3(4), 563–580 (1998)
20. Huggins, J.K., Wallace, C.: An Abstract State Machine Primer. Technical Report 02-04, Computer Science Department, Michigan Technological University (2002)
21. Jackson, D., Wing, J.: Lightweight Formal Methods. IEEE Computer 29(4), 21–22 (1996)
22. Jackson, M.: Problem Frames. Addison-Wesley, Reading (2000)
23. Johnson, R.R.: User Centered Technology: A Rhetorical Theory for Computers and Other Mundane Artifacts. SUNY Press (1998)

24. Newell, A., Simon, H.A.: Human Problem Solving. Prentice-Hall, Englewood Cliffs (1972)
25. Pólya, G.: How to Solve It. Princeton University Press, Princeton (1948)
26. Wallace, C., Wang, X., Bluth, V.: A Course in Problem Analysis and Structuring through Problem Frames. In: Conference on Software Engineering Education and Training (2006)
27. Wright, D.R.: The Decision Pattern: Capturing and Communicating Design Intent. In: ACM International Conference on Design of Communication, pp. 69–74 (2007)

The Algebra of Adjacency Patterns: Rees Matrix Semigroups with Reversion

Marcel Jackson[1,*] and Mikhail Volkov[2,**]

[1] La Trobe University, Victoria 3086, Australia
M.G.Jackson@latrobe.edu.au
[2] Ural State University, Ekaterinburg 620083, Russia
Mikhail.Volkov@usu.ru

For Yuri Gurevich, who built many bridges between logic and algebra, on the occasion of his seventieth birthday.

Abstract. We establish a surprisingly close relationship between universal Horn classes of directed graphs and varieties generated by so-called adjacency semigroups which are Rees matrix semigroups over the trivial group with the unary operation of reversion. In particular, the lattice of subvarieties of the variety generated by adjacency semigroups that are regular unary semigroups is essentially the same as the lattice of universal Horn classes of reflexive directed graphs. A number of examples follow, including a limit variety of regular unary semigroups and finite unary semigroups with NP-hard variety membership problems.

Keywords: Rees matrix semigroup, unary semigroup identity, unary semigroup variety, graph, universal Horn sentence, universal Horn class, variety membership problem, finite basis problem.

1 Introduction and Overview

The aim of this paper is to establish and to explore a new link between graph theory and algebra. Since graphs form a universal language of discrete mathematics, the idea to relate graphs and algebras appears to be natural, and several useful links of this kind can be found in the literature. We mean, for instance, the graph algebras of McNulty and Shallon [20], the closely related flat graph algebras [25], and "almost trivial" algebras investigated in [15,16] amongst other places. While each of the approaches just mentioned has proved to be useful and has yielded interesting applications, none of them seem to share two important features of the present contribution. The two features can be called naturalness and surjectivity.

* The first author was supported by ARC Discovery Project Grant DP0342459. The authors were partially supported also by ARC Discovery Project Grant DP1094578.
** The second author was supported by the Program 2.1.1/3537 of the Russian Education Agency and by the Russian Foundation for Basic Research, grants 09-01-12142 and 10-01-00524.

A. Blass, N. Dershowitz, and W. Reisig (Eds.): Gurevich Festschrift, LNCS 6300, pp. 414–443, 2010.

Speaking about naturalness, we want to stress that the algebraic objects (adjacency semigroups) that we use here to interpret graphs have not been invented for this specific purpose. Indeed, adjacency semigroups belong to a well established class of unary semigroups[1] that have been considered by many authors. We shall demonstrate how graph theory both sheds a new light on some previously known algebraic results and provides their extensions and generalizations. By surjectivity we mean that, on the level of appropriate classes of graphs and unary semigroups, the interpretation map introduced in this paper becomes "nearly" onto; moreover, the map induces a lattice isomorphism between the lattices of such classes provided one excludes just one element on the semigroup side. This implies that our approach allows one to interpret both graphs within unary semigroups and unary semigroups within graphs.

The paper is structured as follows. In Sect. 2 we recall some notions related to graphs and their classes and present a few results and examples from graph theory that are used in the sequel. Section 3 contains our construction and the formulations of our main results: Theorems A and B. These theorems are proved in Sect. 4 and 5 respectively while Sect. 6 collects some of their applications.

We assume the reader's acquaintance with basic concepts of universal algebra and first-order logic such as ultraproducts or the HSP-theorem, see, e.g., [4]. As far as graphs and semigroups are concerned, we have tried to keep the presentation to a reasonable extent self-contained. We do occasionally mention some non-trivial facts of semigroup theory but only in order to place our considerations in a proper perspective. Thus, most of the material should be accessible to readers with very basic semigroup-theoretic background (such as some knowledge of Green's relations and of the Rees matrix construction over the trivial group, cf. [12]).

2 Graphs and Their Classes

In this paper, a *graph* is a structure $\mathsf{G} := \langle V; \sim \rangle$, where V is a set and $\sim\, \subseteq V \times V$ is a binary relation. In other words, **we consider all graphs to be directed**, and do not allow multiple edges (but do allow loops). Of course, V is often referred to as the set of *vertices* of the graph and $\sim$ as the set of *edges*. As is usual, we write $a \sim b$ in place of $(a, b) \in\, \sim$. Conventional undirected graphs are essentially the same as graphs whose edge relation is symmetric (satisfying $x \sim y \to y \sim x$), while a *simple graph* is a symmetric graph without loops. It is convenient for us to allow the empty graph $\underline{\mathbf{0}} := \langle \varnothing; \varnothing \rangle$. On the other hand, when speaking about classes of graphs we always mean **nonempty classes**.

All classes of graphs that come to consideration in this paper are *universal Horn classes*. We recall their definition and some basic properties. Of course, the majority of the statements below are true for arbitrary structures, but our interest is only in the graph case. See Gorbunov [9] for more details.

[1] Here and below the somewhat oxymoronic term "unary semigroup" abbreviates the precise but longer expression "semigroup endowed with an extra unary operation".

Universal Horn classes can be defined both syntactically (via specifying an appropriate sort of first order formulas) and semantically (via certain class operators). We first introduce the operator definition for which we recall notation for a few standard class operators. The operator for taking isomorphic copies is $\mathbb{I}$. We use $\mathbb{S}$ to denote the operator taking a class K to the class of all substructures of structures in K; in the case when K is a class of graphs, substructures are just induced subgraphs of graphs in K. Observe that the empty graph $\underline{\mathbf{0}}$ is an induced subgraph of any graph and thus belongs to any $\mathbb{S}$-closed class of graphs. We denote by $\mathbb{P}$ the operator of taking direct products. For graphs, we allow the notion of an empty direct product, which we identify (as is the standard convention) with the 1-vertex looped graph $\underline{\mathbf{1}} := \langle\{0\}; \{(0,0)\}\rangle$. If we exclude the empty product, we obtain the operator $\mathbb{P}^+$ of taking nonempty direct products. By $\mathbb{P}_{\mathrm{u}}$ we denote the operator of taking ultraproducts. Note that ultrafilters cannot be formed on the empty set, so unlike direct products, ultraproducts cannot be formed over an empty family.

A class K of graphs is an *universal Horn class* if K is closed under each of the operators $\mathbb{I}$, $\mathbb{S}$, $\mathbb{P}^+$, and $\mathbb{P}_{\mathrm{u}}$. In the sequel, we write "uH class" in place of "universal Horn class". It is well known that the least uH class containing a class L of graphs is the class $\mathbb{ISP}^+\mathbb{P}_{\mathrm{u}}(L)$ of all isomorphic copies of induced subgraphs of nonempty direct products of ultraproducts of L; this uH class is referred to as the uH class *generated* by L.

If the operator $\mathbb{P}^+$ in the above definition is extended to $\mathbb{P}$, then one obtains the definition of a *quasivariety* of graphs. The quasivariety *generated* by a given class L is known to be equal to $\mathbb{ISPP}_{\mathrm{u}}(L)$. It is not hard to see that $\mathbb{ISPP}_{\mathrm{u}}(L) = \mathbb{I}(\mathbb{ISP}^+\mathbb{P}_{\mathrm{u}}(L) \cup \{\underline{\mathbf{1}}\})$, showing that there is little or no difference between the uH class and the quasivariety generated by L. However, as examples described later demonstrate, there are many well studied classes of graphs that are uH classes but not quasivarieties.

As mentioned, uH classes also admit a well known syntactic characterization. An *atomic formula* in the language of graphs is an expression of the form $x \sim y$ or $x \approx y$ (where x and y are possibly identical variables). A *universal Horn sentence* (abbreviated to "uH sentence") in the language of graphs is a sentence of one of the following two forms (for some $n \in \omega := \{0, 1, 2, \dots\}$):

$$(\forall x_1 \forall x_2 \dots)\left(\left(\mathop{\&}_{1\le i\le n} \Phi_i\right) \to \Phi_0\right) \quad \text{or} \quad (\forall x_1 \forall x_2 \dots)\left(\bigvee_{0\le i\le n} \neg\Phi_i\right)$$

where the Φ_i are atomic, and $x_1, x_2, \dots$ is a list of all variables appearing. In the case when $n = 0$, a uH sentence of the first kind is simply the universally quantified atomic expression Φ_0. Sentences of the first kind are usually called *quasi-identities*. As is standard, we omit the universal quantifiers when describing uH sentences; also the expressions $x \not\approx y$ and $x \nsim y$ abbreviate $\neg x \approx y$ and $\neg x \sim y$ respectively. Satisfaction of uH sentences by graphs is defined in the obvious way. We write $\mathsf{G} \models \Phi$ ($K \models \Phi$) to denote that the graph G (respectively, each graph in the class K) satisfies the uH sentence Φ.

The Birkhoff theorem identifying varieties of algebras with equationally defined classes has a natural analogue for uH classes, which is usually attributed to Mal'cev. Here we state it in the graph setting.

Lemma 2.1. *A class K of graphs is a uH class if and only if it is the class of all models of some set of uH sentences.*

In particular, the uH class $\mathbb{ISP}^+\mathbb{P}_u(L)$ generated by a class L is equal to the class of models of the uH sentences holding in L.

Recall that we allow the empty graph $\underline{\mathbf{0}} := \langle\varnothing;\varnothing\rangle$. Because there are no possible variable assignments into the empty set, $\underline{\mathbf{0}}$ can fail no uH sentence and hence lies in every uH class. Thus, allowing $\underline{\mathbf{0}}$ brings the advantage that the collection of all uH classes forms a lattice whose meet is intersection: $A \wedge B := A \cap B$ and whose join is the uH class generated by union: $A \vee B := \mathbb{ISP}^+\mathbb{P}_u(A \cup B)$. Furthermore, the inclusion of $\underline{\mathbf{0}}$ allows every set of uH sentences to have a model (for example, the contradiction $x \not\approx x$ axiomatizes the class $\{\underline{\mathbf{0}}\}$). In the world of varieties of algebras, it is the one element algebra that plays these roles.

When $\mathbb{IP}_u(L) = \mathbb{I}(L)$ (such as when L consists of finitely many finite graphs), we have $\mathbb{ISP}^+\mathbb{P}_u(L) = \mathbb{ISP}^+(L)$, and there is a handy structural characterization of the uH class generated by L.

Lemma 2.2. *Let L be an ultraproduct closed class of graphs and let G be a graph. We have $\mathsf{G} \in \mathbb{ISP}^+\mathbb{P}_u(L)$ if and only if there is at least one homomorphism from G into a member of L and the following two* separation conditions *hold:*

1. *for each pair of distinct vertices a, b of G, there is $\mathsf{H} \in L$ and a homomorphism $\phi : \mathsf{G} \to \mathsf{H}$ with $\phi(a) \neq \phi(b)$;*
2. *for each pair of vertices a, b of G with $a \not\sim b$ in G, there is $\mathsf{H} \in L$ and a homomorphism $\phi : \mathsf{G} \to \mathsf{H}$ with $\phi(a) \not\sim \phi(b)$ in H.*

The 1-vertex looped graph $\underline{\mathbf{1}}$ always satisfies the two separation conditions, yet it fails every uH sentence of the second kind; this is why the lemma asks additionally that there be at least one homomorphism from G into some member of L. If $\mathsf{G} = \underline{\mathbf{1}}$ and no such homomorphism exists, then evidently, no member of L contains a loop, and so $L \models x \not\sim x$, a law failing on $\underline{\mathbf{1}}$. Hence $\underline{\mathbf{1}} \notin \mathbb{ISP}^+\mathbb{P}_u(L)$ by Lemma 2.1. Conversely, if there is such a homomorphism, then $\underline{\mathbf{1}}$ is isomorphic to an induced subgraph of some member of L and hence $\underline{\mathbf{1}} \in \mathbb{ISP}^+\mathbb{P}_u(L)$. If the condition that there is at least one homomorphism from G into some member of L is dropped, then Lemma 2.2 instead characterizes membership in the quasivariety generated by L.

We now list some familiar uH sentences.

- reflexivity: $x \sim x$,
- anti-reflexivity: $x \not\sim x$,
- symmetry: $x \sim y \to y \sim x$,
- anti-symmetry: $x \sim y \;\&\; y \sim x \to x \approx y$,
- transitivity: $x \sim y \;\&\; y \sim z \to x \sim y$.

All except anti-reflexivity are quasi-identities.

These laws appear in many commonly investigated classes of graphs. We list a number of examples that are of interest later in the paper (mainly in its application part, see Sect. 6 below).

Example 2.1. Preorders.

This class is defined by reflexivity and transitivity and is a quasivariety. Some well known subclasses are:

- equivalence relations (obtained by adjoining the symmetry law);
- partial orders (obtained by adjoining the anti-symmetry law);
- anti-chains (the intersection of partial orders and equivalence relations);
- complete looped graphs, or equivalently, single block equivalence relations (axiomatized by $x \sim y$).

In fact it is easy to see that, along with the 1-vertex partial orders and the trivial class $\{\underline{\mathbf{0}}\}$, this exhausts the list of all uH classes of preorders, see Fig. 1 (the easy proof is sketched before Corollary 6.4 of [7], for example).

Example 2.2. Simple (that is, anti-reflexive and symmetric) graphs.

Sub-uH classes of simple graphs have been heavily investigated, and include some very interesting families. In order to describe some of these families, we need a series of graphs introduced by Nešetřil and Pultr [21]. For each integer $k \geq 2$, let C_k denote the graph on the vertices $0, \ldots, k+1$ obtained from the complete loopless graph on these vertices by deleting the edges (in both directions) connecting 0 and $k+1$, 0 and k, and 1 and $k+1$. Fig. 2 shows the graphs C_2 and C_3; here and below we adopt the convention that an undirected edge between two vertices, say a and b, represents two directed edges $a \sim b$ and $b \sim a$.

Recall that a simple graph G is said to be *n-colorable* is there exists a homomorphism from G into the complete loopless graph on n vertices.

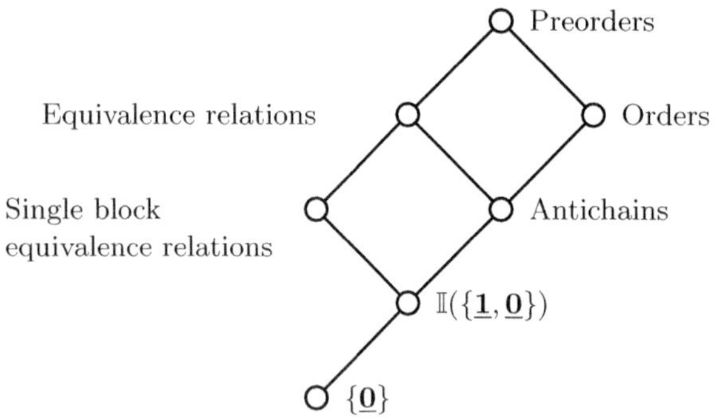

Fig. 1. The lattice of uH classes of preorders

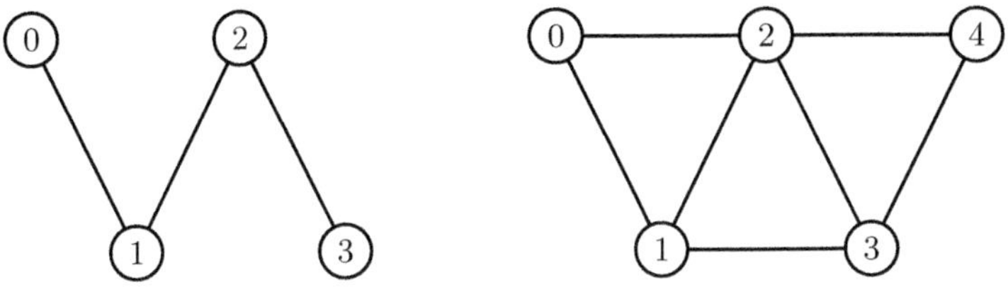

Fig. 2. Graphs C_2 and C_3

Example 2.3. The 2-colorable graphs (equivalently, bipartite graphs).

This is the uH class $\mathbb{ISP}^+(\mathsf{C}_2)$ generated by the graph C_2 (Nešetřil and Pultr [21]) and has no finite axiomatization. Caicedo [5] showed that the lattice of sub-uH classes of $\mathbb{ISP}^+(\mathsf{C}_2)$ is a 6-element chain: besides $\mathbb{ISP}^+(\mathsf{C}_2)$, it contains the class of disjoint unions of complete bipartite graphs, which is axiomatized within simple graphs by the law

$$x_0 \sim x_1 \ \& \ x_1 \sim x_2 \ \& \ x_2 \sim x_3 \rightarrow x_0 \sim x_3\,;$$

the class of disjoint unions of paths of length at most 1 (axiomatized within simple graphs by $x \not\sim y \vee y \not\sim z$); the edgeless graphs (axiomatized by $x \not\sim y$), the class consisting of the 1-vertex edgeless graphs and the empty graph $\underline{\mathbf{0}}$ (axiomatized by $x \approx y$); and the trivial class $\{\underline{\mathbf{0}}\}$.

Every finite simple graph either lies in a sub-uH class of $\mathbb{ISP}^+(\mathsf{C}_2)$ or generates a uH class that: 1) is not finitely axiomatizable, 2) contains $\mathbb{ISP}^+(\mathsf{C}_2)$, and 3) has uncountably many sub-uH classes [10, Theorem 4.7], see also [17].

Example 2.4. The k-colorable graphs.

More generally, Nesetril and Pultr [21] showed that for any $k \geq 2$, the class of all k-colorable graphs is the uH class generated by C_k. These classes have no finite basis for their uH sentences and for $k > 2$ have NP-complete finite membership problem, see [8].

Example 2.5. A generator for the class $\underline{\mathsf{G}}$ of all graphs.

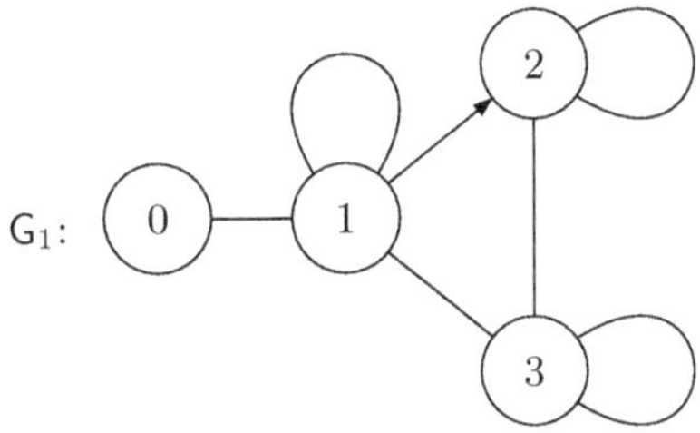

Fig. 3. Generator for the uH class of all graphs

The class of all graphs is generated as a uH class by a single finite graph. Indeed, it is trivial to see that for any graph G, there is a family of 3-vertex graphs such that the separation conditions of Lemma 2.2 hold. Since there are only finitely many non-isomorphic 3-vertex graphs, any graph containing these as induced subgraphs generates the uH class of all graphs. Alternatively, the reader can easily verify using Lemma 2.2 that the graph G_1 in Fig. 3 generates the uH class of all graphs.

Example 2.6. A generator for the class $\underline{\mathsf{G}}_{\mathrm{symm}}$ of all symmetric graphs.

Using Lemma 2.2, it is easy to prove that the class of symmetric graphs is generated as a uH class by the graph S_1 shown in Fig. 4.

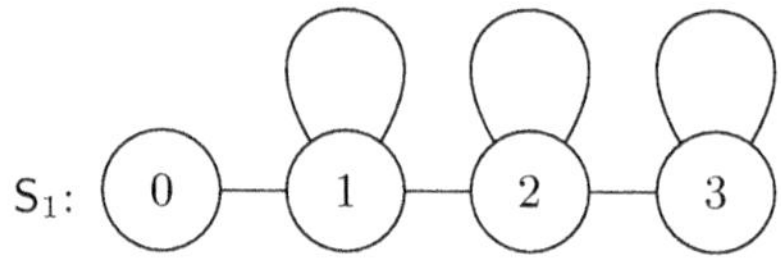

Fig. 4. Generator for the uH class of all symmetric graphs

Example 2.7. The class of simple graphs has no finite generator.

The class of all simple graphs is not generated by any finite graph, since a finite graph on n vertices is n-colorable, while for every positive integer n there is a simple graph that is not n-colorable (the complete simple graph on $n+1$ vertices, for example). However the uH class generated by the following 2-vertex graph S_2 contains all simple graphs (this is well known and follows easily using Lemma 2.2).

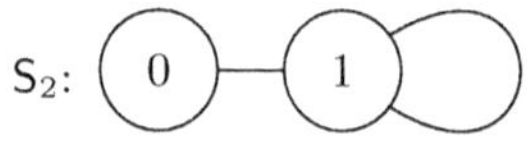

Fig. 5. 2-vertex graph whose uH class contains all simple graphs

Example 2.8. A generator for the class $\underline{\mathsf{G}}_{\mathrm{ref}}$ of all reflexive graphs.

The class of reflexive graphs is generated by the following graph R_1, while the class of reflexive and symmetric graphs is generated by the graph RS_1.

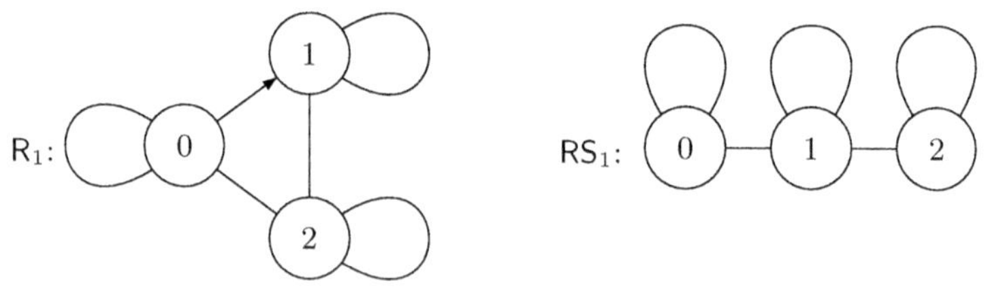

Fig. 6. Generators for reflexive and reflexive-symmetric graphs

3 The Adjacency Semigroup of a Graph

Given a graph $\mathsf{G} = \langle V; \sim \rangle$, its *adjacency semigroup* $\mathrm{A}(\mathsf{G})$ is defined on the set $(V \times V) \cup \{0\}$ and the multiplication rule is

$$(x, y)(z, t) = \begin{cases} (x, t) & \text{if } y \sim z\,, \\ 0 & \text{if } y \nsim z\,; \end{cases}$$
$$a0 = 0a = 0 \text{ for all } a \in \mathrm{A}(\mathsf{G})\,.$$

In terms of semigroup theory, $\mathrm{A}(\mathsf{G})$ is the *Rees matrix semigroup over the trivial group* using the adjacency matrix of the graph G as a sandwich matrix. We describe here the Rees matrix construction in a specific form that is used in the present paper.

Let I, J be nonempty sets and $0 \notin I \cup J$. Let $P = (P_{i,j})$ be a $J \times I$ matrix (the *sandwich matrix*) over the set $\{0, 1\}$. The *Rees matrix semigroup over the trivial group* $M^0[P]$ is the semigroup on the set $(I \times J) \cup \{0\}$ with multiplication

$$a \cdot 0 = 0 \cdot a = 0 \text{ for all } a \in (I \times J) \cup \{0\}, \text{ and}$$
$$(i_1, j_1) \cdot (i_2, j_2) = \begin{cases} 0 & \text{if } P_{j_1,i_2} = 0\,, \\ (i_1, j_2) & \text{if } P_{j_1,i_2} = 1\,. \end{cases}$$

The Rees–Sushkevich Theorem (see [12, Theorem 3.3.1]) states that, up to isomorphism, the completely 0-simple semigroups with trivial subgroups are precisely the Rees matrix semigroups over the trivial group and for which each row and each column of the sandwich matrix contains a nonzero element. If the matrix P has no 0 entries, then the set $M[P] = M^0[P] \setminus \{0\}$ is a subsemigroup. Semigroups of the form $M[P]$ are called *rectangular bands*, and they are precisely the completely simple semigroups with trivial subgroups.

Back to adjacency semigroups, we always think of $\mathrm{A}(\mathsf{G})$ as endowed with an additional unary operation $a \mapsto a'$ which we call *reversion* and define as follows:

$$(x, y)' = (y, x), \quad 0' = 0\,.$$

Notice that by this definition $(a')' = a$ for all $a \in \mathrm{A}(\mathsf{G})$.

The main contribution in this paper is the fact that uH classes of graphs are in extremely close correspondence with unary semigroup varieties generated by adjacency semigroups, and our proof of this will involve a translation of uH sentences of graphs into unary semigroup identities. However, before we proceed with precise formulations and proofs of general results, the reader may find it useful to check that several of the basic uH sentences used in Sect. 2 correspond via the adjacency semigroup construction to rather natural semigroup-theoretic properties. Indeed, all the following are quite easy to verify:

- reflexivity of G is equivalent to $\mathrm{A}(\mathsf{G}) \models xx'x \approx x$;
- anti-reflexivity of G is equivalent to $\mathrm{A}(\mathsf{G}) \models xx'z \approx zxx' \approx xx'$ (these laws can be abbreviated to $xx' \approx 0$);

- symmetry of G is equivalent to $\mathrm{A}(\mathsf{G}) \models (xy)' \approx y'x'$;
- G is empty (satisfies $x \not\approx x$) if and only if $\mathrm{A}(\mathsf{G}) \models x \approx y$;
- G has one vertex (satisfies $x \approx y$) if and only if $\mathrm{A}(\mathsf{G}) \models x \approx x'$; also, G is the one vertex looped graph (satisfies $x \sim y$) if and only if $\mathrm{A}(\mathsf{G})$ additionally satisfies $xx \approx x$.

Observe that the unary semigroup identities that appear in the above examples are in fact used to define the most widely studied types of semigroups endowed with an extra unary operation modelling various notions of an inverse in groups. For instance, a semigroup satisfying the identities

$$x'' \approx x \tag{1}$$

(which always holds true in adjacency semigroups) and

$$(xy)' \approx y'x' \tag{2}$$

(which is a semigroup counterpart of symmetry) is called *involution semigroup* or $*$-*semigroup*. If such a semigroup satisfies also

$$xx'x \approx x \tag{3}$$

(which corresponds to reflexivity), it is called a *regular* $*$-*semigroup*. Semigroups satisfying (1) and (3) are called *I-semigroups* in Howie [12]; note that an I-semigroup satisfies $x'xx' \approx x'x''x' \approx x'$, so that x' is an inverse of x. Semigroups satisfying (3) are often called *regular unary semigroups*. There exists vast literature on all these types of unary semigroups; clearly, the present paper is not a proper place to survey this literature but we just want to stress once more that the range of the adjacency semigroup construction is no less natural than its domain.

When K is a class of graphs, we use the notation $\mathrm{A}(K)$ to denote the class of all adjacency semigroups of members of K. As usual, the operator of taking homomorphic images is denoted by $\mathbb{H}$. We let $\mathcal{A}$ denote the variety $\mathbb{HSP}(\mathrm{A}(\underline{\mathsf{G}}))$ generated by all adjacency semigroups of graphs, and let $\mathcal{A}_{\mathrm{ref}}$ and $\mathcal{A}_{\mathrm{symm}}$ denote the varieties $\mathbb{HSP}(\mathrm{A}(\underline{\mathsf{G}}_{\mathrm{ref}}))$ and $\mathbb{HSP}(\mathrm{A}(\underline{\mathsf{G}}_{\mathrm{symm}}))$ generated by all adjacency semigroups of reflexive graphs and of symmetric graphs respectively.

Our first main result is:

Theorem A. *Let K be any nonempty class of graphs and let G be a graph. The graph G belongs to the uH-class generated by K if and only if the adjacency semigroup $\mathrm{A}(\mathsf{G})$ belongs to the variety generated by the adjacency semigroups $\mathrm{A}(\mathsf{H})$ with $\mathsf{H} \in K$.*

This immediately implies that the assignment $\mathsf{G} \mapsto \mathrm{A}(\mathsf{G})$ induces an injective join-preserving map from the lattice of all uH-classes of graphs to the subvariety lattice of the variety $\mathcal{A}$. The latter fact can be essentially refined for the case of reflexive graphs. In order to describe this refinement, we need an extra definition.

Let I be a nonempty set. We endow the set $B = I \times I$ with a unary semigroup structure whose multiplication is defined by

$$(i, j)(k, \ell) = (i, \ell)$$

and whose unary operation is defined by

$$(i, j)' = (j, i)\,.$$

Observe that while B is not an adjacency semigroup, it is very close to such one. Indeed, B is obtained by removing 0 from the adjacency semigroup of the universal relation on the set I.

It is easy to check that B is a regular $*$-semigroup. We call regular $*$-semigroups constructed this way *square bands*. Clearly, square bands satisfy

$$x^2 \approx x \quad \text{and} \quad xyz \approx xz\,, \tag{4}$$

and in fact it can be shown that the class $\mathcal{SB}$ of all square bands constitutes a variety of unary semigroups defined within the variety of all regular $*$-semigroups by the identities (4).

Let $L(\underline{\mathsf{G}}_{\mathrm{ref}})$ denote the lattice of sub-uH classes of $\underline{\mathsf{G}}_{\mathrm{ref}}$ and let $L(\mathcal{A}_{\mathrm{ref}})$ denote the lattice of subvarieties of $\mathcal{A}_{\mathrm{ref}}$. Let L^+ denote the result of adjoining a new element $\underline{\mathsf{S}}$ to $L(\underline{\mathsf{G}}_{\mathrm{ref}})$ between the class of single block equivalence relations and the class containing the empty graph. (The reader may wish to look at Fig. 1 to see the relative location of these two uH classes.) Meets and joins are extended to L^+ in the weakest way. So L^+ is a lattice in which $L(\underline{\mathsf{G}}_{\mathrm{ref}})$ is a sublattice containing all but one element.

We are now in a position to formulate our second main result.

Theorem B. *Let ι be the map from L^+ to $L(\mathcal{A}_{\mathrm{ref}})$ defined by $\underline{\mathsf{S}} \mapsto \mathcal{SB}$ and $K \mapsto \mathbb{HSP}(\mathrm{A}(K))$ for $K \in L(\underline{\mathsf{G}}_{\mathrm{ref}})$. Then ι is a lattice isomorphism. Furthermore, a variety in $L(\mathcal{A}_{\mathrm{ref}})$ is finitely axiomatized (finitely generated as a variety) if and only if it is the image under ι of either $\underline{\mathsf{S}}$ or a finitely axiomatized (finitely generated, respectively) uH class of reflexive graphs.*

We prove Theorems A and B in the next two sections.

4 Proof of Theorem A

4.1 Equations Satisfied by Adjacency Semigroups

The variety of semigroups generated by the class of Rees matrix semigroups over trivial groups is reasonably well understood: it is generated by a 5-element semigroup usually denoted by A_2 (see [19] for example). (In context of this paper A_2 can be thought as the semigroup reduct of the adjacency semigroup $\mathrm{A}(\mathsf{S}_2)$ where S_2 is the 2-vertex graph from Example 2.7.) This semigroup was shown to have a finite identity basis by Trahtman [23], who gave the following elegant description of the identities: an identity $u \approx v$ (where u and v are semigroup words) holds in A_2 if and only if u and v start with the same letter, end with the same letter and share the same set of two letter subwords. Thus the equational theory of this variety corresponds to pairs of words having the same "adjacency patterns", in the sense that a two letter subword xy records the fact that x occurs next to (and before) y. This adjacency pattern can also be visualized as a graph on the set of letters, with an edge from x to y if xy is a subword, and two distinct markers indicating the first and last letters respectively.

In this subsection we show that the equational theory of $\mathcal{A}$ has the same kind of property with respect to a natural unary semigroup notion of adjacency. The interpretation is that each letter has two sides – left and right – and that the operation $'$ reverses these. A subword xy corresponds to the right side of x matching the left side of y, while $x'y$ or any subword $(x\ldots)'y$ corresponds to the left side of x matching the left side of y. To make this more precise, we give an inductive definition. Under this definition, each letter x in a word will have two associated vertices corresponding to the left and right side. The graph will have an initial vertex, a final vertex as well as a set of (directed) edges corresponding to adjacencies.

Let u be a unary semigroup word, and X be the alphabet of letters appearing in u. We construct a graph $G[u]$ on the set

$$\{\ell_x \mid x \in X\} \cup \{r_x \mid x \in X\}$$

with two marked vertices. If u is a single letter (say x), then the edge set (or *adjacency set*) of $G[u]$ is empty. The *initial vertex* of a single letter x is ℓ_x and the *final* (or *terminal*) *vertex* is r_x.

If u is not a single letter, then it is of the form v' or vw for some unary semigroup words v, w. We deal with the two cases separately. If u is of the form v', where v has set of adjacencies S, initial vertex p_a and final vertex q_b (where $\{p, q\} \subseteq \{\ell, r\}$ and a, b are letters appearing in v), then the set of adjacencies of u is also S, but the initial vertex of u is equal to the final vertex q_b of v and the final vertex of u is equal to the initial vertex p_a of v.

Now say that u is of the form vw for some unary semigroup words v, w, with adjacency set S_v and S_w respectively and with initial vertices p_{a_v}, p_{a_w} respectively and final vertices q_{b_v} and q_{b_w} respectively. Then the adjacency set of $G[u]$ is $S_v \cup S_w \cup \{(q_{b_v}, p_{a_w})\}$, the initial vertex is p_{a_v} and the final vertex is q_{b_w}. Note that the word u may be broken up into a product of two unary words in a number of different ways, however it is reasonably clear that this gives rise to the same adjacency set and initial and final vertices (this basically corresponds to the associativity of multiplication).

For example the word $a'(baa')'$ decomposes as $a' \cdot (baa')'$, and so has initial vertex equal to the initial vertex of a', which in turn is equal to the terminal vertex of a, which is r_a. Likewise, its terminal vertex should be the terminal vertex of $(baa')'$, which is the initial vertex of baa', which is ℓ_b. Continuing, we see that the edge set of the corresponding graph has edges $\{(\ell_a, \ell_a), (r_a, r_a), (r_b, \ell_a)\}$. This graph is the first graph depicted in Fig. 7 (the initial and final vertices are indicated by a sourceless and targetless arrow respectively). The second is the graph

$G[a'(baa')']$ $\qquad$ $G[a(bc)'] = G[(b(ac')')']$

Fig. 7. Two examples of graphs of unary words

of either of the words $a(bc)'$ or $(b(ac')')'$. The fact that $G[a(bc)'] = G[(b(ac')')']$ will be of particular importance in constructing a basis for the identities of $\mathcal{A}$.

We can also construct a second kind of graph from a word w, in which all loops are added to the graph of $G[w]$ (that is, it is the reflexive closure of the edge set), we call this $G_{\text{ref}}[w]$. For example, it is easy to see that $G_{\text{ref}}[a'(baa')'] = G_{\text{ref}}[(ba)']$ (most of the work was done in the previous example). Lastly, we define the graph $G_{\text{symm}}[w]$ corresponding to the symmetric closure of the edge set of $G[w]$.

Notation 1. Let u be a unary semigroup word and let θ be an assignment of the letters of u into nonzero elements of an adjacency semigroup $\mathsf{A}(\mathsf{H})$; say $\theta(x) = (i_x, j_x)$ for each letter x. Note that $\theta(x') = (j_x, i_x)$, so we use the notation $i_{x'} := j_x$ and $j_{x'} = i_x$.

Lemma 4.1. *Let u and θ be as in Notation* 1. *If λ_a is the initial vertex of $G[u]$ and ρ_b is the terminal vertex (so $\lambda, \rho \in \{\ell, r\}$ and a and b are letters in u) and $\theta(u) \neq 0$, then $\theta(u) = (i_{\bar{a}}, j_{\bar{b}})$, where*

$$\bar{a} = \begin{cases} a & \text{if } \lambda = \ell, \\ a' & \text{if } \lambda = r \end{cases} \quad \text{and} \quad \bar{b} = \begin{cases} b & \text{if } \rho = r, \\ b' & \text{if } \rho = \ell. \end{cases}$$

Proof. This follows by an induction following the inductive definition of the graph of u. □

Lemma 4.2. *Let u and θ be as in Notation* 1. *Then $\theta(u) \neq 0$ if and only if the map defined by $\ell_x \mapsto i_x$ and $r_x \mapsto j_x$ is a graph homomorphism from $G[u]$ to H.*

Proof. Throughout the proof we use the notation of Lemma 4.1.

(Necessity.) Say $\theta(u) \neq 0$, and let (ρ_b, λ_a) be an edge in $G[u]$, where $\rho, \lambda \in \{\ell, r\}$ and a and b are letters in u). We use Lemma 4.1 to show that $(j_{\bar{b}}, i_{\bar{a}})$ is an edge of H. Note that in the case where no applications of $'$ are used (so we are dealing in the nonunary case), the edge (ρ_b, λ_a) will necessarily be (r_b, ℓ_a); and we would want (j_b, i_a) to be an edge of H.

Now, since (ρ_b, λ_a) is an edge in $G[u]$, some subwords of u – say u_1 and u_2 – have $u_1 u_2$ a subword of u, and ρ_b the terminal vertex of $G[u_1]$, and λ_a the initial vertex of $G[u_2]$. Applying Lemma 4.1 to both $G[u_1]$ and $G[u_2]$, we find that $\theta(u_1)$ has right coordinate $j_{\bar{b}}$, and $\theta(u_2)$ has left coordinate $i_{\bar{a}}$. But u_1u_2 is a subword, so $\theta(u_1)\theta(u_2) \neq 0$, whence $(j_{\bar{b}}, i_{\bar{a}})$ is an edge of H, as required.

(Sufficiency.) This is easy. □

Lemma 4.2 is easily adapted to the graph $G_{\mathrm{ref}}(u)$ or $G_{\mathrm{symm}}(u)$, where the graph H is assumed to be reflexive or symmetric, respectively.

Proposition 4.1. *An identity $u \approx v$ holds in $\mathcal{A}$ if and only if $G[u] = G[v]$. An identity holds in $\mathcal{A}_{\mathrm{ref}}$ if and only if $G_{\mathrm{ref}}[u] = G_{\mathrm{ref}}[v]$. An identity holds in $\mathcal{A}_{\mathrm{symm}}$ if and only if $G_{\mathrm{symm}}[u] = G_{\mathrm{symm}}[v]$.*

Proof. We prove only the first case; the other two cases are similar.

First we show sufficiency. Let us assume that $G[u] = G[v]$, and consider an assignment θ into an adjacency semigroup $\mathrm{A}(\mathsf{H})$. Now the vertex sets are the same, so u and v have the same alphabet. So we may assume that θ maps the alphabet to nonzero elements of $\mathrm{A}(\mathsf{H})$. By Lemma 4.2, we have $\theta(u) \neq 0$ if and only if $\theta(v) \neq 0$. By Lemma 4.1, we have $\theta(u) = \theta(v)$ whenever both sides are nonzero. Hence $\theta(u) = \theta(v)$ always.

Necessity. Say that $G[u] \neq G[v]$. If the vertex sets are distinct, then $u \approx v$ fails on $\mathrm{A}(\underline{\mathbf{1}})$, which is isomorphic to the unary semigroup formed by the integers 0 and 1 with the usual multiplication and the identity map as the unary operation. Now say that $G[u]$ and $G[v]$ have the same vertices. Without loss of generality, we may assume that either $G[v]$ contains an edge not in $G[u]$, or that the two graphs are identical but have different initial vertices. Let $\mathrm{A}_u := \mathrm{A}(G[u])$ and consider the assignment into A_u that sends each variable x to (ℓ_x, r_x). Observe that the value of u is equal to (λ_a, ρ_b) where λ_a is the initial vertex of $G[u]$ and ρ_b is the final vertex, while the value of v is either 0 (if there is an adjacency not in $G[v]$: we fail to get a graph homomorphism) or has different first coordinate (if $G[v]$ has a different initial vertex). So $u \approx v$ fails in $\mathcal{A}$. □

4.2 A Normal Form

Proposition 4.1 gives a reasonable solution to the word problem in the $\mathcal{A}$-free algebras. In this subsection we go a bit further and show that every unary semigroup word is equivalent in $\mathcal{A}$ to a unary semigroup word of a certain form. Because different forms may have the same adjacency graph, this by itself does not constitute a different solution to the word problem in $\mathcal{A}$-free algebras, however it is useful in analyzing identities of $\mathcal{A}$.

Most of the work in this section revolves around the variety of algebras of type $\langle 2, 1\rangle$ defined by the three laws:

$$\Psi = \{x'' \approx x,\ x(yz)' \approx (y(xz')')',\ (xy)'z \approx ((x'z)'y)'\}$$

as interpreted within the variety of unary semigroups. By examining the adjacency graphs, it is easy to see that these identities are all satisfied by $\mathcal{A}$ (see Fig. 7 for one of these). In fact Ψ defines a strictly larger variety than $\mathcal{A}$ (it contains all groups for example), but they are close enough for us to obtain useful information.

For later reference we refer to the second and third laws in Ψ as the *first associativity across reversion* law (FAAR) and *second associativity across reversion* law (SAAR), respectively. We let $\mathcal{B}$ denote the unary semigroup variety defined by Ψ.

Surprisingly, the laws in Ψ are sufficient to reduce every unary semigroup word to one in which the nesting height of the unary $'$ is at most 2. The proof of this is the main result of this subsection.

Lemma 4.3. *Ψ implies $(a(bcd)'e)' \approx (b'e)'c(ad')'$, where c is possibly empty.*

Proof. We prove the case where c is non-empty only. We have

$$(a(bcd)'e)' \overset{\text{FAAR}}{\approx} ([bc(ad')']'e)' \overset{\text{SAAR}}{\approx} ((b'e)'c(ad')')'' \approx (b'e)'c(ad')' . \qquad \square$$

Let $X := \{x_1, x_2, \ldots\}$. Let $\mathbf{F}(X)$ denote the free unary semigroup freely generated by X and $\mathbf{F}_\Psi(X)$ denote the $\mathcal{B}$-free algebra freely generated by X. We let ψ denote the fully invariant congruence on $\mathbf{F}(X)$ giving $\mathbf{F}_\Psi(X) = \mathbf{F}(X)/\psi$. We find a subset $N \subseteq \mathbf{F}(X)$ with $X \subseteq N$ and show that multiplying two words from N in $\mathbf{F}(X)$, or applying $'$ to a word in N produces a word that is ψ-equivalent to a word in N. It shows that every word in $\mathbf{F}(X)$ is ψ-equivalent to a word in N. In this way the members of N are a kind of weak normal form for terms modulo Ψ (we do not claim that distinct words in N are not ψ-equivalent; for example, Proposition 4.1 shows that $\mathcal{A} \models x(x'y)' \approx x(x'y')'$, but the two words are distinct elements of N).

We let N consist of all (nonempty) words of the form

$$u_1(v_1)'u_2(v_2)' \ldots u_n(v_n)'u_{n+1}$$

for some some $n \in \omega$, where for $i \leq n$,

- the v_i are *semigroup words* in the alphabet

$$X \cup X' = \{x_1, x_1', x_2, x_2', \ldots\},$$

  and all have length at least 2 as semigroup words;
- the u_i are possibly empty semigroup words in the alphabet $X \cup X'$ and if $n = 0$, then u_1 is non-empty.

Notice that $X \subseteq N$ since the case $n = 0$ corresponds to semigroup words over $X \cup X'$. For a member s of N, we refer to the number n in this definition as the *breadth* of s.

The following lemma is trivial.

Lemma 4.4. *If s and t are two words in N, then $s \cdot t$ is ψ-equivalent to a word in N.*

Lemma 4.5. *If s is a word in N, then s' is ψ-equivalent to a word in N.*

Proof. We prove the lemma by induction on the breadth of s. If the breadth of s is 0 then $s' = (u_1)'$ either is in N, or is of the form x'' for some variable x, in which case it reduces to $x \in N$ modulo Ψ. Now say that the result holds for breadth k members of N, and say that the breadth of s is $k+1$. So s can be written in the form $p(y_1 \cdots y_m)'u$ where p is either empty or is a word from N of breadth k, $u = u_{k+2}$ is a possibly empty semigroup word in the alphabet $X \cup X'$ and $y_1, \ldots, y_m$ is a possibly repeating sequence of variables from $X \cup X'$ with $v_{k+1} \equiv y_1 \cdots y_m$ (so $m > 1$). Note that p can be empty only if $k = 0$.

Let us write w for $y_2 \cdots y_{m-1}$ (if $m = 2$, then w is empty). If both p and u are empty, then $s' \in N$ already. If neither p nor u are empty, then by Lemma 4.3 Ψ implies $s' \approx (py_m')'w(y_1'u)'$. The breadth of py_m' is k, so the induction hypothesis and Lemma 4.4 complete the proof.

Now say that p is empty and u is not. We have $((y_1wy_m)'u)' \overset{\text{SAAR}}{\approx} (y_1'u)'wy_m$, and the latter word is contained in N (modulo $x'' \approx x$).

Lastly, if u is empty and p is not, then we have $s' \equiv (p(wy_m)')' \overset{\text{FAAR}}{\approx} w(py_m')'$, and the induction hypothesis applies to $(py_m')'$ since py_m' is of breadth k. By Lemma 4.4, s' is ψ-equivalent to a member of N. □

As explained above, Lemmas 4.4 and 4.5 give us the following result.

Proposition 4.2. *Every unary semigroup word reduces modulo Ψ to a word in N.*

An algorithm for making such a reduction is to iterate the method of proof of Lemmas 4.4 and 4.5; however we will not need this here.

4.3 Subvarieties of $\mathcal{A}$ and Sub-uH Classes of $\underline{\mathsf{G}}$

In this subsection we complete the proof of Theorem A. Recall that the theorem claims that, for any nonempty class K of graphs, any graph G belongs to the uH-class generated by K if and only if the adjacency semigroup $\mathrm{A}(\mathsf{G})$ belongs to the variety generated by the adjacency semigroups $\mathrm{A}(\mathsf{H})$ with $\mathsf{H} \in K$. For the "only if" statement we use a direct argument. For the "if" statement, we use a syntactic argument, translating uH sentences of K into identities of $\mathrm{A}(K)$.

Lemma 4.6. *If* $\mathsf{G} \in \mathbb{ISP}^+\mathbb{P}_\mathrm{u}(K)$, *then* $\mathrm{A}(\mathsf{G}) \in \mathbb{HSP}(\mathrm{A}(K))$.

Proof. First consider a nonempty family $L = \{\mathsf{H}_i \mid i \in I\}$ of graphs from K and an ultraproduct $\mathsf{H} := \prod_U L$ (for some ultrafilter U in 2^I). It is easy to see that the ultraproduct of the family $\{\mathrm{A}(\mathsf{H}_i) \mid i \in I\}$ over the same ultrafilter U is isomorphic to $\mathrm{A}(\mathsf{H})$ (we leave this elementary proof to the reader). Hence, we have $\mathbb{I}(\mathrm{A}\mathbb{P}_\mathrm{u}(K)) = \mathbb{IP}_\mathrm{u}(\mathrm{A}(K))$. Now we have $\mathsf{G} \in \mathbb{ISP}^+(\mathbb{P}_\mathrm{u}(K))$. So it will suffice to prove that $\mathrm{A}(\mathsf{G}) \in \mathbb{HSP}(\mathrm{A}(\mathbb{P}_\mathrm{u}(K)))$, since $\mathbb{HSP}(\mathrm{A}(\mathbb{P}_\mathrm{u}(K))) = \mathbb{HSPP}_\mathrm{u}(\mathrm{A}(K)) = \mathbb{HSP}(\mathrm{A}(K))$. We let P denote $\mathbb{P}_\mathrm{u}(K)$.

Now G is isomorphic to an induced subgraph of the direct product $\prod_{i \in I} \mathsf{H}_i$ with $\mathsf{H}_i \in P$. It does no harm to assume that this embedding is the inclusion map. Let $\pi_i : \mathsf{G} \to \mathsf{H}_i$ denote the projection. Evidently the following properties hold:

(i) if u and v are distinct vertices of G then there is $i \in I$ such that $\pi_i(u) \neq \pi_i(v)$;
(ii) if (u, v) is not an edge of G then there is $i \in I$ with $(\pi_i(u), \pi_i(v))$ not an edge of H_i.

We aim to show that $\mathsf{A}(\mathsf{G})$ is a quotient of a subalgebra of $\prod_{i\in I}\mathsf{A}(\mathsf{H}_i)$. We define a map $\alpha : \mathsf{A}(\mathsf{G}) \to \prod_{i\in I}\mathsf{A}(\mathsf{H}_i)$ by letting $\alpha(0)$ be the constant 0 and $\alpha(u,v)$ be the map $i \mapsto (\pi_i(u), \pi_i(v))$. The map α is unlikely to be a homomorphism. Let B be the subalgebra of $\prod_{i\in I}\mathsf{A}(\mathsf{H}_i)$ generated by the image of $\mathsf{A}(\mathsf{G})$, and let J be the ideal of B consisting of all elements with a 0 coordinate.

Claim 1. Say (u_1, v_1) and (u_2, v_2) are (nonzero) elements of $\mathsf{A}(\mathsf{G})$. If $v_1 \sim u_2$ then $\alpha(u_1, v_1)\alpha(u_2, v_2) = \alpha((u_1, v_1)(u_2, v_2))$.

Proof. Now

$$\alpha((u_1, v_1)(u_2, v_2))[i] = \alpha(u_1, v_2)[i] = (\pi_i(u_1), \pi_i(v_2)),$$

because $v_1 \sim u_2$ in G. Also, for every $i \in I$ we have $\pi_i(v_1) \sim \pi_i(u_2)$, so that

$$\alpha(u_1, v_1)[i]\alpha(u_2, v_2)[i] = (\pi_i(u_1), \pi_i(v_1))(\pi_i(u_2), \pi_i(v_2)) = (\pi_i(u_1), \pi_i(v_2))$$

as required. □

Claim 2. Say (u_1, v_1) and (u_2, v_2) are nonzero elements of $\mathsf{A}(\mathsf{G})$. If $v_1 \not\sim u_2$ then $\alpha(u_1, v_1)\alpha(u_2, v_2) \in J$.

Proof. By the definition of $v_1 \not\sim u_2$ there is $i \in I$ with $\pi_i : \mathsf{G} \to \mathsf{H}_i$ with $\pi_i(v_1) \not\sim \pi_i(u_2)$. Then $(u_1, v_1)[i](u_2, v_2)[i] = 0$. □

Claims 1 and 2 show that α is a semigroup homomorphism from $\mathsf{A}(\mathsf{G})$ onto B/J (at least, if we adjust the co-domain of α to be B/J and identify the constant 0 with J). Now we show that this map is injective. Say $(u_1, v_1) \neq (u_2, v_2)$ in $\mathsf{A}(\mathsf{G})$. Without loss of generality, we may assume that $u_1 \neq u_2$. So there is a coordinate i with $\pi_i(u_1) \neq \pi_i(u_2)$. Then $\alpha(u_1, v_1)$ differs from $\alpha(u_2, v_2)$ on the i-coordinate. So we have a semigroup isomorphism from $\mathsf{A}(\mathsf{G})$ to B/J. Lastly, we observe that α trivially preserves the unary operation, so we have an isomorphism of unary semigroups as well. This completes the proof of Lemma 4.6. □

To prove the other half of Theorem A we take a syntactic approach by translating uH sentences into unary semigroup identities. To apply our technique, we first need to reduce arbitrary uH sentences to logically equivalent ones of a special form.

Our goal is to show that if $\mathsf{G} \notin \mathbb{ISP}^+\mathbb{P}_\mathrm{u}(K)$ then $\mathsf{A}(\mathsf{G}) \notin \mathbb{HSP}(\mathsf{A}(K))$. We first consider some degenerate cases.

If $K = \{\underline{\mathbf{0}}\}$, then $\mathsf{A}(K)$ is the class consisting of the one element unary semigroup and $\mathbb{HSP}(\mathsf{A}(K)) \models x \approx y$. The statement $\mathsf{G} \notin \mathbb{ISP}^+\mathbb{P}_\mathrm{u}(K)$ simply means that $|G| \geq 1$ and so $\mathsf{A}(\mathsf{G}) \not\models x \approx y$. So $\mathsf{A}(\mathsf{G}) \notin \mathbb{HSP}(\mathsf{A}(K))$.

Now we say that K contains a nonempty graph. We can then further assume that the empty graph is not in K. If G is the 1-vertex looped graph $\underline{\mathbf{1}}$, then the statement $\mathsf{G} \notin \mathbb{ISP}^+\mathbb{P}_\mathrm{u}(K)$ simply means that K consists of antireflexive graphs. In this case, $\mathsf{A}(K) \models xx' \approx 0$, while $\mathsf{A}(\mathsf{G}) \not\models xx' \approx 0$. So again, $\mathsf{A}(\mathsf{G}) \notin \mathbb{HSP}(\mathsf{A}(K))$.

So now it remains to consider the case where G is not the 1-vertex looped graph and K does not contain the empty graph. Lemma 2.1 shows that there is some uH sentence Γ holding in each member of K, but failing on G. We now show that Γ can be chosen to be a quasi-identity.

If Γ is a uH sentence of the second kind, say $\bigvee_{1\leq i\leq n} \neg\Phi_i$, then choose some atomic formula Ξ in variables not appearing in any of the Φ_i and that fails on G under some assignment: for example, if $|G| \geq 2$, then a formula of the form $x \approx y$ suffices, while if $|G| = 1$, then G is the one element edgeless graph and $x \sim y$ suffices. Now replace Γ with the quasi-identity $\&_{1\leq i\leq n} \Phi_i \to \Xi$. We need to show that this new quasi-identity holds in K and fails in G. It certainly holds in K, since it is logically equivalent to $\bigvee \neg\Phi_i \vee \Xi$, while $\bigvee \neg\Phi_i$ is constantly true. On the other hand, since $\bigvee_{1\leq i\leq n} \neg\Phi_i$ does not hold on G, there is an assignment θ making $\&_{1\leq i\leq n} \Phi_i$ true, and this assignment can be extended to the variables of Ξ in such a way that Ξ is false under θ. In other words, we have a failing assignment for the new quasi-identity on G.

Next we need to show that the quasi-identity Γ can be chosen to have a particular form. Let us call a quasi-identity *reduced* if the equality symbol $\approx$ does not appear in the premise. One may associate any quasi-identity with a reduced quasi-identity in the obvious way: if $x \approx y$ appears in the premise of the original, then we may replace all occurrences of y in the quasi-identity with x (including in the conclusion) and remove $x \approx y$ from the premise. If a quasi-identity fails on a graph H under some assignment θ, then the corresponding reduced quasi-identity also fails under θ. Conversely, if the reduced quasi-identity fails on H under some assignment θ, then we may extend θ to a failing assignment of the original quasi-identity. This means that we may choose Γ to be reduced.

Let $\Phi := \&_{1\leq i\leq n} u_i \sim v_i$ be a conjunction of adjacencies, where the

$$u_1, \dots, u_n, v_1, \dots, v_n \in \{a_1, \dots, a_m\}$$

are not necessarily distinct variables. For each adjacency $u_i \sim v_i$ in Φ, let w_i denote the word $(u_i v_i)' s_i (u_i v_i')' s_i (u_i' v_i)' s_i (u_i' v_i')'$, where s_i is a new variable. Now let $\sigma : \{1, \dots, m\} \to \{1, \dots, n\}$ be some finite sequence of numbers from $\{1, \dots, n\}$ with the property that for each pair $i, j \in \{1, \dots, n\}$ there is $k < m$ with $\sigma_k = i$ and $\sigma_{k+1} = j$ and such that $\sigma(1) = \sigma(m) = 1$. Define a word D_Φ (depending on σ) as follows:

$$\left(\prod_{1\leq i<m} w_{\sigma(i)} t_{\sigma(i),\sigma(i+1)}\right) w_{\sigma(m)},$$

where the $t_{i,j}$ are new variables. As an example, consider the conjunction $\Phi := x \sim y \ \& \ y \sim z$, where $n = 2$, $u_1 = x$, $v_1 = u_2 = y$ and $v_2 = z$. Using the sequence $\sigma = 1, 2, 2, 1, 1$ we get D_Φ equal to the following expression:

$$(xy)' s_1 (xy')' s_1 (x'y)' s_1 (x'y')' t_{1,2} (yz)' s_2 (yz')' s_2 (y'z)' s_2 (y'z')' t_{2,2} (yz)' s_2 (yz')'$$
$$s_2 (y'z)' s_2 (y'z')' t_{2,1} (xy)' s_1 (xy')' s_1 (x'y)' s_1 (xy)' t_{1,1} (xy)' s_1 (xy')' s_1 (x'y)' s_1 (x'y')'.$$

Lemma 4.7. *Let* H *be a graph,* Φ *be a conjunction of adjacencies in variables* $a_1, \dots, a_m$ *and* θ *be an assignment of the variables of* D_Φ *into* $\mathrm{A}(\mathsf{H})$, *with* $\theta(a_i) =$

(ℓ_i, r_i) say. Let γ be any member of $\{L, R\}^m$. If $\theta(D_\Phi) \neq 0$ then the map ϕ_γ from $a_1, \ldots, a_m$ into the vertices V_H of H defined by

$$\phi_\gamma(a_i) = \begin{cases} \ell_i & \text{if } \gamma(i) = L; \\ r_i & \text{if } \gamma(i) = R \end{cases}$$

satisfies Φ.

Proof. Let $a_i \sim a_j$ be one of the adjacencies in Φ. So all of $a_i a_j$, $a_i' a_j$, $a_i a_j'$ and $a_i' a_j'$ appear in D_Φ and hence are given nonzero values by θ. We have $\ell_i \sim \ell_j$, $\ell_i \sim r_j$, $r_i \sim \ell_j$, $r_i \sim r_j$ in H. So regardless of the choice of γ we have $\phi_\gamma(a_i) \sim \phi_\gamma(a_j)$ in H. □

Lemma 4.8. *Let $\Phi = \&_{1 \leq i \leq n} u_i \sim v_i$ be a nonempty conjunction in the variables $a_1, \ldots, a_m$ and let θ be an assignment of these variables into a graph H such that $\mathsf{H} \models \theta(\Phi)$. Define an assignment θ^+ of the variables of D_Φ into $\mathrm{A}(\mathsf{H})$ by $a_i \mapsto (\theta(a_i), \theta(a_i))$, $\theta^+(t_{i,j}) := (\theta(v_i), \theta(u_j))$ and $\theta^+(s_i) = \theta^+(t_{i,i})$. We have $\theta^+(D_\Phi) = (\theta(v_1), \theta(u_1))$.*

Proof. This is a routine calculation. For each adjacency $u_i \sim v_i$ in Φ (here $\{u_i, v_i\} \subseteq \{a_1, \ldots, a_m\}$) we have

$$\theta^+((u_i v_i)'),\ \theta^+((u_i v_i')'),\ \theta^+((u_i' v_i)'),\ \theta^+((u_i' v_i')')$$

all taking the same nonzero value $(\theta(v_i), \theta(u_i))$. Then we also have $\theta^+(w_i) = (\theta(v_i), \theta(u_i))$ which shows that

$$\begin{aligned} \theta^+(D_\Phi) &= [\theta(u_1), \theta(v_1)] \ldots [\theta(v_i), \theta(u_i)]\ \theta^+(t_{i,j})\ [\theta(v_j), \theta(u_j)] \ldots [\theta(u_1), \theta(v_1)] \\ &= [\theta(u_1), \theta(v_1)] \ldots [\theta(v_i), \theta(u_i)][\theta(v_i), \theta(u_j)][\theta(v_j), \theta(u_j)] \ldots [\theta(u_1), \theta(v_1)] \\ &= [\theta(u_1), \theta(v_1)] \end{aligned}$$

(where the square brackets are used for clarity only). □

Lemma 4.9. *Let H be a nonempty graph and $\Phi \to u \approx v$ be a reduced quasi-identity where Φ is nonempty and one of u or v does not appear in Φ (say it is u). We have $\mathrm{A}(\mathsf{H}) \models uD_\Phi \approx u'D_\Phi$ if and only if $\mathsf{H} \models \Phi \to u \approx v$.*

Proof. First assume that $\mathsf{H} \models \Phi \to u \approx v$, where u does not appear in Φ. Both sides of the identity contain the subword D_Φ and so we may consider an assignment θ sending D_Φ to a nonzero value (if there are none, then we are done). By Lemma 4.7, we have an interpretation of Φ in H. But then, we can choose any value for $\theta(u)$ and find that it is the same value as $\theta(v)$. In other words, H has only one vertex. Also, since D_Φ takes a nonzero value on $\mathrm{A}(\mathsf{H})$, we find that the semigroup reduct of $\mathrm{A}(\mathsf{H})$ is not a null semigroup (that is, a semigroup in which all products are equal to 0). Hence $\mathrm{A}(\mathsf{H})$ is isomorphic to the unary semigroup formed by the integers 0 and 1 with the usual multiplication and the identity map as the unary operation. In this case we have $u \approx u'$ satisfied and the identity holds.

Now say that $\mathsf{H} \not\models \Phi \to u \approx v$, and let θ be a failing assignment. As $\theta(u) \neq \theta(v)$ we can find a vertex a of H such that $a \neq \theta(u_1)$. Extend the assignment θ^+ of Lemma 4.8 by $u \mapsto (a, \theta(u_1))$. Evidently, $\theta^+(uD_\Phi) = (a, \theta(u_1))(\theta(v_1), (\theta(u_1)) = (a, \theta(u_1))$, but $\theta^+(u'D_\Phi)$ either is equal to 0, or is nonzero but has left coordinate different from $\theta^+(uD_\Phi)$. □

Lemma 4.10. *Let* H *be a nonempty graph and* $\&_{1\leq i\leq n} u_i \sim v_i \to u \approx v$ *be a reduced quasi-identity where* Φ *is nonempty and both* u *and* v *appear in* Φ. *If* $u = u_i$ *for some* i *then we have* $\mathrm{A}(\mathsf{H}) \models u_i t_{i,1} D_\Phi \approx v t_{i,1} D_\Phi$ *if and only if* $\mathsf{H} \models \Phi \to u \approx v$. *If* $u = v_i$ *for some* i *then we have* $\mathrm{A}(\mathsf{H}) \models D_\Phi t_{1,i} u_i \approx D_\Phi t_{1,i} v$ *if and only if* $\mathsf{H} \models \Phi \to u \approx v$

Proof. First assume that $\mathsf{H} \models \Phi \to u \approx v$. We consider only the case that $u = u_i$; the other case follows by symmetry. As before, we can consider the case where there is an assignment θ into H satisfying Φ. So we have $\theta(u_i) = \theta(v)$ and hence $\theta^+(u_i) = \theta^+(v)$, in which case both sides of the identity take the same value.

Now say that $\mathsf{H} \not\models \Phi \to u \approx v$ and let θ be a failing assignment. Now, the left side of the identity contains the same adjacencies as D_Φ and so takes a nonzero value in $\mathrm{A}(\mathsf{H})$ under the assignment θ^+; moreover the left coordinate is $\theta(u_i)$. However the right hand side either takes the value 0 (if $\theta(v) \not\sim \theta(v_i)$) or has left coordinate equal to $\theta(v) \neq \theta(u_i)$. In either case, the identity fails. □

Now we come to reduced quasi-identities in which the conclusion is an adjacency $u \sim v$. We consider 9 cases according to whether or not u and v appear in Φ, and if so, whether they appear as the "source" or "target" of an adjacency. The nine identities $\tau_1, \ldots, \tau_9$ are defined in the following table. In this table, the first row corresponds to the situation where neither u nor v appear, while the second corresponds to the situation where u does not appear, but v does appear as some u_j (in other words, as a "source"), and so on. If one of u or v appears as both a source and a target, then there will be choices as to which identity we can choose. The variables z and w are new variables not appearing in D_Φ.

k	$u \sim v$	τ_k
1.	$z \sim w$	$wD_\Phi z \approx (wD_\Phi z)^2$
2.	$z \sim u_j$	$u_j t_{j,1} D_\Phi z \approx (u_j t_{j,1} D_\Phi z)^2$
3.	$z \sim v_j$	$(u_j v_j)' t_{j,1} D_\Phi z \approx ((u_j v_j)' t_{j,1} D_\Phi z)^2$
4.	$u_i \sim w$	$wD_\Phi t_{1,i} (u_i v_i)' \approx (wD_\Phi t_{1,i} (u_i v_i)')^2$
5.	$v_i \sim w$	$wD_\Phi t_{1,i} v_i \approx (wD_\Phi t_{1,i} v_i)^2$
6.	$u_i \sim u_j$	$u_j t_{j,1} D_\Phi t_{1,i} (u_i v_i)' \approx (u_j t_{j,1} D_\Phi t_{1,i} (u_i v_i)')^2$
7.	$u_i \sim v_j$	$(u_j v_j)' t_{j,1} D_\Phi t_{1,i} (u_i v_i)' \approx ((u_j v_j)' t_{j,1} D_\Phi t_{1,i} (u_i v_i)')^2$
8.	$v_i \sim u_j$	$u_j t_{j,1} D_\Phi t_{1,i} v_i \approx (u_j t_{j,1} D_\Phi t_{1,i} v_i)^2$
9.	$v_i \sim v_j$	$(u_j v_j)' t_{j,1} D_\Phi t_{1,i} v_i \approx ((u_j v_j)' t_{j,1} D_\Phi t_{1,i} v_i)^2$

Lemma 4.11. *Let* H *be a graph and* $\Phi \to u \sim v$ *be a quasi-identity where* Φ *is nonempty. Consider the corresponding identity* τ_k. *We have* $\mathrm{A}(\mathsf{H}) \models \tau_k$ *if and only if* $\mathsf{H} \models \Phi \to u \approx v$.

Proof. We prove the case of τ_4 and leave the remaining (very similar) cases to the reader. First assume that $\mathsf{H} \models \Phi \to u_i \sim v$. Consider some assignment θ into $\mathrm{A}(\mathsf{H})$ that gives D_Φ a nonzero value. As w appears on both sides, we may further assume that $\theta(w)$ is nonzero. Observe that the graph of the right hand side of the identity is identical to that of the left side except for the addition of a single edge from ℓ_{u_i} to ℓ_w. Also, the initial and final vertices are the same. So to show that the two sides are equal, it suffices to show that $\theta(u_i')\theta(w)$ is nonzero.

Choose any map γ from the variables of τ_4 to $\{L, R\}$ with $\gamma(u_i) = R$. By Lemma 4.7 we have $\mathsf{H} \models \phi_\gamma(\Phi)$. Using $\Phi \to u_i \sim v$ it follows that for any vertex w we have $\phi_\gamma(u_i) \sim w$. In other words, $\theta(u_i')\theta(w)$ is nonzero as required.

Now say that $\Phi \to u_i \sim v$ fails on H under some assignment θ. Extend θ^+ to w by $w \mapsto (\theta(v), \theta(u_1))$. Under this assignment the left hand side of τ_4 takes the value $(\theta(v), \theta(u_i))$, while the right hand side equals 0. □

Lastly we need to consider the case where Γ has empty premise, that is, where Γ is a universally quantified atomic formula τ. In the language of graphs, there are essentially four different possibilities for τ (up to a permutation of letter names): $x \sim y$, $x \sim x$, $x \approx y$ and $x \approx x$. The last of these is a tautology. The first three are nontautological and correspond to the uH-classes of complete looped graphs, reflexive graphs, and the one element graphs. For Φ one of the three atomic formulas, we let τ_Φ denote the identities $xx \approx x$, $xx'x \approx x$, and $x' \approx x$, respectively.

Lemma 4.12. *Let* H *be a graph and* Φ *be one of the three nontautological atomic formulas in the language of graphs. We have* $\mathsf{H} \models \Phi$ *if and only if* $\mathrm{A}(\mathsf{H}) \models \tau_\Phi$.

Proof. If Φ is $x \sim y$, then it is easy to see that $\mathsf{H} \models \Phi$ if and only if the underlying semigroup of $\mathrm{A}(\mathsf{H})$ satisfies $xx \approx x$. The case of $\Phi = x \sim x$ has been discussed already in Sect. 3. The case of $x \approx y$ corresponds to the 1-vertex graphs, which is clearly equivalent to the property that $\mathrm{A}(\mathsf{H}) \models x' \approx x$. □

Now we can complete the proof of Theorem A. We have a reduced quasi-identity Γ satisfied by K and failing on G. By the appropriate choice out of Lemmas 4.9, 4.10, 4.11 or 4.12 we can construct an identity τ such that $\mathrm{A}(K) \models \tau$ and $\mathrm{A}(\mathsf{G}) \not\models \tau$. Hence $\mathrm{A}(\mathsf{G}) \notin \mathbb{HSP}(\mathrm{A}(K))$. □

5 Proof of Theorem B

In contrast to the proof of Theorem A, this section requires some basic notions and facts from semigroup theory such as Green's relations $\mathscr{J}$, $\mathscr{L}$, $\mathscr{R}$, $\mathscr{H}$ and their manifestation on Rees matrix semigroups. For details, refer to the early chapters of any general semigroup theory text; Howie [12] for example.

The first step to proving Theorem B is the following.

Lemma 5.1. *Let* $\mathcal{V}$ *be a variety of unary semigroups satisfying*

$$xx'x \approx x,\ x'' \approx x,\ (x'x)' \approx x'x,\ (xy)' \approx y'(x'xyy')'x'\,. \tag{5}$$

If $A \in \mathcal{V}$ as a semigroup is a completely 0-simple semigroup with trivial subgroups, then A is of the form $\mathsf{A}(\mathsf{H})$ *for some reflexive graph* H.

Proof. Since A is a completely 0-simple semigroup with trivial subgroups, the Green relation $\mathscr{H}$ is trivial. Now every $a \neq 0$ in A is $\mathscr{L}$-related to $a'a$ (since $a(a'a) = a$) and $\mathscr{R}$-related to aa'. Also, these elements are fixed by $'$ by identity $(x'x)' \approx x'x$ (and $x'' \approx x$). Next we observe that $a \mathrel{\mathscr{L}} b$ if and only if $a' \mathrel{\mathscr{R}} b'$. For this we can use identity $(xy)' \approx y'(x'xyy')'x'$: if $a \mathrel{\mathscr{L}} b$ then $xa = b$ for some x, so $b' = (xa)' = a'z$, for $z = (x'xaa')x'$. So $b' \mathrel{\mathscr{R}} a'$. The other case follows by symmetry (or using $(x')' \approx x$).

This implies that each $\mathscr{L}$-class and each $\mathscr{R}$-class contain precisely one fixed point of $'$ (if $a' = a$, $b' = b$ and $a \mathscr{L} b$, then $a = a' \mathscr{R} b' = b$, so $a \mathscr{H} b$). Represent A as a Rees matrix semigroup (with matrix P) in which fixed points of $'$ correspond to diagonal elements (as xx' is an idempotent, P will have 1 down the diagonal). It is easily seen that this is $\mathsf{A}(\mathsf{H})$ for the graph H with P as adjacency matrix. This graph is reflexive since the identity $xx'x \approx x$ holds. □

In the case where H is a universal relation, the set $\mathsf{A}(\mathsf{H})\backslash\{0\}$ is a subuniverse, and the corresponding subalgebra of $\mathsf{A}(\mathsf{H})$ is a square band.

Lemma 5.2. *Let $\mathcal{V}$ be a variety of unary semigroups satisfying the identities* (5). *If $A \in \mathcal{V}$ as a semigroup is a completely simple semigroup with trivial subgroups, then A is a square band.*

Proof. The proof is basically the same as for Lemma 5.1. □

In order to get a small basis for the identities of $\mathcal{A}_{\text{ref}}$ the following lemma is useful.

Lemma 5.3. *Let Ψ_1 stand for the system of* 5 *identities:*

$$x \approx xx'x,\ (x'x)' \approx x'x,\ x'' \approx x,\ x(yz)' \approx (y(xz')')',\ (xy)'z \approx ((x'z)'y)'.$$

The following laws are consequences of Ψ_1 :

- $(xy)' \approx y'(xyy')' \approx (x'xy)'x' \approx y'(x'xyy')'x'$;
- $(xyz)' \approx (yz)'y(xy)'$.

Proof. For the first item we have Ψ_1 implies $(xy)' \approx ((xx')'xy)' \stackrel{\text{SAAR}}{\approx} (x'xy)'x'$. The other two cases of this item are very similar.

For the second item, first note that using item 1 and Ψ_1, we have $(xyy')' \approx y(xyy'y)' \approx y(xy)'$. Using this we obtain

$$(xyz)' \approx (xy(y'y)'z)' \stackrel{\text{FAAR}}{\approx} ((y'(xyy')')'z)' \stackrel{\text{SAAR}}{\approx} (yz)'(xyy')' \approx (yz)'y(xy)'. \quad \square$$

The second item of Lemma 5.3 enables a refinement of Proposition 4.2.

Corollary 5.1. *The identities Ψ_1 reduce every unary semigroup word to a member of N in which each subword of the form $(v)'$ has the property that v is a semigroup word of length* 1 *or* 2 *(over the alphabet $X \cup X'$).*

We now let Σ_{ref} denote the following set of unary semigroup identities:

$$x'' \approx x,\ x(yz)' \approx (y(xz')')',\ (xy)'z \approx ((x'z)'y)', \tag{Ψ}$$
$$xx'x \approx x, \tag{6}$$
$$(xx')' \approx xx', \tag{7}$$
$$x^3 \approx x^2, \tag{8}$$
$$xyxzx \approx xzxyxzx \approx xzxyx, \tag{9}$$
$$x'yxzx \approx (xzx)'yxzx, \tag{10}$$
$$xyxzx' \approx xyxz(xyx)'. \tag{11}$$

Proposition 4.1 easily shows that all but identity (6) hold in $\mathcal{A}$, while (6) obviously holds in the subvariety $\mathcal{A}_{\mathrm{ref}}$. Hence, to prove that Σ_{ref} is a basis for $\mathcal{A}_{\mathrm{ref}}$, we need to show that every model of Σ_{ref} lies in $\mathcal{A}_{\mathrm{ref}}$. Before we can do this, we need some further consequences of Σ_{ref}.

In the identities that occur in the next lemma we use $\overline{u}$, where u is either x or xyx, to denote either u or u'. We assume that the meaning of the operation $^-$ is fixed within each identity: either it changes nothing or it adds $'$ to all its arguments.

Lemma 5.4. *The following identities all follow from* Σ_{ref}:

- $(\overline{x}u_1)'u_2xyx \approx (\overline{xyx}u_1)'u_2xyx$;
- $xyxu_2(u_1\overline{x})' \approx xyxu_2(u_1\overline{xyx})'$;
- $(u_1\overline{x})'u_2xyx \approx (u_1\overline{xyx})'u_2xyx$;
- $xyxu_2(\overline{x}u_1)' \approx xyxu_2(\overline{xyx}u_1)'$,

where u_1 and u_2 are possibly empty unary semigroup words.

Proof. In each of the eight cases, if u_1 is empty, then the identity is equivalent modulo $x'' \approx x$ to one in Σ_{ref} up to a change of letter names. So we assume that u_1 is non-empty. We can ensure that u_2 is non-empty by rewriting u_2xyx and $xyxu_2$ as $(u_2xx')xyx$ and $xyx(x'xu_2)$ respectively (a process we reverse at the end of each deduction). For the first identity we have Ψ implies $(\overline{x}u_1)'u_2xyx \stackrel{\mathrm{SAAR}}{\approx} ((\overline{x}'u_2xyx)'u_1)'$, and then we use (9) or (10) to replace $\overline{x}$ by $\overline{xyx}$. Reversing the application of SAAR, we obtain the corresponding right hand side.

The second identity is just a dual to the first so follows by symmetry. Similarly, the fourth will follow from the third by symmetry.

For the third identity, Lemma 5.3 can be applied to the left hand side to get $\overline{x}'\overline{x}(u_1\overline{x})'u_2xyx$. Now, the subword $\overline{x}'\overline{x}$ is either $x'x$ or xx'. We will write it as $t(x,x')$ (where $t(x,y)$ is one of the words xy or yx). Using (9), we have $t(x,x')(u_1\bar{x})'u_2xyx \approx t(xyx,x')(u_1\bar{x})'u_2xyx$. But the subword $t(xyx,x')(u_1\bar{x})'$ is of the form required to apply the second identity in the lemma we are proving. Since this second identity has been established, we can use it to deduce $t(xyx,x')(u_1\overline{xyx})'u_2xyx$ and then reverse the procedure to get

$$t(xyx,x')(u_1\overline{xyx})'u_2xyx \approx \overline{x}'\overline{x}(u_1\overline{xyx})'u_2xyx \approx (u_1\overline{xyx})'u_2xyx$$

(the last equality requires a few extra easy steps in the $\overline{u} = u'$ case). □

Recall that a *unary polynomial* $p(x)$ on an algebra S is a function $S \to S$ defined for each $a \in S$ by

$$p(a) = t(a, a_1, \ldots, a_n)$$

where $t(x, x_1, \ldots, x_n)$ is a term, and $a_1, \ldots, a_n$ are elements of S. We let P_x denote the set of all unary polynomials on S. The *syntactic congruence* $\mathrm{Syn}(\theta)$ of an equivalence θ on S is defined to be

$$\mathrm{Syn}(\theta) := \{(a, b) \mid p(a)\, \theta\, p(b) \text{ for all } p(x) \in P_x\}.$$

$\mathrm{Syn}(\theta)$ is known to be the largest congruence of S contained in θ (see [1] or [6]). It is very well known that for standard semigroups, one only needs to consider polynomials $p(x)$ built from the semigroup words x, x_1x, xx_1, x_1xx_2 (see [12] for example). In fact there is a similar – though more complicated – reduction for the variety defined by Σ_{ref} (and more generally still Ψ). This can be gleaned fairly easily from Proposition 4.2 (see [6] for a general approach for establishing this), however we do not need an explicit formulation of it here, and so omit any proof.

We may now prove the key lemma, a variation of [11, Lemma 3.2].

Lemma 5.5. *Every model of* Σ_{ref} *(within the variety of unary semigroups) is a subdirect product of members of* $\mathsf{A}(\mathsf{G}_{\mathrm{ref}}) \cup \mathcal{SB}$.

Proof. Let $S \models \Sigma_{\mathrm{ref}}$. If S is the one element semigroup we are done. Now assume that $|S| > 1$. We need to show that for every pair of distinct elements $a, b \in S$ there is a homomorphism from S onto a square band or an adjacency semigroup $\mathsf{A}(\mathsf{G})$ for some $\mathsf{G} \in \underline{\mathsf{G}}_{\mathrm{ref}}$.

For each element $z \in S$, we let $I_z := \{u \in S \mid z \notin S^1uS^1\}$, in other words, I_z is the ideal consisting of all elements that do not divide z. Note that I_z is closed under the reversion operation (since u' divides u). Define equivalence relations ρ_z and λ_z on S:

$$\begin{aligned}\rho_z :=& \{(x, y) \in S \times S \mid (\forall t \in SzS) \quad xt \equiv yt \mod I_z\};\\ \lambda_z :=& \{(x, y) \in S \times S \mid (\forall t \in SzS) \quad tx \equiv ty \mod I_z\}.\end{aligned}$$

So far the proof is identical to that of [11, Lemma 3.2]. In the semigroup setting, both ρ_z and λ_z are congruences, however this is no longer true in the unary semigroup setting. Instead, we replace ρ_z and λ_z by their syntactic congruences $\mathrm{Syn}(\rho_z)$ and $\mathrm{Syn}(\lambda_z)$.

Let a and b be distinct elements of S. Our goal is to show that one of the congruences $\mathrm{Syn}(\rho_a)$, $\mathrm{Syn}(\rho_b)$, $\mathrm{Syn}(\lambda_a)$ and $\mathrm{Syn}(\lambda_b)$ separates a and b, and that $S/\mathrm{Syn}(\rho_z)$ and $S/\mathrm{Syn}(\lambda_z)$ are isomorphic to a square band or an adjacency semigroup of a reflexive graph. The first part is essentially identical to a corresponding part of the proof of [11, Lemma 3.2]. We include it for completeness only.

First suppose that $a \notin SbS$. So $b \in I_a$. Choose $t = a'a \in SaS$ so that $a = at \not\equiv bt \mod I_a$. Hence $(a, b) \notin \hat{\rho}_a$. Now suppose that $SaS = SbS$, so that a

and b lie in the same $\mathscr{J}$-class $SaS\backslash I_a$ of **S**. One of the following two equalities must fail: $ab'b = b$ or $aa'b = a$ for otherwise $a = aa'b = aa'ab'b = ab'b = b$. Hence as neither a nor b is in $I_a = I_b$, we have either $(a,b) \notin \rho_a \supseteq \hat{\rho}_a$ or $(a,b) \notin \lambda_a \supseteq \hat{\lambda}_a$.

Now it remains to prove that $S/\operatorname{Syn}(\rho_z)$ and $S/\operatorname{Syn}(\lambda_z)$ are adjacency semigroups or square bands. Lemmas 5.1, 5.2 and 5.3 show that it suffices to prove that the underlying semigroup of $S/\operatorname{Syn}(\rho_z)$ is completely 0-simple or completely simple. We look at the $\operatorname{Syn}(\rho_z)$ case only (the $\operatorname{Syn}(\lambda_z)$ case follows by symmetry). Now it does no harm to assume that I_z is empty or $\{0\}$, since $v, w \in I_z$ obviously implies that $(v,w) \in \operatorname{Syn}(\rho_z)$. Hence $K_z := SzS/(I_z \cap SzS)$ is a 0-simple semigroup or a simple semigroup. Since S is periodic (by identity (8) of Σ_{ref}), we have that K_z is completely 0-simple or completely simple. We need to prove that every element of $S\backslash I_z$ is $\operatorname{Syn}(\rho_z)$-related to a member of $SzS\backslash I_z$.

Let $c \in S$. If $c \in SzS$ or $c \subset I_z$ we are done, so let us assume that $c \notin SzS \cup I_z$. So $z = pcq$ for some $p, q \in S^1$. So $z = pcqz'pcq$. Put $w = qz'p$. Note that $w \in SzS$ and $cwc \neq 0$. Our goal is to show that $c \operatorname{Syn}(\rho_z)\, cwc$. Let $s(x, \boldsymbol{y})$ be any unary semigroup word in some variables $x, y_1, \ldots$ and let $t \in SaS$. We need to prove that for any $\boldsymbol{d}$ in S^1 we have $s(c, \boldsymbol{d})t \equiv s(cwc, \boldsymbol{d})t$ modulo I_z. Write t as $ucwcv$, which is possible since both t and cwc are $\mathscr{J}$-related. (Note that modulo the identity $xx'x \approx x$ we may assume both u and v are nonempty.) We want to obtain

$$s(c, \boldsymbol{d})ucwcv = s(cwc, \boldsymbol{d})ucwcv\,. \tag{12}$$

Now using Corollary 5.1, we may rewrite $s(c, \boldsymbol{d})$ as a word in which each application of $'$ covers either a single variable or a word of the form gh where g, h are either letters or $'$ applied to a letter. There may be many occurrences of c in this word. We show how to replace an arbitrary one of these by cwc and by repeating this to each of these occurrences we will achieve the desired equality (12). Let us fix some occurrence of c. So we may consider the expression $s(c, \boldsymbol{d})ucwc$ as being of one of the following forms: w_1cw_2cwc; $w_1c'w_2cwc$; $w_1(cz)'w_2cwc$; $w_1(c'z)'w_2cwc$; $w_1(zc')'w_2cwc$; $w_1(zc)'w_2cwc$. In each case, we can make the required replacement using a single application of Lemma 5.4. This gives equality (12), which completes the proof. □

As an immediate corollary we obtain the following result.

Corollary 5.2. *The identities Σ_{ref} are an identity basis for $\mathcal{A}_{\text{ref}}$.*

Let $\mathcal{SL}$ denote the variety generated by the adjacency semigroup over the 1-vertex looped graph and let $\mathcal{U}$ denote the variety generated by adjacency semigroups over single block equivalence relations (equivalently, $\mathcal{U}$ is the variety generated by the adjacency semigroup over the universal relation on a 2-element set). Recall that $\mathcal{SB}$ denotes the variety of square bands.

Lemma 5.6. $\mathcal{SL} \vee \mathcal{SB} = \mathcal{U}$.

Proof. The direct product of the semigroup $\mathrm{A}(\underline{1})$ with an $I \times I$ square band has a unique maximal ideal and the corresponding Rees quotient is (isomorphic to) the adjacency semigroup over the universal relation on I. So $\mathcal{SL} \vee \mathcal{SB} \supseteq \mathcal{U}$. However if $|I| \geq 2$, and U_I denotes the universal relation on I, then the adjacency semigroup $\mathrm{A}(\mathsf{U}_I) \in \mathcal{U}$ contains as subalgebras both $\mathrm{A}(\underline{1})$ (a generator for $\mathcal{SL}$) and the $I \times I$ square band (a generator for $\mathcal{SB}$). So $\mathcal{SL} \vee \mathcal{SB} \subseteq \mathcal{U}$. □

Lemma 5.7. *Let $\mathcal{V}$ be a subvariety of $\mathcal{A}_{\mathrm{ref}}$ containing the variety $\mathcal{SB}$. Either $\mathcal{V} = \mathcal{SB}$ or $\mathcal{V} \supseteq \mathcal{U}$ and $\mathcal{V} = \mathbb{HSP}(\mathrm{A}(K))$ for some class of (necessarily reflexive) graphs K.*

Proof. Let $\mathbf{A}$ be a nonfinitely generated $\mathcal{V}$-free algebra. If $\mathbf{A} \models xyx \approx x$ then $\mathcal{V}$ is equal to $\mathcal{SB}$. Now say that $xyx \approx x$ fails on $\mathbf{A}$. Lemma 5.5 shows that $\mathbf{A}$ is a subdirect product of some family J of adjacency semigroups and square bands. Note that we have $\mathcal{V} = \mathbb{HSP}(J)$. Our goal is to replace all square bands in J by adjacency semigroups over universal relations.

Since $xyx \approx x$ fails on $\mathbf{A}$, at least one of the subdirect factors of $\mathbf{A}$ is an adjacency semigroup that is not the one element algebra. Hence $\mathcal{V}$ contains the semigroup $\mathrm{A}(\underline{1})$. By Lemma 5.6, $\mathcal{V}$ contains $\mathcal{U}$. Now replace all square bands in J by the adjacency semigroup of a universal relation of some set of size at least 2, and denote the corresponding class by $\bar{J}$; let $\underline{\mathsf{G}}_{\bar{J}}$ denote the corresponding class of graphs. Then $\mathcal{V} = \mathbb{HSP}(J) = \mathbb{HSP}(\bar{J}) = \mathbb{HSP}(\mathrm{A}(\underline{\mathsf{G}}_{\bar{J}}))$. □

Now we may complete the proof of Theorem B.

Proof. Theorem A shows that the map ι described in Theorem A is an order preserving injection from $L(\underline{\mathsf{G}}_{\mathrm{ref}})$ to $L(\mathcal{A}_{\mathrm{ref}})$. Now we show that it is a surjection. That is, every subvariety of $\mathcal{A}_{\mathrm{ref}}$ other than $\mathcal{SB}$ is the image under ι of some uH class of reflexive graphs. Lemma 5.7 shows this is true if $\mathcal{SB} \subseteq \mathcal{V}$. However, if the square bands in $\mathcal{V}$ are all trivial, then Lemma 5.5 shows that either $\mathcal{V}$ is the trivial variety (and equal to $\iota(\{\mathbf{0}\})$) or the ω-generated $\mathcal{V}$-free algebra is a subdirect product of members of $\mathrm{A}(\underline{\mathsf{G}}_{\mathrm{ref}})$. Let F be a set consisting of the subdirect factors and $\underline{\mathsf{G}}_F$ the corresponding graphs. Then $\mathcal{V} = \mathbb{HSP}(F) = \iota(\mathbb{ISP}^+\mathbb{P}_{\mathrm{u}}(\underline{\mathsf{G}}_F))$. To show that ι is a lattice isomorphism, it will suffice to show that ι preserves joins, since meets follow from the fact that ι is an order preserving bijection.

Let $\bigvee_{i \in I} \underline{\mathsf{R}}_i$ be some join in L^+. First assume that $\underline{\mathsf{S}}$ is not amongst the $\underline{\mathsf{R}}_i$. Then

$$\begin{aligned}\mathbb{HSP}(\mathrm{A}(\bigvee_{i\in I} \mathsf{R}_i)) &= \mathbb{HSP}(\mathrm{A}(\mathbb{ISP}^+\mathbb{P}_{\mathrm{u}}(\bigcup_{i\in I} \mathsf{R}_i))) \\ &= \mathbb{HSP}(\mathbb{HSP}(\bigcup_{i\in I} \mathrm{A}(\mathsf{R}_i))) = \bigvee_{i\in I} \mathbb{HSP}(\mathrm{A}(\mathsf{R}_i)).\end{aligned}$$

If $\underline{\mathsf{S}}$ is amongst the $\underline{\mathsf{R}}_i$ then either the join is a join of $\underline{\mathsf{S}}$ with the trivial uH class $\{\mathbf{0}\}$ (and the join is obviously preserved by ι), or using Lemma 5.6, we can replace $\underline{\mathsf{S}}$ by the uH class of universal relations, and proceed as above. This completes the characterization of $\mathcal{L}(\mathcal{A}_{\mathrm{ref}})$.

Next we must show that a class K of graphs generates a finitely axiomatizable uH class if and only if $\mathbb{HSP}(\mathrm{A}(K))$ is finitely axiomatizable. The "only if" case is

Corollary 6.3. Now say that K has a finite basis for its uH sentences. Following the methods of Subsect. 4.3, we may construct a finite set Ξ of identities such that an adjacency semigroup A lies in $\mathbb{HSP}(\mathrm{A}(K))$ if and only if $\mathrm{A} \models \Xi$. We claim that $\Sigma_{\mathrm{ref}} \cup \Xi$ is an identity basis for $\mathbb{HSP}(\mathrm{A}(K))$. Indeed, if S is a unary semigroup satisfying $\Sigma_{\mathrm{ref}} \cup \Xi$, then by Lemma 5.5, S is a subdirect product of adjacency semigroups (or possibly square bands) satisfying Ξ. So these adjacency semigroups lie in $\mathbb{HSP}(\mathrm{A}(K))$, whence so does S.

The proof that ι preserves the property of being finitely generated (and the property of being nonfinitely generated) is very similar and left to the reader. □

6 Applications

The universal Horn theory of graphs is reasonably well developed, and the link to unary Rees matrix semigroups that we have just established provides numerous corollaries. We restrict ourselves to just a few ones which all are based on the examples of uH classes presented in Sect. 2.

We start with presenting finite generators for unary semigroup varieties that we have considered.

Proposition 6.1. *The varieties $\mathcal{A}$, $\mathcal{A}_{\mathrm{symm}}$, and $\mathcal{A}_{\mathrm{ref}}$ are generated by* $\mathrm{A}(\mathsf{G}_1)$, $\mathrm{A}(\mathsf{S}_1)$ *and* $\mathrm{A}(\mathsf{R}_1)$ *respectively.*

Proof. This follows from Theorem A and Examples 2.5, 2.6, and 2.8. □

Observe that the generators are of fairly modest size, with 17, 17 and 10 elements respectively.

Recall that C_3 is a 5-vertex graph generating the uH class of all 3-colorable graphs (Example 2.4, see also Fig. 2).

Proposition 6.2. *The finite membership problem for the variety generated by the 26-element unary semigroup* $\mathrm{A}(\mathsf{C}_3)$ *is* NP*-hard.*

Proof. Let G be a simple graph. By Theorem A the adjacency semigroup $\mathrm{A}(\mathsf{G})$ belongs to $\mathbb{HSP}(\mathrm{A}(\mathsf{C}_3))$ if and only if G is 3-colorable. Thus, we have a reduction to the finite membership problem for $\mathbb{HSP}(\mathrm{A}(\mathsf{C}_3))$ from 3-colorability of simple graphs, a known NP-complete problem, see [8]. Of course, the construction of $\mathrm{A}(\mathsf{G})$ can be made in polynomial time, so this is a polynomial reduction. □

A similar (but more complicated) example in the plain semigroup setting has been found in [14]. Observe that we do not claim that the finite membership problem for $\mathbb{HSP}(\mathrm{A}(\mathsf{C}_3))$ is NP-complete since it is not clear whether or not the problem is in NP.

One can also show that the equational theory of $\mathrm{A}(\mathsf{C}_3)$ is co-NP-complete. (It means that the problem whose instance is a unary semigroup identity $u \approx v$ and whose question is whether or not $u \approx v$ holds in $\mathrm{A}(\mathsf{C}_3)$ is co-NP-complete.) This follows from the construction of identities modelling uH sentences in Subsect. 4.3. The argument is an exact parallel to that associated with [14, Corollary 3.8] and we omit the details.

Proposition 6.3. *If K is a class of graphs without a finite basis of uH sentences, then* $\mathrm{A}(K)$ *is without a finite basis of identities. If K is a class of graphs whose uH class has (infinitely many) uncountably many sub-uH classes, then the variety generated by* $\mathrm{A}(K)$ *has (infinitely many) uncountably many subvarieties.*

Proof. This is an immediate consequence of Theorem A. □

In particular, recall the 2-vertex graph S_2 of Example 2.7, and let K_2 denote the 2-vertex complete simple graph.

Corollary 6.1. *There are uncountably many varieties between the variety generated by* $\mathrm{A}(\mathsf{S}_2)$ *and that generated by* $\mathrm{A}(\mathsf{K}_2)$*. The statement is also true if* S_2 *is replaced by any simple graph that is not* 2*-colorable.*

Proof. The first statement follows from Theorem A and Example 2.7. The second statement follows similarly from statements in Example 2.2. □

Note that the underlying semigroup of $\mathrm{A}(\mathsf{S}_2)$ is simply the familiar semigroup A_2, see Subsection 4.1. The subvariety lattice of the semigroup variety generated by A_2 is reasonably well understood (see Lee and Volkov [19]). This variety contains all semigroup varieties generated by completely 0-simple semigroups with trivial subgroups but has only countably many subvarieties, all of which are finitely axiomatized (see Lee [18]).

Theorem B reduces the study of the subvarieties of $\mathcal{A}_{\mathrm{ref}}$ to the study of uH classes of reflexive graphs. This class of graphs does not seem to have been as heavily investigated as the antireflexive graphs, but contains some interesting examples.

Recently Trotta [24] has disproved a claim made in [22] by showing that there are uncountably many uH classes of reflexive antisymmetric graphs. From this and Theorem B we immediately deduce:

Proposition 6.4. *The unary semigroup variety* $\mathcal{A}_{\mathrm{ref}}$ *has uncountably many subvarieties.*

In contrast, it is easy to check that there are only 6 uH classes of reflexive symmetric graphs, see [3] for example. The lattice they form is shown in Fig. 8 on the left. Theorem B then implies that the subvariety lattice of the corresponding variety of unary semigroups contains 7 elements (it is one of the cases when the "extra" variety $\mathcal{SB}$ of square bands comes into play); the lattice is shown in Fig. 8 on the right. The variety is generated by the adjacency semigroup of the graph RS_1 of Example 2.8 and is nothing but the variety $\mathcal{CSR}$ of so-called combinatorial strict regular *-semigroups which have been one of the central objects of study in [2]. The other join-indecomposable varieties in Fig. 8 are the trivial variety $\mathcal{T}$, the variety $\mathcal{SL}$ of semilattices with identity map as the unary operation, and the variety $\mathcal{BR}$ of combinatorial strict inverse semigroups.

The main results of [2] consisted in providing a finite identity basis for $\mathcal{CSR}$ and determining its subvariety lattice. We see that the latter result is an immediate consequence of Theorem B. A finite identity basis for $\mathcal{CSR}$ can be obtained by adding the involution identity (2) to the identity basis Σ_{ref} of the variety $\mathcal{A}_{\mathrm{ref}}$, see Corollary 5.2. (The basis constructed this way is different from that given in [2].)

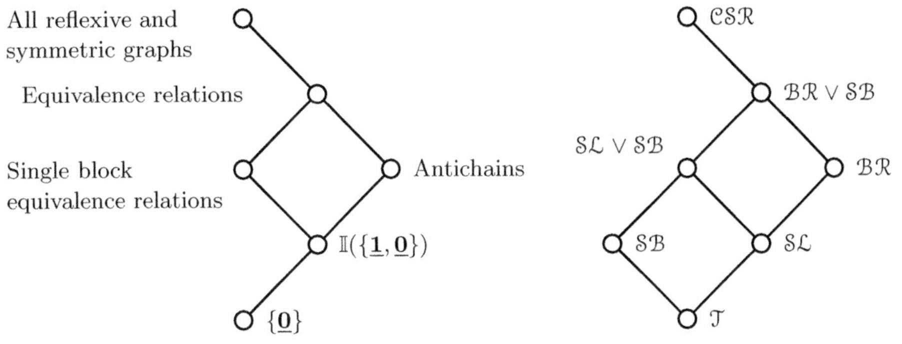

Fig. 8. The lattice of uH classes of reflexive symmetric graphs vs the lattice of varieties of strict regular semigroups

Example 6.1. The adjacency semigroup $\mathrm{A}(\underline{\mathbf{2}})$ of the two element chain $\underline{\mathbf{2}}$ (as a partial order) generates a variety with a lattice of subvarieties isomorphic to the four element chain. The variety is a cover of the variety $\mathcal{BR}$ of combinatorial strict inverse semigroups.

Proof. This follows from Example 2.1, Theorem B and the fact that the uH class of universal relations is not a sub-uH class of the partial orders (so $\mathcal{SB}$ is not a subvariety of $\mathbb{HSP}(\mathrm{A}(\underline{\mathbf{2}}))$. □

The underlying semigroup of $\mathrm{A}(\underline{\mathbf{2}})$ is again the semigroup A_2. Thus, Example 6.1 makes an interesting contrast to Corollary 6.1.

For our final application, consider the 3-vertex graph P shown in Fig. 9.

It is known (see [3]) and easy to verify that the uH-class $\mathbb{ISP}^+\mathbb{P}_\mathrm{u}(\mathsf{P})$ is not finitely axiomatizable and the class of partial orders is the unique maximal sub-uH class of $\mathbb{ISP}^+\mathbb{P}_\mathrm{u}(\mathsf{P})$. Recall that a variety $\mathcal{V}$ is said to be a *limit* variety if $\mathcal{V}$ has no finite identity basis while each of its proper subvarieties is finitely based. The existence of limit varieties is an easy consequence of Zorn's lemma but concrete examples of such varieties are quite rare. We can use the just registered properties of the graph P in order to produce a new example of a finitely generated limit variety of I-semigroups.

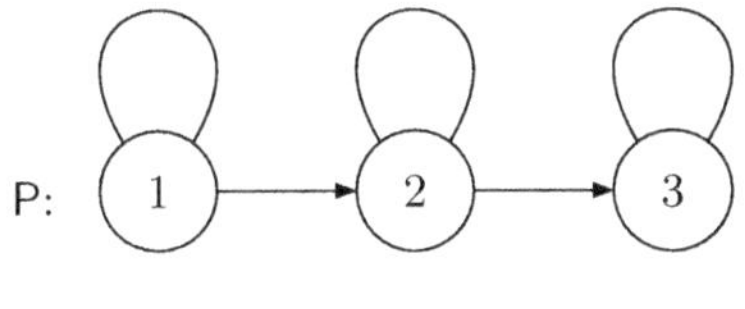

Fig. 9. Generator for a limit uH class

Proposition 6.5. *The variety* $\mathbb{HSP}(\mathsf{A}(\mathsf{P}))$ *is a limit variety whose subvariety lattice is a 5-element chain.*

Proof. This follows from Theorem B and Example 2.1. □

7 Conclusion

We have found a transparent translation of facets of universal Horn logic into the apparently much more restricted world of equational logic. A general translation of this sort has been established for uH classes of arbitrary structures (even partial structures) by the first author [13]. We think however that the special case considered in this paper is of interest because it deals with very natural objects on both universal Horn logic and equational logic sides.

We have shown that the unary semigroup variety $\mathcal{A}_{\mathrm{ref}}$ whose equational logic captures the universal Horn logic of the reflexive graphs is finitely axiomatizable. The question of whether or not the same is true for the variety $\mathcal{A}$ corresponding to all graphs still remains open. A natural candidate for a finite identity basis of $\mathcal{A}$ is the system consisting of the identities (Ψ) and (7)–(11), see Sect. 5.

References

1. Almeida, J.: Finite Semigroups and Universal Algebra. World Scientific, Singapore (1994)
2. Auinger, K.: Strict regular ∗-semigroups. In: Howie, J.M., Munn, W.D., Weinert, H.-J. (eds.) Proceedings of the Conference on Semigroups and Applications, pp. 190–204. World Scientific, Singapore (1992)
3. Benenson, I.E.: On the lattice of quasivarieties of models. Izv. vuzuv. Matematika (12), 14–20 (1979) (Russian); English translation in Soviet Math. (Iz. VUZ) 23(12), 13–21 (1979)
4. Burris, S., Sankappanavar, H.P.: A Course in Universal Algebra. Springer, Heidelberg (1981)
5. Caicedo, X.: Finitely axiomatizable quasivarieties of graphs. Algebra Universalis 34, 314–321 (1995)
6. Clark, D.M., Davey, B.A., Freese, R., Jackson, M.: Standard topological algebras: syntactic and principal congruences and profiniteness. Algebra Universalis 52, 343–376 (2004)
7. Clark, D.M., Davey, B.A., Jackson, M., Pitkethly, J.: The axiomatisability of topological prevarieties. Adv. Math. 218, 1604–1653 (2008)
8. Garey, M.R., Johnson, D.S.: Computers and Intractability: A Guide to the Theory of NP-Completeness. W.H. Freeman and Company, New York (1979)
9. Gorbunov, V.: Algebraic Theory of Quasivarieties. Consultants Bureau, New York (1998)
10. Gorbunov, V., Kravchenko, A.: Antivarieties and colour-families of graphs. Algebra Universalis 46, 43–67 (2001)
11. Hall, T.E., Kublanovsky, S.I., Margolis, S., Sapir, M.V., Trotter, P.G.: Algorithmic problems for finite groups and finite 0-simple semigroups. J. Pure Appl. Algebra 119, 75–96 (1997)

12. Howie, J.M.: Fundamentals of Semigroup Theory. London Mathematical Society Monographs, vol. 12. Clarendon Press, Oxford (1995)
13. Jackson, M.: Flat algebras and the translation of universal Horn logic to equational logic. J. Symbolic Logic 73, 90–128 (2008)
14. Jackson, M., McKenzie, R.: Interpreting graph colourability in finite semigroups. Internat. J. Algebra Comput. 16, 119–140 (2006)
15. Ježek, J., Marković, P., Maróti, M., McKenzie, R.: The variety generated by tournaments. Acta Univ. Carolin. Math. Phys. 40, 21–41 (1999)
16. Ježek, J., McKenzie, R.: The variety generated by equivalence algebras. Algebra Universalis 45, 211–219 (2001)
17. Kravchenko, A.V.: Q-universal quasivarieties of graphs. Algebra Logika 41, 311–325 (2002) (in Russian); English translation in Algebra Logic 41, 173–181 (2002)
18. Lee, E.W.H.: Combinatorial Rees–Sushkevich varieties are finitely based. Internat. J. Algebra Comput. 18, 957–978 (2008)
19. Lee, E.W.H., Volkov, M.V.: On the structure of the lattice of combinatorial Rees–Sushkevich varieties. In: André, J.M., Branco, M.J.J., Fernandes, V.H., Fountain, J., Gomes, G.M.S., Meakin, J.C. (eds.) Proceedings of the International Conference "Semigroups and Formal Languages" in honour of the 65th birthday of Donald B. MCALister, pp. 164–187. World Scientific, Singapore (2007)
20. McNulty, G.F., Shallon, C.R.: Inherently nonfinitely based finite algebras. In: Freese, R., Garcia, O. (eds.) Universal Algebra and Lattice Theory. Lecture Notes in Mathematics, vol. 1004, pp. 206–231. Springer, Heidelberg (1983)
21. Nešetřil, J., Pultr, A.: On classes of relations and graphs determined by subobjects and factorobjects. Discrete Math. 22, 287–300 (1978)
22. Sizyĭ, S.V.: Quasivarieties of graphs. Sib. Matem. Zh. 35, 879–892 (1994) (in Russian); English translation in Sib. Math. J. 35, 783–794 (1994)
23. Trahtman, A.N.: Graphs of identities of a completely 0-simple 5-element semigroup. Preprint, Ural State Technical Institute. Deposited at VINITI on 07.12.81, no.5558-81, 6 pp. (1981) (Russian); Engl. translation (entitled Identities of a five-element 0-simple semigroup) Semigroup Forum 48, 385–387 (1994)
24. Trotta, B.: Residual properties of reflexive antisymmetric graphs. Houston J. Math. (to appear)
25. Willard, R.: On McKenzie's method. Per. Math. Hungar. 32, 149–165 (1996)

Definability of Combinatorial Functions and Their Linear Recurrence Relations

Tomer Kotek* and Johann A. Makowsky**

Department of Computer Science,
Technion–Israel Institute of Technology, Haifa, Israel
{tkotek,janos}@cs.technion.ac.il

For Yuri, on the occasion of his seventieth birthday.

Abstract. We consider functions of natural numbers which allow a combinatorial interpretation as counting functions (speed) of classes of relational structures, such as Fibonacci numbers, Bell numbers, Catalan numbers and the like. Many of these functions satisfy a linear recurrence relation over $\mathbb{Z}$ or $\mathbb{Z}_m$ and allow an interpretation as counting the number of relations satisfying a property expressible in Monadic Second Order Logic (MSOL).

C. Blatter and E. Specker (1981) showed that if such a function f counts the number of binary relations satisfying a property expressible in MSOL then f satisfies for every $m \in \mathbb{N}$ a linear recurrence relation over $\mathbb{Z}_m$.

In this paper we give a complete characterization in terms of definability in MSOL of the combinatorial functions which satisfy a linear recurrence relation over $\mathbb{Z}$, and discuss various extensions and limitations of the Specker-Blatter theorem.

Keywords: Combinatorics, counting functions, monadic second order logic.

1 Introduction

1.1 The Speed of a Class of Finite Relational Structures

Let $\mathcal{P}$ be a graph property, and $\mathcal{P}^n$ be the set of graphs with vertex set $[n] = \{1, \ldots, n\}$ which have property $\mathcal{P}$. We denote by $sp_{\mathcal{P}}(n) = |\mathcal{P}^n|$ the number of labeled graphs in $\mathcal{P}^n$. The function $sp_{\mathcal{P}}(n)$ is called the *speed of* $\mathcal{P}$, or in earlier literature the *counting function of* $\mathcal{P}$[1]. Instead of graph properties we also study

---

* Partially supported by a grant of the Graduate School of the Technion–Israel Institute of Technology.

** Partially supported by a grant of the Fund for Promotion of Research of the Technion–Israel Institute of Technology and grant ISF 1392/07 of the Israel Science Foundation (2007-2010).

[1] In the recent monograph by P. Flajolet and R. Sedgewick [19] the counting function is called "counting sequence of $\mathcal{P}$". Speed is used mostly in case the counting function is monotone increasing.

A. Blass, N. Dershowitz, and W. Reisig (Eds.): Gurevich Festschrift, LNCS 6300, pp. 444–462, 2010.

classes of finite relational structures $\mathcal{K}$ with relations $R_i : i = 1, \ldots, s$ of arity ρ_i. For the case of $s = 1$ and $\rho_1 = 1$ such classes can be identified with binary words over the positions $1, \ldots, n$.

The study of the function $sp_{\mathcal{K}}(n)$ has a rich literature concentrating on two types of behaviours of the sequence $sp_{\mathcal{K}}(n)$:

- Recurrence relations
- Growth rate

Clearly, the existence of recurrence relations limits the growth rate.

(i) In formal language theory it was studied in N. Chomsky and M.P. Schuetzenberger [12] who proved that for $\mathcal{K} = L$, a regular language, the sequence $sp_L(n)$ satisfies a linear recurrence relation over $\mathbb{Z}$. This implies that the formal power series $\sum_n sp_L(n)X^n$ is rational. The paper [12] initiated the field Formal Languages and Formal Power Series.
Furthermore, it is known that L is regular iff L is definable in Monadic Second Order Logic MSOL, [11].

(ii) In C. Blatter and E. Specker [8] the case of $\mathcal{K}$ was studied, where $\rho_i \leq 2$ for all $i \leq s$ and $\mathcal{K}$ definable in MSOL. They showed that in this case for every $m \in \mathbb{N}$, the sequence $sp_{\mathcal{K}}(n)$ is ultimately periodic modulo m, or equivalently, that the sequence $sp_{\mathcal{K}}(n)$ satisfies a linear recurrence relation over $\mathbb{Z}_m$.

(iii) In E.R. Scheinerman and J. Zito [27] the function $sp_{\mathcal{P}}(n)$ was studied for *hereditary* graph properties $\mathcal{P}$, i.e., graph properties closed under induced subgraphs. They were interested in the growth properties of $sp_{\mathcal{P}}(n)$. The topic was further developed by J. Balogh, B. Bollobas and D. Weinreich in a sequence of papers, [10,1,2], which showed that only six classes of growth of $sp_{\mathcal{P}}(n)$ are possible, roughly speaking, constant, polynomial, or exponential growth, or growth in one of three factorial ranges. They also obtained similar results for monotone graph properties, i.e., graph properties closed under subgraphs, [3]. There are early precursors of the study of $sp_{\mathcal{P}}(n)$: for monotone graph properties is [16], and for hereditary graph properties, [25].
We note that hereditary (monotone) graph properties $\mathcal{P}$ are characterized by a countable set $IForb(\mathcal{P})$ ($SForb(\mathcal{P})$) of forbidden induced subgraphs (subgraphs). In the case that $IForb(\mathcal{P})$ is finite, $\mathcal{P}$ is definable in First Order Logic, FOL, and in the case that $IForb(\mathcal{P})$ is MSOL-definable, also $\mathcal{P}$ is MSOL-definable. The same holds also for monotone properties.

(iv) The classification of the growth rate of $sp_{\mathcal{P}}(n)$ was extended to minor-closed classes in [6]. We note that minor-closed classes $\mathcal{P}$ are always MSOL-definable. This is due to the Robertson-Seymour Theorem, which states that they are characterized by a finite set $MForb(\mathcal{P})$ of forbidden minors.

One common theme of all the above cited papers is the connection between the definability properties of $\mathcal{K}$ and the arithmetic properties of the sequence $sp_{\mathcal{K}}(n)$. In this paper we concentrate on the relationship between definability of a class $\mathcal{K}$ of relational structures and the various linear recurrence relations $sp_{\mathcal{K}}(n)$ can satisfy.

1.2 Combinatorial Functions and Specker Functions

We would like to say that a function $f : \mathbb{N} \to \mathbb{N}$ is a combinatorial function if it has a combinatorial interpretation. One way of making this more precise is the following.

Let $\bar{R}$ be a finite set of relation symbols and $\mathcal{K}$ be a class of finite $\bar{R}$-structures. We say that $\mathcal{K}$ is definable in $\mathcal{L}$ if there is a $\mathcal{L}$-sentence ϕ such that for every $\bar{R}$-structure $\mathfrak{A}$, $\mathfrak{A} \in \mathcal{K}$ iff $\mathfrak{A} \models \phi$. Then a function $f : \mathbb{N} \to \mathbb{N}$ is a *combinatorial function* if $f(n) = sp_{\mathcal{K}}(n)$ for some class of finite structures $\mathcal{K}$ definable in a suitable logical formalism $\mathcal{L}$. Here $\mathcal{L}$ could be FOL, MSOL or any interesting fragment of Second Order Logic, SOL. We assume the reader is familiar with these logics, cf. [14].

Definition 1 (Specker[2] function).
*A function $f : \mathbb{N} \to \mathbb{N}$ is called a $\mathcal{L}^k$-*Specker function *if there is a finite set $\bar{R}$ of relation symbols of arity at most k and a class $\mathcal{K}$ of $\bar{R}$-structures definable in $\mathcal{L}$ such that $f(n) = sp_{\mathcal{K}}(n)$.*

A typical non-trivial example is given by A. Cayley's Theorem from 1889, which says that $T(n) = n^{n-2}$ is the number of labeled trees on n vertices. Another example is the sequence of Bell numbers B_n which count the number of equivalence relations on n elements.

In this paper we study under what conditions the Specker function given by the sequence $sp_{\mathcal{K}}(n)$ satisfies a linear recurrence relation.

Example 1

(i) The number of binary relations on $[n]$ is 2^{n^2}, and the number of linear orders on $[n]$ is $n!$. Both are FOL^2-Specker functions. $n!$ satisfies the linear recurrence relation $n! = n \cdot (n-1)!$. We note the coefficient in the recurrence relation is not constant.

(ii) The Stirling numbers of the first kind denoted $\left[{n \atop k}\right]$ are defined as the number of ways to arrange n objects into k cycles. It is well known that for $n > 0$ we have $\left[{n \atop 1}\right] = (n-1)!$. Specker functions are functions in one variable. For fixed k, $\left[{n \atop k}\right]$ is a FOL^2-Specker function. Using our main results, we shall discuss Stirling numbers in more detail in Sect. 4, Proposition 6 and Corollary 2.

(iii) For the functions $2^{n^2}, n^{n-2}$ and $n!$ no linear recurrence relation with constant coefficients exists, because functions defined by linear recurrence relations with constant coefficients grow not faster than $2^{O(n)}$. However, for every $m \in \mathbb{N}$ we have that 2^{n^2} satisfies a linear recurrence relation over $\mathbb{Z}_m$, where the coefficients depend on m.

(iv) The Catalan numbers C_n count the number of valid arrangements of n pairs of parentheses. C_n is even iff n is not of the form $n = 2^k - 1$ for some $k \in \mathbb{N}$ ([21]). Therefore, the sequence C_n cannot be ultimately periodic modulo 2. We discuss the Catalan numbers in Sect. 4.

---

[2] E. Specker studied such functions in the late 1970ties in his lectures on topology at ETH-Zurich.

For R unary we can interpret $\langle [n], R\rangle$ as a binary word where position i is occupied by letter 1 if $i \in R$ and by letter 0 otherwise. Similarly, For $\bar{R} = (R_1, \ldots, R_s)$ which consists of unary relations only we can interpret $\langle [n], R_1, \ldots, R_s\rangle$ as a word over an alphabet of size 2^s. With this way of viewing languages we have the celebrated theorem of R. Büchi (and later but independently of C. Elgot and B. Trakhtenbrot), cf. [22,13] states:

Theorem 1. *Let $\mathcal{K}$ be a language and $\mathcal{K}'$ be the corresponding set of ordered structures with unary predicates for the occurrence of letters. Then $\mathcal{K}$ is regular iff $\mathcal{K}'$ is definable in* MSOL *using the natural order $<_{nat}$ on $[n]$.*

From Theorem 1 and [12] we get immediately:

Proposition 1. *If $f(n) = sp_{\mathcal{K}}(n)$ is definable in* MSOL^1, MSOL *with unary relation symbols only, given the natural order $<_{nat}$ on $[n]$, then it satisfies a linear recurrence relation over $\mathbb{Z}$*

$$sp_{\mathcal{K}}(n) = \sum_{j=1}^{d} a_j \cdot sp_{\mathcal{K}}(n-j)$$

with constant coefficients,

We say a function $f : \mathbb{N} \to \mathbb{N}$ is ultimately periodic over $\mathcal{R} = \mathbb{Z}$ or over $\mathcal{R} = \mathbb{Z}_m$ if there exist $i, n_0 \in \mathbb{N}$ such that for every $n \geq n_0$, $f(n+i) = f(n)$ over $\mathcal{R}$. It is well-known that f is ultimately periodic over $\mathbb{Z}_m$ iff it satisfies a linear recurrence relation with constant coefficients over $\mathbb{Z}_m$. We note that if f satisfies a linear recurrence over $\mathbb{Z}$ then it also satisfies a linear recurrence over $\mathbb{Z}_m$ for every m. C. Blatter and E. Specker proved the following remarkable but little known theorem in [8],[9],[30].

Theorem 2 (Specker-Blatter Theorem). *If $f(n) = sp_{\mathcal{K}}(n)$ is definable in* MSOL^2, MSOL *with unary and binary relation symbols only, then for every $m \in \mathbb{N}$, $f(n)$ satisfies a linear recurrence relation with constant coefficients*

$$sp_{\mathcal{K}}(n) \equiv \sum_{j=1}^{d_m} a_j^{(m)} sp_{\mathcal{K}}(n-j) \pmod{m},$$

and hence is ultimately periodic over $\mathbb{Z}_m$.

In [18] it was shown that in Proposition 1 and in Theorem 2 the logic MSOL can be augmented by modular counting quantifiers.

Furthermore, E. Fischer showed in [17]

Theorem 3. *For every prime $p \in \mathbb{N}$ there is an* FOL^4*-definable function $sp_{\mathcal{K}_p}(n)$, where $\mathcal{K}_p$ consists of finite (E, R)-structures with E binary and R quaternary, which is not ultimately periodic modulo p.*

The definability status of various combinatorial functions from the literature will be discussed in Sect. 4.

1.3 Formal Power Series

Our main result can be viewed as related to the theory of generating functions for formal languages, cf. [26,7] Let A be a commutative semi-ring with unity and denote by $A\langle\langle x\rangle\rangle$ the semi-ring of formal power series F in one variable over A

$$F = \sum_{n=0}^{\infty} f(n)x^n$$

where f is a function from $\mathbb{N}$ to A. A power series F in one variable is an *A-rational series* if it is in the closure of the polynomials over A by the sum, product and star operations, where the star operation F^* is defined as $F^* = \sum_{i=0}^{\infty} F^i$.

We say a function $f : \mathbb{N} \to A$ is A-rational if $\{f(n)\}_{n=0}^{\infty}$ is the sequence of coefficients of an A-rational series F. We will be interested in the cases of $A = \mathbb{N}$ and $A = \mathbb{Z}$. It is trivial that every $\mathbb{N}$-rational function is a $\mathbb{Z}$-rational function. It is well-known that $\mathbb{Z}$-rational functions are exactly those functions $f : \mathbb{N} \to \mathbb{Z}$ which satisfy linear recurrence relations over $\mathbb{Z}$. Furthermore, $\mathbb{Z}$-rational functions can also be characterized as those functions f which are the coefficents of the power series of $P(x)/Q(X)$, where $P, Q \in \mathbb{Z}[x]$ are polynomials and $Q(0) = 1$.

We aim to study Specker functions, which are by definition functions over $\mathbb{N}$. Clearly, every $\mathbb{N}$-rational function is over $\mathbb{N}$, while the $\mathbb{Z}$-rational functions may take negative values. Those non-negative $\mathbb{Z}$-rational functions which are $\mathbb{N}$-rational were characterized by M. Soittola, cf. [29]. However, there are non-negative $\mathbb{Z}$-rational series which are not $\mathbb{N}$-rational, cf. [15,4].

There are strong ties between regular languages and rational series. From Theorem 1 and [26, Thm II.5.1] it follows that:

Proposition 2. *Let $\mathcal{K}$ be a language. If $\mathcal{K}$ is definable in* MSOL *given the natural order $<_{nat}$ on $[n]$, then $sp_{\mathcal{K}}$ is $\mathbb{N}$-rational.*

1.4 Extending MSOL and Order Invariance

In this paper we investigate the existence of linear and modular linear recurrence relations of Specker functions for the case where $\mathcal{K}$ is definable in logics $\mathcal{L}$ which are sublogics of SOL and extend MSOL.

$C_{a,b}$MSOL is the extension of MSOL with modular counting quantifiers "the number of elements x satisfying $\phi(x)$ equals a modulo b". $C_{a,b}$MSOL is a fragment of *SOL* since the modular counting quantifiers are definable in SOL using a linear order of the universe which is existentially quantifed.

Example 2. The Specker function which counts the number of Eulerian graphs with n vertices is not MSOL-definable. It is definable in $C_{a,b}$MSOL and indeed $b = 2$ suffices.

We now look at the case where $[n]$ is equipped with a linear order. We want to count the number of relation without counting different orders.

Definition 2 (Order invariance)

(i) *A class $\mathcal{D}$ of* ordered $\bar{R}$-structures *is a class of $\bar{R} \cup \{<_1\}$-structures, where for every $\mathfrak{A} \in \mathcal{D}$ the interpretation of the relation symbol $<_1$ is always a linear order of the universe of $\mathfrak{A}$.*

(ii) *An $\mathcal{L}$ formula $\phi(\bar{R}, <_1)$ for ordered $\bar{R}$-structures is* truth-value order invariant (t.v.o.i.) *if for any two structures $\mathfrak{A}_i = \langle [n], <_i, \bar{R} \rangle$ $(i = 1, 2)$ we have that $\mathfrak{A}_1 \models \phi$ iff $\mathfrak{A}_2 \models \phi$. Note $\mathfrak{A}_1$ and $\mathfrak{A}_2$ differ only in the linear orders $<_1$ and $<_2$ of $[n]$. We denote by* TV$\mathcal{L}$ *the set of $\mathcal{L}$-formulas for ordered $\bar{R}$-structures which are t.v.o.i. We consider* TV$\mathcal{L}$ *formulas as formulas for $\bar{R}$-structures.*

(iii) *For a class of ordered structures $\mathcal{D}$, let*

$$osp_{\mathcal{D}}(n, <_1) = \left| \{(R_1, \ldots, R_s) \subseteq [n]^{\rho(1)} \times \ldots \times [n]^{\rho(s)} : \langle [n], <_1, R_1, \ldots, R_s \rangle \in \mathcal{D}\} \right| .$$

A function $f : \mathbb{N} \to \mathbb{N}$ is called an $\mathcal{L}^k$-ordered Specker function *if there is a class of ordered $\bar{R}$-structures $\mathcal{D}$ of arity at most k definable in $\mathcal{L}$ such that $f(n) = osp_{\mathcal{D}}$.*

(iv) *A function $f : \mathbb{N} \to \mathbb{N}$ is called a* counting order invariant (c.o.i.) $\mathcal{L}^k$-Specker function *if there is a finite set of relation symbols $\bar{R}$ of arity at most k and a class of ordered $\bar{R}$-structures $\mathcal{D}$ definable in $\mathcal{L}$ such that for all linear orders $<_1$ and $<_2$ of $[n]$ we have $f(n) = osp_{\mathcal{D}}(n, <_1) = osp_{\mathcal{D}}(n, <_2)$.*

Example 3

(i) Every formula $\phi(\bar{R}, <_1) \in \text{TVSOL}^k$ is equivalent to the formula $\psi(\bar{R}) = \exists <_1 \phi(\bar{R}, <_1) \wedge \phi_{linOrd}(<_1) \in \text{SOL}^k$, where $\phi_{linOrd}(<_1)$ says $<_1$ is a linear ordering of the universe.

(ii) Every TVMSOL^k-Specker function is also a counting order invariant MSOL^k-Specker function.

(iii) We shall see in Sect. 3 that there are counting order invariant MSOL^2-definable Specker functions which are not TVMSOL^2-definable.

The following proposition is folklore:

Proposition 3. *Every formula in $C_{a,b}$MSOL^k is equivalent to a formula in* TVMSOL^k.

Proof. We give a sketch of the proof for the $C_{a,b}$MSOL formula $\phi_{even} = C_{0,2}(x = x)$, which says the size of the universe is even. The general proof is similar. ϕ_{even} can be written as $\phi(\bar{R}, <_1) = \exists U \phi_{min}(U) \wedge \forall x \forall y (\phi_{succ}(x, y) \to (x \in U \leftrightarrow y \notin U)) \wedge \forall x (x \in U \to \exists y \; x <_1 y)$ where $\phi_{min}(U) = \forall x ((\neg \exists y \; y <_1 x) \to x \in U)$ says the minimal element x in the order $<_1$ belongs to U, and $\phi_{succ}(x, y) = (x <_1 y) \wedge \neg \exists z (x <_1 z \wedge z <_1 y)$ says y is the successor of x in $<_1$. □

1.5 Main Results

Our first result is a characterization of functions over the natural numbers which satisfy a linear recurrence relation over $\mathbb{Z}$.

Theorem 4. *Let f be a function over $\mathbb{N}$. Then the following are equivalent:*

(i) $f(n)$ *satisfies a linear recurrence relation over* $\mathbb{Z}$*;*
(ii) $f(n) = f_1(n) - f_2(n)$ *is the difference of two counting order invariant* MSOL^1*-Specker functions;*
(iii) $f(n) = d_{L_1}(n) - d_{L_2}(n)$ *is the difference of two counting functions* $d_{L_i}(n)$, $i = 1, 2$, *of some regular languages* L_1 *and* L_2.
(iv) $f(n) = d_{L_3}(n) - c^n$ *for some constant* $c \in \mathbb{N}$ *and some regular language* L_3.

We shall only prove that (i) implies (ii), the other cases are either known or easily derived.

In the terminology of rational functions we get the following corollary:

Corollary 1. *Let f be a function $\mathbb{N} \to \mathbb{N}$. Then f is $\mathbb{Z}$-rational iff f is the difference of two $\mathbb{N}$-rational functions.*

Corollary 1, as stated, is known, cf. [26, Corollary 8.2. in Chapter II]. However, to prove Theorem 4 from Corollary 1, one would need that for every $\mathbb{N}$-rational function $\alpha(x) = \sum_{i=0}^{\infty} a(n)x^n$ the function $a(n)$ is the counting function of some regular language (*). To the best of our knowledge, this is not known to be true. The characterization of regular languages via generating functions uses the multivariate version, see [26].

In the proof of Theorem 4 we introduce the notion of Specker polynomials, which can be thought of as a special case of graph polynomials where graphs are replaced by linear orders.

Next we show that the Specker-Blatter Theorem cannot be extended to counting order invariant Specker functions which are definable in MSOL^2. More precisely:

Proposition 4. *Let $E_{2,=}(n)$ be the number of equivalence relations with two equal-sized equivalence classes. Then $E_{2,=}(2n) = \frac{1}{2}\binom{2n}{n}$, and $E_{2,=}(2n+1) = 0$. $E_{2,=}$ is a counting order invariant MSOL^2-definable. However, it does not satisfy a linear recurrence relation over $\mathbb{Z}_2$, since it is not ultimately periodic modulo 2. To see this note that $E_{2,=}(2n) = 0 \pmod 2$ iff n is an even power of 2.*

In Sect. 4 we shall show in Corollary 3 the same also for the Catalan number.

However, if we require that the defining formula ϕ of a Specker function is itself order invariant, i.e. $\phi \in \mathrm{TVMSOL}^2$, then the Specker-Blatter Theorem still holds.

Theorem 5. *Let f be a TVMSOL^2-Specker function. Then, for all $m \in \mathbb{N}$ the function f satisfies a modular linear recurrence relation modulo m.*

Table 1. Linear recurrences and definability of $\mathcal{L}^k$-Specker functions

k	MSOL^k	$C_{a,b}$MSOL^k	TVMSOL^k	c.o.i.MSOL^k
4	*No MLR*	*No MLR*	*No MLR*	*No MLR*
3	?	?	?	*No MLR*
2	*MLR* **No LR**	*MLR* **No LR**	*MLR* **No LR**	**No MLR**
1	**All functions with LR**			

Table 1 summarizes the relationship between definablity of a $\mathcal{L}^k$-Specker function $f(n)$ and existence of linear recurrence. We denote by MLR that $f(n)$ has a modular linear recurrence (for every $m \in \mathbb{N}$) and by LR that $f(n)$ satisfies a linear recurrence over $\mathbb{Z}$. We write $NO\ LR$ (respectively $NO\ MLR$) to indicate that there is some $\mathcal{L}^k$-Specker function without a linear recurrence over $\mathbb{Z}$ (respectively $\mathbb{Z}_m$, for some $m \in \mathbb{N}$). The entries in bold face are new.

2 Linear Recurrence Relations for $\mathcal{L}^k$-Specker Functions

To prove Theorem 4 we first introduce Specker polynomials and prove a generalized version of one direction of the theorem in Subsect. 2.1. We finish this direction of the proof of Theorem 4 in Subsect. 2.2. The other direction of Theorem 4 is easy and is also given in Subsect. 2.2.

2.1 $\mathcal{L}^k$-Specker Polynomials

Definition 3

(i) A $\mathcal{L}^k$-Specker polynomial $A(n,\bar{x})$ in indeterminate set $\bar{x}$ has the form

$$\sum_{R_1:\Phi_1(R_1)} \cdots \sum_{R_t:\Phi_t(R_1,\ldots,R_t)} \left(\prod_{v_1,\ldots,v_k:\Psi_1(\bar{R},\bar{v})} x_{m_1} \cdots \prod_{v_1,\ldots,v_k:\Psi_l(\bar{R},\bar{v})} x_{m_l} \right)$$

where $\bar{v}$ stands for $(v_1,\ldots,v_k)$, $\bar{R}$ stands for $(R_1,\ldots,R_t)$ and the R_i's are relation variables of arity ρ_i at most k. The R_i's range over relations of arity ρ_i over $[n]$ and the v_i range over elements of $[n]$ satisfying the iteration formulas $\Phi_i, \Psi_i \in \mathcal{L}$.

(ii) Simple ordered $\mathcal{L}^k$-Specker polynomials and order invariance thereof are defined analogously to Specker functions.

Every Specker function can be viewed as a Specker polynomial in zero indeterminates. Conversely, if we evaluate a Specker polynomial at $x = 1$ we get a Specker function.

In this subsection we prove a stronger version of Theorem 4.

Lemma 1. *Let $A(n, \bar{z})$ be a c.o.i.* MSOL[1]*-Specker polynomial with indeterminates $\bar{z} = (z_1, \ldots, z_s)$ and let $h_1(\bar{w}), \ldots, h_s(\bar{w}) \in \mathbb{Z}[\bar{w}]$. Let $A(n, (h_1(\bar{w}), \ldots, h_s(\bar{w})))$ denote the variable subtitution in $A(n, \bar{z})$ where for $i \in [s]$, z_i is substituted to $h_i(\bar{w})$. Then $A\left(n, \bar{h}\right)$ is an integer evaluation of a c.o.i.* MSOL[1]*-Specker polynomial.*

Proof. We look at $A(n, \bar{z})$ with z_1 substituted to the polynomial

$$h_1(\bar{w}) = \sum_{j=1}^{d} c_j w_1^{\alpha_{j1}} \cdots w_t^{\alpha_{jt}}$$

where $d, \alpha_{11}, \ldots, \alpha_{dt} \in \mathbb{N}$ and $c_1, \ldots, c_d \in \mathbb{Z}$. The c.o.i. MSOL[1]–Specker polynomial $A(n, \bar{z})$ is given by

$$\sum_{R_1:\Phi_1(R_1)} \cdots \sum_{R_m:\Phi_m(R_1,\ldots,R_{m-1})} \left(\prod_{v_1:\Psi_1(\bar{R},v_1)} z_1 \cdots \prod_{v_1:\Psi_s(\bar{R},v_1)} z_s \right),$$

so substituting z_1 to $h_1(\bar{w})$ we get $A(n, (h_1(w), z_2, \ldots, z_s)) =$

$$\sum_{R_1:\Phi_1(R_1)} \cdots \sum_{R_m:\Phi_m(R_1,\ldots,R_{m-1})} \left(\prod_{v_1:\Psi_1(\bar{R},v_1)} h_1(\bar{w}) \cdots \prod_{v_1:\Psi_s(\bar{R},v_1)} z_s \right).$$

We note that for every $\alpha(v) \in$ MSOL we can define an MSOL formula with d unary relation variables $\phi_{Part(\alpha)}(U_1 \ldots, U_d)$ which holds iff $U_1, \ldots, U_d$ are a partition of the set of elements of $[n]$ which satisfy $\alpha(v)$. Then $A\left(n, (h_1(\bar{w}), z_2, \ldots, z_s)\right) =$

$$\sum_{R_1:\Phi_1(R_1)} \cdots \sum_{R_m:\Phi_m(R_1,\ldots,R_{m-1})} \sum_{U_1,\ldots,U_d:\phi_{Part(\Psi_1)}(\bar{U})} \left(\prod_{v_1:\Psi_2(\bar{R},v_1)} z_2 \cdots \right.$$
$$\left. \prod_{v_1:\Psi_s(\bar{R},v_1)} z_s \prod_{v_1:v_1\in U_1} c_1 w_1^{\alpha_{11}} \cdots w_t^{\alpha_{1t}} \cdots \prod_{v_1:v_1\in U_d} c_d w_1^{\alpha_{d1}} \cdots w_t^{\alpha_{dt}} \right)$$

Next, we note for any formula θ,

$$\prod_{v_1:\theta} c_j w_1^{\alpha_{j1}} \cdots w_t^{\alpha_{jt}} = \prod_{v_1:\theta} c_j \overbrace{\prod_{v_1:\theta} w_1 \cdots \prod_{v_1:\theta} w_1}^{\alpha_{j1} \text{ times}} \cdots \overbrace{\prod_{v_1:\theta} w_t \cdots \prod_{v_1:\theta} w_t}^{\alpha_{jt} \text{ times}}.$$

We now replace all c_j with new indeterminates w'_j and thus obtain that $A(n, (h_1(\bar{w}), z_2, \ldots, z_s))$ is an evaluation of an c.o.i. MSOL^1-Specker polynomial.

Doing the same for the other z_i, we get that $A(n, (h_1(\bar{w}), \ldots, h_s(\bar{w})))$ is an evaluations of an o.i. MSOL^1-definable Specker polynomial, as required. □

Theorem 6. *Let $A_n(\bar{x})$ be a sequence of polynomials with a finite indeterminate set $\bar{x} = (x_1, \ldots, x_s)$ which satisfies a linear recurrences over $\mathbb{Z}$. Then there exists a c.o.i.* MSOL^1*-Specker polynomial $A'(n, \bar{x}, \bar{y})$ such that $A_n(\bar{x}) = A'(n, \bar{x}, \bar{a})$ where $\bar{a} = (a_1, \ldots, a_l)$ and $a_i \in \mathbb{Z}$ for $i = 1, \ldots, l$.*

Proof. Let $A_n(\bar{x})$ be given by a linear recurrence

$$A_n(\bar{x}) = \sum_{i=1}^{r} f_i(\bar{x}) \cdot A_{n-i}(\bar{x}),$$

where $f_i(\bar{x}) \in \mathbb{Z}[\bar{x}]$ and initial conditions $A_1(\bar{x}), \ldots, A_r(\bar{x}) \in \mathbb{Z}[\bar{x}]$. To write $A_n(\bar{x})$ as a c.o.i. MSOL^1-Specker polynomials, we sum over the paths of the recurrence tree. A path in the recurrence tree corresponds to the successive application of the recurrence

$$A_n(\bar{x}) \to A_{n-i_1}(\bar{x}) \to A_{n-i_1-i_2}(\bar{x}) \to \cdots \to A_{n-i_1-\ldots-i_l}(\bar{x})$$

where $i_1, \ldots, i_l \in [r]$ and $A_{n-i_1-\ldots-i_l}(\bar{x})$ is an initial condition.

In the following, the U_i for $i \in [r]$ stand for the vertices in the path, I_i for $i \in [r]$ stand for initial conditions $A_i(\bar{x})$, and S stands for all those elements of $[n]$ skipped by the recurrence. We may write $A_n(\bar{x})$ as

$$A_n(\bar{x}) = \sum_{\bar{U},\bar{I},S:\phi_{rec}(\bar{U},\bar{I},S)} \prod_{v:v\in U_1} f_1(\bar{x}) \cdots \prod_{v:v\in U_r} f_r(\bar{x}) \prod_{v:v\in I_1} A(1,\bar{x}) \cdots \prod_{v:v\in I_r} A(r,\bar{x}),$$

where $\phi_{rec}(\bar{U}, \bar{I}, S)$ says

- $\phi_{Part}(\bar{U}, \bar{I}, S)$ holds, i.e. $\bar{U}, \bar{I}, S$ is a partition of $[n]$,
- $n \in \bigcup U_i$, i.e. the path in the recurrence tree starts from n,
- $|\bigcup_{i=1}^{r} I_i| = 1$, i.e. the path reaches exactly one initial condition
- if $v \in [n] - [r]$, then $v \notin \bigcup_{i=1}^{r} I_i$, i.e. the path may not reach an initial condition until $v \in [r]$,
- if $v \in [r]$, then $v \notin \bigcup_{i=1}^{r} U_i$, i.e. the path ends when reaching the initial conditions, and
- for every $v \in U_i$, $\{v-1, \ldots, v-(i-1)\} \subseteq S$ and $v - i \in \bigcup_{i=1}^{r} (U_i \cup I_i)$, i.e. the next element in the path is $v - i$.

The formula ϕ_{rec} is MSOL definable using the given order. Let $B(n, \bar{x})$ be

$$B(n, \bar{z}) = \sum_{\bar{U},\bar{I},S:\phi_{rec}(\bar{U},\bar{I},S)} \prod_{v:v\in U_1} z_1 \cdots \prod_{v:v\in U_r} z_r \prod_{v:v\in I_1} z_{r+1} \cdots \prod_{v:v\in I_r} z_{2r}.$$

Then $B(n, \bar{z})$ is a c.o.i. MSOL^1-Specker polynomial. By Lemma 1, substituting z_i to $f_i(\bar{x})$ for $i \in [r]$ and to $A_{i-r}(\bar{x})$ for $i \in [2r]\backslash[r]$, we have that

$$B\left(n, (f_1(\bar{x}), \ldots, f_r(\bar{x}), A(1, \bar{x}), \ldots, A(r, \bar{x}))\right) = A_n(\bar{x})$$

is an evaluation in $\mathbb{Z}$ of a c.o.i. MSOL^1-Specker polynomial. □

2.2 Proof of Theorem 4

Let $f = f_1 - f_2$ and f_1 and f_2 be c.o.i. MSOL^1-Specker functions. By Proposition 1 together with Theorem 1 we have that f_1 and f_2 satisfy linear recurrence relations over $\mathbb{Z}$. It is well known that finite sums, differences and products of functions satisfying a linear recurrence relation again satisfy a linear recurrence relation, cf. [23, Chapter 8] or [28, Chapter 6]. Thus, $f = f_1 - f_2$ satisfies a linear recurrence relation over $\mathbb{Z}$.

Conversely, if f satisfies a linear recurrence relation over $\mathbb{Z}$, then by Theorem 6, f is given by an evaluation $\bar{a} = (a_1, \ldots, a_l)$ where $a_i \in \mathbb{Z}$ for $i = 1, \ldots, l$ of a c.o.i. MSOL^1 Specker polynomial $A(n, \bar{y})$ in variables y_i. We have to show that f is a difference of two c.o.i. MSOL^1-Specker functions. For the sake of simplicity we will show this only for the case of a MSOL^1-Specker polynomial in one indeterminate,

$$A(n, y) = \sum_{R:\Phi(R)} \prod_{v_1:\Psi(R,v_1)} y\,.$$

The general case is similar. We may write $A(n, y)$ as

$$A(n, y) = \sum_{R,Y:\Phi(R)\wedge\Psi'(R,Y)} \prod_{v:v\in Y} y\,,$$

where $\Psi'(Y) = \forall v\,(v \in Y \leftrightarrow \Psi(R, v))$. For $a > 0$ we can write $\prod_{v:v\in Y} a$ as

$$\prod_{v:v\in Y} a = a^{|Y|} = |\{Z_1, \ldots Z_a \mid Z_1, \ldots, Z_a \text{ form a partition of } Y\}|\,.$$

So,

$$A(n, a) = \sum_{R,Y,\bar{Z}:\beta_a(R,Y,\bar{Z})} 1 = \left|\{R, Y, \bar{Z} \mid \beta_a(R, Y, \bar{Z})\}\right|\,,$$

where $\beta_a(R, Y, \bar{Z}) = \Phi(R) \wedge \Psi'(R, Y) \wedge \phi_{part}(Y, \bar{Z})$ and $\phi_{part}(Y, Z_1, \ldots, Z_a)$ says $Z_1, \ldots, Z_a$ form a partition of Y. We note that ϕ_{part} is definable in MSOL. For $a = 0$,

$$A(n, a) = \sum_{R:\gamma(R)} 1 = |\{R \mid \gamma(R)\}|\,,$$

where $\gamma(R) = \Phi(R) \wedge \forall v_1 \neg\Psi(R, v_1)$. Thus, since the constant function 0 is definable in MSOL, we get that if $a \geq 0$ then $A(n, a)$ is the difference of two c.o.i. MSOL^1-Specker functions.

For $a < 0$ we have

$$A(n,a) = \sum_{R,Y:\Phi(R)\wedge\Psi'(R,Y)} \prod_{v:v\in Y} |a| \prod_{v:v\in Y} (-1).$$

As above, we may write $A(n,a)$ as

$$A(n,a) = \sum_{R,Y,\bar{Z}:\beta_{|a|}(R,Y,\bar{Z})} \prod_{v:v\in Y} (-1),$$

and we have

$$A(n,a) = \sum_{R,Y,\bar{Z}:\alpha_{Even}(Y)\wedge\beta_{|a|}(R,Y,\bar{Z})} 1 - \sum_{R,Y,\bar{Z}:\neg\alpha_{Even}(Y)\wedge\beta_{|a|}(R,Y,\bar{Z})} 1,$$

where $\alpha_{Even}(Y)$ says $|Y|$ is even. Thus, $A(n,a)$ is given by $A(n,a) =$

$$\left|\{R,Y,\bar{Z} \mid \alpha_{Even}(Y) \wedge \beta_{|a|}(R,Y,\bar{Z})\}\right| - \left|\{R,Y,\bar{Z} \mid \neg\alpha_{Even}(Y)\wedge\beta_{|a|}(R,Y,\bar{Z})\}\right|.$$

Since α_{Even} is definable in MSOL given an order, as discussed in example 3, we get that $A(n,a)$ is a difference of two c.o.i. MSOL^1-Specker functions for $a < 0$.

3 Modular Linear Recurrence Relations

In this section we prove Theorem 5, the extension of the Specker-Blatter Theorem to TVMSOL^2-Specker functions. We also prove Proposition 4, which shows Theorem 5 cannot be extended to c.o.i. MSOL^2-Specker functions.

3.1 Specker Index

We say a structure $\mathcal{A} = \langle [n], \bar{R}, a\rangle$ is a pointed $\bar{R}$-structure if is consists of a universe $[n]$, relations $R_1, \ldots, R_k$, and an element $a \in [n]$ of the universe. We now define a binary operation on pointed structures. Given two pointed structures $\mathcal{A}_1 = \langle [n_1], \bar{R}^1, a_1\rangle$ and $\mathcal{A}_2 = \langle [n_2], \bar{R}^2, a_2\rangle$, let $Subst(\mathcal{A}_1, \mathcal{A}_2)$ be a new pointed structure $Subst(\mathcal{A}_1, \mathcal{A}_2) = \mathcal{B}$ where $\mathcal{B} = \langle [n_1] \sqcup [n_2] - \{a_1\}, \bar{R}, a_2\rangle$, such that the relations in $\bar{R}$ are defined as follows. For every $R_i \in \bar{R}$ of arity r, $R_i = \left(R_i^1 \cap ([n_1] - \{a_1\})^r\right) \cup R_i^2 \cup I$, where I contains all possibilities of replacing occurrences of a_1 in R_i^1 with members of $[n_2]$.

Similarly, we define $Subst(\mathcal{A}_1, \mathcal{A}_2)$ for a pointed structure $\mathcal{A}_1$ and a structure $\mathcal{A}_2 = \langle [n], \bar{R}\rangle$ (which is not pointed). Let $\mathcal{C}$ be a class of possibly pointed $\bar{R}$-structures. We define an equivalence relation between $\bar{R}$-structures:

- We say $\mathcal{A}_1$ and $\mathcal{A}_2$ are equivalent, denoted $\mathcal{A}_1 \sim_{Su(\mathcal{C})} \mathcal{A}_2$ if for every pointed structure $\mathcal{D}$ we have that $Subst(\mathcal{D}, \mathcal{A}_1) \in \mathcal{C}$ if and only if $Subst(\mathcal{D}, \mathcal{A}_2) \in \mathcal{C}$.
- The *Specker index* of $\mathcal{C}$ is the number of equivalence classes of $\sim_{Su(\mathcal{C})}$.

We use in the next subsection the following lemmas by E. Specker [30]:

Lemma 2. *Let $\mathcal{C}$ be a class of $\bar{R}-$structures of finite Specker index with all relation symbols in $\bar{R}$ at most binary. Then $f_{\mathcal{C}}(n)$ satisfies modular linear recurrence relations for every $m \in \mathbb{N}$.*

Lemma 3. *If $\mathcal{C}$ is a class of $\bar{R}$-structures which is* MSOL^2-*definable, then $\mathcal{C}$ has finite Specker index.*

3.2 Proof of Theorem 5

We prove the following lemma:

Lemma 4. *If $\mathcal{C}$ is a class of $\bar{R}$-structures which is* TVMSOL^2*-definable, then $\mathcal{C}$ has finite Specker index.*

Proof. Let $\mathcal{C}$ be a set of $\bar{R}$-structures defined by the $\mathrm{TVMSOL}(\bar{R})$ formula ϕ. Let $\mathcal{C}'$ be the class of all $\bar{R} \cup R_<$-structures $\langle \mathcal{A}, R_< \rangle$ such that $\mathcal{A} \in \mathcal{C}$ and $R_<$ is a linear ordering of the universe of $\mathcal{A}$. Let ϕ' be the $\mathrm{MSOL}(\bar{R} \cup \{R_<\})$ formula obtained from ϕ by the following changes:

(i) the order used in ϕ, $a <_1 b$, is replaced with the new relation symbol $R_<$
(ii) it is required that $R_<$ is a linear ordering of $[n]$.

We note that ϕ' defines $\mathcal{C}'$, since ϕ is truth-value order invariant and that ϕ' is an MSOL^2-formula.

We will now prove that $\mathcal{C}$ has finite Specker index, by showing that if it does not, then $\mathcal{C}'$ also has infinite Specker index, in contradiction to Lemma 3. Assume $\mathcal{C}$ has infinite Specker index. Then there is an infinite set W of $\bar{R}-$structures, such that for every distinct $\mathcal{A}_1, \mathcal{A}_2 \in W$, $\mathcal{A}_1 \not\sim_{Su(C)} \mathcal{A}_2$. So, for every $\mathcal{A}_1, \mathcal{A}_2 \in W$ there is $\langle \mathcal{G}, s \rangle$ such that

$$Subst(\langle \mathcal{G}, s \rangle, \mathcal{A}_1) \in \mathcal{C} \text{ iff } Subst(\langle \mathcal{G}, s \rangle, \mathcal{A}_2) \notin \mathcal{C}\,.$$

Now look at $W' = \{\langle \mathcal{A}, R_< \rangle \mid \mathcal{A} \in W, R_< \text{ linear order of } [n]\}$, where $[n]$ is the universe of $\mathcal{A}$. We note $Subst(\langle \mathcal{G}, R_{<_\mathcal{G}}, s \rangle, \langle \mathcal{A}_1, R_{<_{\mathcal{A}_1}} \rangle) = \langle Subst(\mathcal{G}, \mathcal{A}_1), R_{<'} \rangle$, where $R_{<'}$ a linear ordering of the universe of $Subst(\mathcal{G}, \mathcal{A}_1)$ which extends $R_{\mathcal{A}_1}$ and $R_\mathcal{G}$, and similarly for $\mathcal{A}_2$. Therefore,

$$Subst(\langle \mathcal{G}, R_{<_\mathcal{G}}, s \rangle, \langle \mathcal{A}_1, R_{<_{\mathcal{A}_1}} \rangle) \in \mathcal{C}' \text{ iff } Subst(\langle \mathcal{G}, R_{<_\mathcal{G}}, s \rangle, \langle \mathcal{A}_2, R_{<_{\mathcal{A}_2}} \rangle) \notin \mathcal{C}'\,.$$

So the Specker index of $\mathcal{C}'$ is infinite, in contradiction. □

Theorem 5 now follows from lemma 2.

3.3 Counting Order Invariant MSOL^2

Here we show the Specker-Blatter Theorem does not hold for c.o.i. MSOL^2-definable Specker functions. We have two such examples, the function $E_{2,=}$, as defined in Proposition 4, and the Catalan numbers, which we discuss in Sect. 4.

More precisely, here we show:

Proposition 5. *$E_{2,=}$, as defined in Proposition 4 is a c.o.i.* MSOL^2*-Specker function.*

Proof. Let $\mathcal{C}$ be defined as follows:

$$\mathcal{C} = \{\langle U, R, F\rangle \mid \langle [n], <_1, U, R, F\rangle \models \Phi\} ,$$

where U and R are unary and F is binary, $<_1$ is a linear order of $[n]$, and Φ is says

(i) F is a function,
(ii) U is the domain of F,
(iii) R is the range of F,
(iv) U and R form a partition of $[n]$,
(v) the first element of $[n]$, is in U,
(vi) $F : U \to R$ is a bijection, and
(vii) F is monotone with respect to $<_1$.

We note $\mathcal{C}$ is MSOL^2 definable. We note also that $ospc_{\mathcal{C}}(n, <_1)$ is counting order invariant. $ospc_{\mathcal{C}}(n, <_1)$ counts the number of partitions of $[n]$ into two equal parts, because there is exactly one monotone bijection between any two subsets of $[n]$ of equal size. The condition that $1 \in U$ assures that we do not count the same partition twice. So $ospc_{\mathcal{C}}(n, <_1) = E_{2,=}(n)$. □

We know that $E_{2,=}$ is not ultimately periodic modulo 2, and hence the Specker-Blatter theorem cannot be extended to c.o.i. MSOL^2-Specker functions.

4 Examples

4.1 Examples of MSOL^k-Specker Functions

Fibonacci and Lucas Numbers. The Fibonacci numbers F_n satisfy the linear recurrence $F_n = F_{n-1} + F_{n-2}$ for $n > 1$, $F_0 = 0$ and $F_1 = 1$. The Lucas numbers L_n, a variation of the Fibonacci numbers, satisfy the same recurrence for $n > 1$, $L_n = L_{n-1} + L_{n-2}$, but have different initial conditions, $L_1 = 1$ and $L_0 = 2$.

It follows from the proof of Theorem 4 that a function which satisfies a linear recurrence relation over $\mathbb{N}$ is a c.o.i. MSOL^1-Specker function. Thus. The Fibonacci and Lucas numbers are natural examples of c.o.i.-MSOL^1-Specker functions.

Stirling Numbers. The Stirling numbers of the first kind, denoted $\left[{n \atop r}\right]$ are defined as the number of ways to arrange n objects into r cycles. For fixed r, this is an MSOL^2-Specker function, since for $E \subseteq [n]^2$ and $U \subseteq E$, the property that U is a cycle in E and the property that E is a disjoint union of cycles are both MSOL^2-definable. Using again the growth argument from Example 1(iii), we can see that the Stirling numbers of the first kind do not satisfy a linear recurrence relation, because $\left[{n \atop 1}\right]$ grows like the factorial $(n-1)!$. However, from the Specker-Blatter Theorem it follows that they satisfy a modular linear recurrence relation for every m.

The Stirling numbers of the second kind, denoted $\left\{{n \atop r}\right\}$, count the number of partitions of a set $[n]$ into r many non-empty subsets. For fixed r, this is $MSOL^2$-definable: We count the number of equivalence relations with r non-empty equivalence classes. From the Specker-Blatter Theorem it follows that they satisfy a modular linear recurrence relation for every m. We did not find in the literature a linear recurrence relation for the Stirling numbers of the second kind which fits our context. But we show below that such a recurrence relation exists.

Proposition 6. *For fixed r, the Stirling numbers of the second kind are c.o.i. $MSOL^1$-Specker functions.*

Proof. We use r unary relations $U_1, \ldots, U_r$ and say that they partition the set $[n]$ into non-empty sets. However, when we permute the indices of the U_i's we count two such partitions twice. To avoid this we use a linear ordering on $[n]$ and require that, with respect to this ordering, the minimal element in U_i is smaller than all the minimal elements in U_j for $j > i$. □

Corollary 2. *For every r there exists a linear recurrence relation with constant coefficients for the Stirling numbers of the second kind $\left\{{n \atop r}\right\}$. Further more there are constants c_r such that $\left\{{n \atop r}\right\} \leq 2^{c_r \cdot n}$.*

Our proof is not constructive, and we did not bother here to calculate the explicit linear recurrence relations or the constants c_r for each r.

Catalan Numbers. Catalan numbers were defined in Sect. 1 Example 1. We already noted that they do not satisfy any modular linear recurrence relation. However, like the example $E_{2,=}$, the functions $f_c(n) = C_n$ is a c.o.i. $MSOL^2$-Specker function. To see this we use the following interpretation of Catalan numbers given in [20].

C_n counts the number of tuples $\bar{a} = (a_0, \ldots, a_{2n-1}) \in [n]^{2n}$ such that

(i) $a_0 = 1$
(ii) $a_{i-1} - a_i \in \{1, -1\}$ for $i = 1, \ldots, 2n-2$
(iii) $a_{2n-1} = 0$

We can express this in $MSOL^2$ using a linear order and two unary functions. The two functions F_1 and F_2 are used to describe $a_0, \ldots, a_{n-1}$ and $a_n, \ldots, a_{2n-1}$ respectively. Let $\Phi_{Catalan}$ be the formula that says:

(i) $F_1, F_2 : [n] \to [n]$
(ii) $F_i(x+1) = F_i(x) \pm 1$ for $i = 1, 2$ and there exists $y = x + 1 \in [n]$
(iii) $F_1(n-1) = F_2(0) \pm 1$.
(iv) $F_1(0) = 1$
(v) $F_2(n-1) = 0$

The resulting formula is not t.vo.i., but $C_n = sp_{\mathcal{C}}(n)$ where

$$\mathcal{C} = |\{(F_1, F_2) \mid \langle [n], <_1, F_1, F_2 \rangle \models \Phi_{Catalan}\}|$$

is a c.o.i. MSOL^2-Specker function.

Corollary 3. *The function $f(n) = C_n$ is a c.o.i. MSOL^2-Specker function and does not satisfy a modular linear recurrence relation modulo 2.*

Bell Numbers. The Bell numbers B_n count the number of equivalence relations on n elements. We note $f(n) = B_n$ is a MSOL^2-Specker function. However, B_n is not c.o.i. MSOL^1-definable due to a growth argument.

4.2 Examples of MSOL^k-Specker Polynomials

Our main interest are $\mathcal{L}^k$-Specker functions, and the $\mathcal{L}^k$-Specker polynomials were introduced as an auxiliary tool. However, there are natural examples in the literature of $\mathcal{L}^k$-Specker polynomials.

Fibonacci, Lucas and Chebyshev Polynomials. The recurrence $F_n(x) = x \cdot F_{n-1}(x) + F_{n-2}(x)$, $F_1(x) = 1$ and $F_2(x) = x$ defines the Fibonacci polynomials. The Fibonacci numbers F_n can be obtained as an evaluation of the Fibonacci polynomial for $x = 1$, $F_n(1) = F_n$. The Lucas polynomials are defined analogously.

The Chebyshev polynomials of the first kind (see [24]) are defined similarly by the recurrence relation $T_{n+1}(x) = 2xT_n(x) - T_{n-1}(x)$, $T_0(x) = 1$, and $T_1(x) = x$. The Fibonacci, Lucas and Chebyshev polynomials are natural examples of Specker polynomials. As they are defined by linear recurrence relations, they are c.o.i. MSOL^1-definable.

Touchard Polynomials. The Touchard polynomials are defined

$$T_n(x) = \sum_{k=1}^{n} \left\{ {n \atop k} \right\} x^k$$

where $\left\{ {n \atop k} \right\}$ is the Stirling number of the second kind. $T_n(x)$ is c.o.i. MSOL^2-definable; To see this we note that it is defined by

$$T_n(x) = \sum_{E:\Phi_{cliques}(E)} \quad \prod_{u:\Phi_{first-in-cc}(E,u)} x \, ,$$

where $\Phi_{cliques}(E)$ says E is a disjoint union of cliques and where

$$\Phi_{first-in-cc}(E, u) = \forall v \left((v <_1 u \wedge v \neq u) \rightarrow (v, u) \notin E \right) ,$$

i.e. it says u is the first vertex in its connected component, with respect to the order (less or equal) of $[n]$. Clearly, $\Phi_{cliques}(E)$ and $\Phi_{first-in-cc}(E, u)$ are in MSOL^2. We note that $\Phi_{first-in-cc}(E, u)$ is not invariant under the order $<_1$. The Bell numbers B_n are given as an evaluation of $T_n(x)$, $B_n = T_n(1)$, which implies $T_n(x)$ is not co.i MSOL^1-definable due to a growth argument.

Mittag-Leffler Polynomials. The Mittag-Leffler polynomial (see [5]) is given by

$$M_n(x) = \sum_{k=0}^{n} \binom{n}{k} (n-1)^{\underline{n-k}} 2^k x^{\underline{k}} .$$

It holds that

$$M_n(x) = \sum_{U \subseteq [n]} (n-1) \cdots k \cdot 2^k \cdot x \cdots (x-(k-1)) = \sum_{U,F,T,S:\Phi_M} 1 ,$$

where $\Phi_M(U, F, T, S)$ says $U \subseteq [n]$, F is an injective function from $[n] - \{n\}$ to $[n]$, T is an injective function from U to $[x]$, and S is a function from U to $\{1, n\}$. So, every evaluation of $M_n(x)$ where $x = m$, $m \in \mathbb{N}$, is a c.o.i. MSOL^2-Specker function.

Note that

$$M_{n+1}(x) = \frac{1}{2} x \left[M_n(x+1) + 2M_n(x) + M_n(x-1) \right] .$$

This looks almost like a linear recurrence relation combined with an interpolation formula, and is not of the kind we are discussing here.

5 Conclusions and Open Problems

We have introduced the notion of one variable $\mathcal{L}^k$-Specker functions $f : \mathbb{N} \to \mathbb{N}$ as the speed of a $\mathcal{L}^k$-definable class of relational structures $\mathcal{K}$, i.e., $f(n) = sp_{\mathcal{K}}(n)$. We have investigated for which fragments $\mathcal{L}$ of SOL the $\mathcal{L}^k$-Specker functions satisfy linear recurrence relations over $\mathbb{Z}$ or $\mathbb{Z}_m$.

We have used order invariance, definability criteria and limitation on the vocabulary to continue the line of study, initiated by C. Blatter and E. Specker [8],[9],[30], as to the kind of linear recurrence relations one can expect from Specker functions. We have completely characterized (Theorem 4) the combinatorial functions which satisfy linear recurrence relations with constant coefficients, and we have discussed (Table 1 in Sect. 1) how far one can extend the Specker-Blatter Theorem in terms of order invariance and MSOL-definability. As a consequence, we obtained (Corollary 1) a new characterization of the $\mathbb{Z}$-rational functions $f : \mathbb{N} \to \mathbb{N}$ as the difference of $\mathbb{N}$-rational functions.

We have not studied many variables $\mathcal{L}^k$-Specker functions arising from many-sorted structures, although this is a natural generalizations: For a class of graphs $\mathcal{K}$, $sp_{\mathcal{K}}(n, m)$ counts the number of graphs with n vertices and m edges which are in $\mathcal{K}$. Even for functions in one variable the following remain open:

(i) Can one prove similar theorems for linear recurrence relations where the coefficients depend on n?
(ii) Can one characterize the $\mathcal{L}^k$-Specker functions which satisfy modular recurrence relations with constant coefficients for each modulus m, i.e., is there some kind of a converse to Theorem 5?
(iii) Does Theorem 5 hold for TVMSOL^3?

Finally, for many-sorted $\mathcal{L}^k$-Specker functions studying both growth rate and recurrence relations seems a promising topic of further research.

Acknowledgements

We are grateful to the anonymous referee for his very careful reading of the manuscript, his stylistic suggestions and pointing out of misprints, and to Daniel Marx for his useful comments.

References

1. Balogh, J., Bollobás, B., Weinreich, D.: The speed of hereditary properties of graphs. J. Comb. Theory, Ser. B 79, 131–156 (2000)
2. Balogh, J., Bollobás, B., Weinreich, D.: The penultimate rate of growth for graph properties. EUROCOMB: European Journal of Combinatorics 22 (2001)
3. Balogh, J., Bollobás, B., Weinreich, D.: Measures on monotone properties of graphs. Discrete Applied Mathematics 116, 17–36 (2002)
4. Barcucci, E., Lungo, A.D., Frosini, A., Rinaldi, S.: A technology for reverse-engineering a combinatorial problem from a rational generating function. Advances in Applied Mathematics 26, 129–153 (2001)
5. Bateman, H.: The polynomial of Mittag-Leffler. Proceedings of the National Academy of Sciences 26, 491–496 (1940)
6. Bernardi, O., Noy, M., Welsh, D.: On the growth rate of minor-closed classes of graphs (2007)
7. Berstel, J., Reutenauer, C.: Rational Series and Their Languages. In: EATCS Monographs on Theoretical Computer Science, vol. 12. Springer, Heidelberg (1988)
8. Blatter, C., Specker, E.: Le nombre de structures finies d'une th'eorie à charactère fin. Sciences Mathématiques, Fonds Nationale de la rechercheScientifique, Bruxelles, pp. 41–44 (1981)
9. Blatter, C., Specker, E.: Recurrence relations for the number of labeled structures on a finite set. In: Börger, E., Hasenjaeger, G., Rödding, D. (eds.) Logic and Machines: Decision Problems and Complexity. LNCS, vol. 171, pp. 43–61. Springer, Heidelberg (1984)
10. Bollobas, B., Thomason, A.: projections of bodies and hereditary properties of hypergraphs. J. London Math. Soc. 27 (1995)
11. Büchi, J.: Weak second-order arithmetic and finite automata. Zeitschrift für mathematische Logik und Grundlagen der Mathematik 6, 66–92 (1960)
12. Chomsky, N., Schützenberger, M.: The algebraic theory of context free languages. In: Brafford, P., Hirschberg, D. (eds.) Computer Programming and Formal Systems, pp. 118–161. North-Holland, Amsterdam (1963)
13. Ebbinghaus, H., Flum, J.: Finite Model Theory. In: Perspectives in Mathematical Logic. Springer, Heidelberg (1995)
14. Ebbinghaus, H., Flum, J., Thomas, W.: Mathematical Logic. Undergraduate Texts in Mathematics, 2nd edn. Springer, Heidelberg (1994)
15. Eilenberg, S.: Automata, Languages, and Machines, Vol. A. Academic Press, London (1974)
16. Erdös, P., Frankl, P., Rödl, V.: The asymptotic enumeration of graphs not containing a fixed subgraph and a problem for hypergraphs having no exponent. Graphs Combin. 2, 113–121 (1986)
17. Fischer, E.: The Specker-Blatter theorem does not hold for quaternary relations. Journal of Combinatorial Theory, Series A 103, 121–136 (2003)

18. Fischer, E., Makowsky, J.: The Specker-Blatter theorem revisited. In: Warnow, T.J., Zhu, B. (eds.) COCOON 2003. LNCS, vol. 2697, pp. 90–101. Springer, Heidelberg (2003)
19. Flajolet, P., Sedgewick, R.: Analytic Combinatorics. Cambridge University Press, Cambridge (2009)
20. Graham, L., Knuth, D.E., Patashnik, O.: Concrete Mathematics: A Foundation for Computer Science. Addison-Wesley Longman Publishing Co., Inc., Boston (1994)
21. Knuth, D.E.: The Art of Computer Programming. Fascicle 4: Generating All Trees–History of Combinatorial Generation (Art of Computer Programming), vol. 4. Addison-Wesley Professional, Reading (2006)
22. Libkin, L.: Elements of Finite Model Theory. Springer, Heidelberg (2004)
23. Lidl, R., Niederreiter, H.: Finite Fields. Encyclopedia of Mathematics and its Applications, vol. 20. Cambridge University Press, Cambridge (1983)
24. Mason, J.C., Handscomb, D.C.: Chebyshev Polynomials. Chapman & Hall/CRC, Boca Raton (2003)
25. Prömel, H., Steger, A.: Excluding induced subgraphs: Quadrilaterals. Random Structures & Algorithms 3, 19–31 (1992)
26. Salomaa, A., Soittola, M.: Automata theoretic aspects of formal power series. Springer, Heidelberg (1978)
27. Scheinerman, E., Zito, J.: On the size of hereditary classes of graphs. J. Comb. Theory, Ser. B 61, 16–39 (1994)
28. Sidi, A.: Practical Extrapolation Methods, Theory and Applications. Cambridge Monographs on Applied and Computational Mathematics, vol. 10. Cambridge University Press, Cambridge (2003)
29. Soittola, M.: Positive rational sequences. Theoretical Computer Science 2, 317–322 (1976)
30. Specker, E.: Application of logic and combinatorics to enumeration problems. In: Börger, E. (ed.): Trends in Theoretical Computer Science. 141–169. Computer Science Press; Reprinted in: Ernst Specker, Selecta, Birkhäuser, pp. 324-350 (1988)

Halting and Equivalence of Program Schemes in Models of Arbitrary Theories

Dexter Kozen

Cornell University, Ithaca, New York 14853-7501, USA
kozen@cs.cornell.edu
http://www.cs.cornell.edu/~kozen

In Honor of Yuri Gurevich on the Occasion of his Seventieth Birthday

Abstract. In this note we consider the following decision problems. Let Σ be a fixed first-order signature.

(i) Given a first-order theory or ground theory T over Σ of Turing degree α, a program scheme p over Σ, and input values specified by ground terms $t_1, \ldots, t_n$, does p halt on input $t_1, \ldots, t_n$ in all models of T?
(ii) Given a first-order theory or ground theory T over Σ of Turing degree α and two program schemes p and q over Σ, are p and q equivalent in all models of T?

When T is empty, these two problems are the classical halting and equivalence problems for program schemes, respectively. We show that problem (i) is Σ_1^α-complete and problem (ii) is Π_2^α-complete. Both problems remain hard for their respective complexity classes even if Σ is restricted to contain only a single constant, a single unary function symbol, and a single monadic predicate. It follows from (ii) that there can exist no relatively complete deductive system for scheme equivalence over models of theories of any Turing degree.

Keywords: dynamic model theory, program scheme, scheme equivalence. Mathematics Subject Classification Codes: 03B70, 3D10, 03D15, 03D28, 03D75, 68Q17.

Let Σ be an arbitrary but fixed first-order signature. A *ground formula* over Σ is a Boolean combination of atomic formulas $P(t_1, \ldots, t_n)$ of Σ, where the t_i are ground terms (no occurrences of variables). A *ground literal* is a ground atomic formula $P(t_1, \ldots, t_n)$ or its negation. A *ground theory* over Σ is a consistent set of ground formulas closed under entailment. A (*first-order*) *theory* over Σ is a consistent set of first-order formulas closed under entailment. A ground theory E is a *complete ground extension* of T if E contains T and every ground formula or its negation appears in E. A first-order theory E is a *complete extension* of T if E contains T and every first-order sentence or its negation appears in E.

A. Blass, N. Dershowitz, and W. Reisig (Eds.): Gurevich Festschrift, LNCS 6300, pp. 463–469, 2010.

Theorem 1. *Let α be an arbitrary Turing degree. The following problems are Σ_1^α-complete and Π_2^α-complete, respectively:*

(i) Given a ground theory T over Σ of Turing degree α, a program scheme p over Σ, and input values specified by ground terms $\bar{t} = t_1, \ldots, t_n$, does p halt on input $\bar{t}$ in all models of T?
(ii) Given a ground theory T over Σ of Turing degree α and two schemes p and q over Σ, are p and q equivalent in all models of T?

Both problems remain hard for their respective complexity classes even if Σ is restricted to contain only a single constant, a single unary function symbol, and a single monadic predicate. Both problems remain complete for their respective complexity classes if "ground theory" in the statement of the problem is replaced by "first-order theory".

Note that for both problems, the theory T is part of the input in the form of an oracle.

If T is the empty ground theory, these are the classical halting and equivalence problems for program schemes, respectively. Classical lower bound proofs (see [7]) establish the r.e. hardness of the two problems for this case. The Π_2^0-hardness of (ii) for this case can also be shown to follow without much difficulty from a result of [4].

Proof. Let T be a first-order theory or ground theory of Turing degree α. For the upper bounds, we do not actually need T to be closed under entailment, but only that ground entailment relative to T is decidable; that is, if ϕ is a ground formula, then $T \vDash \phi$ is decidable with an oracle for T.

We first consider problem (i). For the upper bound, we wish to show that the problem is in Σ_1^α. Given p and $\bar{t}$, we simulate the computation of p on input $\bar{t}$ on all complete extensions of T simultaneously, using the oracle for T to resolve tests. Each branch of the simulation maintains a finite set E of ground literals consistent with T, initially empty. Whenever a test $P(s_1, \ldots, s_k)$ is encountered, we substitute the current values of the program variables, which are ground terms, for the variables to obtain a ground atomic formula $P(u_1, \ldots, u_k)$, then consult T and E to determine which branch to take. If the truth value of the test $P(s_1, \ldots, s_k)$ is determined by T and E, that is, if $T \vDash E \to P(u_1, \ldots, u_k)$ or $T \vDash E \to \neg P(u_1, \ldots, u_k)$, then we just take the appropriate branch. Otherwise, if both $P(u_1, \ldots, u_k)$ and $\neg P(u_1, \ldots, u_k)$ are consistent with $T \cup E$, then the simulation branches, extending E with $P(u_1, \ldots, u_k)$ on one branch and $\neg P(u_1, \ldots, u_k)$ on the other. In each simulation step, all current branches are simulated for one step in a round-robin fashion. We thus simulate the computation of p on all possible complete extensions of T simultaneously. If p halts on all such extensions, then by König's lemma there is a uniform bound on the halting time of all branches of the computation. The simulation halts successfully when that bound is discovered.

We now show that problem (i) is Σ_1^α-hard in the restricted case $\Sigma = \{a, f, P\}$, where a is a constant, f is a unary function symbol, and P is a unary relation symbol. For now we will assume that we have two unary relation symbols Q, R; we can later encode these in a single P by taking $P(f^{2n}(a)) = Q(f^n(a))$ and $P(f^{2n+1}(a)) = R(f^n(a))$.

Let $A \subseteq \mathbb{N}$ be any set. We will show how to encode the halting problem for deterministic oracle Turing machines M with oracle A. This problem is Σ_1^A-complete. Given an input x over M's input alphabet, we will construct a ground theory T_0 Turing-equivalent to A and a scheme p with no input or output such that p halts on all models of T_0 iff M halts on input x. Later, we will extend T_0 to a first-order theory T of the same complexity. The encoding technique used here is fairly standard, but we include the argument for completeness and because we need the resulting scheme p in a certain special form for the proof of (ii).

Consider the Herbrand domain consisting of all terms over a and f. This domain is isomorphic to the natural numbers with 0 and successor under the correspondence $n \mapsto f^n(a)$. An Herbrand model H over this domain is represented by a pair of semi-infinite binary strings representing the truth values of $Q(f^n(a))$ and $R(f^n(a))$ for $n \geq 0$. The correspondence is one-to-one. We will use the string corresponding to Q to encode a computation history of M and the string corresponding to R to encode the oracle A.

Each string x over M's input alphabet determines a unique finite or infinite computation history $\#\alpha_0^x\#\alpha_1^x\#\alpha_2^x\#\cdots$, where α_i^x is a string over a finite alphabet Δ encoding the instantaneous configuration of M on input x at time i consisting of the tape contents, head position, and current state. We also assume that configurations encode all oracle queries and the answers returned by the oracle A (we will be more explicit about the precise format of the encoding below). The configurations α_i^x are separated by a symbol $\# \notin \Delta$. The computation history in turn can be encoded in binary, and this infinite binary string can be encoded by the truth values of $Q(f^n(a))$, $n \geq 0$.

The ground theory T_0 describes the oracle A using R and the starting configuration $\#\alpha_0^x\#$ of M on input x using Q. The description of the starting configuration consists of a finite set S_x of ground literals of the form $Q(f^n(a))$ or $\neg Q(f^n(a))$. The oracle is described by the set

$$\{R(f^n(a)) \mid n \in A\} \cup \{\neg R(f^n(a)) \mid n \notin A\}. \tag{1}$$

In any complete extension of T_0, the infinite string corresponding to Q describes either the unique valid computation history of M on input x or a garbage string. The scheme p can read the n-th bit of this string in the corresponding Herbrand model by testing the value of $Q(f^n(a))$. It starts by scanning the initial part of the string to check that it is of the form $\#\alpha_0^y\#$ for some y. (This step is not strictly necessary for this proof, since we are restricting our attention to models of T_0, in which this step will always succeed; but it will be needed later in the proof of (ii).) Next, p scans the string from left to right to determine whether each successive α_{i+1}^x follows from α_i^x in one step according to the transition rules of M. It does this by comparing corresponding bits in α_i^x and α_{i+1}^x using two variables to simulate pointers into the string. If the current value of the variable

x is $f^n(a)$, then testing $Q(x)$ reads the n-th bit of the string. The pointer is advanced by the assignment $x := f(x)$.

The scheme p must also verify that oracle responses are correct. Without loss of generality, we can assume that M uses the following mechanism to query the oracle. We assume that M has an integer counter initially set to 0. In each step, M may add one to the counter or not, depending on its current state and the tape symbol it is scanning, according to its transition function. It queries the oracle by entering a distinguished oracle query state. If the current value of the counter is n, then M transits to a distinguished "yes" state if $n \in A$ and to a distinguished "no" state if $n \notin A$. The counter is reset to 0.

For p to verify the correctness of the oracle responses, we assume that the format of the encoding of configurations is $\beta\$0^n$, where β is the description of the current state, tape contents, and head position of M and n is the current value of the counter. If p discovers that M is in the oracle query state while scanning β, then after encountering the \$, it sets a variable $z := a$ and executes $z := f(z)$ for each occurrence of 0 after the \$, so that z will have the value $f^n(a)$ when the next # is seen. Then it tests $R(z)$ to determine whether $n \in A$. It then checks that in the subsequent configuration, M is in the "yes" or "no" state according as $R(z)$ is true or false, respectively, and that the counter has been reset.

If p discovers an error, so that the string does not represent a computation history of M on some input, it halts immediately. It also halts if it ever encounters a halting state of M anywhere in the string. Thus the only Herbrand model of T_0 that would cause p not to halt is the one describing the infinite valid computation history of M on x in the case that M does not halt on x. Thus p halts on all Herbrand models of T_0, thus on all models of T_0, iff M halts on x.

We can further restrict the set S_x describing the start configuration of M to be empty by observing that S_x is finite, so it can be hard-wired into the scheme p itself. Thus the initial format check that p performs can be modified to check whether S_x holds and halt immediately if not. This gives a ground theory T_0 consisting of (1) only, independent of the input x, at the expense of coding information about x in the scheme p. However, for purposes of the proof of (ii) below, it will be important that p not depend on the input x but only on the machine M.

Finally, we must produce a first-order theory T extending T_0 such that T is of no higher Turing degree than T_0 (that is, T is still Turing-equivalent to A) and every Herbrand model of T_0 extends to a model of T. Since the halting of p depends only on the Herbrand substructure, p will halt on all models of T iff it halts on all Herbrand models of T_0. The main issue here is that we must be careful to construct a T whose Turing complexity is no greater than that of the ground theory T_0, otherwise the lower bound will not hold. Note that the first-order theory generated by T_0 may not be suitable, because the best we can guarantee is that it is Σ_1 in A.

To construct T, we augment T_0 with all existential formulas of the form

$$\exists x \bigwedge_{n \in C} P(f^n(x)) \wedge \bigwedge_{m \in D} \neg P(f^m(x)), \tag{2}$$

where C and D are disjoint finite subsets of $\mathbb{N}$. We take T to be the set of logical consequences of T_0 and the formulas (2). Every Herbrand model of T_0 extends to a model of T, because new elements outside the Herbrand domain can be freely added as needed to satisfy the existential formulas (2).

To show that T is Turing-equivalent to A, we observe that since the theory is monadic, every first-order sentence reduces effectively via the laws of first-order logic to a Boolean combination of ground formulas $P(f^n(a))$ and existential formulas (2). The latter are all true in T, so every sentence is equivalent modulo T to a Boolean combination of ground formulas $P(f^n(a))$. Any such formula is consistent with T iff it is consistent with T_0, since as previously observed, every Herbrand model of T_0 extends to a model of T. Thus T Turing-reduces to T_0.

This argument shows that the halting problem for program schemes over models of T_0 or T is hard for Σ_1^A. Since A was arbitrary and both T_0 and T are Turing-equivalent to A, we are finished with the proof of (i).

Now we turn to problem (ii). For the upper bound, first we show that equivalence of schemes over models of T is Π_2^T. Equivalently, inequivalence of schemes over models of T is Σ_2^T. It suffices to show that inequivalence of schemes over models of T can be determined by an IND program over $\mathbb{N}$ with oracle T with an $\exists\forall$ alternation structure [5] (see also [6]). As above, we need only that ground entailment relative to T is decidable.

Let p and q be two schemes with input variables $\bar{x} = x_1, \ldots, x_n$. The schemes p and q are not equivalent over models of T iff there exists a complete extension of T with extra constants $\bar{c} = c_1, \ldots, c_n$ in which either

1. both p and q halt on input $\bar{c}$ and produce different output values;
2. p halts on $\bar{c}$ and q does not; or
3. q halts on $\bar{c}$ and p does not.

We start by selecting existentially the alternative 1, 2 or 3 to check.

If alternative 1 was selected, we simulate p and q on input $\bar{c}$, maintaining a finite set E of ground literals and using T and E as in the proof of (i) to resolve tests. Whenever a test is encountered that is not determined by T and E, we guess the truth value and extend E accordingly. Thus we nondeterministically guess a complete extension of T using existential branching in the IND program. We continue the simulation until both p and q halt, then compare output values, accepting if they differ.

If alternative 2 was selected, we simulate p on $\bar{c}$ until it halts, maintaining the guessed truth values of undetermined tests in the set E as above. When p has halted, we have a consistent extension $T \cup E$ of T, where E consists of the finitely many tests that were guessed during the computation of p. So far we have only used existential branching. We must now verify that there exists a complete extension of $T \cup E$ in which q does not halt on input $\bar{c}$. By (i), this problem is $\Pi_1^{T \cup E}$, so we can solve it with a purely universally-branching IND computation.

The argument for alternative 3 is symmetric.

For the lower bound, we reduce the totality problem for oracle Turing machines with oracle A, a well-known Π_2^A-complete problem, to the equivalence problem (ii). The totality problem is to determine whether a given machine halts on all inputs. As above, it will suffice to consider $\Sigma = \{a, f, Q, R\}$.

Given a deterministic oracle Turing machine M with oracle A, let T_0 be the ground theory consisting of the formulas (1). Then T_0 is Turing-equivalent to A. We construct two schemes p and q with no input or output that are equivalent in all complete ground extensions of T_0 iff M halts on all inputs. The scheme p is the one constructed in the proof of (i). As in that proof, each input string x over M's input alphabet determines a unique computation history, and the scheme p checks that the Herbrand model in which it is running encodes a valid computation history of M on some input. As in the proof of (i), the oracle A is encoded by the formulas (1). This allows p to verify responses to the oracle queries in the computation history.

Now unlike the proof of (i), there is an extra source of non-halting. Recall that there is an initial format check in which p checks that the string has a prefix of the form $\#\alpha_0^y\#$ for some y. This check was not really necessary in the proof of (i), since a description of the start configuration on input x could be coded in T_0, but it is necessary here. But if there is no second occurrence of # in the string, then p will loop infinitely looking for it. If it does detect a second occurrence of #, then as before, the only source of non-halting is if M does not halt on x. We therefore build q to simply check for a prefix of the form $\#\alpha_0^x\#$ for some x exactly as p does and halt immediately when it encounters the second occurrence of #. Now p does not halt in the Herbrand model H iff the string represented by the truth values of $Q(f^n(a))$ either

(a) does not have a prefix of the form $\#\alpha_0^x\#$, or
(b) does have a prefix of the form $\#\alpha_0^x\#$ and represents a non-halting computation history of M on x;

and q does not halt in H in case (a) only. Therefore p and q are equivalent iff case (b) never occurs for p; that is, iff M halts on all inputs.

We construct the first-order theory T from T_0 as in the proof of (i) above. Thus the equivalence problem for schemes over models of T_0 or T is Π_2^A-hard. Since A was arbitrary and both T_0 and T are Turing-equivalent to A, we are done. □

In [1], axioms were proposed for reasoning equationally about input-output relations of first-order program schemes over Σ. These axioms have been shown to be adequate for some fairly intricate equivalence arguments arising in program optimization [1,2]. However, unlike the propositional case, it follows from Theorem 1(ii) that there can exist no finite relatively complete axiomatization for first-order scheme equivalence over models of a theory T of any Turing degree. If such an axiomatization did exist, then the scheme equivalence problem over models of T would be r.e. in T.

In the case $\alpha = 0$, it is decidable whether a given first-order sentence ϕ is a consequence of a given finite set E of ground formulas over the signature $\Sigma = \{a, f, P\}$, since $E \models \phi$ iff $E \to \phi$ is a valid sentence of the first-order theory of a one-to-one unary function with monadic predicate, a well-known decidable theory [3] (note that every Σ-structure is elementarily equivalent to one in which the interpretation of f is one-to-one). By Theorem 1(ii), the scheme equivalence problem is Π^0_2-hard relative to E, therefore also relative to the decidable first-order theory generated by E.

Acknowledgments

Thanks to Andreas Blass for insightful comments, which inspired a strengthening of the results. This work was supported in part by NSF grant CCF-0635028.

References

1. Angus, A., Kozen, D.: Kleene algebra with tests and program schematology. Tech. Rep. TR2001-1844, Computer Science Department, Cornell University (July 2001)
2. Barth, A., Kozen, D.: Equational verification of cache blocking in LU decomposition using Kleene algebra with tests. Tech. Rep. TR2002-1865, Computer Science Department, Cornell University (June 2002)
3. Ferrante, J., Rackoff, C.: The computational complexity of logical theories. Lecture Notes in Mathematics, vol. 718. Springer, Heidelberg (1979)
4. Harel, D., Meyer, A.R., Pratt, V.R.: Computability and completeness in logics of programs. In: Proc. 9th Symp. Theory of Comput., pp. 261–268. ACM, New York (1977)
5. Harel, D., Kozen, D.: A programming language for the inductive sets, and applications. Information and Control 63(1-2), 118–139 (1984)
6. Harel, D., Kozen, D., Tiuryn, J.: Dynamic Logic. MIT Press, Cambridge (2000)
7. Manna, Z.: Mathematical Theory of Computation. McGraw-Hill, New York (1974)

Metrization Theorem for Space-Times: From Urysohn's Problem towards Physically Useful Constructive Mathematics

Vladik Kreinovich

Department of Computer Science, University of Texas, El Paso
vladik@utep.edu

To Yuri Gurevich, in honor of his enthusiastic longtime quest for efficiency and constructivity.

Abstract. In the early 1920s, Pavel Urysohn proved his famous lemma (sometimes referred to as "first non-trivial result of point set topology"). Among other applications, this lemma was instrumental in proving that under reasonable conditions, every topological space can be metrized.

A few years before that, in 1919, a complex mathematical theory was experimentally proven to be extremely useful in the description of real world phenomena: namely, during a solar eclipse, General Relativity theory – that uses pseudo-Riemann spaces to describe space-time – was (spectacularly) experimentally confirmed. Motivated by this success, Urysohn started working on an extension of his lemma and of the metrization theorem to (causality-)ordered topological spaces and corresponding pseudo-metrics. After Urysohn's early death in 1924, this activity was continued in Russia by his student Vadim Efremovich, Efremovich's student Revolt Pimenov, and by Pimenov's students (and also by H. Busemann in the US and by E. Kronheimer and R. Penrose in the UK). By the 1970s, reasonably general space-time versions of Urysohn's lemma and metrization theorem have been proven.

However, these 1970s results are not constructive. Since one of the main objectives of this activity is to come up with useful applications to physics, we definitely need constructive versions of these theorems – versions in which we not only claim the theoretical existence of a pseudo-metric, but we also provide an algorithm enabling the physicist to generate such a metric based on empirical data about the causality relation. An additional difficulty here is that for this algorithm to be useful, we need a physically relevant constructive description of a causality-type ordering relation.

In this paper, we propose such a description and show that, for this description, a combination of the existing constructive ideas with the known (non-constructive) proof leads to successful constructive space-time versions of the Urysohn's lemma and of the metrization theorem.

Keywords: Urysohn's lemma, metrization theorem, space-times, constructive mathematics.

A. Blass, N. Dershowitz, and W. Reisig (Eds.): Gurevich Festschrift, LNCS 6300, pp. 470–487, 2010.

1 Introduction

Urysohn's lemma. In the early 1920s, Pavel Urysohn proved his famous lemma (sometimes referred to as "first non-trivial result of point set topology"). This lemma deals with *normal* topological spaces, i.e., spaces in which every two disjoint closed sets have disjoint open neighborhoods; see, e.g., [12]. As the very term "normal" indicates, most usual topological spaces are normal, including the n-dimensional Euclidean space.

Urysohn's lemma states the following:

Lemma 1. *If X is a normal topological space, and A and B are disjoint closed sets in X, then there exists a continuous function $f : X \to [0, 1]$ for which $f(a) = 0$ for all $a \in A$ and $f(b) = 1$ for all $b \in B$.*

Resulting metrization theorem. Urysohn's lemma has many interesting applications. Among other applications, this lemma was instrumental in proving that under reasonable conditions, every topological space X can be metrized, i.e., there exist a *metric* – a function $\rho : X \times X \to \mathbb{R}_0^+$ to the set $\mathbb{R}_0^+$ of all non-negative real numbers for which the following three conditions are satisfied

$$\rho(a, b) = 0 \Leftrightarrow a = b; \quad \rho(a, b) = \rho(b, a); \quad \rho(a, c) \leq \rho(a, b) + \rho(b, c);$$

and for which the original topology on X coincides with the topology generated by the open balls $B_r(x) = \{y : \rho(x, y) < r\}$.

Specifically, from Urysohn's lemma, we can easily conclude that:

Theorem 1. *Every normal space X with countable base is metrizable.*

Comment. It is worth mentioning that the normality condition is too strong for the theorem: actually, it is sufficient to require that the space is:

- *regular*, i.e., for every closed set A and every point $b \notin A$ can be separated by disjoint open neighborhoods, and
- *Hausdorff*, i.e., every two different points have disjoint open neighborhoods.

Space-time geometry and how it inspired Urysohn. A few years before Urysohn's lemma, in 1919, a complex mathematical theory was experimentally proven to be extremely useful in the description of real world phenomena. Specifically, during a solar eclipse, General Relativity theory – that uses pseudo-Riemann spaces to describe space-time – was (spectacularly) experimentally confirmed; see, e.g., [20].

From the mathematical viewpoint, the basic structure behind space-time geometry is not simply a topological space, but a topological space with an *order* $a \preceq b$ whose physical meaning is that the event a can causally influence the event b.

For example, in the simplest case of the Special Relativity theory (see Fig. 1), the event $a = (a_0, a_1, a_2, a_3)$ can influence the event $b = (b_0, b_1, b_2, b_3)$ if we can get from the spatial point (a_1, a_2, a_3) at the moment a_0 to the point (b_1, b_2, b_3)

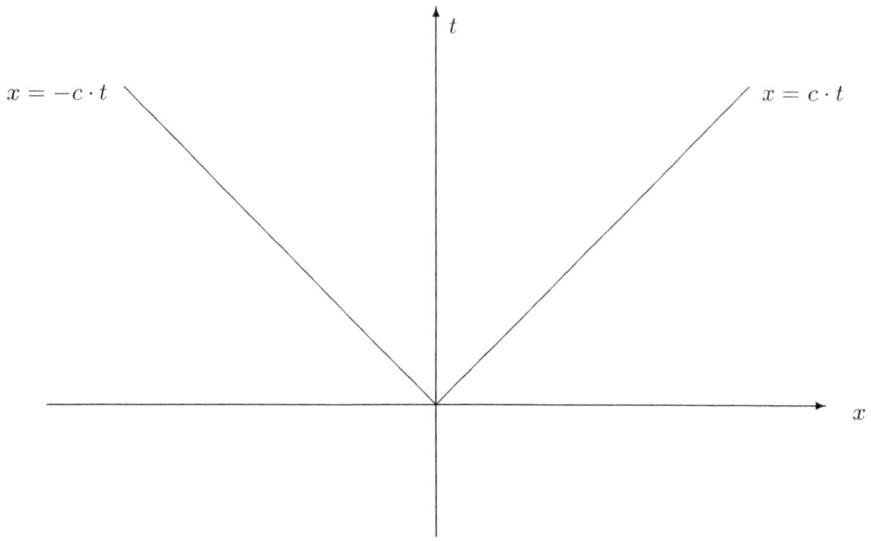

Fig. 1. Causality relation of the Special Relativity theory

at the moment $b_0 > a_0$ while traveling with a speed which is smaller than or equal to the speed of light c:

$$\sqrt{(a_1 - b_1)^2 + (a_2 - b_2)^2 + (a_3 - b_3)^2} \le c \cdot (b_0 - a_0).$$

Motivated by this practical usefulness of ordered topological spaces, Urysohn started working on an extension of his lemma and of the metrization theorem to (causality-)ordered topological spaces and corresponding pseudo-metrics.

Space-time metrization after Urysohn. P. S. Urysohn did not have time to work on the space-time extension of his results, since he died in 1924 at an early age of 26.

After Urysohn's early death, this activity was continued in Russia by his student Vadim Efremovich, by Efremovich's student Revolt Pimenov, and by Pimenov's students – and also by H. Busemann in the US and by E. Kronheimer and R. Penrose in the UK [10,13,22] (see also [16]).

This research actively used the general theory of ordered topological spaces; see, e.g., [21].

By the 1970s, reasonably general space-time versions of Urysohn's lemma and metrization theorem have been proven; see, e.g., [14,15].

Space-time metrization results: main challenge. One of the main objectives of the space-time metrization activity is to come up with useful applications to physics.

From this viewpoint, we definitely need *constructive* versions of these theorems – versions in which we not only claim the theoretical existence of a (pseudo)metric, but we also provide an algorithm enabling a physicist to generate such a metric based on empirical data about the causality relation.

The original 1970s space-time metrization results are not constructive. It is therefore necessary to make them constructive.

An additional difficulty here is that for this algorithm to be useful, we need a physically relevant constructive description of a causality-type ordering relation.

What we do in this paper. In this paper,

- we propose a physically relevant constructive description of a causality-type ordering relation, and
- we show that for this description, a combination of the existing constructive ideas with the known (non-constructive) proof leads to successful constructive space-time versions of the Urysohn's lemma and of the metrization theorem.

2 Known Space-Time Metrization Results: Reminder

Causality relation: the original description. The current formalization of space-time geometry starts with a transitive relation $a \preceq b$ on a topological space X.

The physical meaning of this relation is causality – that an event a can influence the event b. This meaning explains transitivity requirement: if a can influence b and b can influence c, this means that a can therefore (indirectly) influence the event c.

Need for a more practice-oriented definition. On the theoretical level, the causality relation $\preceq$ is all we need to know about the geometry of space-time.

However, from the practical viewpoint, we face an additional problem – that measurements are never 100% accurate and, therefore, we cannot locate events exactly. When we are trying to locate an event a in space and time, then, due to measurement uncertainty, the resulting location $\widetilde{a}$ is only approximately equal to the actual one: $\widetilde{a} \approx a$.

From this viewpoint, when we observe that an event a influences the event b, we record it as a relation between the corresponding approximations – i.e., we conclude that $\widetilde{a} \prec \widetilde{b}$; see Fig. 2. However, this may be a wrong conclusion: for example, if an event b is at the border of the future cone $F_a \stackrel{\text{def}}{=} \{b : a \preceq b\}$ of the event a, then

- we have $a \preceq b$, but
- the approximate location $\widetilde{b}$ may be outside the cone,

so the conclusion $a \preceq \widetilde{b}$ is wrong.

Kinematic causality: a practice-oriented causality relation. To take into account measurement uncertainty, researchers use a different causality relation $a \prec b$, meaning that every event in some small neighborhood of b causally follows a.

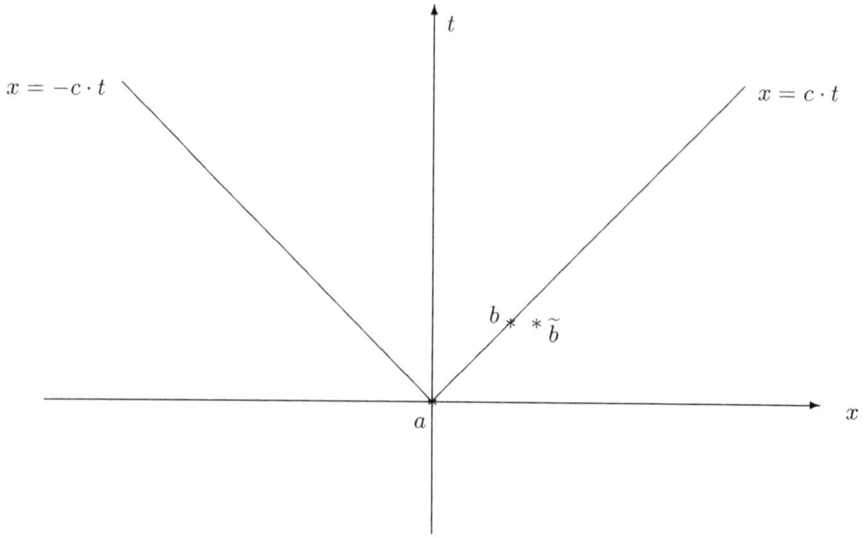

Fig. 2. Need for a more practice-oriented definition of causality

Comment. Since a can only be measured with uncertainty, it is also reasonable to consider a more complex relation: every event in some small neighborhood of b causally follows every element from some small neighborhood of a. Under certain reasonable conditions, however, this more complex definition is equivalent to the above simpler one. Thus, in the following text, we will consider the above simpler definition.

In precise terms, the above definition $a \prec b$ means that b belongs to the interior $\text{Int}(F_a)$ of the future cone F_a.

In the simplest space-time of special relativity, this means that we are excluding the border of the future cones (that corresponds to influencing by photons and other particles traveling at a speed of light c) and only allow causality by particle whose speed is smaller than c. The motion of such particles is known as *kinematics*, hence this new practice-oriented causality relation is called *kinematic causality.*

This definition implies, e.g., that the kinematic casuality relation is transitive, as well as several other reasonable properties. These properties lead to the following formal definition of the kinematic causality relation.

Definition 1. *A relation $\prec$ is called a* kinematic causality *if it is transitive and satisfies the following properties:*

$$a \not\prec a; \quad \forall a\, \exists \underline{a}, \overline{a}\, (\underline{a} \prec a \prec \overline{a}); \quad a \prec b \Rightarrow \exists c\, (a \prec c \prec b);$$

$$a \prec b, c \Rightarrow \exists d\, (a \prec d \prec b, c); \quad b, c \prec a \Rightarrow \exists d\, (b, c \prec d \prec a).$$

Topology and causality can be defined in terms of kinematic causality relation. We started our description with a pre-ordered topological space X with a causality relation $\preceq$. Based on topology and on the causality relation, we defined the kinematic causality $\prec$.

It turns out that in many reasonable cases, it is sufficient to know the kinematic causality relation $\prec$. Based on this relation, we can uniquely reconstruct both the topology and and the original causality relation $\preceq$.

Indeed, as a topology, we can take a so-called *Alexandrov topology* in which intervals

$$(a,b) \stackrel{\text{def}}{=} \{c : a \prec c \prec b\}$$

form the base.

Once the topology is defined, we can now describe causality as

$$a \preceq b \stackrel{\text{def}}{\equiv} b \in \overline{a^+},$$

where $a^+ \stackrel{\text{def}}{=} \{b : a \prec b\}$ and $\overline{S}$ denotes the closure of the set S.

Comment. In principle, we can use a dual definition $a \preceq b \stackrel{\text{def}}{\equiv} a \in \overline{b^-}$, where $b^- \stackrel{\text{def}}{=} \{c : c \prec b\}$. To make sure that these two definitions lead to the same result, the following additional property is usually required:

$$b \in \overline{a^+} \Leftrightarrow a \in \overline{b^-}.$$

Towards a space-time analog of a metric. Traditional metric is defined as a function $\rho : X \times X \to \mathbb{R}_0^+$ to the set to the set $\mathbb{R}_0^+$ of all non-negative real numbers for which the following properties are satisfied:

$$\rho(a,b) = 0 \Leftrightarrow a = b; \quad \rho(a,b) = \rho(b,a); \quad \rho(a,c) \le \rho(a,b) + \rho(b,c).$$

The usual physical meaning of this definition is that $\rho(a,b)$ is the length of the shortest path between a and b. This meaning leads to a natural explanation for the triangle inequality $\rho(a,c) \le \rho(a,b) + \rho(b,c)$. Indeed, the shortest path from a to b (of length $\rho(a,b)$) can be combined with the shortest path from b to c (of length $\rho(b,c)$) into a single combined path from a to c of length $\rho(a,b) + \rho(b,c)$. Thus, the length $\rho(a,c)$ of the shortest possible path between a and c must be smaller than or equal to this combined length: $\rho(a,c) \le \rho(a,b) + \rho(b,c)$.

In space-time, we do not directly measure distances and lengths. The only thing we directly measure is (proper) time along a path. So, in space-time geometry, we talk about times and not lengths.

It is well known that if we travel with a speed close to the speed of light, then the proper travel time (i.e., the time measured by a clock that travels with us) goes to 0. Thus, in space-time, the smallest time does not make sense: it is always 0. What makes sense is the *largest* time. In view of this, we can define a "kinematic metric" $\tau(a,b)$ as the longest (= proper) time along any path from event a to event b.

Of course, such a path is only possible if a kinematically precedes b, i.e., if $a \prec b$.

If $a \prec b$ and $b \prec c$, then the longest path from a to b (of length $\tau(a,b)$) can be combined with the longest path from b to c (of length $\tau(b,c)$) into a single combined path from a to c of length $\tau(a,b)+\tau(b,c)$. Thus, the length $\tau(a,c)$ of the longest possible path between a and c must be larger than or equal to this combined length: $\tau(a,c) \geq \tau(a,b)+\tau(b,c)$. This inequality is sometimes called the *anti-triangle inequality.*

These two properties constitute a formal definition of a kinematic metric.

Definition 2. *By a* kinematic metric *on a set X with a kinematic causality relation $\prec$, we mean a function $\tau : X \times X \to \mathbb{R}_0^+$ to the set $\mathbb{R}_0^+$ of all non-negative real numbers that satisfies the following two properties:*

$$\tau(a,b) > 0 \Leftrightarrow a \prec b; \quad a \prec b \prec c \Rightarrow \tau(a,c) \geq \tau(a,b) + \tau(b,c).$$

Space-time analog of Urysohn's lemma. The main condition under which the space-time analog of Urysohn's lemma is proven is that the space X is *separable*, i.e., there exists a countable dense set $\{x_1, x_2, \ldots, x_n, \ldots\}$.

Lemma 2. *If $\prec$ is a kinematic causality relation on a separable space X, and $a \prec b$, then there exists a continuous $\preceq$-monotonic function $f_{(a,b)} : X \to [0,1]$ for which:*

- $f_{(a,b)}(x) = 0$ *for all x for which $a \not\prec x$, and*
- $f_{(a,b)}(x) = 1$ *for all x for which $b \preceq x$.*

This lemma is *similar* to the original Urysohn's lemma, because it proves the existence of a function $f_{(a,b)}$ that separates two disjoint closed sets:

- the complement $-a^+$ to the set a^+ and
- the set $\overline{b^+}$.

The new statement is *different* from the original Urysohn's lemma, because:

- first, it only considers special closed sets, and
- second, in contrast to the original Urysohn's lemma, the new lemma also requires that the separating function f be monotonic.

Space-time analog of the metrization theorem. Based on the space-time analog of the Urysohn lemma, one can prove the following results:

Theorem 2. *If X is a separable topological space with a kinematic causality relation $\prec$, then there exists a continuous kinematic metric τ which generates the corresponding kinematic causality relation $\prec$ — in the sense that $a \prec b \Leftrightarrow \tau(a,b) > 0$.*

Comment. Since, as we have mentioned, the kinematic causality relation $\prec$ also generates the topology, we can conclude that the kinematic metric τ also determines the corresponding topology.

Proof. First, we prove that for every x, there exists a $\preceq$-monotonic function $f_x : X \to [0,1]$ for which $f_x(b) > 0 \Leftrightarrow x \prec b$.

The proof of this statement is reasonably straightforward.

The only technically cumbersome part of this proof is to show that if a space with a kinematic causality is separable, i.e., if there exists an everywhere dense sequences $\{x_1, \ldots, x_n, \ldots\}$, then there exists a decreasing sequence y_i that converges to x. Moreover, we can select this sequence in such a way that for every i, if $x \prec x_i$ then $y_i \prec x_i$.

Indeed, since the relation $\prec$ is a kinematic causality, there exists a point $\overline{x}$ for which $x \prec \overline{x}$. We then take $y_0 \stackrel{\text{def}}{=} \overline{x}$. By our choice of y_0, we thus have $x \prec y_0$.

Let us assume that we have already selected points $y_0, \ldots, y_k$ for which $x \prec y_k \prec y_{k-1} \prec \ldots \prec y_0$. Let us construct a point y_{k+1} for which, first, $x \prec y_{k+1} \prec y_k$ and, second, if $x \prec x_{k+1}$, then $y_{k+1} \prec x_{k+1}$.

If $x \not\prec x_{k+1}$, then, due to the properties of the kinematic causality, there exists a point c for which $x \prec c \prec y_k$. We will then take $y_{k+1} = c$.

If $x \prec x_{k+1}$, then, due to the properties of the kinematic causality, from $x \prec x_{k+1}$ and $x \prec y_k$, we can conclude that there exists a point d for which $x \prec d \prec x_{k+1}, y_k$. We can then take $y_{k+1} = d$.

Let us now prove that $y_n \to x$, i.e., that for every open neighborhood U of the point x, there exists an index n_0 for which $y_n \in U$ for all $n \geq n_0$. Indeed, let U be such a neighborhood. Since open intervals form a base, there exists an open interval $(a,b) \subseteq U$ that contains the point x. By definition of the interval, $x \in (a,b)$ means that $a \prec x$ and $x \prec b$. By definition of the kinematic causality, there exists a point c for which $x \prec c \prec b$. Thus, the open interval (x,b) is non-empty. Since the sequence $\{x_n\}$ is everywhere dense, it has a point x_{n_0} in this interval, for which $x \prec x_{n_0} \prec b$. By the properties of the sequence y_i, this implies that $x \prec y_{n_0} \prec x_{n_0} \prec b$. Since the sequence $\{y_n\}$ is decreasing, we thus conclude that $x \prec y_n \prec b$ for all $n \geq n_0$. From $a \prec x \prec y_n$, we then deduce that $a \prec y_n$. Hence, $y_n \in (a,b) \subseteq U$ and so, $y_n \in U$ for all $n \geq n_0$. The statement is proven.

Once a decreasing sequence y_i that converges to x is constructed, we can take $f_x(b) = \sum\limits_{i=1}^{\infty} 2^{-i} \cdot f_{(x,y_i)}(b)$.

Next, we prove that for every x, there exists a $\preceq$-decreasing function

$$g_x : X \to [0,1]$$

for which $g_x(a) > 0 \Leftrightarrow a \prec x$. The proof of this second statement is similar to the proof of the first statement.

Once these two auxiliary statements are proven, we can use the countable everywhere dense sequence $\{x_1, x_2, \ldots, x_n, \ldots\}$ to construct the desired kinematic metric as

$$\tau(a,b) = \sum_{i=1}^{\infty} 2^{-i} \cdot \min(g_{x_i}(a), f_{x_i}(b)).$$

It is reasonably easy to prove that thus defined function is indeed a kinematic metric. □

3 Towards a Physically Reasonable Constructive Definition of Causality

Need for a constructive definition of causality. As we have mentioned, in order to provide a physically meaningful constructive version of the space-time metrization theorem, we must come up with a physically meaningful constructive definition of causality.

Towards a constructive definition of casuality: analysis of the physical situation. To come up with a physically meaningful constructive definition of causality, let us recall how causality can be physically detected.

In the ideal world, detecting whether an event a is causally related to the event b (i.e., whether $a \preceq b$) is straightforward. We send a signal at event a in all directions and at all possible speeds, and we check whether this signal was detected at b:

- if this signal is detected at b, we conclude that $a \preceq b$;
- if this signal is not detected at b, we conclude that $a \not\preceq b$.

In practice, we can only locate an event with a certain accuracy. As a result, when we try to detect whether $a \preceq b$, then, instead of two, we now have *three* possible options:

- if the signal is detected in the entire vicinity of b, then we conclude that $a \prec b$;
- if no signal is detected in the entire vicinity of b, then we conclude that $a \not\preceq b$;
- in all other cases, we cannot make any conclusion.

As we increase the location accuracy, we can get more and more information about the casuality. In general, if $a \prec b$, this means that the event a affects all the events in some vicinity of b. Thus, when the location inaccuracy is sufficiently small, we will able to detect that $a \prec b$. In other words, $a \prec b$ if and only we can detect this causality for an appropriate (sufficiently high) level of accuracy.

We can describe this situation by saying that we have a sequence of decidable relations $\prec_n$ corresponding to increasing location accuracy, and

$$a \prec b \Leftrightarrow \exists n\, (a \prec_n b).$$

To detect whether $a \prec b$, we repeat the above experiments with increasing accuracy. If in all these experiments, we do not detect the effect of a on b, this means that a is not in kinematic causality relation with b: $a \not\prec b$. It seems reasonable to argue that if this negative phenomenon does not occur, this means that for some accuracy level n, we will be able to detect the causality. In other words, we require that $\neg(a \not\prec b) \Rightarrow a \prec b$, i.e., that the "Markov principle" $\neg\neg(a \prec b) \Rightarrow a \prec b$ holds for the constructive kinematic causality relation.

As a result, we arrive at the following constructive version of kinematic causality.

Definition 3. *A relation $\prec$ on a set X is called a* constructive kinematic casuality *if it satisfies the following properties:*

- *$\prec$ is transitive: $(a \prec b \,\&\, b \prec c) \Rightarrow a \prec c$.*
- *$\prec$ satisfies the formula $\neg\neg(a \prec b) \Rightarrow a \prec b$.*
- *$\prec$ satisfies the following properties:*

$$a \not\prec a; \quad \forall a\, \exists \underline{a}, \overline{a}\, (\underline{a} \prec a \prec \overline{a}); \quad a \prec b \Rightarrow \exists c\, (a \prec c \prec b);$$

$$a \prec b, c \Rightarrow \exists d\, (a \prec d \prec b, c); \quad b, c \prec a \Rightarrow \exists d\, (b, c \prec d \prec a).$$

- *If $a \prec b$, then $\forall c\, (a \prec c \vee b \not\prec c)$.*
- *There exists a sequence $\{x_i\}$ for which $a \prec b \Rightarrow \exists i\, (a \prec x_i \prec b)$.*
- *There exists a decidable ternary relation $x_i \prec_n x_j$ for which*

$$x_i \prec x_j \Leftrightarrow \exists n\, (x_i \prec_n x_j).$$

A set X with a constructive causality relation is called a constructive space-time.

Constructive meaning: reminder. The main difference between this new definition and the original definition of the kinematic causality is that the existential quantifier $\exists$ (and the disjunction $\vee$) are understood constructively: as the existence of an algorithm that provides the corresponding objects; see, e.g., [1,2,3,4,9,18,19]. In these terms:

- The formula $\forall a\, \exists \underline{a}, \overline{a}\, (\underline{a} \prec a \prec \overline{a})$ means that there exists an algorithm that, given an event a, returns events $\underline{a}$ and $\overline{a}$ for which $\underline{a} \prec a \prec \overline{a}$.
- The formula $a \prec b \Rightarrow \exists c\, (a \prec c \prec b)$ means that there exists an algorithm that, given two events a and b for which $a \prec b$, returns an event c for which $a \prec c \prec b$.
- The formula $a \prec b, c \Rightarrow \exists d\, (a \prec d \prec b, c)$ means that there exists an algorithm that, given events a, b, and c for which $a \prec b, c$, returns an event d for which $a \prec d \prec b, c$.
- The formula $b, c \prec a \Rightarrow \exists d\, (b, c \prec d \prec a)$ means that there exists an algorithm that, given events a, b, and c for which $b, c \prec a$, returns an event d for which $b, c \prec d \prec a$.
- The formula $a \prec b \Rightarrow \forall c\, (a \prec c \vee b \not\prec c)$ means that there exists an algorithm that, given events a, b, and c for which $a \prec b$, returns either a true statement $a \prec c$ or a true statement $b \not\prec c$.
- The formula $a \prec b \Rightarrow \exists i\, (a \prec x_i \prec b)$ means that there exists an algorithm that, given events a and b for which $a \prec b$, returns a natural number i for which $a \prec x_i \prec b$.
- The formula $x_i \prec x_j \Leftrightarrow \exists n\, (x_i \prec_n x_j)$ means that there exists an algorithm that, given natural numbers i and j for which $x_i \prec x_j$, returns a natural number n for which $x_i \prec_n x_j$.

– Finally, the fact that the ternary relation $x_i \prec_n x_j$ is decidable can be described as
$$\forall i\, \forall j\, \forall n\, (x_i \prec_n x_j \vee x_i \not\prec_n x_j).$$

Comment. In strictly constructive terms, we can say that points x_i are simply natural numbers, $x_i \prec_n x_j$ is a ternary relation between natural numbers, and an arbitrary constructive event a can be described by two constructive sequences m_i and M_i for which $x_{m_i} \prec a \prec x_{M_i}$, $x_{m_i} \to x$, and $x_{M_i} \to x$.

In these terms, if an event a is described by sequences m_i and M_i and an event b is described by sequences n_i and N_i, then $a \prec b$ means that there exist i and j for which $x_{M_i} \prec x_{n_j}$.

4 Constructive Space-Time Version of Urysohn's Lemma

Lemma 3. *For every constructive kinematic casuality relation, for every $a \prec b$, there exists a constructive $\preceq$-monotonic function $f_{(a,b)} : X \to [0,1]$ for which $f_{(a,b)}(-a^+) = 0$ and $f_{(a,b)}(b^+) = 1$.*

Comment. This formulation is interpreted constructively.

A real number x is given constructively if we have an algorithm that, given accuracy k, returns a rational number r_k with $|r_k - x| \le 2^{-k}$. A function $f(x)$ is given constructively if, given an input x – i.e., a black box that, given k, returns a 2^{-k}-approximation r_k to the number x – we can compute the value $f(x)$ with a given accuracy.

In this sense, the above formulation means the existence of an algorithm, that, given a, b, x, and a given accuracy k, computes the rational number which is 2^{-k}-close to $f_{(a,b)}(x)$.

Proof.

Part 1. Let us define $\prec$-monotonic values $\gamma(p/2^q)$ for all natural numbers p and q for which $p \le 2^q$. We will define them inductively, first for $q = 0$, then for $q = 1$, etc.

For $q = 0$: We take $\gamma(0) = a$ and $\gamma(1) = b$.

From q to $q+1$: Since $(2p)/2^{q+1} = p/2^q$, the values $\gamma((2p)/2^{q+1})$ are already defined. We just need to define the values $\gamma((2p+1)/2^{q+1})$ corresponding to a midpoint
$$(2p+1)/2^{q+1} = \frac{p/2^q + (p+1)/2^q}{2}$$
between $p/2^q = (2p)/2^{q+1}$ and $(p+1)/2^q = (2p+2)/2^{q+1}$. For each p, since $\gamma(p/2^q) \prec \gamma((p+1)/2^q)$, we can run the algorithm that, given $a \prec b$, returns i for which $a \prec x_i \prec b$; this algorithm is a part of the description of the constructive kinematic causality relation.

By applying this algorithm to $\gamma(p/2^q)$ and $\gamma((p+1)/2^q)$, we get an integer i for which $\gamma(p/2^q) \prec x_i \prec \gamma((p+1)/2^q)$. We then take $\gamma((2p+1)/2^{q+1}) = x_i$.

Comment. In this paper, we operate within an algorithmic approach to constructive mathematics, where existence means the existence of an algorithm. For readers who are more familiar with more general axiomatic approach to constructive mathematics, it is worth mentioning that this construction requires dependent choice.

Part 2. For every x, we now define $f_{(a,b)}(x) \stackrel{\text{def}}{=} \sup\{r : \gamma(r) \prec x\}$. Let us explain how this function can be computed.

Due to the properties of the constructive kinematic causality relation, for each x and for each p and q, we have $\gamma(p/2^q) \prec x \vee \gamma((p+1)/2^q) \not\preceq x$, and hence

$$f_{(a,b)}(x) > p/2^q \vee f_{(a,b)}(x) \leq (p+1)/2^q.$$

In other words, there exist an algorithm that, given x, p, and q, tells us whether $f_{(a,b)}(x) > p/2^q$ or $f_{(a,b)}(x) \leq (p+1)/2^q$. For each q, by applying this algorithm for different $p \leq 2^q$, we can compute the value $f_{(a,b)}(x)$ with accuracy 2^{-q}.

So, the function $f_{(a,b)}(x)$ is indeed computable. □

5 Constructive Space-Time Metrization Theorem

Theorem 3. *For every constructive space-time X with a constructive kinematic causality relation $\prec$, there exists a constructive kinematic metric $\tau(a,b)$ which generates the corresponding kinematic causality relation $\prec$ — in the sense that $a \prec b \Leftrightarrow \tau(a,b) > 0$.*

Comment. This formulation is meant as a constructive one: that there exists an *algorithm* that computes the values of $\tau(a,b)$.

Proof. In this proof, we use the above lemma, that for all $a \prec b$, there exists a $\preceq$-monotonic function $f_{(a,b)} : X \to [0,1]$ for which

$$f_{(a,b)}(-a^+) = 0 \text{ and } f_{(a,b)}(b^+) = 1.$$

Let us define, for every i, the following auxiliary function $f_{x_i} : X \to [0,1]$:

$$f_{x_i}(b) \stackrel{\text{def}}{=} \sum_{j,n:\ x_i \prec_n x_j} 2^{-j} \cdot 2^{-n} \cdot f_{(x_i,x_j)}(b).$$

Since the relation $x_i \prec_n x_j$ is decidable, this function is computable: to compute it with accuracy 2^{-p}, it is sufficient to consider finitely many terms (j,n).

From the $\preceq$-monotonicity of the functions $f_{(x_i,x_j)}(x)$, one can conclude that their linear combination $f_{x_i}(b)$ is also $\preceq$-monotonic.

It is also possible to prove that $f_{x_i}(b) > 0 \Leftrightarrow x_i \prec b$. Indeed:

- If $f_{x_i}(b) > 0$, this means that $f_{(x_i,x_j)}(b) > 0$ for some j. Since $f_{(x_i,x_j)}(-x_i^+) = 0$, this means that b cannot belong to the complement $-x_i^+$, i.e., that

$$\neg\neg(x_i \prec b).$$

  Thus, we have $x_i \prec b$.

- Vice versa, if $x_i \prec b$, then there exists a j for which $x_i \prec x_j \prec b$ and thus, $f_{(x_i,x_j)}(b) = 1$. Since $x_i \prec x_j$, there exists an n for which $x_i \prec_n x_j$ and thus, $f_{x_i}(b) \geq 2^{-j} \cdot 2^{-n} > 0$.

Similarly, we define functions $g_{x_i}(a)$ which are $\preceq$-decreasing and for which $g_{x_i}(a) > 0 \Leftrightarrow a \prec x_i$.

Now, we can define the kinematic metric in the same way as in the non-constructive proof:

$$\tau(a,b) = \sum_{i=1}^{\infty} 2^{-i} \cdot \min(g_{x_i}(a), f_{x_i}(b)).$$

Since $0 \leq g_{x_i}(a) \leq 1$ and $0 \leq f_{x_i}(b) \leq 1$, we have $0 \leq \min(g_{x_i}(a), f_{x_i}(b)) \leq 1$. One can easily check that this formula defines a computable function: to compute it with accuracy 2^{-p}, it is sufficient to compute the sum of the terms $i = 1, \ldots, p$, the remaining terms are bounded from above by the sum

$$2^{-(p+1)} + 2^{-(p+2)} + \ldots = 2^{-p}.$$

So, to complete the proof, we need to prove:

- that the function $\tau(a,b)$ is in correct relation with the kinematic causality relation, and
- that this function satisfies the anti-triangle inequality.

Let us first prove that $a \prec b \Leftrightarrow \tau(a,b) > 0$:

- If $a \prec b$, then there exists i for which $a \prec x_i \prec b$. Thus, by the properties of the functions f_{x_i} and g_{x_i}, we have $g_{x_i}(a) > 0$ and $f_{x_i}(b) > 0$ and thus, $\min(g_{x_i}(a), f_{x_i}(b)) > 0$. Hence, we have $\tau(a,b) > 0$.
- Vice versa, if $\tau(a,b) > 0$, this means that there exists an i for which $\min(g_{x_i}(a), f_{x_i}(b)) > 0$, i.e., for which $g_{x_i}(a) > 0$ and $f_{x_i}(b) > 0$. By the properties of the functions f_{x_i} and g_{x_i}, this means that $a \prec x_i$ and $x_i \prec b$. By transitivity, we can now conclude that $a \prec b$.

To prove the anti-triangle inequality, let us prove that a similar anti-triangle inequality holds for each of the expressions $\min(g_{x_i}(a), f_{x_i}(b))$, i.e., that

$$a \prec b \prec c$$

implies that

$$\min(g_{x_i}(a), f_{x_i}(b)) + \min(g_{x_i}(b), f_{x_i}(c)) \leq \min(g_{x_i}(a), f_{x_i}(c)).$$

Once we prove this, the desired anti-triangle inequality can be obtained by simply multiplying each of these inequalities by 2^{-i} and adding them.

To prove the above inequality, let us take into account that for every real number x, it not possible not to have $x > 0 \vee x \leq 0$: $\neg\neg(x > 0 \vee x \leq 0)$. Thus, we can consider separately

- situations when $\min(g_{x_i}(a), f_{x_i}(b)) > 0$ and
- situations when $\min(g_{x_i}(a), f_{x_i}(b)) = 0$,

and conclude that the double negation of the desired inequality holds. Since for constructive real numbers, $\neg\neg(p \le q)$ is constructively equivalent to $p \le q$, we get the desired inequality.

If $\min(g_{x_i}(a), f_{x_i}(b)) > 0$, this means that $x_i \in (a,b)$. Since $a \prec b \prec c$, this implies that we cannot have $x_i \in (b,c)$, and hence, that $\min(g_{x_i}(b), f_{x_i}(c)) = 0$. Since the function $f_{x_i}(b)$ is $\preceq$-monotonic and $b \prec c$, we have $f_{x_i}(b) \le f_{x_i}(c)$ and thus, $\min(g_{x_i}(a), f_{x_i}(b)) \le \min(g_{x_i}(a), f_{x_i}(c))$. Due to $\min(g_{x_i}(b), f_{x_i}(c)) = 0$, we have $\min(g_{x_i}(a), f_{x_i}(b)) + \min(g_{x_i}(b), f_{x_i}(c)) = \min(g_{x_i}(a), f_{x_i}(b))$ and thus, we get the desired inequality

$$\min(g_{x_i}(a), f_{x_i}(b)) + \min(g_{x_i}(b), f_{x_i}(c)) \le \min(g_{x_i}(a), f_{x_i}(c)).$$

If $\min(g_{x_i}(b), f_{x_i}(c)) > 0$, this means that $x_i \in (b,c)$. Since $a \prec b \prec c$, this implies that we cannot have $x_i \in (a,b)$, and hence, that $\min(g_{x_i}(a), f_{x_i}(b)) = 0$. Since the function $g_{x_i}(b)$ is $\preceq$-decreasing and $a \prec b$, we have $g_{x_i}(b) \le g_{x_i}(a)$ and thus, $\min(g_{x_i}(b), f_{x_i}(c)) \le \min(g_{x_i}(a), f_{x_i}(c))$. Due to $\min(g_{x_i}(a), f_{x_i}(b)) = 0$, we have $\min(g_{x_i}(a), f_{x_i}(b)) + \min(g_{x_i}(b), f_{x_i}(c)) = \min(g_{x_i}(b), f_{x_i}(c))$ and thus, we get the desired inequality

$$\min(g_{x_i}(a), f_{x_i}(b)) + \min(g_{x_i}(b), f_{x_i}(c)) \le \min(g_{x_i}(a), f_{x_i}(c)).$$

Finally, if $\min(g_{x_i}(a), f_{x_i}(b)) = 0$ and $\min(g_{x_i}(b), f_{x_i}(c)) = 0$, then

$$\min(g_{x_i}(a), f_{x_i}(b)) + \min(g_{x_i}(b), f_{x_i}(c)) = 0$$

and hence, since $\min(g_{x_i}(a), f_{x_i}(c)) \ge 0$, we also have the desired anti-triangle inequality. □

6 Additional Results

Similar techniques enable us to prove constructive versions of other results about space-time models.

Definition 4. *By a* time coordinate t *on a space* X *with a kinematic causality relation* $\prec$*, we mean a function* $t : X \to \mathbb{R}$ *for which:*

- $a \prec b \Rightarrow t(a) < t(b)$*; and*
- $a \preceq b \Rightarrow t(a) \le t(b)$.

Proposition 1. *On every constructive space-time* X *with a constructive kinematic causality relation* $\prec$*, there exists a constructive time coordinate.*

Proof. The desired constructive version of a time coordinate can be designed as follows:

$$t(b) \stackrel{\text{def}}{=} \sum_{i=1}^{\infty} 2^{-i} \cdot f_{x_i}(b).$$

Since $f_{x_i}(b) \in [0,1]$, this is constructively defined (computable): to compute $t(b)$ with accuracy 2^{-p}, it is sufficient to add first p terms in the sum.

Let us prove that this function is indeed the time coordinate. Indeed, since each of the functions $f_{x_i}(b)$ is $\preceq$-monotonic, their convex combination $t(b)$ is also $\preceq$-monotonic.

To prove that the function $t(b)$ is $\prec$-monotonic, we can use the fact that $a \prec b$ implies the existence of a natural number i for which $a \prec x_i \prec b$. For this i, we have $f_{x_i}(a) = 0$ and $f_{x_i}(b) > 0$, hence $f_{x_i}(a) < f_{x_i}(b)$. For all other $j \neq i$, due to $a \prec b \Rightarrow a \preceq b$ and $\preceq$-monotonicity of f_{x_j}, we have $f_{x_j}(a) \le f_{x_j}(b)$. Thus, by adding these inequalities, we get $t(a) < t(b)$. □

Comment. This result is similar to the constructive existence of a *utility function* $u(x)$, i.e., a function for which $a \prec b$ implies $u(a) < u(b)$, where $\prec$ is a preference relation; see, e.g., [5,6,7,8].

All possible time coordinates determine the causality relation: non-constructive case. In Newtonian physics, time $t(a)$ is absolute, and

$$a \preceq b \Leftrightarrow t(a) \le t(b).$$

One of the main discoveries that led Einstein to his Special Relativity theory is the discovery that time is relative: a time coordinate corresponding to a moving body is different from the time coordinate corresponding to the stationary one. In general, there are many possible time coordinates t, each of which has the same property:

$$a \preceq b \Rightarrow \forall t\,(t(a) \le t(b)).$$

For each of these time coordinates t, the mere fact that $t(a) \le t(b)$ does not necessarily mean that a causally precedes b: it may happen that in some other time coordinate, we have $t(a) > t(b)$. What *is* true is that if a is *not* causally preceding b, then there exist a time coordinate for which $t(a) > t(b)$:

$$a \not\preceq b \Rightarrow \exists t\,(t(a) > t(b)).$$

In non-constructive space-time geometry, the above two statements simply mean that

$$a \preceq b \Leftrightarrow \forall t\,(t(a) \le t(b)),$$

i.e., that the causality relation is uniquely determined by the class of all possible time coordinates.

All possible time coordinates determine the causality relation: constructive case. Let us show that in the constructive case, under reasonable conditions, we also have the implication

$$a \not\preceq b \Rightarrow \exists t\,(t(a) > t(b)).$$

For that, we will need to impose an additional physically reasonable requirement.

For every event b, the *past cone* $P_b \stackrel{\text{def}}{=} \{c : c \preceq b\}$ is a closed set; thus, classically, its complement $-P_b = \{c : c \preceq b\}$ is an open set. The point a belongs to this set; thus, a whole open neighborhood of a belongs to this set as well. Since the topology is the Alexandrov topology, with intervals as a base, this means that there exist values $\underline{a}$ and $\overline{a}$ which which $\underline{a} \prec a \prec \overline{a}$ and the whole interval $(\underline{a}, \overline{a})$ belongs to the complement $-P_b$.

Since the sequence $\{x_i\}$ is everywhere dense in X, there is a point x_i in the interval $(\underline{a}, a)$, i.e., a point x_i for which $x_i \prec a$ and $x_i \not\preceq b$. By measuring the event locations with higher and higher accuracy, we will be able to detect this relation. Thus, it is reasonable to require that the following additional condition constructively holds:

$$a \not\preceq b \Rightarrow \exists i\, (x_i \prec a \,\&\, x_i \not\preceq b).$$

Let us show that under this condition, the above implication holds.

Proposition 2. *Let X be a constructive space-time with a constructive causality relation $\prec$ for which*

$$a \not\preceq b \Rightarrow \exists i\, (x_i \prec a \,\&\, x_i \not\preceq b).$$

Then, for every $a \not\preceq b$, there exists a constructive time coordinate t for which $t(a) > t(b)$.

Proof. Indeed, let i_0 be an index for which $x_{i_0} \prec a$ and $x_{i_0} \not\preceq b$. For this i_0, we thus have $f_{x_{i_0}}(a) > 0$ and $f_{x_{i_0}}(b) = 0$. Let us now construct the following time coordinate:

$$t(x) = \frac{2}{f_{x_{i_0}}(a)} \cdot f_{x_{i_0}}(x) + \sum_{i \neq i_0}^{\infty} 2^{-i} \cdot f_{x_i}(x).$$

Similar to the above formula, we can check that the function thus defined is indeed a time coordinate. It is therefore sufficient to show that $t(a) > t(b)$. Indeed:

- For $x = a$, the first term in the sum is equal to $\dfrac{2}{f_{x_{i_0}}(a)} \cdot f_{x_{i_0}}(x) = 2$, so $t(a) \geq 2$.
- For $x = b$, the first term is equal to 0. Since $f_{x_i}(x) \leq 1$ for all i, we thus conclude that

$$t(b) = \sum_{i \neq i_0}^{\infty} 2^{-i} \cdot f_{x_i}(x) \leq \sum_{i=1}^{\infty} 2^{-i} = 1.$$

Here, $t(a) \geq 2$ and $t(b) \leq 1$, and hence indeed $t(a) > t(b)$. □

Comment. Without this additional requirement, we can only prove that

$$a \not\preceq b \Rightarrow \neg\neg\exists t\, (t(a) > t(b)).$$

The existence of a standard metric. Another constructive result is the existence of a standard metric on each space-time model.

Proposition 3. *On every constructive space X with a constructive kinematic causality relation $\prec$, there exists a constructive metric $\rho(a, b)$.*

Proof. The desired constructive metric can be defined as follows:

$$\rho(a,b) \stackrel{\text{def}}{=} \sum_{i=1}^{\infty} 2^{-i} \cdot |f_{x_i}(a) - f_{x_i}(b)|.$$

One can easily check that this function is computable, and that it is indeed a metric – i.e., that it is symmetric and satisfies the triangle inequality. □

7 Remaining Challenges

Need to take symmetries into account. In this paper, given space-time X with the kinematic causality relation $\prec$, we designed a kinematic metric τ that is consistent with this relation.

In physics, however, causality is not everything. One of the most important notions of physics is symmetry. If space-time has symmetries – i.e., is invariant with respect to some transformations – it is therefore desirable to find a kinematic metric τ which is invariant with respect to these symmetries.

In the simplest case of a finite symmetry group G, we can explicitly define such a invariant constructive kinematic metric as

$$\tau_{\text{inv}}(a,b) \stackrel{\text{def}}{=} \sum_{g \in G} \tau(g(a), g(b)).$$

An important case is when X is both an ordered group and a space with a kinematic causality relation $\prec$, and the closures of all intervals are compact sets. It known that in the non-constructive case, there exists a left-invariant metric $\tau(a,b)$: namely, $\tau(a,b) = \mu_H(\{c : a \preceq c \preceq b\})$ where μ_H is the (left-invariant) Haar measure; see, e.g., [14]. It is desirable to constructivize this and similar results.

Need for feasible algorithms. In this paper, we have analyzed the existence of algorithms for computing the kinematic metric. From the practical viewpoint, it is important to make sure not only that such algorithms exist, but that they are *feasible* (i.e., can be computed in polynomial time); see, e.g., [11].

Partial analysis of feasibility of different computational problems related to space-time models is given in [17]. It is desirable to extend this analysis to the problem of computing kinematic metric.

Acknowledgements. This work was supported in part by the National Science Foundation grant HRD-0734825, by Grant 1 T36 GM078000-01 from the National Institutes of Health, by the Max Planck Institut für Mathematik, and by Grant MSM 6198898701 from MŠMT of Czech Republic.

The author is thankful to all the participants of Conference on Abelian Groups and on Constructive Mathematics (Boca Raton, Florida, May 9–11, 2008), especially to Douglas S. Bridges, for valuable suggestions, and to the anonymous referees for their help.

References

1. Aberth, O.: Introduction to Precise Numerical Methods. Academic Press, San Diego (2007)
2. Beeson, M.: Foundations of Constructive Mathematics: Metamathematical Studies. Springer, Heidelberg (1985)
3. Beeson, M.: Some relations between classical and constructive mathematics. Journal of Symbolic Logic 43, 228–246 (1987)
4. Bishop, E., Bridges, D.S.: Constructive Analysis. Springer, Heidelberg (1985)
5. Bridges, D.S.: The construction of a continuous demand function for uniformly rotund preferences. J. Math. Economics 21, 217–227 (1992)
6. Bridges, D.S.: The constructive theory of preference orderings on a locally compact space II. Math. Social Sciences 27, 1–9 (1994)
7. Bridges, D.S.: Constructive methods in mathematical economics. Mathematical Utility Theory, J. Econ. (Zeitschrift für Nationalökonomie) (suppl. 8), 1–21 (1999)
8. Bridges, D.S., Mehta, G.B.: Representations of Preference Orderings. Lecture Notes in Economics and Mathematical Systems, vol. 422. Springer, Heidelberg (1996)
9. Bridges, D.S., Vîta, L.: Techniques of Constructive Mathematics. Springer, New York (2006)
10. Busemann, H.: Timelike Spaces. Warszawa, PWN (1967)
11. Gurevich, Y.: Platonism, constructivism, and computer proofs vs. proofs by hand. Bulletin of EATCS (European Association for Theoretical Computer Science) 57, 145–166 (1995)
12. Kelley, J.L.: General Topology. Springer, New York (1975)
13. Kronheimer, E.H., Penrose, R.: On the structure of causal spaces. Proc. Cambr. Phil. Soc. 63, 481–501 (1967)
14. Kreinovich, V.: On the metrization problem for spaces of kinematic type. Soviet Mathematics Doklady 15, 1486–1490 (1974)
15. Kreinovich, V.: Categories of Space-Time Models. Ph.D. dissertation, Soviet Academy of Sciences, Siberian Branch, Institute of Mathematics, Novosibirsk (1979) (in Russian)
16. Kreinovich, V.: Symmetry characterization of Pimenov's spacetime: a reformulation of causality axioms. International J. Theoretical Physics 35, 341–346 (1996)
17. Kreinovich, V., Kosheleva, O.: Computational complexity of determining which statements about causality hold in different space-time models. Theoretical Computer Science 405, 50–63 (2008)
18. Kreinovich, V., Lakeyev, A., Rohn, J., Kahl, P.: Computational Complexity and Feasibility of Data Processing and Interval Computations. Kluwer, Dordrecht (1998)
19. Kushner, B.A.: Lectures on Constructive Mathematical Analysis. American Mathematical Society, Providence (1985)
20. Misner, C.W., Thorne, K.S., Wheeler, J.A.: Gravitation. W.H. Freeman, New York (1973)
21. Nachbin, L.: Topology and Order. Van Nostrand, Princeton (1965); reprinted by R. E. Kreiger: Huntington (1976)
22. Pimenov, R.I.: Kinematic spaces: Mathematical Theory of Space-Time. Consultants Bureau, N.Y (1970)

Thirteen Definitions of a Stable Model

Vladimir Lifschitz

Department of Computer Sciences, University of Texas at Austin, USA
vl@cs.utexas.edu

Abstract. Stable models of logic programs have been studied by many researchers, mainly because of their role in the foundations of answer set programming. This is a review of some of the definitions of the concept of a stable model that have been proposed in the literature. These definitions are equivalent to each other, at least when applied to traditional Prolog-style programs, but there are reasons why each of them is valuable and interesting. A new characterization of stable models can suggest an alternative picture of the intuitive meaning of logic programs; or it can lead to new algorithms for generating stable models; or it can work better than others when we turn to generalizations of the traditional syntax that are important from the perspective of answer set programming; or it can be more convenient for use in proofs; or it can be interesting simply because it demonstrates a relationship between seemingly unrelated ideas.

Keywords: answer set programming, nonmonotonic reasoning, stable models.

1 Introduction

Stable models of logic programs have been studied by many researchers, mainly because of their role in the foundations of answer set programming (ASP) [30,37]. This programming paradigm provides a declarative approach to solving combinatorial search problems, and it has found applications in several areas of science and technology [21]. In ASP, a search problem is reduced to computing stable models, and programs for generating stable models ("answer set solvers") are used to perform search.

This paper is a review of some of the definitions, or characterizations, of the concept of a stable model that have been proposed in the literature. These definitions are equivalent to each other when applied to "traditional rules" – with an atom in the head and a list of atoms, some possibly preceded with the negation as failure symbol, in the body:

$$A_0 \leftarrow A_1, \ldots, A_m, not\ A_{m+1}, not\ A_n. \qquad (1)$$

But there are reasons why each of them is valuable and interesting. A new characterization of stable models can suggest an alternative picture of the intuitive meaning of logic programs; or it can lead to new algorithms for generating stable

A. Blass, N. Dershowitz, and W. Reisig (Eds.): Gurevich Festschrift, LNCS 6300, pp. 488–503, 2010.

models; or it can work better when we turn to generalizations of the traditional syntax that are important from the perspective of answer set programming; or it can be more convenient for use in proofs, such as proofs of correctness of ASP programs; or, quite simply, it can intellectually excite us by demonstrating a relationship between seemingly unrelated ideas.

We concentrate here primarily on programs consisting of finitely many rules of type (1), although generalizations of this syntactic form are mentioned several times in the second half of the paper. Some work on the stable model semantics, for instance [15,23,39,2], is not discussed here because it is about extending, rather than modifying, the definitions proposed earlier; this kind of work does not tell us much new about stable models of traditional programs.

The paper begins with comments on the relevant work that had preceded the invention of stable models – on the semantics of logic programming (Sect. 2) and on formal nomonotonic reasoning (Sect. 3). Early contributions that can be seen as characterizations of the class of stable models in terms of nonmonotonic logic are discussed in Sect. 4. Then we review the "standard" definition of stable models – in terms of reducts (Sect. 5) – and turn to its characterizations in terms of unfounded sets and loop formulas (Sect. 6). After that, we talk about the definitions of stable models in terms of circumscription (Sect. 7) and in terms of support relative to a well-ordering (Sect. 8), discuss two characterizations based on tightening (Sect. 9), and talk about the relationship between stable models and equilibrium logic (Sect. 10).

In recent years, two interesting modifications of the definition of the reduct were introduced (Sect. 11). We have learned also that a simple change in the definition of circumscription can give a characterization of stable models (Sect. 12).

This is an extended version of the conference paper [20].

2 Minimal Models, Completion, and Stratified Programs

2.1 Minimal Models vs. Completion

According to [47], a logic program without negation represents the least (and so, the only minimal) Herbrand model of the corresponding set of Horn clauses. On the other hand, according to [4], a logic program represents a certain set of first-order formulas, called the program's completion.

These two ideas are closely related to each other, but not equivalent. Take, for instance, the program

$$\begin{array}{l} p(a,b). \\ p(X,Y) \leftarrow p(Y,X). \end{array} \tag{2}$$

The minimal Herbrand model

$$\{p(a,b), p(b,a)\} \tag{3}$$

of this program satisfies the program's completion

$$\forall XY(p(X,Y) \leftrightarrow ((X = a \wedge Y = b) \vee p(Y,X))) \wedge a \neq b.$$

But there also other Herbrand interpretations satisfying the program's completion – for instance,

$$\{p(a,b), p(b,a), p(b,b)\}. \tag{4}$$

Another example of this kind, important in applications, is given by the recursive definition of transitive closure:

$$\begin{array}{l} q(X,Y) \leftarrow p(X,Y). \\ q(X,Z) \leftarrow q(X,Y), q(Y,Z). \end{array} \tag{5}$$

The completion of the union of this program with a definition of p has, in many cases, unintended models, in which q is weaker than the transitive closure of p that we want to define.

Should we say then that Herbrand minimal models provide a better semantics for logic programming than program completion? Yes and no. The concept of completion has a fundamental advantage: it is applicable to programs with negation. Such a program, viewed as a set of clauses, usually has several minimal Herbrand models, and some of them may not satisfy the program's completion. Such "bad" models reflect neither the intended meaning of the program nor the behavior of Prolog. For instance, the program

$$\begin{array}{l} p(a). \quad p(b). \quad q(a). \\ r(X) \leftarrow p(X), not\ q(X). \end{array} \tag{6}$$

has two minimal Herbrand models:

$$\{p(a), p(b), q(a), r(b)\} \tag{7}$$

("good") and

$$\{p(a), p(b), q(a), q(b)\} \tag{8}$$

("bad"). The completion of (6)

$$\begin{array}{c} \forall X(p(X) \leftrightarrow (X = a \vee X = b)) \wedge \forall X(q(X) \leftrightarrow X = a) \\ \wedge \forall X(r(X) \leftrightarrow (p(X) \wedge \neg q(X))) \wedge a \neq b \end{array}$$

characterizes the good model.

2.2 The Challenge

In the 1980s, the main challenge in the study of the semantics of logic programming was to invent a semantics that

- in application to a program without negation, such as (2), describes the minimal Herbrand model,
- in the presence of negation, as in example (6), selects a "good" minimal model satisfying the program's completion.

Such a semantics was proposed in two papers presented at the 1986 Workshop on Foundations of Deductive Databases and Logic Programming [1,48]. That approach was not without defects, however. First, it is limited to programs in which recursion and negation "don't mix." Such programs are called stratified. Unfortunately, some useful Prolog programs do not satisfy this condition. For instance, we can say that a position in a two-person game is winning if there exists a move from it to a non-winning position (cf. [46]). This rule is not stratified: it recursively defines winning in terms of non-winning. A really good semantics should be applicable to rules like this.

Second, the definition of the semantics of stratified programs is somewhat complicated. It is based on the concept of the iterated least fixpoint of a program, and to prove the soundness of this definition one needs to show that this fixpoint doesn't depend on the choice of a stratification. A really good semantics should be a little easier to define.

The stable model semantics, as well as the well-founded semantics proposed in [49,50], can be seen as an attempt to generalize and simplify the iterated fixpoint semantics of stratified programs.

3 Nonmonotonic Reasoning

Many events in the history of research on stable models can be only understood if we think of it as part of a broader research effort – the investigation of nonmonotonic reasoning. Early work in this area was motivated primarily by the desire to understand and automate the use of defaults by humans. When commonsense reasoning exploits a default, there is a possibility that taking into account new information may force us to retract the conclusions that we have made. The same kind of nonmonotonicity is observed in the behavior of Prolog programs with negation. For instance, rules (6) warrant the conclusion $r(b)$, but if the fact $q(b)$ is added to the program then this conclusion will be retracted.

As to the stable model semantics, three nonmonotonic formalisms are particularly relevant.

3.1 Circumscription

Circumscription [31,32] is a syntactic transformation that turns a first-order sentence F into the conjunction of F with another formula, which expresses a minimality condition (the exact form of that condition depends on the "circumscription policy"). This additional conjunctive term involves second-order quantifiers.

Circumscription generalizes the concept of a minimal model from [47]. The iterated fixpoint semantics of stratified programs can be characterized in terms of circumscription also [19]. On the other hand, circumscription is similar to program completion in the sense that both are syntactic transformations that make a formula stronger. The relationship between circumscription and program completion was investigated in [43].

3.2 Default Logic

A default theory in the sense of [42] is characterized by a set W of "axioms" – first-order sentences, and a set D of "defaults" – expressions of the form

$$\frac{F \,:\, \mathsf{M}\, G_1, \ldots, \mathsf{M}\, G_n}{H}, \tag{9}$$

where $F, G_1, \ldots, G_n, H$ are first-order formulas. The letter M, according to Reiter, is to be read as "it is consistent to assume." Intuitively, default (9) is similar to the inference rule allowing us to derive the conclusion H from the premise F, except that the applicability of this rule is limited by the justifications $G_1, \ldots, G_n$; deriving H is allowed only if each of the justifications can be "consistently assumed."

This informal description of the meaning of a default is circular: to decide which formulas can be derived using one of the defaults from D we need to know whether the justifications of that default are consistent with the formulas that can be derived from W using the inference rules of classical logic and the defaults from D – including the default that we are trying to understand! But Reiter was able to turn his intuition about M into a precise semantics. His theory of defaults tells us under what conditions a set E of sentences is an "extension" for the default theory with axioms W and defaults D.

In Sect. 4 we will see that one of the earliest incarnations of the stable model semantics was based on treating rules as defaults in the sense of Reiter.

3.3 Autoepistemic Logic

According to [36], autoepistemic logic "is intended to model the beliefs of an agent reflecting upon his own beliefs." Moore's definition of propositional autoepistemic logic builds on the ideas of [35] and [34].

Formulas of this logic are constructed from atoms using propositional connectives and the modal operator L ("is believed"). Its semantics specifies, for any set A of formulas ("axioms"), which sets of formulas are considered "stable expansions" of A. Intuitively, Moore explains, the stable expansions of A are "the possible sets of beliefs that a rational agent might hold, given A as his premises."

In Sect. 4 we will see that one of the earliest incarnations of the stable model semantics was based on treating rules as autoepistemic axioms in the sense of Moore. The term "stable model" is historically related to "stable expansions" of autoepistemic logic.

3.4 Relations between Nonmonotonic Formalisms

The intuitions underlying circumscription, default logic, and autoepistemic logic are different from each other, but related. For instance, circumscribing (that is, minimizing the extent of) a predicate p is somewhat similar to adopting the default

$$\frac{\mathrm{true} \,:\, \mathsf{M}\,\neg p(X)}{\neg p(X)}$$

(if it is consistent to assume that X does not have the property p, conclude that it doesn't). On the other hand, Moore observes that "a formula is consistent if its negation is not believed"; accordingly, Reiter's M is somewhat similar to the combination $\neg\mathsf{L}\neg$ in autoepistemic logic, and default (9), in propositional case, is somewhat similar to the autoepistemic formula

$$F \wedge \neg\mathsf{L}\neg G_1 \wedge \cdots \wedge \neg\mathsf{L}\neg G_n \rightarrow H.$$

However, the task of finding precise and general relationships between these three formalisms turned out to be difficult. Discussing technical work on that topic is beyond the scope of this paper.

4 Definitions A and B, in Terms of Translations into Nonmonotonic Logic

The idea of [13] is to think of the expression *not* A in a logic program as synonymous with the autoepistemic formula $\neg\mathsf{L}A$ ("A is not believed"). Since autoepistemic logic is propositional, the program needs to be grounded before this transformation is applied. After grounding, each rule (1) is rewritten as a formula:

$$A_1 \wedge \cdots \wedge A_m \wedge \neg A_{m+1} \wedge \cdots \wedge \neg A_n \rightarrow A_0, \tag{10}$$

and then L inserted after each negation. For instance, to explain the meaning of program (6), we take the result of its grounding

$$\begin{array}{l} p(a).\quad p(b).\quad q(a). \\ r(a) \leftarrow p(a), not\ q(a). \\ r(b) \leftarrow p(b), not\ q(b). \end{array} \tag{11}$$

and turn it into a collection of formulas:

$$\begin{array}{l} p(a),\quad p(b),\quad q(a), \\ p(a) \wedge \neg\mathsf{L}\, q(a) \rightarrow r(a), \\ p(b) \wedge \neg\mathsf{L}\, q(b) \rightarrow r(b). \end{array}$$

The autoepistemic theory with these axioms has a unique stable expansion, and the atoms from that stable expansion form the intended model (7) of the program.

This epistemic interpretation of logic programs – what we will call *Definition A* – is more general than the iterated fixpoint semantics, and it is much simpler. One other feature of Definition A that makes it attractive is the simplicity of the underlying intuition: negation as failure expresses the absence of belief.

The "default logic semantics" proposed in [3] is translational as well; it interprets logic programs as default theories. The head A_0 of a rule (1) turns into the conclusion of the default, the conjunction $A_1 \wedge \cdots \wedge A_m$ of the positive members of the body becomes the premise, and each negative member *not* A_i turns into the justification $\mathsf{M}\neg A_i$ ("it is consistent to assume $\neg A_i$"). For instance, the last rule of program (6) corresponds to the default

$$\frac{p(X)\ :\ \mathsf{M}\neg q(X)}{r(X)}. \tag{12}$$

There is no need for grounding, because defaults are allowed to contain variables. This difference between the two translations is not essential though, because Reiter's semantics of defaults treats a default with variables as the set of its ground instances. Grounding is simply "hidden" in the semantics of default logic.

This *Definition B* of the stable model semantics stresses an analogy between rules in logic programming and inference rules in logic. Like Definition A, it has an epistemic flavor, because of the relationship between the "consistency operator" M in defaults and the autoepistemic "belief operator" L (Sect. 3.4).

The equivalence between these two approaches to semantics of traditional programs follows from the fact that each of them is equivalent to Definition C of a stable model reviewed in the next section. This was established in [14] for the autoepistemic semantics and in [29] for the default logic approach.

5 Definition C, in Terms of the Reduct

Definitions A and B are easy to understand – assuming that one is familiar with formal nonmonotonic reasoning. Can we make these definitions direct and avoid explicit references to autoepistemic logic and default logic?

This question has led to the most widely used definition of the stable model semantics, *Definition C* [14]. The *reduct* of a program Π relative to a set M of ground atoms is obtained from Π by grounding followed by

(i) dropping each rule (1) containing a term $not\ A_i$ with $A_i \in M$, and
(ii) dropping the negative parts $not\ A_{m+1}, \ldots, not\ A_n$ from the bodies of the remaining rules.

We say that M is a *stable model* of Π if the minimal model of (the set of clauses corresponding to the rules of) the reduct of Π with respect to M equals M. For instance, the reduct of program (6) relative to (7) is

$$\begin{array}{l} p(a).\quad p(b).\quad q(a). \\ r(b) \leftarrow p(b). \end{array} \tag{13}$$

The minimal model of this program is the set (7) that we started with; consequently, that set is a stable model of (6).

Definition C was independently invented in [12].

6 Definitions D and E, in Terms of Unfounded Sets and Loop Formulas

From [45] we learned that stable models can be characterized in terms of the concept of an unfounded set, which was introduced in [49] as part of the definition of the well-founded semantics. Namely, a set M of atoms is a stable model of a (grounded) program Π iff

(i) M satisfies Π,[1] and
(ii) no non-empty subset of M is unfounded for Π with respect to M.[2]

This *Definition D* can be refined using the concept of a loop, introduced many years later in [27]. A *loop* of a grounded program Π is a non-empty set L of ground atoms such that any elements A, A' of L can be connected by a chain

$$A = A_1, A_2, \dots, A_{k-1}, A_k = A' \qquad (k > 1)$$

of elements of L satisfying the following condition: for any $i = 1, \dots, k-1$, the program contains a rule R that such that A_i is the head of R and A_{i+1} is a member of the body of R. For instance, the result of grounding program (2)

$$\begin{array}{l} p(a,b). \\ p(a,a) \leftarrow p(a,a). \\ p(a,b) \leftarrow p(b,a). \\ p(b,a) \leftarrow p(a,b). \\ p(b,b) \leftarrow p(b,b). \end{array} \tag{14}$$

has 3 loops:

$$\{p(a,a)\},\ \{p(b,b)\},\ \{p(a,b), p(b,a)\}. \tag{15}$$

Program (11) has no loops.

According to [17], if we require in condition (i) above that M satisfy the *completion* of the program, rather than the program itself, then it will be possible to relax condition (ii) and require only that no *loop* contained in M be unfounded; there will be no need then to refer to arbitrary non-empty subsets in that condition.

In [27] loops are used in a different way. That paper associates with every loop L of Π a certain propositional formula, called the *loop formula* for L. According to *Definition E*, M is a stable model of Π iff M satisfies the completion of Π conjoined with the loop formulas for all loops of Π. For instance, the loop formulas for the loops (15) of program (14) are

$$\begin{array}{l} p(a,a) \rightarrow \text{false}, \\ p(b,b) \rightarrow \text{false}, \\ (p(a,b) \vee p(b,a)) \rightarrow \text{true}. \end{array}$$

The first two loop formulas eliminate all nonminimal models of the completion of (14).

---

[1] That is, M satisfies the propositional formulas (10) corresponding to the rules of Π.

[2] To be precise, unfoundedness is defined with respect to a partial interpretation, not a set of atoms. But we are only interested here in the special case when the partial interpretation is complete, and assume that complete interpretations are represented by sets of atoms in the usual way.

The invention of loop formulas has led to the creation of systems for generating stable models that use SAT solvers for search ("SAT-based answer set programming"). Several systems of this kind performed well in a recent competition [5].

7 Definition F, in Terms of Circumscription

We saw in Sect. 4 that a logic program can be viewed as shorthand for an autoepistemic theory or a default theory. The characterization of stable models described in [24, Sect. 3.4.1] relates logic programs to the third nonmonotonic formalism reviewed above, circumscription. Like Definitions A and B, it is based on a translation, but the output of that translation is not simply a circumscription formula; it involves also some additional conjunctive terms.

The first step of that translation consists in replacing the occurrences of each predicate symbol p in the negative parts $\neg A_{m+1} \wedge \cdots \wedge \neg A_n$ of the formulas (10) corresponding to the rules of the program with a new symbol p' and forming the conjunction of the universal closures of the resulting formulas. The sentence obtained in this way is denoted by $C(\Pi)$. For instance, if Π is (6) then $C(\Pi)$ is

$$p(a) \wedge p(b) \wedge q(a) \wedge \forall X(p(X) \wedge \neg q'(X) \to r(X)).$$

The translation of Π is a conjunction of two sentences: the circumscription of the old (non-primed) predicates in $C(\Pi)$ and the formulas asserting, for each of the new predicates, that it is equivalent to the corresponding old predicate. For instance, the translation of (6) is

$$\mathrm{CIRC}[C(\Pi)] \wedge \forall X(q'(X) \leftrightarrow q(X)); \tag{16}$$

the circumscription operator CIRC is understood here as the minimization of the extents of p, q, r.

The stable models of Π can be characterized as the Herbrand interpretations satisfying the translation of Π, with the new (primed) predicates removed from them ("forgotten").

An interesting feature of this *Definition F* is that, unlike Definitions A–E, it does not involve grounding. We can ask what non-Herbrand models of the translation of a logic program look like. Can it be convenient in some cases to represent possible states of affairs by such "non-Herbrand stable models" of a logic program? A non-Herbrand model may include an object that is different from the values of all ground terms, or there may be several ground terms having the same value in it; can this be sometimes useful?

We will return to the relationship between stable models and circumscription in Sect. 12.

8 Definition G, in Terms of Support

As observed in [1], an Herbrand model M of a grounded program satisfies the program's completion iff every element A of M is *supported*, in the sense that the program contains a rule (1) such that

$$A_0 = A, \quad A_1, \ldots, A_m \in M, \quad A_{m+1}, \ldots, A_n \notin M. \tag{17}$$

A stronger form of this condition, given in [6], characterizes the class of stable models. According to his *Definition G*, an Herbrand model M of a grounded program is stable if there exists a well-ordering $\leq$ of M such that every element A of M is *supported relative to this well-ordering*, in the sense that the program contains a rule (1) satisfying conditions (17) and

$$A_1, \ldots, A_m < A_0.$$

For instance, the stable model (3) of program (14) is supported relative to the order $p(a,b) < p(b,a)$. On the other hand, model (4) is supported, but it is easy to see that it is not supported relative to any ordering of its elements.[3]

When M is finite, well-orderings of M can be described in terms of functions from atoms to integers, and supportedness relative to an order relation can be described by a formula of difference logic – the extension of propositional logic that includes variables for integers and atomic formulas of the form $x - y \geq c$. This observation suggests the posibility of using solvers for difference logic to generate stable models [38].

9 Definitions H and I, in Terms of Tightening and the Situation Calculus

We will talk now about two characterizations of stable models that are based, like Definition F, on translations into classical logic that use auxiliary predicates.

Programs that have no loops, such as (11), are called *tight*. The stable models of a tight program are identical to the Herbrand models of its completion [8]. *Definition H* [51] is based on a process of "tightening" that makes an arbitrary traditional program tight. This process uses two auxiliary symbols: the object constant 0 and the unary function constant s ("successor"). The tightened program uses also auxiliary predicates with an additional numeric argument. Intuitively, $p(X, N)$ expresses that there exists a sequence of N "applications" of rules of the program that "establishes" $p(X)$. The stable models of a program are described then as Herbrand models of the completion of the result of its tightening, with the auxiliary symbols "forgotten."

We will not reproduce here the definition of tightening, but here is an example: the result of tightening program (6) is

$$\begin{array}{l} p(a, s(N)). \quad p(b, s(N)). \quad q(a, s(N)). \\ r(X, s(N)) \leftarrow p(X, N), \mathit{not}\ q(X). \\ p(X) \leftarrow p(X, N). \\ q(X) \leftarrow q(X, N). \\ r(X) \leftarrow r(X, N). \end{array}$$

[3] To be precise, the characterization of stable models in [6] is limited to finite models, and it uses different terminology.

Rules in line 1 tell us that $p(a)$ can be established in any number of steps that is greater than 0; similarly for $p(b)$ and $q(a)$. According to line 2, $r(X)$ can be established in $N+1$ steps if $p(X)$ can be established in N steps and $q(X)$ cannot be established at all (note that an occurrence of a predicate does not get an additional numeric argument if it is negated). Finally, an atom holds if it can be established by some number N of rule applications.

Definition I [25] treats a rule in a logic program as an abbreviated description of the effect of an action – the action of "applying" that rule – in the situation calculus.[4] For instance, if the action corresponding to the last rule of (6) is denoted by *lastrule*(X) then that rule can be viewed as shorthand for the situation calculus formula

$$p(X,S) \wedge \neg\exists S(q(X,S)) \rightarrow r(X, do(lastrule(X), S))$$

(if $p(X)$ holds in situation S and $q(X)$ does not hold in any situation then $r(X)$ holds after executing action *lastrule*(X) in situation S).

In this approach to stable models, the situation calculus function *do* plays the same role as adding 1 to N in Wallace's theory. Instead of program completion, Lin and Reiter use the process of turning effect axioms into successor state axioms, which is standard in applications of the situation calculus.

10 Definition J, in Terms of Equilibrium Logic

The logic of here-and-there, going back to the early days of modern logic [16], is a modification of classical propositional logic in which propositional interpretations in the usual sense – assignments, or sets of atoms – are replaced by pairs (X, Y) of sets of atoms such that $X \subseteq Y$. (We think of X as the set of the atoms that are true "here", and Y as the set of the atoms that are true "there.") The semantics of this logic defines when (X, Y) *satisfies* a formula F.

In [40], the logic of here-and-there was used as a starting point for defining a nonmonotonic logic closely related to stable models. According to Pearce, a pair (Y, Y) is an *equilibrium model* of a propositional formula F if F is satisfied in the logic of here-and-there by (Y, Y) but is not satisfied by (X, Y) for any proper subset X of Y. Pearce showed that a set M of atoms is a stable model of a program Π iff (M, M) is an equilibrium model of the set of propositional formulas (10) corresponding to the grounded rules of Π.

This *Definition J* is important for two reasons. First, it suggests a way to extend the concept of a stable model from traditional rules – formulas of form (1) – to arbitrary propositional formulas: we can say that M is a stable model of a propositional formula F if (M, M) is an equilibrium model of F. This is valuable from the perspective of answer set programming, because many "nonstandard" constructs commonly used in ASP programs, such as choice rules and weight constraints, can be viewed as abbreviations for propositional formulas [11]. Second, Definition J is a key to the theorem about the relationship between the concept of strong equivalence and the logic of here-and-there [22].

[4] See [44] for a detailed description of the situation calculus [33] as developed by the Toronto school.

11 Definitions K and L, in Terms of Modified Reducts

In [7] the definition of the reduct reproduced in Sect. 5 is modified by including the positive members of the body, along with negative members, in the description of step (i), and by removing step (ii) altogeher. In other words, in the modified process of constructing the reduct relative to M we delete from the program all rules (1) containing in their bodies a term A_i such that $A_i \notin M$ or a term *not* A_i such that $A_i \in M$; the other rules of the program remain unchanged. For instance, the modified reduct of program (6) relative to (7) is

$$\begin{array}{l} p(a). \quad p(b). \quad q(a). \\ r(b) \leftarrow p(b), not\ q(b). \end{array}$$

Unlike the reduct (13), this modified reduct contains negation as failure in the last rule. Generally, unlike the reduct in the sense of Sect. 5, the modified reduct of a program has several minimal models.

According to *Definition K*, M is a stable model of Π if M is a minimal model of the modified reduct of Π relative to M.

In [9] the definition of the reduct is modified in a different way. The reduct of a program Π in the sense of Ferraris is obtained from the formulas (10) corresponding to the grounding rules of Π by replacing every maximal subformula of F that is not satisfied by M with "false". For instance, the formulas corresponding to the grounded rules (11) of (6) are the formulas

$$\begin{array}{l} p(a), \quad p(b), \quad q(a), \\ \text{false} \rightarrow \text{false}, \\ p(b) \wedge \neg\, \text{false} \rightarrow r(b). \end{array}$$

Definition L: M is a stable model of Π if M is a minimal model of the reduct of Π in the sense of Ferraris relative to M.

Definitions K and L are valuable because, like definition J, they can be extended to some nontraditional programs. Definition K was introduced, in fact, in connection with the problem of extending the stable model semantics to programs with aggregates. Definition L provides a satisfactory solution to the problem of aggregates as well. Furthermore, it can be applied in a straightforward way to arbitrary propositional formulas, and this generalization of the stable model semantics turned out to be equivalent to the generalization based on equilibrium logic that was mentioned at the end of Sect. 10.

12 Definition M, in Terms of Modified Circumscription

The authors of [10] defined a modification of circumscription called the *stable model operator*, SM. According to their *Definition M*, an Herbrand interpretation is a stable model of Π iff it satisfies SM[F] for the conjunction F of the universal closures of the formulas (10) corresponding to the rules of Π.

Syntactically, the difference between SM and circumscription is really minor. If F contains neither implications nor negations then SM[F] does not differ from

CIRC[F] at all. If F has "one level of implications" and no negations (as, for instance, when F corresponds to a set of traditional rules without negation, such as (2) and (5)), SM[F] is equivalent to CIRC[F]. But SM becomes essentially different from CIRC as soon as we allow negation in the bodies of rules.

The difference between SM[F] and the formulas used in Definition F is that the former does not involve auxiliary predicates and consequently does not require additional conjunctive terms relating auxiliary predicates to the predicates occurring in the program.

Definition M combines the main attractive feature of Definitions F, H, and I – no need for grounding – with the main attractive feature of Definitions J and L – applicability to formulas of an arbitrarily complex logical form. This fact makes it possible to give a semantics for an ASP language with choice rules and the *count* aggregate without any references to grounding [18].

Among the other definitions of a stable model discussed in this paper, Definition J, based on equilibrium logic, is the closest relative of Definition M. Indeed, in [41] the semantics of equilibrium logic is expressed by quantified Boolean formulas, and we can say that Definition M eliminated the need to ground the program using the fact that the approach of that paper can be easily extended from propositional formulas to first-order formulas.

A characterization of stable models that involves grounding but is otherwise similar to Definition M is given in [28]. It has emerged from research on the nonmonotonic logic of knowledge and justified assumptions [26].

13 Conclusion

Research on stable models has brought us many pleasant surprises.

At the time when the theory of iterated fixpoints of stratified programs was the best available approach to semantics of logic programming, it was difficult to expect that an alternative as general and as simple as Definition C would be found. And prior to the invention of Definition L, who would think that Definition C can be extended to choice rules, aggregates and more without paying any price in terms of the simplicity of the process of constructing the reduct?

A close relationship between stable models and the logic of here-and-there – a nonclassical logic that had been invented decades before the emergence of logic programming – was a big surprise. The possibility of defining stable models by twisting the definition of circumscription just a little was a surprise too.

There was a time when the completion semantics, the well-founded semantics, and the stable model semantics – and a few others – were seen as rivals; every person interested in the semantics of negation in logic programming would tell you then which one was his favorite. Surprisingly, these bitter rivals turned out to be so closely related to each other on a technical level that they eventually became good friends. One cannot study the algorithms used today for generating stable models without learning first about completion and unfounded sets.

And maybe the biggest surprise of all was that an attempt to clarify some semantic issues related to negation in Prolog was destined to be enriched by computational ideas coming from research on the design of SAT solvers and to give rise to a new knowledge representation paradigm, answer set programming.

Acknowledgements

I am grateful to the editors for the invitation to contribute this paper to a volume in honor of my old friend and esteemed colleague Yuri Gurevich.

Many thanks to Michael Gelfond, Tomi Janhunen, Joohyung Lee, Nicola Leone, Yuliya Lierler, Fangzhen Lin, Victor Marek, and Mirek Truszczyński for their comments. This work was partially supported by the National Science Foundation under Grant IIS-0712113.

References

1. Apt, K., Blair, H., Walker, A.: Towards a theory of declarative knowledge. In: Minker, J. (ed.) Foundations of Deductive Databases and Logic Programming, pp. 89–148. Morgan Kaufmann, San Mateo (1988)
2. Balduccini, M., Gelfond, M.: Logic programs with consistency-restoring rules.[5] In: Working Notes of the AAAI Spring Symposium on Logical Formalizations of Commonsense Reasoning (2003)
3. Bidoit, N., Froidevaux, C.: Minimalism subsumes default logic and circumscription in stratified logic programming. In: Proceedings LICS 1987, pp. 89–97 (1987)
4. Clark, K.: Negation as failure. In: Gallaire, H., Minker, J. (eds.) Logic and Data Bases, pp. 293–322. Plenum Press, New York (1978)
5. Denecker, M., Vennekens, J., Bond, S., Gebser, M., Truszczynski, M.: The second answer set programming system competition.[6] In: Erdem, E., Lin, F., Schaub, T. (eds.) LPNMR 2009. LNCS, vol. 5753, pp. 637–654. Springer, Heidelberg (2009)
6. Elkan, C.: A rational reconstruction of nonmonotonic truth maintenance systems. Artificial Intelligence 43, 219–234 (1990)
7. Faber, W., Leone, N., Pfeifer, G.: Recursive aggregates in disjunctive logic programs: Semantics and complexity. In: Alferes, J.J., Leite, J. (eds.) JELIA 2004. LNCS (LNAI), vol. 3229, pp. 200–212. Springer, Heidelberg (2004)
8. Fages, F.: A fixpoint semantics for general logic programs compared with the well–supported and stable model semantics. New Generation Computing 9, 425–443 (1991)
9. Ferraris, P.: Answer sets for propositional theories. In: Baral, C., Greco, G., Leone, N., Terracina, G. (eds.) LPNMR 2005. LNCS (LNAI), vol. 3662, pp. 119–131. Springer, Heidelberg (2005)
10. Ferraris, P., Lee, J., Lifschitz, V.: A new perspective on stable models. In: Proceedings of International Joint Conference on Artificial Intelligence (IJCAI), pp. 372–379 (2007)
11. Ferraris, P., Lifschitz, V.: Weight constraints as nested expressions. Theory and Practice of Logic Programming 5, 45–74 (2005)
12. Fine, K.: The justification of negation as failure. In: Proceedings of the Eighth International Congress of Logic, Methodology and Philosophy of Science, pp. 263–301. North Holland, Amsterdam (1989)

[5] `http://www.krlab.cs.ttu.edu/papers/download/bg03.pdf`
[6] `http://www.cs.kuleuven.be/~dtai/events/asp-competition/paper.pdf`

13. Gelfond, M.: On stratified autoepistemic theories. In: Proceedings of National Conference on Artificial Intelligence (AAAI), pp. 207–211 (1987)
14. Gelfond, M., Lifschitz, V.: The stable model semantics for logic programming. In: Kowalski, R., Bowen, K. (eds.) Proceedings of International Logic Programming Conference and Symposium, pp. 1070–1080. MIT Press, Cambridge (1988)
15. Gelfond, M., Lifschitz, V.: Logic programs with classical negation. In: Warren, D., Szeredi, P. (eds.) Proceedings of International Conference on Logic Programming (ICLP), pp. 579–597 (1990)
16. Heyting, A.: Die formalen Regeln der intuitionistischen Logik. Sitzungsberichte der Preussischen Akademie von Wissenschaften. Physikalisch-mathematische Klasse, pp. 42–56 (1930)
17. Lee, J.: A model-theoretic counterpart of loop formulas. In: Proceedings of International Joint Conference on Artificial Intelligence (IJCAI), pp. 503–508, Professional Book Center (2005)
18. Lee, J., Lifschitz, V., Palla, R.: A reductive semantics for counting and choice in answer set programming. In: Proceedings of the AAAI Conference on Artificial Intelligence (AAAI), pp. 472–479 (2008)
19. Lifschitz, V.: On the declarative semantics of logic programs with negation. In: Minker, J. (ed.) Foundations of Deductive Databases and Logic Programming, pp. 177–192. Morgan Kaufmann, San Mateo (1988)
20. Lifschitz, V.: Twelve definitions of a stable model. In: Garcia de la Banda, M., Pontelli, E. (eds.) ICLP 2008. LNCS, vol. 5366, pp. 37–51. Springer, Heidelberg (2008)
21. Lifschitz, V.: What is answer set programming? In: Proceedings of the AAAI Conference on Artificial Intelligence, pp. 1594–1597. MIT Press, Cambridge (2008)
22. Lifschitz, V., Pearce, D., Valverde, A.: Strongly equivalent logic programs. ACM Transactions on Computational Logic 2, 526–541 (2001)
23. Lifschitz, V., Tang, L.R., Turner, H.: Nested expressions in logic programs. Annals of Mathematics and Artificial Intelligence 25, 369–389 (1999)
24. Lin, F.: A Study of Nonmonotonic Reasoning. PhD thesis, Stanford University (1991)
25. Lin, F., Reiter, R.: Rules as actions: A situation calculus semantics for logic programs. Journal of Logic Programming 31, 299–330 (1997)
26. Lin, F., Zhao, Y.: ASSAT: Computing answer sets of a logic program by SAT solvers. In: Proceedings of National Conference on Artificial Intelligence (AAAI), pp. 112–117. MIT Press, Cambridge (2002)
27. Lin, F., Zhao, Y.: ASSAT: Computing answer sets of a logic program by SAT solvers. Artificial Intelligence 157, 115–137 (2004)
28. Lin, F., Zhou, Y.: From answer set logic programming to circumscription via logic of GK. In: Proceedings of International Joint Conference on Artificial Intelligence (IJCAI) (2007)
29. Marek, V., Truszczyński, M.: Stable semantics for logic programs and default theories. In: Proceedings North American Conf. on Logic Programming, pp. 243–256 (1989)
30. Marek, V., Truszczyński, M.: Stable models and an alternative logic programming paradigm. In: The Logic Programming Paradigm: a 25-Year Perspective, pp. 375–398. Springer, Heidelberg (1999)
31. McCarthy, J.: Circumscription–a form of non-monotonic reasoning. Artificial Intelligence 13, 27–39, 171–172 (1980)
32. McCarthy, J.: Applications of circumscription to formalizing common sense knowledge. Artificial Intelligence 26(3), 89–116 (1986)

33. McCarthy, J., Hayes, P.: Some philosophical problems from the standpoint of artificial intelligence. In: Meltzer, B., Michie, D. (eds.) Machine Intelligence, vol. 4, pp. 463–502. Edinburgh University Press, Edinburgh (1969)
34. McDermott, D.: Nonmonotonic logic II: Nonmonotonic modal theories. Journal of ACM 29(1), 33–57 (1982)
35. McDermott, D., Doyle, J.: Nonmonotonic logic I. Artificial Intelligence 13, 41–72 (1980)
36. Moore, R.: Semantical considerations on nonmonotonic logic. Artificial Intelligence 25(1), 75–94 (1985)
37. Niemelä, I.: Logic programs with stable model semantics as a constraint programming paradigm. Annals of Mathematics and Artificial Intelligence 25, 241–273 (1999)
38. Niemelä, I.: Stable models and difference logic. Annals of Mathematics and Artificial Intelligence 53, 313–329 (2008)
39. Niemelä, I., Simons, P.: Extending the Smodels system with cardinality and weight constraints. In: Minker, J. (ed.) Logic-Based Artificial Intelligence, pp. 491–521. Kluwer, Dordrecht (2000)
40. Pearce, D.: A new logical characterization of stable models and answer sets. In: Dix, J., Przymusinski, T.C., Moniz Pereira, L. (eds.) NMELP 1996. LNCS (LNAI), vol. 1216, pp. 57–70. Springer, Heidelberg (1997)
41. Pearce, D., Tompits, H., Woltran, S.: Encodings for equilibrium logic and logic programs with nested expressions. In: Brazdil, P.B., Jorge, A.M. (eds.) EPIA 2001. LNCS (LNAI), vol. 2258, pp. 306–320. Springer, Heidelberg (2001)
42. Reiter, R.: A logic for default reasoning. Artificial Intelligence 13, 81–132 (1980)
43. Reiter, R.: Circumscription implies predicate completion (sometimes). In: Proceedings of International Joint Conference on Artificial Intelligence (IJCAI), pp. 418–420 (1982)
44. Reiter, R.: Knowledge in Action: Logical Foundations for Specifying and Implementing Dynamical Systems. MIT Press, Cambridge (2001)
45. Saccá, D., Zaniolo, C.: Stable models and non-determinism in logic programs with negation. In: Proceedings of ACM Symposium on Principles of Database Systems (PODS), pp. 205–217 (1990)
46. van Emden, M., Clark, K.: The logic of two-person games. In: Micro-PROLOG: Programming in Logic, pp. 320–340. Prentice-Hall, Englewood Cliffs (1984)
47. van Emden, M., Kowalski, R.: The semantics of predicate logic as a programming language. Journal of ACM 23(4), 733–742 (1976)
48. Van Gelder, A.: Negation as failure using tight derivations for general logic programs. In: Minker, J. (ed.) Foundations of Deductive Databases and Logic Programming, pp. 149–176. Morgan Kaufmann, San Mateo (1988)
49. Van Gelder, A., Ross, K., Schlipf, J.: Unfounded sets and well-founded semantics for general logic programs. In: Proceedings of the Seventh ACM SIGACT-SIGMOD-SIGART Symposium on Principles of Database Systems, Austin, Texas, March 21-23, pp. 221–230. ACM Press, New York (1988)
50. Van Gelder, A., Ross, K., Schlipf, J.: The well-founded semantics for general logic programs. Journal of ACM 38(3), 620–650 (1991)
51. Wallace, M.: Tight, consistent and computable completions for unrestricted logic programs. Journal of Logic Programming 15, 243–273 (1993)

DKAL and Z3: A Logic Embedding Experiment

Sergio Mera and Nikolaj Bjørner

Computer Science Department, University of Buenos Aires,
Ciudad Universitaria, Pabellón 1 (C1428EGA), Buenos Aires, Argentina
smera@dc.uba.ar
Microsoft Research, One Microsoft Way, Redmond, WA, 98074, USA
nbjorner@microsoft.com

For Yuri, on the occasion of his seventieth birthday. The following paper is centered around DKAL, which provides a crisp foundation for pragmatically motivated problems from security. It exemplifies a recurring inspiration of working with Yuri: his ability to crisply and clearly capture the essence of problems and foundations.

Abstract. Yuri Gurevich and Itay Neeman proposed the Distributed Knowledge Authorization Language, DKAL, as an expressive, yet very succinct logic for distributed authorization. DKAL uses a combination of modal and intuitionistic propositional logic. Modalities are used for qualifying assertions made by different principals and intuitionistic logic captures very elegantly assertions about basic information. Furthermore, a non-trivial and useful fragment known as the *primal infon logic* is amenable to efficient linear-time saturation.

In this paper we experiment with an embedding of the full DKAL logic into the state-of-the-art Satisfiability Modulo Theories solver Z3 co-developed by the second author. Z3 supports classical first-order semantics of formulas, so it is not possible to directly embed DKAL into Z3. We therefore use an indirect encoding. The one experimented with in this paper uses the instantiation-based support for quantifiers in Z3.

Z3 offers the feature to return a potential ground counter-model when the saturation procedure ends up with a satisfiable set of ground assertions. We develop an algorithm that extracts a DKAL model from the propositional model, in order to provide root causes for non-derivability.

Keywords: DKAL, Z3, Embedding, Model Extraction.

1 Introduction

DKAL, the *Distributed Knowledge Authorization Language*, has been developed as a foundation for logic-based authorization mechanisms. The formulation of DKAL used in this paper was developed through a sequence of adaptations of logic for authorization, but also a noticeable realization that a combination of modal and intuitionistic propositional logic provide an exceptional match for integrating knowledge information in a distributed system.

A. Blass, N. Dershowitz, and W. Reisig (Eds.): Gurevich Festschrift, LNCS 6300, pp. 504–528, 2010.

We will start out by providing a high-level motivation for DKAL in Section 2 and follow it up by recollecting the logic and main results behind DKAL. These formalities are complemented by a more elaborated case study that illustrates the main DKAL constructs in Section 3. Given that DKAL is based on a core combining propositional modal and intuitionistic logic a simple question arises. Is there a suitable deductive tool to support DKAL? We will here take the approach of embedding the core into the classical first-order theorem prover Z3 (Section 4). We will use a particular approach for the embedding, namely an encoding of DKAL inference rules using first-order axioms (Section 5). The embedding allowed us to develop a prototype for DKAL that we used on the presented case study. The experiment is not intended as a representative approach for implementing DKAL. In fact we here disregard the primal fragment that is amenable to efficient linear-time saturation algorithms. Rather, we see the objective as to study DKAL as a candidate non-classical logic and investigate how such a logic can be encoded using existing features in the context of Z3. Finally, in Section 6, we establish a theoretical property of the embedding: when Z3 completes with a satisfiable ground saturation, it is furthermore possible to extract a Kripke model for the core. The theory was implemented, and we use two models extracted from the case study as an illustration.

2 An Introduction to DKAL

Nowadays, the most widespread method to deal with security policies in systems are access control lists (ACL). This method basically consists of a list of permissions attached to each of the objects that may be accessed. ACLs allow expressing detailed policies at a very granular level, but they have their limitations when it comes to expressing access policies that depend on conditions and that combine policies of potentially multiple parties that don't necessarily trust each other on anything else than a few objects. An approach to authorization based on logic has on the other hand the prospect of expressing policies using expressive conditions and properly combining the intent of multiple parties. Logic based approaches also allow policies to be the object of formal verification, and to thoroughly analyze properties such as security leaks.

In this context, Distributed Knowledge Authorization Language (DKAL) and affiliated logic was proposed in [13] and later DKAL was given a foundation based on a combination of modal and intuitionistic logic in [15].

At the core, this logic deals with pieces of information called *infons*, e.g. John has the right to access the network Corporate. Infons are not required to be true or false, but instead infons can be known or not by relevant principals. For example, we can ask whether the administrator of the network Corporate knows that John has the right to access the network. In this way, each principal has some knowledge (that is, infons that are known to him). DKAL also provides a way in which principals can share the information they have. For example, if Admin knows that John has read access to Corporate, he can communicate this to Patrick and (if Patrick accepts that communication), then Patrick will know that Admin said that John has read access to Corporate.

One may study the logic of infons, and some natural operators arise. You know $a \wedge b$ if you know a and you know b, and the implication connective is also natural, if you know a and $a \to b$, then you know b. In addition to conjunction and implication, infon logic has two unary connectives p *said* and p *implied*, for any principal p. Both connectives represent a way to model the knowledge that are passed from one principal to the other by means of communication assertions. Let us see the intuitive difference between them. A principal p may just say a to q, and then, if the communication is successful, q learns the infon p said a. But p may condition his claim on q knowing b, so then q learns the infon $b \to (p \text{ implied } a)$. So *implied* is a weaker condition, and it will be useful to avoid undesired delegation or probing attacks.

Infons are also partially ordered in terms of information, and the intuitive idea is that $a \leq b$ when the information in a is contained in that of b. In this sense, infons can be studied from the point of view of algebra. For example, $a+b$ is the least upper bound of a and b, and $a \to b = min\{z : a + z \geq b\}$.

There is also interest in studying the logic of infons itself. In this direction, two fragments were identified: the *full infon logic*, and the *primal infon logic*. The latter is weaker, but has nice computational properties and is still expressive enough for modeling security issues. We will say *infon logic* to refer to either of both logics.

2.1 Syntax and Semantics

In this section we present the syntax and semantics of DKAL and the infon logic. We do not give the full definitions, just enough details for the purposes of this article. For a more complete definition see [14].

DKAL. There are three kinds of DKAL assertions: knowledge, communication and filter assertions. Knowledge assertions have the form:

$$A \text{ knows } x$$

where A is a principal and x is an infon term. The intended meaning of this is asserting that A has the knowledge x. The way a principal has to communicate knowledge is through communication assertions. They have the form:

$$A \text{ to } q : [x \leftarrow y] \Leftarrow z$$

where A is a principal, q is either a principal or a variable (that ranges over principals), and x, y and z are infon terms. We use the following shorthands $[x \leftarrow y]$ instead of $[x \leftarrow y] \Leftarrow \text{true}$, $[x] \Leftarrow z$ instead of $[x \leftarrow \text{true}] \Leftarrow z$, and $[x]$ instead of $[x \leftarrow \text{true}] \Leftarrow \text{true}$. We are going to give the precise semantics later, but intuitively, with this assertion the principal A communicates to q the knowledge x, with y as a proviso. This communication takes place only if A already knows z. Filter assertions are used to receive communications. They have the form:

$$B \text{ from } p : [x \leftarrow y] \Leftarrow z$$

where B is a principal, p is either a principal or a variable (that ranges over principals), and x, y and z are infons terms.

The rules that control how knowledge and communications are handled are the following:

Proviso-free scenario

$$\text{(Com1)}\quad \frac{\begin{array}{l} B \text{ to } p : [t_1] \Leftarrow t_3 \\ B \text{ knows } t_3\eta, p\eta = A \\ A \text{ from } q : [s_1] \Leftarrow s_3 \\ A \text{ knows } s_3\theta, q\theta = B \\ s_1\theta = t_1\eta\theta \end{array}}{A \text{ knows } B \text{ said } s_1\theta}$$

Proviso-present scenario

$$\text{(Com2)}\quad \frac{\begin{array}{l} B \text{ to } p : [t_1 \leftarrow t_2] \Leftarrow t_3 \\ B \text{ knows } t_3\eta, p\eta = A \\ A \text{ from } q : [s_1 \leftarrow s_2] \Leftarrow s_3 \\ A \text{ knows } s_3\theta, q\theta = B \\ s_1\theta = t_1\eta\theta \\ s_2\theta = t_2\eta\theta \end{array}}{A \text{ knows } (s_2\theta \to B \text{ implied } s_1\theta)}$$

where η and θ are appropriate substitutions. The knowledge that a principal gets from his knowledge assertions and the communications from other principals gives rise to more knowledge. Here is where the infon logic plays its part.

$$\text{(Ensue)}\quad \frac{\begin{array}{l} A \text{ knows } \Gamma \\ \Gamma \vdash x \end{array}}{A \text{ knows } x}$$

The sequent $\Gamma \vdash x$ represents a derivable sequent in the infon logic, in which Γ is a set of infons, and x is an infon. Note that the derived knowledge is infinite, but A does not compute it all. He will just be interested whether some specific queries follow from his knowledge.

The Infon Logic. We now introduce the syntax and semantics of the infon logic. The syntax is the same for both the full and the primal fragment. Given an infinite set of principals, and a vocabulary of functions names, the set of formulas is defined as:

$$\varphi ::= \text{true} \mid A(t_1, \ldots, t_k) \mid \varphi_1 + \varphi_2 \mid \varphi_1 \to \varphi_2 \mid p \text{ said } \varphi \mid p \text{ implied } \varphi \quad (1)$$

where true and $A(t_1, \ldots, t_k)$ are primitive formulas, A is a function name, $t_1, \ldots, t_k$ are terms, φ, φ_1 and φ_2 range over formulas and p ranges over principals. For the purpose of this article the structure of primitive formulas is of no real importance.

For practical reasons, it is useful to have some fixed built-in functions, such as the nullary function `asInfon` that act as a casting operator that converts the Boolean value `true` into the uninformative infon `true asInfon`. We will also introduce the shortcuts tdonS and tdonI used in DKAL for expressing that principals are trusted on saying or implying certain infons:

p tdonS x abbreviates (p said x) $\to x$,
p tdonI x abbreviates (p implied x) $\to x$.

The infon logic is basically an extension of the intuitionistic propositional system NJp [18], and was introduced is [15]. Here we present the sequent calculus for the full infon logic.

Axioms and Inference Rules

(True) $\vdash$ true (x2x) $x \vdash x$

$$\text{(PI)}\quad \frac{\Gamma \vdash y}{\Gamma, x \vdash y}$$

$$(+\text{E})\quad \frac{\Gamma \vdash x + y}{\Gamma \vdash x} \qquad \frac{\Gamma \vdash x + y}{\Gamma \vdash y} \qquad (+\text{I})\quad \frac{\Gamma \vdash x \quad \Gamma \vdash y}{\Gamma \vdash x + y}$$

$$(\rightarrow\text{E})\quad \frac{\Gamma \vdash x \quad \Gamma \vdash x \rightarrow y}{\Gamma \vdash y} \qquad (\rightarrow\text{I})\quad \frac{\Gamma, x \vdash y}{\Gamma \vdash x \rightarrow y}$$

$$\text{(S)}\quad \frac{\Gamma \vdash y}{q \text{ said } \Gamma \vdash q \text{ said } y} \qquad (I)\quad \frac{\Gamma \vdash y}{q \text{ told } \Gamma \vdash q \text{ implied } y}$$

The rule (PI) is short for *premise inflation.*

If $\Gamma = \{x_1, \ldots, x_n\}$, then q said $\Gamma = \{q \text{ said } x_1, \ldots, q \text{ said } x_n\}$, and q told Γ is any of the 2^n sets $\{q \text{ told}_1\, x_1, \ldots, q \text{ told}_n\, x_n\}$, where each $\text{told}_i \in \{\text{said}, \text{implied}\}$.

Primal infon logic can be given by a sequent calculus that is obtained from the calculus above by replacing ($\rightarrow$I) with the combination of the weaker inference rules ($\rightarrow$IW) and (Trans):

$$(\rightarrow\text{IW})\quad \frac{\Gamma \vdash y}{\Gamma \vdash x \rightarrow y} \qquad \text{(Trans)}\quad \frac{\Gamma \vdash x \quad \Gamma, x \vdash y}{\Gamma \vdash y}$$

Model Theory. A Kripke structure is a nonempty set of elements W, equipped with a binary relation $\leq$ and a collection of binary relations I_q and S_q for every principal p. Additionally a Kripke structure must fulfill the following requirements:

K1. $I_q \subseteq S_q$.
K2. if $u \leq w$ and wI_qv, then uI_qv, and the same for S_q.

The satisfaction relation is defined by induction for every formula φ in terms of a cone $C(\varphi)$ of worlds (that is, an upward closed set). If $u \in C(x)$ we say that x holds in u, and we write $u \models x$. If x is primitive, then $C(x)$ is an arbitrary cone. The satisfaction relation for composite formulas for the full infon logic is the following:

K3. $C(x + y) = C(x) \cap C(y)$.
K4. $C(x \rightarrow y) = \{u : C(x) \cap \{v : v \geq u\} \subseteq C(y)\}$.
K5. $C(q \text{ implied } x) = \{u : \{v : uI_qv\} \subseteq C(x)\}$.
K6. $C(q \text{ said } x) = \{u : \{v : uS_qv\} \subseteq C(x)\}$.
K7. $C(\mathit{true}) = W$.

Let $C(\Gamma) = \bigcap_{x \in \Gamma} C(x)$. A Kripke structure M models a sequent $\Gamma \vdash x$ iff $C(\Gamma) \subseteq C(x)$ in M. A sequent s is *valid* if every structure models s.

The definition for the satisfaction relationship for the primal infon is similar to the above except that K4 is replaced with the following weaker requirement:

K4W. $C(x \rightarrow y)$ is an arbitrary cone such that $C(y) \subseteq C(x \rightarrow y) \subseteq \{u : C(x) \cap \{v : v \geq u\} \subseteq C(y)\}$.

The following theorem can be established:

Theorem 1 **([15]).** *In the case of either full or primal infon logic, the following claims are equivalent for any sequent s:*

1. *s is provable.*
2. *s is valid.*
3. *Every finite Kripke structure models s.*
4. *There is a proof of s that uses only subformulas of s.*

In particular, we are going to see that the subformula property is particularly useful for us. Regarding the complexity of the infon logic, the following two theorems are relevant:

Theorem 2 **([15]).** *The validity problem for the full infon logic is polynomial-space complete.*

For the primal infon logic, the complexity bound can be improved when the maximum depth of a formula is fixed:

Definition 1 (Primal quotation depth). *The primal quotation depth of formulas is defined by induction:*

- $\delta(x) = 0$ *if x is a variable.*
- $\delta(p \textit{ told } x) = 1 + \delta(x)$.
- $\delta(x + y) = max\{\delta(x), \delta(y)\}$.
- $\delta(x \rightarrow y) = \delta(y)$.

Furthermore, $\delta(\Gamma) = max_{x \in \Gamma} \delta(x)$ *for any set Γ of formulas.*

Theorem 3 **([15]).** *For every natural number d, there is a linear time algorithm for the multiple derivability problem for primal infon logic restricted to formulas x, with $\delta(x) \leq d$. Given hypothesis Γ and queries Q, the algorithm determines which of the queries in Q follow from the hypothesis Γ.*

3 DKAL@Starbucks

We will now develop an example using DKAL. The example has been processed mechanically using our Z3 based prototype described in Section 5. We present the example using the syntax used by the front-end to the prototype. The example is extracted from a document that was not specifically aimed at studying authorization but rather various Windows Vista networking user-scenarios, yet several aspects of the scenario can be described directly using DKAL. We include a somewhat detailed excerpt from the document with the intent of giving a sufficient amount of example material for illustrating the different features of DKAL. It models part of the procedure involved to allow Patrick to login remotely to a secure network using a wireless connection. The story begins when Patrick enters in a Starbucks, sits down and opens his laptop. He first has to login to the operating system by introducing his user name and password.

Login Authorization. The entities involved in this part of the process are:

`Patrick`,	the user
`OSLogon`,	the module in charge of the login process
`AuthProvider`,	the module that provides the authentication protocol
`LoginGUI`,	the graphical interface for the login process
`KeyboardAPI`,	the interface that listens to the keyboard events

The primitive functions all take a principal p as argument. They are:

p `hasEnteredUsername`, p `hasValidCredentials`, p `couldLogin`,
p `hasEnteredPassword`, p `hasCredentialsCached`, p `isLoggedIn`

And we are going to use the substrate function p `passesSecurityChecks` that returns a truth value.

Patrick begins by typing his user-name and password. This action is informed by the keyboard API to the login graphical interface:

```
KeyboardAPI to LoginGUI: [Patrick hasEnteredUsername]
KeyboardAPI to LoginGUI: [Patrick hasEnteredPassword]
```

Recall that $[x]$ is an abbreviation for $[x \leftarrow \text{true}] \Leftarrow \text{true}$.

The graphical interface trusts any information, encoded using a variable X, provided by the keyboard API.

`LoginGUI` knows `KeyboardAPI` tdonS X

and the graphical interface also accepts communications Y from any participant X:

`LoginGUI` from X: `[`Y`]`

It therefore follows that the graphical interface learns that Patrick has entered his user-name and password:

```
? LoginGUI knows Patrick hasEnteredUsername
? LoginGUI knows Patrick hasEnteredPassword
```

The graphical interface caches someones credentials when it knows that this person has entered his user-name and password. When that happens, it informs that to the authorization provider. This is encoded using the rule:

```
LoginGUI to AuthProvider: [X hasCredentialsCached] <=
      X hasEnteredUsername + X hasEnteredPassword
```

and the authorization provider accepts all communications from the graphical interface, as the following rule encodes:

`AuthProvider` from `LoginGUI: [`X`]`

Therefore, we can deduce that the provider learns that the interface has cached Patrick's credentials:

(A1) `? AuthProvider` knows `LoginGUI` said
`Patrick hasCredentialsCached`

On the other hand, the authorization provider allows someone to login when the graphical interface has cached his credentials and those credentials are valid. If that is the case, he informs that to the logon module:

`AuthProvider` to `OSLogon: [`X `couldLogin] <=`
`(LoginGUI` said X `hasCredentialsCached) +`
X `hasValidCredentials`

The authorization provider checks Patrick's credentials with its internal database, and concludes that they are valid, so it learns:

(A2) `? AuthProvider` knows `Patrick hasValidCredentials`

Given (A1) and (A2), the authorization provider communicates to the logon module that Patrick can be logged in. Since the logon module accepts communications from the authorization provider regarding a successful login:

`OSLogon` from `AuthProvider: [`X `couldLogin]`

the logon module learns that:

(A3) `? OSLogon` knows `AuthProvider` said `Patrick couldLogin`

The logon module will allow a user X to login when the authorization provider says so, and when it can verify that X passes all the security checks:

`OSLogon` to X`: [`X `isLoggedIn] <= AuthProvider` said
X `couldLogin + asInfon(`X `passesSecurityChecks)`

Given (A3) and the fact that Patrick effectively passes all the security checks, the logon module informs that Patrick is logged in. Patrick accepts any communication from the logon module:

`Patrick` from `OSLogon:[`X`]`

So Patrick learns:

`? Patrick` knows `OSLogon` said `Patrick isLoggedIn`

Patrick trusts the logon module with respect to login issues, so assert the following:

`Patrick` knows `winLogon` tdonS `Patrick isLoggedIn`

Consequently Patrick learns that he is logged in:

`? Patrick` knows `Patrick isLoggedIn`

Patrick Launches IE. A new principal `Windows`, the operating system, is involved in this section. We also use the following primitive functions, where p is a principal and s is a piece of software:

p `wantsToExecute` s, p `isAllowedToRun` s, s `shouldBeLaunched`

We assume there is a constant `IE` that represent the Internet Explorer.

Patrick now continues logging in to the wireless network, and then accessing the secure network. He needs to authenticate with T-Mobile to use hotspot, so he launches IE. He communicates to Windows all the software he wants to execute:

```
Patrick to Windows: [Patrick wantsToExecute X]
                        <= Patrick wantsToExecute X
```

and today he wants to run Internet Explorer:

```
Patrick knows Patrick wantsToExecute IE
```

Windows runs a specific piece of software if the user wants to execute it and he is allowed to do it:

```
                  ((P said P wantsToExecute Software) +
Windows knows      P isAllowedToRun Software)
                     -> Software shouldBeLaunched
```

Patrick is allowed to run IE:

```
Windows knows Patrick isAllowedToRun IE
```

and Windows accepts all communication events related to running software:

```
Windows from P: [P wantsToExecute Software]
```

So Windows learns it should run IE:

```
? Windows knows IE shouldBeLaunched
```

Establishing an SSL Connection. In this section these new principals are involved:

`TMobile`, the wireless router
`TrustedRoot`, the entity that provides validation for certificates

And we use the following functions, where s is a website, c is a certificate and sw is a piece of software:

s `supportsSSL`, s `hasCertificate` c, c `isValid`,
c `isProperlySigned`, sw `shouldShowLock`

We use the constants `ThisCert`, a certificate, `IECurrSite`, a website, and the substrate functions x `currTime`, that returns an integer representing the current time of the running operating system x, y `expTime` that returns an integer representing the expiration time of a certificate y, and the usual comparison function $<$ between integers.

The T-Mobile router redirects IE to the T-Mobile authentication manager web site and it communicates that the authentication site `IECurrSite` supports SSL and the certificate for the authentication site is `ThisCert`:

```
TMobile to Windows: [IECurrSite supportsSSL]
TMobile to Windows: [IECurrSite hasCertificate ThisCert]
```

Windows accepts any communication from T-Mobile:

```
Windows from TMobile:[X]
```

so Windows learns that T-Mobile said that the current site supports SSL and that the certificate is `ThisCert`:

```
? Windows knows TMobile said IECurrSite supportsSSL
? Windows knows TMobile said IECurrSite hasCertificate ThisCert
```

The trusted root communicates that a given certificate is valid when it is properly signed, and with the proviso that the certificate has not expired considering the current time:

```
TrustedRoot to Windows:
   [Y isValid <- asInfon(Windows currTime < Y expTime)]
      <= Y isProperlySigned
```

We will assume that the trusted root knows that `ThisCert` is properly signed. We will also assume that `Windows currTime` is an hour away from `ThisCert expTime`. Therefore:

```
? TrustedRoot knows ThisCert isProperlySigned
? TrustedRoot knows asInfon(
       Windows currTime < ThisCert expTime)
```

Assuming Windows accepts any communication from the trusted root:

```
Windows from TrustedRoot:[X]
```

Windows learns:

```
? Windows knows TrustedRoot implied ThisCert isValid
```

Windows trusts everything the trusted root says:

```
Windows knows TrustedRoot tdonI X
```

so it learns:

```
? Windows knows ThisCert isValid
```

Windows knows that it must display the lock on IE when T-Mobile says that the current site supports SSL, and the site certificate is valid:

```
                 TMobile said IECurrSite supportsSSL +
Windows knows    TMobile said IECurrSite hasCertificate ThisCert +
                 ThisCert isValid
                 -> IE shouldShowLock
```

so it learns:

```
? Windows knows IE shouldShowLock
```

At the T-Mobile Web-site. In this section we add the principal `TMobileHTTP`, the HTTP module of the router, and the function `showThisWelcomePageTo` x, where x is a principal.

At the T-Mobile Authentication web site, Patrick successfully enters his T-Mobile credentials. T-Mobile gets Patrick's credentials and validates them. Then, it communicates to Windows that the credentials are valid:

```
TMobile to Windows: [Patrick hasValidCredentials]
```

Windows is already receiving any communication from T-Mobile, so it learns:

```
? Windows knows TMobile said Patrick hasValidCredentials
```

The T-Mobile HTTP module redirects a user to the welcome page with the proviso that T-Mobile says that the user has valid credentials:

```
TMobileHTTP to Windows: [showThisWelcomePageTo X <-
    TMobile said X hasValidCredentials]
```

Windows receives any communication from the HTTP module:

```
Windows from TMobileHTTP:[X]
```

So Windows learns that:

```
? Windows knows TMobileHTTP implied
    showThisWelcomePageTo Patrick
```

To summarize, the example illustrated the use of DKAL in a distributed scenario involving several parties. The communication assertions were constrained with filters that allowed the parties to exchange just the information they were authorized to. Furthermore, the infon logic was used to express conditional delegation.

4 What Is Z3 Good For?

Z3 [8] is a state-of-the-art Satisfiability Modulo Theories (SMT) solver developed at Microsoft Research by the second author and Leonardo de Moura. It is a theorem prover for first-order logic augmented with various built-in theories, such as the theories of additive arithmetic over reals and integers, bit-vectors and an extension of McCarthy's theory of arrays. Z3 is furthermore a decision procedure for quantifier-free formulas over a combination of the built-in theories. Formulas using quantifiers are handled using an instantiation-based approach that creates ground instances of universal quantifiers based on a search on subterms created during search.

Z3 is currently used in a large array of software analysis, verification and test-case generation tools from Microsoft. These include Spec#/Boogie [2,10], Pex [21], HAVOC [17], PREfix [19], Vigilante [6], a verifying C compiler (VCC) [5], SAGE [11], SLAM/SDV [1], and Yogi [12].

SMT solvers are gaining a distinguished role in the context of software analysis since they offer support for most domains encountered in programs. A well tuned SMT solver that takes into account the state-of-the-art breakthroughs usually scales orders of magnitude beyond custom ad-hoc solvers.

4.1 Embeddings with SMT Solvers and Theorem Provers

The experiment in this paper is for using Z3 to prototype an application it is not directly built for. The underlying combination of modal and intuitionistic logic used in DKAL requires a deep embedding into Z3. We make critical use of the subformula properties of DKAL and the infon logic, and encode both the infon logic and full DKAL. We will be using first-order formulas with quantifiers to encode the DKAL logic and then rely on the instantiation-based quantifier-support in Z3. This strategy is similar to embedding.

The embedding method we pursue here is a variation of deep embeddings of non-classical logics into classical logic; see for instance [20]. We should point out that this is far from being the only way to use a theorem prover for classical logic in the context of non-classical logics. In the context of model checking, where temporal formulas are checked against finite (and even infinite) state systems, the method of *bounded model checking* [3] is used to reduce PSPACE problems into propositional satisfiability problems. The main idea behind bounded model checking is to unfold transition relations a finite number of times, creating a propositional or first-order formula summarizing an n-unfolding of system steps. Given a bound on the diameter required for unfolding, the resulting problem can be solved using classical methods. Another approach for PSPACE complete problems is pursued in [4] where a logic with *linear functional fixed-points* is reduced to propositional linear time temporal logic. The paper develops a hybrid approach using a symbolic model checker and the theorem prover Z3.

Finally, it is entirely possible to use methods, such as Prolog, that are more directly tailored at encoding inference rules. In this context, we also developed an encoding of DKAL into the FORMULA [16] tool. In contrast to Prolog systems, FORMULA performs forward chaining.

4.2 Quantifiers in Z3

There are several motivations for integrating strong quantifier support in the context of SMT solvers. The main motivation stems from program verification applications. In this context, quantified formulas that come from program verification can typically be instantiated based on a local analysis of the ground terms occurring in the input formula. A current main approach to integrating quantifiers with SMT solving is therefore by producing quantifier instantiations. The instantiated quantifiers are then quantifier free formulas that are handled by the main ground SMT solving engine. It is an art and craft to control quantifier instantiations to produce just the useful instantiations.

For controlling which instantiations of the axioms are produced let us consider an annotation of the axioms using *triggers*. A trigger annotated universal formula is a formula of the form

$$\psi_{annot} : \quad \forall \boldsymbol{x} \ . \ \{p_1(\boldsymbol{x}), \ldots, p_k(\boldsymbol{x})\} \ . \ \psi(\boldsymbol{x}),$$

where $k \geq 1, p_1, \ldots, p_k$, are terms that contain all variables from $\boldsymbol{x}$. Given a model $\mathcal{M}$ of a quantifier free formula φ we say that φ is *saturated with respect to* ψ_{annot} if whenever there are subterms $t_1, \ldots, t_k$ in φ and a substitution θ, such that $p_1\theta^{\mathcal{M}} = t_1^{\mathcal{M}}, \ldots, p_k\theta^{\mathcal{M}} = t_k^{\mathcal{M}}$, then φ implies the formula $\psi\theta$. If it is not the case that φ implies $\psi\theta$, then a saturation process adds the conjunct $\psi\theta$ to φ.

We can associate more than one trigger with a quantified formula. In this case we write the set of triggers as different sets of patterns enclosed by $\{\}$. For example,

$$\begin{aligned}
&\forall a, i, j, v\ .\\
&\quad \{\mathit{read}(\mathit{write}(a,i,v),j)\}\\
&\quad \{\mathit{write}(a,i,v), \mathit{read}(a,j)\}\ .\\
&\qquad \mathit{read}(\mathit{write}(a,i,v),j) = \textbf{if } i = j \textbf{ then } v \textbf{ else } \mathit{read}(a,j)\ .
\end{aligned}$$

The patterns instruct the theorem prover to instantiate the axioms for arrays whenever there is a subterm matching the first pattern $\mathit{read}(\mathit{write}(a,i,v),j)$, *or* if there are two subterms. The first is equal to $\mathit{write}(a,i,v)$ and the second is equal to $\mathit{read}(a,j)$ for some instances of the variables a, i, j, v.

E-matching. As mentioned, pattern instantiation is performed using a matching algorithm. To take into account equalities that are assumed during search, the matching algorithm performs E-matching. The problem of E-matching is a special case of rigid E-unification [9]. As with (non-simultaneous rigid E-unification) it remains *NP*-complete, but it is conceptually much simpler. An E-matching algorithm takes as input a set of ground equations E, a ground term t and a term p possibly containing variables. It produces the set of substitutions θ over the variables in p, such that $E \models t \simeq \theta(p)$. In the context of Z3 [7], we found that the *NP*-hardness of E-matching is not the real challenge with integrating good E-matching support. Rather, an emphasis on supporting incremental matching in the context of a backtracking search algorithm is more important for performance.

5 DKAL and Z3

We will now describe an embedding of DKAL into Z3. The embedding encodes inference rules from DKAL as first-order axioms presented to Z3. The encoding furthermore prescribes using *patterns* how the axioms should be instantiated based on ground terms that have already been created. We later show that the encoding preserves decidability of ground DKAL queries as the set of instantiations that can be created based on the patterns is finite.

Terms in Z3 are sorted according to a simple theory of sorts. The universe is assumed partitioned into a set of disjoint sorts and a sort can be introduced by declaring a name identifying the sort. The sorts that are used for DKAL are:

- *Infon* - the sort of infons.
- *InfonSet* - for a set of *Infons*.
- *Principal* - the sort of principals.

The main functions and predicates on these sorts follow the presentation from Section 2 closely.

5.1 The Logic of Infons

To encode infon sequents we introduce the following functions and relations:

- $\in$: *Infon* $\times$ *InfonSet* $\rightarrow \mathcal{B}$. The axiomatization will maintain that $x \in \Gamma$ holds if x is a member of the set Γ

- *insert* : *Infon* $\times$ *InfonSet* $\rightarrow$ *InfonSet*. The operation $insert(x, \Gamma)$ creates the union of the set Γ and $\{x\}$.

The properties that are needed of these functions and relations are similar to the properties useful for the non-extensional theory of arrays. They are:

$$\forall x : Infon,\ \Gamma : InfonSet\ .\ \{insert(x, \Gamma)\}\ .\ x \in insert(x, \Gamma)$$
$$\forall x : Infon,\ y : Infon,\ \Gamma : InfonSet\ .\ \{x \in insert(y, \Gamma)\}\ .\quad x \neq y \ \rightarrow\ (x \in insert(y, \Gamma) \leftrightarrow x \in \Gamma)$$

To ease readability in the following, we will avoid explicit universal quantifiers and simply list the axioms together with their triggers. The above axioms become:

$$\{insert(x, \Gamma)\}\ .\ x \in insert(x, \Gamma) \tag{2}$$
$$\{x \in insert(y, \Gamma)\}\ .\ x \neq y \ \rightarrow\ (x \in insert(y, \Gamma) \leftrightarrow x \in \Gamma) \tag{3}$$

In the following, we let x, y, z, k range over the *Infon* sort, Γ ranges over *InfonSet*, and p, p_1, p_2 have the sort *Principal*.

The Infon Inference Rules. The infons are built according to the BNF grammar outlined in (1). For the embedding we introduce functions that build abstract terms corresponding to each case in the grammar. true : *Infon*, asInfon : $\mathcal{B} \rightarrow$ *Infon*, $\_ + \_$, $\_$ imp $\_$: *Infon* $\times$ *Infon* $\rightarrow$ *Infon* and $\_$ said $\_$, $\_$ implied $\_$: *Principal* $\times$ *Infon* $\rightarrow$ *Infon*.

The entailment relation between infon sets and infons is a binary predicate $\vdash$: *InfonSet* $\times$ *Infon* $\rightarrow \mathcal{B}$. It encodes the entailment relation for sequents over infons. To encode the entailment relation it suffices to follow the inference rules for the infon logic. The subformula property of the infon logic is critical in this context. Rule axioms are only instantiated if there are existing subterms matching the maximal subterms in either the premise or conclusion.

$$\text{asInfon}\quad \{\Gamma \vdash \text{asInfon(true)}\}\ .\ \Gamma \vdash \text{asInfon(true)} \tag{4}$$
$$\text{true}\quad \{\Gamma \vdash \text{true}\}\ .\ \Gamma \vdash \text{true} \tag{5}$$
$$x2x\quad \{x \in \Gamma\},\ \{\Gamma \vdash x\}\ .\ x \in \Gamma \ \rightarrow\ \Gamma \vdash x \tag{6}$$
$$PI\quad \{insert(x, \Gamma) \vdash y\}\ .\ \Gamma \vdash y \ \rightarrow\ insert(x, \Gamma) \vdash y \tag{7}$$
$$+I\quad \{\Gamma \vdash x + y\}\ .\ \Gamma \vdash x \wedge \Gamma \vdash y \ \rightarrow\ \Gamma \vdash x + y \tag{8}$$
$$+E1\quad \{x + y, \Gamma \vdash x\}\ .\ \Gamma \vdash x + y \ \rightarrow\ \Gamma \vdash x \tag{9}$$
$$+E2\quad \{x + y, \Gamma \vdash y\}\ .\ \Gamma \vdash x + y \ \rightarrow\ \Gamma \vdash y \tag{10}$$
$$\rightarrow E\quad \{\Gamma \vdash y,\ (x \text{ imp } y)\}\ .\ (\Gamma \vdash x \wedge \Gamma \vdash x \text{ imp } y) \ \rightarrow\ \Gamma \vdash y \tag{11}$$
$$\rightarrow I\quad \{\Gamma \vdash x \text{ imp } y\}\ .\ insert(x, \Gamma) \vdash y \ \rightarrow\ \Gamma \vdash x \text{ imp } y \tag{12}$$

Encoding Rules for said and implied . The rules S and I require some additional encoding. We will use auxiliary functions saidOf(_, _), toldOf(_, _) : *Principal* × *InfonSet* → *InfonSet* to extract the subset of an infon set Γ where principal p either said or at least implied some infon.

$$S1 \ \{p \text{ said } x \in \Gamma\} \ . \ p \text{ said } x \in \Gamma \ \leftrightarrow \ x \in \text{saidOf}(p, \Gamma) \tag{13}$$

$$S2 \ \{\Gamma \vdash p \text{ said } x\} \ . \ \text{saidOf}(p, \Gamma) \vdash x \ \rightarrow \ \Gamma \vdash p \text{ said } x \tag{14}$$

Extracting implied infons is almost similar, except we take into account that if p said x then p implied x.

$$I1 \ \{p \text{ implied } x \in \Gamma\} \ . \ p \text{ implied } x \in \Gamma \ \leftrightarrow \ x \in \text{toldOf}(p, \Gamma) \tag{15}$$

$$I2 \ \{p \text{ said } x \in \Gamma\} \ . \ p \text{ said } x \in \Gamma \ \leftrightarrow \ x \in \text{toldOf}(p, \Gamma) \tag{16}$$

$$I3 \ \{\Gamma \vdash p \text{ implied } x\} \ . \ \text{toldOf}(p, \Gamma) \vdash x \ \rightarrow \ \Gamma \vdash p \text{ implied } x \tag{17}$$

Ground Completeness. The main property of the pattern-based encoding of the infon logic is the following Theorem.

Theorem 4 (Ground Infon Logic Completeness). *Let $\mathcal{A}$ be the axioms given by equations (2)-(17) and let $\Gamma \vdash \varphi$ be a ground sequent, then $\Gamma \vdash \varphi$ is derivable in the infon logic if and only if the (pattern-guided) saturation of $\mathcal{A} \wedge \neg(\Gamma \vdash \varphi)$ contains a subset of ground inconsistent formulas.*

Proof: It is very simple to observe that the encoding of the infon logic is sound, so we need only to consider the case where $\Gamma \vdash \varphi$ is derivable, and we wish to show that the saturation $\mathcal{A} \wedge \neg(\Gamma \vdash \varphi)$ contains a subset of ground inconsistent formulas. If $\Gamma \vdash \varphi$ is derivable, then there is a proof in the infon logic satisfying the *subformula property*: only subformulas of the original sequent are used in the proof. In particular, consider each proof step. We outline a justification that a corresponding axiom from $\mathcal{A}$ gets instantiated.

The rules $+I, +E1, +E2, \rightarrow E, \rightarrow I$ are similar in that if the last step in the subformula preserving proof is one of these, then the pattern annotation matches the conclusion and one of the existing subformulas. The corresponding axiom is instantiated. Since we assume that the current saturation satisfies the negation of the conclusion of the rule, it will also have to falsify at least one of the premises.

The axioms for $x2x$, true and asInfon are more liberal from their inference rule counter-parts, as they apply in the context of an arbitrary context Γ, but the rule *PI* ensures that this difference is benign.

Finally, the rules I and S are simulated using axioms $S1-2, I1-3$ since they encode a sequence of *PI* rules followed by either S or I. These axioms introduce the auxiliary sets saidOf(p, Γ) and toldOf(p, Γ). On the other hand $x2x$ and (2) and (3) take care of extracting premises when they become relevant. □

For good order we should note that axiom instantiation also terminates. In other words, it is not possible to repeatedly instantiate the axioms without re-introducing an already existing instance.

5.2 DKAL Knowledge and Communication Rules

In DKAL, knowledge (using the predicate _ knows _ : $\mathit{Principal} \times \mathit{Infon} \to \mathcal{B}$) is accumulated through derivations on infons and through communication. We model the communication assertions as predicates with two principals and a knowledge filter. Their signatures are: _ to _ _, _ from _ _ : $\mathit{Principal} \times \mathit{Principal} \times \mathit{KnowledgeFilter} \to \mathcal{B}$. The different filters are presented as auxiliary functions $[\_] \Leftarrow \_ : \mathit{Infon} \times \mathit{Infon} \to \mathit{KnowledgeFilter}$, $[\_ \leftarrow \_] \Leftarrow \_ : \mathit{Infon} \times \mathit{Infon} \times \mathit{Infon} \to \mathit{KnowledgeFilter}$.

One difficulty arises when encoding the communication rules. The original formulation of the rules (Com1) and (Com2) delay instantiating premises, but we will have to present the rules with eagerly instantiated premises in order to present the rules with ground instances. We use a predicate T(_) : $\mathit{Infon} \to \mathcal{B}$ to capture premises that are tautologies. The predicate satisfies T(true). We use the following encoding of the communication rules:

(Com1)
Proviso-free scenario
$\left\{ \begin{array}{l} p_1 \text{ to } p_2 : [x] \Leftarrow y, \\ p_2 \text{ from } p_1 : [x] \Leftarrow z \end{array} \right\}$
$\{p_2 \text{ knows } (p_1 \text{ said } x), \mathrm{T}(y), \mathrm{T}(z)\}$
$\{p_1 \text{ to } p_2\, [x] \Leftarrow y,\ \mathrm{T}(z)\}$
$\{p_2 \text{ from } p_1\, [x] \Leftarrow z,\ \mathrm{T}(y)\}$

$$\begin{array}{l} \quad p_1 \text{ to } p_2\, [x] \Leftarrow y \\ \wedge\ p_1 \text{ knows } y \\ \wedge\ p_2 \text{ from } p_1\, [x] \Leftarrow z \\ \wedge\ p_2 \text{ knows } z \\ \to\ p_2 \text{ knows } (p_1 \text{ said } x) \end{array}$$

(Com2)
Proviso-present scenario
$\left\{ \begin{array}{l} p_1 \text{ to } p_2\, [x \leftarrow y] \Leftarrow z, \\ p_2 \text{ from } p_1\, [x \leftarrow y] \Leftarrow k \end{array} \right\},$
$\{p_2 \text{ knows } (y \text{ imp } p_1 \text{ implied } x), \mathrm{T}(z), \mathrm{T}(k)\}$
$\{p_1 \text{ to } p_2\, [x \leftarrow y] \Leftarrow z, \mathrm{T}(k)\}$
$\{p_2 \text{ from } p_1\, [x \leftarrow y] \Leftarrow k,\ \mathrm{T}(z)\}$

$$\begin{array}{l} \quad p_1 \text{ to } p_2\, [x \leftarrow y] \Leftarrow z \\ \wedge\ p_1 \text{ knows } z \\ \wedge\ p_2 \text{ from } p_1\, [x \leftarrow y] \Leftarrow k \\ \wedge\ p_2 \text{ knows } k \\ \to\ p_2 \text{ knows } (y \text{ imp } p_1 \text{ implied } x) \end{array}$$

We split the Ensue rule into two parts. The first part applies the Ensue rule assuming x is derivable from some set knowsOf(p) comprising of the infons known to x. The second defines the set knowsOf(p).

(Ensue1)
$\{p \text{ knows } x\}$
$\{\text{knowsOf}(p) \vdash x\}$
$\text{knowsOf}(p) \vdash x \ \to\ (p \text{ knows } x)$

(Ensue2)
$\{p \text{ knows } x\}$
$\{x \in \text{knowsOf}(p)\}$
$(p \text{ knows } x) \ \leftrightarrow\ x \in \text{knowsOf}(p)$

We make a note that our treatment of the communication rules is specialized to our embedding into an instantiation-based procedure. The encoding ensures that the rules are not instantiated indiscriminately, but on the other hand, it restricts rule applications to a subset of the general formulation from Section 2.1. The main sanity check of this limited encoding has been through experimental evaluation through case studies, including the one presented in Section 3.

6 Derivations and Model Extraction

The embedding of DKAL into first-order axioms with patterns allows Z3 to answer DKAL queries as a first order theorem proving task. This process takes a query Q and checks it relative to a collection of knowledge and communication assertions $\mathcal{A}$. Z3 checks if $\mathcal{A} \wedge \neg Q$ is consistent, and if that is the case, then $\mathcal{A}$ implies Q is a DKAL theorem, as reflected in Theorem 4. When a proof cannot be found, a useful feature is to have a way to obtain a counter-model as a justification for the failure of finding a proof. Having a counter-model at hand may help the user who wrote the specification to understand the reasons of why the theory $\mathcal{A}$ does not implies the query Q. In our context, for counter-model we mean a Kripke model $\mathcal{M}$ in which there is a state $w \in \mathcal{M}$ where $\mathcal{M}, w \models \mathcal{A}$ but $\mathcal{M}, w \not\models Q$. In this section we present a procedure to extract a counter-model when Z3 fails to find a proof for $\mathcal{A} \wedge \neg Q$.

Our algorithm is based on a Z3 feature that returns a potential ground counter-model when the saturation procedure does not succeed in finding a proof. The algorithm we propose takes that model and extract a Kripke model from it. Let's see first a sketch of how Z3 outputs the ground model. Z3 *saturates* the assertions $\mathcal{A}$ with respect to the negated query Q (we denote this $Sat(\mathcal{A} \wedge \neg Q)$). The pattern annotations ensures that the saturation takes into account the subterms in Q. The saturation works by instantiating the universally quantified assertions from $\mathcal{A}$ and the rules encoding the infon logic, communication and knowledge rules. This saturation procedure ends when no new instantiations can be produced. Then ground assertions are checked for satisfiability. If they are unsatisfiable, then a proof for $(\mathcal{A} \Rightarrow Q)$ is found. If they are satisfiable, a (finite) propositional model $\mathcal{M}$ is produced, in which each ground assertion is assigned to true or false.

Recall that saturation is complete for establishing ground facts for the infon logic as is reflected by Theorem 4. This of course limits the procedure to work only with ground queries, leaving outside parametrized queries. These limitations also are reflected in the encoding of the Com rules. As we said before, the original formulation of these rules delay instantiating premises, but the encoding forces the rules to behave eagerly in order to use the pattern-based instantiation.

We will here show how a Kripke model according to Section 2.1 can be extracted. In overview, the main algorithm for extracting a model is as follows:

1. Given a saturated conjunction $Sat(\mathcal{A} \wedge \neg Q)$ let $\mathcal{M}$ be an evaluation that satisfies all ground conjuncts in the saturation.
2. We collect the set of atomic formulas of the form $\Gamma \vdash \varphi$ in $Sat(\mathcal{A} \wedge \neg Q)$, such that $\mathcal{M}$ assigns $\Gamma \vdash \varphi$ to *false*. These sequents are *underivable*.
3. The Kripke model $K = \langle W, \leq, S_q, I_q : W \times W, V : \text{PROP} \rightarrow \wp(W) \rangle$ is built using these underivable sequents as the building blocks. An element w_Γ is

the set of all underivable sequents in $\mathcal{M}$ with the same antecedent ($w_\Gamma = \{\Gamma \vdash \varphi \mid \mathcal{M} \not\models \Gamma \vdash \varphi$, for fixed Γ and arbitrary $\varphi\}$)

(a) The domain of K is the set of all sets w_Γ.
(b) The valuation function V is the following: $w_\Gamma \in V(p) \iff \Gamma \vdash p \notin w_\Gamma$
(c) For every pair of elements $w_{\Gamma_1}, w_{\Gamma_2}$ in K, $w_{\Gamma_1} \leq w_{\Gamma_2}$ iff $\Gamma_1 \subseteq \Gamma_2$.
(d) For every pair of elements $w_{\Gamma_1}, w_{\Gamma_2} \in K, said_p(w_{\Gamma_1}, w_{\Gamma_2}) \iff \{\varphi \mid p \text{ said } \varphi \in \Gamma_1\} \subseteq \Gamma_2$, and for all ψ in the Sub(Γ_1), $\Gamma_2 \vdash \psi \in w_{\Gamma_2}$ implies that $\Gamma_1 \vdash p$ said $\psi \in w_{\Gamma_1}$.
(e) For every pair of elements $w_{\Gamma_1}, w_{\Gamma_2} \in K, implied_p(w_{\Gamma_1}, w_{\Gamma_2}) \iff \{\varphi \mid p \text{ told } \varphi \in \Gamma_1\} \subseteq \Gamma_2$, and for all ψ in the Sub(Γ_1), $\Gamma_2 \vdash \psi \in w_{\Gamma_2}$ implies that $\Gamma_1 \vdash p$ implied $\psi \in w_{\Gamma_1}$.

We will now justify the model construction with more rigorous detail.

Definition 2 (Ω_Γ). *Given a set Γ of formulas, we define Ω_Γ as any set of sequents of the shape $\Gamma \vdash \varphi$, with φ arbitrary.*

Definition 3 (Complete sets). *A set Ω_Γ is* complete *if each $\Gamma \vdash \varphi \in \Omega_\Gamma$ is underivable and for each $\psi \in Sub(\Gamma \cup \{\varphi\})$, either*

1. $\psi \in \Gamma$ or
2. $\Gamma \vdash \psi \in \Omega_\Gamma$ or
3. both $\Gamma \vdash \psi$ and $\Gamma, \psi \vdash \varphi$ are derivable.

Definition 4 (Saturation). *A set Ω_Γ is* saturated for invertible rules *if every sequent $\Gamma \vdash \varphi \in \Omega_\Gamma$ is underivable and the following conditions are satisfied for any $\psi_1, \psi_2 \in Sub(\Gamma)$:*

1) if $\Gamma \vdash \psi_1 + \psi_2 \in \Omega_\Gamma$ then $\Gamma \vdash \psi_1 \in \Omega_\Gamma$ or $\Gamma \vdash \psi_2 \in \Omega_\Gamma$
2) Let $\psi_1 + \psi_2 \in Sub(\Gamma)$. Then if $\Gamma \vdash \psi_1 \in \Omega_\Gamma$ or $\Gamma \vdash \psi_2 \in \Omega_\Gamma$ then $\Gamma \vdash \psi_1 + \psi_2 \in \Omega_\Gamma$
3) if $\psi_1 + \psi_2 \in \Gamma$ then $\Gamma \vdash \psi_1 \notin \Omega_\Gamma$ and $\Gamma \vdash \psi_2 \notin \Omega_\Gamma$
4) if $\psi_1 \to \psi_2 \in \Gamma$ then $\Gamma \vdash \psi_1 \in \Omega_\Gamma$ or $\Gamma \vdash \psi_2 \notin \Omega_\Gamma$

Lemma 1. *If Ω_Γ is complete, then it is saturated for invertible rules.*

Proof: We say that ψ *clashes* with $\Gamma \vdash \varphi$ if both $\Gamma, \psi \vdash \varphi$ and $\Gamma \vdash \psi$ are derivable.

1) Note that if

$$\Gamma \vdash \psi_1 \text{ and } \Gamma \vdash \psi_2 \tag{18}$$

are derivable, then $\Gamma \vdash \psi_1 + \psi_2$ is derivable by $(+I)$ rule, contradicting completeness of Ω_Γ. Hence one of ψ_1, ψ_2 (say ψ_1) is not in Γ, and one of ψ_1, ψ_2 does not clash with $\Gamma \vdash \psi_1 + \psi_2$. If $\Gamma \vdash \psi_1 \notin \Omega_\Gamma$, then ψ_1 clashes with $\Gamma \vdash \psi_1 + \psi_2$ by completeness. Hence the sequent $\Gamma \vdash \psi_1$ is derivable, and ψ_2 does not clash with $\Gamma \vdash \psi_1 + \psi_2$. Also $\psi_2 \notin \Gamma$, since otherwise sequents in (18) are derivable. Hence $\Gamma \vdash \psi_2 \in \Omega_\Gamma$ as required.

2) We know that one of the two sequents (say $\Gamma \vdash \psi_1$) is in Ω_Γ. If $\psi_1 + \psi_2 \in \Gamma$, then $\Gamma \vdash \psi_1 + \psi_2$ is derivable, and using $(+E)$, $\Gamma \vdash \psi_1$ is derivable. This is not possible since $\Gamma \vdash \psi_1 \in \Omega_\Gamma$, and hence is underivable. On the other hand, because Ω_Γ is complete and $\psi_1 + \psi_2 \in \text{Sub}(\Gamma)$, if $\psi_1 + \psi_2 \notin \Gamma$ and $\Gamma \vdash \psi_1 + \psi_2 \notin \Omega_\Gamma$, then $\psi_1 + \psi_2$ clashes with $\Gamma \vdash \psi_1 + \psi_2$, and again by the same argument, $\Gamma \vdash \psi_1$ is derivable. This is again absurd, and therefore $\Gamma \vdash \psi_1 + \psi_2 \in \Omega_\Gamma$.
3) Let $\psi_1 + \psi_2 \in \Gamma$. Then $\Gamma \vdash \psi_1 + \psi_2$ is derivable using $(x2x)$. Then, by $(+E)$, both $\Gamma \vdash \psi_1$ and $\Gamma \vdash \psi_2$ are derivable. Then none of them can be in Ω_Γ.
4) Let $\psi_1 \to \psi_2 \in \Gamma$. By $(x2x)$, we know that $\Gamma \vdash \psi_1 \to \psi_2$ is derivable. If $\psi_1 \in \Omega_\Gamma$, again by $(x2x)$ the sequent $\Gamma \vdash \psi_1$ is derivable. Applying $(\to E)$ we derive $\Gamma \vdash \psi_2$, and therefore $\Gamma \vdash \psi_2 \notin \Omega_\Gamma$. The other possibility is, $\psi_1 \notin \Gamma$ and $\Gamma \vdash \psi_1 \notin \Omega_\Gamma$, and given that Ω_Γ is complete, we know that $\Gamma \vdash \psi_1$ is derivable, and we can apply $(\to E)$ again to derive $\Gamma \vdash \psi_2$, and therefore $\Gamma \vdash \psi_2 \notin \Omega_\Gamma$. □

Lemma 2 (Completion). *Given two set of formulas, Γ and Δ, such that $\Sigma = \{\Gamma \vdash \varphi \mid \varphi \in \Delta\}$ is a set of underivable sequents, then Σ can be extended to a complete set consisting of subformulas of $\Gamma \cup \Delta$.*

Proof: Consider an enumeration $\psi_0, \psi_1, \ldots$ of all formulas in $\text{Sub}(\Gamma, \Delta)$. Define the sequences $\Gamma \subset \Gamma_1 \subset \ldots$, and $\Delta \subset \Delta_1 \subset \ldots$ of finite sets of formulas, and $\Omega_{\Gamma_i} = \{\Gamma_i \vdash \psi \mid \psi \in \Delta_i\}$ such that Ω_{Γ_i} is complete for all formulas ψ_j, $j < i$, that is to say, for each sequent $\Gamma_i \vdash \varphi \in \Omega_{\Gamma_i}$ either $\psi_j \in \Gamma_i$, or $\Gamma_i \vdash \psi_j \in \Omega_{\Gamma_i}$, or both $\Gamma_i \vdash \psi_j$ and $\Gamma_i, \psi_j \vdash \varphi$ are derivable. Let $\Gamma_{i+1} := \Gamma_i \cup \{\psi_i\}$ if all the sequents $\{\Gamma_i, \psi_i \vdash \varphi \mid \varphi \in \Delta_i\}$ are underivable; otherwise $\Gamma_{i+1} := \Gamma$. Then let $\Delta_{i+1} := \Delta_i \cup \{\psi_i\}$ if $\Gamma_{i+1} \vdash \psi_i$ is underivable, and otherwise let $\Delta_{i+1} := \Delta_i$. Finally, let $\Gamma^+ := \bigcup \Gamma_i$ and $\Delta^+ := \bigcup \Delta_i$. The completeness of $\Omega_{\Gamma^+} = \{\Gamma^+ \vdash \varphi \mid \varphi \in \Delta^+\}$ easily follows. □

Definition 5. *Consider the Kripke model $K = \langle W, \leq, S_q, I_q, V\rangle$, over the signature $\langle \text{PROP}, \text{REL}\rangle$, where W is a nonempty set, $\leq$, S_q and I_q are binary relations over W, $V : \text{PROP} \to \wp(W)$ and*

- *W is the set of all sets of complete sequents.*
- *$\Omega_\Gamma \leq \Omega_{\Gamma'}$ iff $\Gamma \subseteq \Gamma'$.*
- *$S_q(\Omega_\Gamma, \Omega_{\Gamma'})$ iff $\{\varphi \mid p \text{ said } \varphi \in \Gamma\} \subseteq \Gamma'$ and for all $\psi \in \text{Sub}(\Gamma)$, $\Gamma' \vdash \psi \in \Omega_{\Gamma'}$ implies $\Gamma \vdash p \text{ said } \psi \in \Omega_\Gamma$.*
- *$I_q(\Omega_\Gamma, \Omega_{\Gamma'})$ iff $\{\varphi \mid p \text{ told } \varphi \in \Gamma\} \subseteq \Gamma'$ and for all $\psi \in \text{Sub}(\Gamma)$, $\Gamma' \vdash \psi \in \Omega_{\Gamma'}$ implies $\Gamma \vdash p \text{ implied } \psi \in \Omega_\Gamma$.*
- *$\Omega_\Gamma \in V(p)$ iff $\Gamma \vdash p \notin \Omega_\Gamma$.*

Let's see that the model K is a Kripke model for the infon logic.

- $\leq$ is reflexive and transitive given that $\subseteq$ is reflexive and transitive.
- We have to check that, if $\Omega_\Gamma \leq \Omega_{\Gamma'}$ and $S_p(\Omega_{\Gamma'}, \Omega_{\Gamma''})$, then $S_p(\Omega_\Gamma, \Omega_{\Gamma''})$ (and the same for I_p). If $\Gamma \subseteq \Gamma'$, and $\{\varphi \mid p \text{ said } \varphi \in \Gamma'\} \subseteq \Gamma''$, then

$\{\varphi \mid p \text{ said } \varphi \in \Gamma\} \subseteq \Gamma''$. If $\Gamma'' \vdash \psi \in \Omega_{\Gamma''}$, then $\Gamma' \vdash p \text{ said } \psi \in \Omega_{\Gamma'}$. Thus, $\Gamma' \vdash p \text{ said } \psi$ is underivable, and because, $\Gamma \subseteq \Gamma'$, $\Gamma \vdash p \text{ said } \psi$ is also underivable. Therefore, $p \text{ said } \psi \notin \Gamma$, and by completeness of Ω_Γ, $\Gamma \vdash p \text{ said } \psi \in \Omega_\Gamma$. Therefore $S_p(\Omega_\Gamma, \Omega_{\Gamma''})$ as desired. The same reasoning can be applied to I_p.

- To see that the valuation function V is monotonic, we have to prove that given Ω_Γ and $\Omega_{\Gamma'}$ such that $\Gamma \subseteq \Gamma'$, then if $\Gamma \vdash p \notin \Omega_\Gamma$ then $\Gamma' \vdash p \notin \Omega_{\Gamma'}$. Assume that $\Gamma \vdash p \notin \Omega_\Gamma$. If $p \in \Gamma$, then $p \in \Gamma'$, therefore $\Gamma' \vdash p$ is derivable, and therefore $\Gamma' \vdash p \notin \Omega_{\Gamma'}$. On the other hand, if $p \notin \Gamma$, and knowing that Ω_Γ is complete, then $\Gamma \vdash p$ is derivable. That means that $\Gamma' \vdash p$ is also derivable, and again we have that $\Gamma' \vdash p \notin \Omega_{\Gamma'}$.

Definition 6. *A set M of sets Ω_Γ is* saturated for non-invertible rules *if the following conditions are satisfied for every $\Omega_\Gamma \in M$:*

1) *If $\Gamma \vdash \psi_1 \to \psi_2 \in \Omega_\Gamma$ then there is a $\Omega_{\Gamma'} \in M$ such that $\Gamma \cup \{\psi_1\} \subseteq \Gamma'$ and $\Gamma' \vdash \psi_2 \in \Omega_{\Gamma'}$.*
2) *If $\Gamma \vdash p$ said $\varphi \in \Omega_\Gamma$ then there is a $\Omega_{\Gamma'} \in M$ such that $S_p(\Omega_\Gamma, \Omega_{\Gamma'})$ and $\Gamma' \vdash \varphi \in \Omega_{\Gamma'}$.*
3) *If $\Gamma \vdash p$ implied $\varphi \in \Omega_\Gamma$ then there is a $\Omega_{\Gamma'} \in M$ such that $I_p(\Omega_\Gamma, \Omega_{\Gamma'})$ and $\Gamma' \vdash \varphi \in \Omega_{\Gamma'}$.*

A set M of sets Ω_Γ is saturated *if every $\Omega_\Gamma \in M$ is saturated for invertible rules, and M is saturated for non-invertible rules.*

Lemma 3. *The set W of all sets of complete sequents Ω_Γ is saturated.*

Proof: We check the three conditions:

1) If $\Gamma \vdash \psi_1 \to \psi_2 \in \Omega_\Gamma$, then the sequent $\Gamma, \psi_1 \vdash \psi_2$ is underivable (we can apply rule $(\to I)$ otherwise). By the Completion Lemma 2, we can extend the singleton set $\{\Gamma, \psi_1 \vdash \psi_2\}$ to a complete set $\Omega_{\Gamma'}$ such that $\Gamma \subseteq \Gamma'$ and $\Gamma' \vdash \psi_2 \in \Omega_{\Gamma'}$. Therefore $\Omega_{\Gamma'} \in W$.
2) Let $\Delta = \{\varphi \mid \Gamma \vdash p \text{ said } \varphi \in \Omega_\Gamma\}$. Let $\Gamma' = \Gamma'_1 \cup \Gamma'_2$, where $\Gamma'_1 = \{\psi \mid p \text{ said } \psi \in \Gamma\}$ and $\Gamma'_2 = \{\beta \mid \Gamma \vdash p \text{ said } \beta \text{ is derivable}\}$. Let's check first that $\{\Gamma' \vdash \varphi \mid \varphi \in \Delta\}$ is a set of underivable sequents. If that is not the case, then there is some sequent $\Gamma' \vdash \varphi_i$ that is derivable. Let us restrict to the premises that are used in the derivation of the sequent, so $\psi_1, \ldots, \psi_n, \beta_1, \ldots, \beta_m \vdash \varphi_i$ is derivable, where $\psi_i \in \Gamma'_1$ and $\beta_i \in \Gamma'_2$. Using rule (S), we can derive $p \text{ said } \{\psi_1, \ldots, \psi_n, \beta_1, \ldots, \beta_m\} \vdash p \text{ said } \varphi_i$. Applying rule $(\to I)$ m times, we get $p \text{ said } \{\psi_1, \ldots, \psi_n\} \vdash p \text{ said } \beta_1 \to \cdots \to p \text{ said } \beta_m \to p \text{ said } \varphi_i$. We know that $p \text{ said } \{\psi_1, \ldots, \psi_n\} \subseteq \Gamma$, so $\Gamma \vdash p \text{ said } \beta_1 \to \ldots p \text{ said } \beta_m \to p \text{ said } \varphi_i$ is also derivable. Because each $\Gamma \vdash p \text{ said } \beta_i$, $1 \leq i \leq m$, is derivable, we can apply $(\to E)$ m times, and conclude $\Gamma \vdash p \text{ said } \varphi_i$ is derivable. This is absurd, since $\Gamma \vdash p \text{ said } \varphi_i \in \Omega_\Gamma$.

   Therefore we can use Completion Lemma 2 and extend $\{\Gamma' \vdash \varphi \mid \varphi \in \Delta\}$ to a complete set Ω_{Γ^+}, and therefore $\Omega_{\Gamma^+} \in M$. Let us now check that Ω_{Γ^+} satisfies the conditions we want. By construction, we know that for each

$\Gamma \vdash p$ said $\varphi \in \Omega_\Gamma$, $\Gamma^+ \vdash \varphi \in \Omega_{\Gamma^+}$. Furthermore, let's see that for every ψ such that $\Gamma^+ \vdash \psi \in \Omega_{\Gamma^+}$, $\Gamma \vdash p$ said $\psi \in \Omega_\Gamma$. If we assume the contrary, there is a ψ such that $\Gamma^+ \vdash \psi \in \Omega_{\Gamma^+}$, but $\Gamma \vdash p$ said $\psi \notin \Omega_\Gamma$. Observe that if p said $\psi \in \Gamma$, then by construction $\psi \in \Gamma^+$, and $\Gamma^+ \vdash \psi$ would be derivable. Since $\Gamma^+ \vdash \psi \in \Omega_{\Gamma^+}$, this is absurd. So, by completeness of Ω_Γ, $\Gamma \vdash p$ said ψ is derivable, but in that case $\psi \in \Gamma^+$ again by construction. This implies that $\Gamma^+ \vdash \psi$ is derivable, which is an absurd. Therefore, the desired condition holds. Finally, it is easy to see that $\{\varphi \mid p \text{ said } \varphi \in \Gamma\} \subseteq \Gamma^+$, and therefore $S_p(\Omega_\Gamma, \Omega_{\Gamma^+})$.

3) The proof for this condition is similar to 2). □

Theorem 5. *Let K be as in Definition 5. Then for $w \equiv \Omega_\Gamma$, $w \in M$, the following holds for all $\varphi \in Sub(\Gamma)$:*

$$\varphi \in \Gamma \textit{ implies } K, w \models \varphi \tag{19}$$

$$\Gamma \vdash \varphi \in \Omega_\Gamma \textit{ iff } K, w \not\models \varphi \tag{20}$$

Proof: We prove both claims by simultaneous induction on φ. For the base case, let $\varphi = p$. To see claim (19), if $p \in \Gamma$, then $\Gamma \vdash p$ is derivable, and therefore $\Gamma \vdash p \notin \Omega_\Gamma$. That means by definition of V that $K, w \models p$. Claim (20) follows directly by definition of the valuation function V.

For the case $\varphi = \psi_1 + \psi_2$. If $\psi_1 + \psi_2 \in \Gamma$, by Lemma 1 (3), both $\Gamma \vdash \psi_1$ and $\Gamma \vdash \psi_2$ do not belong to Ω_Γ. Therefore by inductive hypothesis $K, w \models \psi_1$ and $K, w \models \psi_2$, and by the truth definition, $K, w \models \psi_1 + \psi_2$. To see the other claim, $\Gamma \vdash \psi_1 + \psi_2 \in \Omega_\Gamma$ iff (by Lemma 1, 1) and 2)) one of $\Gamma \vdash \psi_1$, $\Gamma \vdash \psi_2$ (say $\Gamma \vdash \psi_1$) is in Ω_Γ. By inductive hypothesis, $K, w \not\models \psi_1$, and by the definition of truth, $K, w \not\models \psi_1 + \psi_2$.

For the case $\varphi = \psi_1 \rightarrow \psi_2$. If $\psi_1 \rightarrow \psi_2 \in \Gamma$, then for every $w' \equiv \Omega_{\Gamma'}$ such that $\Omega_\Gamma \leq \Omega_{\Gamma'}$, we have $\psi_1 \rightarrow \psi_2 \in \Gamma'$. By Lemma 1 (4), this implies that $\Gamma' \vdash \psi_1 \in \Omega_{\Gamma'}$ or $\Gamma' \vdash \psi_2 \notin \Omega_{\Gamma'}$. By inductive hypothesis, $K, w' \not\models \psi_1$ or $K, w' \models \psi_2$. This implies $K, w \models \psi_1 \rightarrow \psi_2$ as desired. For the second claim, if $\Gamma \vdash \psi_1 \rightarrow \psi_2 \in \Omega_\Gamma$, by the saturation condition we know that there is a $w' \equiv \Omega_{\Gamma'} \in M$ such that $\Omega_\Gamma \leq \Omega_{\Gamma'}$, $\psi_1 \in \Gamma'$ and $\Gamma' \vdash \psi_2 \in \Omega_{\Gamma'}$. By the inductive hypothesis, $K, w' \models \psi_1$ and $K, w' \not\models \psi_2$. Therefore $K, w \not\models \psi_1 \rightarrow \psi_2$. For the other direction, let us suppose that $K, w \not\models \psi_1 \rightarrow \psi_2$. By the truth definition, that means that there is a $w' \equiv \Omega_{\Gamma'} \in M$ such that $\Omega_\Gamma \leq \Omega_{\Gamma'}$, $K, w' \models \psi_1$ and $K, w' \not\models \psi_2$. By inductive hypothesis, this implies that $\Gamma' \vdash \psi_1 \notin \Omega_{\Gamma'}$ and $\Gamma' \vdash \psi_2 \in \Omega_{\Gamma'}$. First note that $\Gamma' \vdash \psi_1$ is derivable (this is because either $\psi_1 \in \Gamma'$, and therefore $\Gamma' \vdash \psi_1$ is derivable, or because $\psi_1 \notin \Gamma'$ and then the completeness condition imposes derivability). That means that $\Gamma' \vdash \psi_1 \rightarrow \psi_2$ cannot be derivable, because in that case we could apply ($\rightarrow E$) and conclude that $\Gamma' \vdash \psi_2$ is derivable, which would contradict the fact that $\Gamma' \vdash \psi_2 \in \Omega_{\Gamma'}$. Because $\Gamma \subseteq \Gamma'$, we know that $\Gamma \vdash \psi_1 \rightarrow \psi_2$ is not derivable either. Therefore $\psi_1 \rightarrow \psi_2 \notin \Gamma$, and by the completeness condition, the only possibility is $\Gamma \vdash \psi_1 \rightarrow \psi_2 \in \Omega_\Gamma$.

For the case $\varphi = p$ said ψ. If p said $\psi \in \Gamma$, then for every $w' \equiv \Omega_{\Gamma'}$ such that $S_p(\Omega_\Gamma, \Omega_{\Gamma'})$ we have $\psi \in \Gamma'$. By inductive hypothesis, $K, w' \models \psi$. Using the truth definition, this implies $K, w \models p$ said ψ as desired. For the second claim, if $\Gamma \vdash p$ said $\psi \in \Omega_\Gamma$, then by the saturation condition, there is a $\Omega_{\Gamma'} \in M$ such that $S_p(\Omega_\Gamma, \Omega_{\Gamma'})$ and $\Gamma' \vdash \psi \in \Omega_{\Gamma'}$. By the inductive hypothesis, $K, w' \not\models \psi$. Therefore $K, w \not\models p$ said ψ. For the other direction, if $K, w \not\models p$ said ψ, there is a $w' \equiv \Omega_{\Gamma'} \in M$ such that $S_p(w, w')$ and $K, w' \not\models \psi$. By inductive hypothesis, we know that $\Gamma' \vdash \psi \in \Omega_{\Gamma'}$. Because $S_p(\Omega_\Gamma, \Omega_{\Gamma'})$ and $\Gamma' \vdash \psi \in \Omega_{\Gamma'}$, $\Gamma \vdash$ p said $\psi \in \Omega_\Gamma$.

The case for $\varphi = p$ implied ψ is equivalent to the previous one. □

Corollary 1. *$K, w \not\models \Gamma \vdash \varphi$ for every $\Gamma \vdash \varphi \in \Omega_\Gamma$.*

Proof: This is an immediate consequence of Theorem 5. □

Corollary 2 (Completeness). *Each underivable sequent in the infon calculus is falsified in a state of the canonical model K. Therefore, every valid sequence is derivable in the infon calculus.*

Proof: By the Completion Lemma 2, any underivable sequent $\Gamma \vdash \varphi$ can be extended to a complete set $w \equiv \Omega_{\Gamma'}$ such that $\Gamma \subseteq \Gamma'$ and $\Gamma' \vdash \varphi \in \Omega_{\Gamma'}$. By Theorem 5, there is a $w \equiv \Omega_{\Gamma'} \in M$ such that $K, w \not\models \Gamma' \vdash \varphi$. Therefore $K, w \not\models \Gamma \vdash \varphi$. □

Corollary 3 (Z3oundness). *The model extraction algorithm is complete for ground sequents when using the pattern-based encoding.*

Proof: Theorem 4 establishes that the pattern-based encoding is complete for ground sequents. In particular, every sequent that evaluates to *false* in the model produced by saturation is underivable. Lemma 2 establishes that a subset of these underivable sequents can be extended to a complete set. Corollary 2 therefore applies and allows extracting a model K for the underivable ground sequent. □

6.1 DKAL Models@Starbucks

Here we show examples of some models that have been extracted using our prototype. In order to force the desired query to be underivable we feed the tool with an incomplete specification. Recall the Starbucks example presented in Section 3, in which an SSL connection is established. When the trusted root informs Windows that the certificate is valid (and Windows accepts the communication) it learns:

? `Windows` knows `TrustedRoot` implied `ThisCert isValid`

But in this example we will remove the fact that says that Windows trusts on the trusted root:

`/* Windows` knows `TrustedRoot` tdonI X`*/`

So now the query:

```
? Windows knows ThisCert isValid
```

will not be derivable. The Kripke model K returned by the prototype in this case is the following (for the knowledge relevant to Windows):

where the valuation function is the following: $V($`ThisCert isValid`$) = \emptyset$, and the value for the rest of the atoms is the set $\{w_0, w_1\}$. In the picture the relation $\leq$ is shown, and the other relations are empty. Note that

$$K, w_0 \not\models \texttt{ThisCert isValid},$$

but in K, w_0 the rest of the assertions are satisfied. So K is effectively a counter-model. The next query that is tested relates to the process of establishing an SSL connection:

`? Windows` knows `IE shouldShowLock`

It is not derivable either, since knowing that the certificate is valid is a condition for Windows to know that it should show the lock on IE. The output in this case is the following Kripke model K:

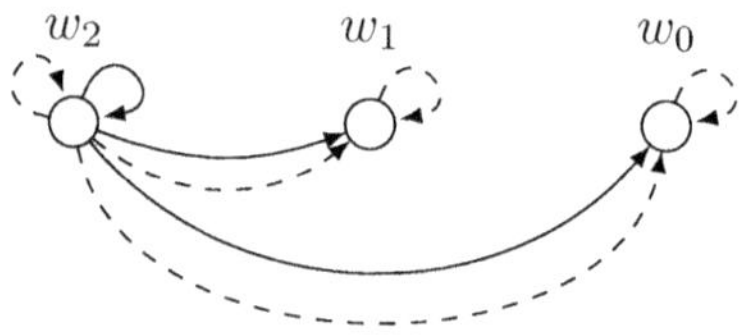

In the picture, the dashed edges represent the $\leq$ relationship. All the other relations ($said_{patrick}$, $said_{windows}$, etc.) are the same and they are represented by the solid edges. The valuation function V is the following:

$V($`ThisCert isValid`$) = \emptyset$,
$V($`IECurrSite supportsSSL`$) = \{w_0, w_1, w_2\}$
$V($`ThisCert isProperlySigned`$) = \{w_0, w_1, w_2\}$
$V($`IE shouldShowLock`$) = \{w_1, w_2\}$.

We can see again that

$$K, w_0 \not\models \texttt{ThisCert isValid},$$

but the rest of the assertions are satisfied.

7 Conclusion

We experimented with embedding the DKAL logic into the theorem prover Z3 for classical first-order logic and interpreted theories. More specifically, we used the instantiation-based support for quantifiers in Z3. We also established how ground counter-models produced by Z3 can be mapped back to Kripke models for the infon logic. A prototype embedding into Z3 was implemented and checked on smaller illustrative case studies. This paper includes one such case study that exercised various features available in DKAL. We do not claim to have produced any dedicated, high-performance, DKAL engine, but the embedding into Z3 allows re-using the extensible support for theories in Z3 to handle auxiliary logical constraints, such as comparing timeouts symbolically.

References

1. Ball, T., Rajamani, S.K.: The SLAM project: debugging system software via static analysis. SIGPLAN Not. 37(1), 1–3 (2002)
2. Barnett, M., Leino, K.R.M., Schulte, W.: The Spec# Programming System: An Overview. In: Barthe, G., Burdy, L., Huisman, M., Lanet, J.-L., Muntean, T. (eds.) CASSIS 2004. LNCS, vol. 3362, pp. 49–69. Springer, Heidelberg (2005)
3. Biere, A., Cimatti, A., Clarke, E., Zhu, Y.: Symbolic model checking without BDDs. In: Cleaveland, W.R. (ed.) TACAS 1999. LNCS, vol. 1579, pp. 193–207. Springer, Heidelberg (1999)
4. Bjørner, N., Hendrix, J.: Linear functional fixed-points. In: Bouajjani, A., Maler, O. (eds.) CAV 2009. LNCS, vol. 5643, pp. 124–139. Springer, Heidelberg (2009)
5. Cohen, E., Moskal, M., Schulte, W., Tobies, S.: A Precise Yet Efficient Memory Model For C. In: Proceedings of Systems Software Verification Workshop (SSV 2009) (2009) (to appear)
6. Costa, M., Crowcroft, J., Castro, M., Rowstron, A.I.T., Zhou, L., Zhang, L., Barham, P.: Vigilante: end-to-end containment of internet worms. In: Herbert, A., Birman, K.P. (eds.) SOSP, pp. 133–147. ACM, New York (2005)
7. de Moura, L., Bjørner, N.: Efficient E-Matching for SMT Solvers. In: Pfenning, F. (ed.) CADE 2007. LNCS (LNAI), vol. 4603, pp. 183–198. Springer, Heidelberg (2007)
8. de Moura, L., Bjørner, N.: Z3: An Efficient SMT Solver. In: Ramakrishnan, C.R., Rehof, J. (eds.) TACAS 2008. LNCS, vol. 4963, pp. 337–340. Springer, Heidelberg (2008)
9. Degtyarev, A., Gurevich, Y., Narendran, P., Veanes, M., Voronkov, A.: Decidability and complexity of simultaneous rigid E-unification with one variable and related results. Theoretical Computer Science 243, 167–184 (2000)
10. DeLine, R., Leino, K.R.M.: BoogiePL: A typed procedural language for checking object-oriented programs. Technical Report 2005-70, Microsoft Research (2005)
11. Godefroid, P., Levin, M., Molnar, D.: Automated Whitebox Fuzz Testing. Technical Report 2007-58, Microsoft Research (2007)
12. Gulavani, B.S., Henzinger, T.A., Kannan, Y., Nori, A.V., Rajamani, S.K.: Synergy: a new algorithm for property checking. In: Young, M., Devanbu, P.T. (eds.) SIGSOFT FSE, pp. 117–127. ACM, New York (2006)
13. Gurevich, Y., Neeman, I.: Dkal: Distributed-knowledge authorization language. In: CSF, pp. 149–162. IEEE Computer Society, Los Alamitos (2008)

14. Gurevich, Y., Neeman, I.: DKAL 2 — A Simplified and Improved Authorization Language. Technical Report 2009-11, Microsoft Research (2009)
15. Gurevich, Y., Neeman, I.: The Infon Logic. Bulletin of European Association for Theoretical Computer Science (2009)
16. Jackson, E.K., Schulte, W.: Model Generation for Horn Logic with Stratified Negation. In: Suzuki, K., Higashino, T., Yasumoto, K., El-Fakih, K. (eds.) FORTE 2008. LNCS, vol. 5048, pp. 1–20. Springer, Heidelberg (2008)
17. Lahiri, S.K., Qadeer, S.: Back to the Future: Revisiting Precise Program Verification using SMT Solvers. In: POPL 2008 (2008)
18. Mints, G.: Grigori. In: A short introduction to intuitionistic logic. Kluwer Academic, New York (2000)
19. Moy, Y., Bjørner, N., Sielaff, D.: Modular Bug-finding for Integer Overflows in the Large: Sound, Efficient, Bit-precise Static Analysis. Technical Report MSR-TR-2009-57, Microsoft Research (2009)
20. Ohlbach, H.J., Nonnengart, A., de Rijke, M., Gabbay, D.M.: Encoding two-valued nonclassical logics in classical logic. In: Robinson, J.A., Voronkov, A. (eds.) Handbook of Automated Reasoning, pp. 1403–1486. Elsevier, MIT Press (2001)
21. Tillmann, N., Schulte, W.: Unit Tests Reloaded: Parameterized Unit Testing with Symbolic Execution. IEEE software 23, 38–47 (2006)

Decidability of the Class E by Maslov's Inverse Method

Grigori Mints

Stanford University, Stanford, CA
gmints@stanford.edu

Dedicated to Yuri Gurevich on the occasion of his 70th birthday

Abstract. We present a simple formulation of S. Maslov's inverse method of proof (and proof search) for first order predicate logic and illustrate it by proving decidability of validity for formulas with one existential quantifier and arbitrary function symbols.

Keywords: decidable classes, automated deduction, inverse method.

1 Introduction

This paper contains a proof of a result obtained independently by two quite different methods by Yuri Gurevich [6] and V. Orevkov and S. Maslov [14], [9] in the sixties. These two methods made it possible to give definitive treatments of decidable fragments for classical predicate logic in [9] and [6]. It was the time of the frequent and fruitful contacts of Yuri with the Leningrad group of mathematical logic headed by N. Shanin which included (among others) S. Maslov, V. Orevkov and the present author. The proof in the present paper is very close to one presented [in Russian] in [8].

An important reason for the present publication is the need to reintroduce into the literature on and around automated deduction the basic ideas of the inverse method proposed by to S. Maslov [7]. This powerful extension of Gentzen-style treatment for classical predicate logic allows for very smooth extensions to almost any non-classical logic admitting Gentzen-style treatment. Unfortunately the untimely death of S. Maslov in July 1982 at the age of 43 years prevented him from developing his ideas in that direction. His publications on inverse method are notoriously brief which may have been an obstacle for their assimilation into logic and computer science, and the vast potential of the suggested methods and results have not been realized. S. Maslov also developed versions of the resolution method having close connections to Gentzen-style systems [10]. These ideas were taken up, presented in detail and developed by the present author [11], [12], by A. Voronkov and Degtyarev [4] as well as by other authors. In this later work the term *inverse method* is applied to resolution-like developments, while S. Maslov used it primarily for his enhancement of Gentzen-type methods. The authors of [4] present an inverse method with a lot of technical detail, then switch to

A. Blass, N. Dershowitz, and W. Reisig (Eds.): Gurevich Festschrift, LNCS 6300, pp. 529–537, 2010.

resolution-like methods and explain the reason for the switch at the beginning of their Sect. 4.

In Sect. 1 we present S. Maslov's formulation of his inverse method for skolemized formulas which avoids many technical details and give a short proof of its completeness which shows a close connection with Gentzen-type derivations. Then we discuss two notions most prominent in S. Maslov's approach to decidable classes and automated deduction: those of a favorable free clause and of factorization. They make it possible to simplify a formula to be tested for validity "on the fly", in the process of testing.

S. Maslov proved in [7] decidability of a class K containing essentially all decidable classes of predicate formulas known by the time [7] was written. That proof used a combination of Lemmas 1 and 2 with much more careful analysis of tautological disjunction than in the present paper. It would be interesting to give decision procedures using inverse method for decidable classes that were discovered after that, for example guarded formulas [2]. Another possible application that requires development of the inverse method for intuitionistic logic may be a new more efficient decision algorithm for the formulas of intuitionistic predicate logic that do not contain negative quantifiers [13]. Negative quantifiers are ones that become $\exists$ in the prenex normal form. The need for such an algorithm in computer science is noticed in [5] where an algorithm close to [13] is provided for an extension of this class.

I thank Elena Maslova for help in editing this paper.

2 Inverse Method for Skolemized Formulas

2.1 Calculus U_F for Deriving Clauses

A *clause* is a disjunction of *literals* $P(t_1, \ldots, t_n)$, $\neg P(t_1, \ldots, t_n)$ where t_i are terms possibly containing function symbols. As usual, clauses are treated up to order of literals and contraction of identical literals. For exmple $L \vee K \vee L$ is identified with $K \vee L$.

We consider closed prenex first order formulas of the form

$$\exists x_1 \ldots \exists x_n \&_{1 \leq i \leq \delta} D_i \equiv \exists \mathbf{x} M \equiv F \tag{1}$$

where D_i are clauses.

An *F-substitution* is an expression of the form $(t_1/x_1, \ldots, t_n/x_n)$ where t_i are terms.

Derivable objects of the calculus U_F constructed for the formula F are clauses

$$D_{i_1}\sigma_1 \vee \ldots \vee D_{i_m}\sigma_m \qquad (m \geq 0;\ 1 \leq i_1, \ldots i_m \leq \delta)$$

where $\sigma_1, \ldots, \sigma_m$ are F-substitutions.

Rule A: C, if C is a tautology.

Rule B.

$$\frac{C_1 \vee (D_1\sigma_1) \quad \ldots \quad C_\delta \vee (D_\delta\sigma_\delta)}{C_1\sigma^1 \vee \ldots C_\delta\sigma^\delta}\, B$$

provided $\sigma_1\sigma^1 = \ldots = \sigma_\delta\sigma^\delta$.

A clause is F-favorable iff it is derivable in U_F.

Comment. A simple (but sufficiently general) instance of the rule B is

$$\frac{C \vee (D_1\sigma) \quad \ldots \quad C \vee (D_\delta\sigma)}{C}\ B$$

which can be seen as

$$\frac{C \vee (\& D_i\sigma)}{C}\ B$$

Here a substitution instance of the whole formula F is cut out of the derived clauses to infer the clause C.

Example 1. Consider

$$F \equiv \exists x(\neg Px \vee Pf(x)).$$

Here we have $\delta = 1$ hence just one premise in the rule B. The following figure is a derivation of the the empty clause $\emptyset$ in U_F.

$$\dfrac{\dfrac{\begin{array}{c}\text{axiom}\\ D_1 \vee D_1(fx/x)\end{array}}{D_1}\ B}{\emptyset}\ B$$

2.2 Completeness of U_F

Theorem 1. *A clause C is F-favorable iff $C \vee F$ is valid, that is, derivable in predicate logic.*

Proof. If $C \vee F$ is valid, there are F-substitutions $\tau_1, \ldots, \tau_p$ such that

$$C \vee M\tau_1 \vee \ldots \vee M\tau_p$$

is derivable (in fact is a tautology). We use induction on p to construct a derivation (of height p) of C in U_F.

Induction base. $p = 0$, that is C is a tautology. Then C is an axiom of U_F.

Induction step. If

$$C \vee M\tau_1 \vee \ldots \vee M\tau_p \vee M\theta$$

is derivable, then distributing the last conjunction

$$M\theta \equiv \&_i D_i\theta$$

over disjunction we get δ disjunctions

$$(C \vee D_1\theta) \vee M\tau_1 \vee \ldots \vee M\tau_p, \ldots (C \vee D_\delta\theta) \vee M\tau_1 \vee \ldots \vee M\tau_p.$$

By induction hypothesis we have in U_F

$$C \vee D_1\theta, \ldots, C \vee D_\delta\theta,$$

and hence one application of the rule B derives the clause C.

Only if. Use induction on the derivation in U_F. One application of the rule B is simulated by derivable implications:

$$(C_1 \vee D_1\sigma \vee F)\& \ldots \&(C_\delta \vee D_\delta\sigma \vee F) \rightarrow$$

$$(C_1 \vee \ldots \vee C_\delta) \vee (D_1 \vee \ldots \vee D_\delta)\sigma \vee F \equiv$$

$$(C_1 \vee \ldots \vee C_\delta) \vee M\sigma \vee F \rightarrow$$

$$(C_1 \vee \ldots \vee C_\delta) \vee F \vee F \rightarrow (C_1 \vee \ldots \vee C_\delta) \vee F$$

where $\sigma = \sigma_1\sigma^1 = \ldots = \sigma_\delta\sigma^\delta$. □

Corollary 1. *A formula F is derivable iff the empty clause $\emptyset$ is F-favorable.*

Proof. Let $C = \emptyset$ in the theorem; $\emptyset \vee F$ is $\bot \vee F \iff F$. □

2.3 Free Favorable Clauses, Factorization, General Proof Search Method

The main tool of decidability proofs by inverse method is the following simplification lemma.

Lemma 1. *If D_α is F-favorable, then*

$$F \iff \exists \mathbf{x} \&_{i \neq \alpha} D_i$$

(abbreviated $F \iff F^-$).

Proof. $F \rightarrow F^-$ is obvious. By Theorem 1 $D_\alpha \vee F$ is derivable in the predicate logic, and hence

$$\forall \mathbf{x} D_\alpha \vee F.$$

Hence in the proof of $F^- \rightarrow F$ one can use $\forall \mathbf{x} D_\alpha$, but then the task is trivial. □

Definition 1. *A favorable clause $D_{i_1}\sigma_1 \vee \ldots D_{i_p}\sigma_p$ is* free *if each of the substitutions $\sigma_1, \ldots, \sigma_p$ is the identity $(x_1/x_1, \ldots, x_n/x_n)$.*

A favorable clause D_α is a unit free clause.

We shall see below how Lemma 1 decides the class E.

Another source of simplification is the notion of *factorization*.

Definition 2. *A clause $C \vee D$ is* factored *into clauses C and D (and C, D are called its factors) if C and D have no common variables.*

Notation $C \vdash E$ for derivability of a clause E from a clause C in U_F is self-explaining.

Lemma 2. *If an F-favorable clause $C \vee D$ is factored into C and D and a clause E does not have common variables with $C \vee D$ then*

$$C \vee D \vdash E \text{ if and only if } (C \vdash E \text{ and } D \vdash E).$$

Proof. If $C \vee D \vdash E$ then erasing all traces of D from the given derivation of E we get $C \vdash E$ and similarly for $D \vdash E$, even if E has common variables with $C \vee D$.

Let $C \vdash E$. Since D has no common variables with C, E, we can rename variables in D so that they do not occur at all in the given derivation. Then adding the resulting clause (say D') to all clauses in the derivation gives us

$$(C \vee D') \vdash E \vee D'. \tag{2}$$

Renaming variables from D in the derivation $D \vdash E$ we get $D' \vdash E$. Adding this to (2) we get $C \vee D' \vee D' \vdash E \vee E$, and after contraction and renaming $C \vee D \vdash E$ as required. □

A general proof search algorithm for the first order logic based on the inverse method [3] generates favorable clauses for the goal formula F from the clauses given by the rule A by the application of the rule B combined with the unification as it is done in the resolution method. The proof search terminates when an empty clause is generated.

If at some moment a free unit clause is derived, it is deleted from the formula (cf. Lemma 1) and the proof search is continued with a simplified formula. If a factorable clause $C \vee D$ is generated, the search branches according to Lemma 2. If in addition one of the clauses C, D is a unit free clause, the formula F can be simplified in corresponding branch.

Example. $F \equiv \exists x \exists y[(\neg Px \vee Pf(x))\& (\neg Qy \vee Qg(y))]$
$\equiv \exists x \exists y[D_1(x) \& D_2(y)]$.

Rule A gives two favorable clauses:

$$D_1(x) \vee D_1(f(x)); \qquad D_2(y) \vee D_2(g(y)).$$

Rule B gives (with substitutions $\sigma_1 = (x/f(x)),\ \sigma_2 = (y/g(y))$)

$$D_1(x) \vee D_2(y).$$

This clause is factored into two unit free clauses $D_1(x)$ and $D_2(y)$. Taking first the factor $D_2(y)$ (to save notation), we simplify F to $D_1(x)$ and have $\emptyset$ by Example 1. The factor $D_1(x)$ is treated similarly. □

3 Decidability of the Class *E*

From now on we assume that $n = 1$ in (1) that is

$$F \equiv \exists x M \equiv \exists x \&_{i \leq \delta} D_i. \tag{3}$$

Writing substitution (t/x) we shall drop x, so that $E(t)$ is the result of substituting t for all occurrencies of x in E.

Definition 3. *Define the* depth $d(t)$ *to be the nesting of function symbols in a term* t*:* $d(a) = 0$ *for variables and constants* a*, and*

$$d(f(t_1, \ldots, t_n)) = \max(d(t_1), \ldots, d(t_n)) + 1.$$

Let $d(F)$ be the maximum of $d(t)$ for terms t occurring in F.

The Herbrand universe *of the formula F is the least set of terms containing all constants occurring in F and closed under substitutions into terms occurring in F.*

$H^m(F)$ is the Herbrand expansion *of the depth m for F, that is*

$$M(t_1) \vee \ldots \vee M(t_k) \tag{4}$$

where $\{t_1, \ldots t_k\}$ is the list of all terms of the depth $\leq m$ from the Herbrand universe of F.

F-terms are terms of the depth $\leq d(F)$ constructed from constants and variable x by means of function symbols in F.

*F-*atoms *are constructed from F-terms by means of predicates in F.*

Formula F is saturated *if each clause D_i contains each F-atom (possibly with negation).*

Obviously each formula can be made saturated by using

$$D \iff (D \vee A)\&(D \vee \neg A).$$

The next four lemmas are used in the proof of the main Theorem 2.

Lemma 3. *Let E, E' be terms or literals, r be a term not occurring in E, E' and $E(r) \equiv E'(r)$. Then $E \equiv E'$.*

Proof. Induction on E, E'. □

Lemma 4. *Let T, t, s, t', s' be terms, $d(T) > \max(d(t), d(t'))$ and $T \equiv t(s) \equiv t'(s')$. Then s contains s' or s' contains s.*

Proof. Induction on $\max(d(t), d(t'))$. □

Lemma 5. *Let T, r, s, r', s' be terms, $T \equiv r(s) \equiv r'(s')$, $m = \max(d(r), d(r'))$, $d(T) \geq 2m$, $d(s') \geq d(s)$. Then $d(r) \geq d(r')$.*

Proof. Induction on m. □

Lemma 6. *Let u, u' be literals or terms, s, T terms, $u'(s) \equiv u(T)$, $d(T) \geq \max(d(u), d(u'), d(s))$, and x occurs in u. Then there is a term r such that*

$$u' \equiv u(r) \text{ and } r(s) \equiv T.$$

Proof. Induction on $d(u')$. □

Now we prove the basic result.

Theorem 2. *Let F be a derivable formula of the form (3) and $H^{2d(F)}(F)$ be underivable. Then one of the clauses D_i ($1 \leq i \leq \delta$) is derivable in U_F by at most one application of the rule B.*

Proof. Case 1. One of D_i does not contain x. If this D_i is a tautology, it is derivable by the rule A. Otherwise F is underivable, since D_i is refutable.

Case 2. All D_i contain x. Find Herbrand expansion (4) which is a tautology. Adding to (4) new clauses and changing the order of formulas $M(t_i)$ we can achieve the following:

a If some t_j $(j = 1, \ldots k)$ contains a term t then $M(t)$ is also contained in (4),
b $t_j \neq t_l$ for $j \neq l$,
c if $d(t_j) < d(t_l)$ then $j < l$.

Distributing & over $\vee$ in (4), we see that (4) is a tautology iff all clauses

$$D_{\alpha_1}(t_1) \vee \ldots \vee D_{\alpha_k}(t_k) \; (1 \leq \alpha_1, \ldots, \alpha_k \leq \delta) \tag{5}$$

are tautologies. Reducing k if necessary we can assume that

$$\text{Disjunction } M(t_1) \vee \ldots \vee M(t_{k-1}) \text{ is not a tautology.} \tag{6}$$

So there is a non-tautological disjunction

$$\Gamma \equiv \; D_{\alpha_1}(t_1) \vee \ldots \vee D_{\alpha_{k-1}}(t_{k-1})$$

such that all clauses

$$\Gamma \vee D_1[T], \ldots, \Gamma \vee D_k[T] \tag{7}$$

are tautologies. By (3), T has the maximal depth among all t_j, and since $H^{2d(F)}(F)$ is underivable, we have

$$d(T) > 2d(F) \geq d(F) \tag{8}$$

so that T does not occur in F.

Case 2.1. One of the clauses $D_i(T)$ is a tautology. This means that $L_1(T) \equiv \neg L_2(T)$ for some literals L_1, L_2 in D_i. By (8) and Lemma 3 this implies $L_1 \equiv \neg L_2$, that is, D_i is a tautology, and hence derivable in U_F by the rule A.

Case 2.2. None of $D_i(T)$ is a tautology. Since Γ is not a tautology, all $D_{\alpha_j}(t_j)$ for $j = 1, \ldots, k-1$ are not tautologies. From the fact that all clauses (7) are tautologies it follows that for every $\beta = 1, \ldots, \delta$ there is a j, $1 \leq j \leq k-1$ and a literal L_j in a clause D_{α_j} such that $L_j(t_j)$ is the negation of some literal from $D_\beta(T)$, that is of some literal $L^\beta(T)$ where L^β occurs in D_β.

Assume (for contradiction) that each of these literals L^β, $\beta \in \{1, \ldots, \delta\}$ does not contain x. Then $L^\beta(t) \equiv L^\beta$ for any term t, in particular for $t \equiv t_j$, and hence the clause $L_j(t_j) \vee D_\beta(t)$ contains a pair of complementary literals. Setting $\beta := \alpha_j$ we see that $\Gamma \vee D_{\alpha_j}(t_j) \; \equiv \Gamma$ is a tautology, contrary to the assumption concerning Γ. So one of the literals L^β has to contain x. Using equality $L_j(t_j) \equiv \neg L^\beta(T)$ and Lemma 6 (which is applicable in view of (3)) we obtain

$$L_j \equiv \neg L^\beta(r) \text{ and } T \equiv r(t_j) \tag{9}$$

for some F-term r. In view of (8) and Lemma 4 there is a maximal term t_l $(1 \leq l \leq k-1)$ such that

$$T \equiv t(t_l) \tag{10}$$

for some F-term t. In other words, from (9,10) for some F-terms r and t it follows that t_j is contained in t_l, in particular $d(t_l) \geq d(t_j)$. By (8) and Lemma 5 this implies

$$d(r) \geq d(t). \tag{11}$$

Now we shall prove that it is the clause D_{α_l} that is derivable in U_F by one application of the rule B. For this it is sufficient to establish that

$$D_{\alpha_l} \vee D_\beta(t) \tag{12}$$

is a tautology for every $\beta = 1, \ldots, \delta$. For this it is sufficient to establish that

$$D_{\alpha_l} \text{ contains } \neg L^\beta(t). \tag{13}$$

where L^β is the literal occurring as a disjunctive term in D_β and mentioned in (9). By (9,11) we have

$$d(L^\beta(t)) \leq d(L^\beta(r)) = d(L_j) \leq d(F),$$

the latter because L_j belongs to D_{α_j}. So $L^\beta(t)$ is an F-atom, and since F is saturated, the clause D_{α_l} has to contain either this literal or its negation. But D_{α_l} cannot contain $L^\beta(t)$ as a disjunctive term, since in that case Γ and even $D_{\alpha_j}(t_j) \vee D_{\alpha_l}(t_l)$ would be tautologies in view of the equality $L_j(t_j) \equiv \neg L^\beta(t(t_l))$. This concludes the proof of the theorem. □

Now the decision algorithm for the class E is given by Lemma 1 and Theorem 2.

References

1. Chang, C., Lee, R.: Symbolic Logic and Mechanical Theorem Proving. Academic Press, New York (1973)
2. Andreka, H., van Benthem, J., Nemeti, I.: Modal Logics and Bounded Fragments of Predicate Logic. J. Philos. Log. 27, 217–230 (1998)
3. Davydov, G., Maslov, S., Mints, G., Orevkov, V., Slisenko, A.: A Computer Algorithm for Establishing Deducibility Based on Inverse Method. In: Seminars in Math., V.A. Steklov Math. Inst., vol. 16, pp. 1–16 (1971)
4. Degtyarev, A., Voronkov, A.: The Inverse Method. In: Robinson, A., Voronkov, A. (eds.) Handbook of Automated Reasoning, pp. 179–272. Elsevier, Amsterdam (2001)
5. Dowek, G., Jiang, Y.: Eigenvariables, Bracketing and the Decidability of Positive Minimal Predicate Logic. Theor. Comput. Sci. 360, 193–208 (2006)
6. Gurevich, Y.: Decision Problem for the Logic of Predicates and Operations. Algebra and Log. 8(3), 284–308 (1968)
7. Maslov, S.: The Inverse Method for Logical Calculi. Trudy Mat. Inst. Steklov 98, 26–87 (1968)
8. Maslov, S., Mints, G.: Proof Search Theory and the Inverse Method (Russian). Supplement to the Russian Translation of the book by Chang and Lee [1], Nauka, Moscow, pp. 310–340 (1983)
9. Maslov, S., Orevkov, V.: Decidable Classes Reducible to the One-quantifier Class. Trudy Inst. Steklov 121, 57–66 (1972)

10. Maslov, S.: Connection between the Strategies of the Inverse Method and the Resolution Method Seminars in Math., vol. 16, pp. 48–54. Plenum Press, New York (1971)
11. Mints, G.: Gentzen-type Systems and Resolution Rules. Part I. Propositional Logic. In: Martin-Löf, P., Mints, G. (eds.) COLOG 1988. LNCS, vol. 417, pp. 198–231. Springer, Heidelberg (1990)
12. Mints, G.: Gentzen-type Systems and Resolution Rule. Part II. In: Logic Colloquium 1990. Lecture Notes in Logic, vol. 2, pp. 163–190 (1994)
13. Mints, G.: Solvability of the Problem of Deducibility in LJ for the Class of Formulas not Containing Negative Occurrences of Quantifiers. Proc. Steklov Inst. of Mathematics 98, 135–145 (1971)
14. Orevkov, V.: One Decidable Class of Formulas of the Predicate Calculus with Function Symbols. In: Proceedings of the II-nd Symposium in Cybernetics, Tbilisi, p. 176 (1965)

Logics for Two Fragments beyond the Syllogistic Boundary

Lawrence S. Moss

Department of Mathematics, Indiana University, Bloomington, IN, USA 47405
lsm@cs.indiana.edu

Dedicated to Yuri Gurevich

Abstract. This paper is a contribution to *natural logic*, the study of logical systems for linguistic reasoning. We construct a system with the following properties: its syntax is closer to that of a natural language than is first-order logic; it can faithfully represent simple sentences with standard quantifiers, subject relative clauses (a recursive construct), and negation on nouns and verbs. We also give a proof system which is complete and has the finite model property. We go further by adding comparative adjective phrases, assuming interpretations by transitive relations. This last system has all the previously-mentioned properties as well.

The paper was written for theoretical computer scientists and logicians interested in areas such as decidability questions for fragments of first-order logic, modal logic, and natural deduction.

Keywords: natural logic, natural deduction, relative clause, comparative adjective, decidability.

1 Introduction

Q: Hello, I'm wondering if you can help me.

A: Sure, but who *are* you? Your manner is familiar, but I don't recognize your face. I think I need an Introduction.

Q: Well, my name is Quisani, and ...

A: Quisani! It's a pleasure to meet you. I almost feel that I know you already, since I've been following your discussions with Yuri Gurevich for so many years. What brings you here?

Q: I have heard that there is a resurgence of interest in logical systems that deal with inference in natural language. Is this so? And can you explain a bit of it to me, an open-minded theoretical computer scientist?

A: This is a tall order of business for a short chat, but I'll try. It is true that there has been activity in the last few years that could be of interest. One area coming from AI and natural language processing has been concerned with getting

A. Blass, N. Dershowitz, and W. Reisig (Eds.): Gurevich Festschrift, LNCS 6300, pp. 538–564, 2010.

programs to carry out the inferences from te that people do, or usually do, when they read some text. This is an active area with connections to Information Extraction. But there is not much logic there, at least not yet. Another area, one probably closer to your interests, has been an exploration of fragments of natural language that correspond to complexity classes. This work has been carried out by Ian Pratt-Hartmann [13,14]. I think it could be interesting to complexity theorists. And then there is the matter of giving logical systems in which one can carry out as much simple reasoning in language as possible. This has been going on for some time (see van Benthem [1] for one early reference), and I could tell you more about it.

Q: But surely the general problem is undecidable. Actually, when I think about matters like vagueness, incomplete sentences, failures of reference, figurative language, and more and more problems ..., I'm not even sure it makes much sense to talk about what you call "simple reasoning in language".

A: To quote your teacher [8] on the Entscheidungsproblem,

> The ambitious attempt to mechanize mathematics via a decision algorithm for first-order logic failed. Does this mean that the field should be abandoned? Of course not. One should try to see what can be done. It is natural to try to isolate special cases of interest where mechanization is possible. There are many ways to define syntactic complexity of logic formulas. It turns out that, with respect to some natural definitions, sentences of low syntactic complexity suffice to express many mathematical problems. For example, many mathematical problems can be formulated with very few quantifier alternations. The decision problem for such classes of sentences is of interest.

If we replace "mathematics" by "language", and "mathematical problems" by "simple natural language arguments," you'll get my point.

Q: All right, I would like to know more about this. But first, could you please give me an example of some non-trivial inference carried out in natural language?

A: Sure, I'd be happy to. Let's say we're talking about a big table full of fruit. (*He writes on the board.*)

$$\frac{\begin{array}{l}\textsf{Every sweet fruit is bigger than every ripe fruit}\\ \textsf{Every pineapple is bigger than every kumquat}\\ \textsf{Every non-pineapple is bigger than every unripe fruit}\end{array}}{\textsf{Every fruit bigger than some sweet fruit is bigger than every kumquat}} \qquad (1)$$

I say that if we assume (or believe, or know) all the sentences above the line, then we should do the same for the sentence below the line.

Q: "non-pineapple"?! I thought this was supposed to be natural language.

A: Take it as a shorthand for "piece of fruit which is not a pineapple".

Q: Ok, I get it. But is everything either a pineapple or a non-pineapple?

A: You bet. You'll need to use this when you convince yourself that (1) is valid.

Q: Do I need to use anything else?

A: Yes, you need to know that bigger than is a transitive relation.

Q: I'll have to think about the reasoning. But anyhow, why don't you just type (1) into a theorem prover? It is in first-order logic. Or rather, it translates easily into FOL.

A: I'd rather translate to a decidable system. The full expressive power of first-order logic does not seem appropriate either to modeling how people reason, or for the matter of getting a computer to carry out the reasoning in examples like (1).

Q: Well, looking more carefully, I can see that your argument can be translated into the two-variable fragment FO^2. Then it really would fit into a decidable logic, by Mortimer's Theorem [11].

A: Not so fast. Although what I wrote translates to FO^2, the fact that bigger than is transitive means that we go beyond it. As you know, three variables is enough to have undecidability. You might try to study FO^2 on transitive relations, but this too is undecidable [7].

Q: But are there other logics that are on the one hand big enough to express the sentences in (1) and yet are decidable?

A: Yes, there is at least one that I can think of: *Boolean modal logic*, again with the stipulation that the semantics lives on transitive relations. This is known to be decidable [9].

Q: I thought that modal logic was about possible worlds, that sort of thing. It's hard to see why it's rearing its ugly head here. But again, I ask: if this Boolean modal logic is so great, why not translate (1) into it, and be done with it?

A: If I was only interested in the complexity of the example, or similar ones, then I'd agree. But with my logical baggage, I mean background, I feel that I should be looking for general principles and not just algorithms and complexity results. Think about Aristotle. As it happens, his system can be reformulated a little, and it turns out to be *complete*. But do you think he would have been happy if someone came and told him to forget the syllogisms because they had an algorithm to tell whether a conclusion followed, without telling him the reasons?

Q: Aristotle, ..., hmm. You know, I once heard that the three greatest logicians of all time are Aristotle, Frege, and Gödel. I know a bit about the last two, but I have almost no idea what Aristotle did. Something about *all men are mortal*?

A: Aristotle raised the matter of *inference* in the first place, with no precedent. And then for the fragment he was concerned with he provided a complete answer. I can't imagine a bigger project.

Q: I can see that you're really taken with Aristotle. But okay, let's go back to what's on the blackboard (1). Do you have a logical system in which we can prove (1), and which is still decidable?

A: Yes. Want to see it?

Q: Sure, but before you start in, I have a last question. You started out mentioning that there is some interest in natural logic from people in much more applied areas of computer science. Are they going to be interested in logical systems for cases like (1), the kind of thing that you are going to show me?

A: Probably not. Presumably they would be much more interested in an algorithm that worked quickly and correctly on 90% of the real-world inferences that come up in practice, than on a logical system that was complete in the logician's sense but was only good for a small amount of real-world inference. On the other hand, knowing what a complete system looked like could be an inspiration, or at least a comfort.

Q: Can you give me an example? Something that a logical system is not likely to get, but which is an inference from text in the sense of current work in natural language processing?

A: (*Again writing on the board.*)

$$\frac{\textsf{Frege's favorite food was sushi}}{\textsf{Frege ate sushi at least once}} \tag{2}$$

Q: Are you sure you don't mean Russell?

A: Oh yes. By the way, nearly everyone would agree to (2), but almost nobody will get (1) on their own.

Q: Yes, but real-world knowledge ... that's just not my cup of tea. Actually, formal systems and completeness proofs are not really my cup of tea either, but I'm curious enough to want to see yours.

A: Speaking of tea, can I show you the logical system for (1) and related matters over a cup?

Q: Sure, but now that you bring up eating, I'd prefer sushi and pineapples.

2 Logic for a Fragment $\mathcal{L}$ with Set Terms and Negation

There are only two languages in this paper (actually they are families of languages parameterized by sets of basic symbols): the language $\mathcal{L}$ of this section, and the extension $\mathcal{L}(adj)$ studied in Section 3. $\mathcal{L}$ is based on three pairwise disjoint sets called **P**, **R**, and **K**. These are called *unary atoms*, *binary atoms*, and *constant symbols*. (The reason we use the letters **P** and **R** is that they remind us of *predicate symbols* and *relation symbols*.)

Expression	Variables	Syntax
unary atom	p, q	
binary atom	s	
constant	j, k	
unary literal	l	$p \mid \bar{p}$
binary literal	r	$s \mid \bar{s}$
set term	b, c, d	$l \mid \exists(c, r) \mid \forall(c, r)$
sentence	φ, ψ	$\forall(c, d) \mid \exists(c, d) \mid c(j) \mid r(j, k)$

Fig. 1. Syntax of sentences of $\mathcal{L}$

2.1 Syntax and Semantics

We present the syntax of $\mathcal{L}$ in Figure 1. Sentences are built from constant symbols, unary and binary atoms using an involutive symbol for negation, a formation of set terms, and also a form of quantification. The second column indicates the variables that we shall use in order to refer to the objects of the various syntactic categories. Because the syntax is not standard, it will be worthwhile to go through it slowly and to provide glosses in English for expressions of various types.

One might think of the constant symbols as proper names such as John and Mary. The unary atoms may be glossed as one-place predictates such as boys, girls, etc. And the relation symbols correspond to transitive verbs (that is, verbs which take a direct object) such as likes, sees, etc. They also correspond to comparative adjective phrases such as is bigger than. (However, later on in Section 3, we introduce a new syntactic primitive for the adjectives.)

Unary atoms *appear to be* one-place relation symbols, especially because we shall form sentences of the form $p(j)$. However, we do not have sentences $p(x)$, since we have no variables at this point in the first place. Similar remarks apply to binary atoms and two-place relation symbols. So we chose to change the terminology from *relation symbols* to *atoms.*

We form unary and binary *literals* using the bar notation. We think of this as expressing classical negation. So we take it to be involutive, so that $\overline{\overline{p}} = p$ and $\overline{\overline{s}} = s$.

The set terms in this language are the only recursive construct. If b is read as boys and s as sees, then one should read $\forall(b, s)$ as sees all boys, and $\exists(b, s)$ as sees some boys. Hence these set terms correspond to simple verb phrases. We also allow negation on the atoms, so we have $\forall(b, \overline{s})$; this can be read as fails to see all boys, or (better) sees no boys or doesn't see any boys. We also have $\exists(b, \overline{s})$, fails to see some boys. But the recursion allows us to embed set terms, and so we have set terms like

$$\exists(\forall(\forall(b, \overline{s}), h), a)$$

which may be taken to symbolize a verb phrase such as admires someone who hates everyone who does not see any boy.

We should note that the relative clauses which can be obtained in this way are all "missing the subject", never "missing the object". The language is too poor to express predicates like $\lambda x.$all boys see x.

The main sentences in the language are of the form $\forall(b,c)$ and $\exists(b,c)$; they can be read as statements of the inclusion of one set term extension in another, and of the non-empty intersection. We also have sentences using the constants, such as $\forall(g,s)(m)$, corresponding to Mary sees all girls. But we are not able to say all girls see Mary; the syntax again is too weak. (However, in our Conclusion we shall see how to extend our system to handle this.) This weakness in expressive power corresponds to a less complex decidability result, as we shall see.

Semantics. A *structure* (for this language $\mathcal{L}$) is a pair $\mathcal{M} = \langle M, [\![\ ]\!]\rangle$, where M is a non-empty set, $[\![p]\!] \subseteq M$ for all $p \in \mathbf{P}$, $[\![r]\!] \subseteq M^2$ for all $r \in \mathbf{R}$, and $[\![j]\!] \in M$ for all $j \in \mathbf{K}$.

Given a model $\mathcal{M}$, we extend the interpretation function $[\![\ ]\!]$ to the rest of the language by setting

$$\begin{array}{ll} [\![\overline{p}]\!] & = M \setminus [\![p]\!] \\ [\![\overline{r}]\!] & = M^2 \setminus [\![r]\!] \\ [\![\exists(l,t)]\!] & = \{x \in M : \text{ for some } y \text{ such that } [\![l]\!](y),\ [\![t]\!](x,y)\} \\ [\![\forall(l,t)]\!] & = \{x \in M : \text{ for all } y \text{ such that } [\![l]\!](y),\ [\![t]\!](x,y)\} \end{array}$$

We define the truth relation $\models$ between models and sentences by:

$$\begin{array}{lll} \mathcal{M} \models \forall(c,d) & \text{iff} & [\![c]\!] \subseteq [\![d]\!] \\ \mathcal{M} \models \exists(c,d) & \text{iff} & [\![c]\!] \cap [\![d]\!] \neq \emptyset \\ \mathcal{M} \models c(j) & \text{iff} & [\![c]\!]([\![j]\!]) \\ \mathcal{M} \models r(j,k) & \text{iff} & [\![r]\!]([\![j]\!],[\![k]\!]) \end{array}$$

If Γ is a set of formulas, we write $\mathcal{M} \models \Gamma$ if for all $\varphi \in \Gamma$, $\mathcal{M} \models \varphi$.

Example 1. We consider a simple case, with one unary atom p, one binary atom s, and two constants j and k. Consider the following model. We take $M = \{w,x,y,z\}$, and $[\![p]\!] = \{w,x,y\}$. For the relation symbol, s, we take the arrows below:

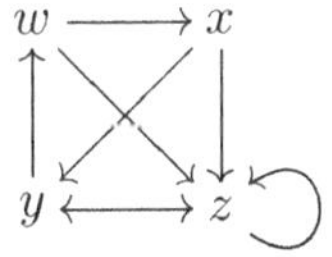

For example, $[\![\overline{p}]\!] = \{z\}$, $[\![\forall(p,s)]\!] = \emptyset$, $[\![\exists(\overline{p},s)]\!] = M$, and $[\![\exists(\forall(p,\overline{s}),s)]\!] = \emptyset$. Here are two $\mathcal{L}$-sentences true in $\mathcal{M}$: $\forall(p,\exists(\overline{p},s))$ and $\forall(\exists(\forall(p,\overline{s}),s),\overline{p})$.

Now set $[\![j]\!] = w$ and $[\![k]\!] = x$. We get additional sentences true in $\mathcal{M}$ such as $s(j,k)$, $\overline{s}(k,j)$, and $\exists(\overline{p},s)(k)$.

Here is a point that will be important later. For all terms c, $\mathcal{M} \models c(j)$ iff $\mathcal{M} \models c(k)$. (The easiest way to check this is to show that for all set terms c, $[\![c]\!]$ is one of the following four sets: $\emptyset, M, \{w,x,y\}$, or $\{z\}$.) However, $\mathcal{M} \models s(j,k)$ and $\mathcal{M} \models \overline{s}(k,j)$.

Satisfiability. A sentence φ is *satisfiable* if there exists $\mathcal{M}$ such that $\mathcal{M} \models \varphi$; satisfiability of a set of formulas Γ is defined similarly. We write $\Gamma \models \varphi$ to mean that every model of every sentence in Γ is also a model of φ.

The satisfiability problem for the language is decidable for a very easy reason: the language $\mathcal{L}$ translates to the *two-variable fragment* FO^2 of first-order logic. (We shall see this shortly.) Thus we have the finite model property (by Mortimer [11]) and decidability of satisfiability in non-deterministic exponential time (Grädel et al [6]). It might therefore be interesting to ask whether the smaller fragment $\mathcal{L}$ is of a lower complexity. As it happens, it is. Pratt-Hartmann [14] showed that the satisfiability problem for a certain fragment $\mathcal{E}_2$ of FO^2 can be decided in EXPTIME in the length of the input Γ, and his fragment was essentially the same as the one in this paper.

The bar notation. We have already seen that our unary and binary atoms come with negative forms. We extend this notation to all sentences in the following ways: $\overline{\overline{p}} = p$, $\overline{\overline{s}} = s$, $\overline{\exists(l,r)} = \forall(l,\overline{r})$, $\overline{\forall(l,r)} = \exists(l,\overline{r})$, $\overline{\forall(c,d)} = \exists(c,\overline{d})$, $\overline{\exists(c,d)} = \forall(c,\overline{d})$, $\overline{c(j)} = \overline{c}(j)$, and $\overline{r(j,k)} = \overline{r}(j,k)$.

Translation of the syllogistic into $\mathcal{L}$. We indicate briefly a few translations to orient the reader. First, the classical syllogistic translates into $\mathcal{L}$:

All p are q	$\mapsto \forall(p,q)$	No p are q	$\mapsto \forall(p,\overline{q})$
Some p are q	$\mapsto \exists(p,q)$	Some p aren't q	$\mapsto \exists(p,\overline{q})$

The same is true of the relational syllogistic; cf. [15]. In the other direction, we translate $\mathcal{L}$ to FO^2, the fragment of first order logic using only the variables x and w. We do this by mapping the set terms two ways, called $c \mapsto \varphi_{c,x}$ and $c \mapsto \varphi_{c,y}$. Here are the recursion equations for $c \mapsto \varphi_{c,x}$:

$$\begin{array}{ll} p \mapsto P(x) & \forall(c,r) \mapsto (\forall y)(\varphi_{c,y}(y) \rightarrow r(x,y)) \\ \overline{p} \mapsto \neg P(x) & \exists(c,r) \mapsto (\exists y)(\varphi_{c,y}(y) \wedge r(x,y)) \end{array}$$

The equations for $c \mapsto \varphi_{c,y}$ are similar. Then the translation of the sentences into FO^2 follows easily.

Translation of $\mathcal{L}$ into Boolean modal logic. We shall write $\hat{\mathcal{L}}$ for the following version of Boolean modal logic. $\hat{\mathcal{L}}$ has each $p \in \mathbf{P}$ as an *atomic proposition*, and it has two *modal operators*, $\Box_s$ and $\blacksquare_s$, one for each $s \in \mathbf{R}$. The syntax of $\hat{\mathcal{L}}$ is given by

$$\varphi \quad := \quad p \mid \neg\varphi \mid \varphi \wedge \psi \mid \Box_s\varphi \mid \blacksquare_s\varphi$$

The language is interpreted on the same kind of structures that we have been using for $\mathcal{L}$. Then $[\![p]\!]$ is given for all atoms p, and we also set $[\![\neg\varphi]\!] = M \setminus [\![\varphi]\!]$, $[\![\varphi \wedge \psi]\!] = [\![\varphi]\!] \cap [\![\psi]\!]$, and

$$\begin{array}{l} [\![\Box_s\varphi]\!] = \{x \in M : \text{for all } y \text{ such that } [\![s]\!](x,y),\ y \in [\![\varphi]\!]\} \\ [\![\blacksquare_s\varphi]\!] = \{x \in M : \text{for all } y \text{ such that not } [\![s]\!](x,y),\ y \in [\![\varphi]\!]\} \end{array}$$

We write $\Gamma \models \varphi$ to mean that for all structures $\mathcal{M}$ and all $x \in M$, if $x \in [\![\psi]\!]$ for all $\psi \in \Gamma$, then again $x \in [\![\varphi]\!]$.

Let $\mathcal{L}_0$ be the set of sentences of $\mathcal{L}$ which do not involve constants. We translate $\mathcal{L}_0$ into $\hat{\mathcal{L}}$. First, for each set term c, we define a sentence c^* of $\hat{\mathcal{L}}$. The definition is: $p^* = p$, $\overline{p}^* = \neg p$,

$$\begin{array}{ll} \forall(c,s)^* = \blacksquare_s \neg c^* & \forall(c,\overline{s})^* = \Box_s \neg c^* \\ \exists(c,s)^* = \neg\Box_s \neg c^* & \exists(c,\overline{s})^* = \neg\blacksquare_s \neg c^* \end{array}$$

An easy induction shows that $[\![c]\!] = [\![c^*]\!]$ for all set terms c. Then we translate $\forall(c,d)$ to $\forall(c,d)^*$ and $\exists(c,d)$ to $\exists(c,d)^*$:

$$\begin{aligned} \forall(c,d)^* &= \Box_s(c^* \to d^*) \wedge \blacksquare_s(c^* \to d^*) \\ \exists(c,d)^* &= \neg\forall(c,\overline{d})^* \end{aligned}$$

where s is an arbitrary element of $\mathbf{R}$. Then for all sentences φ of $\mathcal{L}_0$, and all models $\mathcal{M}$,

$$\mathcal{M} \models \varphi \quad \text{iff} \quad [\![\varphi^*]\!] = M \quad \text{iff} \quad [\![\varphi^*]\!] \neq \emptyset.$$

It follows easily from this that for $\Gamma \cup \{\varphi\}$ a set of $\mathcal{L}_0$ sentences, $\Gamma \models \varphi$ in the semantics of $\mathcal{L}$ iff $\Gamma^* \models \varphi^*$ in the semantics of $\hat{\mathcal{L}}$.

2.2 Proof System

Pratt-Hartmann and Moss [15] investigated several logical systems for the *relational syllogistic* and asked whether they were axiomatizable in a purely syllogistic fashion. We shall not enter into their definition of *syllogistic proof system* except to say that it is an adequate formalization of the concept. It comes in two flavors, depending on whether one permits *reductio ad absurdum* or not. It turns out that the consequence relation for some some logical languages can be captured by syllogistic systems even without *reductio*, some can be captured with *reductio* but (provably) not without it, and some are so strong that they cannot be captured even with *reductio*. The language of this paper would be one example of this latter phenomenon. Theorem 6.12 of [15] shows that for the language $\mathcal{L}$ of this paper (but without constant symbols), there indeed is no finite complete syllogistic system. It is therefore of interest to build a proof system which goes beyond syllogistic logic. This is the main technical goal of this paper.

We present our system in natural-deduction style in Figure 3. It makes use of *introduction* and *elimination* rules, and more critically of *variables*. For a textbook account of a proof system for first-order logic presented in this way, see van Dalen [3].

General sentences in this fragment are what usually are called *formulas*. We prefer to change the standard terminology to make the point that here, sentences are not built from formulas by quantification. In fact, sentences in our sense do not have variable occurrences. But general sentences do include *variables*. They are only used in our proof theory.

Expression	Variables	Syntax
individual variable	x, y	
individual term	t, u	$x \mid j$
general sentence	α	$\varphi \mid c(x) \mid r(x, y) \mid \bot$

Fig. 2. Syntax of general sentences of $\mathcal{L}$, with φ ranging over sentences

The syntax of general sentences is given in Figure 2. What we are calling *individual terms* are just variables and constant symbols. (There are no function symbols here.) Using terms allows us to shorten the statements of our rules, but this is the only reason to have terms.

An additional note: we don't need general sentences of the form $r(j, x)$ or $r(x, j)$. In larger fragments, we would expect to see general sentences of these forms, but our proof theory will not need these.

The bar notation, again. We have already seen the bar notation $\overline{c}$ for set terms c, and $\overline{\varphi}$ for sentences φ. We extend this to formulas $\overline{b(x)} = \overline{b}(x)$, $\overline{r(x, y)} = \overline{r}(x, y)$. We technically have a general sentence $\overline{\bot}$, but this plays no role in the proof theory.

We write $\Gamma \vdash \varphi$ if there is a proof tree conforming to the rules of the system with root labeled φ and whose axioms are labeled by elements of Γ. (Frequently we shall be sloppy about the labeling and just speak, e.g, of the root as if it *were* a sentence instead of being *labeled by* one.) Instead of giving a precise definition here, we shall content ourselves with a series of examples in Section 2.3 just below.

The system has two rules called $(\forall E)$, one for deriving general sentences of the form $c(x)$ or $c(j)$, and one for deriving general sentences $r(x, y)$ or $r(j, k)$. (Other rules are doubled as well, of course.) It surely looks like these should be unified, and the system would of course be more elegant if they were. But given the way we are presenting the syntax, there is no way to do this. That is, we do not have a concept of *substitution*, and so rules like $(\forall E)$ cannot be formulated in the usual way. Returning to the two rules with the same name, we could have chosen to use different names, say $(\forall E1)$ and $(\forall E2)$. But the result would have been a more cluttered notation, and it is always clear from context which rule is being used.

Although we are speaking of trees, we don't distinguish left from right. This is especially the case with the $(\exists E)$ rules, where the canceled hypotheses may occur in either order.

Side Conditions. As with every natural deduction system using variables, there are some side conditions which are needed in order to have a *sound* system.

In $(\forall I)$, x must not occur free in any uncanceled hypothesis. For example, in the version whose root is $\forall(c, d)$, one must cancel all occurrences of $c(x)$ in the leaves, and x must not appear free in any other leaf.

$$\frac{c(t) \quad \forall(c,d)}{d(t)}\ \forall E \qquad\qquad \frac{c(u) \quad \forall(c,r)(t)}{r(t,u)}\ \forall E$$

$$\frac{c(t) \quad d(t)}{\exists(c,d)}\ \exists I \qquad\qquad \frac{r(t,u) \quad c(u)}{\exists(c,r)(t)}\ \exists I$$

$$\frac{\begin{array}{c}[c(x)]\\ \vdots \\ d(x)\end{array}}{\forall(c,d)}\ \forall I \qquad\qquad \frac{\begin{array}{c}[c(x)]\\ \vdots \\ r(t,x)\end{array}}{\forall(c,r)(t)}\ \forall I$$

$$\frac{\exists(c,d) \quad \begin{array}{c}[c(x)] \quad [d(x)]\\ \vdots \\ \alpha\end{array}}{\alpha}\ \exists E \qquad\qquad \frac{\exists(c,r)(t) \quad \begin{array}{c}[c(x)] \quad [r(t,x)]\\ \vdots \\ \alpha\end{array}}{\alpha}\ \exists E$$

$$\frac{\alpha \quad \overline{\alpha}}{\bot}\ \bot I \qquad\qquad \frac{\begin{array}{c}[\overline{\varphi}]\\ \vdots \\ \bot\end{array}}{\varphi}\ RAA$$

Fig. 3. Proof rules. See the text for the side conditions in the ($\forall I$) and ($\exists E$) rules.

In ($\exists E$), the variable x must not occur free in the conclusion α or in any uncanceled hypothesis in the subderivation of α.

In contrast to usual first-order natural deduction systems, there are *no side conditions* on the rules ($\forall E$) and ($\exists I$). The usual side conditions are phrased in terms of concepts such as *free substitution*, and the syntax here has no substitution to begin with. To be sure on this point, one should check the soundness result of Lemma 1.

Formal proofs in the Fitch style. Textbook presentations of logic overwhelmingly use natural deduction. Pelletier [12] discusses the history of this in texts used in philosophy classes. In such books, by far the most common style of presentation is via Fitch diagrams. Pelletier was not concerned with books for computer science students, and here the situation is more mixed, I believe. I have chosen to present the system in a more "classical" Gentzen-style format. But the system may easily be re-formatted to look more like a Fitch system, as we shall see in Example 3 and Figure 5. These examples might give the impression that we have merely re-presented Fitch-style natural deduction proofs. The difference is that our syntax is not a special case of the syntax of first-order logic. Corresponding to this, our proof rules our rather restrictive, and the system cannot be used for much of anything beyond the language $\mathfrak{L}$. However, the fact that our Fitch-style proofs *look like* familiar formal proofs is a virtue: for example, it means that one could teach logic using this material.

1	$\forall(c,d)$		hyp
2	x	$\exists(c,r)(x)$	hyp
3		$c(y)$	$\exists E$, 2
4		$r(x,y)$	$\exists E$, 2
5		$d(y)$	$\forall E$, 1, 3
6		$\exists(d,r)(x)$	$\exists I$, 4, 5
7	$\forall(\exists(c,r),\exists(d,r))$		$\forall I$, 1–6

$$\dfrac{\dfrac{[\exists(c,r)(x)]^2 \qquad \dfrac{[r(x,y)]^1 \qquad \dfrac{[c(y)]^1 \quad \forall(c,d)}{d(y)}\,\forall E}{\exists(d,r)(x)}\,\exists I}{\exists(d,r)(x)}\,\exists E^1}{\forall(\exists(c,r),\exists(d,r))}\,\forall I^2$$

Fig. 4. Derivations in Example 3

2.3 Examples

We present a few examples of the proof system at work, along with comments pertaining to the side conditions. Many of these are taken from the proof system $\mathbf{R}^*$ for the language $\mathcal{R}^*$ of [15]. That system $\mathbf{R}^*$ is among the strongest of the known syllogistic systems, and so it is of interest to check the current proof system is at least as strong.

Example 2. Here is a proof of the classical syllogism *Darii*: $\forall(b,d), \exists(c,b) \vdash \exists(c,d)$:

$$\dfrac{\exists(c,b) \qquad \dfrac{\dfrac{[b(x)]^1 \quad \forall(b,d)}{d(x)}\,\forall E \qquad [c(x)]^1}{\exists(c,d)}\,\exists I}{\exists(c,d)}\,\exists E^1$$

Example 3. Next we study a principle called (K) in [15]. Intuitively, if all watches are expensive items, then everyone who owns a watch owns an expensive item. The formal statement in our language is $\forall(c,d) \vdash \forall(\exists(c,r),\exists(d,r))$. See Figure 4. We present a Fitch-style proof on the left and the corresponding one in our formalism on the right. One aspect of the Fitch-style system is that $(\exists E)$ gives two lines; see lines 3 and 4 on the left in Figure 4.

Example 4. Here is an example of a derivation using (RAA). It shows $\forall(c,\overline{c}) \vdash \forall(d,\forall(c,r))$.

$$\dfrac{[d(x)]^2 \qquad \dfrac{\dfrac{\dfrac{\dfrac{[c(y)]^1 \quad \forall(c,\overline{c})}{\overline{c}(y)}\,\forall E \qquad [c(y)]^1}{\bot}\,\bot I}{r(x,y)}\,RAA}{\forall(c,r)(x)}\,\forall I^1}{\forall(d,\forall(c,r))}\,\forall I^2$$

Example 5. Here is a statement of the *rule of proof by cases*: If $\Gamma + \varphi \vdash \psi$ and $\Gamma + \overline{\varphi} \vdash \psi$, then $\Gamma \vdash \psi$. (Here and below, $\Gamma + \varphi$ denotes $\Gamma \cup \{\varphi\}$.) Instead of giving a derivation, we only indicate the ideas. Since $\Gamma + \varphi \vdash \psi$, we have $\Gamma + \varphi + \overline{\psi} \vdash \bot$ using $(\bot I)$. From this and (RAA), $\Gamma, \overline{\psi} \vdash \overline{\varphi}$. Take a derivation showing $\Gamma + \overline{\varphi} \vdash \psi$, and replace the labeled $\overline{\varphi}$ with derivations from $\Gamma + \overline{\psi}$. We thus see that $\Gamma + \overline{\psi} \vdash \psi$. Using $(\bot I)$, $\Gamma + \overline{\psi} \vdash \bot$. And then using (RAA) again, $\Gamma \vdash \psi$. (This point is from [15].)

Example 6. The example at the beginning of this paper cannot be formalized in this fragment because the correct reasoning uses the transitivity of is bigger than. However, we can prove a result which may itself be used in a formal proof of (1):

$$\frac{\begin{array}{l}\text{Every sweet fruit is bigger than every ripe fruit}\\ \text{Every pineapple is bigger than every kumquat}\\ \text{Every non-pineapple is bigger than every unripe fruit}\end{array}}{\text{Every sweet fruit is bigger than every kumquat}} \qquad (3)$$

To discuss this, we take the set $\mathbf{P}$ of unary atoms to be

$$\mathbf{P} \quad = \quad \{\text{sweet}, \text{ripe}, \text{pineapple}, \text{kumquat}\}.$$

We also take $\mathbf{R} = \{\text{bigger}\}$ and $\mathbf{K} = \emptyset$. Figure 5 contains a derivation showing (3), done in the manner of Fitch [4]. The main way in which we have bent the English in the direction of our formalism is to use the bar notation on the nouns. The main reason for presenting the derivation as a Fitch diagram is that the derivation given as a tree (as demanded by our definitions) would not fit on a page. This is because the cases rule is not a first-class rule in the system, it is a derived rule (see Example 5 above). Our Fitch diagram pretends that the system has a rule of cases. Another reason to present the derivation as in Figure 5 is to make the point that the treatment in this paper is a beginning of a formalization of the work that Fitch was doing.

2.4 Soundness

Before presenting a soundness result, it might be good to see an improper derivation. Here is one, purporting to infer some men see some men from some men see some women:

$$\dfrac{\exists(m,\exists(w,s)) \qquad \dfrac{[\exists(w,s)(x)]^2 \qquad \dfrac{\dfrac{\dfrac{[s(x,x)]^1 \quad [m(x)]^2}{\exists(s,m)(x)}\,\exists I \qquad [m(x)]^2}{\exists(m,\exists(m,s))}\,\exists I}{}}{\exists(m,\exists(m,s))}\,\exists E^1}{\exists(m,\exists(m,s))}\,\exists E^2$$

The specific problem here is that when $[s(x,x)]$ is withdrawn in the application of $\exists I^1$, the variable x is free in the as-yet-uncanceled leaves labeled $m(x)$.

1		Every sweet fruit is bigger than every ripe fruit	hyp
2		Every pineapple is bigger than every kumquat	hyp
3		Every $\overline{\text{pineapple}}$ is bigger than every $\overline{\text{ripe fruit}}$	hyp
4	x	x is a sweet fruit	hyp
5		x is bigger than every ripe fruit	$\forall E$, 1, 4
6		x is a pineapple	hyp
7		x is bigger than every kumquat	$\forall E$, 2, 6
8		x is a $\overline{\text{pineapple}}$	hyp
9		x is bigger than every $\overline{\text{ripe fruit}}$	$\forall E$, 3, 8
10	y	y is a kumquat	hyp
11		y is a ripe fruit	hyp
12		x is bigger than y	$\forall E$, 5, 11
13		y is a $\overline{\text{ripe fruit}}$	hyp
14		x is bigger than y	$\forall E$, 9, 13
15		x is bigger than y	cases, 13–14, 11–12
16		x is bigger than every kumquat	$\forall I$, 10–15
17		x is bigger than every kumquat	cases, 6–7, 8–16
18		Every sweet fruit is bigger than every kumquat	$\forall I$, 4–17

Fig. 5. A derivation corresponding to the argument in (3)

To state a result pertaining to the soundness of our system, we need to define the truth value of a general sentence under a variable assignment. First, a *variable assignment* in a model $\mathcal{M}$ is a function $v : V \to M$, where V is the set of variable symbols and M is the universe of $\mathcal{M}$. We need to define $\mathcal{M} \models \alpha[v]$ for general sentences α. If α is a sentence, then $\mathcal{M} \models \alpha[v]$ iff $\mathcal{M} \models \alpha$ in our earlier sense. If α is $b(x)$, then $\mathcal{M} \models \alpha[v]$ iff $[\![b]\!](v(x))$. If α is $r(x,y)$, then $\mathcal{M} \models \alpha[v]$ iff $[\![r]\!](v(x), v(y))$. If α is $\bot$, then $\mathcal{M} \not\models \bot$ for all models $\mathcal{M}$.

Lemma 1. *Let Π be any proof tree for this fragment all of whose nodes are labeled with $\mathcal{L}$-formulas, let φ be the root of Π, let $\mathcal{M}$ be a structure, let $v : X \to M$ be a variable assignment, and assume that for all uncanceled leaves ψ of Π, $\mathcal{M} \models \psi[v]$. Then also $\mathcal{M} \models \varphi[v]$.*

Proof. By induction on Π. We shall only go into details concerning two cases. First, consider the case when the root of Π is

$$\dfrac{\exists(c,r)(t) \qquad \begin{array}{c}[c(x)] \quad [r(t,x)] \\ \vdots \\ \alpha\end{array}}{\alpha}\ \exists E$$

To simplify matters further, let us assume that t is a variable. Let v be a variable assignment making true all of the leaves of the tree, except possibly $c(x)$ and $r(t,x)$. By induction hypothesis, $\mathcal{M} \models \exists(c,r)(t)[v]$. Let $a \in A$ witness this assertion. In the obvious notation, $[\![c]\!](a)$ and $[\![r]\!](t^{\mathcal{M},v},a)$. Let w be the same variable assignment as v, except that $w(x) = a$. Then since x is not free in any leaves except those labeled $c(x)$ and $r(t,x)$, we have $\mathcal{M} \models \psi[w]$ for all those ψ. And so $\mathcal{M} \models \alpha[w]$, using the induction hypothesis applied to the subtree on the right. And since x is not free in the conclusion α, we also have $\mathcal{M} \models \alpha[v]$, as desired.

Second, let us consider the case when the root is

$$\frac{c(y) \quad \forall(c,r)(x)}{r(x,y)}\ \forall E$$

(That is, we are considering an instance of $(\forall E)$ when the terms t and u are variables.) The variables x and y might well be the same. Let $\mathcal{M}$ be a structure, and v be a variable assignment making true the leaves of the tree. By induction hypothesis, $[\![c]\!](v(y))$ and also $[\![r]\!](v(x),m))$ for all $m \in [\![c]\!]$. In particular, $[\![r]\!](v(x),v(y))$.

The remaining cases are similar. □

2.5 The Henkin Property

The completeness of the logic parallels the Henkin-style completeness result for first-order logic. Given a consistent theory Γ, we get a model of Γ in the following way: (1) take the underlying language $\mathcal{L}$, add constant symbols to the language to witness existential sentences; (2) extend Γ to a maximal consistent set in the larger language; and then (3) use the set of constant symbols as the carrier of a model in a canonical way. In the setting of this paper, the work is in some ways easier than in the standard setting, and in some ways harder. There are more details to check, since the language has more basic constructs. But one doesn't need to take a quotient by equivalence classes, and in other ways the work here is easier.

Given two languages $\mathcal{L}$ and $\mathcal{L}'$, we say that $\mathcal{L}' \supseteq \mathcal{L}$ if every symbol (of any type) in $\mathcal{L}$ is also a symbol (of the same type) in $\mathcal{L}'$. In this paper, the main case is when $\mathbf{P}(\mathcal{L}) = \mathbf{P}(\mathcal{L}')$, $\mathbf{R}(\mathcal{L}) = \mathbf{R}(\mathcal{L}')$, and $\mathbf{K}(\mathcal{L}) \subseteq \mathbf{K}(\mathcal{L}')$; that is, $\mathcal{L}'$ arises by adding constants to $\mathcal{L}$.

A *theory* in a language is just a set of sentences in it. Given a theory Γ in a language $\mathcal{L}$, and a theory Γ^* in an extension $\mathcal{L}' \supseteq \mathcal{L}$, we say that Γ^* is a *conservative extension* of Γ if for every $\varphi \in \mathcal{L}$, if $\Gamma^* \vdash \varphi$, then $\Gamma \vdash \varphi$.

Lemma 2. *Let Γ be a consistent $\mathcal{L}$-theory, and let $j \notin \mathbf{K}(\mathcal{L})$.*

1. *If $\exists(c,d) \in \Gamma$, then $\Gamma + c(j) + d(j)$ is a conservative extension of Γ.*
2. *If $\exists(c,r)(j) \in \Gamma$, then $\Gamma + r(j,k) + c(k)$ is a conservative extension of Γ.*

Proof. For (1), suppose that Γ contains $\exists(c, d)$ and that $\Gamma + c(j) + d(j) \vdash \varphi$. Let Π be a derivation tree. Replace the constant j by an individual variable x which does not occur in Π. The result is still a derivation tree, except that the leaves are not labeled by *sentences*. (The reason is that our proof system has no rules specifically for constants, only for terms which might be constants and also might be individual variables.) Call the resulting tree Π'. Now the following proof tree shows that $\Gamma \vdash \varphi$:

$$\dfrac{\exists(c,d) \qquad \begin{array}{c}[c(x)] \quad [d(x)] \\ \vdots \\ \varphi\end{array}}{\varphi}\ \exists E$$

The subtree on the right is Π'. The point is that the occurrences of $c(x)$ and $d(x)$ have been canceled by the use of $\exists E$ at the root.

This completes the proof of the first assertion, and the proof of the second is similar. □

Definition 1. *An* $\mathfrak{L}$-theory Γ *has the* Henkin property *if the following hold:*

1. *If* $\exists(c, d) \in \Gamma$, *then for some constant* j, $c(j)$ *and* $d(j)$ *belong to* Γ.
2. *If* r *is a literal of* $\mathfrak{L}$ *and* $\exists(c, r)(j) \in \Gamma$, *then for some constant* k, $r(j, k)$ *and* $c(k)$ *belong to* Γ.

Lemma 3. *Let* Γ *be a consistent* $\mathfrak{L}$*-theory. Then there is some* $\mathfrak{L}^* \supset \mathfrak{L}$ *and some* $\mathfrak{L}^*$*-theory* Γ^* *such that* Γ^* *is a maximal consistent theory with the Henkin property. Moreover, if* $s \in \mathbf{R}(\mathfrak{L})$, $j \in \mathbf{K}(\mathfrak{L}^*)$ *and* $k \in \mathbf{K}(\mathfrak{L})$, *and if* $s(j, k) \in \Gamma^*$, *then* $j \in \mathbf{K}(\mathfrak{L})$.

Proof. This is a routine argument, using Lemma 2. One dovetails the addition of constants which is needed for the Henkin property together with the addition of sentences needed to insure maximal consistency. The formal details would use Lemma 2 for steps of the first kind, and for the second kind we need to know that if Γ is consistent, then for all φ, either $\Gamma + \varphi$ or $\Gamma + \overline{\varphi}$ is consistent. This follows from the derivable rule of proof by cases; see Example 5 in Section 2.3. □

The last point in Lemma 3 states a technical property that will be useful in Section 3.1.

It might be worthwhile noting that the extensions produced by Lemma 3 add *infinitely many constants* to the language.

2.6 Completeness via Canonical Models

In this section, fix a language $\mathfrak{L}$ and a maximal consistent Henkin $\mathfrak{L}$-theory Γ. We construct a *canonical model* $\mathcal{M} = \mathcal{M}(\Gamma)$ as follows: $M = \mathbf{K}(\mathfrak{L})$; $[\![p]\!](j)$ iff $p(j) \in \Gamma$; $[\![s]\!](j, k)$ iff $s(j, k) \in \Gamma$; and $[\![j]\!] = j$. That is, we take the constant symbols of the language to be the points of the model, and the interpretations of the atoms are the natural ones. Each constant symbol is interpreted by itself.

Lemma 4. *For all set terms c, $[\![c]\!] = \{j : c(j) \in \Gamma\}$.*

Proof. By induction on c. The base case of unary atoms p is by definition of $\mathcal{M}$.

Before we turn to the induction proper, here is a preliminary point. Assuming that $[\![c]\!] = \{j : c(j) \in \Gamma\}$, we check that $[\![\overline{c}]\!] = \{j : \overline{c}(j) \in \Gamma\}$:

$$j \in [\![\overline{c}]\!] \quad \text{iff} \quad j \notin [\![c]\!] \quad \text{iff} \quad c(j) \notin \Gamma \quad \text{iff} \quad \overline{c}(j) \in \Gamma.$$

The last point uses the maximal consistency of Γ.

Turning to the inductive steps, assume our result for c; we establish it for $\forall(c, s)$ and $\exists(c, s)$; it then follows from the preliminary point that we have the same fact for $\forall(c, \overline{s})$ and $\exists(c, \overline{s})$.

Let $j \in [\![\forall(c, s)]\!]$. We claim that $\forall(c, s)(j) \in \Gamma$. For if not, then $\exists(c, \overline{s})(j) \in \Gamma$. By the Henkin property, let k be such that Γ contains $c(k)$ and $\overline{s}(j, k)$. By the induction hypothesis, $k \in [\![c]\!]$, and by the definition of $\mathcal{M}$, $[\![s]\!](j, k)$ is false. Thus $j \notin [\![\forall(c, s)]\!]$. This is a contradiction.

In the other direction, assume that $\forall(c, s)(j) \in \Gamma$; this time we claim that $j \in [\![\forall(c, s)]\!]$. Let $k \in [\![c]\!]$. By induction hypothesis, Γ contains $c(k)$. By $(\forall E)$, we see that $\Gamma \vdash s(j, k)$. Hence Γ contains $s(j, k)$. So in $\mathcal{M}$, $[\![s]\!](j, k)$. Since k was arbitrary, we see that indeed $j \in [\![\forall(c, s)]\!]$.

The other induction step is for $\exists(c, s)$. Let $j \in [\![\exists(c, s)]\!]$. We thus have some $k \in [\![c]\!]$ such that $[\![s]\!](j, k)$. That is, $s(j, k) \in \Gamma$. Using $(\exists I)$, we have $\Gamma \vdash \exists(c, s)(j)$; from this we see that $\exists(c, s)(j) \in \Gamma$, as desired.

Finally, assume that $\exists(c, s)(j) \in \Gamma$. By the Henkin condition, let k be such that Γ contains $c(k)$ and $s(j, k)$. Using the derivation above, we have the desired conclusion that $j \in [\![\exists(c, s)]\!]$.

This concludes the proof. □

Lemma 5. *$\mathcal{M} \models \Gamma$.*

Proof. We check the sentence types in turn. Throughout the proof, we shall use Lemma 4 without mention.

First, let Γ contain the sentence $\forall(c, d)$. Let $j \in [\![c]\!]$, so that $c(j) \in \Gamma$. We have $d(j) \in \Gamma$ using $(\forall E)$. This for all j shows that $\mathcal{M} \models \forall(c, d)$.

Second, let $\exists(c, d) \in \Gamma$. By the Henkin condition, let j be such that both $c(j)$ and $d(j)$ belong to Γ. This element j shows that $[\![c]\!] \cap [\![d]\!] \neq \emptyset$. That is, $\mathcal{M} \models \exists(c, d)$.

Continuing, consider a sentence $c(j) \in \Gamma$. Then $j \in [\![c]\!]$, so that $\mathcal{M} \models c(j)$.

Finally, the case of sentences $r(j, k) \in \Gamma$ is immediate from the structure of the model. □

Theorem 1. *If $\Gamma \models \varphi$, then $\Gamma \vdash \varphi$.*

Proof. We rehearse the standard argument. Due to the classical negation, we need only show that consistent sets Γ are satisfiable. Let $\mathcal{L}$ be the language of Γ, Let $\mathcal{L}' \supseteq \mathcal{L}$ be an extension of $\mathcal{L}$, and let $\Gamma^* \supseteq \Gamma$ be a maximal consistent theory in $\mathcal{L}'$ with the Henkin property (see Lemma 3). Consider the canonical model $\mathcal{M}(\Gamma^*)$ as defined in this section. By Lemma 5, $\mathcal{M}(\Gamma^*) \models \Gamma^*$. Thus Γ^* is satisfiable, and hence so is Γ. □

2.7 The Finite Model Property

Let Γ be a consistent finite theory in some language $\mathcal{L}$. As we now know, Γ has a model. Specifically, we have seen that there is some $\Gamma^* \supseteq \Gamma$ which is a maximal consistent theory with the Henkin property in an extended language $\mathcal{L}^* \supseteq \mathcal{L}$. Then we may take the set of constant symbols of $\mathcal{L}^*$ to be the carrier of a model of Γ^*, hence of Γ. The model obtained in this way is infinite. It is of interest to build a finite model, so in this section Γ must be finite. The easiest way to see that Γ has a finite model is to recall that our overall language is a sub-language of the two variable fragment FO^2 of first-order logic. And FO^2 has the finite model property by Mortimer's Theorem [11].

However, it is possible to give a direct argument for the finite model property, along the lines of filtration in modal logic (but with some differences). We sketch the result here because we shall use the same method in Section 3.1 below to prove a finite model property for our second logical system $\mathcal{L}(adj)$ with respect to its natural semantics; that result does not follow from others in the literature.

Let $\mathcal{M} = \mathcal{M}(\Gamma^*)$ be the canonical model as defined in Section 2.6. Let $Sub(\Gamma)$ be the collection of set terms occurring in any sentence in the original finite theory Γ. So $Sub(\Gamma)$ is finite, and if $\forall(c, r) \in Sub(\Gamma)$ or $\exists(c, r) \in Sub(\Gamma)$, then also $c \in Sub(\Gamma)$. For constant symbols j and k of $\mathcal{L}^*$, write $j \equiv k$ iff the following conditions hold:

1. If either j or k is a constant of $\mathcal{L}$, then $k = j$.
2. For all $c \in Sub(\Gamma)$, $c(j) \in \Gamma$ iff $c(k) \in \Gamma$.

Remark 1. The equivalence relation $\equiv$ may be defined on any structure. It is not necessarily a congruence, as Example 1 shows. Specifically, we had constant symbols j and k such that $j \equiv k$, and yet in our structure $s(j, k)$ and $\overline{s}(k, j)$. In the case of $\mathcal{M}(\Gamma^*)$, we have no reason to think that $\equiv$ is a congruence. That is, the construction in Section 2.6 did not arrange for this.

Let $N = \{[k] : k \in \mathbf{K}(\mathcal{L})\} \times \{\forall, \exists\}$. (We use $\forall$ and $\exists$ as tags to give two copies of the quotient $\mathbf{K}/\equiv$.) We endow N with an $\mathcal{L}$-structure as follows:

$[\![p]\!] = \{([j], Q) : p(j) \in \Gamma^* \text{ and } Q \in \{\forall, \exists\}\}$.
$[\![s]\!](([j], Q), ([k], Q'))$ iff one of the following two conditions holds:

1. There is a set term c such that Γ^* contains $c(k)$ and $\forall(c, s)(j)$.
2. $Q' = \exists$, and for some $j_* \equiv j$ and $k_* \equiv k$, Γ^* contains $s(j_*, k_*)$.

For a constant j of $\mathcal{L}$, $[\![j]\!] = ([j], \exists)$. (Of course, $[j]$ is the singleton set $\{j\}$.)

Before going on, we note that the first of the two alternatives in the definition of $[\![s]\!](([j], Q), ([k], Q'))$ is independent of the choice of representatives of equivalence classes. And clearly so is the second alternative.

We shall write $\mathcal{N}$ for the resulting $\mathcal{L}$-structure, hiding the dependence on Γ and Γ^*.

Lemma 6. *For all* $c \in \mathrm{Sub}(\Gamma)$, $[\![c]\!] = \{([j], Q) : c(j) \in \Gamma^* \textit{ and } Q \in \{\forall, \exists\}\}$.

Proof. By induction on set terms c. We are not going to present any of the details here because in Lemma 10 below, we shall see all the details on a more involved result. □

Lemma 7. $\mathcal{N} \models \Gamma$.

Proof. Again we are only highlighting a few details, since the full account is similar to what we saw in Lemma 5, and to what we shall see in Lemma 11. One would check the sentence types in turn, using Lemma 6 frequently. We want to go into details concerning sentences in Γ of the form $s(j,k)$ or $\overline{s}(j,k)$. Recall that we are dealing in this result with sentences of $\mathcal{L}$, and so j and k are constant symbols of that language. Also recall that $[\![j]\!] = ([j], \exists)$, and similarly for k.

First, consider sentences in Γ of the form $s(j,k)$. By the definition of $[\![s]\!]$, we have

$$[\![s]\!](([j], \exists), ([k], \exists)).$$

By the way binary atoms and constants are interpreted in $\mathcal{N}$, we have $\mathcal{N} \models s(j,k)$, as desired.

We conclude with the consideration of a sentence in Γ of the form $\overline{s}(j,k)$. We wish to show that $\mathcal{N} \models \overline{s}(j,k)$. Suppose towards a contradiction that $\mathcal{N} \models s(j,k)$. Then we have $[\![s]\!](([j], \exists), ([k], \exists))$. There are two possibilities, corresponding to the alternatives in the semantics of s. The first is when there is a set term c such that Γ^* contains $c(k)$ and $\forall(c,s)(j)$. By $(\forall E)$, Γ^* then contains $s(j,k)$. But recall that Γ contains $\overline{s}(j,k)$. So in this alternative, $\Gamma^* \supseteq \Gamma$ is inconsistent. In the second alternative, there are $j_* \equiv j$ and $k_* \equiv k$ such that $s(j_*, k_*) \in \Gamma^*$. But recall that the equivalence classes of constant symbols from the base language $\mathcal{L}$ are singletons. Thus in this alternative, $j_* = j$ and $k_* = k$; hence $s(j,k) \in \Gamma^*$. But then again Γ^* is inconsistent, a contradiction. □

Theorem 2 (Finite Model Property). *If Γ is consistent, then Γ has a model of size at most 2^{2n}, where n is the number of set terms in Γ.*

Complexity notes. Theorem 2 implies that the satisfiability problem for our language is in NEXPTIME. We can improve this to an EXPTIME-completeness result by quoting the work of others. Pratt-Hartmann [14] define a certain logic $\mathcal{E}_2$ and showed that the complexity of its satisfiability problem is EXPTIME-complete. $\mathcal{E}_2$ corresponds to a fragment of first-order logic, and it is somewhat bigger than the language $\mathcal{L}$. (It would correspond to adding converses to the binary atoms in $\mathcal{L}$, as we mention at the very end of this paper.) Since satisfiability for $\mathcal{E}_2$ is EXPTIME-complete, the same problem for $\mathcal{L}$ is in EXPTIME.

A different way to obtain this upper bound is via the embedding into Boolean modal logic which we saw in Section 2.1. For this, see Theorem 7 of Lutz and Sattler [9]. We shall use an extension of that result below in connection with an extension $\mathcal{L}(adj)$ of $\mathcal{L}$.

The EXPTIME-hardness for $\mathcal{L}$ follows from Lemma 6.1 in [15]. That result dealt with a language called $\mathcal{R}^\dagger$, and $\mathcal{R}^\dagger$ is a sub-language of $\mathcal{L}$.

3 Adding Transitivity: The Language $\mathcal{L}(adj)$

Before going further, let us briefly recapitulate the overall problem of this paper and point out where we are and what remains to be done. We aim to formalize a fragment of first-order logic in which one may represent arguments as complex as that in (1) in the Introduction. We are especially interested in decidable systems, and so the systems must be weaker than first-order logic. We presented in Section 2 a language $\mathcal{L}$ and a proof system for it. Validity in the logic cannot be captured by a purely syllogistic proof system, and so our proof system uses variables. But the use is very special and restricted. The proof system is complete and decidable in exponential time. To our knowledge, it is the first system with these properties.

There are a number of ways in which one can go further. In this paper, we want to explore one such way, connected to our example in (1). One key feature of this example is that comparative adjectives such as bigger than are transitive. This is true for all comparative adjectives. (Another point of interest is that comparative adjectives are typically irreflexive. We are going to ignore that in the present paper.)

We extend our language $\mathcal{L}$ to a language $\mathcal{L}(adj)$ by taking a basic set $\mathbf{A}$ of comparative adjective phrases in the base. The proof system simply extends the one we have already seen with a rule corresponding to the transitivity of comparatives. Our completeness result, Theorem 1, extends to the new setting. The next section does this. The decidability of the language is a more delicate matter than before, since it does not follow from Mortimer's Theorem [11] on the finite model property for FO^2. Indeed, adding transitivity statements to FO^2 renders the logic undecidable, as shown in Grädel, Otto, and Rosen [7]. Instead, one could use Theorem 12 of Lutz and Sattler [9] on the decidability of a variant on Boolean modal logic in which some of the relations are taken to be transitive. This would indeed give the ExpTime-completeness of $\mathcal{L}(adj)$ with our semantics. However, we have decided to present a direct proof for several reasons. First, Lutz and Sattler's result does not give a finite model property, and our result does do this. Second, our argument is shorter. Finally, our treatment connects to modal filtration arguments and is therefore different; [9] uses automata on infinite trees and is based on Vardi and Wolper [16].

I do not wish to treat the transitivity of comparison with adjectives as an enthymeme (missing premise) because the transitivity seems more fundamental, more 'logical' somehow. Hence it should be treated on a deeper level. The decidability considerations give a supporting argument: if we took the transitivity to be a *meaning postulate*, then it would seem that the underlying language would have to be rich enough to state transitivity. This requires three universal quantifiers. For other reasons, we want our languages to be closed under negation. It thus seems very likely that any logical system with these properties is going to be undecidable. The upshot is a system in which the transitivity turns out to be a *proof postulate* rather than a *meaning postulate*. We turn to the system itself.

Syntax and semantics. We start with four pairwise disjoint sets $\mathbf{A}$ (for *comparative adjective phrases*) and the three that we saw before: $\mathbf{P}$, $\mathbf{R}$, and $\mathbf{K}$. We use a as a variable to range over $\mathbf{A}$ in our statement of the syntax and the rules.

For the syntax, we take elements $a \in \mathbf{A}$ to be binary atoms, just as the elements $s \in \mathbf{R}$ are. Thus, the binary literals are the expressions of the form s, $\overline{s}$, a, or $\overline{a}$.

The syntax is the same as before, except that we allow the binary atoms to be elements of $\mathbf{A}$ in addition to elements of $\mathbf{R}$. So in a sense, we have the same syntax is before, except that some of the binary atoms are taken to render transitive verbs, and some are taken to render comparative adjective phrases. The only difference is in the semantics. Here, we require that (in every model $\mathcal{M}$) for an adjective $a \in \mathbf{A}$, $[\![a]\!]$ must be a transitive relation.

Proof system. We adopt the same proof system as in Figure 3, but with one addition. This addition is the rule for transitivity:

$$\frac{a(t_1, t_2) \quad a(t_2, t_3)}{a(t_1, t_3)}\ \text{trans}$$

This rule is added for all $a \in \mathbf{A}$.

Example 7. We have seen an informal example in (1) at the beginning of this paper. At this point, we can check that our system does indeed have a derivation corresponding to this. We need to check that $\Gamma \vdash \varphi$, where Γ contains

$$\forall(\mathsf{sw}, \forall(\mathsf{ripe}, \mathsf{bigger})), \forall(\mathsf{pineapple}, \forall(\mathsf{kq}, \mathsf{bigger})), \forall(\overline{\mathsf{pineapple}}, \forall(\overline{\mathsf{ripe}}, \mathsf{bigger})),$$

and φ is

$$\forall(\exists(\mathsf{sw}, \mathsf{bigger}), \forall(\mathsf{kq}, \mathsf{bigger})).$$

(We are going to use kq as an abbreviation of $\mathsf{kumquat}$ for typographical convenience, and similarly for sw and sweet.) In Example 6, we saw that $\Gamma \vdash \forall(\mathsf{sweet}, \forall(\mathsf{kq}, \mathsf{bigger}))$. (Recall that we saw this with a derivation in a different format, in Figure 5. This could be converted to our official format of natural deduction trees.) That work used $\mathbf{R} = \{\mathsf{bigger}\}$, but here we want $\mathbf{R} = \emptyset$ and $\mathbf{A} = \{\mathsf{bigger}\}$. The same derivation works, of course. Transitivity enables us to obtain a derivation for (1):

$$\dfrac{\dfrac{[\exists(\mathsf{sw}, \mathsf{bigger})(x)]^3 \qquad \dfrac{\dfrac{[\mathsf{bigger}(x,y)]^2 \qquad \dfrac{[\mathsf{kq}(z)]^1 \qquad \dfrac{[\mathsf{sw}(y)]^2 \quad \overset{\vdots}{\forall(\mathsf{sw}, \forall(\mathsf{kq}, \mathsf{bigger}))}}{\forall(\mathsf{kq}, \mathsf{bigger})(y)}\ \forall E}{\mathsf{bigger}(y,z)}\ \forall E}{\mathsf{bigger}(x,z)}\ \text{trans}}{\forall(\mathsf{kq}, \mathsf{bigger})(x)}\ \forall I^1}{\forall(\mathsf{kq}, \mathsf{bigger})(x)}\ \exists E^2}{\forall(\exists(\mathsf{sw}, \mathsf{bigger}), \forall(\mathsf{kq}, \mathsf{bigger}))}\ \forall I^3$$

Adding the transitivity rule gives a sound and complete proof system for the semantic consequence relation $\Gamma \models \varphi$. The soundness is easy, and so we only sketch the completeness. We must show that a set Γ which is consistent in the new logic has a transitive model. The canonical model $\mathcal{M}(\Gamma)$ as defined in Section 2.6 is automatically transitive; this is immediate from the transitivity rule. And as we know, it satisfies Γ.

3.1 $\mathcal{L}(adj)$ Has the Finite Model Property

Our final result is that $\mathcal{L}(adj)$ has the finite model property. We extend the work in Section 2.7. The inspiration for our definitions comes from the technique of *filtration* in modal logic, but we shall not refer explicitly to this area.

We again assume that Γ is consistent, and Γ^* has the properties of Lemma 3.

Definition 2. *For $a \in \mathbf{A}$, we say that j* reaches k *(*by a chain of $\equiv$ and a statements*) if there is a sequence*

$$j = j_0 \equiv k_0, \quad j_1 \equiv k_1, \quad \ldots, \quad j_n \equiv k_n = k \tag{4}$$

such that $n \geq 1$, and Γ^* contains $a(k_0, j_1)$, …, $a(k_{n-1}, j_n)$.

Lemma 8. *Assume that j reaches k by a chain of $\equiv$ and a statements.*

1. *If $c(k) \in \Gamma^*$, then Γ^* contains $\exists(c, a)(j)$.*
2. *If $j, k \in \mathbf{K}(\mathcal{L})$, then Γ^* contains $a(j, k)$.*

Proof. By induction on $n \geq 1$ in (4). For $n = 1$, we have essentially seen the argument as a step in Lemma 6. Here it is again. Since $c(k_1)$ and $j_1 \equiv k_1$, we see that $c(j_1)$. Together with $s(k_0, j_1)$, we have $\exists(c, a)(k_0)$. And as $j_0 \equiv k_0$, we see that $\exists(c, a)(j_0)$.

Assume our result for n, and now consider a chain as in (4) of length $n + 1$. The induction hypothesis applies to

$$j = j_1 \equiv k_1, \quad j_2 \equiv k_2, \quad \ldots, \quad j_{n+1} \equiv k_{n+1} = k$$

and so we have $\exists(c, a)(j_1)$. Since $a(k_0, j_1)$, we easily have $\exists(c, a)(k_0)$ by transitivity. And as $j_0 \equiv k_0$, we have $\exists(c, a)(j_0)$.

The second assertion is also proved by induction on $n \geq 1$. For $n = 1$, we have $j = j_0 \equiv k_0$, Γ^* contains $a(k_0, j_1)$; and $j_1 \equiv k_1 = k$. Then since the $\equiv$ is the identity on $\mathbf{K}(\mathcal{L})$, $j = j_0 = k_0$, and $j_1 = k_1 = k$. Hence Γ^* contains $s(j, k)$. Assuming our result for n, we again consider a chain as in (4) of length $n + 1$. Just as before, $j = j_0 = k_0$, and so Γ^* contains $a(j, j_1)$. By induction hypothesis, Γ^* contains $a(j_1, k)$. By transitivity, Γ^* contains $a(j, k)$. □

We endow N with an $\mathcal{L}$-structure as follows:

$[\![p]\!] = \{([j], Q) : p(j) \in \Gamma^* \text{ and } Q \in \{\forall, \exists\}\}$.
$[\![s]\!](([j], Q), ([k], Q'))$ iff one of the following two conditions holds:

1. There is a set term c such that Γ^* contains $c(k)$ and $\forall(c, s)(j)$.
2. $Q' = \exists$, and for some $j_* \equiv j$ and $k_* \equiv k$, Γ^* contains $s(j_*, k_*)$.

$[\![a]\!](([j], Q), ([k], Q'))$ iff

1. If $\forall(c, a)(k) \in \Gamma^*$, then also $\forall(c, a)(j) \in \Gamma^*$.
2. In addition, either (a) or (b) below holds:
   (a) There is a set term c such that Γ^* contains $c(k)$ and $\forall(c, a)(j)$.
   (b) $Q' = \exists$, and j reaches k by a chain of $\equiv$ and a statements.

(Notice that this definition is independent of the representatives in $[j]$ and $[k]$.)

For a constant j of $\mathfrak{L}$, $[\![j]\!] = ([j], \exists)$.

Once again, we suppress Γ and Γ^* and simply write $\mathcal{N}$ for the resulting $\mathfrak{L}$-structure.

Lemma 9. *For $a \in \mathbf{\Lambda}$, each relation $[\![a]\!]$ is transitive in $\mathcal{N}$.*

Proof. In this proof and the next, we are going to use l to stand for a constant symbol, even though earlier in the paper we used it for a literal. Assume that

$$([j], Q)\ [\![a]\!]\ ([k], Q')\ [\![a]\!]\ ([l], Q''). \tag{5}$$

Clearly we have the first requirement concerning $[\![a]\!]$: if $\forall(c, a)(l) \in \Gamma^*$, then also $\forall(c, a)(j) \in \Gamma^*$.

We have four cases, depending on the reasons for the two assertions in (5).

Case 1 There is a set term b such that Γ^* contains $b(k)$ and $\forall(b, a)(j)$, and there is also a set term c such that Γ^* contains $c(l)$ and $\forall(c, a)(k)$. By (1), Γ^* contains $c(l)$ and $\forall(c, a)(j)$. And so we have requirement (2a) concerning $[\![a]\!]$ for $([j], Q)$ and $([l], Q'')$.

Case 2 There is a set term b such that Γ^* contains $b(k)$ and $\forall(b, a)(j)$, and k reaches l. Note that $a(j, k)$. So j reaches l.

Case 3 j reaches k by a chain of $\equiv$ and a statements, and there is a set term c such that Γ^* contains $c(l)$ and $\forall(c, a)(k)$. Then $a(k, l)$. And so j reaches l.

Case 4 j reaches k, and k reaches l. Then concatenating the chains shows that j reaches l. □

Lemma 10. *For all $c \in \mathrm{Sub}(\Gamma)$, $[\![c]\!] = \{([j], Q) : c(j) \in \Gamma^*$ and $Q \in \{\forall, \exists\}\}$.*

Proof. We argue by induction on c. Much of the proof is as in Lemma 6, For c a unary atom, the result is obvious. Also, assuming that $[\![c]\!] = \{([j], Q) : c(j) \in \Gamma^*\}$ we easily have the same result for $\overline{c}$ using the maximal consistency of Γ^*:

$$([j], Q) \in [\![\overline{c}]\!] \quad \text{iff} \quad ([j], Q) \notin [\![c]\!] \quad \text{iff} \quad c(j) \notin \Gamma^* \quad \text{iff} \quad \overline{c}(j) \in \Gamma^*.$$

Assume about c that if $c \in Sub(\Gamma)$, then $[\![c]\!] = \{([j], Q) : c(j) \in \Gamma^*\}$. In view of what we just saw, we only need to check the same result for $\forall(c, s)$, $\exists(c, s)$, $\forall(c, a)$, and $\exists(c, \overline{a})$.

$\underline{\forall(c,s)}$ Suppose that $\forall(c,s) \in \mathit{Sub}(\Gamma)$, so that $c \in \mathit{Sub}(\Gamma)$ as well. We prove that

$$[\![\forall(c,s)]\!] \quad = \quad \{([j],Q) : \forall(c,s)(j) \in \Gamma^*\}.$$

Let $([j],Q) \in [\![\forall(c,s)]\!]$. We shall show that $\forall(c,s)(j) \in \Gamma^*$. If not, then by maximal consistency, $\exists(c,\overline{s})(j) \in \Gamma^*$. By the Henkin property, let k be such that Γ^* contains $c(k)$ and $\overline{s}(j,k)$. By induction hypothesis, $([k],\forall) \in [\![c]\!]$. And so $([j],\forall)[\![s]\!]([k],\forall)$. Thus there is a set term b such that Γ^* contains $b(k)$ and $\forall(b,s)(j)$. From these, Γ^* contains $s(j,k)$. And thus Γ^* is inconsistent. This contradiction shows that indeed $\forall(c,s)(j) \in \Gamma^*$.

In the other direction, suppose that $([j],Q)$ is such that $\forall(c,s)(j) \in \Gamma^*$. Let $([k],Q') \in [\![c]\!]$, so by induction hypothesis, $c(k) \in \Gamma^*$. By the way we interpret binary relations in $\mathcal{N}$, $[\![s]\!](([j],Q),([k],Q'))$. This for all $([k],Q') \in [\![c]\!]$ shows that $([j],Q) \in [\![\forall(c,s)]\!]$.

$\underline{\exists(c,s)}$ Suppose that $\exists(c,s) \in \mathit{Sub}(\Gamma)$, so that $c \in \mathit{Sub}(\Gamma)$ as well. Let $([j],Q) \in [\![\exists(c,s)]\!]$. Let k and Q' be such that $[\![c]\!]([k],Q')$ and $[\![s]\!](([j],Q),([k],Q'))$. By induction hypothesis, $c(k) \in \Gamma^*$. First, let us consider the case when $Q' = \forall$. Let b be such that Γ^* contains $b(k)$ and $\forall(b,s)(j)$. Using $(\forall E)$, we have $\Gamma^* \vdash \exists(c,s)(j)$. And as Γ^* is closed under deduction, $\exists(c,s)(j) \in \Gamma^*$ as desired. The more interesting case is when $Q' = \exists$, so that for some $j_* \equiv j$ and $k_* \equiv k$, Γ^* contains $s(j_*,k_*)$. Since $c(k)$ and $k \equiv k_*$, we have $c(k_*) \in \Gamma^*$. Then using $(\exists I)$, we see that $\exists(c,s)(j_*) \in \Gamma^*$. Since $j \equiv j_*$, once again we have $\exists(c,s)(j) \in \Gamma^*$.

Conversely, suppose that $\exists(c,s)(j) \in \Gamma^*$. By the Henkin property, let k be such that $c(k)$ and $s(j,k)$ belong to Γ^*. Then $[\![s]\!](([j],Q),([k],\exists))$, and by induction hypothesis, $[\![c]\!](k)$. Hence $([j],Q) \in [\![\exists(c,s)]\!]$.

$\underline{\forall(c,a)}$ Suppose that $\forall(c,a) \in \mathit{Sub}(\Gamma)$, so that $c \in \mathit{Sub}(\Gamma)$ as well. We prove that

$$[\![\forall(c,a)]\!] \quad = \quad \{([j],Q) : \forall(c,a)(j) \in \Gamma^*\}.$$

The first part argument is the left-to-right inclusion. It is exactly the same as what we saw above for the sentences of the form $\forall(c,s)$.

In the other direction, suppose that $\forall(c,a)(j) \in \Gamma^*$; we show that $([j],Q) \in [\![\forall(c,a)]\!]$. For this, let $([k],Q') \in [\![c]\!]$. By induction hypothesis, $c(k) \in \Gamma^*$. We must verify that if $\forall(b,a)(k) \in \Gamma^*$, then also $\forall(b,a)(j) \in \Gamma^*$. This is shown in the derivation below:

$$\dfrac{\dfrac{\dfrac{c(k) \quad \forall(c,a)(j)}{a(j,k)}\,\forall E \qquad \dfrac{[b(x)]^1 \quad \forall(b,a)(k)}{a(k,x)}\,\forall E}{a(j,x)}\,\text{trans}}{\forall(b,a)(j)}\,\forall I^1$$

Since Γ^* is closed under deduction, we see that indeed $\forall(b,a)(j) \in \Gamma^*$. Going on, we see from the structure of $\mathcal{N}$ that $[\![s]\!](([j],Q),([k],Q'))$. This for all $([k],Q') \in [\![c]\!]$ shows that $([j],Q) \in [\![\forall(c,a)]\!]$.

$\underline{\exists(c,a)}$ Suppose that $\exists(c,a) \in Sub(\Gamma)$, so that $c \in Sub(\Gamma)$ as well.

Let $([j],Q) \in [\![\exists(c,a)]\!]$. Let k and Q' be such that the following two assertions hold: $[\![a]\!](([k],Q')$ and $[\![a]\!](([j],Q),([k],Q'))$. By induction hypothesis, $c(k) \in \Gamma^*$. There are two cases depending on whether $Q' = \forall$ or $Q' = \exists$. The argument for $Q' = \forall$ is the same as the one we saw in our work on sentences $\exists(c,s)$ above. The more interesting case is when $Q' = \exists$. This time, j reaches k. By Lemma 8, $\exists(c,a)(j) \in \Gamma^*$.

Conversely, suppose that $\exists(c,a)(j) \in \Gamma^*$. By the Henkin property, let k be such that $c(k)$ and $a(j,k)$ belong to Γ^*. The derivation below shows that if $\forall(d,a)(k) \in \Gamma^*$, then $\forall(d,a)(j) \in \Gamma^*$ as well:

$$\dfrac{\dfrac{a(j,k) \qquad \dfrac{[d(x)]^1 \quad \forall(d,a)(k)}{a(k,x)}\ \forall E}{a(j,x)}\ \text{trans}}{\forall(d,a)(j)}\ \forall I^1$$

So $[\![a]\!](([j],Q),([k],\exists))$, and by induction hypothesis, $[\![c]\!](k)$. Hence $([j],Q) \in [\![\exists(c,a)]\!]$.

This completes the induction. □

Lemma 11. $\mathcal{N} \models \Gamma$.

Proof. We check the sentence types in turn, using Lemma 10 without mention.

First, let Γ contain the sentence $\forall(b,c)$. Then b and c belong to $Sub(\Gamma)$. Let $([j],Q) \in [\![b]\!]$, so that $b(j) \in \Gamma^*$. We have $d(j) \in \Gamma^*$ using $(\forall E)$. This for all $([j],Q)$ shows that $\mathcal{N} \models \forall(b,c)$.

Second, let $\exists(c,d) \in \Gamma$. By the Henkin property, let j be such that both $c(j)$ and $d(j)$ belong to Γ^*. The element $([j],\forall)$ shows that $[\![c]\!] \cap [\![d]\!] \neq \emptyset$. That is, $\mathcal{N} \models \exists(c,d)$.

Continuing, consider a sentence $b(j) \in \Gamma$. As $b \in Sub(\Gamma)$, we have $([j],\exists) \in [\![b]\!]$, so that $\mathcal{N} \models b(j)$.

The work for sentences of the forms $s(j,k)$ and $\overline{s}(j,k)$ was done in Lemma 7.

The most intricate part of this proof concerns sentences $a(j,k), \overline{a}(j,k) \in \Gamma$. Recall that we are dealing in this result with sentences of $\mathcal{L}$, and so j and k are constant symbols of that language. Also recall that $[\![j]\!] = ([j],\exists)$, and similarly for k.

Consider sentences in Γ of the form $a(j,k)$. It is easy to see that if $\exists(c,a)(k)$ belongs to Γ, then so does $\exists(c,a)(j)$. (See the $\exists(c,a)$ case in Lemma 10.) From this it follows easily that $[\![a]\!]([\![j]\!],[\![k]\!])$. And so $\mathcal{N} \models a(j,k)$ in this case. We conclude with the consideration of a sentence in Γ of the form $\overline{a}(j,k)$. We wish to show that $\mathcal{N} \models \overline{a}(j,k)$. Suppose towards a contradiction that $\mathcal{N} \models a(j,k)$. Then we have $[\![a]\!](([j],\exists),([k],\exists))$. There are two possibilities, corresponding to the alternatives in the semantics of a. The first is when there is a set term c such that Γ^* contains $c(k)$ and $\forall(c,a)(j)$. Using $(\forall E)$, Γ^* then contains $a(j,k)$. But recall that Γ contains $\overline{a}(j,k)$. So in this alternative, $\Gamma^* \supseteq \Gamma$ is inconsistent. In the second alternative, j reaches k by a chain of $\equiv$ and a statements. By Lemma 8, $a(j,k) \in \Gamma^*$. So Γ^* is inconsistent, and we have our contradiction. □

Once again, this gives us the finite model property for $\mathcal{L}(adj)$. The result is not interesting from a complexity-theoretic point of view, since we already could see from Lutz and Sattler [9] that the logic had an ExpTime satisfiability problem.

4 Conclusion and Future Work

This paper has provided two logical systems, $\mathcal{L}$ and $\mathcal{L}(adj)$, along with semantics. We presented proof systems in the format of natural deduction, and in both cases we have completeness theorems and the finite model property. The semantics of the language allows us to translate some natural language sentences into the languages faithfully.

Set terms in this sense of this paper come from McAllester and Givan [10], where they are called *class terms*. That paper was probably the first to present an infinite fragment relevant to natural language and to study its logical complexity. The language of [10] did not have negation, and they showed that satisfiability is NP-complete. The language of [10] is included in the language $\mathcal{R}^*$ of Pratt-Hartmann and Moss [15]; the difference is that $\mathcal{R}^*$ has "a small amount" of negation. Yet more negation is found in the language $\mathcal{R}^{*\dagger}$ of [15]. This fragment has binary and unary atoms and negation. It is equivalent in most respects to the language $\mathcal{L}$ of this paper, but there are two small differences. First, here we have added constant symbols. In addition to making the system more expressive, the reason for adding constants is in order to present the Henkin-style completeness proof in Sections 2.5. The other change is that $\mathcal{R}^{*\dagger}$ does not allow recursively defined set terms, only "flat" terms. However, from the point of view of decidability and complexity, this change is really minor: one may add new symbols to flatten a sentence, at the small cost of adding new sentences. The flat version is also essentially the same as the language $\mathcal{E}_2$ of Pratt-Hartmann [14].

The decidability of $\mathcal{L}(adj)$ follows from known results on Boolean modal logics [9], but the finite model property appears to be new here.

Proof systems for fragments that are weaker than $\mathcal{L}$ appear in [15]. These proof systems are syllogistic; there are no variables or what we have in this paper called general sentences. Modulo complexity hypotheses, the proof systems of this paper are the first ones which are complete and go beyond the capabilities of syllogistic proof systems. At the same time, they are decidable and useable. For example, we have seen how the inference in (1) in the Introduction is handled in our system; see Examples 6 and 7.

The use of natural deduction proofs in connection with natural language is very old, going back to Fitch [4]. Fitch's paper does not deal with a formalized fragment, and so it is not possible to even ask about questions like completeness and decidability. Also, the phenomena of interest in the paper went beyond what we covered here. We would like to think that the methods of this paper could eventually revive interest in Fitch's proposal by giving it a foundation.

Francez and Dyckhoff [5] propose a *proof-theoretic semantics* for natural language. Their far-ranging proposal goes beyond what we can discuss in this paper. We only want to mention that our proof rules bear some similarity to theirs.

Their system had no recursive constructs and also no negative determiners, but it went beyond ours in covering both readings of scope-ambiguous simple sentences. Since our motivation was not proof-theoretic in this paper, we did not investigate proof-theoretic properties of our system. But it would be interesting to do so.

It is of interest to go further in order to render more of natural language inference in complete and decidable logical systems. One next step would be to add converses to the binary atoms in order to express simple sentences beyond what we have seen. For example, writing sees^{-1} for the inverse of see, we could render $\mathsf{Every\ girl\ sees\ Mary}$ as $\forall(\mathsf{girl}, \mathsf{see}^{-1})(\mathsf{Mary})$. It is possible to extend our work in such a way as to incorporate these converses. The logic would axiomatized on top of $\mathcal{L}$ by adding the rule deriving $r^{-1}(t, u)$ from $r(u, t)$. But this is only one of the many things to do.

Acknowledgments

My thanks to Andreas Blass and Ian Pratt-Hartmann for useful discussions of these topics and for corrections at various stages.

References

1. van Benthem, J.: Essays in Logical Semantics. Reidel, Dordrecht (1986)
2. Böerger, E., Grädel, E., Gurevich, Y.: The Classical Decision Problem. In: Perspectives in Mathematical Logic, p. 1197. Springer, Heidelberg
3. van Dalen, D.: Logic and Structure, 4th edn. Springer, Berlin (2004)
4. Fitch, F.B.: Natural Deduction Rules for English. Philosophical Studies 24(2), 89–104 (1973)
5. Francez, N., Dyckhoff, R.: Proof-Theoretic Semantics for a Natural Language Fragment. In: Jaeger, G., Ebert, C., Michaelis, J. (eds.) Proceedings, MoL 10/11. LNCS (LNAI). Springer, Heidelberg (2010) (to appear)
6. Grädel, E., Kolaitis, P., Vardi, M.: On the Decision Problem for Two-Variable First-Order Logic. Bulletin of Symbolic Logic 3(1), 53–69 (1997)
7. Grädel, E., Otto, M., Rosen, E.: Undecidability Results on Two-Variable Logics. Archive for Mathematical Logic 38, 313–354 (1999)
8. Gurevich, Y.: On the Classical Decision Problem. Logic in Computer Science Column. The Bulletin of the European Association for Theoretical Computer Science (October 1990)
9. Lutz, C., Sattler, U.: The Complexity of Reasoning with Boolean Modal Logics. In: Wolter, F., et al. (eds.) Advances in Modal Logics, vol. 3. CSLI Publications, Stanford (2001)
10. McAllester, D., Givan, R.: Natural Language Syntax and First Order Inference. Artificial Intelligence 56 (1992)
11. Mortimer, M.: On Languages with Two Variables. Zeitschrift für Mathematische Logik und Grundlagen der Mathematik 21, 135–140 (1975)
12. Pelletier, F.J.: A Brief History of Natural Deduction. History and Philosophy of Logic 20, 1–31

13. Pratt-Hartmann, I.: A Two-Variable Fragment of English. Journal of Logic, Language and Information 12(1), 13–45 (2003)
14. Pratt-Hartmann, I.: Fragments of Language. Journal of Logic, Language and Information 13, 207–223 (2004)
15. Pratt-Hartmann, I., Moss, L.S.: Logics for the Relational Syllogistic. Review of Symbolic Logic 2(4), 647–683 (2009)
16. Vardi, M.Y., Wolper, P.: Automata-Theoretic Techniques for Modal Logics of Programs. Journal of Computer and System Sciences 32(2), 183–221 (1986)

Choiceless Computation and Symmetry

Benjamin Rossman*

Computer Science and Artificial Intelligence Laboratory, MIT
brossman@theory.csail.mit.edu

Abstract. Many natural problems in computer science concern structures like graphs where elements are not inherently ordered. In contrast, Turing machines and other common models of computation operate on strings. While graphs may be *encoded* as strings (via an adjacency matrix), the encoding imposes a linear order on vertices. This enables a Turing machine operating on encodings of graphs to *choose* an arbitrary element from any nonempty set of vertices at low cost (the Augmenting Paths algorithm for Bipartite Matching being an example of the power of choice). However, the outcome of a computation is liable to depend on the external linear order (i.e., the choice of encoding). Moreover, isomorphism-invariance/encoding-independence is an undecidable property of Turing machines. This trouble with encodings led Blass, Gurevich and Shelah [3] to propose a model of computation known as *BGS machines* that operate directly on structures. BGS machines preserve symmetry at every step in a computation, sacrificing the ability to make arbitrary choices between indistinguishable elements of the input structure (hence "choiceless computation"). Blass et al. also introduced a complexity class CPT+C (Choiceless Polynomial Time with Counting) defined in terms of polynomially bounded BGS machines. While every property finite structures in CPT+C is polynomial-time computable in the usual sense, it is open whether conversely every isomorphism-invariant property in P belongs to CPT+C. In this paper we give evidence that CPT+C $\neq$ P by proving the separation of the corresponding classes of *function problems*. Specifically, we show that there is an isomorphism-invariant polynomial-time computable function problem on finite vector spaces ("given a finite vector space V, output the set of hyperplanes in V") that is not computable by any CPT+C program. In addition, we give a new simplified proof of the Support Theorem, which is a key step in the result of [3] that a weak version of CPT+C absent counting cannot decide the parity of sets.

Keywords: choiceless polynomial time, descriptive complexity, finite model theory.

1 Introduction

Is there a "logic" capturing exactly the polynomial-time computable properties of finite structures? This question was raised by Gurevich [9] in the mid 80s,

* Supported by the NSF Graduate Research Fellowship.

A. Blass, N. Dershowitz, and W. Reisig (Eds.): Gurevich Festschrift, LNCS 6300, pp. 565–580, 2010.

nearly a decade after Fagin [8] showed that the NP properties of finite structures are precisely what can be defined in existential second-order logic. Today this question remains a central open problem in finite model theory.

Addressing this question, Blass, Gurevich and Shelah [3,4] introduced an logic/complexity class known as CPT+C or Choiceless Polynomial Time with Counting. CPT+C is based on a model of computation known as *BGS machines* (after the inventors). BGS machines operate directly on structures in a manner that preserves symmetry at every step of a computation. By contrast, Turing machines encode structures as strings. This encoding violates the symmetry of structures like graphs (which might possess nontrivial automorphisms) by imposing a linear order on elements. Note that Turing machines are able to exploit this linear order to efficiently *choose* an element from any set constructed in the course of a computation. Thus, it is not uncommon in the high-level description of an algorithm (say, the well-known Augmenting Paths algorithm for BIPARTITE MATCHING) to read something along the lines of "let w be any unmatched neighbor of the vertex v". A description like this (implicit) carries a claim that the ultimate outcome of the computation will not dependent on the choice of w. However, the validity of such claims cannot be taken for granted: by Rice's Theorem, encoding-invariance is an undecidable property of Turing machines.

The BGS machine model of computation is said to be "choiceless" because it disallows choices which violate the inherent symmetry of the input structure. Pseudo-instructions of the form "let i be an arbitrary element of the set I" is forbidden. Similarly, "let w be the first neighbor of v" is meaningless (unless referring to an explicitly constructed linear order on vertices). The inability of BGS machines to choose is compensated by *parallelism* (the power to explore all choices in parallel) and the machinery of set theory (the power to build sets using comprehension).

BGS machines may in fact be viewed as the syntactic elements (i.e., formulas) of a logic, whose semantics is well-defined on any structure. One rough description BGS logic is:

BGS logic = propositional logic +

- *a least-fixed-point operator*
- *a cardinality operator*
- *basic set-theoretic predicates* ($\in$, $\cup$, *etc.*) *and*
- *comprehension terms* $\{\mathsf{term}_1(x) : x \in \mathsf{term}_2 : \mathsf{formula}(x)\}$

evaluated in the domain $\mathrm{HF}(\mathcal{A}) = A \cup \wp(A) \cup \wp(A \cup \wp(A)) \cup \cdots$
of hereditarily finite objects over a structure $\mathcal{A}$ *with universe* A.

The complexity class CPT+C or Choiceless Polynomial Time with Counting (itself also a logic) is obtained by imposing polynomial bounds on the running time and number of processors required to run a BGS machine. (Somewhat more precisely: this involves requiring that fixed-points converge in polynomially many steps and that only polynomial many elements of $\mathrm{HF}(\mathcal{A})$ participate in the evaluation of a formula.) Every CPT+C computable property of structures like graphs can be implemented on a polynomial-time Turing machine (on encodings on such structures). Thus, CPT+C $\subseteq$ P. However, it is open whether, conversely, every isomorphism-invariant polynomial-time property of finite structures is computable in CPT+C.

The main result of this paper (Theorem 6) suggests that CPT+C $\neq$ P. We show that there is an isomorphism-invariant polynomial-time computable function problem on finite vector spaces ("given a finite vector space V, output the set of hyperplanes in V") that is not computable by any CPT+C program. An additional result of this paper (Theorem 5) is a new simplified proof of the Support Theorem of [3], which is a key step in the result of [3] that a weak version of CPT+C absent counting cannot decide the parity of sets.

Outline of the paper. In §2 we present all relevant definitions. §3 contains a brief summary of the key results on CPT and CPT+C from previous work of Blass et al. [3] and others. In §4 we introduce some new notions relating BGS machines to the automorphism groups of structures. In §5 we prove that the symmetric group has a certain "support property", leading to a simplified proof of the Support Theorem from the original paper of Blass et al. [3]. In §6 we establish a similar result for the general linear group; as a corollary, we show that there is a polynomial-time computable *function problem* on finite vector spaces that is not computable by any CPT+C program.

2 Definitions

We begin by defining *hereditarily finite expansions* of structures in §2.1. We then define BGS logic and classes CPT and CPT+C in §2.2. The definition of BGS logic presented here differs from the BGS machines of Blass, et al. [3], but classes CPT and CPT+C are exactly the same. BGS logic has a bare-bones syntax that is well-suited for induction, whereas the BGS machines of [3] have an intuitive and attractive syntax (borrowing from Abstract State Machines) that is recommended for the actual description of CPT+C algorithms (see [3] for examples). Let us also mention that CPT(+C) is elsewhere [3,4,7] written as $\widetilde{\mathrm{C}}$PT(+C), the tilde over C evoking the "less" in "choiceless".

2.1 Hereditarily Finite Expansion

For every structure $\mathcal{A}$ in a signature σ, we define a structure HF($\mathcal{A}$) in an enlarged signature σ^{HF} called the *hereditarily finite expansion* of $\mathcal{A}$.

Definition 1. [Hereditarily Finite Objects] Let A be for a set (a special case being the universe of a structure $\mathcal{A}$) whose elements we call *atoms* and assume to be non-set entities (or "ur-elements").

1. HF(A) denote the (unique) smallest set such that $A \subseteq \mathrm{HF}(A)$ and $B \in \mathrm{HF}(A)$ for every finite subset B of HF(A). Elements of HF(A) are called *hereditarily finite objects* (*h.f. objects*) *over* A and elements of $\mathrm{HF}(A) \setminus A$

(i.e., elements of $\mathrm{HF}(A)$ which are sets) are called *hereditarily finite sets* (*h.f. sets*) *over* A. (Note that if A is finite, then $\mathrm{HF}(A)$ is the countable union $A \cup \wp(A) \cup \wp(A \cup \wp(A)) \cup \wp(A \cup \wp(A) \cup \wp(A \cup \wp(A))) \cup \cdots$.)
2. The *rank* of a h.f. object x is defined as 0 if $x \in A \cup \{\emptyset\}$ and $1 + \max_{y \in x} \mathrm{rank}(y)$ otherwise.
3. A h.f. set x is *transitive* if $y \subseteq x$ for all $y \in x$ such that y is a set. The *transitive closure* $\mathrm{TC}(x)$ of a h.f. object x is the (unique) smallest transitive set containing x as an element.
4. The *finite von Neumann ordinals* κ_i for $i < \omega$ are elements of $\mathrm{HF}(\emptyset)$ defined by $\kappa_0 = \emptyset$ and $\kappa_{i+1} = \{\kappa_0, \ldots, \kappa_i\}$.

Definition 2. [Group Actions] For a group G acting on A (a special case being the action of $\mathrm{Aut}(\mathcal{A})$ on the universe of a structure $\mathcal{A}$), we shall consider the extension of this action to $\mathrm{HF}(A)$ defined for $g \in G$ and $x \in \mathrm{HF}(A) \backslash A$ inductively by $gx = \{gy : y \in x\}$.

Definition 3. [Signatures and Structures] A *signature* σ consists of relation symbols R_i and function symbols f_j (with designated arities) as well as constant symbols c_k. A σ*-structure* $\mathcal{A}$ (or *structure with signature* σ) consists of a set A (called the *universe* of $\mathcal{A}$) together with interpretations for the various symbols in σ, that is, relations $R_i^{\mathcal{A}} \subseteq A^{\mathrm{arity}(R_i)}$, functions $f_j^{\mathcal{A}} : A^{\mathrm{arity}(f_j)} \to A$ and constants $c_k^{\mathcal{A}} \in A$ (or simply R_i, f_j and c_k when $\mathcal{A}$ is known from context).

Definition 4. [Hereditarily Finite Expansion] σ^{HF} denotes the disjoint union of σ and $\{\mathsf{In}, \mathsf{EmptySet}, \mathsf{Atoms}, \mathsf{Union}, \mathsf{Pair}, \mathsf{TheUnique}, \mathsf{Card}\}$ where In is a binary relation symbol, $\mathsf{EmptySet}$ and Atoms are constant symbols, Union, $\mathsf{TheUnique}$ and Card unary function symbols, and Pair is a binary function symbol. For a finite σ-structure $\mathcal{A}$, let $\mathrm{HF}(\mathcal{A})$ denote the σ^{HF}-structure with universe $\mathrm{HF}(A)$ where:

- symbols from σ have the same interpretation in $\mathrm{HF}(\mathcal{A})$ as in $\mathcal{A}$, with the convention that functions from σ take value $\emptyset$ whenever any coordinate of the input is not an atom,
- $(x, y) \in \mathsf{In}$ if and only if $x \in y$,
- $\mathsf{EmptySet} = \emptyset$, $\mathsf{Atoms} = A$, $\mathsf{Pair}(x, y) = \{x, y\}$,
- $\mathsf{Union}(x) = \bigcup_{y \in x} y$ (in particular, $\mathsf{Union}(x) = \emptyset$ if $x \in \mathsf{Atoms} \cup \{\emptyset\}$),
- $\mathsf{TheUnique}(x) = \begin{cases} y & \text{if } x = \{y\}, \\ \emptyset & \text{if } x \text{ is not a singleton,} \end{cases}$
- $\mathsf{Card}(x) = \begin{cases} |x| \text{ as a von Neumann ordinal} & \text{if } x \text{ is a set,} \\ \emptyset & \text{if } x \text{ is an atom.} \end{cases}$

$\mathrm{HF}(\mathcal{A})$ is called the *hereditarily finite expansion* of $\mathcal{A}$. Structures of the form $\mathrm{HF}(\mathcal{A})$ are called *hereditarily finite structures.*

2.2 BGS Logic and CPT+C

Just like first-order logic, BGS logic (and its weaker cousin BGS$^-$ logic) are defined with respect to a fixed signature σ. Similarly, BGS logic has terms and formulas. However, BGS logic has an additional syntactic element called *programs* (which compute a term $t(t(t(...t(\emptyset)\dots)))$ iteratively using a "step" term $t(\cdot)$ until some "halting" formula is satisfied, whereupon an "output" term is computed).

Definition 5. [BGS Logic]
BGS logic over a signature σ consists of *terms*, *formulas* and *programs*, defined below.

1. ***Terms*** and ***formulas*** are defined inductively:
   - (base case) variables are terms;
   - (base case) constant symbols in σ^{HF} are terms;
   - if f is an r-ary function symbol in σ^{HF} and $t_1, \dots, t_r$ are terms, then $f(t_1, \dots, t_r)$ is a term;
   - if R is an r-ary relation symbol in σ^{HF} and $t_1, \dots, t_r$ are terms, then $R(t_1, \dots, t_r)$ is a formula;
   - if t_1 and t_2 are terms, then $t_1 = t_2$ is a formula;
   - if ϕ_1 and ϕ_2 are formulas, then so are $\neg\phi_1$ and $\phi_1 \wedge \phi_2$ and $\phi_1 \vee \phi_2$;
   - if s and t are terms, v is a variable (which is not free in t) and ϕ is a formula, then $\{s(v) : v \in t : \phi(v)\}$ is a term ("the set of $s(v)$ for $v \in t$ such that $\phi(v)$ is true").

   Terms of the form $\{s(v) : v \in t : \phi(v)\}$ are called *comprehension terms.*
2. Each occurrence of a variable in a term or formula is either *free* or *bound*, with the comprehension construct $\{s(v) : v \in t : \phi(v)\}$ binding the free occurrences of v within s and ϕ. As a matter of notation, we write $t(v_1, \dots, v_\ell)$ or $\phi(v_1, \dots, v_\ell)$ for a term or formula whose free variables are contained among $v_1, \dots, v_\ell$. A term or formula with no free variables is said to be *ground.*

   The *variable rank* of a term t (resp. formula φ) is the maximum number of free variables in any subterm/formula of t (resp. φ).
3. Terms and formulas have the obvious semantics when evaluated on hereditary finite structures with free variables assigned to h.f. objects. For a term $t(v_1, \dots, v_\ell)$ and elements $x_1, \dots, x_\ell \in \mathrm{HF}(A)$, the value of t when free variables $v_1, \dots, v_\ell$ are assigned to $x_1, \dots, x_\ell$ is denoted by $[\![t(\bar{x})]\!]^{\mathcal{A}} \in \mathrm{HF}(A)$ (or simply $[\![t(\bar{x})]\!]$ if $\mathcal{A}$ is known from context). For a formula $\varphi(v_1, \dots, v_\ell)$, the value $[\![\varphi(\bar{x})]\!]^{\mathcal{A}}$ is an element of $\{\mathsf{True}, \mathsf{False}\}$ where (following [3]) we identify $\mathsf{True} = \{\emptyset\}$ and $\mathsf{False} = \emptyset$.

   We omit a rigorous definition of the semantic operator $[\![\cdot]\!]$. Let us however explicitly state the semantics of comprehension terms (since this construct may be less familiar). For a comprehension term $r(\bar{v}) = \{s(\bar{v}, w) : w \in t(\bar{v}) : \varphi(\bar{v}, w)\}$ and parameters $\bar{x}$ from $\mathrm{HF}(A)$, the value $[\![r(\bar{x})]\!]$ is defined (in the obvious way) as the set of $[\![s(\bar{x}, y)]\!]$ for $y \in [\![t(\bar{x})]\!]$ such that $[\![\varphi(\bar{x}, y)]\!] = \mathsf{True}$.

4. A ***program*** $\Pi = (\Pi_{\text{step}}, \Pi_{\text{halt}}, \Pi_{\text{out}})$ consists of a term $\Pi_{\text{step}}(v)$, a formula $\Pi_{\text{halt}}(v)$, and a term or formula $\Pi_{\text{out}}(v)$ (depending whether Π computes a decision problem or produces an output) all with a single free variable v. If Π_{out} is a formula, then Π is said to be *Boolean*.

   The *variable rank* of Π is the maximum of the variable ranks of Π_{step}, Π_{halt} and Π_{out}.
5. A program Π is executed on an input structure $\mathcal{A}$ as follows. Let $x_0 = \emptyset$ and $x_{i+1} = [\![\Pi_{\text{step}}(x_i)]\!]$ for all $i < \omega$. In the event that $[\![\Pi_{\text{halt}}(x_i)]\!] = \mathsf{False}$ for all i, let $\Pi(\mathcal{A}) = \bot$ (and the computation is said to *diverge*). Otherwise, let $\Pi(\mathcal{A}) = [\![\Pi_{\text{out}}(x_i)]\!]$ for the minimal i such that $[\![\Pi_{\text{halt}}(x_i)]\!] = \mathsf{True}$.
6. ***BGS$^-$ logic*** consists of all terms, formulas and programs of BGS logic which exclude unary function symbol Card.

We remark that BGS logic has the ability to carry out bounded existential and universal quantification in $\text{HF}(\mathcal{A})$ (and thus subsumes first-order logic over the base structure $\mathcal{A}$). To see this, note that $(\exists v \in t)\, \phi(v)$ is equivalent to the formula $\{\emptyset : v \in t : \phi(v)\} = \{\emptyset\}$ (pedantically, we should write $\mathsf{EmptySet}$ instead of $\emptyset$ and $\mathsf{Pair}(\mathsf{EmptySet}, \mathsf{EmptySet})$ instead of $\{\emptyset\}$). Similarly, $(\forall v \in t)\, \phi(v)$ is equivalent to the formula $\{\emptyset : v \in t : \neg\phi(v)\} = \emptyset$.

We now define the crucial resource by which we measure the complexity of BGS programs. Informally, a h.f. object $x \in \text{HF}(A)$ is *active* for the operation of a program Π on a structure $\mathcal{A}$ if x is the value of any term involved in the computation of Π on $\mathcal{A}$ (until a halt state is reached). By setting a polynomial bound on the number of active objects, as well as requiring programs to halt on every input, we arrive at classes CPT and CPT+C.

Definition 6. [Active Objects and Classes CPT and CPT+C] As in the definition of $[\![\cdot]\!]^{\mathcal{A}}$, we omit the superscript from the *active-element operator* $\langle\!\langle\cdot\rangle\!\rangle^{\mathcal{A}}$ (defined below) when the structure $\mathcal{A}$ is clear from context (as below).

1. For every term $t(v_1, \ldots, v_\ell)$ and assignment of free variables to values $x_1, \ldots, x_\ell \in \text{HF}(A)$, we define $\langle\!\langle t(x_1, \ldots, x_\ell)\rangle\!\rangle \in \text{HF}(A)$ inductively as follows:
   - (base case) if term $t(v)$ is precisely the variable v, then $\langle\!\langle t(x)\rangle\!\rangle = \{x\}$ for every $x \in \text{HF}(A)$;
   - (base case) if c is a constant symbol in σ^{HF}, then $\langle\!\langle c\rangle\!\rangle = \{c^{\text{HF}(\mathcal{A})}\}$;
   - if R is an r-ary relation symbol in σ^{HF} and $t_1, \ldots, t_r$ are terms, then
   $$\langle\!\langle R(t_1, \ldots, t_r)\rangle\!\rangle = \textstyle\bigcup_{i=1}^r \langle\!\langle t_i\rangle\!\rangle;$$
   - if f is an r-ary function symbol in σ^{HF} and $t_1, \ldots, t_r$ are terms, then
   $$\langle\!\langle f(t_1, \ldots, t_r)\rangle\!\rangle = \big\{[\![f(t_1, \ldots, t_r)]\!]\big\} \cup \textstyle\bigcup_{i=1}^r \langle\!\langle t_i\rangle\!\rangle;$$
   - for logical connectives,
   $$\langle\!\langle \neg\phi\rangle\!\rangle = \langle\!\langle \phi\rangle\!\rangle, \quad \langle\!\langle \phi \wedge \psi\rangle\!\rangle = \langle\!\langle \phi \vee \psi\rangle\!\rangle = \langle\!\langle \phi\rangle\!\rangle \cup \langle\!\langle \psi\rangle\!\rangle, \quad \langle\!\langle t_1 = t_2\rangle\!\rangle = \langle\!\langle t_1\rangle\!\rangle \cup \langle\!\langle t_2\rangle\!\rangle;$$

∘ for comprehension terms,[1]

$$\begin{aligned}&\langle\!\langle \{s(v) : v \in t : \phi(v)\} \rangle\!\rangle \\ &\qquad = \{[\![\{s(v) : v \in t : \phi(v)\}]\!]\} \cup \langle\!\langle t \rangle\!\rangle \cup \textstyle\bigcup_{x \in [\![t]\!]} \left(\langle\!\langle \phi(x) \rangle\!\rangle \cup \langle\!\langle s(x) \rangle\!\rangle\right).\end{aligned}$$

2. The set $\mathrm{Active}(\Pi, \mathcal{A}) \subseteq \mathrm{HF}(A)$ of *active objects of* Π *on* $\mathcal{A}$ is defined as:

$$\begin{cases} \bigcup_{i<\omega} \langle\!\langle \Pi_{\text{step}}(x_i) \rangle\!\rangle \cup \langle\!\langle \Pi_{\text{halt}}(x_i) \rangle\!\rangle & \text{if } \Pi(\mathcal{A}) = \bot, \\ \langle\!\langle \Pi_{\text{halt}}(x_t) \rangle\!\rangle \cup \langle\!\langle \Pi_{\text{out}}(x_t) \rangle\!\rangle \cup \bigcup_{i=0}^{t-1} \left(\langle\!\langle \Pi_{\text{step}}(x_i) \rangle\!\rangle \cup \langle\!\langle \Pi_{\text{halt}}(x_i) \rangle\!\rangle\right) & \text{otherwise,} \end{cases}$$

where $x_0 = \emptyset$, $x_{i+1} = [\![\Pi_{\text{step}}(x_i)]\!]$ and t is the least nonnegative integer such that $[\![\Pi_{\text{halt}}(x_t)]\!] = \mathsf{True}$.

3. For a function $f(n)$, we denote by $\mathrm{BGS}^{(-)}(f(n))$ the class of $\mathrm{BGS}^{(-)}$ programs Π such that $\Pi(\mathcal{A}) \neq \bot$ and $|\mathrm{Active}(\Pi, \mathcal{A})| \leqslant f(|A|)$ for all finite structures $\mathcal{A}$.

4. Classes CPT and CPT+C are defined by

$$\mathrm{CPT} = \mathrm{BGS}^{-}(n^{O(1)}), \qquad \mathrm{CPT{+}C} = \mathrm{BGS}(n^{O(1)}).$$

We also denote by CPT(+C) the "complexity class" consisting of classes ("languages") of finite structures recognized by a Boolean program $\Pi \in$ CPT(+C). That is, for a class $\mathscr{C}$ of finite structures, we write $\mathscr{C} \in$ CPT(+C) if and only if $\mathscr{C} = \{\mathcal{A} : \Pi(\mathcal{A}) = \mathsf{True}\}$ for some CPT(+C) program Π.

3 Brief Survey of Results on CPT and CPT+C

For background we give a brief and partial survey of results on CPT and CPT+C. Theorems 1, 2, 3 are from the original paper [3] of Blass, Gurevich and Shelah.

Theorem 1. CPT+C $\subseteq$ P.

The idea behind the proof is that a CPT+C program Π can be simulated by a polynomial-time dynamic programming algorithm. The "subproblems" in the dynamic program correspond to terms and formulas occurring in the evaluation of Π, together with assignments of free variables to h.f. objects. The key observation is that, while these terms and formulas like $\Pi_{\text{step}}(\Pi_{\text{step}}(\dots \Pi_{\text{step}}(v) \dots))$ may grow polynomially long, there can only be polynomial many such terms and formulas and, moreover, the number of free variables in any subformula is bounded by a constant (the variable rank of Π).

---

[1] There is a reasonable alternative:

$$\begin{aligned}&\langle\!\langle \{s(v) : v \in t : \phi(v)\} \rangle\!\rangle \\ &\qquad = \{[\![\{s(v) : v \in t : \phi(v)\}]\!]\} \cup \langle\!\langle t \rangle\!\rangle \cup \textstyle\bigcup_{x \in [\![t]\!]} \langle\!\langle \phi(x) \rangle\!\rangle \cup \bigcup_{x \in [\![t]\!] \,:\, [\![\phi(x)]\!] = \mathsf{True}} \langle\!\langle s(x) \rangle\!\rangle.\end{aligned}$$

For the purposes of this paper, either definition is fine (i.e., all results hold just the same).

Theorem 2. CPT = CPT+C = P *on structures with a built-in linear order.*

Theorem 2 uses the fact that least-fixed-point logic LFP is a subclass of CPT (this is shown in [3]) and that LFP = P on structures with a built-in linear order [10,12].

Theorem 3. *The class* PARITY *of finite sets of even cardinality is not definable in* CPT.

Theorem 3 moreover shows that CPT $\neq$ CPT+C, since PARITY is clearly definable in CPT+C. In a significantly strengthening of Theorem 3, Shelah [11] proved that CPT has a "zero-one law" (see [1] for an alternative exposition of this result):

Theorem 4. *For every relational signature σ and every* CPT*-definable property $\mathcal{P}$ of σ-structures, the probability that a uniform random σ-structure of cardinality n has property $\mathcal{P}$ tends (as a function of n) to* 0 *or* 1.

Although it is open whether CPT+C = P, a number of "choiceless" algorithms have been devised which solve some particular P problems on unordered structures in new and surprising ways.

- The paper [4] contains a CPT+C algorithm for Bipartite Matching among other problems.
- In [2] it is explained how "choicelessly" to implement the Csanky algorithm for computing the determinant of an *unordered matrix*, that is, a function $I \times I \to F$ where I is a finite (unordered) set and F is a finite field.
- The paper [7] presents a CPT algorithm that solves a tractable special case of Graph Isomorphism due to Cai, Fürer and Immerman [5]. (The Cai-Fürer-Immerman problem was used to show that LFP+C $\neq$ P, that is, first-order logic with least-fixed-point and counting operators does not capture P.)

Among polynomial-time problems for which no CPT+C algorithm is known is Perfect Matching (see [4]).

4 The Role of Symmetry

This section introduces a tool for showing that certain h.f. objects in certain structures (with rich automorphism groups) cannot be activated by any CPT+C program.

Definition 7. Let G be a group acting faithfully on a finite set A of cardinality n (a special case is $\mathrm{Aut}(\mathcal{A})$ acting on the universe of a structure $\mathcal{A}$). Recall that the action of G extends to $\mathrm{HF}(A)$ via $gx = \{gy : y \in x\}$ for sets $x \in \mathrm{HF}(A) \setminus A$.

1. For $x \in \mathrm{HF}(A)$, the *stabilizer* of x is the subgroup of G defined by $\mathrm{Stab}(x) = \{g \in G : gx = x\}$. If x is a set, the *pointwise stabilizer* of x is the subgroup of G defined by $\mathrm{Stab}^\bullet(x) = \{g \in G : gy = y \text{ for all } y \in x\}$.

2. A set of atoms $B \subseteq A$ is a *support* for a subgroup $H \subseteq G$ if $\mathrm{Stab}^\bullet(B) \subseteq H$. If H has a support of size $\leqslant k$, then H is said to be *k-supported.*
3. For all k and r, the set of *(k,r)-constructible subgroups* of G is the minimal family of subgroups of G such that
   - every k-supported subgroup is (k,r)-constructible, and
   - if $H_1,\dots,H_r$ are (k,r)-constructible, $H_1 \cap \dots \cap H_r \subseteq H$ and $[G:H] \leqslant n^k$, then H is (k,r)-constructible.[2]
4. G has the *(k,r)-support property* if every (k,r)-constructible subgroup is k-supported.

We extend the notion of (k,r)-constructibility from subgroups of G to elements of $\mathrm{HF}(A)$.

5. A h.f. object $x \in \mathrm{HF}(A)$ is *(k,r)-constructible* if $\mathrm{Supp}(y)$ is a (k,r)-constructible subgroup of G for all $y \in \mathrm{TC}(x)$.

Note that (k,r)-constructibility is a transitive property: if a h.f. set x is (k,r)-constructible, then so are all $y \in x$. Whenever we speak about "(k,r)-constructible" elements of $\mathrm{HF}(A)$ in the context of a structure $\mathcal{A}$ without mentioning G, let it be understood that G is the automorphism group $\mathrm{Aut}(\mathcal{A})$.

The following lemma gives a condition equivalent to the (k,r)-support property.

Lemma 1. *G has the (k,r)-support property if, and only if, every kr-supported subgroup with index $\leqslant n^k$ is k-supported.*

Proof. ($\Longrightarrow$) Suppose G has the (k,r)-support property and H is kr-supported and $[G:H] \leqslant n^k$. Let $B \subseteq A$ be a support for H of size $|B| \leqslant kr$. Fix an arbitrary partition $B = B_1 \cup \dots \cup B_r$ into sets of size $|B_i| \leqslant k$. For $i \in \{1,\dots,r\}$, let $H_i = \mathrm{Stab}^\bullet(B_i)$ and note that H_i is (k,r)-constructible (since every k-supported subgroup is (k,r)-constructible). Since $H_1 \cap \dots \cap H_r = \mathrm{Stab}^\bullet(B) \subseteq H$ and $[G:H] \leqslant n^k$, it follows that H is (k,r)-constructible.

($\Longleftarrow$) Suppose that every kr-supported subgroup with index $\leqslant n^k$ is k-supported. To show that every (k,r)-constructible subgroup is k-supported, assume $H_1,\dots,H_r$ are k-supported and $H_1 \cap \dots \cap H_r \subseteq H$ and $[G:H] \leqslant n^k$. It suffices to show that H is k-supported. But note that H is kr-supported (the union of the supports for $H_1,\dots,H_r$ of size $\leqslant k$ is a support for H). So H is k-supported by assumption.

The following proposition links these concepts to CPT+C.

Proposition 1. *Suppose Π is a non-Boolean program in* $\mathrm{BGS}(n^k)$ *with variable rank $\leqslant r$. Then $\Pi(\mathcal{C})$ is (k,r)-constructible for every finite structure $\mathcal{A}$.*

The proof follows after two lemma. The first lemma states that the semantics of terms and formulas respects automorphisms of $\mathcal{A}$ in the expected way.

[2] Recall that the *index of H in G* is defined by $[G:H] = |G|/|H|$.

Lemma 2. *Let $\gamma(v_1, \dots, v_\ell)$ be any term or formula of BGS logic. For every structure $\mathcal{A}$, automorphism $\alpha \in \mathrm{Aut}(\mathcal{A})$ and elements $x_1, \dots, x_\ell \in \mathrm{HF}(A)$,*

$$\llbracket\gamma(\alpha x_1, \dots, \alpha x_\ell)\rrbracket = \alpha\llbracket\gamma(x_1, \dots, x_\ell)\rrbracket,$$
$$\langle\!\langle\gamma(\alpha x_1, \dots, \alpha x_\ell)\rangle\!\rangle = \alpha\langle\!\langle\gamma(x_1, \dots, x_\ell)\rangle\!\rangle.$$

In particular, $\mathrm{Stab}(\llbracket\gamma\rrbracket) = \mathrm{Stab}(\langle\!\langle\gamma\rangle\!\rangle) = \mathrm{Aut}(\mathcal{A})$ *for every ground term or formula γ.*

The proof (omitted) is a straightforward induction on terms and formulas.

Lemma 3. *Suppose $t(v_1, \dots, v_\ell)$ is a term with variable rank $\leqslant r$ and $x_1, \dots, x_\ell$ are (k, r)-constructible elements of* $\mathrm{HF}(A)$ *such that $|\{\alpha y : y \in \langle\!\langle t(x_1, \dots, x_\ell)\rangle\!\rangle, \alpha \in \mathrm{Aut}(\mathcal{A})\}| \leqslant n^k$. Then $\llbracket t(x_1, \dots, x_\ell)\rrbracket$ is (k, r)-constructible.*

Proof. The proof is by induction on terms. The bases cases are when t is a constant symbol or a variable; both cases are trivial. For the induction step, we consider the various types of term constructs (see Definition 5(1)), namely when t is:

(i) $f(t_1(\bar{v}), \dots, t_m(\bar{v}))$ where f is an m-ary function symbol f in the signature of $\mathcal{A}$,
(ii) $\mathsf{Pair}(t_1(\bar{v}), t_2(\bar{v}))$,
(iii) $\mathsf{TheUnique}(t_1(\bar{v}))$,
(iv) $\mathsf{Union}(t_1(\bar{v}))$, or
(v) $\{s(\bar{v}, w) : w \in t_1(\bar{v}) : \varphi(\bar{v}, w)\}$ (i.e., a comprehension term with subterms $t_1(\bar{v})$ and $s(\bar{v}, w)$ and subformula $\varphi(\bar{v}, w)$).

That is, in each case we assume that the lemma holds for subterms $t_1, t_2, \dots$ (as well as s in case (v)) and prove that $\llbracket t(\bar{x})\rrbracket$ is (k, r)-constructible. For this, it is sufficient to show: first, that every element of $\llbracket t(\bar{x})\rrbracket$ is (k, r)-constructible; and second, that $\mathrm{Stab}(\llbracket t(\bar{x})\rrbracket)$ is a (k, r)-constructible subgroup of $\mathrm{Aut}(\mathcal{A})$ (using Lemma 2).

As for the first claim that every element of $\llbracket t(\bar{x})\rrbracket$ is (k, r)-constructible, we consider cases (i)–(v) separately. Note that every subterm t_i of t has variable rank $\leqslant r$ and satisfies $|\{\alpha y : y \in \langle\!\langle t_i(\bar{x})\rangle\!\rangle, \alpha \in \mathrm{Aut}(\mathcal{A})\}| \leqslant n^k$ since $\langle\!\langle t_i(\bar{x})\rangle\!\rangle \subseteq \langle\!\langle t(\bar{x})\rangle\!\rangle$; therefore, $\llbracket t_i(\bar{x})\rrbracket$ is (k, r)-constructible by the induction hypothesis.

- Case (i): $\llbracket f(t_1(\bar{x}), \dots, t_m(\bar{x}))\rrbracket$ is either an atom (if $\llbracket t_1(\bar{x})\rrbracket, \dots, \llbracket t_m(\bar{x})\rrbracket$ are all atoms) or $\emptyset$ (otherwise). In either case, $\llbracket f(t_1(\bar{x}), \dots, t_m(\bar{x}))\rrbracket$ is (k, r)-constructible.
- Case (ii): By the induction hypothesis, $\llbracket t_1(\bar{x})\rrbracket$ and $\llbracket t_2(\bar{x})\rrbracket$ are both (k, r)-constructible.
- Case (iii): If $\llbracket t_1(\bar{x})\rrbracket$ is not a singleton, then $\llbracket\mathsf{TheUnique}(t_1(\bar{x}))\rrbracket = \emptyset$ and hence is (k, r)-constructible. So assume $\llbracket t_1(\bar{x})\rrbracket$ is a singleton $\{y\}$. By the induction hypothesis, $\llbracket t_1(\bar{x})\rrbracket$ is (k, r)-constructible. By transitivity of (k, r)-constructibility, y is (k, r)-constructible.

- Case (iv): By the induction hypothesis, $[\![t_1(\bar{x})]\!]$ is (k,r)-constructible. By transitivity of (k,r)-constructibility, all elements of $\bigcup[\![t_1(\bar{x})]\!]$ are (k,r)-constructible.
- Case (v): Suppose $t(\bar{v})$ is a comprehension term $\{t_2(\bar{v},w) : w \in t_1(\bar{v}) : \varphi(\bar{v},w)\}$. Recall that

$$[\![t(\bar{x})]\!] = \{[\![s(\bar{x},y)]\!] : y \in [\![t_1(\bar{x})]\!] \text{ such that } [\![\varphi(\bar{x},y)]\!] = \mathsf{True}\}.$$

By the induction hypothesis, $[\![t_1(\bar{x})]\!]$ is (k,r)-constructible. Therefore, every $y \in [\![s(\bar{x})]\!]$ is (k,r)-constructible (by transitivity of (k,r)-constructibility); it follows that $[\![s(\bar{x},y)]\!]$ is (k,r)-constructible (by the induction hypothesis on s, noting that s has variable rank $\leqslant r$ and $\langle\!\langle s(\bar{x},y)\rangle\!\rangle \subseteq \langle\!\langle t(\bar{x})\rangle\!\rangle$ so $|\{\alpha z : z \in \langle\!\langle s(\bar{x},y)\rangle\!\rangle, \alpha \in \mathrm{Aut}(\mathcal{A})\}| \leqslant n^k$).

To finish the proof, we prove that $\mathrm{Stab}([\![t(x_1,\dots,x_\ell)]\!])$ is a (k,r)-constructible subgroup of $\mathrm{Aut}(\mathcal{A})$ (in all cases (i)–(v)). Because t has variable rank $\leqslant r$, at most r of the variables $v_1,\dots,v_\ell$ occur free in t (this is obvious if $\ell \leqslant r$, but we allow $\ell > r$). Let $j_1,\dots,j_r \in \{1,\dots,\ell\}$ be such that $v_{j_1},\dots,v_{j_r}$ are the only variables which occur free in t. Let $H_i = \mathrm{Stab}(x_{j_i})$ for $i = 1,\dots,r$. Lemma 2 implies that

$$H_1 \cap \dots \cap H_r \subseteq \mathrm{Stab}([\![t(\bar{x})]\!]),$$

that is, every automorphism of $\mathrm{Aut}(\mathcal{A})$ which fixes each of $x_{j_1},\dots,x_{j_r}$ also fixes $[\![t(\bar{x})]\!]$. Since $H_1,\dots,H_r$ are (k,r)-constructible subgroups of $\mathrm{Aut}(\mathcal{A})$ (by the assumption that $x_{j_1},\dots,x_{j_r}$ are (k,r)-constructible elements of $\mathrm{HF}(\mathcal{A})$) and their intersection is contained in $\mathrm{Stab}([\![t(\bar{x})]\!])$, it suffices to show that $[\mathrm{Aut}(\mathcal{A}) : \mathrm{Stab}([\![t(\bar{x})]\!])] \leqslant n^k$. This follows from our assumption that $|\{\alpha y : y \in \langle\!\langle t(\bar{x})\rangle\!\rangle, \alpha \in \mathrm{Aut}(\mathcal{A})\}| \leqslant n^k$, as we have

$$\begin{aligned}
[\mathrm{Aut}(\mathcal{A}) : \mathrm{Stab}([\![t(\bar{x})]\!])] &= |\{\alpha[\![t(\bar{x})]\!] : \alpha \in \mathrm{Aut}(\mathcal{A})\}| \\
&\leqslant |\{\alpha y : y \in \langle\!\langle t(\bar{x})\rangle\!\rangle, \alpha \in \mathrm{Aut}(\mathcal{A})\}| \\
&\quad \text{(since } [\![t(\bar{x})]\!] \in \langle\!\langle t(\bar{x})\rangle\!\rangle\text{)} \\
&\leqslant n^k \quad \text{(by assumption).}
\end{aligned}$$

Finally, we prove Proposition 1 using Lemma 3.

Proof (Proof of Proposition 1). Let Π be a non-Boolean program $\mathrm{BGS}(n^k)$ with variable rank $\leqslant r$. For any finite structure $\mathcal{A}$, note that

$$\Pi(\mathcal{A}) = [\![\Pi_{\mathrm{out}}(\underbrace{\Pi_{\mathrm{step}}(\dots\Pi_{\mathrm{step}}(\emptyset)\dots)}_{m \text{ times}})]\!]$$

for some finite m. Let t denote this term $\Pi_{\mathrm{out}}(\Pi_{\mathrm{step}}(\dots\Pi_{\mathrm{step}}(\emptyset)\dots))$. Whatever m happens to be, t is a ground term with variable rank $\leqslant r$. By Lemma 2, $\langle\!\langle t\rangle\!\rangle$ is fixed by all automorphisms of $\mathcal{A}$ (i.e., $\mathrm{Stab}(\langle\!\langle t\rangle\!\rangle) = \mathrm{Aut}(\mathcal{A})$). Thus,

$$\{\alpha y : y \in \langle\!\langle t\rangle\!\rangle, \alpha \in \mathrm{Aut}(\mathcal{A})\} = \{y : y \in \langle\!\langle t\rangle\!\rangle\} \subseteq \mathrm{Active}(\Pi,\mathcal{A}).$$

Since $|\mathrm{Active}(\Pi,\mathcal{A})| \leqslant n^k$ (by definition of $\mathrm{BGS}(n^k)$), we have $|\{\alpha y : y \in \langle\!\langle t\rangle\!\rangle, \alpha \in \mathrm{Aut}(\mathcal{A})\}| \leqslant n^k$. Therefore, $\Pi(\mathcal{A})$ is (k,r)-constructible by Lemma 3.

In the next two sections, we will use Proposition 1 to prove that CPT+C programs cannot activate certain h.f. objects over "naked" sets and vector spaces.

5 PARITY $\notin$ CPT

We denote by $[n]$ the "naked" set $\{1,\ldots,n\}$ viewed as structure in the empty signature. Let PARITY denote the class of naked sets with even cardinality (i.e., the "language" of empty sets). Earlier we stated the result of Blass et al. from [3] that PARITY $\notin$ CPT (Theorem 3). A key step in the proof is the following so-called Support Theorem (Theorem 24 of [3]).

Theorem 5. *For every $\Pi \in$ CPT, there is a constant c such that for all sufficiently large n, every object in $\mathrm{Active}(\Pi,[n])$ has a support of cardinality $\leqslant c$.*

The original proof of Theorem 5 in [3] involves a fairly intricate combinatorial argument. We give an alternative and simpler proof using the support property defined in the previous section. Theorem 5 follows directly from Proposition 1 and the following proposition.

Proposition 2. *For $n > 2kr$, the symmetric group S_n has the (k,r)-support property.*

We remark that S_{kr+1} fails to have the (k,r)-support property, as the alternating subgroup is (k,r)-constructible but not k-supported (its smallest support has size kr). The following lemma and corollary from [3] are also used in the original proof of Theorem 5. We include proofs for completeness.

Lemma 4. *Let $H \subseteq S_n$ and suppose $U, V \subset [n]$ such that $\mathrm{Stab}^\bullet(U), \mathrm{Stab}^\bullet(V) \subseteq H$ and $U \cup V \neq [n]$. Then $\mathrm{Stab}^\bullet(U \cap V) \subseteq H$.*

Proof. $\mathrm{Stab}^\bullet(U\cap V)$ is generated by transpositions $(i\ j)$ where $i,j \in [n]\setminus(U\cap V)$. Therefore, it suffices to show that $(i\ j) \in H$ for all $i,j \in [n] \setminus (U \cap V)$. By assumption, there exists $k \in [n]\setminus(U\cup V)$. Since $(i\ j) = (i\ k)(j\ k)(i\ k)$, it suffices to show that $(i\ k) \in H$ for all $i \in [n]\setminus(U\cap V)$. We consider two cases depending whether $i \notin U$ or $i \notin V$: if $i \notin U$, then $(i\ k) \in \mathrm{Stab}^\bullet(U)$ and hence $(i\ k) \in H$ as $\mathrm{Stab}^\bullet(U) \leqslant H$; if $i \notin V$, then $(i\ k) \in \mathrm{Stab}^\bullet(V)$ and hence $(i\ k) \in H$ as $\mathrm{Stab}^\bullet(V) \leqslant H$.

Corollary 1. *If $H \subseteq S_n$ has a support of size $< n/2$, then H has a unique minimal support of size $< n/2$.* □

We now give our proof of Proposition 2 (which bypasses the lengthy combinatorial argument in [3]).

Proof (Proof of Proposition 2). Suppose $H_1,\ldots,H_r$ are k-supported subgroups of S_n. Let H be another subgroup of S_n such that $H_1 \cap \cdots \cap H_r \subseteq H$ and $[S_n : H] \leqslant n^k$. By Lemma 1, it suffices to show that H is also k-supported. For each $i \in [r]$, fix $U_i \subset [n]$ such that $|U_i| \leqslant k$ and $\mathrm{Stab}^\bullet(U_i) \subseteq H_i$. Let

$U = U_1 \cup \cdots \cup U_r$. Note that U is a support for H, as $\mathrm{Stab}^\bullet(U) = \mathrm{Stab}^\bullet(U_1) \cap \cdots \cap \mathrm{Stab}^\bullet(U_r) \subseteq H_1 \cap \cdots \cap H_r \subseteq H$. Also note that $|U| \leqslant rk < n/2$. So by Corollary 1, H has a unique minimal support V of size $< n/2$.

We claim that $H \subseteq \mathrm{Stab}(V)$. For contradiction, assume otherwise. Then there exists $h \in H$ such that $hV \neq V$. Note that hV is a support for $hHh^{-1} = H$. Since $V \cup hV \neq [n]$ (as $|V \cup hV| \leqslant 2|V| < n$), the intersection $V \cap hV$ is a support for H by Lemma 4. But $V \cap hV \subset V$, which contradicts the minimality of V. Therefore, $H \subseteq \mathrm{Stab}(V)$ as claimed. It follows that $[S_n : H] \geqslant [S_n : \mathrm{Stab}(V)] = \binom{n}{|V|}$. Since $[S_n : H] \leqslant n^k$, we conclude that $|V| \leqslant k$. Therefore, H is k-supported.

6 CPT+C Cannot Construct the Dual of a Finite Vector Space

Let V be a finite vector space over a fixed finite field F. We view V as a structure with binary operation $+$ and unary operations for scalar multiplication by each element of F. Let $\mathcal{H}(V)$ denote the set of hyperplanes in V. Note that $\mathcal{H}(V)$ is an element of $\mathrm{HF}(V)$ (in particular, $\mathcal{H}(V)$ is a set of subsets of V).

The task of computing $\mathcal{H}(V)$ given V is a polynomial-time *function problem* (as opposed to *decision problem*) in the usual sense of complexity theory ($\mathcal{H}(V)$ has a polynomial-size description as a hereditary finite object, i.e., $|\mathrm{TC}(\mathcal{H}(V))| = O(\mathrm{poly}(|V|))$. Moreover, $\mathcal{H}(V)$ is an invariant of V (i.e., not depending on any extrinsic linear order on V). It is thus reasonable to ask whether any CPT+C program computes the operation $V \longmapsto \mathcal{H}(V)$.

We remark the results of this section hold just the same for the operation $V \longmapsto V^*$ of computing from V the *dual space* V^* of linear functions $V \longrightarrow F$ (suitably represented as an element of $\mathrm{HF}(V \cup F)$).

Theorem 6. *No program in* CPT+C *program computes the operation* $V \longmapsto \mathcal{H}(V)$ *over finite vector spaces over a fixed finite field* F.

Noting that a hyperplane in an n-dimensional vector space has smallest support size $n-1$, Theorem 6 follows from the following vector-space analogue of Proposition 2.

Proposition 3. *If* V *is a finite vector space of dimension* $> r^2k^2$*, then the group* $\mathrm{GL}(V)$ *of linear automorphisms of* V *has the* (k,r)*-support property.*

To prove Theorem 6 from Proposition 3, note that if $\Pi \in$ CPT+C, then $\Pi \in \mathrm{BGS}(n^k)$ for some k. Let r be the variable rank Π. Consider a finite vector space V on dimension $> r^2k^2$. By Proposition 1, $\Pi(V)$ (the output of Π on V) is (k,r)-constructible. By Proposition 3, $\mathrm{GL}(V)$ ($= \mathrm{Aut}(V)$) has the (k,r)-support property. Therefore, $\Pi(V)$ is k-supported. Since $\mathcal{H}(V)$ is not k-supported for any $k < \dim(V)$, we conclude that $\Pi(V) \neq \mathcal{H}(V)$.

The proof of Proposition 3 proceeds along similar lines as the proof of Proposition 2. We have the following vector-space analogues of Lemma 4 and Corollary 1.

Lemma 5. *If subspaces $W, W' \subseteq V$ both support a subgroup $H \subseteq \mathrm{GL}(V)$ and if $W + W' \neq V$, then the intersection $W \cap W'$ supports H.*

Proof. Let P, P' and Q denote the pointwise stabilizers of W, W' and $W \cap W'$, respectively. It suffices to show that Q is the subgroup of $\mathrm{GL}(V)$ generated by $P \cup P'$. This is a simple exercise in linear algebra. Choose a basis $v_1, ..., v_n$ for V such that for some $1 \leqslant i \leqslant j \leqslant j' < n = \dim(V)$,

- $v_1, ..., v_j$ span W,
- $v_i, ..., v_{j'}$ span W',
- $v_i, ..., v_j$ span $W \cap W'$.

We now identify particular sets of generators for P, P' and Q. For all $r, s \in \{1, \ldots, n\}$ and $\lambda \in F^{\times}$, define $n \times n$ matrix $\sigma_{r,s,\lambda}$ by

$$\sigma_{r,s,\lambda} = \begin{cases} \begin{pmatrix} I_{r-1} & & & & \\ & 1 & & \lambda & \\ & & I_{s-r-1} & & \\ & 0 & & 1 & \\ & & & & I_{n-s} \end{pmatrix} & \text{if } r < s, \\ \begin{pmatrix} I_{r-1} & & \\ & \lambda & \\ & & I_{n-r} \end{pmatrix} & \text{if } r = s, \\ \begin{pmatrix} I_{s-1} & & & & \\ & 1 & & 0 & \\ & & I_{r-s-1} & & \\ & \lambda & & 1 & \\ & & & & I_{n-r} \end{pmatrix} & \text{if } r > s. \end{cases}$$

For all $r, s \in \{1, \ldots, n\}$ with $r < s$, define $n \times n$ matrix $\tau_{r,s}$ by

$$\tau_{r,s} = \begin{pmatrix} I_{r-1} & & & & \\ & 0 & & 1 & \\ & & I_{s-r-1} & & \\ & 1 & & 0 & \\ & & & & I_{n-s} \end{pmatrix}.$$

Linear transformations $\sigma_{r,s,\lambda}$ and $\tau_{r,s}$ are the familiar "row reduction" generators of $\mathrm{GL}(V)$ with respect to the basis $v_1, ..., v_n$. Note that P (respectively: P', Q) is generated by the set of all $\sigma_{r,s,\lambda}$ such that $r \notin \{1, ..., j\}$ (respectively: $r \notin \{i, ..., j'\}$, $r \notin \{i, ..., j\}$), together with all $\tau_{r,s}$ such that $r, s \notin \{1, \ldots, j\}$ (respectively: $r, s \notin \{i, ..., j'\}$, $r, s \notin \{i, ..., j\}$) .

The only generators of Q which are not also generators of P or P' are those of the form $\tau_{r,s}$ where $r \in \{1, ..., i-1\}$ and $s \in \{j+1, ..., j'\}$. Note that $\tau_{r,s} = \tau_{r,n}\tau_{s,n}\tau_{r,n}$ and $\tau_{r,n} \in P'$ and $\tau_{s,n} \in P$. Therefore, $\tau_{r,s}$ is in the subgroup of $\mathrm{GL}(V)$ generated by $P \cup P'$. We conclude that Q is the subgroup of $\mathrm{GL}(V)$ generated by $P \cup P'$.

Corollary 2. *If a subgroup $H \subseteq \mathrm{GL}(V)$ is supported by a subspace of dimension $< n/2$, then H is supported by a unique minimal subspace of dimension $< n/2$.* □

With this corollary, we are ready to prove Proposition 3.

Proof (Proof of Proposition 3). Let $H_1, \ldots, H_r$ be k-supported subgroups of $\mathrm{GL}(V)$. Let H be another subgroup of $\mathrm{GL}(V)$ such that $H \supseteq H_1 \cap \cdots \cap H_r$ and $[\mathrm{GL}(V) : H] \leqslant |V|^k = q^{nk}$ where q is the size of the field F. By Lemma 1, it suffices show that H is also k-supported. For each $i \in [r]$, fix $U_i \subset [n]$ such that $|U_i| \leqslant k$ and $\mathrm{Stab}^{\bullet}(U_i) \subseteq H_i$. Let $U = U_1 \cup \cdots \cup U_r$. Note that U is a support for H, as $\mathrm{Stab}^{\bullet}(U) = \mathrm{Stab}^{\bullet}(U_1) \cap \cdots \cap \mathrm{Stab}^{\bullet}(U_r) \subseteq H_1 \cap \cdots \cap H_r \subseteq H$. Also note that $|U| \leqslant rk < n/2$. So by Corollary 1, H is supported by a unique minimal subspace W of dimension $\leqslant rk$.

We claim that $H \subseteq \mathrm{Stab}(W)$. For contradiction, assume otherwise. Then there exists $h \in H$ such that $hW \neq W$. Note that hW is a support for $hHh^{-1} = H$. Since $W + hW \neq V$ (as $\dim(W + hW) \leqslant 2\dim(W) < n$), the intersection $W \cap hW$ is a support for H by Lemma 4. But $W \cap hW \subset W$, which contradicts the minimality of W. Therefore, $H \subseteq \mathrm{Stab}(W)$ as claimed. It follows that $[\mathrm{GL}(V) : H] \geqslant [\mathrm{GL}(V) : \mathrm{Stab}(W)] = \#\{\dim(W)\text{-dimensional subspaces of } V\}$. For all $d \leqslant rk\ (= \sqrt{n})$, we have

$$\begin{aligned}
\#\{d\text{-dimensional subspaces of } V\} = \prod_{i=0}^{d-1} \frac{q^{n-i} - 1}{q^{i+1} - 1} &\geqslant \prod_{i=0}^{d-1} q^{n-2i-2} \\
&= q^{dn - 2\binom{d-1}{2} - 2} \\
&\geqslant q^{dn - (\sqrt{n}-1)(\sqrt{n}-2) - 2} \\
&= q^{(d-1)n + 3\sqrt{n} - 4} \\
&> |V|^{d-1}.
\end{aligned}$$

Since $[\mathrm{GL}(V) : H] \leqslant |V|^k$, it follows that $\dim(W) \leqslant k$. Because every basis for W is a support for H, it follows that H is k-supported.

Acknowledgements. My thanks to Swastik Kopparty and an anonymous referee for their helpful comments.

References

1. Blass, A., Gurevich, Y.: Strong extension axioms and Shelah's zero-one law for choiceless polynomial time. Journal of Symbolic Logic 68(1), 65–131 (2003)
2. Blass, A., Gurevich, Y.: A quick update on the open problems in Blass-Gurevich-Shelah's article. On polynomial time computations over unordered structures (Decembmer 2005), `http://research.microsoft.com/en-us/um/people/gurevich/Opera/150a.pdf`
3. Blass, A., Gurevich, Y., Shelah, S.: Choiceless polynomial time. Annals of Pure and Applied Logic 100(1–3), 141–187 (1999)

4. Blass, A., Gurevich, Y., Shelah, S.: On polynomial time computation over unordered structures. Journal of Symbolic Logic 67(3), 1093–1125 (2002)
5. Cai, J.-Y., Fürer, M., Immerman, N.: An optimal lower bound on the number of variables for graph identification. Combinatorica 12(4), 389–410 (1992)
6. Chandra, A., Harel, D.: Structure and complexity of relational queries. Journal of Computer and System Sciences 25, 99–128 (1982)
7. Dawar, A., Richerby, D., Rossman, B.: Choiceless polynomial time, counting and the Cai-Fürer-Immerman graphs. Annals of Pure and Applied Logic 152, 31–50 (2008)
8. Fagin, R.: Generalized first-order spectra and polynomial-time recognizable sets. In: Karp, R.M. (ed.) Complexity of Computation. SIAM-AMS Proceedings, vol. 7, pp. 43–73 (1974)
9. Gurevich, Y.: Toward logic tailored for computational complexity. In: Richter, M.M., et al. (eds.) Computation and Proof Theory. Springer Lecture Notes in Mathematics, pp. 175–216. Springer, Heidelberg (1984)
10. Immerman, N.: Relational queries computable in polynomial time. Information and Control 68(1–3), 86–104 (1986)
11. Shelah, S.: Choiceless polynominal time logic: Inability to express. In: Clote, P.G., Schwichtenberg, H. (eds.) CSL 2000. LNCS, vol. 1862, pp. 72–125. Springer, Heidelberg (2000)
12. Vardi, M.Y.: The complexity of relational query languages. In: Proc. 14th ACM Symp. on Theory of Computing, pp. 137–146 (1982)

Hereditary Zero-One Laws for Graphs

Saharon Shelah and Mor Doron*

Department of Mathematics, The Hebrew University of Jerusalem, Israel
{shelah,mord}@math.huji.ac.il

This article is dedicated to Yuri Gurevich, on the occasion of his seventieth birthday.

Abstract. We consider the random graph $M^n_{\bar{p}}$ on the set $[n]$, where the probability of $\{x, y\}$ being an edge is $p_{|x-y|}$, and $\bar{p} = (p_1, p_2, p_3, ...)$ is a series of probabilities. We consider the set of all $\bar{q}$ derived from $\bar{p}$ by inserting 0 probabilities into $\bar{p}$, or alternatively by decreasing some of the p_i. We say that $\bar{p}$ hereditarily satisfies the 0-1 law if the 0-1 law (for first order logic) holds in $M^n_{\bar{q}}$ for every $\bar{q}$ derived from $\bar{p}$ in the relevant way described above. We give a necessary and sufficient condition on $\bar{p}$ for it to hereditarily satisfy the 0-1 law.

Keywords: random graphs, zero-one laws.

1 Introduction

In this paper we will investigate the random graph on the set $[n] = \{1, 2, ..., n\}$ where the probability of a pair $i \neq j \in [n]$ being connected by an edge depends only on their distance $|i - j|$. Let us define:

Definition 1. *For a sequence $\bar{p} = (p_1, p_2, p_3, ...)$ where each p_i is a probability, i.e. a real in $[0, 1]$, let $M^n_{\bar{p}}$ be the random graph defined by:*

- *The set of vertices is $[n] = \{1, 2, ..., n\}$.*
- *For $i, j \leq n$, $i \neq j$, the probability of $\{i, j\}$ being an edge is $p_{|i-j|}$.*
- *All the edges are drawn independently.*

Convention 1. *Formally speaking Definition 1 defines a probability on the space of subsets of $G^n := \{G : G$ is a graph with vertex set $[n]\}$. If H is a subset of G^n we denote its probability by $Pr[M^n_{\bar{p}} \in H]$. If ϕ is a sentence in some logic we write $Pr[M^n_{\bar{p}} \models \phi]$ for the probability of $\{G \in G^n : G \models \phi\}$. Similarly if A_n is some property of graphs on the set of vertices $[n]$, then we write $Pr[A_n]$ or $Pr[A_n$ holds in $M^n_{\bar{p}}]$ for the probability of the set $\{G \in G^n : G$ has the property $A_n\}$.*

* The authors would like to thank the Israel Science Foundation for partial support of this research (Grant no. 242/03). Publication no. 953 on Saharon Shelah's list.

A. Blass, N. Dershowitz, and W. Reisig (Eds.): Gurevich Festschrift, LNCS 6300, pp. 581–614, 2010.

If $\mathfrak{L}$ is some logic, we say that $M^n_{\bar{p}}$ satisfies the 0-1 law for the logic $\mathfrak{L}$ if for each sentence $\psi \in \mathfrak{L}$ the probability that ψ holds in $M^n_{\bar{p}}$ tends to 0 or 1, as n approaches ∞. The relations between properties of $\bar{p}$ and the asymptotic behavior of $M^n_{\bar{p}}$ were investigated in [2]. It was proved there that for L, the first order logic in the vocabulary with only the adjacency relation, we have:

Theorem 2. *1. Assume $\bar{p} = (p_1, p_2, ...)$ is such that $0 \leq p_i < 1$ for all $i > 0$ and let $f_{\bar{p}}(n) := \log(\prod_{i=1}^{n}(1 - p_i))/\log(n)$. If $\lim_{n\to\infty} f_{\bar{p}}(n) = 0$ then $M^n_{\bar{p}}$ satisfies the 0-1 law for L.*
2. The demand above on $f_{\bar{p}}$ is the best possible. Formally for each $\epsilon > 0$, there exists some $\bar{p}$ with $0 \leq p_i < 1$ for all $i > 0$ such that $|f_{\bar{p}}(n)| < \epsilon$ but the 0-1 law fails for $M^n_{\bar{p}}$.

Part (1) above gives a sufficient condition on $\bar{p}$ for the 0-1 law to hold in $M^n_{\bar{p}}$, but the condition is not necessary and a full characterization of $\bar{p}$ seems to be harder. However we give below a complete characterization of $\bar{p}$ in terms of the 0-1 law in $M^n_{\bar{q}}$ for all $\bar{q}$ "dominated by $\bar{p}$", in the appropriate sense. Alternatively one may ask which of the asymptotic properties of $M^n_{\bar{p}}$ are kept under some operations on $\bar{p}$. The notion of "domination" or the "operations" are taken from examples of the failure of the 0-1 law, and specifically the construction for part (2) above. Those are given in [2] by either adding zeros to a given sequence or decreasing some of the members of a given sequence. Formally define:

Definition 3. *For a sequence $\bar{p} = (p_1, p_2, ...)$:*

1. *$Gen_1(\bar{p})$ is the set of all sequences $\bar{q} = (q_1, q_2, ...)$ obtained from $\bar{p}$ by adding zeros to $\bar{p}$. Formally $\bar{q} \in Gen_1(\bar{p})$ iff for some increasing $f : \mathbb{N} \to \mathbb{N}$ we have for all $l > 0$*
$$q_l = \begin{cases} p_i & f(i) = l \\ 0 & l \notin Im(f). \end{cases}$$
2. *$Gen_2(\bar{p}) := \{\bar{q} = (q_1, q_2, ...) : l > 0 \Rightarrow q_l \in [0, p_l]\}$.*
3. *$Gen_3(\bar{p}) := \{\bar{q} = (q_1, q_2, ...) : l > 0 \Rightarrow q_l \in \{0, p_l\}\}$.*

Definition 4. *Let $\bar{p} = (p_1, p_2, ...)$ be a sequence of probabilities and $\mathfrak{L}$ be some logic. For a sentence $\psi \in \mathfrak{L}$ denote by $Pr[M^n_{\bar{p}} \models \psi]$ the probability that ψ holds in $M^n_{\bar{p}}$.*

1. *We say that $M^n_{\bar{p}}$ satisfies the 0-1 law for $\mathfrak{L}$, if for all $\psi \in \mathfrak{L}$ the limit $\lim_{n\to\infty} Pr[M^n_{\bar{p}} \models \psi]$ exists and belongs to $\{0, 1\}$.*
2. *We say that $M^n_{\bar{p}}$ satisfies the convergence law for $\mathfrak{L}$, if for all $\psi \in \mathfrak{L}$ the limit $\lim_{n\to\infty} Pr[M^n_{\bar{p}} \models \psi]$ exists.*
3. *We say that $M^n_{\bar{p}}$ satisfies the weak convergence law for $\mathfrak{L}$, if for all $\psi \in \mathfrak{L}$, $\limsup_{n\to\infty} Pr[M^n_{\bar{p}} \models \psi] - \liminf_{n\to\infty} Pr[M^n_{\bar{p}} \models \psi] < 1$.*
4. *For $i \in \{1, 2, 3\}$ we say that $\bar{p}$ i-hereditarily satisfies the 0-1 law for $\mathfrak{L}$, if for all $\bar{q} \in Gen_i(\bar{p})$, $M^n_{\bar{q}}$ satisfies the 0-1 law for $\mathfrak{L}$.*
5. *Similarly to (4) for the convergence and weak convergence law.*

The main theorem of this paper is the following strengthening of Theorem 2:

Theorem 5. *Let $\bar{p} = (p_1, p_2, ...)$ be such that $0 \leq p_i < 1$ for all $i > 0$, and $j \in \{1, 2, 3\}$. Then $\bar{p}$ j-hereditarily satisfies the 0-1 law for L iff*

$$(*) \qquad \lim_{n \to \infty} \log(\prod_{i=1}^{n}(1 - p_i))/\log n = 0.$$

Moreover we may replace above the "0-1 law" by the "convergence law" or "weak convergence law".

Note that the 0-1 law implies the convergence law which in turn implies the weak convergence law. Hence it is enough to prove the "if" direction for the 0-1 law and the "only if" direction for the weak convergence law. Also note that the "if" direction is an immediate consequence of Theorem 2 (in the case $j = 1$ it is stated in [2] as a corollary at the end of section 3). The case $j = 1$ is proved in section 2, and the case $j \in \{2, 3\}$ is proved in section 3. In section 4 we deal with the case $U^*(\bar{p}) := \{i : p_i = 1\}$ is not empty. We give an almost full analysis of the hereditary 0-1 law in this case as well. The only case which is not fully characterized is the case $j = 1$ and $|U^*(\bar{p})| = 1$. We give some results regarding this case in section 5. The case $j = 1$ and $|U^*(\bar{p})| = 1$ and the case that the successor relation belongs to the dictionary, will be dealt with in [3]. Table 1 summarizes the results in this article regarding the j-hereditary laws.

Table 1. Summation of Results

<table>
<tr><th></th><th>$|U^*| = \infty$</th><th>$2 \leq |U^*| < \infty$</th><th>$|U^*| = 1$</th><th>$|U^*| = 0$</th></tr>
<tr><td>$j = 1$</td><td rowspan="3">The weak convergence law fails</td><td>The 0-1 law holds
$\Updownarrow$
$\{l : 0 < p_l < 1\} = \emptyset$</td><td>See section 5</td><td rowspan="3">$\lim_{n\to\infty} \frac{\log(\prod_{i=1}^{n}(1-p_i))}{\log n} = 0$
$\Updownarrow$
The 0-1 law holds
$\Updownarrow$
The convergence law holds
$\Updownarrow$
The weak convergence law holds</td></tr>
<tr><td>$j = 2$</td><td colspan="2">The 0-1 law holds
$\Updownarrow$
$|\{l : p_l > 0\}| \leq 1$</td></tr>
<tr><td>$j = 3$</td><td colspan="2">The 0-1 law holds
$\Updownarrow$
$\{l : 0 < p_l < 1\} = \emptyset$</td></tr>
</table>

Notation 1. *1. $\mathbb{N}$ is the set of natural numbers (including 0), $\mathbb{N}^*$ denotes the set $\mathbb{N} \setminus \{0\}$.*

2. n, m, r, i, j and k will denote natural numbers. l will denote a member of $\mathbb{N}^*$ (usually an index).

3. p, q and similarly p_l, q_l will denote probabilities, i.e. reals in $[0, 1]$.

4. ϵ, ζ and δ will denote positive reals.

5. $L = \{\sim\}$ is the vocabulary of graphs, i.e. $\sim$ is a binary relation symbol. All L-structures are assumed to be graphs, i.e. $\sim$ is interpreted by a symmetric irreflexive binary relation.

6. If $x \sim y$ holds in some graph G, we say that $\{x, y\}$ is an edge of G or that x and y are "connected" or "neighbors" in G.

2 Adding Zeros

In this section we prove Theorem 5 for $j = 1$. As the "if" direction is immediate from Theorem 2 it remains to prove that if $(*)$ of 5 fails then the 0-1 law for L fails for some $\bar{q} \in Gen_1(\bar{p})$. In fact we will show that it fails "badly" i.e. for some $\psi \in L$, $Pr[M^n_{\bar{q}} \models \psi]$ has both 0 and 1 as limit points. Formally:

Definition 6. *1. Let ψ be a sentence in some logic $\mathfrak{L}$, and $\bar{q} = (q_1, q_2, ...)$ be a series of probabilities. We say that ψ holds infinitely often in $M^n_{\bar{q}}$ if $\limsup_{n\to\infty} Prob[M^n_{\bar{q}} \models \psi] = 1$.*
2. We say that the 0-1 law for $\mathfrak{L}$ strongly fails in $M^n_{\bar{q}}$, if for some $\psi \in \mathfrak{L}$ both ψ and $\neg\psi$ hold infinitely often in $M^n_{\bar{q}}$.

Obviously the 0-1 law strongly fails in some $M^n_{\bar{q}}$ iff $M^n_{\bar{q}}$ does not satisfy the weak convergence law. Hence in order to prove Theorem 5 for $j = 1$ it is enough if we prove:

Lemma 7. *Let $\bar{p} = (p_1, p_2, ...)$ be such that $0 \leq p_i < 1$ for all $i > 0$, and assume that $(*)$ of 5 fails. Then for some $\bar{q} \in Gen_1(\bar{p})$ the 0-1 law for L strongly fails in $M^n_{\bar{q}}$.*

In the remainder of this section we prove Lemma 7. We do so by inductively constructing $\bar{q}$, as the limit of a series of finite sequences. Let us start with some basic definitions:

Definition 8. *1. Let $\mathfrak{P}$ be the set of all, finite or infinite, sequences of probabilities. Formally each $\bar{p} \in \mathfrak{P}$ has the form $\langle p_l : 0 < l < n_{\bar{p}}\rangle$ where each $p_l \in [0,1]$ and $n_{\bar{p}}$ is either ω (the first infinite ordinal) or a member of $\mathbb{N} \setminus \{0,1\}$. Let $\mathfrak{P}^{inf} = \{\bar{p} \in \mathfrak{P} : n_{\bar{p}} = \omega\}$, and $\mathfrak{P}^{fin} := \mathfrak{P} \setminus \mathfrak{P}^{inf}$.*
2. For $\bar{q} \in \mathfrak{P}^{fin}$ and increasing $f : [n_{\bar{q}}] \to \mathbb{N}$, define $\bar{q}^f \in \mathfrak{P}^{fin}$ by $n_{\bar{q}^f} = f(n_{\bar{q}})$, $(\bar{q}^f)_l = q_i$ if $f(i) = l$ and $(\bar{q}^f)_l = 0$ if $l \notin Im(f)$.
3. For $\bar{p} \in \mathfrak{P}^{inf}$ and $r > 0$, let $Gen^r_1(\bar{p}) := \{\bar{q} \in \mathfrak{P}^{fin} :$ for some increasing $f : [r+1] \to \mathbb{N}, (\bar{p}|_{[r]})^f = \bar{q}\}$.
4. For $\bar{p}, \bar{p}' \in \mathfrak{P}$ denote $\bar{p} \lhd \bar{p}'$ if $n_{\bar{p}} < n_{\bar{p}'}$ and for each $l < n_{\bar{p}}$, $p_l = p'_l$.
5. If $\bar{p} \in \mathfrak{P}^{fin}$ and $n > n_{\bar{p}}$, we can still consider $M^n_{\bar{p}}$ by putting $p_l = 0$ for all $l \geq n_{\bar{p}}$.

Observation 1. *1. Let $\langle \bar{p}_i : i \in \mathbb{N}\rangle$ be such that each $\bar{p}_i \in \mathfrak{P}^{fin}$, and assume that $i < j \in \mathbb{N} \Rightarrow \bar{p}_i \lhd \bar{p}_j$. Then $\bar{p} = \cup_{i\in\mathbb{N}}\bar{p}_i$ (i.e. $p_l = (p_i)_l$ for some $\bar{p}_i$ with $n_{\bar{p}_i} > l$) is well defined and $\bar{p} \in \mathfrak{P}^{inf}$.*
2. Assume further that $\langle r_i : i \in \mathbb{N}\rangle$ is non-decreasing and unbounded, and that $\bar{p}_i \in Gen^{r_i}_1(\bar{p}')$ for some fixed $\bar{p}' \in \mathfrak{P}^{inf}$, then $\cup_{i\in\mathbb{N}}\bar{p}_i \in Gen_1(\bar{p}')$.

We would like our graphs $M^n_{\bar{q}}$ to have a certain structure, namely that the number of triangles in $M^n_{\bar{q}}$ is $o(n)$ rather than say $o(n^3)$. We can impose this structure by making demands on $\bar{q}$. This is made precise by the following:

Definition 9. *A sequence $\bar{q} \in \mathfrak{P}$ is called proper (for l^*), if:*

1. *l^* and $2l^*$ are the first and second members of $\{0 < l < n_{\bar{q}} : q_l > 0\}$.*
2. *Let $l^{**} = 4l^* + 2$. If $l < n_{\bar{q}}$, $l \notin \{l^*, 2l^*\}$ and $q_l > 0$, then $l \equiv 1 \pmod{l^{**}}$.*

For $\bar{q}, \bar{q}' \in \mathfrak{P}$ we write $\bar{q} \lhd^{prop} \bar{q}'$ if $\bar{q} \lhd \bar{q}'$, and both $\bar{q}$ and $\bar{q}'$ are proper.

Observation 2. *1. If $\langle \bar{p}_i : i \in \mathbb{N} \rangle$ is such that each $\bar{p}_i \in \mathfrak{P}$, and $i < j \in \mathbb{N} \Rightarrow \bar{p}_i \lhd^{prop} \bar{p}_j$, then $\bar{p} = \cup_{i \in \mathbb{N}} \bar{p}_i$ is proper.*
2. Assume that $\bar{q} \in \mathfrak{P}$ is proper for l^* and $n \in \mathbb{N}$. Then the following event holds in $M^n_{\bar{q}}$ with probability 1:
$(*)_{\bar{q},l^*}$ *If $m_1, m_2, m_3 \in [n]$ and $\{m_1, m_2, m_3\}$ is a triangle in $M^n_{\bar{q}}$, then $\{m_1, m_2, m_3\} = \{l, l + l^*, l + 2l^*\}$ for some $l > 0$.*

We can now define the sentence ψ for which we have failure of the 0-1 law.

Definition 10. *Let k be an even natural number. Let ψ_k be the L sentence "saying": There exists $x_0, x_1, ..., x_k$ such that:*

- *$(x_0, x_1, ..., x_k)$ is without repetitions.*
- *For each even $0 \leq i < k$, $\{x_i, x_{i+1}, x_{i+2}\}$ is a triangle.*
- *The valency of x_0 and x_k is 2.*
- *For each even $0 < i < k$ the valency of x_i is 4.*
- *For each odd $0 < i < k$ the valency of x_i is 2.*

If the above holds (in a graph G) we say that $(x_0, x_1, ..., x_k)$ is a chain of triangles (in G).

Definition 11. *Let $n \in \mathbb{N}$, $k \in \mathbb{N}$ be even and $l^* \in [n]$. For $1 \leq m < n - k \cdot l^*$ a sequence $(m_0, m_1, ..., m_k)$ is called a candidate of type (n, l^*, k, m) if it is without repetitions, $m_0 = m$ and for each even $0 \leq i < k$, $\{m_i, m_{i+1}, m_{i+2}\} = \{l, l + l^*, l + 2l^*\}$ for some $l > 0$. Note that for given (n, l^*, k, m), there are at most 4 candidates of type (n, l^*, k, m) (and at most 2 if $k > 2$).*

claim 1. Let $n \in \mathbb{N}$, $k \in \mathbb{N}$ be even, and $\bar{q} \in \mathfrak{P}$ be proper for l^*. For $1 \leq m < n - k \cdot l^*$ let $E^n_{\bar{q},m}$ be the following event (on the probability space $M^n_{\bar{q}}$): "No candidate of of type (n, l^*, k, m) is a chain of triangles." Then $M^n_{\bar{q}}$ satisfies with probability 1: $M^n_{\bar{q}} \models \neg\psi_k$ iff $M^n_{\bar{q}} \models \bigwedge_{1 \leq m < n - k \cdot l^*} E^n_{\bar{q},m}$

Proof. The "only if" direction is immediate. For the "if" direction note that by Observation 2(2), with probability 1, only a candidate can be a chain of triangles, and the claim follows immediately. □

The following claim shows that by adding enough zeros at the end of $\bar{q}$ we can make sure that ψ_k holds in $M^n_{\bar{q}}$ with probability close to 1. Note that we do not make a "strong" use of the properness of $\bar{q}$, i.e we do not use item (2) of Definition 9.

claim 2. Let $\bar{q} \in \mathfrak{P}^{fin}$ be proper for l^*, $k \in \mathbb{N}$ be even, and $\zeta > 0$ be some rational. Then there exists $\bar{q}' \in \mathfrak{P}^{fin}$ such that $\bar{q} \lhd^{prop} \bar{q}'$ and $Pr[M^{n_{\bar{q}'}}_{\bar{q}'} \models \psi_k] \geq 1 - \zeta$.

Proof. For $n > n_{\bar{q}}$ denote by $\bar{q}^n$ the member of $\mathfrak{P}$ with $n_{\bar{q}^n} = n$ and $(q^n)_l$ is q_l if $l < n_{\bar{q}}$ and 0 otherwise. Note that $\bar{q} \lhd^{prop} \bar{q}^n$, hence if we show that for n

large enough we have $Pr[M^n_{\bar{q}^n} \models \psi_k] \geq 1 - \zeta$ then we will be done by putting $\bar{q}' = \bar{q}^n$. Note that (recalling Definition 8(5)) $M^n_{\bar{q}} = M^n_{\bar{q}^n}$ so below we need not distinguish between them. Now set $n^* = \max\{n_{\bar{q}}, k \cdot l^*\}$. For any $n > n^*$ and $1 \leq m \leq n - n^*$ consider the sequence $s(m) = (m, m + l^*, m + 2l^*, ..., m + k \cdot l^*)$ (note that $s(m)$ is a candidate of type (n, l^*, k, m)). Denote by E_m the event that $s(m)$ is a chain of triangles (in $M^n_{\bar{q}}$). We then have:

$$Pr[M^n_{\bar{q}} \models E_m] \geq (q_{l^*})^k \cdot (q_{2l^*})^{k/2} \cdot (\prod_{l=1}^{n_{\bar{q}}-1} (1 - q_l))^{2(k+1)}.$$

Denote the expression on the right by $p^*_{\bar{q}}$ and note that it is positive and depends only on k and $\bar{q}$ (but not on n). Now assume that $n > 6 \cdot n^*$ and that $1 \leq m < m' \leq n - n^*$ are such that $m' - m > 2 \cdot n^*$. Then the distance between the sequences $s(m)$ and $s(m')$ is larger than $n_{\bar{q}}$ and hence the events E_m and $E_{m'}$ are independent. We conclude that $Pr[M^n_{\bar{q}} \not\models \psi_k] \leq (1 - p^*_{\bar{q}})^{n/(2 \cdot n^* + 1)} \to_{n \to \infty} 0$ and hence by choosing n large enough we are done. □

The following claim shows that under our assumptions we can always find a long initial segment $\bar{q}$ of some member of $Gen_1(\bar{p})$ such that ψ_k holds in $M^n_{\bar{q}}$ with probability close to 0. This is where we make use of our assumptions on $\bar{p}$ and the properness of $\bar{q}$.

claim 3. Let $\bar{p} \in \mathfrak{P}^{inf}$, $\epsilon > 0$ and assume that for an unbounded set of $n \in \mathbb{N}$ we have $\prod_{l=1}^{n}(1 - p_l) \leq n^{-\epsilon}$. Let $k \in \mathbb{N}$ be even such that $k \cdot \epsilon > 2$. Let $\bar{q} \in Gen_1^r(\bar{p})$ be proper for l^*, and $\zeta > 0$ be some rational. Then there exists $r' > r$ and $\bar{q}' \in Gen_1^{r'}(\bar{p})$ such that $\bar{q} \lhd^{prop} \bar{q}'$ and $Pr[M^{n_{\bar{q}'}}_{\bar{q}'} \models \neg\psi_k] \geq 1 - \zeta$.

Proof. First recalling Definition 9 let $l^{**} = 3l^* + 2$, and for $l \geq n_{\bar{q}}$ define $r(l) := \lceil (l - n_{\bar{q}} + 1)/l^{**} \rceil$. Now for each $n > n_{\bar{q}} + l^{**}$ denote by $\bar{q}_n$ the member of $\mathfrak{P}$ defined by:

$$(q_n)_l = \begin{cases} q_l & 0 < l < n_{\bar{q}} \\ 0 & n_{\bar{q}} \leq l < n \text{ and } l \not\equiv 1 \mod l^{**} \\ p_{r+r(l)} & n_{\bar{q}} \leq l < n \text{ and } l \equiv 1 \mod l^{**}. \end{cases}$$

Note that $n_{\bar{q}_n} = n$, $\bar{q}_n \in Gen_1^{r'}(\bar{p})$ where $r' = r + r(n-1) > r$ and $\bar{q} \lhd^{prop} \bar{q}_n$. Hence if we show that for some n large enough we have $Pr[M^n_{\bar{q}_n} \models \neg\psi_k] \geq 1 - \zeta$ then we will be done by putting $\bar{q}' = \bar{q}_n$. As before let $n^* := \max\{kl^*, n_{\bar{q}} + l^*\}$. Now fix some $n > n^*$ and for $1 \leq m < n - k \cdot l^*$ let $s(m)$ be some candidate of type (n, l^*, k, m). Denote by $E = E(s(m))$ the event that $s(m)$ is a chain of triangles in $M^n_{\bar{q}_n}$. We then have:

$$Pr[M^n_{\bar{q}_n} \models E] \leq (q_{l^*})^k \cdot (q_{2l^*})^{k/2} \cdot (\prod_{n^*+1}^{\lfloor (n-n^*)/2 \rfloor} (1 - (q_i)_l))^k.$$

Now denote:

$$p^*_{\bar{q}} := (q_{l^*})^k \cdot (q_{2l^*})^{k/2} \cdot (\prod_{l=1}^{n^*}(1 - (q_i)_l))^{-k}$$

and note that it is positive and does not depend on n. Together we get:

$$Pr[M^n_{\bar{q}_n} \models E] \leq p^*_{\bar{q}} \cdot (\prod_{l=1}^{\lfloor (n-n^*)/2 \rfloor} (1-(q_i)_l))^k \leq p^*_{\bar{q}} \cdot (\prod_{l=1}^{\lfloor (n-n^*)/(2l^{**}) \rfloor} (1-p_l))^k.$$

For each m in the range $1 \leq m < n - k \cdot l^*$ the number of candidates of type (n, l^*, k, m) is at most 4, hence the total number of candidates is no more then $4n$. We get that the expected number (in the probability space $M^n_{\bar{q}_n}$) of candidates which are a chain of triangles is at most $p^*_{\bar{q}} \cdot (\prod_{l=1}^{\lfloor (n-n^*)/(2l^{**}) \rfloor} (1-p_l))^k \cdot 4n$. Let E^* be the following event: "No candidate is a chain of triangles". Then using Claim 1 and Markov's inequality we get:

$$Pr[M^n_{\bar{q}} \models \psi_k] = Pr[M^n_{\bar{q}} \not\models E^*] \leq p^*_{\bar{q}} \cdot (\prod_{l=1}^{\lfloor (n-n^*)/(2l^{**}) \rfloor} (1-p_l))^k \cdot 4n.$$

Finally by our assumptions, for an unbounded set of n we have $\prod_{l=1}^{\lfloor (n-n^*)/(2l^{**}) \rfloor} (1-p_l) \leq (\lfloor (n-n^*)/(2l^{**}) \rfloor)^{-\epsilon}$, and note that for n large enough we have $(\lfloor (n-n^*)/(2l^{**}) \rfloor)^{-\epsilon} \leq n^{-\epsilon/2}$. Hence for infinitely many $n \in \mathbb{N}$ we have $Pr[M^n_{\bar{q}} \models \psi_k] \leq p^*_{\bar{q}} \cdot 4 \cdot n^{1-\epsilon \cdot k/2}$, and as $\epsilon \cdot k > 2$ this tends to 0 as n tends to ∞, so we are done. □

We are now ready to prove Lemma 7. First as (∗) of 5 does not hold we have some $\epsilon > 0$ such that for an unbounded set of $n \in \mathbb{N}$, we have $\prod_{l=1}^{n}(1-p_l) \leq n^{-\epsilon}$. Let $k \in \mathbb{N}$ be even such that $k \cdot \epsilon > 2$. Now for each $i \in \mathbb{N}$ we will construct a pair $(\bar{q}_i, r_i)$ such that the following holds:

1. For $i \in \mathbb{N}$, $\bar{q}_i \in Gen_1^{r_i}(\bar{p})$ and put $n_i := n_{\bar{q}_i}$.
2. For $i \in \mathbb{N}$, $\bar{q}_i \lhd^{prop} \bar{q}_{i+1}$.
3. For each odd $i > 0$, $Pr[M^{n_i}_{\bar{q}_i} \models \psi_k] \geq 1 - \frac{1}{i}$ and $r_i = r_{i-1}$.
4. For each even $i > 0$, $Pr[M^{n_i}_{\bar{q}_i} \models \neg\psi_k] \geq 1 - \frac{1}{i}$ and $r_i > r_{i-1}$.

Clearly if we construct such $\langle (\bar{q}_i, r_i) : i \in \mathbb{N} \rangle$ then by taking $\bar{q} = \cup_{i \in \mathbb{N}} \bar{q}_i$ (recall Observation 1), we have $\bar{q} \in Gen_1(\bar{p})$ and both ψ_k and $\neg\psi_k$ holds with probability approaching 1 in $M^n_{\bar{q}}$, thus finishing the proof. We turn to the construction of $\langle (\bar{q}_i, r_i) : i \in \mathbb{N} \rangle$, and naturally we use induction on $i \in \mathbb{N}$.

Case 1: $i = 0$. Let $l_1 < l_2$ be the first and second indexes such that $p_{l_i} > 0$. Put $r_0 :- l_2$. If $l_2 \leq 2l_1$ define $\bar{q}_0$ by:

$$(q_0)_l = \begin{cases} p_l & l \leq l_1 \\ 0 & l_1 \leq l < 2l_1 \\ p_{l_2} & l = 2l_1. \end{cases}$$

Otherwise if $l_2 > 2l_1$ define $\bar{q}_0$ by:

$$(q_0)_l = \begin{cases} 0 & l < \lceil l_2/2 \rceil \\ p_{l_1} & l = \lceil l_2/2 \rceil \\ 0 & \lceil l_2/2 \rceil < l < 2\lceil l_2/2 \rceil \\ p_{l_2} & l = 2\lceil l_2/2 \rceil. \end{cases}$$

clearly $\bar{q}_0 \in Gen_1^{r_0}(\bar{p})$ as desired, and note that $\bar{q}_0$ is proper (for either l_1 or $\lceil l_2/2 \rceil$).

Case 2: $i > 0$ is odd. First set $r_i = r_{i-1}$. Next we use Claim 2 where we set: $\bar{q}_{i-1}$ for $\bar{q}$, $\frac{1}{i}$ for ζ and $\bar{q}_i$ is the one promised by the claim. Note that indeed $\bar{q}_{i-1} \lhd^{prop} \bar{q}_i$, $\bar{q}_i \in gen^{r_i}(\bar{p})$ and $Pr[M^{n_i}_{\bar{q}_i} \models \psi_k] \geq 1 - \frac{1}{i}$.

Case 3: $i > 0$ is even. We use Claim 3 where we set: $\bar{q}_{i-1}$ for $\bar{q}$, $\frac{1}{i}$ for ζ and $(r_i, \bar{q}_i)$ are $(r', \bar{q}')$ promised by the claim. Note that indeed $\bar{q}_{i-1} \lhd^{prop} \bar{q}_i$, $\bar{q}_i \in Gen_1^{r_i}(\bar{p})$ and $Pr[M^{n_i}_{\bar{q}_i} \models \psi_k] \geq 1 - \frac{1}{i}$. This completes the proof of Lemma 7.

3 Decreasing Coordinates

In this section we prove Theorem 5 for $j \in \{2, 3\}$. As before, the "if" direction is an immediate consequence of Theorem 2. Moreover as $Gen_3(\bar{p}) \subseteq Gen_2(\bar{p})$ it remains to prove that if $(*)$ of 5 fails then the 0-1 law strongly fails for some $\bar{q} \in Gen_3(\bar{p})$. We divide the proof into two cases according to the behavior of $\sum_{l=1}^{n} p_i$, which is an approximation of the expected number of neighbors of a given node in $M^n_{\bar{p}}$. Define:

$$(**) \qquad \iff \qquad \lim_{n\to\infty} \log(\sum_{i=1}^{n} p_i)/\log n = 0.$$

Assume that $(**)$ above fails. Then for some $\epsilon > 0$, the set $\{n \in \mathbb{N} : \sum_{i=1}^{n} p_i \geq n^{\epsilon}\}$ is unbounded, hence we finish by Lemma 12. On the other hand if $(**)$ holds then $\sum_{i=1}^{n} p_i$ increases slower than any positive power of n; formally for all $\delta > 0$ for some $n_\delta \in \mathbb{N}$ we have that $n > n_\delta$ implies $\sum_{i=1}^{n} p_i \leq n^{\delta}$. As we assume that $(*)$ of Theorem 5 fails we have that for some $\epsilon > 0$ the set $\{n \in \mathbb{N} : \prod_{i=1}^{n}(1 - p_i) \leq n^{-\epsilon}\}$ is unbounded. Together (with $-\epsilon/6$ as δ) we have that the assumptions of Lemma 13 hold, hence we finish the proof.

Lemma 12. *Let $\bar{p} \in \mathfrak{P}^{inf}$ be such that $p_l < 1$ for $l > 0$. Assume that for some $\epsilon > 0$ we have for an unbounded set of $n \in \mathbb{N}$: $\sum_{l \leq n} p_l \geq n^{\epsilon}$. Then for some $\bar{q} \in Gen_3(\bar{p})$ and $\psi = \psi_{isolated} := \exists x \forall y \neg x \sim y$, both ψ and $\neg\psi$ holds infinitely often in $M^n_{\bar{q}}$.*

Proof. We construct a series, $(\bar{q}_1, \bar{q}_2, ...)$ such that for $i > 0$: $\bar{q}_i \in \mathfrak{P}^{fin}$, $\bar{q}_i \lhd \bar{q}_{i+1}$ and $\cup_{i>0}\bar{q}_i \in Gen_3(\bar{p})$. For $i \geq 1$ denote $n_i := n_{\bar{q}_i}$. We will show that:

$*_{even}$ For even $i > 1$: $Pr[M^{n_i}_{\bar{q}_i} \models \psi] \geq 1 - \frac{1}{i}$.
$*_{odd}$ For odd $i > 1$: $Pr[M^{n_i}_{\bar{q}_i} \models \neg\psi] \geq 1 - \frac{1}{i}$.

Taking $\bar{q} = \cup_{i>0}\bar{q}_i$ will then complete the proof. We construct $\bar{q}_i$ by induction on $i > 0$:

Case 1: $i = 1$: Let $n_1 = 2$ and $(q_1)_1 = p_1$.

Case 2: even $i > 1$: As $(\bar{q}_{i-1}, n_{i-1})$ is given, let us define $\bar{q}_i$ where $n_i > n_{i-1}$ is to be determined later: $(q_i)_l = (q_{i-1})_l$ for $l < n_{i-1}$ and $(q_i)_l = 0$ for $n_{i-1} \leq$

$l < n_i$. For $x \in [n_i]$ let E_x be the event: "x is an isolated point". Denote $p' := (\prod_{0<l<n_{i-1}}(1-(q_{i-1})_l)^2$ and note that $p' > 0$ and does not depend on n_i. Now for $x \in [n_i]$, $Pr[M^{n_i}_{\bar{q}_i} \models E_x] \geq p'$, furthermore if $x, x' \in [n_i]$ and $|x - x'| > n_{i-1}$ then E_x and $E_{x'}$ are independent in $M^{n_i}_{\bar{q}_i}$. We conclude that $Pr[M^{n_i}_{\bar{q}_i} \models \neg\psi] \leq (1-p')^{\lfloor n_i/(n_{i-1}+1)\rfloor}$ which approaches 0 as $n_i \to \infty$. So by choosing n_i large enough we have $*_{even}$.

Case 3: odd $i > 1$: As in case 2 let us define $\bar{q}_i$ where $n_i > n_{i-1}$ is to be determined later: $(q_i)_l = (q_{i-1})_l$ for $l < n_{i-1}$ and $(q_i)_l = p_l$ for $n_{i-1} \leq l < n_i$. Let $n' = \max\{n < n_i/2 : n = 2^m$ for some $m \in \mathbb{N}\}$, so $n_i/4 \leq n' < n_i/2$. Denote $a = \sum_{0<l\leq n'}(q_i)_l$ and $a' = \sum_{0<l\leq\lfloor n_i/4\rfloor}(q_i)_l$. Again let E_x be the event: "x is isolated". Now as $n' < n_i/2$, $Pr[M^{n_i}_{\bar{q}_i} \models E_x] \leq \prod_{0<l\leq n'}(1-(q_i)_l)$. By a repeated use of: $(1-x)(1-y) \leq (1-\frac{x+y}{2})^2$ we get $Pr[M^{n_i}_{\bar{q}_i} \models E_x] \leq (1-\frac{a}{n'})^{n'}$ which for n' large enough is smaller then $2 \cdot e^{-a}$, and as $a' \leq a$, we get $Pr[M^{n_i}_{\bar{q}_i} \vdash E_x] \leq 2 \cdot e^{-a'}$. By the definition of a' and $\bar{q}_i$ we have $a' = \sum_{l=1}^{\lfloor n_1/4\rfloor} p_l - \sum_{l<n_{i-1}}(p_l - (q_{i-1})_l)$. By our assumption for an unbounded set of $n_i \in \mathbb{N}$ we have $a' \geq (\lfloor n_i/4\rfloor)^{\epsilon} - \sum_{l<n_{i-1}}(p_l - (q_{i-1})_l)$. But as the sum on the right is independent of n_i we have (again for n_i large enough): $a' \geq (n_i/5)^{\epsilon}$. Consider the expected number of isolated points in the probability space $M^{n_i}_{\bar{q}_i}$, denote this number by $X(n_i)$. By all the above we have:

$$X(n_i) \leq n_i \cdot 2 \cdot e^{-a} \leq n_i \cdot 2 \cdot e^{-a'} \leq 2n_i \cdot e^{-(n_i/5)^{\epsilon}}.$$

The last expression approaches 0 as $n_i \to \infty$. So by choosing n_i large enough (while keeping $a' \geq (n_i/5)^{\epsilon}$ we have $*_{odd}$.

Finally notice that indeed $\cup_{i>0}\bar{q}_i \in Gen_3(\bar{p})$, as the only change we made in the inductive process is decreasing p_l to 0 when $n_{i-1} < l \leq n_i$ and i is even. □

Lemma 13. *Let $\bar{p} \in \mathfrak{P}^{inf}$ be such that $p_l < 1$ for $l > 0$. Assume that for some $\epsilon > 0$ we have for an unbounded set of $n \in \mathbb{N}$:*

(α) $\sum_{l\leq n} p_l \leq n^{\epsilon/6}$.
(β) $\prod_{l\leq n}(1-p_l) \leq n^{-\epsilon}$.

Let $k = \lceil\frac{6}{\epsilon}\rceil + 1$ and $\psi = \psi_k$ be the sentence "saying" there exists a connected component which includes a path of length k, formally:

$$\psi_k := \exists x_1 ... \exists x_k$$
$$\bigwedge_{1\leq i\neq j\leq k} x_i \neq x_j \wedge \bigwedge_{1\leq i<k} x_i \sim x_{i+1} \wedge \forall y[(\bigwedge_{1\leq i\leq k} x_i \neq y) \to (\bigwedge_{1\leq i\leq k} \neg x_i \sim y)].$$

Then for some $\bar{q} \in Gen_3(\bar{p})$, each of ψ and $\neg\psi$ holds infinitely often in $M^n_{\bar{q}}$.

Proof. The proof follows the same line as the proof of 12. We construct an increasing series, $(\bar{q}_1, \bar{q}_2, ...)$, and demand $*_{even}$ and $*_{odd}$ as in 12. Taking $\bar{q} = \cup_{i>0}\bar{q}_i$ will then complete the proof. We construct $\bar{q}_i$ by induction on $i > 0$:

Case 1: $i = 1$: Let $l(*) := \min\{l > 0 : p_l > 0\}$ and define $n_1 = l(*) + 1$ and $(q_1)_l = p_l$ for $l < n_1$.

Case 2: even $i > 1$: As before, for $n_i > n_{i-1}$ define: $(q_i)_l = (q_{i-1})_l$ for $l < n_{i-1}$ and $(q_i)_l = 0$ for $n_{i-1} \leq l < n_i$. For $1 \leq x < n_i - k \cdot l(*)$ let E^x be the event: "$(x, x + l(*), ..., x + l(*)(k-1))$ exemplifies ψ." Formally E^x holds in $M^{n_i}_{\bar{q}_i}$ iff $\{(x, x + l(*), ..., x + l(*)(k-1))\}$ is isolated and for $0 \leq j < k-1$, $\{x + jl(*), x + (j+1)l(*)\}$ is an edge of $M^{n_i}_{\bar{q}_i}$. The remainder of this case is similar to case 2 of Lemma 12 so we will not go into details. Note that $Pr[M^{n_i}_{\bar{q}_i} \models E^x] > 0$ and does not depend on n_i, and if $|x - x'|$ is large enough (again not depending on n_i) then E^x and $E^{x'}$ are independent in $M^{n_i}_{\bar{q}_i}$. We conclude that by choosing n_i large enough we have $*_{even}$.

Case 3: odd $i > 1$: In this case we make use of the fact that almost always, no $x \in [n]$ has to many neighbors. Formally:

claim 4. Let $\bar{q} \in \mathfrak{P}^{inf}$ be such that $q_l < 1$ for $l > 0$. Let $\delta > 0$ and assume that for an unbounded set of $n \in \mathbb{N}$ we have, $\sum_{l=1}^{n} q_l \leq n^{\delta}$. Let E^n_{δ} be the event: "No $x \in [n]$ has more than $8n^{2\delta}$ neighbors". Then we have:

$$\limsup_{n \to \infty} Pr[E^n_{\delta} \text{ holds in } M^n_{\bar{q}}] = 1.$$

Proof. First note that the size of the set $\{l > 0 : q_l > n^{-\delta}\}$ is at most $n^{2\delta}$. Hence by ignoring at most $2n^{2\delta}$ neighbors of each $x \in [n]$, and changing the number of neighbors in the definition of E^n_{δ} to $6n^{2\delta}$ we may assume that for all $l > 0$, $q_l \leq n^{-\delta}$. The idea is that the number of neighbors of each $x \in [n]$ can be approximated (or in our case only bounded from above) by a Poisson random variable with parameter close to $\sum_{i=l}^{n} q_l$. Formally, for each $l > 0$ let B_l be a Bernoulli random variable with $Pr[B_l = 1] = q_l$. For $n \in \mathbb{N}$ let X^n be the random variable defined by $X^n := \sum_{l=1}^{n} B_l$. For $l > 0$ let Po_l be a Poisson random variable with parameter $\lambda_l := -\log(1 - q_l)$ that is for $i = 0, 1, 2, ...$ $Pr[Po_l = i] = e^{-\lambda_l} \frac{(\lambda_l)^i}{i!}$. Note that $Pr[B_l = 0] = Pr[Po_l = 0]$. Now define $Po^n := \sum_{i=1}^{n} Po_l$. By the last sentence we have $Po^n \geq_{st} X^n$ (Po^n is stochastically larger than X^n) that is, for $i = 0, 1, 2, ...$ $Pr[Po^n \geq i] \geq Pr[X^n \geq i]$. Now Po^n (as the sum of Poisson random variables) is a Poisson random variable with parameter $\lambda^n := \sum_{l=1}^{n} \lambda_l$. Let $n \in \mathbb{N}$ be such that $\sum_{l=1}^{n} q_l \leq n^{\delta}$, and define $n' = n'(n) := min\{n' \geq n : n' = 2^m \text{ for some } m \in \mathbb{N}\}$, so $n \leq n' < 2n$. For $0 < l \leq n'$ let q'_l be q_l if $l \leq n$ and 0 otherwise, so we have: $\prod_{l=1}^{n}(1 - q_l) = \prod_{l=1}^{n'}(1 - q'_l)$ and $\sum_{l=1}^{n} q_l = \sum_{l=1}^{n'} q'_l$. Note that if $0 \leq p, q \leq 1/4$ then $(1-p)(1-q) \geq (1 - \frac{p+q}{2})^2 \cdot \frac{1}{2}$. By a repeated use of the last inequality we get that $\prod_{i=l}^{n'}(1 - q'_l) \geq (1 - \frac{\sum_{i=l}^{n'} q'_l}{n'})^{n'} \cdot \frac{1}{n'}$. We can now evaluate λ^n:

$$\begin{aligned}
\lambda^n = \sum_{l=1}^{n} \lambda_l = \sum_{l=1}^{n} -\log(1 - q_l) &= -\log(\prod_{l=1}^{n}(1 - q_l)) = -\log(\prod_{l=1}^{n'}(1 - q'_l)) \\
&\leq -\log[(1 - \frac{\sum_{l=1}^{n'} q'_l}{n'})^{n'} \cdot \frac{1}{n'}] = -\log[(1 - \frac{\sum_{l=1}^{n} q_l}{n'})^{n'} \cdot \frac{1}{n'}] \\
&\approx -\log[e^{-\sum_{l=1}^{n} q_l} \cdot \frac{1}{n'}] \leq -\log[e^{-n^{\delta}} \cdot \frac{1}{2n}] \leq -\log[e^{-n^{2\delta}}] = n^{2\delta}.
\end{aligned}$$

Hence by choosing $n \in \mathbb{N}$ large enough while keeping $\sum_{l=1}^{n} q_l \leq n^{\delta}$ (which is possible by our assumption) we have $\lambda^n \leq n^{2\delta}$. We now use the Chernoff bound for Poisson random variables: If Po is a Poisson random variable with parameter λ and $i > 0$ we have $Pr[Po \geq i] \leq e^{\lambda(i/\lambda - 1)} \cdot (\frac{\lambda}{i})^i$. Applying this bound to Po^n (for n as above) we get:

$$Pr[Po^n \geq 3n^{2\delta}] \leq e^{\lambda^n(3n^{2\delta}/\lambda^n - 1)} \cdot (\frac{\lambda^n}{3n^{2\delta}})^{3n^{2\delta}} \leq e^{3n^{2\delta}} \cdot (\frac{\lambda^n}{3n^{2\delta}})^{3n^{2\delta}} \leq (\frac{e}{3})^{3n^{2\delta}}.$$

Now for $x \in [n]$ let X_x^n be the number of neighbors of x in $M_{\bar{q}}^n$ (so X_x^n is a random variable on the probability space $M_{\bar{q}}^n$). By the definition of $M_{\bar{q}}^n$ we have $X_x^n \leq_{st} 2 \cdot X^n \leq_{st} 2 \cdot Po^n$. So for unboundedly many $n \in \mathbb{N}$ we have for all $x \in [n]$, $Pr[X_x^n \geq 6n^{2\delta}] \leq (\frac{e}{3})^{3n^{2\delta}}$. Hence by the Markov inequality for unboundedly many $n \in \mathbb{N}$ we have,

$$Pr[E^n \text{ does not hold in } M_{\bar{q}}^n] = Pr[\text{for some } x \in [n], X_x^n \geq 3n^{2\delta}] \leq n \cdot (\frac{e}{3})^{6n^{2\delta}}.$$

But the last expression approaches 0 as n approaches ∞, Hence we are done proving the claim. □

We return to **Case 3:** of the proof of 13, and it remains to construct $\bar{q}_i$. As before for $n_i > n_{i-1}$ define: $(q_i)_l = (q_{i-1})_l$ for $l < n_{i-1}$ and $(q_i)_l = p_l$ for $n_{i-1} \leq l < n_i$. By the claim above and (α) in our assumptions, for n_i large enough we have $Pr[E_{\epsilon/6}^{n_i}$ holds in $M_{\bar{q}_i}^{n_i}] \geq 1/2i$, so assume in the rest of the proof that n_i is indeed large enough, and assume that $E_{\epsilon/6}^{n_i}$ holds in $M_{\bar{q}_i}^{n_i}$, and all the probabilities on the space $M_{\bar{q}_i}^{n_i}$ will be conditioned to $E_{\epsilon/6}^{n_i}$ (even if not explicitly said so). A k-tuple $\bar{x} = (x_1, ..., x_k)$ of members of $[n_i]$ is called a k-path (in $M_{\bar{q}_i}^{n_i}$) if it is without repetitions and for $0 < j < k$ we have $M_{\bar{q}_i}^{n_i} \models x_j \sim x_{j+1}$. A k-path is isolated if in addition no member of $\{x_1, ..., x_k\}$ is connected to a member of $[n_i] \setminus \{x_1, ..., x_k\}$. Now (recall we assume $E_{\epsilon/6}^{n_i}$) with probability 1: the number of k-paths in $M_{\bar{q}_i}^{n_i}$ is at most $8^k \cdot n^{1+k\epsilon/3}$. For each $(x_1, ..., x_k)$ without repetitions we have:

$$Pr[(x_1, ..., x_k) \text{ is isolated in } M_{\bar{q}_i}^{n_i}] = \prod_{j=1}^{k} \prod_{y \neq x_j} (1 - (q_i)_{|x_j - y|}) \leq (\prod_{l=1}^{\lfloor n_i/2 \rfloor} (1 - (q_i)_l))^k.$$

By assumption (β) we have for an unbounded set of $n_i \in \mathbb{N}$:

$$\prod_{l=1}^{\lfloor n_i/2 \rfloor} (1 - (q_i)_l) \leq \prod_{l=n_i-1}^{\lfloor n_i/2 \rfloor} (1 - p_l) \leq \prod_{l<n_i} (1 - q_l) \cdot (\lfloor n_i/2 \rfloor)^{-\epsilon} \leq (n_i)^{-\epsilon/2}.$$

Together letting $Y(n_i)$ be the expected number of isolated k tuples in $M_{\bar{q}_i}^{n_i}$ we have:

$$Y(n_i) \leq 8^k \cdot (n_i)^{1+k\epsilon/3} \cdot (n_i)^{-k\epsilon/2} = 8^k \cdot (n_i)^{1-k\epsilon/6} \to_{n_i \to \infty} 0.$$

So by choosing n_i large enough and using Markov's inequality, we have $*_{odd}$, and we are done. □

4 Allowing Some Probabilities to Equal 1

In this section we analyze the hereditary 0-1 law for $\bar{p}$ where some of the p_i-s may equal 1. For $\bar{p} \in \mathfrak{P}^{inf}$ let $U^*(\bar{p}) := \{l > 0 : p_l = 1\}$. The situation $U^*(\bar{p}) \neq \emptyset$ was discussed briefly at the end of section 4 of [2], and an example was given there of some $\bar{p}$ consisting of only ones and zeros with $|U^*(\bar{p})| = \infty$ such that the 0-1 law fails for $M^n_{\bar{p}}$. We follow the lines of that example and prove that if $|U^*(\bar{p})| = \infty$ and $j \in \{1, 2, 3\}$, then the j-hereditary 0-1 law for L fails for $\bar{p}$. This is done in 14. The case $0 < |U^*(\bar{p})| < \infty$ is also studied and a full characterization of the j-hereditary 0-1 law for L is given in Conclusion 1 for $j \in \{2, 3\}$, and for $j = 1$, $1 < |U^*(\bar{p})|$. The case $j = 1$ and $1 = |U^*(\bar{p})|$ is discussed in section 5.

Theorem 14. *Let $\bar{p} \in \mathfrak{P}^{inf}$ be such that $U^*(\bar{p})$ is infinite, and j be in $\{1, 2, 3\}$. Then $M^n_{\bar{p}}$ does not satisfy the j-hereditary weak convergence law for L.*

Proof. We start with the case $j = 1$. The idea here is similar to that of section 2. We show that some $\bar{q} \in Gen_1(\bar{p})$ has a structure (similar to the "proper" structure defined in 9) that allows us to identify the sections "close" to 1 or n in $M^n_{\bar{q}}$. It is then easy to see that if $\bar{q}$ has infinitely many ones and infinitely many "long" sections of consecutive zeros, then the sentence saying: "there exists an edge connecting vertices close to the the edges", will exemplify the failure of the 0-1 law for $M^n_{\bar{q}}$. This is formulated below. Consider the following demands on $\bar{q} \in \mathfrak{P}^{inf}$:

1. Let $l^* < l^{**}$ be the first two members of $U^*(\bar{q})$; then l^* is odd and $l^{**} = 2 \cdot l^*$.
2. If l_1, l_2, l_3 all belong to $\{l > 0 : q_l > 0\}$ and $l_1 + l_2 = l_3$ then $l_1 = l_2 = l^*$.
3. The set $\{n \in \mathbb{N} : n - 2l^* < l < n \Rightarrow q_l = 0\}$ is infinite.
4. The set $U^*(\bar{q})$ is infinite.

We first claim that some $\bar{q} \in Gen_1(\bar{p})$ satisfies the demands (1)-(4) above. This is straightforward. We inductively add enough zeros before each nonzero member of $\bar{p}$ guaranteeing that it is larger than the sum of any two (not necessarily different) nonzero members preceding it. We continue until we reach l^*, then by adding zeros either before l^* or before l^{**} we can guarantee that l^* is odd and that $l^{**} = 2 \cdot l^*$, and hence (1) holds. We then continue the same process from l^{**}, adding at least $2l^*$ zero's at each step. This guarantees (2) and (3). (4) follows immediately from our assumption that $U^*(\bar{p})$ is infinite. Assume that $\bar{q}$ satisfies (1)-(4) and $n \in \mathbb{N}$. With probability 1 we have:

$\{x, y, z\}$ is a triangle in $M^n_{\bar{q}}$ iff $\{x, y, z\} = \{l, l + l^*, l + l^{**}\}$ for some $0 < l \leq n$.

To see this use (1) for the "if" direction and (2) for the "only if" direction. We conclude that letting $\psi_{ext}(x)$ be the L formula saying that x belongs to exactly one triangle, for each $n \in \mathbb{N}$ and $m \in [n]$ with probability 1 we have:

$$M^n_{\bar{q}} \models \psi_{ext}[m] \text{ iff } m \in [1, l^*] \cup (n - l^*, n].$$

We are now ready to prove the failure of the weak convergence law in $M^n_{\bar{q}}$, but in the first stage let us only show the failure of the convergence law. This will be useful for other cases (see Remark 15 below). Define

$$\psi := (\exists x \exists y)\psi_{ext}(x) \wedge \psi_{ext}(y) \wedge x \sim y.$$

Recall that l^* is the *first* member of $U^*(\bar{q})$, and hence for some $p > 0$ (not depending on n) for any $x, y \in [1, l^*]$ we have $Pr[M^n_{\bar{q}} \models \neg x \sim y] \geq p$ and similarly for any $x, y \in (n - l^*, n]$. We conclude that:

$$Pr[(\exists x \exists y)(x, y \in [1, l^*] \text{ or } x, y \in (n - l^*, n]) \text{ and } x \sim y] \leq 1 - p^{2\binom{l^*}{2}} < 1.$$

By all the above, for each l such that $q_l = 1$ we have $Pr[M^{l+1}_{\bar{q}} \models \psi] = 1$, as the pair $(1, l+1)$ exemplifies ψ in $M^{l+1}_{\bar{q}}$ with probability 1. On the other hand if n is such that $n - 2l^* < l < n \Rightarrow q_l = 0$ then $Pr[M^n_{\bar{q}} \models \psi] \leq 1 - p^{2\binom{l^*}{2}}$. Hence by (3) and (4) above, ψ exemplifies the failure of the convergence law for $M^n_{\bar{q}}$ as required.

We return to the proof of the failure of the weak convergence law. Define:

$$\psi' = \exists x_0 ... \exists x_{2l^*-1}[\bigwedge_{0 \leq i < i' < 2l^*} x_i \neq x_{i'} \wedge \forall y((\bigwedge_{0 \leq i < 2l^*} y \neq x_i) \to \neg \psi_{ext}(y))$$
$$\wedge \bigwedge_{0 \leq i < 2l^*} \psi_{ext}(x_i) \wedge \bigwedge_{0 \leq i < l^*} x_{2i} \sim x_{2i+1}].$$

We will show that each of ψ' and $\neg\psi'$ holds infinitely often in $M^n_{\bar{q}}$. First let $n \in \mathbb{N}$ be such that $q_{n-l^*} = 1$. Then by choosing for each i in the range $0 \leq i < l^*$, $x_{2i} := i+1$ and $x_{2i+1} := n-l^*+1+i$, we will get that the sequence $(x_0, ..., x_{2l^*-1})$ exemplifies ψ' in $M^n_{\bar{q}}$ (with probability 1). As by assumption (4) above the set $\{n \in \mathbb{N} : q_{n-l^*} = 1\}$ is unbounded we have $\limsup_{n\to\infty}[M^n_{\bar{q}} \models \psi'] = 1$. For the other direction let $n \in \mathbb{N}$ be such that for each l in the range $n - 2l^* < l < n$, $q_l = 0$. Then $M^n_{\bar{q}}$ satisfies (again with probability 1) for each $x, y \in [1, l^*] \cup (n - l^*, n]$ such that $x \sim y$: $x \in [1, l^*]$ iff $y \in [1, l^*]$. Now assume that $(x_0, ..., x_{2l^*-1})$ exemplifies ψ' in $M^n_{\bar{q}}$. Then for each i in the range $0 \leq i < l^*$, $x_{2i} \in [1, l^*]$ iff $x_{2i+1} \in [1, l^*]$. We conclude that the set $[1, l^*]$ is of even size, thus contradicting (1). So we have $Pr[M^n_{\bar{q}} \models \psi'] = 0$. But by assumption (3) above the set of natural numbers, n, for which we have $n - 2l^* < l < n$ implies $q_l = 0$ is unbounded, and hence we have $\limsup_{n\to\infty}[M^n_{\bar{q}} \models \neg\psi'] = 1$ as desired.

We turn to the proof of the case $j \in \{2, 3\}$, and as $Gen_3(\bar{p}) \subseteq Gen_2(\bar{p})$ it is enough to prove that for some $\bar{q} \in Gen_3(\bar{p})$ the 0-1 law for L strongly fails in $M^n_{\bar{q}}$. Motivated by the example mentioned above appearing in the end of section 4 of [2], we let ψ be the sentence in L implying that each edge of the graph is contained in a cycle of length 4. Once again we use an inductive construction of $(\bar{q}_1, \bar{q}_2, \bar{q}_3, ...)$ in $\mathfrak{P}^{fin}$ such that $\bar{q} = \bigcup_{i>0} \bar{q}_i \in Gen_3(\bar{p})$ and both ψ and $\neg\psi$ hold infinitely often in $M^n_{\bar{q}}$. For $i = 1$ let $n_{\bar{q}_1} = n_1 := \min\{l : p_l = 1\} + 1$ and define $(q_1)_l = 0$ if $0 < l < n_1 - 1$ and $(q_1)_{n_1-1} = 1$. For even $i > 1$ let $n_{\bar{q}_i} = n_i := \min\{l > 4n_{i-1} : p_l = 1\} + 1$ and define $(q_i)_l = (q_{i-1})_l$ if

$0 < l < n_{i-1}$, $(q_i)_l = 0$ if $n_{i-1} \leq l < n_i - 1$ and $(q_1)_{n_1-1} = 1$. For odd $i > 1$ recall $n_1 = \min\{l : p_l = 1\} + 1$ and let $n_{\bar{q}_i} = n_i := n_{i-1} + n_1$. Now define $(q_i)_l = (q_{i-1})_l$ if $0 < l < n_{i-1}$ and $(q_i)_l = 0$ if $n_{i-1} \leq l < n_i$. Clearly we have for even $i > 1$, $Pr[M^{n_i+1}_{\bar{q}_{n_i+1}} \models \psi] = 0$ and for odd $i > 1$ $Pr[M^{n_i}_{\bar{q}_{n_i}} \models \psi] = 1$. Note that indeed $\bigcup_{i>0} \bar{q}_i \in Gen_3(\bar{p})$, and hence we are done. □

Remark 15. In the proof of the failure of the convergence law in the case $j = 1$ the assumption $|U^*(\bar{p})| = \infty$ is not needed, our proof works under the weaker assumption $|U^*(\bar{p})| \geq 2$ and for some $p > 0$, $\{l > 0 : p_l > p\}$ is infinite. See below more on the case $j = 1$ and $1 < |U^*(\bar{p})| < \infty$.

Lemma 16. *Let $\bar{q} \in \mathfrak{P}^{inf}$ and assume:*

1. *Let $l^* < l^{**}$ be the first two members of $U^*(\bar{q})$ (in particular assume $|U^*(\bar{q})| \geq 2$); then $l^{**} = 2 \cdot l^*$.*
2. *If l_1, l_2, l_3 all belong to $\{l > 0 : q_l > 0\}$ and $l_1 + l_2 = l_3$ then $\{l_1, l_2, l_3\} = \{l, l + l^*, l + l^{**}\}$ for some $l \geq 0$.*
3. *Let l^{***} be the first member of $\{l > 0 : 0 < q_l < 1\}$ (in particular assume $|\{l > 0 : 0 < q_l < 1\}| \geq 1$); then the set $\{n \in \mathbb{N} : n \leq l \leq n + l^{**} + l^{***} \Rightarrow q_l = 0\}$ is infinite.*

Then the 0-1 law for L fails for $M^n_{\bar{q}}$.

Proof. The proof is similar to the case $j = 1$ in the proof of Theorem 14, so we will not go into detail. Below n is some large enough natural number (say larger than $3 \cdot l^{**} \cdot l^{***}$) such that (3) above holds, and if we say that some property holds in $M^n_{\bar{q}}$ we mean it holds there with probability 1. Let $\psi^1_{ext}(x)$ be the formula in L implying that x belongs to at most two distinct triangles. Then for all $m \in [n]$:

$$M^n_{\bar{q}} \models \psi^1_{ext}[m] \text{ iff } m \in [1, l^{**}] \cup (n - l^{**}, n].$$

Similarly for any natural $t < n/3l^{**}$ define (using induction on t):

$$\psi^t_{ext}(x) := (\exists y \exists z) x \sim y \wedge x \sim z \wedge y \sim z \wedge (\psi^{t-1}_{ext}(y) \vee \psi^{t-1}_{ext}(z))$$

we then have for all $m \in [n]$:

$$M^n_{\bar{q}} \models \psi^t_{ext}[m] \text{ iff } m \in [1, tl^{**}] \cup (n - tl^{**}, n].$$

Now for $1 \leq t < n/3l^{**}$ let $m^*(t)$ be the minimal number of edges in $M^n_{\bar{q}}|_{[1,t \cdot l^{**}] \cup (n - t \cdot l^{**}, n]}$ i.e only edges with probability one and within one of the intervals are counted, formally

$$m^*(t) := 2 \cdot |\{(m, m') : m < m' \in [1, t \cdot l^{**}] \text{ and } q_{m'-m} = 1\}|.$$

Let $1 \leq t^* < n/3l^{**}$ be such that $l^{***} < l^{**} \cdot t^*$ (it exists as n is large enough). Note that $m^*(t^*)$ depends only on $\bar{q}$ and not on n and hence we can define

$\psi :=$ "There exist exactly $m^*(t^*)$ couples $\{x, y\}$ s.t. $\psi^{t^*}_{ext}(x) \wedge \psi^{t^*}_{ext}(y) \wedge x \sim y$."

We then have $Pr[M^n_{\bar{q}} \models \psi] \leq (1 - q_{l^{***}})^2 < 1$ as we have $m^*(t^*)$ edges on $[1, t^* l^{**}] \cup (n - t^* l^{**}, n]$ that exist with probability 1, and at least two additional edges (namely $\{1, l^{***} + 1\}$ and $\{n - l^{***}, n\}$) that exist with probability $q_{l^{***}}$ each. On the other hand if we define:

$$p' := \prod \{1 - q_{m'-m} : m < m' \in [1, t^* \cdot l^{**}] \text{ and } q_{m'-m} < 1\}$$

and note that p' does not depend on n, then (recalling assumption (3) above) we have $Pr[M^n_{\bar{q}} \models \psi] \geq (p')^2 > 0$ thus completing the proof. □

Lemma 17. *Let $\bar{q} \in \mathfrak{P}^{inf}$ be such that for some $l_1 < l_2 \in \mathbb{N} \setminus \{0\}$ we have: $0 < p_{l_1} < 1$, $p_{l_2} = 1$ and $p_l = 0$ for all $l \notin \{l_1, l_2\}$. Then the 0-1 law for L fails for $M^n_{\bar{q}}$.*

Proof. Let ψ be the sentence in L "saying" that some vertex has exactly one neighbor and this neighbor has at least three neighbors. Formally:

$$\psi := (\exists x)(\exists! y) x \sim y \wedge (\forall z)[x \sim z \rightarrow (\exists u_1 \exists u_2 \exists u_3) \bigwedge_{0<i<j\leq 3} u_i \neq u_j \wedge \bigwedge_{0<i\leq 3} z \sim u_i].$$

We first show that for some $p > 0$ and $n_0 \in \mathbb{N}$, for all $n > n_0$ we have $Pr[M^n_{\bar{q}} \models \psi] > p$. To see this simply take $n_0 = l_1 + l_2 + 1$ and $p = (1 - p_{l_1})(p_{l_1})$. Now for $n > n_0$ in $M^n_{\bar{q}}$, with probability $1 - p_{l_1}$ the node $1 \in [n]$ has exactly one neighbor (namely $1 + l_2 \in [n]$) and with probability at least p_{l_1}, $1 + l_2$ is connected to $1 + l_1 + l_2$, and hence has three neighbors (1, $1 + 2l_2$ and $1 + l_1 + l_2$). This yields the desired result. On the other hand for some $p' > 0$ we have for all $n \in \mathbb{N}$, $Pr[M^n_{\bar{q}} \models \neg\psi] > p'$. To see this note that for all n, only members of $[1, l_2] \cup (n - l_2, n]$ can possibly exemplify ψ, as all members of $(l_2, n - l_2]$ have at least two neighbors with probability one. For each $x \in [1, l_2] \cup (n - l_2, n]$, with probability at least $(1 - p_1)^2$, x does not exemplify ψ (since the unique neighbor of x has less then three neighbors). As the size of $[1, l_2] \cup (n - l_2, n]$ is $2 \cdot l_2$ we get $Pr[M^n_{\bar{q}} \models \neg\psi] > (1 - p_1)^{2l_2} := p' > 0$. Together we are done. □

Lemma 18. *Let $\bar{p} \in \mathfrak{P}^{inf}$ be such that $|U^*(\bar{p})| < \infty$ and $p_i \in \{0, 1\}$ for $i > 0$. Then $M^n_{\bar{p}}$ satisfies the 0-1 law for L.*

Proof. Let S^n be the (not random) structure in vocabulary $\{Suc\}$, with universe $[n]$ and Suc is the successor relation on $[n]$. It is straightforward to see that any sentence $\psi \in L$ has a sentence $\psi^S \in \{Suc\}$ such that

$$Pr[M^n_{\bar{p}} \models \psi] = \begin{cases} 1 & S^n \models \psi^S \\ 0 & S^n \not\models \psi^S. \end{cases}$$

Also by a special case of Gaifman's result from [1] we have: for each $k \in \mathbb{N}$ there exists some $n_k \in \mathbb{N}$ such that if $n, n' > n_k$ then S^n and $S^{n'}$ have the same first order theory of quantifier depth k. Together we are done. □

Conclusion 1. *Let $\bar{p} \in \mathfrak{P}^{inf}$ be such that $0 < |U^*(\bar{p})| < \infty$.*

1. *The 2-hereditary 0-1 law holds for* $\bar{p}$ *iff* $|\{l > 0 : p_l > 0\}| \leq 1$.
2. *The 3-hereditary 0-1 law holds for* $\bar{p}$ *iff* $\{l > 0 : 0 < p_l < 1\} = \emptyset$.
3. *If furthermore* $1 < |U^*(\bar{p})|$ *then the 1-hereditary 0-1 law holds for* $\bar{p}$ *iff* $\{l > 0 : 0 < p_l < 1\} = \emptyset$.

Proof. For (1) note that if indeed $|\{i > 0 : p_l > 0\}| > 1$ then some $\bar{q} \in Gen_2(\bar{p})$ is as in the assumption of Lemma 17; otherwise any $\bar{q} \in Gen_2(\bar{p})$ has at most 1 nonzero member and hence $M^n_{\bar{q}}$ satisfies the 0-1 law by either 18 or 2.

For (2) note that if $\{i > 0 : 0 < p_l < 1\} \neq \emptyset$ then some $\bar{q} \in Gen_3(\bar{p})$ is as in the assumption of Lemma 17; otherwise any $\bar{q} \in Gen_3(\bar{p})$ is as in the assumption of Lemma 18 and we are done.

Similarly for (3) note that if $1 < |U^*(\bar{p})|$ and $\{l > 0 : 0 < p_l < 1\} \neq \emptyset$ then some $\bar{q} \in Gen_1(\bar{p})$ satisfies assumptions (1)-(3) of Lemma 16; otherwise any $\bar{q} \in Gen_1(\bar{p})$ is as in the assumption of Lemma 18 and we are done. □

5 When Exactly One Probability Equals 1

In this section we assume:

Assumption 1. *$\bar{p}$ is a fixed member of $\mathfrak{P}^{inf}$ such that $|U^*(\bar{p})| = 1$ and we write $U^*(\bar{p}) = \{l^*\}$, and assume*

$$(*)' \qquad \lim_{n\to\infty} \log\Big(\prod_{l\in[n]\setminus\{l^*\}} (1 - p_l)\Big)/\log(n) = 0.$$

We try to determine when the 1-hereditary 0-1 law holds. The assumption of $(*)'$ is justified as the proof in section 2 works also in this case and in fact in any case that $U^*(\bar{p})$ is finite. To see this replace in section 2 products of the form $\prod_{l<n}(1 - p_l)$ by $\prod_{l<n, l\notin U^*(\bar{p})}(1 - p_l)$, sentences of the form "x has valency m" by "x has valency $m+2|U^*(\bar{p})|$", and similar simple changes. So if $(*)'$ fails then the 1-hereditary weak convergence law fails, and we are done. It seems that our ability to "identify" the l^*-boundary (i.e. the set $[1, l^*] \cup (n - l^*, n])$ in $M^n_{\bar{p}}$ is closely related to the holding of the 0-1 law. In Conclusion 2 we use this idea and give a necessary condition on $\bar{p}$ for the 1-hereditary weak convergence law. The proof uses methods similar to those of the previous sections. Finding a sufficient condition for the 1-hereditary 0-1 law seems to be harder. It turns out that the analysis of this case is, in a way, similar to the analysis when we add the successor relation to our vocabulary. This is because the edges of the form $\{l, l+l^*\}$ appear with probability 1 similarly to the successor relation. There are, however, some obvious differences. Let L^+ be the vocabulary $\{\sim, S\}$, and let $(M^+)^n_{\bar{p}}$ be the random L^+ structure with universe $[n]$, $\sim$ is the same as in $M^n_{\bar{p}}$, and $S^{(M^+)^n_{\bar{p}}}$ is the successor relation on $[n]$. Now if for some $l^{**} > 0$, $0 < p_{l^{**}} < 1$ then $(M^+)^n_{\bar{p}}$ does not satisfy the 0-1 law for L^+. This is because the elements 1 and $l^{**} + 1$ are definable in L^+ and hence some L^+ sentence holds in $(M^+)^n_{\bar{p}}$ iff $\{1, l^{**} + 1\}$ is an edge of $(M^+)^n_{\bar{p}}$ which holds with probability $p_{l^{**}}$. In our case, as in L we can not distinguish edges of the form $\{l, l + l^*\}$ from the rest of the edges, the

0-1 law may hold even if such l^* exists. In Lemma 24 below we show that if, in fact, we can not "identify the edges" in $M^n_{\bar{p}}$ then the 0-1 law holds in $M^n_{\bar{p}}$. This is translated in Theorem 27 to a sufficient condition on $\bar{p}$ for the 0-1 law holding in $M^n_{\bar{p}}$, but not necessarily for the 1-hereditary 0-1 law. The proof uses "local" properties of graphs. It seems that some form of "1-hereditary" version of 27 is possible. In any case we could not find a necessary and sufficient condition for the 1-hereditary 0-1 law, and the analysis of this case is not complete.

We first find a necessary condition on $\bar{p}$ for the 1-hereditary weak convergence law. Let us start with a definition of a structure on a sequence $\bar{q} \in \mathfrak{P}$ that enables us to "identify" the l^*-boundary in $M^n_{\bar{q}}$.

Definition 19. *1. A sequence $\bar{q} \in \mathfrak{P}$ is called nice if:*
- *(a) $U^*(\bar{q}) = \{l^*\}$.*
- *(b) If $l_1, l_2, l_3 \in \{l < n_{\bar{q}} : q_l > 0\}$ then $l_1 + l_2 \neq l_3$.*
- *(c) If $l_1, l_2, l_3, l_4 \in \{l < n_{\bar{q}} : q_l > 0\}$ then $l_1 + l_2 + l_3 \neq l_4$.*
- *(d) If $l_1, l_2, l_3, l_4 \in \{l < n_{\bar{p}} : q_l > 0\}$, $l_1 + l_2 = l_3 + l_4$ and $l_1 + l_2 < n_{\bar{q}}$ then $\{l_1, l_2\} = \{l_3, l_4\}$.*

2. Let ϕ^1 be the following L-formula:

$$\phi^1(y_1, z_1, y_2, z_2) := y_1 \sim z_1 \wedge z_1 \sim z_2 \wedge z_2 \sim y_2 \wedge y_2 \sim y_1 \wedge y_1 \neq z_2 \wedge z_1 \neq y_2.$$

3. For $k \geq 0$ define by induction on k the L-formula $\phi^1_k(y_1, z_1, y_2, z_2)$ by:
- $\phi^1_0(y_1, z_1, y_2, z_2) := y_1 = y_2 \wedge z_1 = z_2 \wedge y_1 \neq z_1$.
- $\phi^1_1(y_1, z_1, y_2, z_2) := \phi^1(y_1, z_1, y_2, z_2)$.
- $\phi^1_{k+1}(y_1, z_1, y_2, z_2) :=$ $(\exists y \exists z)[(\phi^1_k(y_1, z_1, y, z) \wedge \phi^1(y, z, y_2, z_2)) \vee (\phi^1_k(y_2, z_2, y, z) \wedge \phi^1(y_1, z_1, y, z))]$.

4. For $k_1, k_2, \in \mathbb{N}$ let $\phi^2_{k_1,k_2}(y, z)$ be the following L-formula:

$$(\exists x_1 \exists x_2 \exists x_3 \exists x_4)[\phi^1_{k_1}(y, z, x_2, x_3) \wedge \phi^1_{k_2}(x_2, x_1, x_4, x_3) \wedge \neg x_1 \sim x_4].$$

5. For $k_1, k_2, \in \mathbb{N}$ let $\phi^3_{k_1,k_2}$ be the following L formula:

$$\phi^3_{k_1,k_2}(x) := (\exists! y)[x \sim y \wedge \neg \phi^2_{k_1,k_2}(x, y)].$$

Observation 3. *Let $\bar{q} \in \mathfrak{P}$ be nice and $n \in \mathbb{N}$ be such that $n < n_{\bar{q}}$. Then the following holds in $M^n_{\bar{q}}$ with probability 1:*

1. *For $y_1, z_1, y_2, z_2 \in [n]$, if $M^n_{\bar{q}} \models \phi^1[y_1, z_1, y_2, z_2]$ then $y_1 - z_1 = y_2 - z_2$. (Use (d) in the definition of nice).*
2. *For $k \in \mathbb{N}$ and $y_1, z_1, y_2, z_2 \in [n]$, if $M^n_{\bar{q}} \models \phi^1_k[y_1, z_1, y_2, z_2]$ then $y_1 - z_1 = y_2 - z_2$. (Use (1) above and induction on k).*
3. *For $k_1, k_2 \in \mathbb{N}$ and $y, z \in [n]$, if $M^n_{\bar{q}} \models \phi^2_{k_1,k_2}[y, z]$ then $|y - z| \neq l^*$. (Use (2) above and the definition of $\phi^2_{k_1,k_2}(y, z)$).*
4. *For $k_1, k_2 \in \mathbb{N}$ and $x \in [n]$, if $M^n_{\bar{q}} \models \phi^3_{k_1,k_2}[x]$ then $x \in [1, l^*] \cup (n - l^*, n]$. (Use (3) above).*

The following claim shows that if $\bar{q}$ is nice (and has a certain structure) then, with probability close to 1, $\phi^3_{3,0}[y]$ holds in $M^n_{\bar{q}}$ for all $y \in [1, l^*] \cup (n - l^*, n]$. This, together with (4) in the observation above gives us a "definition" of the l^*-boundary in $M^n_{\bar{q}}$.

claim 5. Let $\bar{q} \in \mathfrak{P}^{fin}$ be nice and denote $n = n_{\bar{q}}$. Assume that for all $l > 0$, $q_l > 0$ implies $l < \lfloor n/3 \rfloor$. Assume further that for some $\epsilon > 0$, $0 < q_l < 1 \Rightarrow \epsilon < q_l < 1 - \epsilon$. Let $y_0 \in [1, l^*] \cup (n - l^*, n]$. Denote $m := |\{0 < l < n_{\bar{p}} : 0 < q_l < 1\}|$. Then:

$$Pr[M^n_{\bar{q}} \models \neg\phi^3_{3,0}[y_0]] \leq (\sum_{\{y \in [n] : |y_0 - y| \neq l^*\}} q_{|y_0 - y|})(1 - \epsilon^{11})^{m/2-1}.$$

Proof. We deal with the case $y_0 \in [1, l^*]$; the case $y_0 \in (n - l^*, n]$ is symmetric. Let $z_0 \in [n]$ be such that $l_0 := z_0 - y_0 \in \{0 < l < n : 0 < q_l < 1\}$ (so $l_0 \neq l^*$ and $l_0 < \lfloor n/3 \rfloor$), and assume that $M^n_{\bar{q}} \models y_0 \sim z_0$. For any $l_1, l_2 < \lfloor n/3 \rfloor$ denote (see Figure 1): $y_1 := y_0 + l_1$, $y_2 := y_0 + l_2$, $y_3 := y_2 + l_1 = y_1 + l_2 = y_0 + l_1 + l_2$ and symmetrically for z_1, z_2, z_3 (so y_i and z_i for $i \in \{0, 1, 2, 3\}$ all belong to $[n]$).

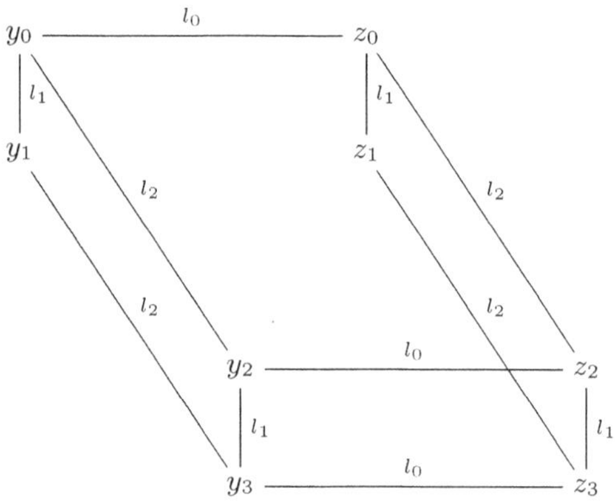

Fig. 1.

The following holds in $M^n_{\bar{q}}$ with probability 1: <u>If</u> for some $l_1, l_2 < \lfloor n/3 \rfloor$ such that (l_0, l_1, l_2) is without repetitions, we have:

$(*)_1$ (y_0, y_1, y_3, y_2), (z_0, z_1, z_3, z_2) and (y_2, y_3, z_3, z_2) are all cycles in $M^n_{\bar{q}}$.
$(*)_2$ $\{y_1, z_1\}$ is *not* an edge of $M^n_{\bar{q}}$.

<u>Then</u> $M^n_{\bar{q}} \models \phi^2_{0,3}[y_0, z_0]$. Why? As (y_1, y_0, z_0, z_1), in the place of (x_1, x_2, x_3, x_4), exemplifies $M^n_{\bar{p}} \models \phi^2_{0,3}[y_0, z_0]$. Let us fix $z_0 = y_0 + l_0$ and assume that $M^n_{\bar{q}} \models y_0 \sim z_0$. (Formally we condition the probability space $M^n_{\bar{q}}$ on the event $y_0 \sim z_0$.) Denote

$$L^{y_0, z_0} := \{(l_1, l_2) : q_{l_1}, q_{l_2} > 0, l_0 \neq l_1, l_0 \neq l_2, l_1 \neq l_2\}.$$

For $(l_1, l_2) \in L^{y_0,z_0}$, the probability that $(*)_1$ and $(*)_2$ holds, is $(1 - q_{l_0})(q_{l_0})^2(q_{l_1})^4(q_{l_2})^4$. Denote the event that $(*)_1$ and $(*)_2$ holds by $E^{y_0,z_0}(l_1, l_2)$. Note that if $(l_1, l_2), (l'_1, l'_2) \in L^{y_0,z_0}$ are such that (l_1, l_2, l'_1, l'_2) is without repetitions and $l_1 + l_2 \neq l'_1 + l'_2$ then the events $E^{y_0,z_0}(l_1, l_2)$ and $E^{y_0,z_0}(l'_1, l'_2)$ are independent. Now recall that $m := |\{l > 0 : \epsilon < q_l < 1 - \epsilon\}|$. Hence we have some $L' \subseteq L^{y_0,z_0}$ such that: $|L'| = \lfloor m/2 - 1 \rfloor$, and if $(l_1, l_2), (l'_1, l'_2) \in L'$ then the events $E^{y_0,z_0}(l_1, l_2)$ and $E^{y_0,z_0}(l'_1, l'_2)$ are independent. We conclude that

$$Pr[M^n_{\bar{q}} \models \neg\phi^2_{0,3}[y_0, z_0] | M^n_{\bar{q}} \models y_0 \sim z_0] \leq$$

$$(1 - (1 - q_{l_0})(q_{l_0})^2(q_{l_1})^4(q_{l_2})^4)^{m/2-1} \leq (1 - \epsilon^{11})^{m/2-1}.$$

This is a common bound for all $z_0 = y_0 + l_0$, and the same bound holds for all $z_0 = y_0 - l_0$ (whenever it belongs to $[n]$). We conclude that the expected number of $z_0 \in [n]$ such that: $|z_0 - y_0| \neq l^*$, $M^n_{\bar{q}} \models y_0 \sim z_0$ and $M^n_{\bar{q}} \models \neg\phi^2_{0,3}[y_0, z_0]$ is at most $(\sum_{\{y \in [n]: |y_0 - y| \neq l^*\}} q_{|y_0 - y|})(1 - \epsilon^{11})^{m/2-1}$. Now by (3) in Observation 3, $M^n_{\bar{q}} \models \phi^2_{0,3}[y_0, y_0 + l^*]$. By Markov's inequality and the definition of $\phi^3_{0,3}(x)$ we are done.

We now prove two lemmas which allow us to construct a sequence $\bar{q}$ such that for $\varphi := \exists x \phi^3_{0,3}(x)$ each of φ and $\neg\varphi$ will hold infinitely often in $M^n_{\bar{q}}$.

Lemma 20. *Assume $\bar{p}$ satisfies $\sum_{l>0} p_l = \infty$, and let $\bar{q} \in Gen^r_1(\bar{p})$ be nice. Let $\zeta > 0$ be some rational number. Then there exists some $r' > r$ and $\bar{q}' \in Gen^{r'}_1(\bar{p})$ such that: $\bar{q}'$ is nice, $\bar{q} \lhd \bar{q}'$ and $Pr[M^{n_{\bar{q}'}}_{\bar{q}'} \models \varphi] \leq \zeta$.*

Proof. Define $p^1 := (\prod_{l \in [n_{\bar{q}}] \setminus \{l^*\}} (1 - p_l))^2$, and choose $r' > r$ large enough so that $\sum_{r < l \leq r'} p_l \geq 2l^* \cdot p^1/\zeta$. Now define $\bar{q}' \in Gen^{r'}_1(\bar{p})$ in the following way:

$$q'_l = \begin{cases} q_l & 0 < l < n_{\bar{q}} \\ 0 & n_{\bar{q}} \leq l < (r' - r) \cdot n_{\bar{q}} \\ p_{r+i} & l = (r' - r + i) \cdot n_{\bar{q}} \text{ for some } 0 < i \leq (r' - r) \\ 0 & (r' - r) \cdot n_{\bar{q}} \leq l < 2(r' - r) \cdot n_{\bar{q}} \text{ and } l \not\equiv 0 \pmod{n_{\bar{q}}}. \end{cases}$$

Note that indeed $\bar{q}'$ is nice and $\bar{q} \lhd \bar{q}'$. Denote $n := n_{\bar{q}'} = 2(r' - r) \cdot n_{\bar{q}}$. Note further that every member of $M^n_{\bar{q}'}$ have at most one neighbor of distance more more than $n/2$, and all the rest of its neighbors are of distance at most $n_{\bar{q}}$. We now bound from above the probability of $M^n_{\bar{q}'} \models \exists x \phi^3_{0,3}(x)$. Let x be in $[1, l^*]$. For each i in the range $0 < i \leq (r' - r)$ denote $y_i := x + (r' - r + i) \cdot n_{\bar{q}}$ (hence $y_i \in [n/2, n]$) and let E_i be the following event: "$M^n_{\bar{q}'} \models y_i \sim z$ iff $z \in \{x, y_i + l^*, y_i - l^*\}$". By the definition of $\bar{q}'$, each y_i can only be connected either to x or to members of $[y - n_{\bar{q}}, y + n_{\bar{q}}]$, and hence we have

$$Pr[E_i] = q'_{(r'-r+i) \cdot n_{\bar{q}}} \cdot p^1 = p_{r+i} \cdot p^1.$$

As $i \neq j \Rightarrow n/2 > |y_i - y_j| > n_{\bar{q}}$ we have that the E_i-s are independent events. Now if E_i holds then by the definition of $\phi^2_{0,3}$ we have $M^n_{\bar{q}'} \models \neg\phi^2_{0,3}[x, y_i]$, and as $M^n_{\bar{q}'} \models \neg\phi^2_{0,3}[x, x+l^*]$ this implies $M^n_{\bar{q}'} \models \neg\phi^3_{0,3}[x]$. Let the random variable X denote the number of i in the range $0 < i \leq (r'-r)$ such that E_i holds in $M^n_{\bar{q}'}$. Then by Chebyshev's inequality we have:

$$Pr[M^n_{\bar{q}'} \models \phi^3_{0,3}[x]] \leq$$
$$Pr[X=0] \leq \frac{Var(X)}{Exp(X)^2} \leq \frac{1}{Exp(X)} \leq \frac{p^1}{\sum\limits_{0<i\leq(r'-r)} p_{r+i}} \leq \frac{\zeta}{2l^*}.$$

This is true for each $x \in [1, l^*]$ and the symmetric argument gives the same bound for each $x \in (n-l^*, n]$. Finally note that if $x, x+l^*$ both belong to $[n]$ then $M^n_{\bar{q}'} \models \neg\phi^2_{0,3}[x, x+l^*]$ (see Observation 3(4)). Hence if $x \in (l^*, n-l^*]$ then $M^n_{\bar{q}'} \models \neg\phi^3_{0,3}[x]$. We conclude that:

$$Pr[M^n_{\bar{q}'} \models \exists x \phi^3_{0,3}(x)] = Pr[M^n_{\bar{q}'} \models \phi] \leq \zeta$$

as desired. □

Lemma 21. *Assume $\bar{p}$ satisfies $0 < p_l < 1 \Rightarrow \epsilon < p_l < 1-\epsilon$ for some $\epsilon > 0$, and $\sum_{n=1}^{\infty} p_n = \infty$. Let $\bar{q} \in Gen^r_1(\bar{p})$ be nice, and $\zeta > 0$ be some rational number. Then there exist some $r' > r$ and $\bar{q}' \in Gen^{r'}_1(\bar{p})$ such that: $\bar{q}'$ is nice, $\bar{q} \lhd \bar{q}'$ and $Pr[M^{n_{\bar{q}'}}_{\bar{q}'} \models \varphi] \geq 1-\zeta$.*

Proof. This is a direct consequence of Claim 5. For each $r' > r$ denote $m(r') := |\{0 < l \leq r' : 0 < p_l < 1\}|$. Trivially we can choose $r' > r$ such that $m(r')(1-\epsilon^{11})^{m(r')/2-1} \leq \zeta$. As $\bar{q}$ is nice there exists some nice $\bar{q}' \in Gen^{r'}_1(\bar{p})$ such that $\bar{q} \lhd \bar{q}'$. Note that

$$\sum_{\{y\in[n]:|1-y|\neq l^*\}} q'_{|1-y|} \leq \sum_{\{0<l<n_{\bar{q}'}:l\neq l^*\}} q'_l \leq m(r')$$

and hence by 5 we have:

$$Pr[M^n_{\bar{q}'} \models \neg\phi] \leq Pr[M^n_{\bar{q}'} \models \neg\phi^3_{2,0}[1]] \leq m(r')(1-\epsilon^{11})^{m(r')/2-1} \leq \zeta$$

as desired. □

From the last two lemmas we conclude:

Conclusion 2. *Assume that $\bar{p}$ satisfies $0 < p_l < 1 \Rightarrow \epsilon < p_l < 1-\epsilon$ for some $\epsilon > 0$, and $\sum_{n=1}^{\infty} p_n = \infty$. Then $\bar{p}$ does not satisfy the 1-hereditary weak convergence law for L.*

The proof is by inductive construction of $\bar{q} \in Gen_1(\bar{p})$ such that for $\varphi := \exists x \phi^3_{0,3}(x)$ both φ and $\neg\varphi$ hold infinitely often in $M^n_{\bar{q}}$, using Lemmas 20, 21 as done in previous proofs.

From Conclusion 2 we have a necessary condition on $\bar{p}$ for the 1-hereditary weak convergence law. We now find a sufficient condition on $\bar{p}$ for the (not necessarily 1-hereditary) 0-1 law. Let us start with definitions of distance in graphs and of local properties in graphs.

Definition 22. *Let G be a graph on vertex set $[n]$.*

1. *For $x, y \in [n]$ let*
$$dist^G(x, y) := \min\{k \in \mathbb{N} : G \text{ has a path of length } k \text{ from } x \text{ to } y\}.$$
Note that for each $k \in \mathbb{N}$ there exists some L-formula $\theta_k(x, y)$ such that for all G and $x, y \in [n]$:
$$G \models \theta_k[x, y] \quad \text{iff} \quad dist^G(x, y) \leq k.$$
2. *For $x \in [n]$ and $r \in \mathbb{N}$ let $B^G(r, x) := \{y \in [n] : dist^G(x, y) \leq r\}$ be the ball with radius r and center x in G.*
3. *An L-formula $\phi(x)$ is called r-local if every quantifier in ϕ is restricted to the set $B^G(r, x)$. Formally each appearance of the form $\forall y...$ in ϕ is of the form $(\forall y)[\theta_r(x, y) \rightarrow ...]$, and similarly for $\exists y$ and other variables. Note that for any G, $x \in [n]$, $r \in \mathbb{N}$ and an r-local formula $\phi(x)$ we have:*
$$G \models \phi[x] \quad \text{iff} \quad G|_{B(r,x)} \models \phi[x].$$
4. *An L-sentence is called* local *if it has the form*
$$\exists x_1 ... \exists x_m \bigwedge_{1 \leq i \leq m} \phi(x_i) \wedge \bigwedge_{1 \leq i < j \leq m} \neg\theta_{2r}(x_i, x_j)$$
where $\phi = \phi(x)$ is an r-local formula for some $r \in \mathbb{N}$.
5. *For $l, r \in \mathbb{N}$ and an L-formula $\phi(x)$ we say that the l-boundary of G is r-indistinguishable by $\phi(x)$ if for all $z \in [1, l] \cup (n - l, n]$ there exists some $y \in [n]$ such that $B^G(r, y) \cap ([1, l] \cup (n - l, n]) = \emptyset$ and $G \models \phi[z] \leftrightarrow \phi[y]$*

We can now use the following famous result from [1]:

Theorem 23 (Gaifman's Theorem). *Every L-sentence is logically equivalent to a boolean combination of local L-sentences.*

We will use Gaifman's theorem to prove:

Lemma 24. *Assume that for all $k \in \mathbb{N}$ and k-local L-formulas $\varphi(z)$ we have:*
$$\lim_{n \to \infty} Pr[\text{The } l^*\text{-boundary of } M^n_{\bar{p}} \text{ is } k\text{-indistinguishable by } \varphi(z)] = 1.$$
Then the 0-1 law for L holds in $M^n_{\bar{p}}$.

Proof. By Gaifman's theorem it is enough if we prove that the 0-1 law holds in $M^n_{\bar{p}}$ for local L-sentences. Let
$$\psi := \exists x_1 ... \exists x_m \bigwedge_{1 \leq i \leq m} \phi(x_i) \quad \bigwedge_{1 \leq i < j \leq m} \neg\theta_{2r}(x_i, x_j)$$
be some local L-sentence, where $\phi(x)$ is an r-local formula.

Define $\mathfrak{H}$ to be the set of all 4-tuples (l, U, u_0, H) such that: $l \in \mathbb{N}$, $U \subseteq [l]$, $u_0 \in U$ and H is a graph with vertex set U. We say that some $(l, U, u_0, H) \in \mathfrak{H}$ is r-proper for $\bar{p}$ (but as $\bar{p}$ is fixed we usually omit it) if it satisfies:

($*_1$) For all $u \in U$, $dist^H(u_0, u) \leq r$.
($*_2$) For all $u \in U$, if $dist^H(u_0, u) < r$ then $u + l^*, u - l^* \in U$.
($*_3$) $Pr[M^l_{\bar{p}}|_U = H] > 0$.

We say that a member of $\mathfrak{H}$ is proper if it is r-proper for some $r \in \mathbb{N}$.

Let H be a graph on vertex set $U \subseteq [l]$ and G be a graph on vertex set $[n]$. We say that $f : U \to [n]$ is a strong embedding of H in G if:

- f in one-to one.
- For all $u, v \in U$, $H \models u \sim v$ iff $G \models f(u) \sim f(v)$.
- For all $u, v \in U$, $f(u) - f(v) = u - v$.
- If $i \in Im(f)$, $j \in [n] \setminus Im(f)$ and $|i - j| \neq l^*$ then $G \models \neg i \sim j$.

We make two observations which follow directly from the definitions:

1. If $(l, U, u_0, H) \in \mathfrak{H}$ is r-proper and $f : U \to [n]$ is a strong embedding of H in G then $Im(f) = B^G(r, f(u_0))$. Furthermore for any r-local formula $\phi(x)$ and $u \in U$ we have, $G \models \phi[f(u)]$ iff $H \models \phi[u]$.
2. Let G be a graph on vertex set $[n]$ such that $Pr[M^n_{\bar{p}} = G] > 0$, and $x \in [n]$ be such that $B^G(r - 1, x)$ is disjoint to $[1, l^*] \cup (n - l^*, n]$. Denote by m and M the minimal and maximal elements of $B^G(r, x)$ respectively. Denote by U the set $\{i - m + 1 : i \in B^G(r, x)\}$ and by H the graph on U defined by $H \models u \sim v$ iff $G \models (u + m - 1) \sim (v + m - 1)$. Then the 4-tuple $(M - m + 1, U, x - m + 1, H)$ is an r-proper member of $\mathfrak{H}$. Furthermore for any r-local formula $\phi(x)$ and $u \in U$ we have $G \models \phi[u - m + 1]$ iff $H \models \phi[u]$.

We now show that for any proper member of $\mathfrak{H}$ there are many disjoint strong embeddings into $M^n_{\bar{p}}$. Formally:

claim 6. Let $(l, U, u_0, H) \in \mathfrak{H}$ be proper, and $c > 1$ be some fixed real. Let E^n_c be the following event on $M^n_{\bar{p}}$: "For any interval $I \subseteq [n]$ of length at least n/c there exists some $f : U \to I$ a strong embedding of H in $M^n_{\bar{p}}$". Then

$$\lim_{n \to \infty} Pr[E^n_c \text{ holds in } M^n_{\bar{p}}] = 1.$$

We skip the proof of this claim because an almost identical lemma is proved in [2] (see Lemma at page 8 there).

We can now finish the proof of Lemma 24. Recall that $\phi(x)$ is an r-local formula. We consider two possibilities. First assume that for some r-proper $(l, U, u_0, H) \in \mathfrak{H}$ we have $H \models \phi[u_0]$. Let $\zeta > 0$ be some real. Then by the claim above, for n large enough, with probability at least $1 - \zeta$ there exist $f_1, ..., f_m$ strong embeddings of H into $M^n_{\bar{p}}$ such that $\langle Im(f_i) : 1 \leq i \leq m \rangle$ are pairwise disjoint. By observation (1) above we have:

- For $1 \leq i < j \leq m$, $B^{M^n_{\bar{p}}}(r, f_i(u_0)) \cap B^{M^n_{\bar{p}}}(r, f_j(u_0)) = \emptyset$.
- For $1 \leq i \leq m$, $M^n_{\bar{p}} \models \phi[f_i(u_0)]$.

Hence $f_1(u_0), ..., f_m(u_0)$ exemplifies ψ in $M^n_{\bar{p}}$, so $Pr[M^n_{\bar{p}} \models \psi] \geq 1 - \zeta$ and as ζ was arbitrary we have $\lim_{n \to \infty} Pr[M^n_{\bar{p}} \models \psi] = 1$ and we are done.

Otherwise assume that for all r-proper $(l, U, u_0, H) \in \mathfrak{H}$ we have $H \models \neg\phi[u_0]$. We will show that $\lim_{n\to\infty} Pr[M^n_{\bar{p}} \models \psi] = 0$ which will finish the proof. Towards contradiction assume that for some $\epsilon > 0$ for unboundedly many $n \in \mathbb{N}$ we have $Pr[M^n_{\bar{p}} \models \psi] \geq \epsilon$. Define the L-formula:

$$\varphi(z) := (\exists x)(\theta_{r-1}(x, z) \wedge \phi(x)).$$

Note that $\varphi(z)$ is equivalent to a k-local formula for $k = 2r - 1$. Hence by the assumption of our lemma for some (large enough) $n \in \mathbb{N}$ we have with probability at least $\epsilon/2$: $M^n_{\bar{p}} \models \psi$ and the l^*-boundary of $M^n_{\bar{p}}$ is k-indistinguishable by $\varphi(z)$. In particular for some $n \in \mathbb{N}$ and G a graph on vertex set $[n]$ we have:

(α) $Pr[M^n_{\bar{p}} = G] > 0$.
(β) $G \models \psi$.
(γ) The l^*-boundary of G is k indistinguishable by $\varphi(z)$.

By (β) for some $x_0 \in [n]$ we have $G \models \phi[x_0]$. If x_0 is such that $B^G(r - 1, x_0)$ is disjoint to $[1, l^*] \cup (n - l^*, n]$ then by (α) and observation (2) above we have some r-proper $(l, U, u_0, H) \in \mathfrak{H}$ such that $H \models \phi[u_0]$ in contradiction to our assumption. Hence assume that $B^G(r-1, x_0)$ is not disjoint to $[1, l^*] \cup (n-l^*, n]$ and let $z_0 \in [n]$ belong to their intersection. So by the definition of $\varphi(z)$ we have $G \models \varphi[z_0]$ and by (γ) we have some $y_0 \in [n]$ such that $B^G(k, y_0) \cap ([1, l^*] \cup (n - l^*, n]) = \emptyset$ and $G \models \varphi[y_0]$. Again by the definition of $\varphi(z)$, and recalling that $k = 2r-1$ we have some $x_1 \in [n]$ such that $B^G(r - 1, x_1) \cap ([1, l^*] \cup (n - l^*, n]) = \emptyset$ and $G \models \phi[x_1]$. So again by (α) and observation (2) we get a contradiction. □

Remark 25. Lemma 24 above gives a sufficient condition for the 0-1 law. If we are only interested in the convergence law, then a weaker condition is sufficient; all we need is that the probability of any local property holding in the l^*-boundary converges. Formally:

Assume that for all $r \in \mathbb{N}$ and r-local L-formulas, $\phi(x)$, and for all $1 \leq l \leq l^*$ we have: Both $\langle Pr[M^n_{\bar{p}} \models \phi[l] : n \in \mathbb{N}\rangle$ and $\langle Pr[M^n_{\bar{p}} \models \phi[n - l + 1] : n \in \mathbb{N}\rangle$ converge to a limit. Then $M^n_{\bar{p}}$ satisfies the convergence law.

The proof is similar to the proof of Lemma 24. A similar proof on the convergence law in graphs with the successor relation is Theorem 2(i) in [2].

We now use 24 to get a sufficient condition on $\bar{p}$ for the 0-1 law holding in $M^n_{\bar{p}}$. Our proof relies on the assumption that $M^n_{\bar{p}}$ contains few cycles, and only those that are "unavoidable". We start with a definition of such cycles:

Definition 26. *Let $n \in \mathbb{N}$.*

1. *For a sequence $\bar{x} = (x_0, x_1, ..., x_k) \subseteq [n]$ and $0 \leq i < k$ denote $l^{\bar{x}}_i := x_{i+1} - x_i$.*
2. *A sequence $(x_0, x_1, ..., x_k) \subseteq [n]$ is called possible for $\bar{p}$ (but as $\bar{p}$ is fixed we omit it and similarly below) if for each i in the range $0 \leq i < k$, $p_{|l^{\bar{x}}_i|} > 0$.*
3. *A sequence $(x_0, x_1, ..., x_k)$ is called a cycle of length k if $x_0 = x_k$ and $\langle\{x_i, x_{i+1}\} : 0 \leq i < k\rangle$ is without repetitions.*

4. *A cycle of length k, is called simple if $(x_0, x_1, ..., x_{k-1})$ is without repetitions.*
5. *For $\bar{x} = (x_0, x_1, ..., x_k) \subseteq [n]$, a pair $(S \dot{\cup} A)$ is called a symmetric partition of $\bar{x}$ if:*
   - $S \dot{\cup} A = \{0, ..., k-1\}$.
   - *If $i \neq j$ belong to A then $l_i^{\bar{x}} + l_j^{\bar{x}} \neq 0$.*
   - *The sequence $\langle l_i^{\bar{x}} : i \in S\rangle$ can be partitioned into two sequences of length $r = |S|/2$: $\langle l_i : 0 \leq i < r\rangle$ and $\langle l'_i : 0 \leq i < r\rangle$ such that $l_i + l'_i = 0$ for each i in the range $0 \leq i < r$.*
6. *For $\bar{x} = (x_0, x_1, ..., x_k) \subseteq [n]$ let $(Sym(\bar{x}), Asym(\bar{x}))$ be some symmetric partition of $\bar{x}$ (say the first in some prefixed order). Denote $Sym^+(\bar{x}) := \{i \in Sym(\bar{x}) : l_i^{\bar{x}} > 0\}$.*
7. *We say that $\bar{p}$ has no unavoidable cycles if for all $k \in \mathbb{N}$ there exists some $m_k \in \mathbb{N}$ such that if $\bar{x}$ is a* possible *cycle of length k then for each $i \in Asym(\bar{x})$, $|l_i^{\bar{x}}| \leq m_k$.*

Theorem 27. *Assume that $\bar{p}$ has no unavoidable cycles, $\sum_{l=1}^{\infty} p_l = \infty$ and $\sum_{l=1}^{\infty} (p_l)^2 < \infty$. Then $M_{\bar{p}}^n$ satisfies the 0-1 law for L.*

Proof. Let $\phi(x)$ be some r-local formula, and j^* be in $\{1, 2, ..., l^*\} \cup \{-1, -2, ..., -l^*\}$. For $n \in \mathbb{N}$ let $z_n^* = z^*(n, j^*)$ equal j^* if $j^* > 0$ and $n - j^* + 1$ if $j^* < 0$ (so z_n^* belongs to $[1, l^*] \cup (n - l^*, n]$). We will show that with probability approaching 1 as $n \to \infty$ there exists some $y^* \in [n]$ such that $B^{M_{\bar{p}}^n}(r, y^*) \cap ([1, l^*] \cup (n - l^*, n]) = \emptyset$ and $M_{\bar{p}}^n \models \phi[z_n^*] \leftrightarrow \phi[y^*]$. This will complete the proof by Lemma 24. For simplicity of notation assume $j^* = 1$ hence $z_n^* = 1$ (the proof of the other cases is similar). We use the notations of the proof of 24. In particular recall the definition of the set $\mathfrak{H}$ and of an r-proper member of $\mathfrak{H}$. Now if for two r-proper members of $\mathfrak{H}$, (l^1, x^1, U^1, H^1) and (l^2, x^2, U^2, H^2) we have $H^1 \models \phi[x^1]$ and $H^2 \models \neg\phi[x^2]$ then by Claim 6 we are done. Otherwise all r-proper members of $\mathfrak{H}$ give the same value to $\phi[x]$ and without loss of generality assume that if $(l, x, U, H) \in \mathfrak{H}$ is a r-proper then $H \models \phi[x]$ (the dual case is identical). If $\lim_{n\to\infty} Pr[M_{\bar{p}}^n \models \phi[1]] = 1$ then again we are done by 6. Hence we may assume that:

$\odot$ For some $\epsilon > 0$, for an unbounded set of $n \in \mathbb{N}$, $Pr[M_{\bar{p}}^n \models \neg\phi[1]] \geq \epsilon$.

In the construction below we use the following notations: 2 denotes the set $\{0, 1\}$. ${}^k 2$ denotes the set of sequences of length k of members of 2, and if η belongs to ${}^k 2$ we write $|\eta| = k$. ${}^{\leq k}2$ denotes $\bigcup_{0 \leq i \leq k} {}^k 2$ and similarly ${}^{<k}2$. $\langle\rangle$ denotes the empty sequence, and for $\eta, \eta' \in {}^{\leq k}2$, $\eta\hat{}\eta'$ denotes the concatenation of η and η'. Finally for $\eta \in {}^k 2$ and $k' < k$, $\eta|_{k'}$ is the initial segment of length k' of η.

Call $\bar{y}$ a saturated tree of depth k in $[n]$ if:

- $\bar{y} = \langle y_\eta \in [n] : \eta \in {}^{\leq k}2\rangle$.
- $\bar{y}$ is without repetitions.
- $\{y_{\langle 0\rangle}, y_{\langle 1\rangle}\} = \{y_{\langle\rangle} + l^*, y_{\langle\rangle} - l^*\}$.
- If $0 < l < k$ and $\eta \in {}^l 2$ then $\{y_\eta + l^*, y_\eta - l^*\} \subseteq \{y_{\eta\hat{}\langle 0\rangle}, y_{\eta\hat{}\langle 1\rangle}, y_{\eta|_{l-1}}\}$.

Let G be a graph with set of vertices $[n]$, and $i \in [n]$. We say that $\bar{y}$ is a cycle free saturated tree of depth k for i in G if:

(i) $\bar{y}$ is a saturated tree of depth k in $[n]$.
(ii) $G \models i \sim y_{\langle\rangle}$ but $|i - y_{\langle\rangle}| \neq l^*$.
(iii) For each $\eta \in {}^{<k}2$, $G \models y_\eta \sim y_{\hat{\eta}\langle 0\rangle}$ and $G \models y_\eta \sim y_{\hat{\eta}\langle 1\rangle}$.
(iv) None of the edges described in (ii),(iii) belongs to a cycle of length $\leq 6k$ in G.
(v) Recalling that $\bar{p}$ has no unavoidable cycles let m_{2k} be the one from Definition 26(7). For all $\eta \in {}^{\leq k}2$ and $y \in [n]$ if $G \models y_\eta \sim y$ and $y \notin \{y_{\hat{\eta}\langle 0\rangle}, y_{\hat{\eta}\langle 1\rangle}, y_{\eta|_{l-1}}, i\}$ then $|y - y_\eta| > m_{2k}$.

For $I \subseteq [n]$ we say that $\langle \bar{y}^i : i \in I\rangle$ is a cycle free saturated forest of depth k for I in G if:

(a) For each $i \in I$, $\bar{y}^i$ is a cycle free saturated tree of depth k for i in G.
(b) As sets $\langle \bar{y}^i : i \in I\rangle$ are pairwise disjoint.
(c) If $i_1, i_2 \in I$ and $\bar{x}$ is a path of length $k' \leq k$ in G from $y^{i_1}_{\langle\rangle}$ to i_2, then for some $j < k'$, $(x_j, x_{j+1}) = (y^{i_1}_{\langle\rangle}, i_1)$.

claim 7. For $n \in \mathbb{N}$ and G a graph on $[n]$ denote by $I^*_k(G)$ the set $([1, l^*] \cup (n - l^*, n]) \cap B^G(1, k)$. Let $E^{n,k}$ be the event: "There exists a cycle free saturated forest of depth k for $I^*_k(G)$". Then for each $k \in \mathbb{N}$:

$$\lim_{n\to\infty} Pr[E^{n,k} \text{ holds in } M^n_{\bar{p}}] = 1.$$

Proof. Let $k \in \mathbb{N}$ be fixed. The proof proceeds in six steps:

Step 1. We observe that only a bounded number of cycles starts in each vertex of $M^n_{\bar{p}}$. Formally: For $n, m \in \mathbb{N}$ and $i \in [n]$ let $E^1_{n,m,i}$ be the event: "More than m different cycles of length at most $12k$ include i". Then for all $\zeta > 0$ for some $m = m(\zeta)$ (m depends also on $\bar{p}$ and k but as those are fixed we omit them from the notation and similarly below) we have:

$\circledast_1$ For all $n \in \mathbb{N}$ and $i \in [n]$, $Pr_{M^n_{\bar{p}}}[E^1_{n,m,i}] \leq \zeta$.

To see this note that if $\bar{x} = (x_0, ..., x_{k'})$ is a possible cycle in $[n]$, then

$$Pr[\bar{x} \text{ is a weak cycle in } M^n_{\bar{p}}] := p(\bar{x}) = \prod_{i\in Asym(\bar{x})} p_{|l^{\bar{x}}_i|} \cdot \prod_{i\in Sym^+(\bar{x})} (p_{l^{\bar{x}}_i})^2.$$

Now as $\bar{p}$ has no unavoidable, cycles let m_{12k} be as in 26(7). Then the expected number of cycles of length $\leq 12k$ starting in $i = x_0$ is

$$\sum_{\substack{k'\leq 12k, \bar{x}=(x_0,\ldots,x_{k'})\\ \text{is a possible cycle}}} p(\bar{x}) \leq$$

$$(m_{12k})^{12k} \cdot \sum_{0<l_1,\ldots,l_{6k}<n} \prod_{i=1}^{6k} (p_{l_i})^2 \leq (m_12k)^{12k} \cdot (\sum_{0<l<n} (p_l)^2)^{6k}.$$

But as $\sum_{0<l<n} (p_l)^2$ is bounded by $\sum_{l=1}^{\infty} (p_l)^2 := c^* < \infty$, if we take $m = (m_{12k})^{12k} \cdot (c^*)^{6k}/\zeta$ then we have $\circledast_1$ as desired.

Step 2. We show that there exists a positive lower bound on the probability that a cycle passes through a given edge of $M^n_{\bar{p}}$. Formally: Let $n \in \mathbb{N}$ and $i, j \in [n]$ be such that $p_{|i-j|} > 0$. Denote By $E^2_{n,i,j}$ the event: "There does not exists a cycle of length $\leq 6k$ containing the edge $\{i, j\}$". Then there exists some $q_2 > 0$ such that:

⊛$_2$ For any $n \in \mathbb{N}$ and $i, j \in [n]$ such that $p_{|i-j|} > 0$, $Pr_{M^n_{\bar{p}}}[E^2_{n,i,j} | i \sim j] \geq q_2$.

To see this call a path $\bar{x} = (x_0, ..., x_{k'})$ good for $i, j \in [n]$ if $x_0 = j$, $x_{k'} = i$, $\bar{x}$ does not contain the edge $\{i, j\}$ and does not contain the same edge more than once. Let $E'^2_{n,i,j}$ be the event: "There does not exists a path good for i, j of length $< 6k$". Note that for $i, j \in [n]$ and G a graph on $[n]$ such that $G \models i \sim j$ we have: $(i, j, x_2, ..., x_{k'})$ is a cycle in G iff $(j, x_2, ..., k_{k'})$ is a path in G good for i, j. Hence for such G we have: $E^2_{n,i,j}$ holds in G iff $E'^2_{n,i,j}$ holds in G. Since the events $i \sim j$ and $E'^2_{n,i,j}$ are independent in $M^n_{\bar{p}}$ we conclude:

$$Pr_{M^n_{\bar{p}}}[E^2_{n,i,j} | i \sim j] = Pr_{M^n_{\bar{p}}}[E'^2_{n,i,j} | i \sim j] = Pr_{M^n_{\bar{p}}}[E'^2_{n,i,j}].$$

Next recalling Definition 26(7) let m_k be as there. Since $\sum_{l>0}(p_l)^2 < \infty$, $(p_l)^2$ converges to 0 as l approaches infinity, and hence so does p_l. Hence for some $m^0 \in \mathbb{N}$ we have that $l > m^0$ implies $p_l < 1/2$. Let $m^*_k := \max\{m_{6k}, m^0\}$. We now define for a possible path $\bar{x} = (x_0, ...x_{k'})$, $Large(\bar{x}) = \{0 \leq r < k' : |l^{\bar{x}}_r| > m^*_k\}$. Note that as $\bar{p}$ has no unavoidable cycles we have for any possible cycle $\bar{x}$ of length $\leq 6k$, $Large(\bar{x}) \subseteq Sym(\bar{x})$, and $|Large(\bar{x})|$ is even. We now make the following claim: For each k^* in the range $0 \leq k^* \leq \lfloor k/2 \rfloor$ let $E'^{2,k^*}_{n,i,j}$ be the event: "There does not exists a path, $\bar{x}$, good for i, j of length $< 6k$ with $|Large(\bar{x})| = 2k^*$". Then there exists a positive probability q_{2,k^*} such that for any $n \in \mathbb{N}$ and $i, j \in [n]$ we have:

$$Pr_{M^n_{\bar{p}}}[E'^{2,k^*}_{n,i,j}] \geq q_{2,k^*}.$$

Then by taking $q_2 = \prod_{0 \leq k^* \leq \lfloor k/2 \rfloor} q_{2,k^*}$ we will have ⊛$_2$. Let us prove the claim. For $k^* = 0$ we have (recalling that no cycle consists only of edges of length l^*):

$$Pr_{M^n_{\bar{p}}}[E'^{2,0}_{n,i,j}] = \prod_{\substack{k' \leq 6k,\ \bar{x}=(i=x_0, j=x_1, ..., x_{k'}) \\ \text{is a possible cycle, } |Large(\bar{x})|=0}} (1 - \prod_{r=1}^{k'-1} p_{|l^{\bar{x}}_r|})$$

$$\geq (1 - \max\{p_l : 0 < l \leq m^*_k, l \neq l^*\})^{6k \cdot (m^*_k)^{6k-1}}.$$

But as the last expression is positive and depends only on $\bar{p}$ and k we are done. For $k^* > 0$ we have:

$$Pr_{M^n_{\bar{p}}}[E'^{2,k^*}_{n,i,j}] = \prod_{\substack{k' \leq 6k,\ \bar{x}=(i=x_0, j=x_1, ..., x_{k'}) \\ \text{is a possible cycle, } |Large(\bar{x})|=k^*}} (1 - \prod_{m=1}^{k'-1} p_{|l^{\bar{x}}_m|})$$

$$= \prod_{\substack{k' \leq 6k,\ \bar{x}=(i=x_0,j=x_1,\ldots,x_{k'}) \\ \text{is a possible cycle,} \\ |Large(\bar{x})|=k^*\ ,0 \notin Large(\bar{x})}} (1 - \prod_{m=1}^{k'-1} p_{|l_m^{\bar{x}}|}) \cdot$$

$$\prod_{\substack{k' \leq 6k,\ \bar{x}=(i=x_0,j=x_1,\ldots,x_{k'}) \\ \text{is a possible cycle,} \\ |Large(\bar{x})|=k^*\ ,0 \in Large(\bar{x})}} (1 - \prod_{m=1}^{k'-1} p_{|l_m^{\bar{x}}|}).$$

But the product on the second line is at least

$$[\prod_{l_1,\ldots,l_{k^*} > m_k^*} (1 - \prod_{m=1}^{k^*} (p_{l_m})^2)]^{(m_k^*)^{(6k-2k^*)} \cdot (6k)^{2k^*}},$$

and as $\sum_{l > m_k^*} (p_l)^2 \leq c^* < \infty$ we have $\sum_{l_1,\ldots,l_{k^*} > m_k^*} \prod_{m=1}^{k^*} (p_{l_m})^2 \leq (c^*)^{k^*} < \infty$ and hence $\prod_{l_1,\ldots,l_{k^*} > m_k^*} (1 - \prod_{m=1}^{k^*} (p_{l_m})^2) > 0$ and we have a bound as desired. Similarly the product on the third line is at least

$$[\prod_{l_1,\ldots,l_{k^*-1} > m_k^*} (1 - \prod_{m=1}^{k^*-1} (p_{l_m})^2) \cdot 1/2]^{(m_k^*)^{(6k-2k^*-1)} \cdot (6k)^{2k^*}},$$

and again we have a bound as desired.

Step 3. Denote

$$E^3_{n,i,j} := E^2_{n,i,j} \wedge \bigwedge_{r=1,\ldots,k} (E^2_{n,j+(r-1)l^*,j+rl^*} \wedge E^2_{n,j,j-(r-1)l^*,j-rl^*})$$

and let $q_3 = q_2^{(2l^*+1)}$. We then have:

⊛$_3$ For any $n \in \mathbb{N}$ and $i, j \in [n]$ such that $p_{|i-j|} > 0$ and $j + kl^*, j - kl^* \in [n]$, $Pr_{M^n_{\bar{p}}}[E^3_{n,i,j} | i \sim j] \geq q_3$.

This follows immediately from ⊛$_2$, and the fact that if i, i', j, j' all belong to $[n]$ then the probability $Pr_{M^n_{\bar{p}}}[E^2_{n,i,j} | E^2_{n,i',j'}]$ is no smaller then the probability $Pr_{M^n_{\bar{p}}}[E^2_{n,i,j}]$.

Step 4. For $i, j \in [n]$ such that $j + kl^*, j - kl^* \in [n]$ denote by $E^4_{n,i,j}$ the event: "$E^3_{n,i,j}$ holds and for $x \in \{j + rl^* : r \in \{-k, -k+1, \ldots, k\}\}$ and $y \in [n] \setminus \{i\}$ we have $x \sim y \Rightarrow (|x - y| = l^* \vee |x - y| > m_{2k})$". Then for some $q_4 > 0$ we have:

⊛$_4$ For any $n \in \mathbb{N}$ and $i, j \in [n]$ such that $p_{|i-j|} > 0$ and $j + kl^*, j - kl^* \in [n]$, $Pr_{M^n_{\bar{p}}}[E^4_{n,i,j} | i \sim j] \geq q_4$.

To see this simply take $q_4 = q_3 \cdot (\prod_{l \in \{1,\ldots,m_{2k}\} \setminus \{l^*\}} (1 - p_l))^{2k+1}$, and use ⊛$_3$.

Step 5. For $n \in \mathbb{N}$, $S \subseteq [n]$, and $i \in [n]$ let $E^5_{n,S,i}$ be the event: "For some $j \in [n] \setminus S$ we have $i \sim j$, $|i - j| \neq l^*$ and $E^4_{n,i,j}$". Then for each $\delta > 0$ and $s \in \mathbb{N}$, for $n \in \mathbb{N}$ large enough (depending on δ and s) we have:

$\circledast_5$ For all $i \in [n]$ and $S \subseteq [n]$ with $|S| \leq s$, $Pr_{M^n_{\bar{p}}}[E^5_{n,S,i}] \geq 1 - \delta$.

First let $\delta > 0$ and $s \in \mathbb{N}$ be fixed. Second for $n \in \mathbb{N}$, $S \subseteq [n]$ and $i \in [n]$ denote by $J^{n,S}_i$ the set of all possible candidates for j, namely $J^{n,S}_i := \{j \in (kl^*, n-kl^*] \backslash S : |i-j| \neq l^*\}$. For $j \in J^{n,\emptyset}_i$ let $U_j := \{j + rl^* : r \in \{-k, -k+1, ..., k\}\}$. For $m \in \mathbb{N}$ and G a graph on $[n]$ call $j \in J^{n,S}_i$ a candidate of type (n, m, S, i) in G, if each $j' \in U(j)$, belongs to at most m different cycles of length at most $6k$ in G. Denote the set of all candidates of type (n, m, S, i) in G by $J^{n,S}_i(G)$. Now let $X^{n,m}_i$ be the random variable on $M^n_{\bar{p}}$ defined by:

$$X^{n,m}_i(M^n_{\bar{p}}) = \sum\{p_{|i-j|} : j \in J^{n,S}_i(M^n_{\bar{p}})\}.$$

Denote $R^{n,S}_i := \sum\{p_{|i-j|} : j \in J^{n,S}_i\}$. Trivially for all n, m, S, i as above, $X^{n,m}_i \leq R^{n,S}_i$. On the other hand, by $\circledast_1$ and the definition of a candidate, for all $\zeta > 0$ we can find $m = m(\zeta) \in \mathbb{N}$ such that for all n, S, i as above and $j \in J^{n,S}_i$, the probability that j is a candidate of type (n, m, S, i) in $M^n_{\bar{p}}$ is at least $1 - \zeta$. Then for such m we have: $Exp(X^{n,m}_i) \geq R^{n,S}_i(1 - \zeta)$. Hence we have $Pr_{M^n_{\bar{p}}}[X^{n,m}_i \leq R^{n,S}_i/2] \leq 2\zeta$. Recall that $\delta > 0$ was fixed, and let $m^* = m(\delta/4)$. Then for all n, S, i as above we have with probability at least $1 - \delta/2$, $X^{n,m^*}_i(M^n_{\bar{p}}) \geq R^{n,S}_i/2$. Now denote $m^{**} := (2l^* + 1)(m^* + 2m_{2k})6k(m^* + 1)$, and fix $n \in \mathbb{N}$ such that $\sum_{0<l<n} p_l > 2 \cdot ((m^{**}/(q_4 \cdot \delta) \cdot 2m_{2k}(2l^* + 1) + (s + 2kl^* + 2))$. Let $i \in [n]$ and $S \subseteq [n]$ be such that $|S| \leq s$. We relativize our probability space $M^n_{\bar{p}}$ to the event $X^{n,m^*}_i(M^n_{\bar{p}}) \geq R^{n,S}_i/2$, and all probabilities until the end of Step 5 will be conditioned to this event. If we show that under this assumption we have $Pr_{M^n_{\bar{p}}}[E^5_{n,S,i}] \geq 1 - \delta/2$, then we will have $\circledast_5$.

Let G be a graph on $[n]$ such that, $X^{n,m^*}_i(G) \geq R^{n,S}_i/2$. For $j \in J^{n,S}_i$ let $C_j(G)$ denote the set of all the pairs of vertices which are relevant for the event $E^4_{n,i,j}$. Namely $C_j(G)$ will contain: $\{i, j\}$, all the edges $\{u, v\}$ such that : $u \in U(j)$, $v \neq i$ and $|u - v| < m_{2k}$, and all the edges that belong to a cycle of length $\leq 6k$ containing some member of $U(j)$. We make some observations:

1. $X^{n,m^*}_i(G) \geq (m^{**}/(q_4 \cdot \delta)) \cdot 2m_{2k}(2l^* + 1)$.
2. There exists $J^1(G) \subseteq J^{n,S}_i$ such that:
   (a) The sets $U(j)$ for $j \in J^1(G)$ are pairwise disjoint. Moreover if $j_1, j_2 \in J^1(G)$, $u_l \in U(j_l)$ for $l \in \{1, 2\}$ and $j_1 \neq j_2$ then $|u_1 - u_2| > m_{2k}$.
   (b) Each $j \in J^1(G)$ is a candidate of type (n, m^*, S, i) in G.
   (c) The sum $\sum\{p_{|i-j|} : j \in J^1(G)\}$ is at least $m^{**}/(q_4 \cdot \delta)$.
   [To see this use (1) and construct J^1 by adding the candidate with the largest $p_{|i-j|}$ that satisfies (a). Note that each new candidate excludes at most $m_{2k}(2l^* + 1)$ others.]
3. Let j belong to $J^1(G)$. Then the set $\{j' \in J^1(G) : C_j(G) \cap C_{j'}(G) \neq \emptyset\}$ has size at most m^{**}. [To see this use (2)(b) above, the fact that two cycles of length $\leq 6k$ that intersect in an edge give a cycle of length $\leq 12k$ and similar trivial facts.]

4. From (3) we conclude that there exists $J^2(G) \subseteq J^1(G)$ and $\langle j_1, ... j_r \rangle$ an enumeration of $J^2(G)$ such that:
   (a) For any $1 \le r' \le r$ the sets $C(j_{r'})$ and $\cup_{1 \le r'' < r'} C(j_{r''})$ are disjoint.
   (b) The sum $\sum\{p_{|i-j|} : j \in J^2(G)\}$ is greater or equal $1/(q_4 \cdot \delta)$.

Now for each $j \in J_i^{n,S}$ let E_j^* be the event: "$i \sim j$ and $E^4_{n,i,j}$". By $\circledast_4$ we have for each $j \in J_i^{n,S}$, $Pr_{M^n_{\bar{p}}}[E^*_j] \ge q_4 \cdot p_{|i-j|}$. Recall that we condition the probability space $M^n_{\bar{p}}$ to the event $X_i^{n,m^*}(M^n_{\bar{p}}) \ge R_i^{n,S}/2$, and let $\langle j_1, ... j_r \rangle$ be the enumeration of $J^2(M^n_{\bar{p}})$ from (4) above. (Formally speaking r and each $j_{r'}$ is a function of $M^n_{\bar{p}}$). We then have for $1 \le r' < r'' \le r$, $Pr_{M^n_{\bar{p}}}[E^*_{j_{r'}} | E^*_{j_{r''}}] \ge Pr_{M^n_{\bar{p}}}[E^*_{j_{r'}}]$, and $Pr_{M^n_{\bar{p}}}[E^*_{j_{r'}} | \neg E^*_{j_{r''}}] \ge Pr_{M^n_{\bar{p}}}[E^*_{j_{r'}}]$. To see this use (2)(a) and (4)(a) above and the definition of $C_j(G)$.

Let the random variables X and X' be defined as follows. X is the number of $j \in J^2(M^n_{\bar{p}})$ such that E^*_j holds in $M^n_{\bar{p}}$. In other words X is the sum of r random variables $\langle Y_1, ..., Y_r \rangle$, where for each r' in the range $1 \le r' \le r$, $Y_{r'}$ equals 1 if $E^*_{j_{r'}}$ holds, and 0 otherwise. X' is the sum of r *independent* random variables $\langle Y'_1, ..., Y'_r \rangle$, where for each r' in the range $1 \le r' \le r$ $Y'_{r'}$ equals 1 with probability $q_4 \cdot p_{|i-j_{r'}|}$ and 0 with probability $1 - q_4 \cdot p_{|i-j_{r'}|}$. Then by the last paragraph for any $0 \le t \le r$,

$$Pr_{M^n_{\bar{p}}}[X \ge t] \ge Pr[X' \ge t].$$

But $Exp(X') = Exp(X) = q_4 \cdot \sum_{1 \le r' \le r} p_{|i-j_{r'}|}$ and by (4)(b) above this is greater than or equal to $1/\delta$. Hence by Chebyshev's inequality we have:

$$Pr_{M^n_{\bar{p}}}[\neg E^5_{n,S,i}] \le Pr_{M^n_{\bar{p}}}[X = 0] \le Pr[X' = 0] \le \frac{Var(X')}{Exp(X')^2} \le \frac{1}{Exp(X')} \le \delta$$

as desired.

Step 6. We turn to the construction of the cycle free saturated forest. Let $\epsilon > 0$, and we will prove that for $n \in \mathbb{N}$ large enough we have $Pr[E^{n,k}$ holds in $M^n_{\bar{p}}] \ge 1 - \epsilon$. Let $\delta = \epsilon/(l^* 2^{k+2})$ and $s = 2l^*((k + 2^k)(2l^* k + 1))$. Let $n \in \mathbb{N}$ be large enough so that $\circledast_5$ holds for n, k, δ and s. We now choose (formally we show that with probability at least $1 - \epsilon$ such a choice exists) by induction on $(i, \eta) \in I^*_k(M^n_{\bar{p}}) \times {}^{\le k}2$ (ordered by the lexicographic order) $y^i_\eta \in [n]$ such that:

1. $\langle y^i_\eta \in [n] : (i, \eta) \in I^*_k(M^n_{\bar{p}}) \times {}^{\le k}2 \rangle$ is without repetitions.
2. If $\eta = \langle\rangle$ then $M^n_{\bar{p}} \models i \sim y^i_\eta$, but $|i - y^i_\eta| \ne l^*$.
3. If $\eta \ne \langle\rangle$ then $M^n_{\bar{p}} \models y^i_\eta \sim y^i_{\eta|_{|\eta|-1}}$.
4. If $\eta = \langle\rangle$ then $M^n_{\bar{p}}$ satisfies E^4_{n,i,y^i_η}; else, denoting $\rho := \eta|_{|\eta|-1}$, $M^n_{\bar{p}}$ satisfies $E^4_{n,y^i_\rho,y^i_\eta}$.

Before we describe the choice of y^i_η, we need to define sets $S^i_\eta \subseteq [n]$. For a graph G on $[n]$ and $i \in I^*_k(G)$ let $S^*_i(G)$ be the set of vertices in the first (in some fixed order) path of length $\leq k$ from 1 to i in G. Now let $S^*(G) = \bigcup_{i \in I^*_k(G)} S^*_i(G)$. For $(i, \eta) \in I^*_k(M^n_{\bar{p}}) \times {}^{\leq k}2$ and $\langle y^{i'}_{\eta'} \in [n] : (i', \eta') <_{lex} (i, \eta)\rangle$ define:

$$S^i_\eta(G) = S^*(G) \cup \{[y^{i'}_{\eta'} - kl^*, y^{i'}_{\eta'} + kl^*] : (i'\eta') <_{lex} (i, \eta)\}.$$

Note that indeed $|S^*(G)| \leq s$ for all G. In the construction below when we write S^i_η we mean $S^i_\eta(M^n_{\bar{p}})$ where $\langle y^{i'}_{\eta'} \in [n] : (i', \eta') <_{lex} (i, \eta)\rangle$ were already chosen. Now the choice of y^i_η is as follows:

- If $\eta = \langle\rangle$ by $\circledast_5$ with probability at least $1 - \delta$, $E^5_{n, S^i_\eta, i}$ holds in $M^n_{\bar{p}}$ and hence we can choose y^i_η that satisfies (1)-(4).
- If $\eta = \langle 0 \rangle$ (resp. $\eta = \langle 1 \rangle$) choose $y^i_\eta = y^i_{\langle\rangle} - l^*$ (resp. $y^i_\eta = y^i_{\langle\rangle} + l^*$). By the induction hypothesis and the definition of $E^4_{n,i,j}$ this satisfies (1)-(4) above.
- If $|\eta| > 1$, $|y^i_{\eta\restriction|\eta|-1} - y^i_{\eta\restriction|\eta|-2}| \neq l^*$ and $\eta(|\eta|) = 0$ (resp. $\eta(|\eta|) = 1$) then choose $y^i_\eta = y^i_{\eta\restriction|\eta|-1} - l^*$ (resp. $y^i_\eta = y^i_{\eta\restriction|\eta|-1} + l^*$). Again by the induction hypothesis and the definition of $E^4_{n,i,j}$ this satisfies (1)-(4).
- If $|\eta| > 1$, $y^i_{\eta\restriction|\eta|-1} - y^i_{\eta\restriction|\eta|-2} = l^*$ (resp. $y^i_{\eta\restriction|\eta|-1} - y^i_{\eta\restriction|\eta|-2} = -l^*$) and $\eta(|\eta|) = 0$, then choose $y^i_\eta = y^i_{\eta\restriction|\eta|-1} - l^*$ (resp. $y^i_\eta = y^i_{\eta\restriction|\eta|-1} + l^*$).
- If $|\eta| > 1$, $|y^i_{\eta\restriction|\eta|-1} - y^i_{\eta\restriction|\eta|-2}| = l^*$ and $\eta(|\eta|) = 1$, then by $\circledast_5$ with probability at least $1 - \delta$, $E^5_{n, S^i_\eta, y^i_{\eta\restriction|\eta|-1}}$ holds in $M^n_{\bar{p}}$, and hence we can choose y^i_η that satisfies (1)-(4).

At each step of the construction above the probability of "failure" is at most δ; hence with probability at least $1-(l^*2^{k+2})\delta = 1-\epsilon$ we complete the construction. It remains to show that indeed $\langle y^i_\eta : i \in I^n, \eta \in {}^{\leq k}2\rangle$ is a cycle free saturated forest of depth k for I^*_k in $M^n_{\bar{p}}$. This is straightforward from the definitions. First each $\langle y^i_\eta : \eta \in {}^{\leq k}2\rangle$ is a saturated tree of depth k in $[n]$ by its construction. Second (ii) and (iii) in the definition of a saturated tree holds by (2) and (3) above (respectively). Third note that by (4) each edge (y, y') of our construction satisfies $E^2_{n,y,y'}$ and $E^4_{n,y,y'}$ and hence (iv) and (v) (respectively) in the definition of a saturated tree follows. Lastly we need to show that (c) in the definition of a saturated forest holds. To see this note that if $i_1, i_2 \in i^*_k(M^n_{\bar{p}})$ then by the definition of $S^i_\eta(M^n_{\bar{p}})$ there exists a path of length $\leq 2k$ from i_1 to i_2 with all its vertices in $S^i_\eta(M^n_{\bar{p}})$. Now if $\bar{x}$ is a path of length $\leq k$ from $y^{i_1}_{\langle\rangle}$ to i_2 and $(y^{i_1}_{\langle\rangle}, i_1)$ is not an edge of $\bar{x}$, then necessarily $\{y^{i_1}_{\langle\rangle}, i_1\}$ is included in some cycle of length $\leq 3k + 2$. This is a contradiction to the choice of $y^{i_1}_{\langle\rangle}$. This completes the proof of the claim. □

By $\odot$ and the claim above we conclude that, for some large enough $n \in \mathbb{N}$, there exists a graph $G = ([n], \sim)$ such that:

1. $G \models \neg\phi[1]$.
2. $Pr[M^n_{\bar{p}} = G] > 0$.

3. There exists $\langle \bar{y}^i : i \in I_r^*(G) \rangle$, a cycle free saturated forest of depth r for $I_r^*(G)$ in G.

Denote $B = B^G(1,r)$, $I = I_r^*(G)$, and we will prove that for some r-proper $(l, u_0, U, H) \in \mathfrak{H}$ we have $(B,1) \cong (H, u_0)$ (i.e. there exists a graph isomorphism from $G|_B$ to H mapping 1 to u_0). As ϕ is r-local we will then have $H \models \neg\phi[u_0]$ which is a contradiction of our assumption and we will be done. We turn to the construction of (l, u_0, U, H). For $i \in I$ let $r(i) = r - dist^G(1,i)$. Denote

$$Y := \{y^i_\eta : i \in I, \eta \in {}^{<r(i)}2\}.$$

Note that by (ii)-(iii) in the definition of a saturated tree we have $Y \subseteq B$. We first define a one-to-one function $f : B \to \mathbb{Z}$ in three steps:

Step 1. For each $i \in I$ define

$$B_i := \{x \in B : \text{ there exists a path of length } \leq r(i) \text{ from } x \text{ to } i \text{ disjoint to } Y\}$$

and $B^0 := I \cup \bigcup_{i \in I} B_i$. Now define for all $x \in B^0$, $f(x) = x$. Note that:

- $\bullet_1$ $f|_{B^0}$ is one-to-one (trivially).
- $\bullet_2$ If $x \in B^0$ and $dist^G(1,x) < r$ then $x + l^* \in [n] \Rightarrow x + l^* \in B^0$ and $x - l^* \in [n] \Rightarrow x - l^* \in B^0$ (use the definition of a saturated tree).

Step 2. We define $f|_Y$. We start by defining $f(y)$ for $y \in \bar{y}^1$, so let $\eta \in {}^{\leq r}2$ and denote $y = y^1_\eta$. We define $f(y)$ using induction on η where ${}^{\leq r}2$ is ordered by the lexicographic order. First if $\eta = \langle\rangle$ then define $f(y) = 1 - l^*$. If $\eta \neq \langle\rangle$ let $\rho : \eta|_{|\eta|-1}$, and consider $u := f(y^1_\rho)$. Denote $F = F_\eta := \{f(y^1_{\eta'}) : \eta' <_{lex} \eta\}$. Now if $u - l^* \notin F$ define $f(y) = u - l^*$. If $u - l^* \in F$ but $u + l^* \notin F$ define $f(y) = u + l^*$. Finally, if $u - l^*, u + l^* \in F$, choose some $l = l_\eta$ such that $p_l > 0$ and $u - l < \min F - rl^* - n$, and define $f(y) = u - l$. Note that by our assumptions $\{l : p_l > 0\}$ is infinite so we can always choose l as desired. Note further that we chose $f(y)$ so that $f|_{\bar{y}^1}$ is one-to-one. Now for each $i \in I \cap [1, l^*]$ and $\eta \in {}^{<r(i)}2$, define $f(y^i_\eta) = f(y^1_\eta) + (f(i) - 1)$ (recall that $f(i) = i$ was defined in Step 1, and that $k(i) \leq k(1)$ so $f(y^i_\eta)$ is well defined). For $i \in I \cap (n - l^*, n]$ perform a similar construction in "reversed directions". Formally define $f(y^i_{\langle\rangle}) = i + l^*$, and the induction step is similar to the case $i = 1$ above only now choose l such that $u + l > \max F + rl^* + n$, and define $f(y) = u + l$. Note that:

- $\bullet_3$ $f|_Y$ is one-to-one.
- $\bullet_4$ $f(Y) \cap f(B^0) = \emptyset$. In fact:
- $\bullet_4^+$ $f(Y) \cap [n] = \emptyset$.
- $\bullet_5$ If $i \in I \cap [1, l^*]$ then $i - l^* \in f(Y)$ (namely $i - l^* = f(y^i_{\langle\rangle})$).
- $\bullet_5'$ If $i \in I \cap (n - l^*, n]$ then $i + l^* \in f(Y)$ (namely $i + l^* = f(y^i_{\langle\rangle})$).
- $\bullet_6$ If $y \in Y \setminus \{y^i_{\langle\rangle} : i \in I\}$ and $dist^G(1,y) < r$ then $f(y) + l^*, f(y) - l^* \in f(Y)$. (Why? If $dist^G(1, y^i_\eta) < r$ then $|\eta| < r(i)$, and the construction of **Step 2**).

Step 3. For each $i \in I$ and $\eta \in {}^{<r(i)}2$, define B^i_η by

$$\{x \in B : \text{there exists a path of length } \leq r(i) \text{ from } x \text{ to } y^i_\eta \text{ disjoint to } Y \setminus \{y^i_\eta\}\}$$

and $B^1 := \bigcup_{i \in I, \eta \in {}^{<r(i)}2} B^i_\eta$.

We now make a few observations:

(α) If $i_1, i_2 \in I$ then, in G there exists a path of length at most $2r$ from i_1 to i_2 disjoint to Y. Why? By the definition of I and (c) in the definition of a saturated forest.

(β) B^0 and B^1 are disjoint and cover B. Why? Trivially they cover B, and by (α) and (iv) in the definition of a saturated tree they are disjoint.

(γ) $\langle B^i_\eta : i \in I, \eta \in {}^{<r(i)}2\rangle$ is a partition of B^1. Why? Again trivially they cover B^1, and by (iv) in the definition of a saturated tree they are disjoint.

(δ) If $\{x, y\}$ is an edge of $G|_B$ then either $x, y \in B^0$, $\{x, y\} = \{i, y^i_{\langle\rangle}\}$ for some $i \in I$, $\{x, y\} \subseteq Y$ or $\{x, y\} \subseteq B^i_\eta$ for some $i \in I$ and $\eta \in {}^{<r(i)}2$. (Use the properties of a saturated forest.)

We now define $f|_{B^1}$. Let $\langle (B_j, y_j) : j < j^* \rangle$ be some enumeration of $\langle (B^i_\eta, y^i_\eta) : i \in I, \eta \in {}^{<r(i)}2\rangle$. We define $f|_{B_j}$ by induction on $j < j^*$ so assume that $f|_{(\cup_{j'<j} B_{j'})}$ is already defined, and denote: $F = F_j := f(B^0) \cup f(Y) \cup f(\cup_{j'<j} B_{j'})$. Our construction of $f|_{B_j}$ will satisfy:

- $f|_{B_j}$ is one-to-one.
- $f(B_j)$ is disjoint to F_j.
- If $y \in B_j$ then either $f(y) = y$ or $f(y) \notin [n]$.

Let $\langle z^j_s : s < s(j)\rangle$ be some enumeration of the set $\{z \in B_j : G \models y_j \sim z\}$. For each $s < s(j)$ choose $l(j, s)$ such that $p_{l(j,s)} > 0$ and:

⊗ If $k \leq 4r$, $(m_1, ..., m_k)$ are integers with absolute value not larger than $4r$ and not all equal 0, and $(s_1, ...s_k)$ is a sequence of natural numbers smaller than $j(s)$ without repetitions, then $|\sum_{1 \leq i \leq m}(m_i \cdot l(j, s_i))| > n + \max\{|x| : x \in F_j\}$.

Again as $\{l : p_l > 0\}$ is infinite we can always choose such $l(j, s)$. We now define $f|_{B_j}$. For each $y \in B_j$ let $\bar{x} = (x_0, ...x_k)$ be a path in G from y to y_j, disjoint to $Y \setminus \{y_j\}$, such that k is minimal. So we have $x_0 = y$, $x_k = y_j$, $k \leq r$ and $\bar{x}$ is without repetitions. Note that by the definition of B_j such a path exists. For each t in the range $0 \leq t < k$ define

$$l_t = l_t(\bar{x}) = \begin{cases} l(j,s) & l^{\bar{x}}_t = |y_j - z^j_s| \text{ for some } s < s(j) \\ -l(j,s) & l^{\bar{x}}_t = -|y_j - z^j_s| \text{ for some } s < s(j) \\ l^{\bar{x}}_t & \text{otherwise.} \end{cases}$$

Now define $f(y) = f(y_j) + \sum_{0 \leq t < k} l_t$. We have to show that $f(y)$ is well defined. Assume that both $\bar{x}_1 = (x_0, ...x_{k_1})$ and $\bar{x}_2 = (x'_0, ...x'_{k_1})$ are paths as above. Then $k_1 = k_2$ and $\bar{x} = (x_0, ..., x_{k_1}, x'_{k_2-1}, ..., x'_0)$ is a cycle of length $k_1 + k_2 \leq 2r$. By (v) in the definition of a saturated tree we know that for each $s < s(j)$, $|y_j - z^j_s| > m_{2r}$. Hence as $\bar{p}$ is without unavoidable cycles we have for each $s < s(j)$ and $0 \leq t < k_1 + k_2$, if $|l^{\bar{x}}_t| = |y_j - z^j_s|$ then $t \in Sym(\bar{x})$. (see Definition 26(6,7)). Now put for $w \in \{1,2\}$ and $s < s(j)$, $m^+_w(s) := |\{0 \leq t < k_w : l^{\bar{x}_w}_t = y_j - z^j_s\}|$ and similarly $m^-_w(s) := |\{0 \leq t < k_w : -l^{\bar{x}_w}_t = y_j - z^j_s\}|$. By the definition of $\bar{x}$ we have, $m^+_1(s) - m^-_1(s) = m^+_2(s) - m^-_2(s)$. But from the definition of $l_t(\bar{x})$ we have for $w \in \{1,2\}$,

$$\sum_{0\leq t<k_w} l_t(\bar{x}_w) = \sum_{0\leq t<k_w} l^{\bar{x}_w}_t + \sum_{s<s(j)} (m^+_w(s) - m^-_w(s))(l(j,s) - (y_j - z^j_s)).$$

Now as $\sum_{0\leq t<k_1} l^{\bar{x}_1}_t = \sum_{0\leq t<k_2} l^{\bar{x}_2}_t$ we get $\sum_{0\leq t<k_1} l_t(x_1) = \sum_{0\leq t<k_2} l_t(x_2)$ as desired.

We now show that $f|_{B_j}$ is one-to-one. Let $y^1 \neq y^2$ be in B_j. So for $w \in \{1,2\}$ we have a path $\bar{x}_w = (x^w_0, ...x^w_{k_w})$ from y^w to y_j. As before, for $s < s(j)$ denote $m^+_w(s) := |\{0 \leq t < k_w : l^{\bar{x}_w}_t = y_j - z^j_s\}|$ and similarly $m^-_w(s)$. By the definition of f_{B_j} we have

$$f(y^1) - f(y^2) = y^1 - y^2 + \sum_{s<s(j)} [(m^+_1(s) - m^-_1(s)) - (m^+_2(s) - m^-_2(s))] \cdot l(j,s).$$

Now if for each $s < s(j)$, $m^+_1(s) - m^-_1(s) = m^+_2(s) - m^-_2(s)$ then we are done as $y^1 \neq y^2$. Otherwise note that for each $s < s(j)$, $|m^+_1(s) - m^-_1(s)| = |m^+_2(s) - m^-_2(s)| \leq 4r$. Note further that $|\{s < s(j) : m^+_1(s) - m^-_1(s) = m^+_2(s) - m^-_2(s) \neq 0\}| \leq 4r$. Hence by $\otimes$, and as $|y^1 - y^2| \leq n$ we are done.

Next let $y \in B_j$ and $\bar{x} = (x_0, ..., x_k)$ be a path in G from y to y_j. For each $s < s(j)$ define $m^+(s)$ and $m^-(s)$ as above, hence we have $f(y) = y_j + \sum_{s<s(j)}(m^+(s) - m^-(s))l(j,s)$. Consider two cases. First if $(m^+(s) - m^-(s)) = 0$ for each $s < s(j)$ then $f(y) = y$. Hence $f(y) \notin f(B^0) = B^0$ (by (β) above), $f(y) \notin f(Y)$ (as $f(Y) \cap [n] = \emptyset$) and $f(y) \notin f(\cup_{j'<j} B_{j'})$ (by (γ) and the induction hypothesis). So $f(y) \notin F_j$. Second assume that for some $s < s(j)$, $(m^+(s) - m^-(s)) \neq 0$. Then by the $\otimes$ we have $f(y) \notin [n]$ and furthermore $f(y) \notin F_j$. In both cases the demands for $f|_{B_j}$ are met and we are done. After finishing the construction for all $j < j^*$ we have $f|_{B^1}$ such that:

- $\bullet_7$ $f|_{B^1}$ is one-to-one.
- $\bullet_8$ $f(B^1)$ is disjoint to $f(B^0) \cup f(Y)$.
- $\bullet_9$ If $y \in B^1$ and $dist^G(1,y) < r$ then $f(y) + l^*, f(y) - l^* \in f(B^1)$. In fact $f(y + l^*) = f(y) + l^*$ and $f(y - l^*) = f(y) - l^*$. (By the construction of Step 3.)

Putting $\bullet_1 - \bullet_9$ together we have constructed $f : B \to \mathbb{Z}$ that is one-to-one and satisfies:

($\circ$) If $y \in B$ and $dist^G(1,y) < r$ then $f(y) + l^*, f(y) - l^* \in f(B)$. Furthermore:
($\circ\circ$) $\{y, f^{-1}(f(y) - l^*)\}$ and $\{y, f^{-1}(f(y) + l^*)\}$ are edges of G.

For ($\circ\circ$) use: $\bullet_2$ with the definition of $f|_{B^0}$, $\bullet_5 + \bullet'_5$ with the fact that $G \models i \sim y^i_{\langle\rangle}$, $\bullet_6$ with the construction of Step 2 and $\bullet_9$.

We turn to the definition of (l, u_0, U, H) and the isomorphism $h : B \to H$. Let $l_{min} = \min\{f(b) : b \in B\}$ and $l_{max} = \max\{f(b) : b \in B\}$. Define:

- $l = l_{min} + l_{max} + 1$.
- $u_0 = l_{min} + 2$.
- $U = \{z + l_{min} + 1 : z \in Im(f)\}$.
- For $b \in B$, $h(b) = f(b) + l_{min} + 1$.
- For $u, v \in U$, $H \models u \sim v$ iff $G \models h^{-1}(u) \sim h^{-1}(v)$.

As f was one-to-one so is h, and trivially it is onto U and maps 1 to u_0. Also by the definition of H, h is a graph isomorphism. So it remains to show that (l, u_0, U, H) is r-proper. First $(*)_1$ in the definition of proper is immediate from the definition of H. Second for $(*)_2$ in the definition of proper let $u \in U$ be such that $dist^H(u_0, u) < r$. Denote $y := h^{-1}(u)$; then by the definition of H we have $dist^G(1, y) < r$, hence by ($\circ$), $f(y) + l^*, f(y) - l^* \in f(B)$ and hence by the definition of h and U, $u + l^*, u - l^* \in U$ as desired. Lastly to see $(*)_3$ let $u, u' \in U$ and denote $y = h^{-1}(u)$ and $y' = h^{-1}(u')$. Assume $|u - u'| = l^*$; then by ($\circ\circ$) we have $G \models y \sim y'$ and by the definition of H, $H \models u \sim u'$. Now assume that $H \models u \sim u'$; then $G \models y \sim y'$. Using observation (δ) above and rereading 1-3 we see that $|u - u'|$ is either l^*, $|y - y'|$, l_η for some $\eta \in {}^{<r}2$ (see Step 2) or $l(j, s)$ for some $j < j^*, s < s(j)$ (see step 3). In all cases we have $P_{|u-u'|} > 0$. Together we have $(*)_3$ as desired. This completes the proof of Theorem 27. $\square$

References

1. Gaifman, H.: On local and nonlocal properties. In: Proceedings of the Herbrand symposium (Marseilles, 1981). Stud. Logic Found. Math., vol. 107, pp. 105–135. North-Holland, Amsterdam (1982)
2. Łuczak, T., Shelah, S.: Convergence in homogeneous random graphs. Random Structures Algorithms 6(4), 371–391 (1995)
3. Shelah, S.: Hereditary convergence laws with successor (in preparation)

On Monadic Theories of Monadic Predicates

Wolfgang Thomas

RWTH Aachen University, Lehrstuhl Informatik 7, 52056 Aachen, Germany
thomas@informatik.rwth-aachen.de

For Yuri Gurevich on the occasion of his 70th birthday

Abstract. Pioneers of logic, among them J.R. Büchi, M.O. Rabin, S. Shelah, and Y. Gurevich, have shown that monadic second-order logic offers a rich landscape of interesting decidable theories. Prominent examples are the monadic theory of the successor structure $\mathcal{S}_1 = (\mathbb{N}, +1)$ of the natural numbers and the monadic theory of the binary tree, i.e., of the two-successor structure $\mathcal{S}_2 = (\{0,1\}^*, \cdot 0, \cdot 1)$. We consider expansions of these structures by a monadic predicate P. It is known that the monadic theory of $(\mathcal{S}_1, P)$ is decidable iff the weak monadic theory is, and that for recursive P this theory is in Δ^0_3, i.e. of low degree in the arithmetical hierarchy. We show that there are structures $(\mathcal{S}_2, P)$ for which the first result fails, and that there is a recursive P such that the monadic theory of $(\mathcal{S}_2, P)$ is Π^1_1-hard.

Keywords: monadic second-order logic, tree automata, decidable theories.

1 Introduction

Over the past century, starting with Löwenheim [16] in 1915, monadic second-order logic has been developed as a framework in which decision procedures can be provided for interesting theories of high expressive power. In building this rich domain of effective logic, two techniques were crucial. The first was based on the correspondence between monadic second-order formulas and finite automata. This "match made in heaven" (cf. Vardi [28]) was first established for weak monadic second-order logic over the successor structure $\mathcal{S}_1 = (\mathbb{N}, +1)$ by Büchi, Elgot, and Trakhtenbrot. Büchi [2] and Rabin [19] extended this to the full monadic second-order theory of $\mathcal{S}_1$ and of the binary tree $\mathcal{S}_2 = (\{0,1\}^*, \cdot 0, \cdot 1)$. The logic-automata connection first led to the decidability of $\mathrm{MT}(\mathcal{S}_1)$ and $\mathrm{MT}(\mathcal{S}_2)$, the monadic second-order theories of $\mathcal{S}_1$ and $\mathcal{S}_2$, respectively (or shorter: the "monadic theory" of these structures). The results were extended to many further logical systems and led to new approaches in verification, data base theory, and further areas of computer science.

A. Blass, N. Dershowitz, and W. Reisig (Eds.): Gurevich Festschrift, LNCS 6300, pp. 615–626, 2010.

The second technique, technically more demanding but more general in its scope, is the "composition method" as developed by Shelah [24] (building on earlier work by Ehrenfeucht, Fraïssé, Läuchli, and others). The idea here is to consider finite fragments of a theory and to compose such theory-fragments according to the combination of models. The method has been applied successfully over orderings, trees, and graphs. Over orderings, the "combination" is concatenation. Shelah's work provided a deep analysis of monadic theories of orderings where automata do not help (or at least are hard to imagine), for example over dense orderings.

In both approaches, Yuri Gurevich has played a central role and contributed most influential papers. For the automata theoretic approach, it might suffice to recall his path-breaking work with Harrington [12] on the monadic second-order theory of the binary tree. As an example of his papers involving the composition method, we mention the work [13,14] which explains over which "short" orderings (neither embedding ω_1 nor its reverse) the monadic theory is decidable. For the reader who wants to enter the field, Yuri's survey *Monadic second-order theories* [11] is still the first choice.

In the present paper, a very small mosaic piece is added to this rich picture. We consider the expansions of the binary tree $\mathcal{S}_2$ by recursive monadic predicates P. We study which complexity (on the scale of recursion theory) the monadic second-order theory of such an expansion $(\mathcal{S}_2, P)$ can have, and we compare the weak and the strong monadic second-order theory of the structures $(\mathcal{S}_2, P)$.

As a starting point we take the corresponding results on expansions of the successor structure $\mathcal{S}_1$ by recursive predicates. We recall (in Sect. 2) that for recursive $P \subseteq \mathbb{N}$, the monadic theory of $(\mathcal{S}_1, P)$ belongs to a low level of the arithmetical hierarchy, namely to the class Δ^0_3. It is also known that for any monadic predicate P, the unrestricted monadic theory of $(\mathcal{S}_1, P)$ is decidable iff the weak monadic theory is (where set quantification is restricted to finite sets). In contrast, we show in Sect. 3 that for recursive P the monadic theory of $(\mathcal{S}_2, P)$, which in general is confined to the analytical class Δ^1_2, can be Π^1_1-hard. In Sections 4 and 5 we prove that there is a predicate P such that the weak monadic theory of $(\mathcal{S}_2, P)$ is decidable but the full monadic theory is undecidable. For the proofs, both the automata theoretic and the composition method are useful.[1]

We assume that the reader is familiar with the basics of the subject. We use standard terminology on monadic theories, automata, and recursion theory (see, e.g., [10,11,21,27]).

[1] The second result should be attributed to the late Andrei Muchnik; it is stated in a densely written abstract *Automata on infinite objects, monadic theories, and complexity* of the Dagstuhl seminar report [7] of 1992. This abstract, written jointly by A. Muchnik and A.L. Semenov, lists – in a dozen of lines – ten topics and results, among them "an example of predicate on tree for which the weak monadic theory is decidable and the monadic theory undecidable". A manuscript with Muchnik's proof does not seem to exist. The talk itself, which was a memorable scientific event appreciated by all who attended (among them the present author), dealt with a different result, the "Muchnik tree iteration theorem"; see for example [1].

2 The Monadic Theory of Structures $(\mathcal{S}_1, P)$

Let us recall some well-known facts on structures $(\mathcal{S}_1, P)$. First we remark that for recursive P the theory $\mathrm{MT}(\mathcal{S}_1, P)$ may be undecidable:

Proposition 1. *There is a recursive predicate $P \subseteq \mathbb{N}$ such that* $\mathrm{MT}(\mathcal{S}_1, P)$ *(and even the first-order theory* $\mathrm{FT}(\mathcal{S}_1, P)$*) is undecidable.*

Proof. Let Q be a non-recursive, recursively enumerable set of natural numbers with effective enumeration $j_0, j_1, j_2, \ldots$. From this enumeration we define P. We present the characteristic sequence χ_P of P (with $\chi_P(i) = 1$ iff $i \in P$, else $\chi_P(i) = 0$):

$$\chi_P \quad = \quad 1\ 0^{j_0}\ 1\ 0^{j_1}\ 1\ 0^{j_2}\ 1\ \ \ldots .$$

Clearly χ_P (and hence P) is recursive. We have

$$n \in Q \ \text{ iff } \ (\mathcal{S}_1, P) \models \exists x (P(x) \ \wedge \ \bigwedge_{i=1}^{n} \neg P(x+i) \ \wedge \ P(x+n+1)) \ ,$$

where $x+i$ indicates the i-fold application of "+1" to x. So Q is 1-reducible even to the first-order order theory of $(\mathcal{S}_1, P)$. Hence also $\mathrm{MT}(\mathcal{S}_1, P)$ is undecidable. □

The set $\mathbb{P}$ of prime numbers gives an interesting example of a predicate P where the status of $\mathrm{MT}(\mathcal{S}_1, P)$ is unknown. Observing that we can express the order relation $<$ over $\mathbb{N}$ in monadic logic over $\mathcal{S}_1$, we note that the (open) twin prime hypothesis is expressible by the sentence

$$\forall x \exists y \ (x < y \ \wedge \ \mathbb{P}(y) \ \wedge \ \mathbb{P}(y+2)).$$

Hence it will be hard to show decidability of $\mathrm{MT}(\mathcal{S}_1, \mathbb{P})$; for a detailed analysis see [4]. On the other hand, no "natural" examples of predicates P are known such that $\mathrm{MT}(\mathcal{S}_1, P)$ is undecidable. The known undecidability results rely on predicates built for the purpose, as in Proposition 1 above.

The conversion of monadic formulas into automata provides nice examples of predicates P where $\mathrm{MT}(\mathcal{S}_1, P)$ is decidable. We use the results of Büchi [2] and McNaughton [17] which together yield a transformation from monadic formulas to deterministic ω-automata: *For each monadic second-order formula $\varphi(X)$ in the monadic second-order language of $\mathcal{S}_1 = (\mathbb{N}, +1)$ one can construct a deterministic Muller automaton $\mathcal{A}_\varphi$ such that for each predicate Q*

$$\mathcal{S}_1 \models \varphi[Q] \ \text{ iff } \ \mathcal{A}_\varphi \text{ accepts } \chi_Q.$$

We can use the left-hand side for a *fixed* predicate P, replacing in $\varphi(X)$ each occurrence of X by the predicate constant P. Then we have for each *sentence* φ of the monadic second-order language of the structure $(\mathcal{S}_1, P)$:

$$(\mathcal{S}_1, P) \models \varphi \ \text{ iff } \ \mathcal{A}_\varphi \text{ accepts } \chi_P.$$

This reduces the decision problem for the theory $\mathrm{MT}(\mathcal{S}_1, P)$ to the following acceptance problem Acc_P: *Given a Muller automaton $\mathcal{A}$ over the input alphabet $\{0,1\}$, does $\mathcal{A}$ accept χ_P?*

This reduction can be exploited in a concrete way, regarding example predicates P, and also in a general way, regarding the recursion theoretic complexity of theories $\mathrm{MT}(\mathcal{S}_1, P)$.

Concrete examples of predicates P such that $\mathrm{MT}(\mathcal{S}_1, P)$ is decidable were first proposed by Elgot and Rabin [9], namely, the set of factorial numbers, the set of k-th powers and the set of powers of k, for each $k > 1$. The idea is to solve the acceptance problem Acc_P as follows: A given automaton $\mathcal{A}$ accepts χ_P iff $\mathcal{A}$ accepts a modified sequence χ' where the distances between successive letters 1 are contracted below a certain length (a contracted 0-segment just should induce the same state transformation as the original one and should cause the automaton to visit the same states as the original one). In each of the cases mentioned above (factorials, k-th powers, powers of k), the contracted sequence χ' turns out to be ultimately periodic (where phase and period depend on $\mathcal{A}$). So one can decide whether $\mathcal{A}$ accepts χ' and hence whether it accepts χ_P. The method has been extended to further predicates (see e.g. [8]), and criteria for the decidability of $\mathrm{MT}(\mathcal{S}_1, P)$ have been developed in [23,4,22].

For the general aspect we analyze the acceptance problem Acc_P for a Muller automaton $\mathcal{A} = (S, \Sigma, s_0, \delta, \mathcal{F})$ in more detail. As usual, we write S for the set of states, Σ for the input alphabet, s_0 for the initial state, δ for the transition function from $S \times \Sigma$ to S, and $\mathcal{F} \subseteq 2^S$ for the acceptance component; recall that $\mathcal{A}$ accepts an input word α if the set of states visited infinitely often in the unique run of $\mathcal{A}$ on α coincides with a set in $\mathcal{F}$. Let us write $\delta(s_0, \alpha[0,j])$ for the state reached by $\mathcal{A}$ after processing the initial sement $\alpha(0)\dots\alpha(j)$. Then, taking $\alpha = \chi_P$, the automaton $\mathcal{A}$ accepts χ_P iff the following condition holds:

$$(*)_{\mathcal{A},P} \qquad \bigvee_{F\in\mathcal{F}} \Big(\bigwedge_{s\in F} (\forall i \exists j > i \; \delta(s_0, \chi_P[0,j]) = s) \\ \wedge \bigwedge_{s\in S\setminus F} (\neg \forall i \exists j > i \; \delta(s_0, \chi_P[0,j]) = s) \Big) .$$

Assuming that P is recursive, we obtain a reduction of the decision problem for $\mathrm{MT}(\mathcal{S}_1, P)$ to Boolean combinations of conditions that are in Π^0_2; note that the condition $\delta(s_0, \chi_P[0,j]) = s$ can be decided if P is recursive. By relativization, and using recursion theoretic terminology, we obtain for arbitrary $P \subseteq \mathbb{N}$:

$$\mathrm{MT}(\mathcal{S}_1, P) \;\leq_{\mathrm{tt}}\; P''.$$

Here $\leq_{\mathrm{tt}}$ is truth-table reducibility and P'' is the second jump of P. (In [25] it is shown that the slightly sharper bounded truth-table reducibility does not suffice.) We conclude the following fact, first noted in [3]:

Proposition 2. ([3]) *For each recursive $P \subseteq \mathbb{N}$, the theory $\mathrm{MT}(\mathcal{S}_1, P)$ belongs to the class Δ^0_3 of the arithmetical hierarchy.*

In particular, it is not possible to show the undecidability of a theory $\mathrm{MT}(\mathcal{S}_1, P)$ by a reduction of true first-order arithmetic to it.

A second consequence of the formulation $(*)_{\mathcal{A},P}$ is a reduction of the strong monadic language over $(\mathcal{S}_1, P)$ to the weak monadic language. For this we observe that the condition $(*)_{\mathcal{A},P}$ from above can be formalized in the weak monadic language over $(\mathcal{S}_1, P)$; note that the statement "$\delta(s_0, \chi_P[0,x]) = s$" involves only a finite run (up to position x) and hence can be expressed by a weak monadic formula $\psi_s(y)$. This shows that for each monadic sentence φ one can construct an equivalent weak monadic sentence φ' such that $(\mathcal{S}_1, P) \models \varphi$ iff $(\mathcal{S}_1, P) \models \varphi'$ (in φ' we use a definition of $<$ in weak monadic logic over $\mathcal{S}_1$). So we obtain:

Proposition 3. *For each $P \subseteq \mathbb{N}$: $\mathrm{MT}(\mathcal{S}_1, P)$ is decidable iff $\mathrm{WMT}(\mathcal{S}_1, P)$ is decidable.*[2]

Our aim in the subsequent sections is to show that both propositions fail when we consider the binary tree $\mathcal{S}_2$ instead of the successor structure $\mathcal{S}_1$.

3 A Recursive Predicate Where $\mathrm{MT}(\mathcal{S}_2, P)$ Is Π^1_1-Hard

In the same way as described above for theories $\mathrm{MT}(\mathcal{S}_1, P)$, the automata theoretic approach can be applied to study the complexity of the monadic theory of an expansion $(\mathcal{S}_2, P)$ of the binary tree. Here we identify a structure $(\mathcal{S}_2, P)$ with a $\{0,1\}$-labelled tree t_P which has label 1 at node u iff $u \in P$. We know from Rabin's Tree Theorem [19] that for each monadic sentence φ in the language of $(\mathcal{S}_2, P)$ one can construct a Rabin tree automaton $\mathcal{A}_\varphi$ such that

$$(\mathcal{S}_2, P) \models \varphi \quad \text{iff} \quad \mathcal{A}_\varphi \ \text{ accepts } \ t_P.$$

For recursive P, the right-hand side is a Σ^1_2-statement of the form $\exists X \forall Y \psi(X, Y)$ with first-order formula ψ, namely, "there is an $\mathcal{A}_\varphi$-run on t_P such that each infinite path of this run satisfies the Rabin acceptance condition". Since Rabin automata are closed under complement, the statement can also be phrased in Π^1_2-form. This proves the first statement of the following result:

Theorem 1. *For recursive $P \subseteq \{0,1\}^*$, the theory $\mathrm{MT}(\mathcal{S}_2, P)$ belongs to the class Δ^1_2, and there is a recursive $P \subseteq \{0,1\}^*$ such that $\mathrm{MT}(\mathcal{S}_2, P)$ is Π^1_1-hard.*

For the proof of the second statement we have to find a recursive P such that a known Π^1_1-complete set is reducible to $\mathrm{MT}(\mathcal{S}_2, P)$. As Π^1_1-complete set we use a coding of *finite-path trees* (cf. [21, Ch. 16.3]). We work with the infinitely branching tree $\mathcal{S}_\omega$ whose nodes are sequences $(n_1, \ldots n_k)$ of natural numbers. The empty sequence is the root, and the nodes $(n_1, \ldots, n_k, i)$ are the successors of $(n_1, \ldots, n_k)$. Paths in $\mathcal{S}_\omega$ are defined accordingly. We say that a subset S of $\mathcal{S}_\omega$ *defines a finite-path tree* if S is closed under taking predecessors and if it does not contain an infinite path. For a recursion theoretic treatment, we use a computable bijective coding of the finite sequences over $\mathbb{N}$ by natural

[2] Although this Proposition is very close to Proposition 2, a result of [3], it was left as an open problem in [3]. In a more general context an answer was then given in [26].

numbers, writing $\langle n_1, \ldots, n_k \rangle$ for the code of $(n_1, \ldots, n_k)$. Furthermore, we refer to a standard numbering of the partial recursive functions; we write f_e for the function with number e. A function $f : \mathbb{N} \to \mathbb{N}$ is the characteristic function of a finite-path tree if

1. f is total and has only values 0 or 1,
2. the set $\{(n_1, \ldots, n_k) \mid f(\langle n_1, \ldots, n_k \rangle) = 1\}$ defines a finite-path tree.

Let

$$\mathrm{FPT} = \{e \in \mathbb{N} \mid f_e \text{ is characteristic function of a finite-path tree}\}.$$

We use the following fact (see [21, Ch. 16.3]):

Proposition 4. FPT *is a* Π^1_1*-complete set of natural numbers.*

Proof of Theorem 1: It suffices to define a recursive set P of nodes of the *binary* tree $\mathcal{S}_2$ such that for each number e we can construct a monadic second-order sentence φ_e with

$$e \in \mathrm{FPT} \text{ iff } (\mathcal{S}_2, P) \models \varphi_e.$$

We build the structure $(\mathcal{S}_2, P)$ as a sequence of $\{0,1\}$-labelled trees $t_0, t_1, \ldots$ attached to the rightmost branch of $\mathcal{S}_2$. So the root of t_e is the node $r_e := 1^e0$. In the tree t_e we obtain a copy of $\mathcal{S}_\omega$: Its node $(n_1, \ldots, n_k)$ is coded by $r_e 1^{n_1+1} 0 1^{n_2+1} \ldots 1^{n_k+1} 0$. The predicate P will only apply to nodes of the leftmost branch starting in such a node. We define P by attaching labels 0 and 1 to the nodes $r_e 1^{n_1+1} 0 1^{n_2+1} \ldots 1^{n_k+1} 0^i$ for $i = 1, 2, 3 \ldots$. All other nodes get label 0 by default.

In order to define the labelling, we imagine an effective procedure $\mathcal{P}$ that computes, in a dovetailed fashion, the values $f_e(\langle n_1, \ldots, n_k \rangle)$ of all functions f_e simultaneously. So the procedure treats each pair $(e, \langle n_1, \ldots, n_k \rangle)$ again and again, and when dealing with this pair it progresses with the computation of $f_e(\langle n_1, \ldots, n_k \rangle)$ for one further step (unless a value has been computed already). Consider the i-th step of $\mathcal{P}$ ($i = 1, 2, 3, \ldots$). It will determine the bit label attached to all the nodes $r_e 1^{n_1+1} 0 1^{n_2+1} \ldots 1^{n_k+1} 0^i$, reporting on the current status of the computation of $f_e(\langle n_1, \ldots, n_k \rangle)$ at $\mathcal{P}$-step i. If the i-th $\mathcal{P}$-step produces the value $f_e(\langle n_1, \ldots, n_k \rangle)$ then we attach label 1 to the node $r_e 1^{n_1+1} 0 \ldots 1^{n_k+1} 0^i$, for all other nodes $r_{e'} 1^{m_1+1} 0 1^{m_2+1} \ldots 1^{m_{k'}+1} 0^i$ we attach label 0. In fact, when we find a value for $f_e(\langle n_1, \ldots, n_k \rangle)$, we attach to the nodes $r_e 1^{n_1+1} 0 \ldots 1^{n_k+1} 0^j$ for $j = i, i+1, i+2$ the labels 100, respectively 110, respectively 111, depending on whether the computation of $f_e(\langle n_1, \ldots, n_k \rangle)$ produced value 0, 1, or > 1, respectively. After such a block of letters 1 on the path $r_e 1^{n_1+1} 0 \ldots 1^{n_k+1} 0^\omega$, all subsequent labels will be 0.

Clearly this attachment of labels defines a recursive predicate over $\mathcal{S}_2$. ¿From the labels on the 0^ω-parts of the paths $r_e 1^{n_1+1} 0 \ldots 1^{n_k+1} 0^\omega$ (for fixed e) we can infer whether f_e is a characteristic function, i.e., whether for all tuples $(n_1, \ldots, n_k)$ the value $f_e(\langle n_1, \ldots, n_k \rangle)$ is defined and either 0 or 1: This happens if for all $(n_1, \ldots, n_k)$, on the 0^ω-part of $r_e 1^{n_1+1} 0 \ldots 1^{n_k+1} 0^\omega$ precisely one or two labels 1 occur. (Let us call such a path associated to $(n_1, \ldots, n_k)$ "once 1-labelled", respectively "twice 1-labelled".) So, using P, we can easily express in monadic logic for any given e whether f_e is a characteristic function. The function f_e is the characteristic function *of a finite-path tree* if moreover the nodes $r_e 1^{n_1+1}0 \ldots 1^{n_k+1}0$ whose associated path is twice 1-labelled form a set that that is closed under prefixes (i.e., there is no prefix whose associated path is only once 1-labelled), and that each path through $r_e(1^+0)^\omega$ eventually hits a node outside the coded tree, i.e., a node whose associated path is only once 1-labelled. All these conditions can be expressed by a monadic sentence φ_e. Hence we have $e \in$ FPT iff $(\mathcal{S}_2, P) \models \varphi_e$, as desired. □

4 Some Background on Types and Tree Automata

For the comparison between the weak and the strong monadic theory of structures $(\mathcal{S}_2, P)$, we need some preparations concerning "types" (i.e., finite theory fragments) and concerning tree automata. For a more detailed treatment, the reader can consult [11] or [27].

For the analysis of weak monadic logic over structures $(\mathcal{S}_2, P)$, it is convenient to use a syntax in which only second-order variables $X, Y, Z, \ldots$ are present. As atomic formulas we use $X \subseteq Y$, Sing(X) ("X is a singleton"), $S_i(X, Y)$ for $i = 0, 1$ ("X, Y are singletons, and the element of X has the element of Y as the i-th successor"), and $X \subseteq P$. Formulas are built up from atomic formulas by means of Boolean connectives and the (weak monadic) quantifiers $\exists, \forall$. It is clear that this relational language is equivalent in expressive power to the original one with first-order and weak monadic second-order quantifiers and the (functional) signature with symbols for the functions $\cdot 0$ and $\cdot 1$.

As in the previous section, we identify a structure $(\mathcal{S}_2, P)$ with a $\{0,1\}$-labelled tree t_P, i.e. with a mapping $t_P : \{0,1\}^* \to \{0,1\}$. Conversely, each $\{0,1\}$-labelled infinite binary tree t induces a structure $(\mathcal{S}_2, P_t)$; we freely use this correspondence and mean by "tree" always a $\{0,1\}$-labelled infinite tree. The set of all these trees is denoted by $T_{\{0,1\}}$. A tree t is *regular* if it has only finitely many non-isomorphic subtrees (or equivalently, if a finite Moore automaton generates t by producing the label $t(u)$ after processing the input word u). It is well-known that a regular tree is definable in the weak monadic language over $\mathcal{S}_2$; so its (weak and strong) monadic theory is decidable.

Let $m > 1$. Two trees s, t are *m-equivalent* (short: $s \equiv_m t$) if they satisfy the same weak monadic sentences (of the relational signature just introduced) of quantifier depth $\leq m$. There are finitely many equivalence classes, called *m-types*. Each m-type τ is definable by a weak monadic sentence φ_τ which again is of quantifier depth m. As finite representations of an m-type τ we use such a sentence φ_τ defining it.

In the sequel we shall work with natural compositions of trees and corresponding compositions of m-types. First we consider the combination of two trees via a 0-labelled or 1-labelled root: For two trees s, t let $0 \cdot \langle s, t\rangle$, respectively $1 \cdot \langle s, t\rangle$, be the tree with a 0-, respectively 1-labelled root and s, t as its left and right subtree. Next, we consider the composition of a given infinite sequence $t_0, t_1, t_2, \ldots$

of trees or of a sequence $(s_0, t_0), (s_1, t_1), \ldots$ of pairs of trees. In the first case we attach the trees $t_0, t_1, \ldots$ along the 0-labelled right-hand branch of the binary tree: We insert the tree t_i at the node 1^i0; i.e., the root of t_0 is node 0, the root of t_1 is 10, etc., and – as mentioned – the right-hand branch 1^ω is labelled 0. The resulting tree we denote as $[t_0, t_1, \ldots]$. In the second case we consider the two sons of the nodes $0, 10, 110$ etc. and insert s_i at the left son of 1^i0 and t_i at the right son of 1^i0. The nodes 1^i and 1^i0 are all labelled 0. We denote the tree obtained in this way as $[(s_0, t_0), (s_1, t_1), \ldots]$.

A simple Ehrenfeucht-Fraïssé type argument now shows the following lemma:

Lemma 1. *Let* $m > 1$.

(a) *The m-types* σ *of* s *and* τ *of* t *determine the m-types of* $0 \cdot \langle s, t \rangle$ *and* $1 \cdot \langle s, t \rangle$ *and these types are computable from* σ, τ.
(b) *If* $t_i \equiv_m t'_i$ *for* $i > 0$ *then* $[t_1, t_2, \ldots] \equiv_m [t'_1, t'_2, \ldots]$. *Similarly, if* $s_i \equiv_m s'_i$ *and* $t_i \equiv_m t'_i$, *then* $[(s_0, t_0), (s_1, t_1), \ldots] \equiv_m [(s'_0, t'_0), (s'_1, t'_1), \ldots]$.
(c) *If the sequence* $\tau_0, \tau_1, \ldots$ *of m-types of* $t_0, t_1, \ldots$ *is ultimately perdiodic, say of the form* $\tau_0 \ldots \tau_{k-1}(\tau_k \ldots \tau_{\ell-1})^\omega$, *then the m-type of* $[t_0, t_1, \ldots]$ *is determined by the types* $\tau_1, \ldots, \tau_{\ell-1}$ *and computable from them.*

Next we turn to prerequisites from tree automata theory, mainly using the concept of Büchi tree automaton (see e.g. [27] for details) and a fundamental example due to Rabin [20] which shows their expressive weakness in comparison with Rabin tree automata. Rabin presented a tree language T_0 which is definable in monadic logic (or by a Rabin tree automaton) but which is *not* recognizable by a Büchi tree automaton. It is a variant of the language of finite-path trees:

$$T_0 = \{t \in T_{\{0,1\}} \mid \text{on each path of } t \text{ there are only finitely many letters } 1\}.$$

We have to recall the construction of Rabin since we exploit it below. For $n \geq 0$ define the tree t_n inductively as follows:

1. t_0 has a 1-labelled root and is otherwise labelled 0.
2. t_{n+1} has a 1-labelled root, otherwise a 0-labelled right-hand branch 1^ω, a 0-labelled left subtree, and a copy of t_n inserted at each node in 1^+0.

So

$$t_n(u) = 1 \text{ iff } (u = \varepsilon \text{ or } u \in 1^+0 + (1^+01^+0) + \ldots + (1^+0)^n).$$

Let us verify that the m-type of t_n determines the m-type of t_{n+1} (and that the latter can be computed from the former): By Lemma 1 (c) we can compute the m-type of the right-hand subtree of the root of t_{n+1} from the m-type of t_n (note that the copies of t_n give a constant and hence periodic sequence of m-types). The left-hand subtree of the root of t_{n+1} is labelled 0; we can compute its m-type (since it is regular). Now Lemma 1 (a) yields the claim. So there is a map F over the finite domain of m-types that produces the m-type of t_{n+1} from the m-type of t_n. Starting with the m-type τ_0 of t_0, we obtain with the values $F^{(i)}(\tau_0)$ an ultimately periodic sequence. We summarize:

Lemma 2. *The m-types of the trees* $t_0, t_1, t_2, \ldots$ *form a computable ultimately periodic sequence* $\tau_0 \ldots \tau_{k-1}(\tau_k \ldots \tau_{\ell-1})^\omega$.

Clearly, each tree t_n belongs to T_0. We use the following lemma shown in [20] (see also [27]):

Lemma 3. *For each Büchi tree automaton $\mathcal{A}$ with $< n$ states accepting t_n one can construct a regular tree $t'_n \notin T_0$ which is again accepted by $\mathcal{A}$.*

Let us sketch the proof. Assume that the Büchi tree automaton $\mathcal{A}$ with $< n$ states and the set F of final states accepts t_n. Then one can construct a regular run ϱ of $\mathcal{A}$ on t_n (since t_n is regular and accepted). We define a path in ϱ as follows: Pick a node $u_1 = 1^{k_1}$ on the right-hand branch where $\varrho(1^{k_1}) \in F$. Pick a node $u_2 = 1^{k_1}01^{k_2}$ on the right-hand branch starting in $1^{k_1}0$ where again $\varrho(u_2) \in F$, and so on until such a node $u_n = 1^{k_1}01^{k_2} \ldots 1^{k_n}$ with $\varrho(u_n) \in F$ is chosen. These nodes exist since on each path of ϱ infinitely many visits of F occur. Now $t_n(u_i0) = 1$ for $i = 1, \ldots, n$ by definition of t_n. Since $\mathcal{A}$ has $< n$ states, there are u_i, u_j with $i < j$ such that $\varrho(u_i) = \varrho(u_j)$; observe that between these nodes a 1-labelled node of t_n occurs (for example at u_i0). Repeating the t_n-segment determined by the path segment from u_i (included) to u_j (excluded) indefinitely, we obtain a regular tree t'_n which is accepted by $\mathcal{A}$ and which has a path with infinitely many labels 1.

A set of trees definable in weak monadic logic is easily seen to be recognized by a Büchi tree automaton. So the lemma also shows that T_0 is not definable in weak monadic logic.

5 Comparing Weak and Strong Monadic Logic

The aim of this section is to show the following:

Theorem 2. *There is a predicate $P \subseteq \{0,1\}^*$ such that* $\mathrm{WMT}(\mathcal{S}_2, P)$ *is decidable and* $\mathrm{MT}(\mathcal{S}_2, P)$ *undecidable.*

We shall start with a tree t_ω which for each given quantifier depth m is m-equivalent to an effectively constructible regular tree. This gives us the decidability of the weak monadic theory of t_ω. Then we modify t_ω first to a tree s_ω and then to a tree t'_ω such that for each quantifier depth m the trees t_ω, s_ω, and t'_ω cannot be distinguished by m-types from some computable level onwards (which ensures that the weak monadic theory of t'_ω is also decidable). However, t'_ω will be constructed such that in the full monadic theory an undecidability proof as for Proposition 1 can be carried through.

Proof of Theorem 2: Define, using the trees t_i of the previous section,

$$t_\omega := [(t_0, t_0), (t_1, t_1), (t_2, t_2), \ldots].$$

By Lemma 2, for each given m the tree t_ω is m-equivalent to an effectively constructible regular tree; just take a fixed representative for each m-type τ that appears in the ultimately periodic sequence of m-types of the trees $0 \cdot \langle t_0, t_0 \rangle, 0 \cdot \langle t_1, t_1 \rangle, \ldots$ (use Lemma 2 and Lemma 1 (a)). Hence the weak monadic theory of t_ω is decidable.

As a next step we now construct a tree s_ω from t_ω. First we pick, for each m-type τ ($m = 1, 2, \ldots$), a Büchi tree automaton $\mathcal{A}_\tau$ that defines τ. Let n_τ be the number of states of $\mathcal{A}_\tau$. Define

$$N_m := \max\{n_\tau \mid \tau \text{ is } m\text{-type}\} + 1.$$

These numbers N_m will be called *special* below.

Consider t_{N_m}; denote by τ its m-type. Then $\mathcal{A}_\tau$ accepts t_{N_m}. The number of states of $\mathcal{A}_\tau$ is $< N_m$. By Lemma 3, we can construct a tree $t'_{N_m} \notin T_0$ that is again accepted by $\mathcal{A}_\tau$; so its m-type is τ. We conclude

$$(*) \qquad t_{N_m} \equiv_m t'_{N_m} \text{ and also } t_{N_i} \equiv_m t'_{N_i} \quad \text{for } i > m.$$

Now let s_ω be obtained from $t_\omega = [(t_0, t_0), (t_1, t_1), (t_2, t_2), \ldots]$ by replacing, for each $m > 1$, the pair (t_{N_m}, t_{N_m}) of subtrees by (t'_{N_m}, t_{N_m}). By Lemma 1 (b), for each m, the subtree of s_ω with root 1^{N_m} is m-equivalent to the corresponding subtree of t_ω (note that $t_{N_i} \equiv_m t'_{N_i}$ for $i \geq m$). Hence also s_ω is m-equivalent to an effectively constructible regular tree, and thus its weak monadic theory is decidable.

We now focus on the "special numbers" N_m (including N_0 which is set to 0). A tree node 1^n is called special if n is special. It is worth noting that the set of special tree nodes 1^n of s_ω is definable in monadic logic: A node 1^n is special iff the subtree with root 1^n00 does not belong to the tree language T_0, which in turn is definable in (strong!) monadic logic.

Now, copying Proposition 1, we code a non-recursive, recursively enumerable set Q with enumeration $j_0, j_1, \ldots$ on the domain S of special numbers. We introduce a *marker* on a special number N_i when in the proof of Proposition 1 the value 1 was chosen for i. So the number N_0 is marked, the next j_0 special numbers are unmarked, the special number N_{j_0+1} is marked, the next j_1 special numbers are unmarked, and so on. For each *marked* N_i we modify the entry (t'_{N_i}, t_{N_i}) of s_ω to (t'_{N_i}, t'_{N_i}), thus obtaining the desired tree t'_ω, or in other words, the desired predicate P over $\mathcal{S}_2$. Again we call a node 1^n marked if n is marked.

We finish the proof by verifying that the weak monadic theory of t'_ω is decidable and that the strong monadic theory of t'_ω is undecidable.

For the first claim, one observes, using $(*)$, that exactly as for the tree s_ω, also t'_ω is m-equivalent to an effectively constructible regular tree, for each given $m > 0$. For the second claim we use the following equivalence, regarding the considered non-recursive set Q: $n \in Q$ iff there are two marked and special nodes $1^k, 1^{k'}$ in t'_ω such that there are exactly n special nodes between them, all of them unmarked. Clearly this condition is expressible in monadic logic. Hence Q is 1-reducible to the monadic theory of t'_ω $(=: (\mathcal{S}_2, P))$. □

6 Conclusion

The study of the monadic theory of structures $(\mathcal{S}_2, P)$ with monadic predicate P seems far from finished. Let us list three open problems.

1. More examples of predicates P shold be found such that $\mathrm{MT}(\mathcal{S}_2, P)$ is decidable. The "contraction method" of Elgot and Rabin [9] has been transferred to the binary tree by Montanari and Puppis [18], but it seems that not many interesting predicates are (as yet) manageable by this approach. For example, consider predicates induced by the binary representations (or inverse binary representations) of numbers of interesting sets $S \subseteq \mathbb{N}$. For the powers of 2, the corresponding predicate P with the nodes $0, 10, 100, \ldots$ is a definable set, whence the monadic theory of $(\mathcal{S}_2, P)$ is decidable. What about the corresponding predicate for the set of squares?
2. The lack of a deterministic automaton model over trees (capturing monadic logic over the binary tree) may be considered as the deeper reason for the result of Sect. 3 that the theory $\mathrm{MT}(\mathcal{S}_2, P)$ can be non-arithmetical for recursive P. However, this leaves open the question whether an undecidability proof for such a theory can be done via a reduction of true first-order arithmetic. A partial (negative) answer follows from work of Gurevich and Shelah [15] on the uniformization problem for monadic logic over $\mathcal{S}_2$; it is shown there that no well-ordering of $\mathcal{S}_2$ exists that is definable in monadic logic. A more recent treatment, also covering definability in structures $(\mathcal{S}_2, P)$ with monadic P, is given in [5,6]: The structure $(\mathbb{N}, <)$ (and hence $(\mathbb{N}, +, \cdot)$) is not monadic second-order interpretable in a structure $(\mathcal{S}_2, P)$ (even with non-recursive P) when the universe $\mathbb{N}$ is represented by the full domain of the binary tree.
3. A natural question, already raised at the end of Rabin's paper [20] (and attributed there to H. Gaifman), is concerned with decidability of weak definability: Can one decide for a monadic formula $\varphi(X_1, \ldots, X_n)$ interpreted over $\mathcal{S}_2$ whether it is equivalent to a formula of weak monadic logic?

Acknowledgment

Many thanks are due to Nachum Dershowitz for his patience and help and to Christof Löding and Alex Rabinovich for their comments.

References

1. Berwanger, D., Blumensath, A.: The monadic theory of tree-like structures. In: Grädel, E., Thomas, W., Wilke, T. (eds.) Automata, Logics, and Infinite Games. LNCS, vol. 2500, pp. 285–302. Springer, Heidelberg (2002)
2. Büchi, J.R.: On a decision method in restricted second-order arithmetic. In: Nagel, E., et al. (eds.) Logic, Methodology, and Philosophy of Science: Proceedings of the 1960 International Congress, pp. 1–11. Stanford Univ. Press, Stanford (1962)
3. Büchi, J.R., Landweber, L.H.: Definability in the monadic second-order theory of successor. J. Symb. Logic 34, 166–170 (1969)
4. Bateman, P.T., Jockusch, C.G., Woods, A.R.: Decidability and undecidability of theories with a predicate for the primes. J. Symb. Logic 58, 672–687 (1993)
5. Carayol, A., Löding, C.: MSO on the infinite binary tree: Choice and order. In: Duparc, J., Henzinger, T.A. (eds.) CSL 2007. LNCS, vol. 4646, pp. 161–176. Springer, Heidelberg (2007)

6. Carayol, A., Löding, C., Niwiński, D., Walukiewicz, I.: Choice functions and well-orderings over the infinite binary tree. Central Europ. J. of Math. (to appear)
7. Compton, K., Pin, J.E., Thomas, W. (eds.): Automata Theory: Infinite Computations. Dagstuhl Seminar Report 9202 (1992)
8. Carton, O., Thomas, W.: The monadic theory of morphic infinite words and generalizations. Information and Computation 176, 51–76 (2002)
9. Elgot, C.C., Rabin, M.O.: Decidability and undecidability of extensions of second (first) order theory of (generalized) successor. J. Symb. Logic 31, 169–181 (1966)
10. Grädel, E., Thomas, W., Wilke, T. (eds.): Automata, Logics, and Infinite Games. LNCS, vol. 2500. Springer, Heidelberg (2002)
11. Gurevich, Y.: Monadic theories. In: Barwise, J., Feferman, S. (eds.) Model-Theoretic Logics, pp. 479–506. Springer, Berlin (1985)
12. Gurevich, Y., Harrington, L.: Trees, automata, and games. In: Proc. 14th STOC, pp. 60–65 (1982)
13. Gurevich, Y.: Modest theory of short chains. J. Symb. Logic 44, 481–490 (1979)
14. Gurevich, Y., Shelah, S.: Modest theory of short chains II. J. Symb. Logic 44, 491–502 (1979)
15. Gurevich, Y., Shelah, S.: Rabin's uniformization problem. J. Symb. Logic 48, 1105–1119 (1983)
16. Löwenheim, L.: Über Möglichkeiten im Relativkalkül. Math. Ann. 76, 447–470 (1915)
17. McNaughton, R.: Testing and generating infinite sequences by a finite automaton. Inf. Contr. 9, 521–530 (1966)
18. Montanari, A., Puppis, G.: A contraction method to decide MSO theories of deterministic trees. In: Proc. 22nd IEEE Symposium on Logic in Computer Science (LICS), pp. 141–150
19. Rabin, M.O.: Decidability of second-order theories and automata on infinite trees. Trans. Amer. Math. Soc. 141, 1–35 (1969)
20. Rabin, M.O.: Weakly definable relations and special automata. In: Bar-Hillel, Y. (ed.) Math. Logic and Foundations of Set Theory, pp. 1–23. North-Holland, Amsterdam (1970)
21. Rogers, H.: The Theory of Recursive Functions and Effective Computability. McGraw-Hill, New York (1967)
22. Rabinovich, A., Thomas, W.: Decidable theories of the ordering of natural numbers with unary predicates. In: Ésik, Z. (ed.) CSL 2006. LNCS, vol. 4207, pp. 562–574. Springer, Heidelberg (2006)
23. Semenov, A.: Decidability of monadic theories. In: Chytil, M.P., Koubek, V. (eds.) MFCS 1984. LNCS, vol. 176, pp. 162–175. Springer, Heidelberg (1984)
24. Shelah, S.: The monadic theory of order. Ann. Math. 102, 379–419 (1975)
25. Thomas, W.: The theory of successor with an extra predicate. Math. Ann. 237, 121–132 (1978)
26. Thomas, W.: On the bounded monadic theory of well-ordered structures. J. Symb. Logic 45, 334–338 (1980)
27. Thomas, W.: Languages, automata and logic. In: Rozenberg, G., Salomaa, A. (eds.) Handbook of Formal Language Theory, vol. 3. Springer, New York (1997)
28. Vardi, M.Y.: Logic and Automata: A match made in heaven. In: Baeten, J.C.M., Lenstra, J.K., Parrow, J., Woeginger, G.J. (eds.) ICALP 2003. LNCS, vol. 2719, pp. 64–65. Springer, Heidelberg (2003)

Author Index

GPSR Compliance

The European Union's (EU) General Product Safety Regulation (GPSR) is a set of rules that requires consumer products to be safe and our obligations to ensure this.

If you have any concerns about our products, you can contact us on ProductSafety@springernature.com

In case Publisher is established outside the EU, the EU authorized representative is:

Springer Nature Customer Service Center GmbH
Europaplatz 3
69115 Heidelberg, Germany

Batch number: 10844611

Printed by Printforce, the Netherlands